ANALOG
AND
ANALOG/HYBRID
COMPUTER PROGRAMMING

PRENTICE-HALL ELECTRICAL ENGINEERING SERIES

PRENTICE-HALL INTERNATIONAL, INC. *London*
PRENTICE-HALL OF AUSTRALIA, PTY. LTD. *Sydney*
PRENTICE-HALL OF CANADA, LTD. *Toronto*
PRENTICE-HALL OF INDIA PRIVATE LIMITED *New Delhi*
PRENTICE-HALL OF JAPAN, INC. *Tokyo*

ANALOG
AND
ANALOG / HYBRID
COMPUTER PROGRAMMING

ARTHUR HAUSNER

Harry Diamond Laboratories

PRENTICE-HALL, INC. *Englewood Cliffs, N. J.*

Current printing (last digit):
10 9 8 7 6 5 4 3 2 1

13-032607-0

Library of Congress Catalog Card Number: 74-130835

Printed in the United States of America

To Geneve, Robert, and Bethany

PREFACE

I felt the need for a book of this kind when I taught a course in analog computation at the National Bureau of Standards' graduate school in 1965. Each student, whether a mathematician, physicist, chemist, or engineer, was actively engaged in scientific research or engineering development and wanted to know how to use an analog computer to solve his own problems. I endeavored to teach this and the more general problems that can be attacked by such a computer.

The emphasis is on techniques of using a general-purpose analog/hybrid computer for solving classes of mathematical problems, with minimal discussions on equipment design or on special-purpose computers. Equipment discussions that are included are based on the computers with which I am most familiar—an Electronic Associates, Inc. PACE 231R-V with an MLG logic board and electronic mode control, together with their digital expansion system, the DES-30. But the techniques described are applicable to all computers.

The chapter headings reflect the fact that no one field of science, except mathematics, is particularly emphasized because the techniques to solve a set of differential equations are virtually independent of whether the problem originated in nuclear physics, chemistry, or other areas. Exceptions, perhaps, are the chapters on Lagrange's equations and electrical network analysis because of the special methods developed for these problems.

Throughout the book, the parametric nature of the computer is stressed and is perhaps the central mathematical theme. All techniques are referred to the "parametric technique" for obtaining computer solutions, in precisely the same way that we at the Harry Diamond Laboratories have been utilizing our computer for over a decade. This orientation is a thoroughly new feature in books on analog computation. Judging from the success we have had in solving some of our own technical problems, such an approach is long overdue.

This book could probably be used as early as the junior year for a general course in analog computation, the only prerequisite being a scant knowledge of differential equations. On the other hand, many of the more modern concepts in analog/hybrid computer usage involving memory and automatic iteration in optimization techniques make the book suitable for a graduate course. Other features, such as graphs of step responses to Chebyshev and Padé approximants for a time delay and many previously unpublished techniques, should also serve the experienced worker in the field.

This book is written for a two-semester course, where an analog computer is available for laboratory sessions. In the first semester the basic material of Chapters 1–11 can be covered. In the last nine chapters more advanced material, representative applications, error analysis, and iterative techniques are meant for a second semester.

Many problems of varying degrees of difficulty are provided. These can be used both for homework and laboratory problems. Since the skill of a programmer depends to a large extent upon laboratory experience, this aspect of any course should not be minimized.

I am indebted to many people who assisted, directly and indirectly, in the writing of this book. Special thanks must be given to my wife, Geneve, who typed the manuscript; to my colleagues, R. F. Butler, A. I. Talkin, C. F. Ravilious, P. G. Hershall, C. M. Furlani, and H. M. Bloom, who each reviewed much technical material; to H. Ogata, I. Rotkin, and R. R. Vorkink, who all provided editorial reviews; and to B. M. Kurkjian, chief of the computation and analysis branch, R. D. Hatcher, chief of the systems research laboratory, M. Apstein, associate technical director, and B. M. Horton, technical director, all of Harry Diamond Laboratories, who gave much-needed encouragement.

ARTHUR HAUSNER

CONTENTS

ANALOG AND ANALOG/HYBRID COMPUTER PROGRAMMING

Chapter 1

INTRODUCTION TO AN ANALOG COMPUTER

1.1 Prologue

A simple mechanical system consisting of a mass, linear spring, and damper exerting a force in proportion to its velocity (Fig. 1.1a) is analogous to the electrical circuit consisting of an inductance, resistance, and capacitance (Fig. 1.1b). Both are describable by a second-order differential equation of the form

$$a_1 \frac{d^2y}{dt^2} + a_2 \frac{dy}{dt} + a_3 y = f(t) \tag{1.1}$$

We could construct a list of analogous quantities [such as mass m and inductance L, both represented by a_1 in (1.1)]. Either system can therefore be used

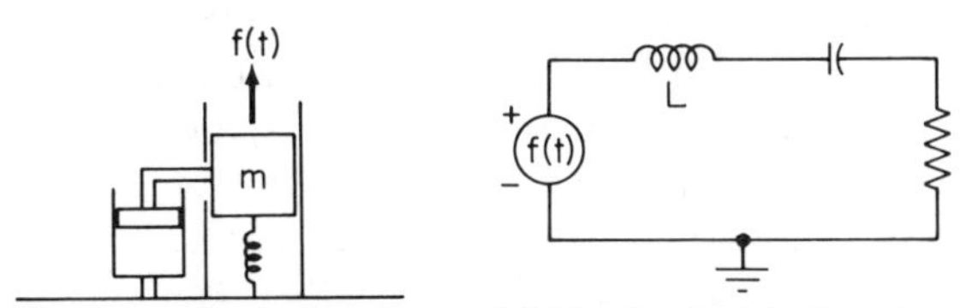

(a) Simple mechanical system (b) Simple electrical system

FIG. 1.1 Analogous systems.

to study the other system because their mathematical descriptions are identical. Such a method of study uses a *direct* analog, which has a long and interesting history [1] and is still being used today.

This book, however, will be concerned instead with a convenient *machine* to solve differential equations—a general-purpose *indirect* or *functional dc electronic analog computer*, or simply an analog computer. With an indirect computer of any type, an equation such as (1.1) is studied by means of *computing elements* in which particular mathematical operations, such as addition, multiplication, integration, etc., are performed on continuous time-varying physical quantities, e.g., distances, angles of shaft rotations, currents, or voltages. Both inputs and outputs may be measured and recorded as a function of time. The elements may be connected in a way satisfying (1.1). There is usually a one-to-one correspondence between the physical quantities and equation variables, making an interpretation of the computer outputs quite simple.

A modern dc electronic analog computer uses computing elements whose outputs are voltages varying with time. A general-purpose computer must also have means to interconnect these elements conveniently, control the solution process, and record solution outputs. Because the main purpose of an analog computer is to solve differential equations, many refer to it as an *electronic differential analyzer* or *EDA*.

Analog computers are widely used as tools to study systems involving differential equations. This includes simulation of engineering, physical, physiological, and mathematical systems [2]. The results may produce such insight into systems that many use an analog computer as a teaching aid [3]–[8].

1.2 Computer Symbols

An analog computer is a parallel machine, i.e., all computing elements operate *simultaneously*. A *computer program* is a diagram showing how the various elements are to be interconnected. This is entirely different in concept from a digital computer program which concerns itself with the logic of *sequential* operations.

Symbols that represent computer elements have necessarily evolved as an adjunct to programming. The Institute of Radio Engineers (IRE), now incorporated into the Institute of Electrical and Electronics Engineers (IEEE), proposed a set of standard symbols in February 1962 for use by analog computer programmers [9]. Similarly, the Simulation Councils, Inc., proposed "uniform graphics" for computer diagrams appearing in their journal *Simulation* [10]. All symbols used in this text will be first specifically defined and based on these standards. One should be aware, however, that standards were proposed late enough so that one will find different programmers using different symbols.

Figure 1.2 shows the symbols used for the *linear computing elements*, so-called because they are the main elements that are required to investigate linear

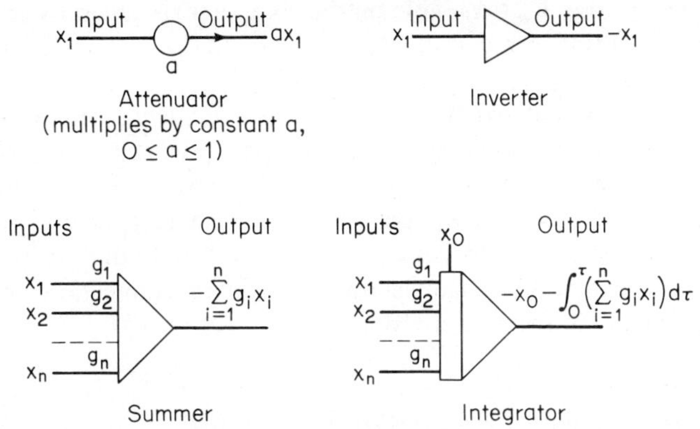

Notes: (1) All x_i are voltage functions of time τ.
(2) All $g_i \geq 0$.

FIG. 1.2 Linear computing elements.

systems of differential equations with constant coefficients. We note that:

1. Summers and integrators in common use negate the outputs and are so shown. New programmers must get used to this arrangement.
2. Provision is made for an initial condition voltage x_0, also to be negated, for the integrator. Integration is with respect to computer time τ measured in seconds.
3. When the symbols are shaped like arrowheads, they indicate the direction of the process. If any confusion is possible, arrowheads are drawn.
4. The g_i are called *gains*. When $g_i = 1$ (a unity gain), it is usually omitted. All gains are otherwise drawn outside the symbol.

Typical symbols for nonlinear computing elements are shown in Fig. 1.3.

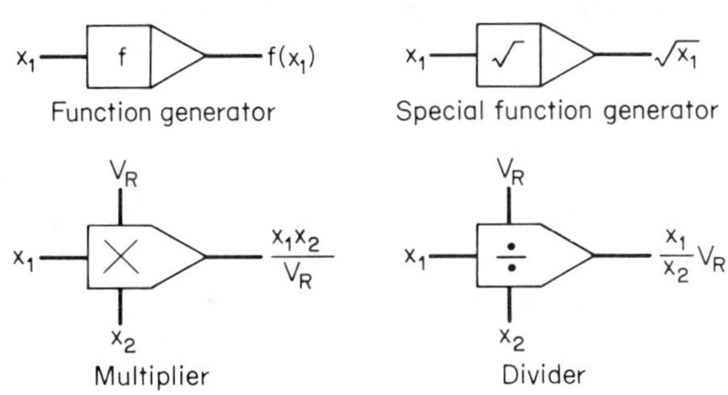

Note: V_R = reference voltage.

FIG. 1.3 Some symbols for nonlinear elements.

The *reference voltage* V_R (available on the computer) is generally considered to be an input to multipliers and dividers as a normalizing factor. To simplify the computer diagrams presented in this book V_R will not be shown.

It should be emphasized that the element symbols are useful in a general discussion to indicate the signal flow and for writing the computer program, but may not be so useful as a working diagram for interconnecting the elements. For example, many computers have no divider element, and a multiplier is connected in a special way to perform division. A working diagram of interconnecting elements would usually show the electrical connections actually necessary; in this case, to set up a division.

1.3 Measuring and Recording Equipment

To enable more meaningful interpretations of the voltage outputs of each computing element, means to measure and record them are necessary. Not all the kinds of equipment described in this section are available at every analog computer installation. On the other hand, some installations are so large that more than one of some are available for use in solving a problem. The list is in no particular order.

1.3.1 Digital voltmeter (DVM)

A digital voltmeter displays digitally the static voltage output of any computing element quite accurately. In most cases, four digits are displayed, which is roughly the maximum feasible above the noise level of equipment. Since it is used chiefly to set attenuation constants accurately, some computer manufacturers provide a *ratiometer* which displays the setting directly. On application of the reference voltage V_R at the input to an attenuator, a ratiometer reads the setting a instead of a voltage output aV_R. This makes it easy to set the value a, particularly when V_R is not an integral power of 10.

Selector pushbuttons on the main computer console can be manually operated to select the computing element output (sometimes the input) to be read.

1.3.2 Vacuum-tube voltmeter (VTVM)

This voltmeter usually has high input impedance and is used to track near-static voltages. Not being particularly accurate, it is useful only to check approximate values of slowly changing voltages, and to read certain high power voltages outside the range of the DVM.

1.3.3 Strip-chart recorder

A strip-chart recorder, or recorder, usually has many channels (typically 6 or 8) for plotting voltage-time graphs. The graph paper is moved at convenient constant speeds controllable by the user. Very sophisticated designs have evolved including such features as rectangular grid lines, pressurized instant-dry ink, automatic timing marks, pushbutton control of paper speeds in convenient steps for ranges of 10,000 to 1 mm/s, etc. The frequency response is usually such that a full amplitude sine wave can be recorded with no appreciable error for frequencies as high as 60 Hz. Since the graph is small, its resolution is generally poor. This unit is the bulwark of all recording devices.

1.3.4 X-Y variplotter

This versatile unit, also called an x-y plotter or plotter, records any one (sometimes two) voltage as a function of another voltage on rectangular graph paper (typically 11 by 17 in.). Two servo mechanisms position the x and y axes, so that the variplotter's frequency response is poor. The accuracy, however, is usually between 0.1 and 1.0 per cent of full scale, making this unit extremely useful for obtaining accurate graphs.

1.3.5 Oscilloscopes

Cathode ray oscilloscopes, or scopes, are indispensable for observing high-speed repetitive voltage traces. Although any scope can be used, many manufacturers provide large-screen (17 in.) multi-input (4 to 8) oscilloscopes, specifically designed for computer use. Many models provide cameras for recording the light traces. In some, the major grid lines are flashed on the screen in a manner similar to the voltage traces, so that any nonlinearity of the electron gun is taken into account. The resolution is typically 1 per cent of full scale.

Some oscilloscopes can be used in the way a variplotter is used; i.e., the horizontal sweep can be controlled by an arbitrary voltage rather than by time to allow cross plotting several y_i versus x.

1.3.6 Printers

Some computers contain an arrangement to print static input and output voltage values on the computer. This provides a permanent record of test cases and is also useful for checking results. In most cases, a sequencing circuit enables automatic printing of the voltage outputs of all or a group of similar elements, e.g., amplifiers.

We will have occasion to use plotters, recorders, and oscilloscopes while solving problems. Symbols used for these devices are shown in Fig. 1.4. If

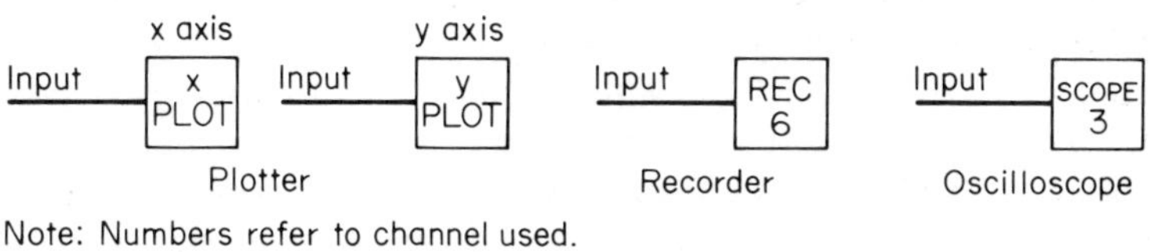

FIG. 1.4 Symbols for recording devices.

more than one are available and connected to the computer, an additional identifying number within the box is used.

1.4 Interconnections on a Patch Panel with Patch Cords

Inputs and outputs of computing elements and inputs to recording equipment are generally electrically wired to a central location in the computer console called the *patch bay.* In the patch bay, there usually is a *removable patch panel* (Fig. 1.5). Interconnections can be made with *patch cords* on the patch

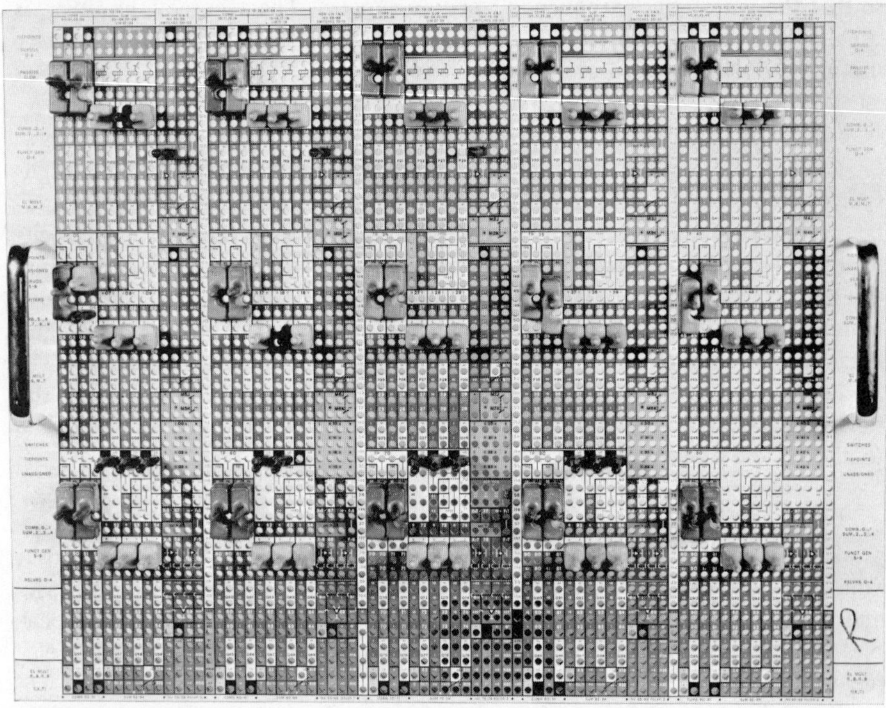

FIG. 1.5 Patch panel (Courtesy Electronic Associates, Inc.).

panel in any desired manner away from the computer. A panel may be inserted and locked into the patch bay at a moment's notice (Fig. 1.6). All electrical connections are made instantly by the patch cord tips wiping on protruding contacts in the patch bay. A research center may have 10 or more patch panels, each patched to solve or investigate a specific problem. This arrangement permits investigation of many problems in rapid succession so that computer time may be used efficiently.

FIG. 1.6 Inserting a wired panel into the patch bay.

Modern patch panel designs provide for ease of checking connections as well as ease of patching. Areas are color-coded (yellow for attenuator areas, green for inputs, red for outputs, etc.) and are clearly labeled. In addition, sections of the board contain terminations for a variety of equipment, eliminating the need for long patch cords (modular design).

The patch cords are generally shielded coaxial cable, color-coded by length. *Bottle plugs* are also provided for connecting adjacent terminations.

1.5 Computer Control

The computer controls provided vary considerably, from simple manual switches to fully automated switching by a digital computer. In any case, the computer can be placed in several different modes. The most important of these are:

RESET or INITIAL CONDITION (IC) mode—Places initial voltages on every integrator. This mode must be selected before any integrations are performed.

OPERATE mode—Starts integrations of inputs to all integrators. Computer time τ, initially zero, is measured from the instant the computer is placed in the OPERATE mode. Recording and plotting equipment also can be started automatically when the computer is placed in OPERATE. This mode is selected when differential equations are being solved.

HOLD mode—Stops the integration. The output voltages of all integrators at the instant the computer is placed in the HOLD mode remain constant. This mode is selected to stop a problem before completion when intermediate results are needed. The operator can then continue the solution by placing the computer in the OPERATE mode.

If the operator places the computer in any one of the above modes, appropriate relays or electronic switches are activated. Details of the switching can be determined by study of the computer manuals.

Many computer systems contain a master clock controlling signals that periodically place the computer into the RESET and OPERATE modes. Such a computer is called a *repetitive* computer. One advantage is that the solutions obtained during the operate period can be displayed on an oscilloscope where effects of parameter changes can be instantly observed. Repetition frequencies are limited if the computer modes are controlled by relays. Even so, 10 solutions per second can be obtained, and this is often adequate to determine the effect of parameter changes.

More sophisticated arrangements are also available in some computers. Individual integrators can be placed in the various computing modes by means of *programmed logic*. In such a case, several integrators can be in the OPERATE mode with others in the HOLD mode and still others in the RESET mode. Very complicated programs can then be programmed, even ones that can alter themselves. Integrators can then be patched as *memory* units. Computers in which this is the case are frequently called *iterative analog computers* or *analog/hybrid* computers, because the computer may then be programmed to do iterative calculations.

An analog/hybrid computer is a repetitive computer by its very nature. Where many iterations per second are needed, electronic switching of the integrator modes (the mode of individual integrators) is used instead of relay switching. Some modern computers can produce 1000, or even more, solutions per second, each one giving an entire solution to a set of differential equations, and each one being slightly different from the next because parameters are changed between solutions.

A computer with this capability is generally equipped with a separate patch panel where the logic circuitry can be patched. The outputs here are *digital* binary signals (0 or 1, off or on, etc.) which control the state of electronic switches, and are produced by digital elements such as comparators, flip-flops, counters, etc. The use of these elements is discussed in Chapters 6, 18–20. The latter three chapters also describe how such flexibility can be applied in the field of optimization.

Finally, there are analog computers that are linked with a digital computer in what is called a *hybrid* computer. The digital computer provides much memory and does much calculation as well as control the modes of the analog integrators. This book will not be concerned with such hybrid computers.

In the early chapters, we assume that a computer with manual mode control is available, and that all integrators are simultaneously placed in similar modes.

1.6 The Kelvin Feedback Method

Let us consider just how we could interconnect computer elements of the types shown in Figs. 1.2 and 1.3 to investigate the solution of (1.1). Lord Kelvin (W. Thomson) originated the method for doing this in a series of papers concerning a mechanical wheel-and-disc integrator invented by his brother, J. Thomson [11]–[14]. Before presenting the argument, (1.1) is rearranged:

$$\frac{d^2y}{dt^2} = \frac{1}{a_1}\left[-a_2\frac{dy}{dt} - a_3y + f(t)\right] \tag{1.2}$$

Now,

(1) Consider that y, dy/dt, and f are voltages, and that a_1, a_2, and a_3 are dimensionless numbers with $a_1 > 1$, $a_2 > 0$, and $a_3 > 0$. Assume a variable voltage d^2y/dt^2 is somehow generated on the computer at point P (Fig. 1.7).

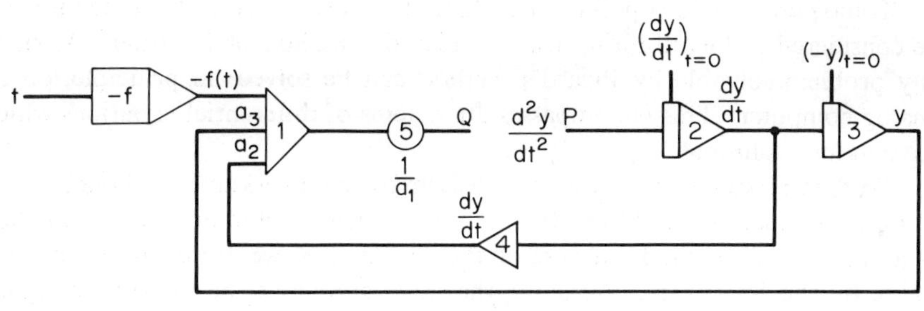

FIG. 1.7 The Kelvin feedback method.

(2) Let $t = \tau$ (t is now a voltage numerically equal to τ), and successively integrate d^2y/dt^2 with integrators 2 and 3 to produce $-dy/dt$ and y.

(3) Construct the right-hand side of (1.2) at point Q continuously, by using summer 1, which adds the terms $-f(t) + a_2(dy/dt) + a_3y$ individually, and attenuator 5 which multiplies by $1/a_1$. Note that inverter 4 is required because a_2 was assumed positive.

(4) The voltage at P should equal the voltage at Q. *Agreement is forced by electrically connecting these two wires.* Figure 1.8 shows the final computer

diagram with the feedback loop closed. The solution $y(t)$ is produced continuously in a single process.

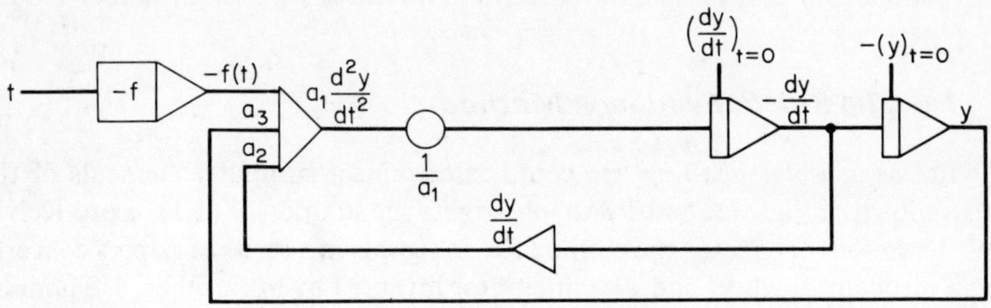

FIG. 1.8 Feedback loop closed.

Lord Kelvin was excited and surprised that a machine could solve differential equations in a single process, particularly since there is no analytical equivalent of the method. He first had envisioned using the integrators in a manner which he knew converged to the true solutions under very general conditions: generate the function at Q in terms of y, dy/dt, and $f(t)$ assuming an arbitrary function $F_k(t)$ is fed in at P. Use the generated function as a new input function $F_{k+1}(t)$ to be fed in at P. In this way the sequence of functions $F_k(t)$, $k = 1, 2, \ldots,$ is formed, with $F_{k+1} - F_k \rightarrow 0$ as $k \rightarrow \infty$. This process is known as Picard's method of iteration for solving differential equations.

Tomovich [15]–[16] pointed out that the Kelvin feedback procedure can be considered to be a limiting case of Picard's method of iteration. As such, any problem solvable by Picard's method can be solved, in principle, on an analog computer. This encompasses all systems of differential equations which have unique solutions.

To the uninitiated, the process may seem to be unworkable, as though something is obtained for nothing. In fact, some writers refer to Kelvin's method as a "bootstrap" method, because of the feeling that we are trying to lift ourselves by the bootstraps. Actually, the integrators must be carefully designed to prevent the outputs from improperly interacting with the inputs. This usually means amplification of some sort, independent of the type of integrator designed. (Mechanical integrators were not particularly successful in solving differential equations requiring feedback, until a torque amplifier was invented.) The basis of the voltage amplifier in the electronic analog computer is the *dc feedback*, or *operational amplifier*, to be discussed in Chapter 2.

In practice, before Kelvin's feedback method can be applied, *scaling* is required because of several reasons to be discussed in Chapter 4.

PROBLEMS

1-1. Use the Kelvin feedback method to draw a computer diagram to graph y versus y_1 on a variplotter from the following pair of first-order equations:

$$\frac{dy}{dt} = y_1$$

$$\frac{dy_1}{dt} = \frac{1}{a_1}[-a_2 y_1 - a_3 y + f(t)]$$

with initial conditions of $y_1 = (y_1)_0$ and $y = (y)_0$ at $t = 0$. Assume $a_1 > 1, a_2$ and $a_3 > 0$, and that a voltage equal to t can feed a function generator. How does your diagram compare with Fig. 1.8?

1-2. In one of the first papers describing electronic analog computing principles, Ragazzini *et al.* [17] set up *differentiators* to solve (1.1). The symbol sometimes used for a differentiator is shown in Fig. 1.9. Using this symbol and the reasoning in Kelvin's feedback technique, draw a circuit diagram to solve (1.1). Again, assume that $f(t)$ can be generated. *Note:* Differentiation is a noise amplifying process and is not used in analog computing unless it cannot be avoided.

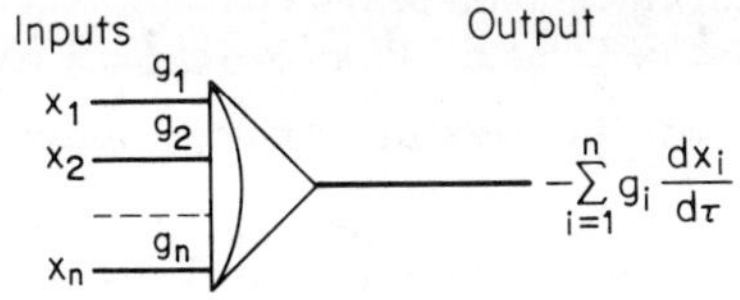

FIG. 1.9 Differentiator symbol.

1-3. Draw a computer diagram for the solution of

$$\frac{dx}{dt} = -\frac{t}{x} + xt + 3$$

Hint: Integrate -1 V to obtain a voltage numerically equal to t, and let $t = \tau$.

REFERENCES

[1] Roedel, J., "An Introduction to Analog Computers," *ISA Journal*, vol. 1, no. 8, Aug. 1954.

[2] McLeod, J., ed., *Simulation* (New York: McGraw-Hill Book Co., 1968).

[3] Kovach, L. D., and W. Comley, "The Analog Computer as a Teaching Aid in Differential Equations," *AICA*, vol. 3, April 1961.

[4] Karplus, W, J., and L. D. Kovach, "The Use of Computers in Analysis," *Proc. Spring Joint Comp. Conference*, 1962.

[5] Rasof, B., and R. Tomović, "The Analog Computer as an Aid in Teaching Mathematics," *AICA*, vol. 4, Oct. 1962.

[6] Betchov R., and L.D. Kovach, "Teaching Mathematics with Analog Computers," *Simulation*, vol. 3, Nov. 1964.

[7] Berry, F. G., and A. C. Soudack, "The Use of the Analog Computer in Teaching Engineering Systems," *AICA*, vol. 7, April 1965.

[8] James, M. L., "The Analog Computer as a Teaching Aid," *Simulation*, vol. 7, Sept. 1966.

[9] "Proposed IRE Standards for Analog Computers," *IRETEC*, vol. 11, Feb. 1962.

[10] "Uniform Graphics for SIMULATION," *Simulation*, vol. 6, March 1966. Revised in vol. 9, Dec. 1967.

[11] Thomson, J., "On an Integrating Machine Having a New Kinetic Principle," *Proc. Royal Soc.*, vol. 24, Feb. 3, 1876.

[12] Thomson, W., "On an Instrument for Calculating $\int \phi(x)\psi(x)\,dx$, the Integral of the Product of Two Given Functions," *Proc. Royal Soc.*, vol. 24, Feb. 3, 1876.

[13] ______, "Mechanical Integration of the Linear Differential Equation of the Second Order with Variable Coefficients," *Proc. Royal Soc.*, vol. 24, Feb. 3, 1876.

[14] ______, "Mechanical Integration of the General Differential Equation of Any Order with Variable Coefficients," *Proc. Royal Soc.*, vol. 24, Feb. 3, 1876.

[15] Tomovich, R., "Picard's Method and Analog Computation," *IRETEC*, vol. 6, March 1957.

[16] Tomović, R. "Method of Iteration and Analog Computation," *AICA*, vol. 1, Sept. 1958.

[17] Ragazzini, J. R., *et al.*, "Analysis of Problems in Dynamics by Electronic Circuits," *Proc. IRE*, vol. 35, May 1947.

Chapter 2

POTENTIOMETERS, OPERATIONAL AMPLIFIERS, AND MULTIPLIERS

In this chapter we examine the equipment commonly used in an analog computer to perform integration and the basic arithmetic operations—addition, subtraction, multiplication, and divison. With this equipment we can solve the large class of differential equations involving rational functions. The design of the equipment varies among manufacturers, but the basic principles of operation are usually similar. Therefore, we shall use typical designs and introduce the symbols used in diagrams and thus prepare the reader to use such equipment.

2.1 Potentiometers

Potentiometers (attenuators, coefficient potentiometers, or simply pots) are used to multiply voltages by constants between 0 and 1. They are voltage dividers consisting of a resistance with one end grounded and a sliding contact (usually fused) as shown in Fig. 2.1a. The total resistance R_P is typically 30 kΩ with linearity of about 0.05 per cent and resolution of 0.01 per cent. Such a potentiometer can therefore be set to four decimal digits. In the working diagram symbol, it is customary to place the pot number in the circle and the value to which it is set near, usually below, the circle.

Hand-set pots are placed near the computer operator controls, so that they

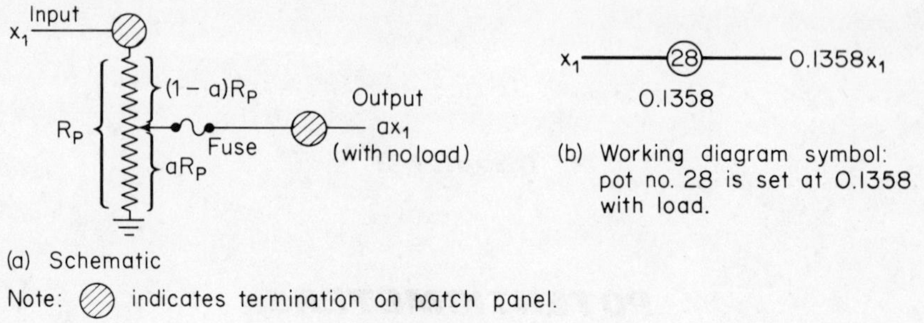

FIG. 2.1 Potentiometer principles.

may be changed rapidly between computer runs. This is a convenience when pot values are proportional to parameters of a problem and must be reset several times in trial-and-error procedures.

2.1.1 Loading corrections

Most commercial computers use multiturn, wirewound pots with dial indicators exposed and with a locking device. The dial setting will correspond to the pot setting within the specified linearity provided there is no loading error. Loading errors are introduced when current is drawn by the wiping arm because of an equivalent resistive load. The equivalent circuit with dial setting a is shown in Fig. 2.2 with pot resistance R_P, and load resistance R_L, usually an

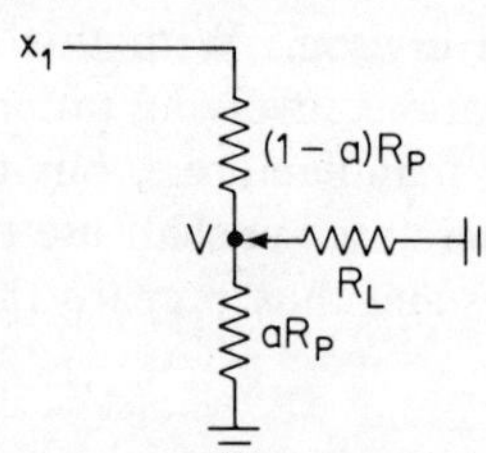

FIG. 2.2 Equivalent circuit with load.

input resistor to an operational amplifier. The equivalent resistance from V to ground is given by

$$R_e = \frac{1}{(1/R_L) + (1/R_P)}$$

with the voltage V now given by

$$V = x_1 \frac{R_e}{R_e + (1 - a)R_P}$$

By appropriate substitution,

$$V = x_1 \frac{a}{1 + a(R_P/R_L) - a^2(R_P/R_L)} \tag{2.1}$$

If the desired multiplier of x_1 were a^*, i.e., $V = a^* x_1$, then

$$a^* = \frac{a}{1 + a(R_P/R_L) - a^2(R_P/R_L)} \tag{2.2}$$

so that the dial setting should be

$$\begin{aligned} a &= \frac{a^*(R_P/R_L) - 1 + \sqrt{[1 - a^*(R_P/R_L)]^2 + 4a^{*2}(R_P/R_L)}}{2a^*(R_P/R_L)} \\ &= \frac{2a^*}{1 - a^*(R_P/R_L) + \sqrt{[1 - a^*(R_P/R_L)]^2 + 4a^{*2}(R_P/R_L)}} \end{aligned} \tag{2.3}$$

Loading correction curves showing how much of a loading correction ϵ should be added to the desired multiplier a^* can be prepared as function of a^*. We have

$$a = a^* + \epsilon(a^*) \tag{2.4}$$

and from (2.3)

$$\epsilon = \frac{2a^*}{1 - a^*(R_P/R_L) + \sqrt{[1 - a^*(R_P/R_L)]^2 + 4a^{*2}(R_P/R_L)}} - a^* \tag{2.5}$$

2.1.2 Setting pots

Correction curves are unnecessary if a digital voltmeter can be used to read the voltage output of potentiometers directly, as is the case in most installations. A switch that automatically places the reference voltage V_R at the input is generally located beside the pot to permit the operator to set it precisely under the proper load. There is a special computer mode to do this called POTSET, which should be used to protect amplifiers and insure correct settings. Most potentiometers are set while the computer is in POTSET prior to operating the computer (Fig. 2.3).

Large installations are generally equipped with automatic, servo-set potentiometers with only a few that can be hand-set. This relieves the operator of the tedious job of pot setting in large problems, where 100 or more potentiometers may have to be set. If these must be hand-set, a good method is to set the dial of each pot to the desired reading before resetting it accurately with the voltmeter. This greatly reduces the tedium of holding down the spring-loaded switch which applies the reference voltage while adjusting the pot.

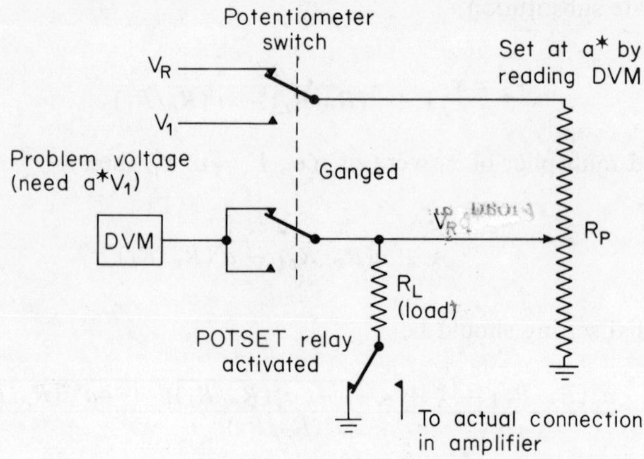

FIG 2.3 Potentiometer setting circuit.

An important use of potentiometers is to obtain a constant voltage. Both the reference voltage V_R and its negative $-V_R$ are available on the patch panel. Any constant between these extremes can therefore be obtained.

2.1.3 Ungrounded pots

Computers also contain ungrounded pots (Fig. 2.4) for use in special circuits involving diodes (Chapter 6). If there is a grounded resistive output load R_L, it can be shown that the output is

$$V_0 = \frac{x_2 + a(x_1 - x_2)}{1 + a(1 - a)(R_P/R_L)} \tag{2.6}$$

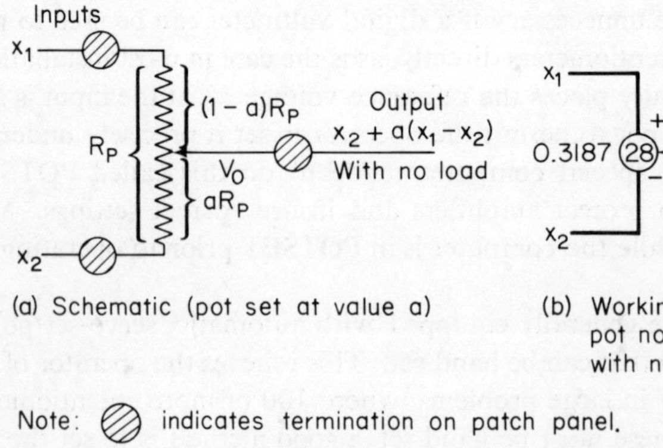

FIG. 2.4 Ungrounded potentiometer.

There is now no one setting a to produce a desired output $x_2 + a^*(x_1 - x_2)$, since it requires

$$x_2 + a^*(x_1 - x_2) = \frac{x_2 + a(x_1 - x_2)}{1 + a(1 - a)(R_P/R_L)}$$

which cannot be solved for a independent of x_1 and x_2. Thus, a has meaning only to define the resistances on either side of the wiper arm. As such, it can be set only with no load. To set an ungrounded pot, simply ground it temporarily, remove any load, and treat it as a normal pot. Note that the working diagram should indicate the + and − sides of the pot, also called the HI and LO sides, and show the no-load (infinite R_L) DVM setting.

The low end of an ungrounded pot can be grounded to convert it to a normal pot.

2.1.4 Ganged pots

Some computers provide ganged pots, where one knob controls the setting of several potentiometers. This is generally more useful in repetitive computers when a parameter appears in more than one equation. Changing the parameter without a set of ganged pots can otherwise disturb the smooth information flow between the computer and operator, since more than one knob must be turned and properly set. The accuracy of ganged pots, however, is limited by misalignments and nonlinearities, but settings are at least within 0.001 of each other. For more accurate settings, ganged pots should not be used.

2.1.5 Potentiometer errors

Like every computing element, a potentiometer deviates from an ideal element. There are end resistances which prohibit exact settings of zero and one. At very high frequencies, parasitic capacitances produce undesirable phase shifts, although special capacitors can be used to compensate for these shifts. Even at low frequencies, a source of difficulty is the possible setting error of $\pm$ 0.00005 due to finite resolution of the DVM reading. This may be critical for lower settings since the percentage error (or relative error) of the setting increases when the setting decreases. This often motivates programmers to arrange a computation with large pot settings, and sometimes to use two pots in series to improve results.

Let ϵ be the maximum absolute error in the setting of a potentiometer. The maximum relative error E_1 when setting it to the value a is given by

$$E_1 = \frac{\epsilon}{a}$$

If two pots are used and are set to the values of a_1 and a_2, where $a = a_1 a_2$, then the maximum absolute error of each setting remains ϵ. Thus, the true setting is at worst $(a_1 + \epsilon)(a_2 + \epsilon)$, so that the new maximum relative error E_2 becomes

$$E_2 = \frac{\epsilon(a_1 + a_2) + \epsilon^2}{a} = \frac{\epsilon[a_1 + (a/a_1)] + \epsilon^2}{a}$$

To find the proper settings of a_1 and a_2 to minimize E_2, we have

$$\frac{dE_2}{da_1} = \frac{\epsilon}{a}\left(1 - \frac{a}{a_1^2}\right) = 0$$

so that

$$a_1 = a_2 = \sqrt{a} \tag{2.7}$$

Note now that

$$E_2 \simeq \frac{2\epsilon}{\sqrt{a}}$$

which is smaller than E_1 only when $a < 0.2500$. In practice, two pots usually are used in series only when $a \leq 0.1000$ and is critical in the solution of a problem. Even then, it is not usually convenient to set each pot at $\sqrt{a}$, but at convenient values which are close to one another, and whose product is a. Care should be taken to set the pots in proper sequence and to insure that the voltage source can supply the current required by two pots in series.

2.2 Operational Amplifiers

The heart of an analog computer is the *direct-coupled feedback amplifier*, often called an operational amplifier because its output V_0 can be determined by the use of operational calculus. Howe [1], Korn and Korn [2], Fifer [3], or computer manuals provide detailed discussions on their design. It suffices here to examine the ideal performance of an amplifier, since this is usually closely realized.

Operational amplifiers are designed to work satisfactorily between two output voltages, $-V_R$ and $+V_R$, where V_R is the reference voltage (typically 100, 50, or 10V). Networks of linear components are connected to the amplifier to obtain various responses.

The concept of impedance simplifies the derivation of the equations involving these networks. To calculate the impedance of a network, the Heaviside operator $p \equiv d/d\tau$ is a convenient shortcut. Thus, $pI \equiv dI/d\tau$. Its inverse is $p^{-1}I \equiv \int I\,d\tau$, integration being the inverse of differentiation. For example, the impedance of a capacitor can now be derived from fundamental considerations. If E is the voltage drop across a capacitor C, and I is the current, then $E = (1/C)\int I\,d\tau$. We may therefore write $E = (1/C)p^{-1}I$ from which, by Ohm's law, p^{-1}/C is the impedance of the capacitor. It follows from the nature of linear networks that the impedance may be manipulated algebraically, just as we manipulate resistances. This is shown in Chapter 10.

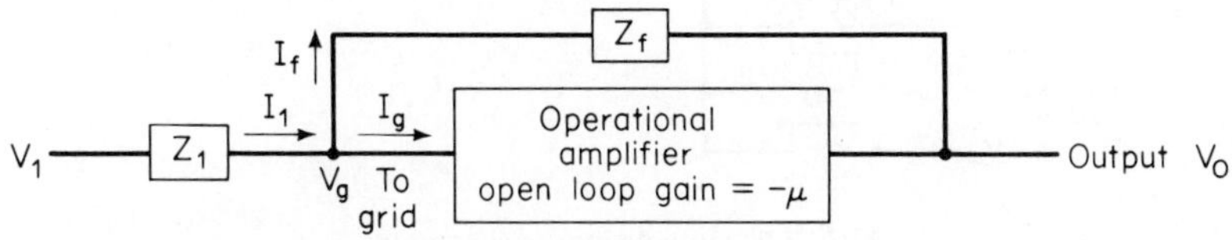

FIG. 2.5 Operational amplifier schematic.

An operational amplifier with input and feedback networks having impedances Z_1 and Z_f, respectively, connected to the grid of the first stage is shown in Fig. 2.5. From basic concepts, the operational equations are simply

$$V_1 - V_g = Z_1 I_1, \qquad V_g - V_0 = Z_f I_f, \\ I_g = I_1 - I_f \tag{2.8}$$

Since an operational amplifier is one with an extremely large open-loop gain $-\mu$, we can replace the value of V_g by

$$V_g = -\frac{V_0}{\mu} \tag{2.9}$$

by definition. Now we must make an assumption closely realized in practice, namely,

$$I_g = 0 \tag{2.10}$$

so that by substitution of (2.9) and (2.10) into (2.8), we can show

$$V_0 = -\frac{Z_f}{Z_1} V_1 \left[\frac{1}{1 + (Z_1 + Z_f)/(\mu Z_1)} \right] \tag{2.11}$$

Since μ is of the order of 10^8, (2.11) reduces to

$$V_0 \simeq - Z_f \frac{V_1}{Z_1} \tag{2.12}$$

with extremely small error.

In a similar manner we can derive the operational output for the case where n inputs are connected through networks to the grid (Fig. 2.6):

$$V_0 \simeq - Z_f \sum_{i=1}^{n} \frac{V_i}{Z_i} \tag{2.13}$$

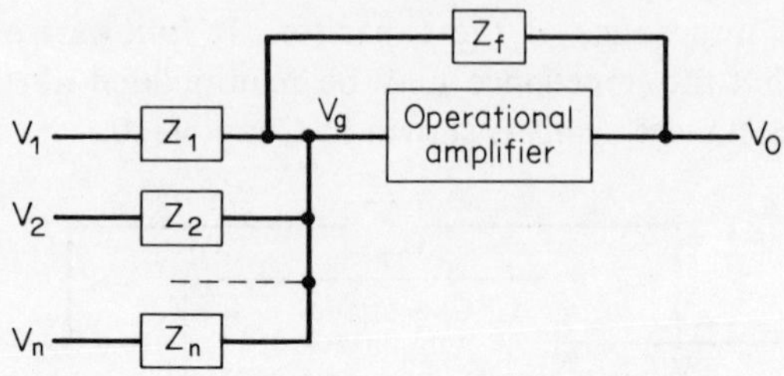

FIG. 2.6 Multiple inputs schematic.

This is the fundamental equation governing the output of an operational amplifier. By choosing various input and feedback impedances, the output is forced to behave in some desired way.

In all cases, the high open-loop gain forces the grid to be maintained at virtual ground when operating satisfactorily. If the amplifier output is not what it should be, it is said to be *saturated*, and this condition implies bad scaling (Chapter 4) or malfunctioning equipment. Most computers provide for both audio and visual indication of saturation (by speaker signal and light).

2.2.1 Inverters

Let $n = 1$, $Z_1 = Z_f =$ resistance R. Then

$$V_0 = -R\frac{V_1}{R} = -V_1$$

which is the equation for an inverter. Since an inverter is a special case of a summer (one unity-gain input) some computers do not have inverters. Others provide inverters with special nonlinear equipment requiring inverters. Still others provide inverters for all amplifiers (bipolar amplifiers). The schematic of an inverter and the symbol to be used in working diagrams are shown in Fig.

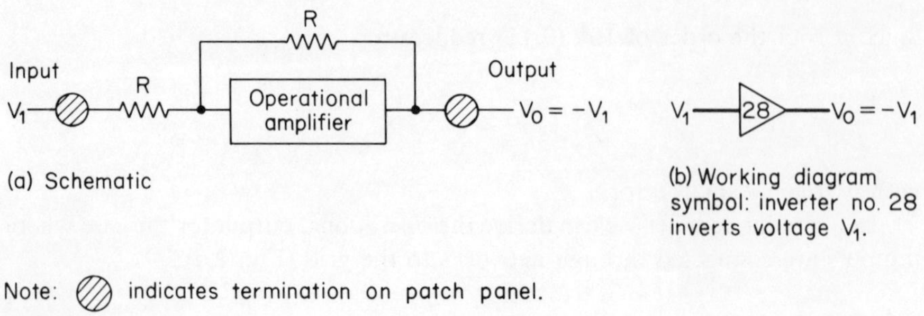

FIG. 2.7 Inverter.

2.7. Typical values for R are 0.1 or 1MΩ. Inverters are used for subtraction; a voltage is first inverted and then added to other voltages.

2.2.2 Summers

Let Z_f = resistance R_f, and let multiple input impedances Z_i be resistances R_i. Then

$$V_0 = -R_f \sum_{i=1}^{n} \frac{V_i}{R_i} \tag{2.14}$$

A simplified schematic of a summer in a typical general-purpose computer and a symbol for use in working diagrams are shown in Fig. 2.8 without the control circuitry required for the POTSET mode. It is customary for R_f/R_i to be integral when the R_i are permanently wired in, providing fixed gains. There are typically six input resistances, three each providing gains of one and ten.

The patch panel termination usually gives access to the summing junction (SJ), grid (G), feedback resistor (F), as well as input holes (V_i) and output holes

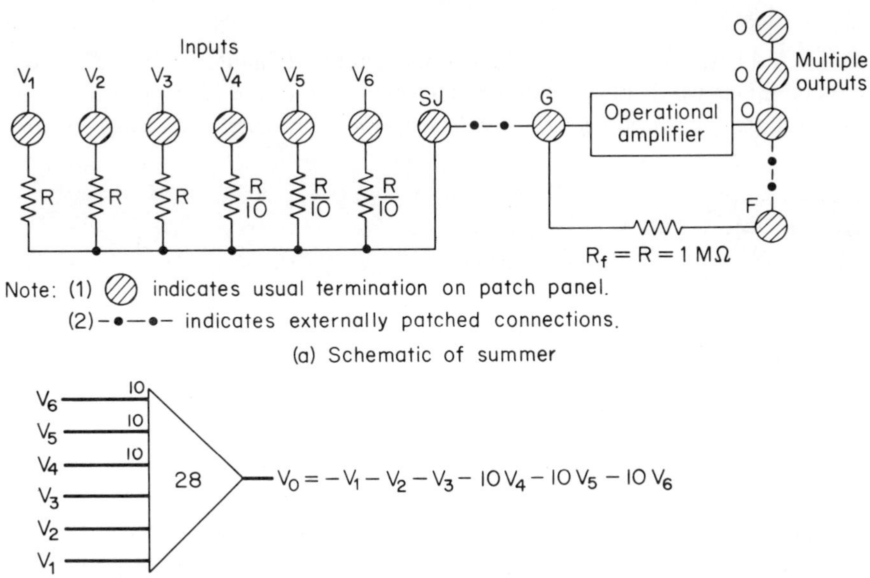

FIG. 2.8 Typical summer information.

(O). In this way other external components can be patched to the summing junction if needed. Also, special circuits may require access to the grid. The SJ-G and F-O connections are usually not wired permanently but must be patched on the patch panel. Bottle plugs are provided to make these connections simultaneously and therefore rapidly, and to allow the feedback resistor to be disconnected for special circuits.

In this book we assume the fixed feedback resistor for inverters and summers to be one megohm for simplicity. Gains of 10 are provided by using 0.1-MΩ input resistances.

In general, components are controlled to be within 0.01 per cent of their nominal values to maintain a four-figure accuracy. The input and feedback resistances are often kept in an oven to provide for temperature stability.

2.2.3 High-gain amplifiers

A summing amplifier with its feedback resistance removed is called a *high-gain* amplifier. Figure 2.9 shows the symbol for the high-gain amplifier. Its schematic is identical with Fig. 2.8a without the O-F connection. Since this is effectively an infinite feedback resistance, from (2.14) we see that a finite voltage can exist at the output only if

$$\sum_{i=1}^{n} \frac{V_i}{R_i} = 0 \tag{2.15}$$

This is the basis of many implicit function circuits (Chapter 9) and diode circuits (Chapter 6). If the inputs do not sum to zero, the high open-loop gain will saturate the amplifier. *High-gain amplifiers are used only when external*

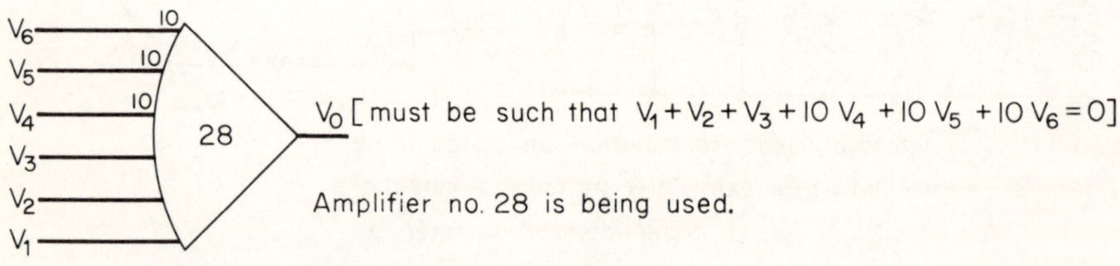

Note: This symbol is often used for an operational amplifier (input resistances are externally drawn).

FIG. 2.9 High-gain amplifier working diagram symbol.

feedback is provided, since the chance of (2.15) being satisfied is essentially zero. We may use (2.15) to determine the output of the amplifier, as illustrated immediately below.

A frequent use of the high-gain amplifier is to provide gains of more than 10 by using a pot in the feedback loop (Fig. 2.10). Thus, from (2.15),

$$\sum_{i=1}^{n} g_i V_i + a g_0 V_0 = 0$$

from which

$$V_0 = -\frac{1}{ag_0}\sum_{i=1}^{n} g_i V_i$$

If $ag_0 = 1$, then the arrangement is equivalent to a summer. Large gains are possible when a is set near zero. (The output, however, must still not be permitted to saturate.)

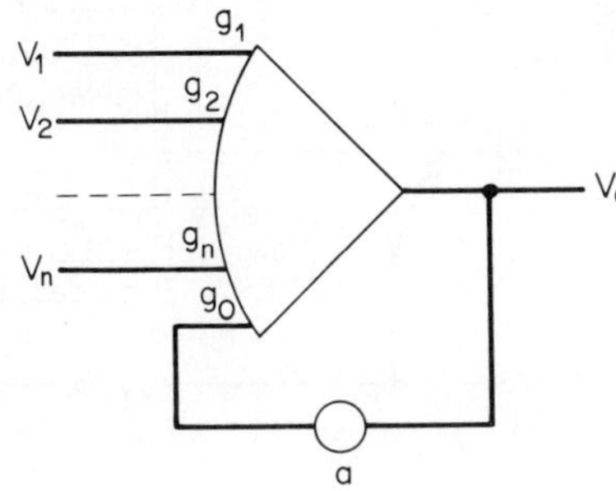

FIG. 2.10 Using a potentiometer in the feedback loop.

2.2.4 Integrators

If the feedback component is a capacitor with impedance p^{-1}/C, then from (2.13)

$$V_0 = -\int_0^{\tau}\left(\sum_{i=1}^{n}\frac{V_i}{R_i C}\right)d\tau \tag{2.16}$$

Capacitor values are fixed during a solution, but elaborate arrangements for easily changing the value between solutions, usually by factors of 10, are often employed. It is not uncommon for computers to provide for four different capacitor values ranging from 1 to 0.001 μF; some manufacturers provide for six different ones. These are usually oven-mounted when the input and feedback resistors are.

The rate of integration depends inversely on the value of C; it is convenient to write (2.16) as

$$V_0 = -\frac{1}{10^6 C}\int_0^{\tau}\left(\sum_{i=1}^{n}\frac{V_i}{10^{-6}R_i}\right)d\tau$$

and express resistors in units of megohms and capacitors in units of microfarads. The V_i can then be patched into gains labeled $1/(10^{-6}R_i)$, while the capacitor chosen determines the rate. Most integrator designs are arranged in this way. Figure 2.11 is a schematic of a typical integrator (shown without any control circuitry) and its symbol. Although relay switches are shown, modern computers use electronic switches for fast operation. As in the summer, a connection must be made on the patch panel; in this case, the G-C connection.

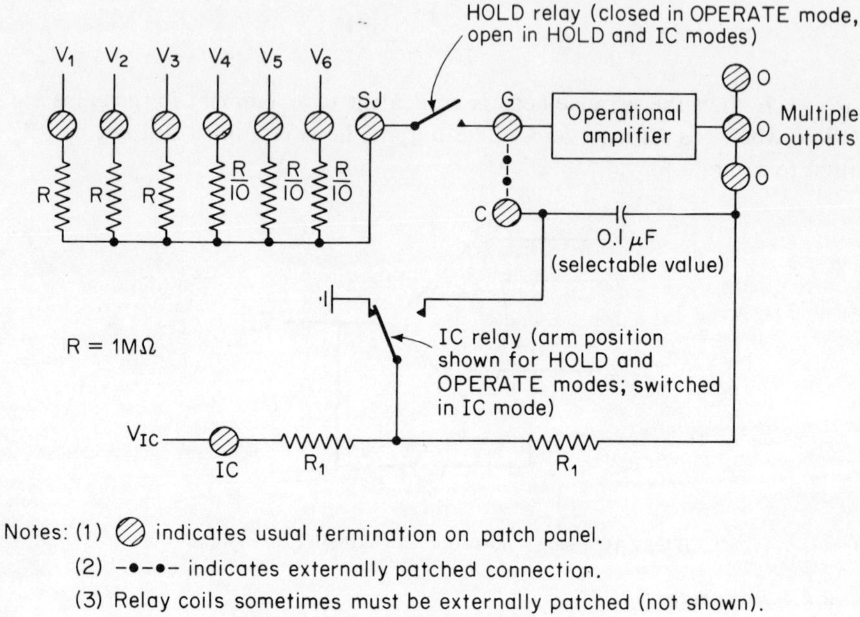

(a) Schematic of integrator

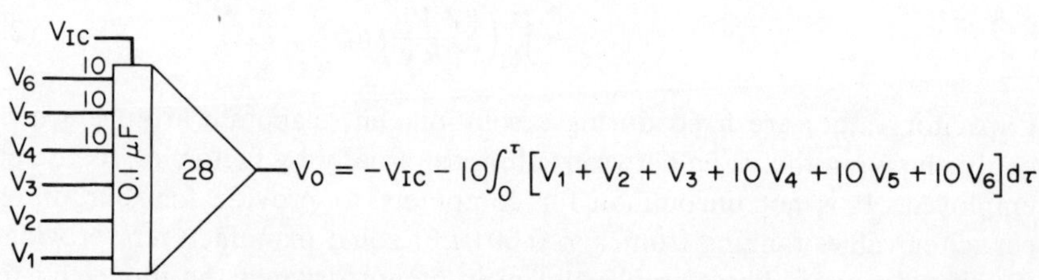

(b) Working diagram symbol: integrator no. 28 is used with a 0.1-μF capacitor

FIG. 2.11 Typical integrator information.

The integrator is very similar to the summer, and virtually all of them are designed to be patchable as summers by replacing the feedback capacitor with a feedback resistor. Such an amplifier is called a *combination amplifier* and makes for better utilization of the operational amplifiers. The G-C plug can be removed easily so that this amplifier can be converted to a summer with another bottle plug by connecting in a feedback resistor that is not shown in Fig. 2.11.

When the integrator is in the IC mode, the feedback capacitor is charged to the initial voltage V_{IC} by a circuit characterized by two matched resistors. These age and tend to become mismatched with time. Initial condition voltages, therefore, are best placed on capacitors by adjusting the IC pot while reading the output of the integrator on a DVM with the computer in the IC mode.

The capacitance of any capacitor used should be indicated inside or below the rectangular portion of the symbol, unless it is the standard one at the installation (usually 1 μF). If all integrators use the same nonstandard capacitance, then a note on the computer diagram would suffice.

We shall assume throughout this book that input resistors are fixed (gains

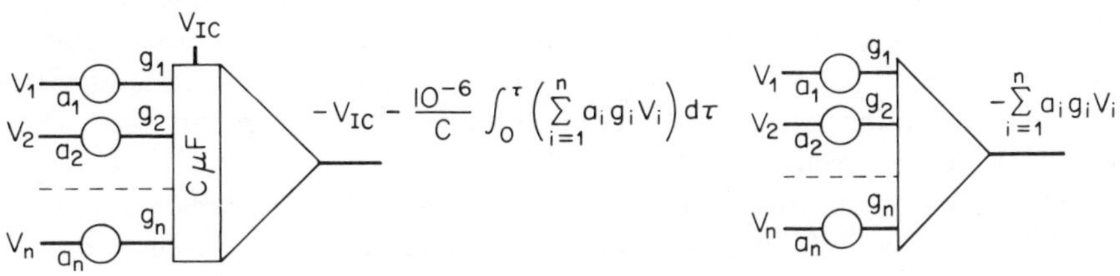

FIG. 2.12 Obtaining variable gains.

g_i of 1 or 10) and only a limited number of capacitor values can be used. As such, variable gains must be obtained with potentiometers for both the summer and integrator as shown in Fig. 2.12.

2.2.5 *Other possible input and feedback impedance combinations*

Any series-parallel combination of resistors and capacitors can theoretically be used as input and/or feedback elements. In fact, early users of operational amplifiers wired or plugged the resistors and capacitors into operational amplifiers externally. It was advantageous to use a minimum number of amplifiers so that techniques were developed to enable a programmer to set up common transfer functions quickly, [4], [5], using one operational amplifier. With general-purpose computers using fixed, oven-mounted, precision resistors and capacitors, these techniques are almost never used. Any transfer function set up with one amplifier has an equivalent representation using integrators, summers, and inverters (Chapter 10). In view of this, we shall not consider complex input and feedback impedances.

There is one exception to the above generalization. A circuit to differentiate can only be approximated with integrators whereas a true *differentiator* requires an input capacitor and feedback resistor as shown in Fig. 2.13. Because differentiators are rarely needed in practice, general-purpose computers do not provide fixed capacitor inputs to operational amplifiers. When needed, external capacitors must be patched into the summing junction (SJ). Most computers

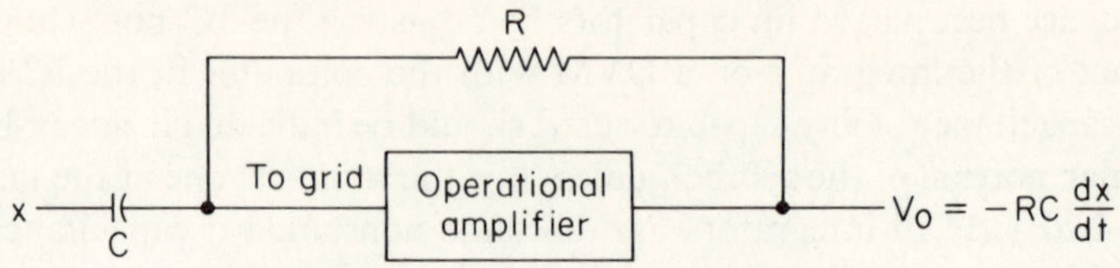

FIG. 2.13 True differentiator.

provide oven-mounted precision capacitors (and resistors) for this purpose in order to preserve accuracy. Both inputs and outputs of such capacitors are available on the patch panel. The circuit of Fig. 2.13 is not without its difficulties, so that modifications are actually required for successful differentiation. These modifications are discussed in Chapter 10.

Theoretically, inductors could be used as elements in an impedance network, but since accurate inductors are difficult to obtain, this is almost never done.

2.3 Multipliers

There are very many ways to perform analog multiplication ([6], [7], [9]), all of which have both accuracy and bandwidth limitations. The two most common methods used in general-purpose computers are *servo* multiplication and *quarter-square*, *biased-diode* multiplication. Both methods are capable of good absolute accuracy (0.01 to 0.1 per cent of full scale) but differ from each other appreciably in bandwidth limitation. A servo multiplier is limited to full-scale frequencies of no more than 10 Hz, whereas the quarter-square multiplier (or electronic multiplier) operates well with signals having frequencies of 1000 Hz and even greater. The speed advantage of the electronic multiplier is somewhat offset by errors occurring with near zero inputs. In some applications when one factor is zero, it is important to obtain a zero product. This can be done with servo multipliers, but not electronic ones. For most applications, however, electronic multiplication is adequate and preferred because of the high frequencies allowable.

2.3.1 Servo multiplier

A servo multiplier is an electromechanical feedback system that controls the positions of the wiper arms of a set of ganged potentiometers mounted in cup-shaped containers called *cups*. A servo multiplier is shown schematically in Fig. 2.14 in a normal connection. The difference between the output of the *F*-cup (follow-up cup of feedback potentiometer) and the input x_1 is amplified and applied to a servo motor. The motor turns the wiper of the *F*-cup in a direction to reduce the input error voltage. Thus, the error voltage is nulled automatically. Since the wiper arm of the multiplying cup is similarly positioned,

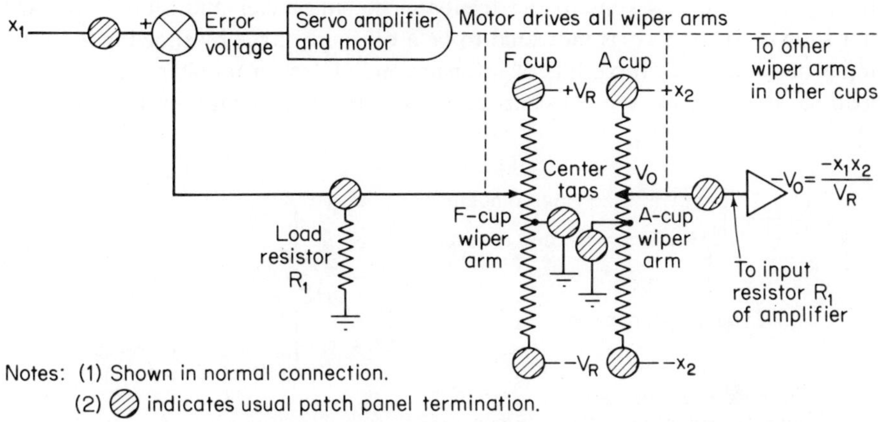

FIG. 2.14 Servo multiplier schematic.

its output voltage is equal to x_1x_2/V_R provided the potentiometers are linear and the cups are aligned properly. By grounding the center taps of both pots, errors due to nonlinearities are minimized when x_1 (the driving voltage) is zero. The alignment of a multiplying cup is made with $x_1 = 0$; its position is adjusted until the output is zero in order to minimize per cent errors when x_1 is small. When $x_1 = 0$, some error is unavoidable because of backlash. If $x_2 = 0$, the output is zero regardless of the F-cup position—an important consideration in automatic nulling problems. The product voltage V_0 can therefore be expressed as

$$V_0 = \frac{x_2(x_1 + \epsilon)}{V_R} \tag{2.17}$$

where ϵ represents an error due to nonlinearity and misalignment.

2.3.2 Loading error and compensation of servo multipliers

If the output is used as an input to an operational amplifier, a loading error results just as with a normal potentiometer [Eq. (2.5)]. To compensate for loading error, it is customary to load the feedback potentiometer and multiplying cup equally. Since the same positioning error now appears for both wiper arms, the voltage error cancels so that the output is exact. In practice, nonprecision external load resistors may be used. One per cent load resistors produce errors well below the error specifications due to backlash and nonlinearity.

2.3.3 Working diagram for servo multipliers

We will use the Electronic Associates, Inc., symbols for servo multipliers in working diagrams in this book because they are less confusing. These are shown in Fig. 2.15. In this system, the feedback potentiometer and servomotor are not shown. Instead, x_1 is indicated to be a driving voltage by showing it entering a box that gives three bits of information: (1) servo number, (2) value of load resistor, and (3) type of connection. For the multiplying cup, we use the

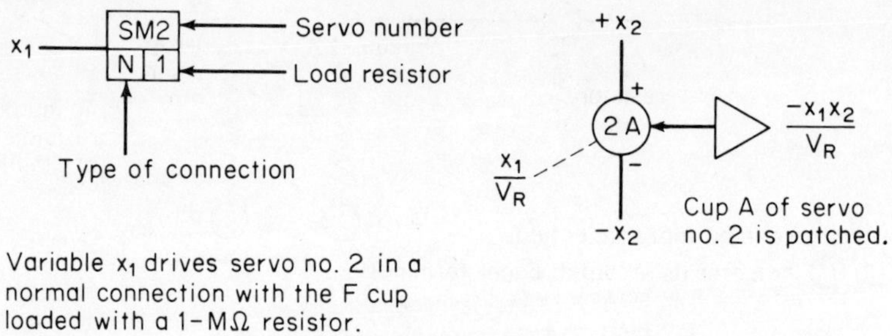

FIG. 2.15 Working diagram symbols for a servo multiplier.

symbol of an ungrounded pot in working diagrams with the designation showing the servo number and cup letter. There are normally several cups labeled A, B, C, etc. in order to permit forming several products having x_1 as a factor. No confusion can result from this notation, particularly since the dashed line indicates that the pot is automatically set to the value of x_1/V_R. Its advantage is that the multiplying cups need not be located close to the driving voltage source in the diagram.

2.3.4 Other connections

There are two other standard types of connections besides the normal one shown in Fig. 2.15. In a plus connection (Fig. 2.16), the driving variable is known to be positive only. To improve accuracy, the *lower* (LO) ends of both pots are grounded with the center taps left floating. The minus connection (Fig. 2.17) requires grounding the *upper* (HI) ends of both pots when x_1 is known to be negative. The lower end of the feedback pot is patched with $-V_R$. (The HI end of the feedback pot must always have greater potential than the LO end; otherwise the servo system is unstable.)

In all cases, interchanging x_2 with $-x_2$ simply inverts the product sign. In Fig. 2.17, the output of the C-cup is fed into a 10 gain (0.1-MΩ input resistor) to illustrate loading with 0.1 MΩ. In practice, this can only be done if the product $x_1 x_2/V_R < 0.1 V_R$, or if other inputs are patched to the summer to keep its output below the reference voltage.

When more than one multiplying cup is used, each must be fed into the same load to compensate for the loading error on the follow-up cup. Figure 2.18 shows three cups used in a plus connection with each cup and the follow-up cup loaded identically. Note that the symbol S (for Special) can be used to

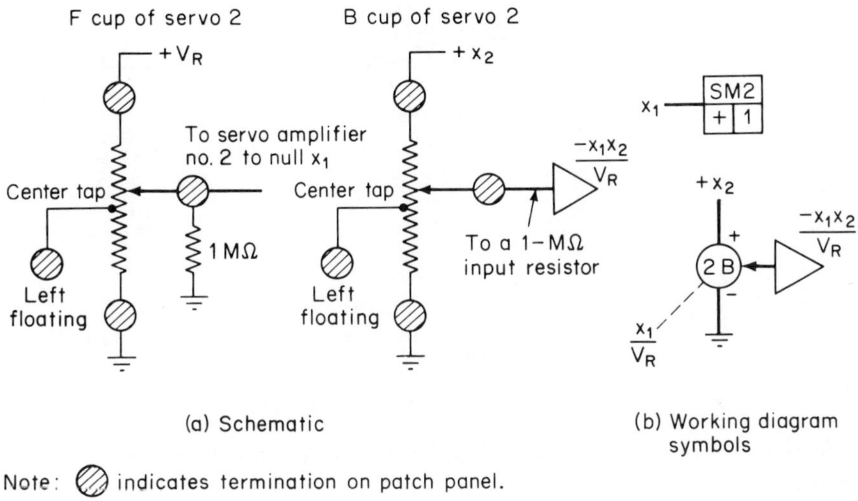

FIG. 2.16 Plus connection when $x_1 \geq 0$.

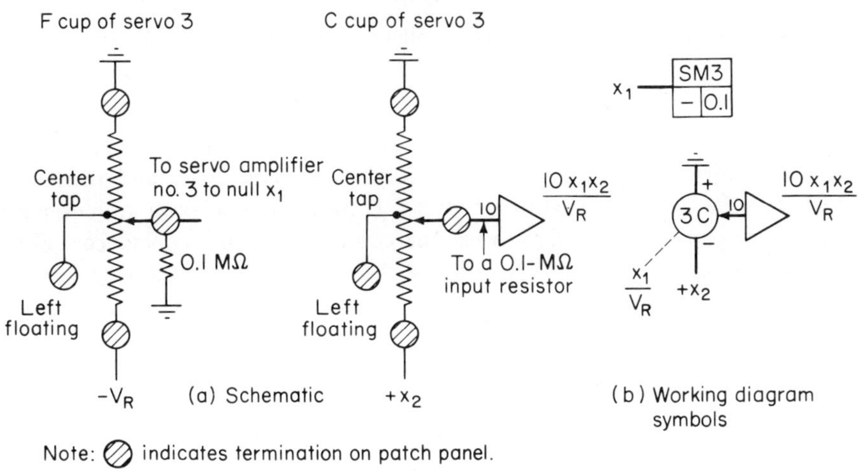

FIG. 2.17 Minus connection when $x_1 \leq 0$.

show a parallel arrangement of resistors instead of calculating the actual load of $\frac{1}{13}$MΩ. It then becomes easier to load the other outputs using the standard 1-MΩ and 0.1-MΩ resistors available.

When no load is needed, the symbol ∞ is used. It is also possible to arrange circuitry to require no load (see Chapter 8). Much important information concerning the design of any computer, its maintenance, operation, limitations, etc., is contained in the manuals supplied by the manufacturer. For proper operation, gain and damping controls must usually be manually adjusted; overload

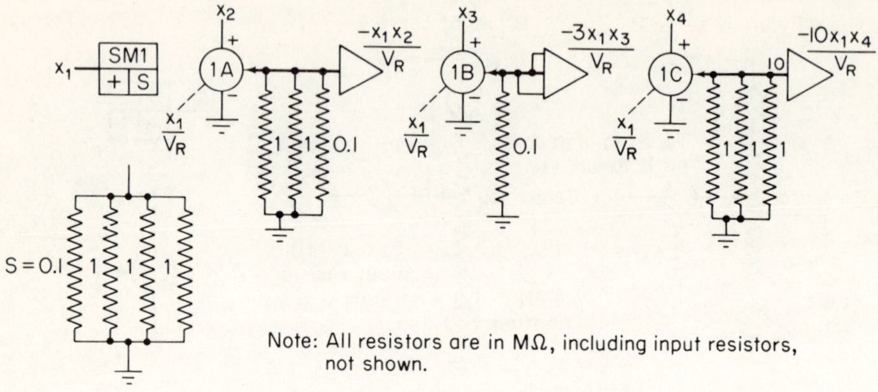

FIG. 2.18 Arrangement for varying input loads.

indications are often provided; mechanical and/or electrical limit stops are used for protection; etc. A programmer quickly learns these factors through practice with the help of the manuals.

2.3.5 Quarter-square electronic multipliers

This name is derived from an old well-known fundamental identity

$$xy \equiv \tfrac{1}{4}[(x + y)^2 - (x - y)^2]$$

which reduces the multiplication process to one of addition, subtraction, and squaring. To square a voltage, a fixed-function, biased-diode and resistance network of the type discussed in Chapter 8 is used, but the programmer need know nothing about the design to use the equipment with competence. Four networks are usually used, one for each of the possible combinations of signs of $x + y$ and $x - y$ ($++$, $--$, $-+$, and $+-$), thus resulting in four-quadrant multiplication. Each network produces a voltage only approximately proportional to the square of the input voltage. The functional relationship is actually

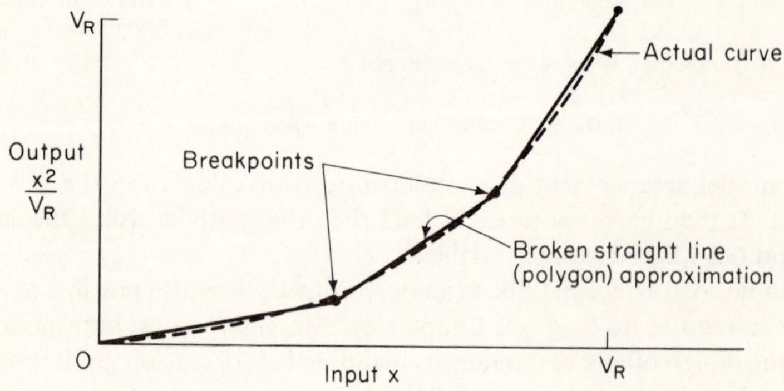

FIG. 2.19 Piecewise linear approximation to a squaring function.

a piecewise linear approximation, i. e., a set of connected straight line segments of the type shown in Fig. 2.19. This results in a product V_0 with an error function ϵ,

$$V_0 = xy + \epsilon(x, y) \tag{2.18}$$

Much work has been done to determine the number of segments N required and the breakpoint locations to minimize the error, notably by Fisher [8]. He considers minimizing mean-square-error integrals of the type

$$\bar{E}^2 = \frac{\iint \rho^2(x, y)\epsilon^2(x, y)\,dx\,dy}{\iint \rho^2(x, y)\,dx\,dy}$$

where ρ is a weighting factor. One conclusion of this investigation is that we cannot optimize the design to minimize the per cent error [using $\rho = 1/(xy)$] because the required integrals diverge. This is often the cause of a large percentage of error in computation. A programmer must be careful about using the electronic multiplier when both inputs are near zero.

Despite these difficulties near zero, the high bandwidth makes electronic multiplication very attractive. It is virtually mandatory for high-speed computation, when both input variables can be changing rapidly.

Manufacturer's designs vary greatly as to the required patching to obtain a product. Ideally, the patching would require the two inputs x and y, and yield the product xy/V_R, thus requiring just three terminations on the patch panel. In practice, special arrangements for division, squaring, and sometimes square-rooting are often provided. Sometimes the chassis containing the squaring networks is interchangeable with a servo multiplier chassis, but output amplifiers must be patched in order to obtain the product. Because of these variations, illustrations using electronic multipliers will be drawn with the standard multiplying symbol (Fig. 1.3).

The *class* of an electronic multiplier refers to the number of external amplifiers required to produce a product. Our symbol for the electronic multiplier, therefore, assumes it is of class 0.

2.4 Dividers

No divider unit with sufficient accuracy has been developed to satisfy the requirements of analog computation. To perform division, a product is used as a feedback path into a high-gain amplifier (Fig. 2.20). From (2.15)

$$\frac{yV_0}{V_R} - x = 0$$

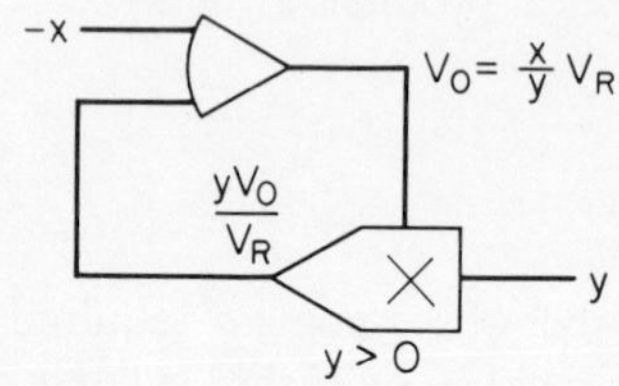

FIG. 2.20 Standard division circuit.

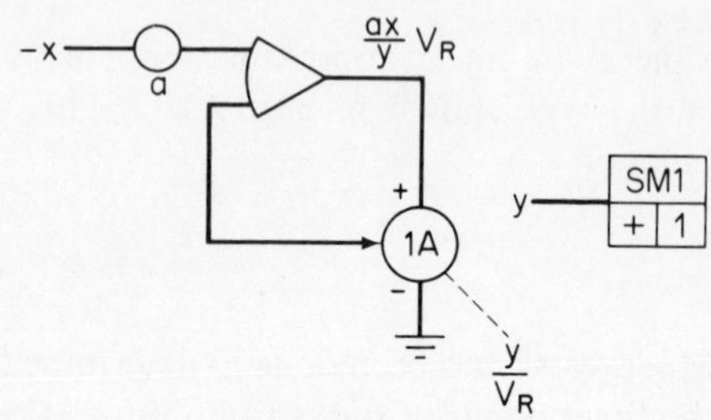

FIG. 2.21 Division circuit using a servo multiplier showing attenuation.

from which $V_0 = V_R(x/y)$. This is actually an implicit function computation (Chapter 9), given here to complete the circuits necessary to perform the basic arithmetic operations. As discussed in Chapter 9, it is necessary that $y > 0$ for stability. Because of amplifier restrictions, the programmer must also arrange

$$x \leq y \tag{2.19}$$

If (2.19) does not hold, it is necessary to attenuate the voltage x, as shown in Fig. 2.21, where a circuit using a servo multiplier is also shown. The factor a must then be accounted for by scaling other equations (Chapter 4).

PROBLEMS

2-1. Assume $R_P/R_L = 0.03$. Draw a loading correction curve $\epsilon(a^*)$ from (2.5). Also for $R_P/R_L = 0.3$.

2-2. Derive an expression for $\epsilon(a^*)$ in terms of a using (2.2) and (2.4). Because of its simplicity, many prefer to work with this formula rather than (2.5). The setting a, however, must then be determined by *iteration* (usually one is sufficient), because a is unknown initially.

2-3. Using the same assumptions for which (2.12) was derived, prove (2.13).

2-4. Given the voltages A, B, C, and D, draw a computer diagram to obtain the voltage $0.7A - 0.2B + C - D + 32$, assuming $V_R = 100$ V.

2-5. Given the voltages $-A$, $-B$, and $-C$, draw a computer diagram to obtain $(A+B+10C)/10$ using a high-gain amplifier.

2-6. Given the voltages $-A$, $-B$, and $-C$, draw a computer diagram to obtain the voltage $32.17 + \int_0^\tau (9.731A + 0.7812B + 10.8391C)\,d\tau$.

2-7. Using (2.13), determine the differential equation governing the output V_0 of the circuit in Fig. 2.22.

2-8. Let $V \geq 0$ drive servo number 5. Given the voltages $-X$ and $-Y$, draw a computer diagram showing the simultaneous products XV/V_R and $10\,YV/V_R$ accounting for load.

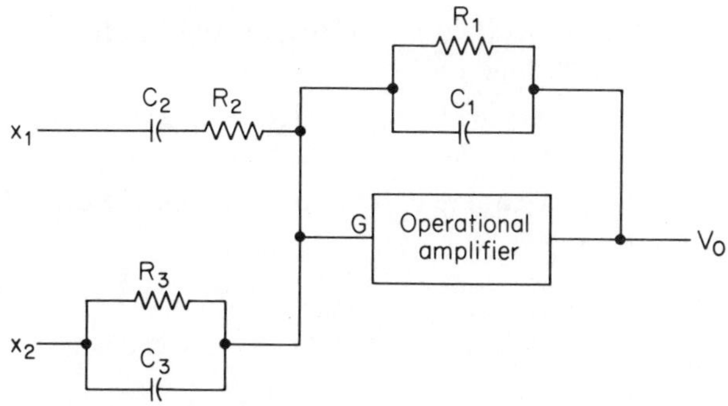

FIG. 2.22 Problem 2-7.

2-9. Let Z drive servo number 6. Draw a computer diagram showing the function Z^3/V_R^2.

2-10. Given the voltage $V \geq 0$ and the voltages A, B, and C, draw a computer diagram to generate the voltage $AV^2/V_R^2 + BV/V_R + C$, using a servo multiplier.

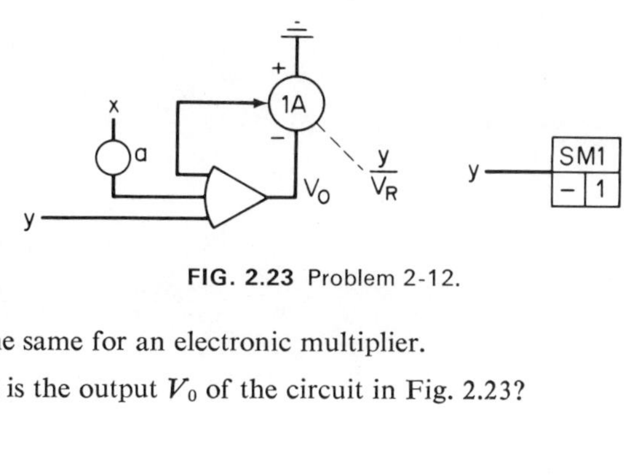

FIG. 2.23 Problem 2-12.

2-11. Do the same for an electronic multiplier.

2-12. What is the output V_0 of the circuit in Fig. 2.23?

REFERENCES

[1] Howe, R. M., *Design Fundamentals of Analog Computer Components* (Princeton, N. J.: Van Nostrand, 1961).

[2] Korn, G. A., and T. M. Korn, *Electronic Analog and Hybrid Computers* (New York: McGraw-Hill Book Co., 1964).

[3] Fifer, S., *Analogue Computation*, vol. 2 (New York: McGraw-Hill Book Co., 1961).

[4] Heizer, L. E., and S. J. Abraham, "Transfer Function Simulation by Means of Amplifiers and Potentiometers," *Journal ACM*, vol. 3, July 1956.

[5] Bridgman, A., and R. Brennan, "Simulation of Transfer Functions Using Only One Operational Amplifier," *IRE Wescon Convention Record*, 1957.

[6] Maslov, A. A., "Survey and Classification of Multiplying Devices," *Automatika i Telemekhanika*, vol. 21, Oct. 1960.

[7] Celinski, O., and I. H. Rimawi, "Classification of Analog Multipliers," *Instr. and Control Systems*, vol. 37, June 1964.

[8] Fisher, M. E., "The Optimum Design of Quarter-Square Multipliers with Segmented Characteristics, *J. Sci. Instr.*, vol. 34, Aug. 1957.

[9] Davis, S. A., "31 Ways to Multiply," *Control Eng.*, vol. 1, Nov. 1954.

Chapter 3

PROBLEM PREPARATION I: OBTAINING COMPUTER EQUATIONS

In solving a problem on an analog computer, programmers will approach a given problem in different ways. Nevertheless, the same factors require consideration: the required accuracy, equipment availability and limitations, convenience, ease of generating certain functions, simplification, etc. The programmer eventually must:

(1) Obtain a set of *computer* equations, i.e., the equations that will be mechanized on the computer.
(2) Scale the computer equations.
(3) Determine how to generate explicit functions needed.
(4) Draw a working diagram to follow when patching the computer.
(5) Patch the problem on the patchboard.
(6) Check for patching errors and equipment malfunctions.
(7) Plan the computer runs for efficient computer use.
(8) Run the problem.
(9) Estimate errors and decide if they are acceptable. If not, rearrange the program, if possible, to obtain acceptable results.

The above steps are not so clearly separate in practice as they are on paper nor are they always taken in the order listed. Feedback from one step may affect previous steps.

We will concern ourselves only with step (1) in this chapter. We will discuss the process of formulating an appropriate set of computer equations that can be mechanized. Our interest is in obtaining equations that involve the rational operations of arithmetic and integration. The problem of generating other algebraic and transcendental functions, when needed, will be treated later.

3.1 Parametric Technique

There are an infinite number of ways of mechanizing a computer to solve a set of differential equations. A method known as the *parametric technique* for setting up the equations has been developed to help choose an intelligent way. Using this technique, a programmer can more easily choose a proper mechanization while obtaining more insight into computer processes.

The parametric technique originates from attempts to eliminate division ([1]–[3]), often a source of difficulty in any mechanization. In essence, it treats computer time τ as a parametric variable. To motivate the discussion, we consider the elementary equation

$$\frac{dy}{dx} = -y - x \tag{3.1}$$

for solution on the computer. In (3.1), initial conditions (x_0, y_0) are given. To use the Kelvin feedback method (see Chapter 1), τ must be introduced, since all integrations must be performed with respect to τ. Intuition suggests replacing the problem independent variable x by the computer independent variable τ:

$$\frac{dy}{d\tau} = -y - \tau, \qquad \text{IC} = (\tau_0, y_0)$$

But this is of no immediate help, since a voltage numerically equal to $\tau_0 + \tau$ must be generated for feedback. To do this, we could integrate 1 V with an initial condition of τ_0 on the integrator. Excluding any scaling considerations, the computer diagram could appear as in Fig. 3.1.

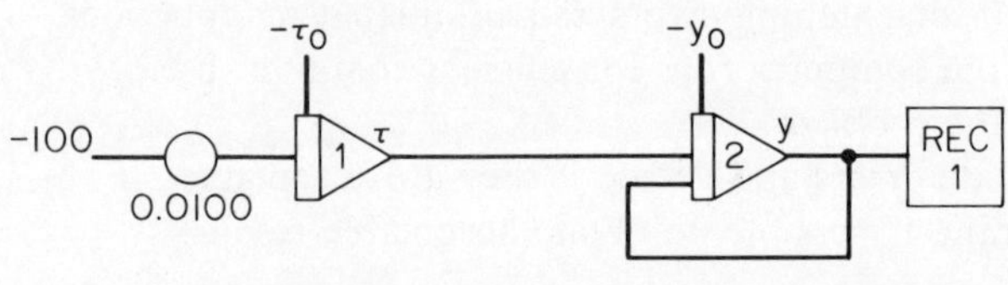

FIG. 3.1 Diagram for solving $dy/d\tau = -y - \tau$.

We should recognize, however, that two first-order equations are being solved on the computer because two integrators are used. There is really no advantage in substituting τ for x. Figure 3.2 shows the same circuit (which therefore means the same problem is being solved) except that the more convenient problem notation is kept. From the diagram, we must be solving

$$\frac{dy}{d\tau} = -y - x, \qquad \text{IC: } y_{(\tau=0)} = y_0 \tag{3.2}$$

$$\frac{dx}{d\tau} = 1, \qquad \text{IC: } x_{(\tau=0)} = x_0 \tag{3.3}$$

It is clear that the computer can solve only *autonomous* equations, i.e., equations of the type

$$\frac{dy_i}{d\tau} = f_i(y_1, y_2, \ldots, y_n)$$

for $i = 1, 2, \ldots, n$, where computer time does not appear explicitly in the right-hand-side expressions. Only *functions* of computer time can so appear, and these may be considered to be problem dependent variables.

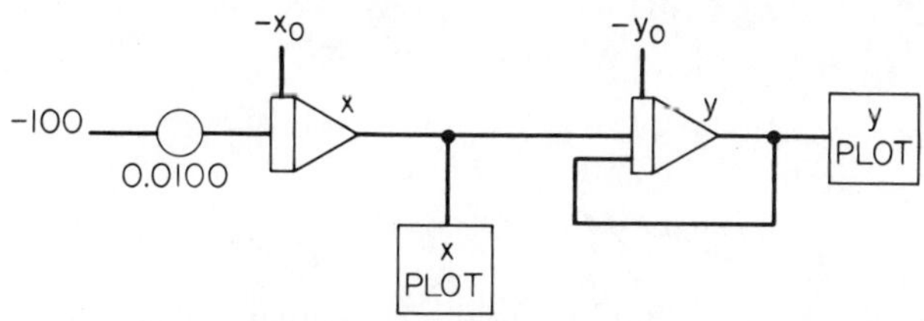

FIG. 3.2 Solution of (3.2) and (3.3).

We shall hereafter use the Newtonian symbolism of placing a dot above a variable when speaking of its derivative with respect to computer time τ. Also, the subscript 0 will refer to the initial value of the subscripted variable.

The most convenient way of arriving at the equations (3.2) and (3.3) is simply to replace dy/dx by

$$\frac{dy}{dx} = \frac{\dot{y}}{\dot{x}} = -y - x$$

cross-multiply by $\dot{x}$,

$$\dot{y} = (-y - x)\dot{x} \tag{3.4}$$

and *choose*

$$\dot{x} = 1, \qquad \text{IC: } x_0$$

Then

$$\dot{y} = -y - x, \qquad \text{IC: } y_0$$

The programmer must choose the value for $\dot{x} \neq 0$ (from the chain-rule for calculating derivatives). This method has other advantages. If solutions for $x < x_0$ were desired, a good choice of $\dot{x}$ would be

$$\dot{x} = -1, \qquad \text{IC: } x_0$$

Then

$$\dot{y} = y + x, \qquad \text{IC: } y_0$$

Considering x to be computer time involves awkward phrases such as "running the problem backwards," "negative time," or "making time go backwards" whereas it is clear that time goes only forward, but the value of x decreases on the computer. In fact, we are able to introduce the concept of *time scaling* by letting

$$\dot{x} = k, \qquad \text{IC: } x_0$$

from which

$$\dot{y} = k(-x - y), \qquad \text{IC: } y_0$$

for any constant $k \neq 0$. But it should be equally clear that any function $g(x, y) \neq 0$ would be suitable for $\dot{x}$: The equations

$$\dot{x} = kg(x, y), \qquad \text{IC: } x_0$$
$$\dot{y} = -k(x + y)g(x, y), \qquad \text{IC: } y_0$$

also result in solutions of (3.1). Now, however, graphs *must* be obtained with an x-y plotter; a recorder simply gives $x(\tau)$ and $y(\tau)$ which are of no direct interest. For this specific example, the choice of $\dot{x} = k =$ constant results in the simplest possible mechanization, as is directly obvious from inspection of (3.4).

Note that in situations involving infinite slopes or rational functions, this method has great advantages. Consider, for example, obtaining graphs of solutions of

$$\frac{dy}{dx} = \frac{ax + by + c}{ex + fy + g} \tag{3.5}$$

with initial conditions (x_0, y_0). From (3.5)

$$\dot{y} = \frac{ax + by + c}{ex + fy + g}\dot{x} = (ax + by + c)\left(\frac{\dot{x}}{ex + fy + g}\right)$$

and it is instantly clear that an excellent mechanization results from the choice

$$\dot{x} = ex + fy + g, \qquad \text{IC: } x_0$$

from which

$$\dot{y} = ax + by + c, \qquad \text{IC: } y_0$$

Not only is the quotient now eliminated (this could become infinite and unscalable), but an important property of the parametric representation of the solution is evident. If $ex_1 + fy_1 + g = 0$ for some point (x_1, y_1) where $ax_1 + by_1 + c \neq 0$, the resulting singularity occurring in (3.5) is eliminated and the graph of the solution is readily continued through the point (x_1, y_1). It becomes possible to obtain graphs of solutions to differential equations using bounded functions only.

The process of replacing first-order problem equations of the form

$$\frac{dy_i}{dx} = f_i(x, y_1, y_2, \ldots, y_n) \tag{3.6}$$

for $i = 1, 2, \ldots, n$, by

$$\dot{y}_i = f_i(x, y_1, y_2, \ldots, y_n)\dot{x}$$

and selecting a convenient $\dot{x}$ to obtain *computer* equations is called the *parametric technique*. Computer time is used strictly as a parameter with which to change computer variables according to the original set of differential equations.

Before leaving this aspect of problem preparation, a remark on the choice of $\dot{x}$ is appropriate. In many simulations investigated on the computer, physical time t is the independent variable, with many equations to mechanize. In such a case, the choice of $\dot{t}$ = constant is appropriate, for many multiplications are required otherwise. For smaller systems, however, a few extra multiplications may offer a great convenience in not having to worry about the scaling of certain quotients. Although some programmers may differ on the proper way to mechanize a given set of equations, a clear choice is indicated most of the time.

3.2 Obtaining First-Order Equations

In order to employ the parametric technique, we attempt to reduce all ordinary differential equations to the canonical form of (3.6)

$$\frac{dy_i}{dx} = f_i(x, y_1, y_2, \ldots, y_n)$$

for $i = 1, 2, \ldots, n$, where each y_i is then called a *state* variable. There are other reasons for doing this. Scaling techniques (Chapter 4) are simplified. We are also quickly able to see which variables are to be generated by integration, and which are not, for there is always a one-to-one correspondence between the final number of equations and the number of integrators to be used. Finally, other scientific disciplines employ canonical form (3.6), so we provide a simple means for these to use an analog computer. For example, control engineers consistently use form (3.6) because it is essentially a vector notation. (Chapter 19 will discuss the maximum principle which is most easily stated with first-order equations.) Digital computer programs to solve differential equations are invariably written for first-order equations. The transition to and from digital computer programs is smoother. Note that the subscript notation is not particularly important for our applications: A system such as

$$\frac{dy}{dx} = f(x, y, z), \qquad y(x_0) = y_0$$

$$\frac{dz}{dx} = g(x, y, z), \qquad z(x_0) = z_0$$

is considered to be in form (3.6) with $y = y_1$ and $z = y_2$. The emphasis is strictly on the first-order form.

Under most circumstances, obtaining equations in first-order form is very straightforward. Let us examine typical situations which occur in practice.

A single equation of nth order, which can be solved for the highest derivative,

$$\frac{d^n y}{dx^n} = f\left(x, y, \frac{dy}{dx}, \ldots, \frac{d^{n-1}y}{dx^{n-1}}\right) \tag{3.7}$$

can be brought to first-order form with the transformation

$$\begin{aligned} y = y_0, \qquad & \frac{dy}{dx} = y_1 \\ \frac{d^2y}{dx^2} = y_2, \qquad \ldots, \qquad & \frac{d^{n-1}y}{dx^{n-1}} = y_{n-1} \end{aligned} \tag{3.8}$$

The n first-order equations are found to be

$$\begin{aligned} &\frac{dy_0}{dx} = y_1, \qquad \frac{dy_1}{dx} = y_2, \qquad \ldots, \\ &\frac{dy_{n-1}}{dx} = f(x, y_0, y_1, \ldots, y_{n-1}) \end{aligned} \tag{3.9}$$

with appropriate initial conditions. It should be emphasized that (3.8) is not the only transformation to bring (3.7) into first-order form, although it is certainly the simplest, the most general, and the one most frequently used. Other ways will be discussed in connection with specific examples.

Simultaneous systems of high-order equations, each of which are solvable for the highest derivative, can be treated similarly.

Example 3.1. Determine a set of first-order equations for the problem

$$\frac{d^2y}{dt^2} = -g + A(y)\left(\frac{dy}{dt}\right)^2$$

which describes approximately the height $y(t)$ of a falling ball under the influence of gravity g, and in an atmosphere which resists the fall with a force proportional to $A(y)(dy/dt)^2$. Here $A(y)$ is a function of y, due primarily to density variations of the atmosphere with height, and t is *not* computer time, but the problem independent variable. We let

$$\frac{dy}{dt} = v$$

so that

$$\frac{dv}{dt} = -g + A(y)v^2$$

Note that the choice of a symbol can remind us of the role of the variable. In this case, the symbol v was chosen to remind us that dy/dt is a velocity.

Example 3.2. Reduce to first-order form

$$m_1\frac{d^2x_1}{dt^2} = -k_1x_1 + F(t) + k_2(x_2 - x_1) + C_2\left(\frac{dx_2}{dt} - \frac{dx_1}{dt}\right) - C_1\frac{dx_1}{dt}$$

$$m_2\frac{d^2x_2}{dt^2} = -k_2(x_2 - x_1) - C_2\left(\frac{dx_2}{dt} - \frac{dx_1}{dt}\right)$$

which are derivable from Newton's laws and describe the coupled mass-spring-damper system shown in Fig. 3.3. Let

$$\frac{dx_1}{dt} = v_1, \qquad \frac{dx_2}{dt} = v_2$$

Then

$$\frac{dv_1}{dt} = -k_1x_1 + F(t) + k_2(x_2 - x_1) + C_2(v_2 - v_1) - C_1v_1$$

$$\frac{dv_2}{dt} = -k_2(x_2 - x_1) - C_2(v_2 - v_1)$$

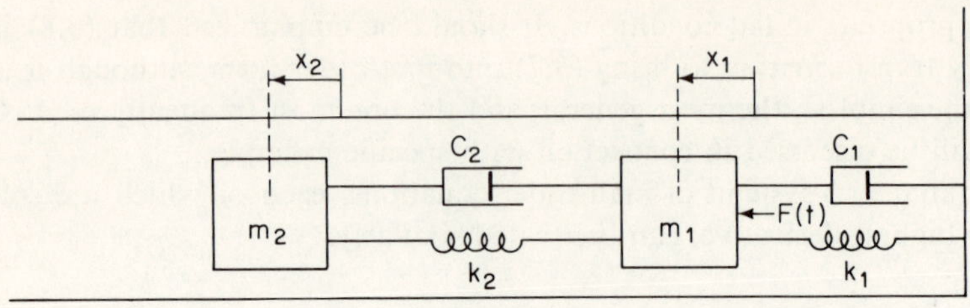

Note: $x_i = 0$ when both springs are in equilibrium.

FIG. 3.3 Coupled mass-spring-damper system.

This is a fourth-order system shown without any initial conditions. Simultaneous systems of equations can thus be easily treated when all highest-order derivatives can be found explicitly.

Another ordinary differential equation that can be put into first-order form describes the steady-state sag of a beam. It is a boundary-value problem (the boundary conditions depend on how the beam is hinged) taking the form

$$\frac{d^2}{dx^2}\left[A(x)\frac{d^2 y}{dx^2}\right] = Q(x)$$

Rather than differentiate the bracketed expression, we define

$$y_3 = \frac{d}{dx}\left[A(x)\frac{d^2 y}{dx^2}\right]$$

from which

$$\frac{dy_3}{dx} = Q(x)$$

Continuing, we define

$$y_2 = A(x)\frac{d^2 y}{dx^2}, \qquad y_1 = \frac{dy}{dx}$$

so that

$$\frac{dy_2}{dx} = y_3, \qquad \frac{dy_1}{dx} = \frac{y_2}{A(x)}, \qquad \frac{dy}{dx} = y_1$$

represent the rest of the system.

The programmer must manipulate some systems of equations more to get the canonical form. Honnell and Horn [4] show, for example, that

$$\frac{d^2x}{dt^2} + \frac{d^2y}{dt^2} - \frac{d^2z}{dt^2} + \frac{dy}{dt} + x + y = 0 \tag{3.10}$$

$$\frac{dy}{dt} + z = 0 \tag{3.11}$$

$$2\frac{dx}{dt} + \frac{dz}{dt} + z = 0 \tag{3.12}$$

define a fourth-order system, requiring four initial conditions for unique solutions. One method of obtaining a canonical form is to differentiate (3.11) and (3.12) and substitute directly into (3.10). Thus

$$\frac{d^2y}{dt^2} = -\frac{dz}{dt} \tag{3.13}$$

$$\frac{d^2z}{dt^2} = -2\frac{d^2x}{dt^2} - \frac{dz}{dt} \tag{3.14}$$

results in

$$3\frac{d^2x}{dt^2} + \frac{dy}{dt} + x + y = 0$$

By letting

$$\frac{dx}{dt} = x_1$$

we have

$$\frac{dx_1}{dt} = \frac{1}{3}(-x - y + z), \qquad \frac{dy}{dt} = -z, \qquad \frac{dz}{dt} = -z - 2x_1$$

Note that we could have solved for d^2y/dt^2, d^2x/dt^2, and d^2z/dt^2 from (3.13), (3.14), and (3.10) and obtained a sixth-order system. However, it is possible to introduce errors [4] in such a procedure, unless the lower-order equations (3.11) and (3.12) are treated as *constraints* (Chapter 13).

If much algebraic manipulation is required, such as can be involved in the solution of Lagrange's equations (Chapter 12), it may not be worthwhile to get a canonical form. In such cases, we may use *implicit* function techniques (Chapter 9) which involve the solution of algebraic equations (Chapter 11). For example, systems such as

$$\sum_{j=1}^{n} a_{i,j}(x, y_1, y_2, \ldots, y_n)\frac{dy_j}{dx} = f_i(x, y_1, y_2, \ldots, y_n)$$

for $i = 1, 2, \ldots, n$, can theoretically be explicitly solved for dy_i/dx in terms of $x, y_1, y_2, \ldots, y_n$, but it can be tedious to do so. Computer time τ can still be introduced as required by the parametric technique, but it may be quite difficult to make an intelligent choice of $\dot{x}$.

There is also a large class of problems which cannot be solved explicity for the highest derivatives. For example, the general first-order equation

$$f\left(x, y, \frac{dy}{dx}\right) = 0 \tag{3.15}$$

sometimes cannot be solved for dy/dx, although there could exist a (possibly multivalued) functional relationship

$$\frac{dy}{dx} = \phi(x, y)$$

Here, we are forced to use other methods for obtaining solutions on the computer. One such method is to raise the order of the equation by differentiating (3.15) and then solve for the highest derivative. Thus, with $y' = dy/dx$,

$$\frac{\partial f}{\partial x} + \frac{\partial f}{\partial y}\frac{dy}{dx} + \frac{\partial f}{\partial (y')}\frac{d^2 y}{dx^2} = 0$$

results in

$$\frac{d^2 y}{dx^2} = -\frac{\left(\dfrac{\partial f}{\partial x} + \dfrac{\partial f}{\partial y}\dfrac{dy}{dx}\right)}{\dfrac{\partial f}{\partial (y')}}$$

This will prove to be a powerful technique in connection with methods of satisfying constraints (Chapter 13). Another method will involve solving the equation by implicit methods (Chapter 9).

3.3 Avoiding Explicit Unbounded Functions

It is extremely important to avoid generating functions which can become unbounded during a solution process. If a function explicitly involved in the expression of the differential equation can be expected to become unbounded, it can be replaced by the quotient of two bounded functions. For example, the function $\ln|x|$ is unbounded for a domain including $x = 0$, but it can be replaced by $x \ln|x|/x$ because $x \ln|x|$ is bounded in an interval containing zero. The parametric technique will permit the choice of $\dot{x}$ to avoid generating the actual quotient required, and the only functions needed will be bounded. All elementary unbounded trigonometric and hyperbolic functions can be treated in this way. Other functions can usually be so treated too.

Example 3.3. Find a suitable first-order equation for solving

$$\frac{dy}{dx} = \frac{\tan y + \ln x}{x^2 + 1}$$

for $0 < x \leq 1$ with initial conditions: at $x = 1$, $y = \pi/2$. Replace $\tan y$ by $(\sin y)/(\cos y)$, and $\ln x$ by $(x \ln x)/x$ and clear fractions:

$$\frac{dy}{dx} = \frac{x \sin y + (\cos y)(x \ln x)}{x(x^2 + 1) \cos y}$$

Actual methods of generating these functions will be discussed in subsequent chapters. If x could be expected not to approach zero, then the function $\ln x$ could be generated directly.

3.4 The Use of Transformations

Transformations of variables are so basic in all of mathematics, that a programmer should be constantly aware of the possible need or desire for such changes in the basic equations. In analog computer practice, transformations on original problem equations have been used primarily to simplify the mechanization ordinarily required for solution. The motivation of a programmer is therefore similar to that of a mathematician, who uses transformations to simplify the path to an analytical result.

Let us consider first the question of unbounded high-order derivatives which may occur, even when low-order derivatives are bounded. Equations such as these cannot be programmed effectively using (3.8). For example, consider programming the equation

$$\frac{d^2 y}{dx^2} = -\frac{\left[1 + \left(\frac{dy}{dx}\right)^2\right]}{y} \tag{3.16}$$

With the transformation (3.8), only partial solutions are obtained because dy/dx will eventually become infinite. For, if we let

$$\frac{dy}{dx} = y_1$$

so that

$$\frac{dy_1}{dx} = -\frac{(1 + y_1^2)}{y}$$

then the computer can solve the equations

$$\dot{y}_1 = -(1 + y_1^2), \qquad \dot{y} = y_1 y, \qquad \dot{x} = y$$

There is now no direct worry about dividing by y, but we see that y is a factor in the expressions for $\dot{y}$ and $\dot{x}$, while y_1 continually decreases because its derivative is negative. If y approaches zero, we can expect to be forced to terminate the computation before y actually becomes zero. To continue the solution, it would be necessary to transform the original equation into expressions for d^2x/dy^2 and dx/dy, and program a new set of equations using as initial conditions the final conditions of the old set. Thus at least two sets of equations would need to be programmed.

For this example, however, the simple transformation

$$u = y\frac{dy}{dx}$$

reduces the original equation to the point where it can be solved analytically. For, by differentiating and substituting (3.16)

$$\frac{du}{dx} = \left(\frac{dy}{dx}\right)^2 + y\frac{d^2y}{dx^2} = -1$$

so that

$$u = -x + C_1$$

From the definition of u,

$$\frac{dy}{dx} = \frac{u}{y} = \frac{-x + C_1}{y}$$

which is separable and by integration results in

$$(x - C_1)^2 + y^2 = C_2^2$$

If we solve the transformed equations on the computer, the bounded function u replaces the unbounded dy/dx as one of the functions generated:

$$\frac{dy}{dx} = \frac{u}{y}, \qquad \text{IC: } y_0$$

$$\frac{du}{dx} = -1, \qquad \text{IC: } u_0 = y_0\left(\frac{dy}{dx}\right)_0$$

It is always possible to generate on the computer some function related to the derivative dy/dx rather than the derivative itself. Transformations achieving simplicity are not generally this obvious, nor will they always result in an analytical integration of the given differential equation. Certainly an equation of the form

$$\frac{d^2y}{dx^2} = -\frac{\left[f(x, y) + \left(\frac{dy}{dx}\right)^2\right]}{y}$$

can be simplified to

$$\frac{du}{dx} = -f(x, y), \qquad \frac{dy}{dx} = \frac{u}{y}$$

with the transformation shown, but only experience and ingenuity on the part of the programmer can bring good results, if possible, for a specific problem.

A transformation which can always eliminate an unbounded dy/dx can be obtained by defining *two* functions, p and q, such that

$$\frac{dy}{dx} = \frac{p(x)}{q(x)} \tag{3.17}$$

with the constraint

$$p^2 + q^2 = 1 \tag{3.18}$$

By differentiating (3.17) and (3.18) with respect to x, and solving for both dp/dx and dq/dx, we may show that

$$\frac{dp}{dx} = q^3\frac{d^2y}{dx^2}, \qquad \frac{dq}{dx} = -pq^2\frac{d^2y}{dx^2}$$

An equation of the form

$$\frac{d^2y}{dx^2} = f\left(x, y, \frac{dy}{dx}\right) \tag{3.19}$$

can then be transformed to the system

$$\begin{aligned} \frac{dy}{dx} &= \frac{p}{q} \\ \frac{dp}{dx} &= q^3 f\left(x, y, \frac{p}{q}\right) \\ \frac{dq}{dx} &= -pq^2 f\left(x, y, \frac{p}{q}\right) \end{aligned} \tag{3.20}$$

which will allow computing the solution at least until an infinite slope is reached ($q = 0$). This transformation is equivalent to the elementary transformation

$$z = \arctan\left(\frac{dy}{dx}\right)$$

with $p = \sin z$ and $q = \cos z$, which simplifies expressions of the form

$$\left[1 + \left(\frac{dy}{dx}\right)^2\right]^{n/2}$$

for odd n.

Example 3.4. Apply (3.17) and (3.18) to simplify the equation

$$\frac{d^2y}{dx^2} = g\left(x, y, \frac{dy}{dx}\right)\left[1 + \left(\frac{dy}{dx}\right)^2\right]^{3/2}$$

and obtain a system in canonical form. We have

$$\left[1 + \left(\frac{dy}{dx}\right)^2\right]^{3/2} = \left[1 + \left(\frac{p}{q}\right)^2\right]^{3/2} = \left(\frac{q^2 + p^2}{q^2}\right)^{3/2} = \frac{1}{q^3}$$

Using (3.20)

$$\frac{dy}{dx} = \frac{p}{q}, \qquad \frac{dp}{dx} = g\left(x, y, \frac{p}{q}\right), \qquad \frac{dq}{dx} = -\frac{p}{q}g\left(x, y, \frac{p}{q}\right)$$

If a function y_i has a lower limit, $y_i \geq k$, and it is part of a system of computer equations,

$$\dot{y}_i = f(x, y_1, y_2, \ldots, y_i, \ldots, y_n) \tag{3.21}$$

then the transformation

$$z = \frac{1}{y_i - k + 1}$$

is a new function with $0 \leq z \leq 1$. Also the computer equation becomes

$$\dot{z} = -z^2 f\left(x, y_1, y_2, \ldots, \left[\frac{1}{z} + k - 1\right], \ldots, y_n\right)$$

This may require more equipment than (3.21), but can result in more convenient mechanizations.

Tomovic and Karplus [5] suggest another idea suitable for n linear equations of the form

$$\frac{dx_i}{dt} = \sum_{j=1}^{n} a_{i,j}x_j + b_i(t) \tag{3.22}$$

for constant $a_{i,j}$, which may contain divergent solutions as $t \to \infty$. If we desire to investigate solutions of (3.22) for large t, we may use the transformation

$$y_i = x_i e^{-\lambda t}$$

for some $\lambda > 0$. By differentiation and substitution, the system

$$\frac{dy_i}{dt} = \sum_{j=1}^{n} a_{i,j} y_j + e^{-\lambda t} b_i(t) - \lambda y_i$$

is obtained. For many functions $b_i(t)$, this new system can be made to contain convergent solutions for a large enough λ.

White [6] describes a transformation useful for saving n multiplications for a system of n equations of the form

$$\frac{dy_i}{dx} = f_i(y_i) g_i(x, y_1, y_2, \ldots, y_n)$$

for $i = 1, 2, \ldots, n$, which arises in connection with various physical systems. By introducing n functions z_i such that

$$\frac{dy_i}{dz_i} = f_i(y_i) \tag{3.23}$$

the equations

$$\frac{dz_i}{dx} = g_i(x, y_1, y_2, \ldots, y_n) \tag{3.24}$$

hold. The functions y_i can now be found by integrating (3.23) which is separable, yielding

$$z_i = \int \frac{dy_i}{f_i(y_i)} = h_i(y_i) \tag{3.25}$$

and generating the inverse functions

$$y_i = h_i^{-1}(z_i) \tag{3.26}$$

Since both (3.25) and (3.26) can be found before (3.24) is mechanized on the computer, the final system solved is

$$\frac{dz_i}{dx} = g_i[x, h_1^{-1}(z_1), h_2^{-1}(z_2), \ldots, h_n^{-1}(z_n)]$$

Thus, n function generators are still required [generating $h_i^{-1}(z_i)$ instead of $f_i(y_i)$], but the original multiplications $f_i g_i$ are not needed. In this case, the transformations conserve equipment.

PROBLEMS

3-1. Reduce the following equation to first-order form, pick a set of computer equations, and draw an unscaled computer diagram to solve them.

$$\frac{d^3y}{dx^3} + 3y^2\frac{d^2y}{dx^2} + 4x\frac{dy}{dx} + x^2 + y^2 = 0$$

3-2. The study of a parametron involves the equation [7]

$$C\frac{d^2(LI)}{dt^2} + \frac{1}{R}\frac{d(LI)}{dt} + I = 0$$

where $L = L_0(1 + 2\Gamma \sin \omega t + \beta I^2)$, and $C, R, L_0, \Gamma, \omega$, and β are constants. Reduce to first-order form, choose a set of computer equations, and draw an unscaled computer diagram assuming $\sin \omega t$ is generated with a function generator.

3-3. Reduce the following fifth-order system to a canonical form:

$$\frac{d^2x}{dt^2} + \frac{d^2y}{dt^2} - \frac{d^2z}{dt^2} + \sin y = 0 \tag{1}$$

$$\frac{d^2x}{dt^2} + \frac{d^2z}{dt^2} + \sin(x + t) = 0 \tag{2}$$

$$\frac{dx}{dt} + \frac{dy}{dt} - xe^z = 0 \tag{3}$$

in two different ways:

(a) By integrating (1) and (2), defining

$$w = \int_{t_1}^{t} \sin y\, dt + C_1, \qquad v = \int_{t_1}^{t} \sin(x + t)\, dt + C_2$$

and obtaining equations for dw/dt, dv/dt, dx/dt, dy/dt, dz/dt.

(b) By differentiating (3) and solving for d^2x/dt^2 and dz^2/dt^2. Which is easier? Assuming the initial conditions are given for $x, y, z, dx/dt$, and dz/dt, indicate the initial conditions for the variables in (a).

3-4. Solve for dx/dt and dy/dt draw an unscaled computer diagram to solve

$$x\frac{dx}{dt} + y\frac{dy}{dt} = a, \qquad -y\frac{dx}{dt} + x\frac{dy}{dt} = b$$

where a and b are constants.

3-5. If

$$\frac{d^2y}{dt^2} = f\left(\frac{dy}{dt}, y\right)$$

then much information can be obtained from the phase-plane trajectories of dy/dt versus y, because this graph is independent of t. Let $dy/dt = v$; then the differential equation of the trajectories is

$$\frac{dv}{dy} = \frac{f(v, y)}{v}$$

Obtain the phase-plane differential equation for the equation

$$\frac{d^2y}{dt^2} = \frac{ay + b(dy/dt)}{1 - y^2},$$

select a set of computer equations, and draw an unscaled computer diagram to obtain the graphs.

3-6. Parametric techniques play an important role in the analytical solution of implicit differential equations. Solve the equation

$$\frac{dy}{dx} + \frac{1}{2} \sin\left(\frac{dy}{dx}\right) = x$$

with $y = y_0$ when $x = x_0$ by introducing the parameter t. [Let $y' = t$ and obtain $y = \int t\left(\frac{dx}{dt}\right) dt$].

3-7. Prepare an unscaled circuit diagram for the solution of

$$\frac{dy}{dx} = \cot y + \sec x$$

assuming all functions needed are obtained from function generators.

3-8. The equation

$$\frac{d^2y}{dx^2} = 2(h - y)\left[1 + \left(\frac{dy}{dx}\right)^2\right]^{3/2} - \frac{\frac{dy}{dx}\left[1 + \left(\frac{dy}{dx}\right)^2\right]}{x}$$

where h is constant and describes the shape of the molten zone of a cylindrical rod is an aid in making transistors. Apply transformation (3.17) and (3.18) by using (3.20) directly. Simplify and choose a reasonable $\dot{x}$. Draw the unscaled circuit diagram for solution on the computer [2].

3-9. Show that the equation

$$\frac{d^2y}{dx^2} = \frac{g(x, y)}{f(x, y)}\left[1 + \left(\frac{dy}{dx}\right)^n\right]^{3/n}$$

can be solved on the computer by the set of computer equations

$$\dot{y} = kpf(x, y)$$

$$\dot{x} = kqf(x, y)$$

$$\dot{p} = kq^{n-1}g(x, y)$$

$$\dot{q} = -kp^{n-1}g(x, y)$$

with the transformation

$$p^n + q^n = 1, \qquad \frac{dy}{dx} = \frac{p}{q}$$

REFERENCES

[1] Fisher, M. E., "Avoiding the Need for Dividing Units in Setting Up Electronic Differential Analysers," *J. Sci. Instr.*, vol. 34, Aug. 1957.

[2] Hausner, A., "Parametric Techniques for Eliminating Division and Treating Singularities in Computer Solutions of Differential Equations," *IRETEC*, vol. 11, Feb. 1962.

[3] Taylor, D. K., "Elimination of Division Circuits by Variable Time Scale," *Electronic Eng.*, vol. 37, May 1965.

[4] Horn, R. E., and P. M. Honnell, "Matrix Programming of Electronic Analog Computers," *Trans AIEE, Part I* (*Comm. and Elect.*), vol. 77, Sept. 1958.

[5] Tomovic, R., and W. J. Karplus, *High-Speed Analog Computers* (New York: John Wiley & Sons, Inc., 1962), p. 28.

[6] White, M. E., "An Analog Computer Technique for Solving a Class of Non-linear Ordinary Differential Equations," *IEEETEC*, vol. 15, April 1966.

[7] Randery, V. K., "Study of a Parametron on an Analog Computer," *IEEETEC*, vol. 13, Oct. 1964.

See also:

[8] Hammond, J. L., Jr., "Application of State Variable Techniques in Analog Computer Programming," *Simulation*, vol. 4, May 1965.

[9] Just, L. C., and N. F. Morehouse, Jr., "The Simulation of Large Transients in Neutron Reactors," *Simulation*, vol. 3, July 1964.

[10] Bingulac, S., *et al.*, "Elimination of Undesired Sensitivity in the Analog Solution of Diffusion Equations," *Proceedings of the 4th Inter. Analogue Comp. Meetings* (Brussels: Presses Académiques Européennes, 1966).

Chapter 4

PROBLEM PREPARATION II: COMPUTER DIAGRAMS AND SCALING

4.1 The Need for Scaling

Analog computing elements are designed to operate satisfactorily within particular voltage and frequency ranges. The voltage limits require amplitude scaling prior to putting a problem on the computer, while the frequency limits require time scaling. The scaling process is very straightforward but is not totally independent of the computer diagram which shows how the functions are to be produced.

A good general rule is to *scale variables so that outputs have a value no greater than the reference voltage V_R in absolute value*. There are minor exceptions—amplifiers and potentiometers can generally be allowed to exceed V_R slightly depending on the load they must drive. However, the importance of this rule is enhanced by the fact that the inputs to some nonlinear equipment (servo multipliers, for example) should never exceed V_R in absolute value.

It is generally easy to transform variables so that all amplifier output voltages V_i are such that $|V_i| \leq V_R$, but there is actually a more stringent requirement governing amplitude scaling. To minimize the problem of noise, drift, offset, etc., *all V_i should also be as large as is practicable*. Consider, for example, integrating a voltage V, obtained by inverting the voltage $-V$, as shown in

FIG. 4.1 Effect of offset.

Fig. 4.1. Because of offset, the voltage is actually $V + \epsilon$, resulting in an error in the desired integral. Note that the error term has a factor of 10. If V were less than $0.1V_R$ volts it would be possible to arrange to integrate ten times V in a one gain, thus reducing errors due to offset and drift by a factor of 10, as shown in Fig. 4.2. Proper scale factors can therefore contribute materially to overall accuracy of results [1].

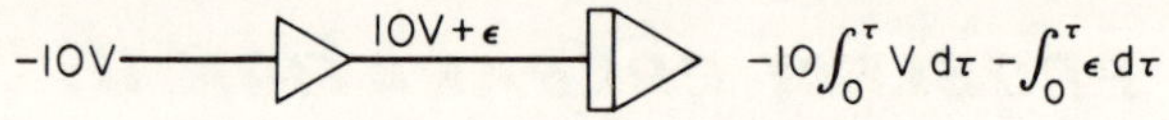

FIG. 4.2 Minimizing the offset and drift effects.

This problem of minimizing errors is much more serious with multipliers, which, after all, only approximate a product. Error specifications of electronic multipliers, for example, are generally given as a per cent of full scale (defined as the difference of the possible output maximum and minimum; $2V_R$ in this case,) so that the actual product voltage P of the two voltages V_1 and V_2 is

$$P = \frac{V_1V_2}{V_R} + \epsilon$$

where ϵ is much larger than the usual amplifier offset voltage. The per cent error of P is now inversely proportional to V_1V_2 and given by

$$\text{Per cent error} = \frac{100\epsilon V_R}{V_1V_2}$$

a value which can be significant if both V_1 and V_2 are small. If both V_1 and V_2 can be increased by a factor of 10 without overloads, then the maximum per cent error is reduced by a factor of 100. For triple products or ones which involve more than 3 factors, it is imperative to insure that factors are reasonably large, or else meaningless results can be obtained.

Other nonlinear equipment contains similar pitfalls. This problem can be so severe that some computers have devices which indicate not only when voltages are too large but also when their levels remain too small throughout the solution of a problem.

There may arise situations where voltage levels are too low in a portion of the problem, seriously affecting the accuracy of the results. In such cases, it is necessary to change the scaling during the problem, either by putting the computer in the HOLD mode and manually making the change, or by automatic means, with relays or other circuitry. This will be easy to do once the principles of scaling are mastered.

The process of controlling the rate at which a computer solves a set of differential equations is called *time scaling*. We prefer to get a problem finished quickly because we wish to have results quickly and computer time is usually at a premium. Another reason for speed is the finite leakage resistances of integrating capacitors which cause their outputs to drift with time, resulting in error. On the other hand, amplifiers and other computing elements have frequency limitations that can cause large errors if the frequencies are too high. Servo multipliers are limited in operation to a speed of no more than several Hertz. Curve followers (Chapter 8) and x-y plotters, also servo-controlled, are similarly limited. Time scaling must achieve an optimum combination of speed and accuracy.

4.2 Planning a Computer Diagram

To scale a set of computer equations properly, we must know how the problem is to be mechanized. We begin by drawing an unscaled computer diagram.

There is never a unique mechanization to a set of equations. Each mechanization will require a separate set of scaled equations. To illustrate the factors considered in arriving at a given scaled set, let us consider how to solve the following elementary problem:

Example 4.1. Using a variplotter, draw a graph $y(x)$ of the solution of

$$\frac{dy}{dx} = x^2 - xy + 30$$

for $y(0) = 0$ and $0 \leq x \leq 10$.

Good computer equations are simply

$$\begin{aligned} \dot{x} &= k, & x_0 &= 0 \\ \dot{y} &= k(x^2 - xy + 30), & y_0 &= 0 \end{aligned}$$

for constant $k > 0$. The computer equations should include values for initial conditions, and these are most conveniently written next to each equation for the derivative. The subscript 0 indicates the value at $\tau = 0$.

To draw a computer diagram, we might begin by drawing the integrators, labeling one x and one y. The Kelvin feedback method requires that all right-hand expressions now be generated. What is the best way to generate the function $k(x^2 - xy + 30)$?

The *equipment rule for programming* which is supported by experience and often by an error analysis helps in this determination. This rule states that *a mechanization that minimizes amplifiers and multipliers probably produces the most accurate results.* The equipment rule will intentionally be broken on occasion for convenience, when error characteristics will hardly be affected, or when it is not true (Chapter 14.) It should be pointed out that since multipliers have much larger errors than amplifiers, it is more important to minimize multiplications by factoring polynomials, even if this means more amplifiers. The function $k(x^2 - xy + 30)$ can be factored to require one multiplication as follows: $k[x(x - y) + 30]$. We now have the option of summing in a summer or at the input to the integrator as can be seen by tracing through the

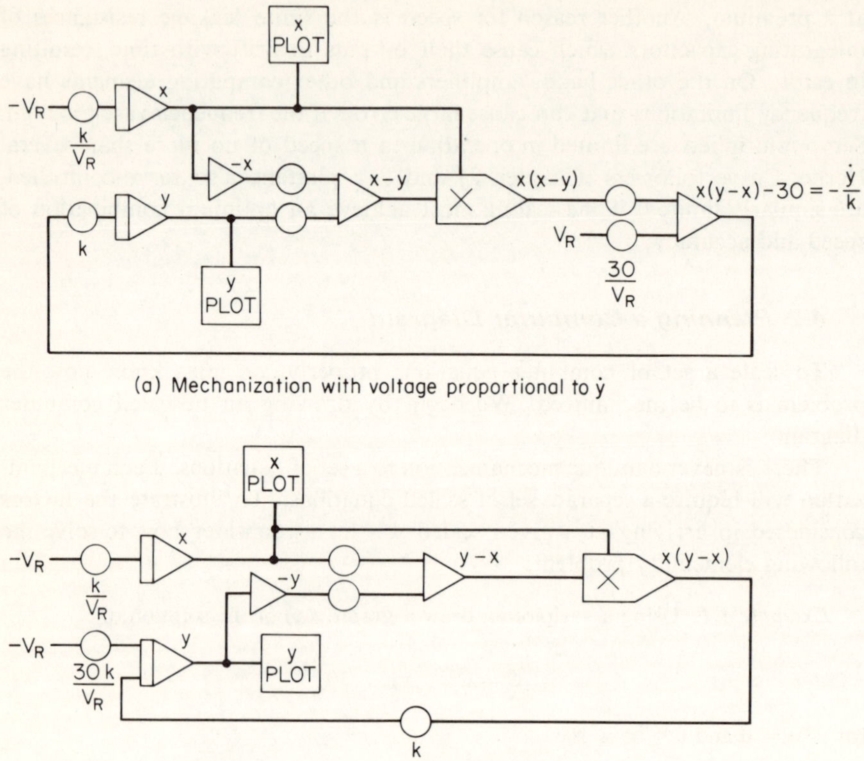

(a) Mechanization with voltage proportional to $\dot{y}$

(b) Summing at the integrator input

FIG. 4.3 Two mechanizations for $dy/dx = x^2 - xy + 30$.

diagrams in Fig. 4.3. If $\dot{y}$ is needed, Fig. 4.3a produces a voltage proportional to $\dot{y}$; otherwise the mechanization in Fig. 4.3b is preferable. It is both easier to scale, not requiring the amplifier labeled $-\dot{y}/k$, and contains less equipment.

In either mechanization, the voltages are sometimes attenuated with potentiometers, and sometimes not. These decisions follow from another simple pair of *potentiometer rules*, rarely broken:

1. *Attenuate all voltages entering summers and integrators to account for possible scale factors.* If a required gain can be obtained without a potentiometer, as determined from the final scaled equations, the pot need not be patched.

2. *Never attenuate inputs to multipliers or inverters,* on the principle that they are properly scaled when the variables feeding them are properly scaled. In the case of multipliers, it is also possible that it may be difficult to compensate for the pot loading.

Although Example 4.1 is elementary, there is very little difference in the actual procedure when there are many equations and each variable interacts with all other variables. It may then be necessary to arrange for a shortcut notation to simplify a complicated diagram. For example, Johnson [2] omits some feedback lines when they tend to interfere with a clear understanding

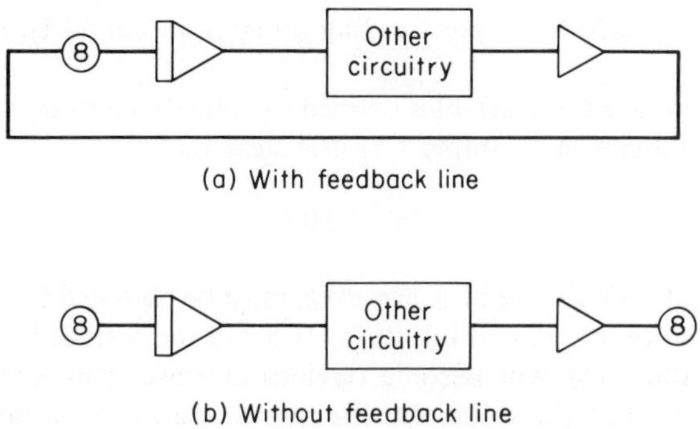

FIG. 4.4 Johnson's method for avoiding feedback lines.

and interpretation of the diagram by drawing the potentiometer circle twice (Fig. 4.4). Others [3] have found it convenient to simply indicate the functional expression feeding the pot as in Fig. 4.5. In any case, a programmer learns to draw diagrams so that he does not confuse himself. If possible, he also tries to make them understandable to others.

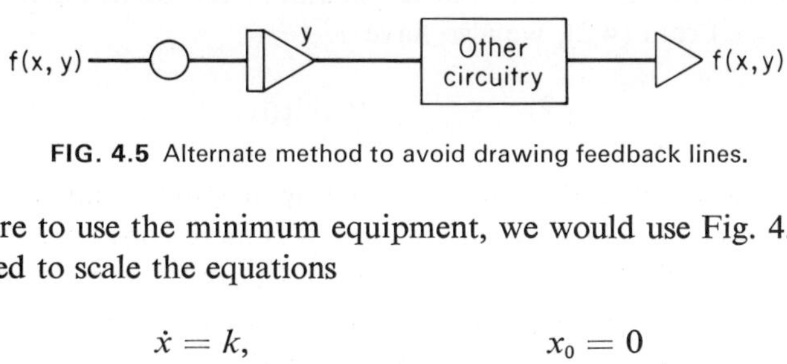

FIG. 4.5 Alternate method to avoid drawing feedback lines.

If we are to use the minimum equipment, we would use Fig. 4.3b, and we may proceed to scale the equations

$$\begin{aligned} \dot{x} &= k, & x_0 &= 0 \\ \dot{y} &= kx(x - y) + 30k, & y_0 &= 0 \end{aligned} \tag{4.1}$$

4.3 Amplitude Scaling in Volts

The process of amplitude scaling is straightforward requiring basically three steps:

(1) Estimate maximums of all variables defined by a time derivative to permit a suitable scaling transformation. (These variables appear at integrator outputs.)

(2) Rewrite the equations using the new scaled computer variables.

(3) Account for the scale factors of quotients and products. Redistribute constants to insure that all other amplifiers are scaled properly.

Since all terms in the right-hand expressions of the equations in canonical form contain the factor k, all pots feeding integrators should be made proportional to k.

In (4.1), there are two variables defined by time derivatives, x and y. From the problem statement of Example 4.1, it is clear that

$$|x| \leq 10$$

The maximum absolute value of y, however, must be obtained from calculation, knowledge of the problem, and intuition. It is not too serious if we err in our guess, because the error will become obvious in subsequent computer results, and we can then revise the value. In this case we may draw a tangent field to approximate the maximum value of y. Let us assume we arrive at $|y| \leq 20$. Assuming $V_R = 100$ volts, we let

$$Y = 5y, \qquad X = 10x \tag{4.2}$$

where X and Y are computer variables. The scale factors 5 and 10 have such units that X and Y have units of volts, and are selected so that $|X|_{\max}$ and $|Y|_{\max} \leq V_R$. (In practice, only convenient scale factors such as $1, 2, 4, 5 \times 10^k$, k integral, would be chosen. The scaled maximum can be less than but should be close to V_R.) From (4.2), we also have

$$\dot{Y} = 5\dot{y}, \qquad \dot{X} = 10\dot{x}$$

We now proceed to effect the transformation by substituting into (4.1). Instead of using the actual variables (capital letters), it is more convenient to keep the problem variable notation by writing $(5y)$ and $(10x)$ where y and x appear, using parentheses or brackets to remind us that these are computer variables. This procedure simplifies rescaling when necessary. Thus, we have at the conclusion of step 2,

$$\begin{aligned} (10\dot{x}) &= 10k, & (10x)_0 &= 0 \\ (5\dot{y}) &= 5k\frac{(10x)}{10}[0.1(10x) - 0.2(5y)] + 150k, & (5y)_0 &= 0 \end{aligned} \tag{4.3}$$

Constants have been inserted to balance the equations. Cancellation of all these constants should transform the equation back to the original computer equation. *This is the best practical way to determine if an error was made in the transformation.*

To redistribute the constants properly, we must remember the normalizing factor of 0.01 for the product and also scale the variable $y - x$, since it appears as an amplifier output. In (4.3), the expression $0.1(10x) - 0.2(5y)$ can be multiplied by at least 5 since $0.5(10x) - (5y)$ will not exceed 100 in absolute value if neither $(10x)$ nor $(5y)$ do. The factor $\frac{1}{5}$ is therefore used as another factor of k. These modifications result in

$$(10\dot{x}) = 10k, \qquad (10x)_0 = 0$$

$$(5\dot{y}) = 10k\,\frac{(10x)[0.5(10x) - (5y)]}{100} + 150k, \qquad (5y)_0 = 0$$

The constant voltages of $10k$ and $150k$ must also be scaled. These are obtained from the reference and are therefore written $0.1k(100)$ and $1.5k(100)$, respectively. The equations solved on the computer now appear as

$$\begin{aligned} (10\dot{x}) &= 0.1k(100), & (10x)_0 &= 0 \\ (5\dot{y}) &= 10k\,\frac{(10x)[0.5(10x) - (5y)]}{100} + 1.5k(100), & (5y)_0 &= 0 \end{aligned} \tag{4.4}$$

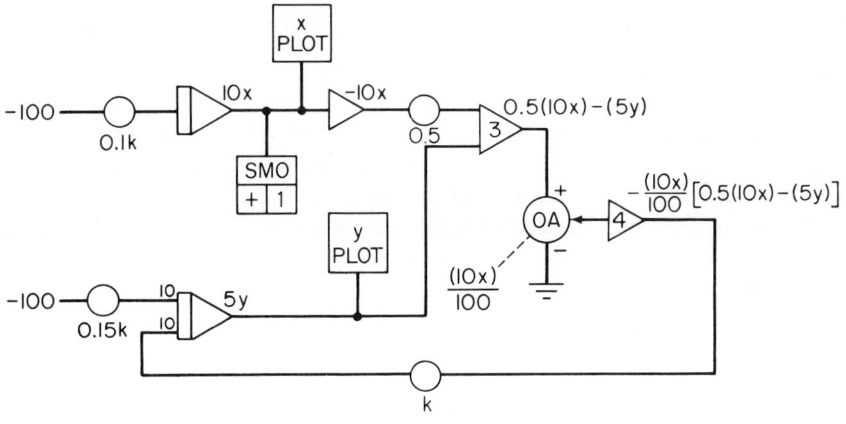

FIG. 4.6 Amplitude scaled diagram.

Remember to think of the variable $(10x)$ as a computer variable in volts, rather than 10 times the problem variable x. Figure 4.6 shows a computer diagram using a servo multiplier before time scaling. One input pot was removed at the input of summer 3 because of the required unity gain.

4.4 Time Scaling

The factor k appearing in all right-hand expressions in (4.4), for which pots were judiciously supplied for all inputs to every integrator, is a *time scale* factor.

This factor in no way affects amplitudes in the solution, since $d(5y)/d(10x) = 5\dot{y}/(10\dot{x})$ is independent of k. Only a rate of solution is determined by the choice of k.

As a good general rule, solving problems on a one-shot, nonrepetitive, or slow computer should take 5–60 s. In Example 4.1, the time of solution is determined by the constant voltage integrated to produce $(10x)$. Since the total excursion of $(10x)$ is 100 V (from 0 to 100 V), a reasonable choice of k is 1, requiring 10 V at the input of this integrator, which in turn results in a 10-s solution. It is important, however, to determine if this speed produces any high frequencies in the other equation. Since this may be difficult to determine, some computer manufacturers provide a capacitor selection scheme that can be used to slow down a solution automatically. If the programmer is to have any confidence in the solution, the results at both speeds should be identical. This subject belongs in the area of error analysis and checking (Chapter 17), but some foresight is generally attained after some experience.

With $k = 1$, the equations become

$$
\begin{aligned}
(10\dot{x}) &= 0.1(100), & (10x)_0 &= 0 \\
(5\dot{y}) &= 10\,\frac{(10x)[0.5(10x) - (5y)]}{100} + 1.5(100), & (5y)_0 &= 0
\end{aligned}
\tag{4.5}
$$

and the corresponding diagram is changed as in Fig. 4.7. Note that amplifier 4 in Fig. 4.6 has been eliminated by shifting the inverting amplifier from $(10x)$ to $(5y)$. This was possible because no pot was needed to feed the lead from the multiplier directly to the ten gain of the integrator.

If the problem independent variable is time and $k = 1$, the problem is said to be solved in *real time*. If $k > 1$ (or $0 < k < 1$), the problem is solved faster (or slower) than real time.

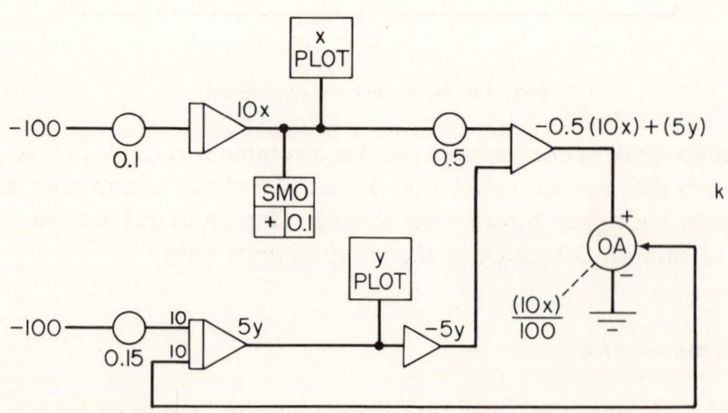

FIG. 4.7 Final diagram.

Other methods of scaling have been used, particularly when the computer equations are not first order, ([8]–[13]). They are essentially all equivalent to each other and to the method given here. Examples in this book will consistently use the state variable approach.

4.5 Rescaling

If this problem were run, we would find that $(5y)$ exceeds 100 V and should have been scaled $(4y)$. Then (4.5) must be rescaled, and the corresponding diagram changed. Rescaling is a simple process of readjusting the constants once again. We quickly see that the proper equations are

$$
\begin{aligned}
(10\dot{x}) &= 0.1(100), & (10x)_0 &= 0 \\
(4\dot{y}) &= 10\,\frac{(10x)[0.4(10x) - (4y)]}{100} + 1.2(100), & (4y)_0 &= 0
\end{aligned}
\tag{4.6}
$$

In the corrected diagram (Fig. 4.8), the method of obtaining 1.2(100) or 120 V is more accurate, by a factor of about 10, than using a pot set at 0.12 in a 10 gain. In this problem, the difference will probably not be readily detectable, but a programmer learns to be on guard for more accurate arrangements when possible. In many cases the improvements possible are great indeed.

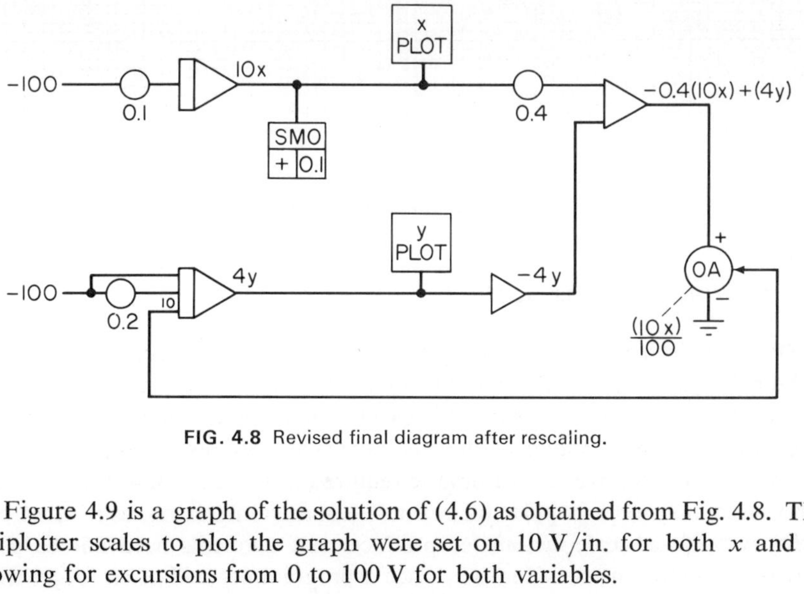

FIG. 4.8 Revised final diagram after rescaling.

Figure 4.9 is a graph of the solution of (4.6) as obtained from Fig. 4.8. The variplotter scales to plot the graph were set on 10 V/in. for both x and y, allowing for excursions from 0 to 100 V for both variables.

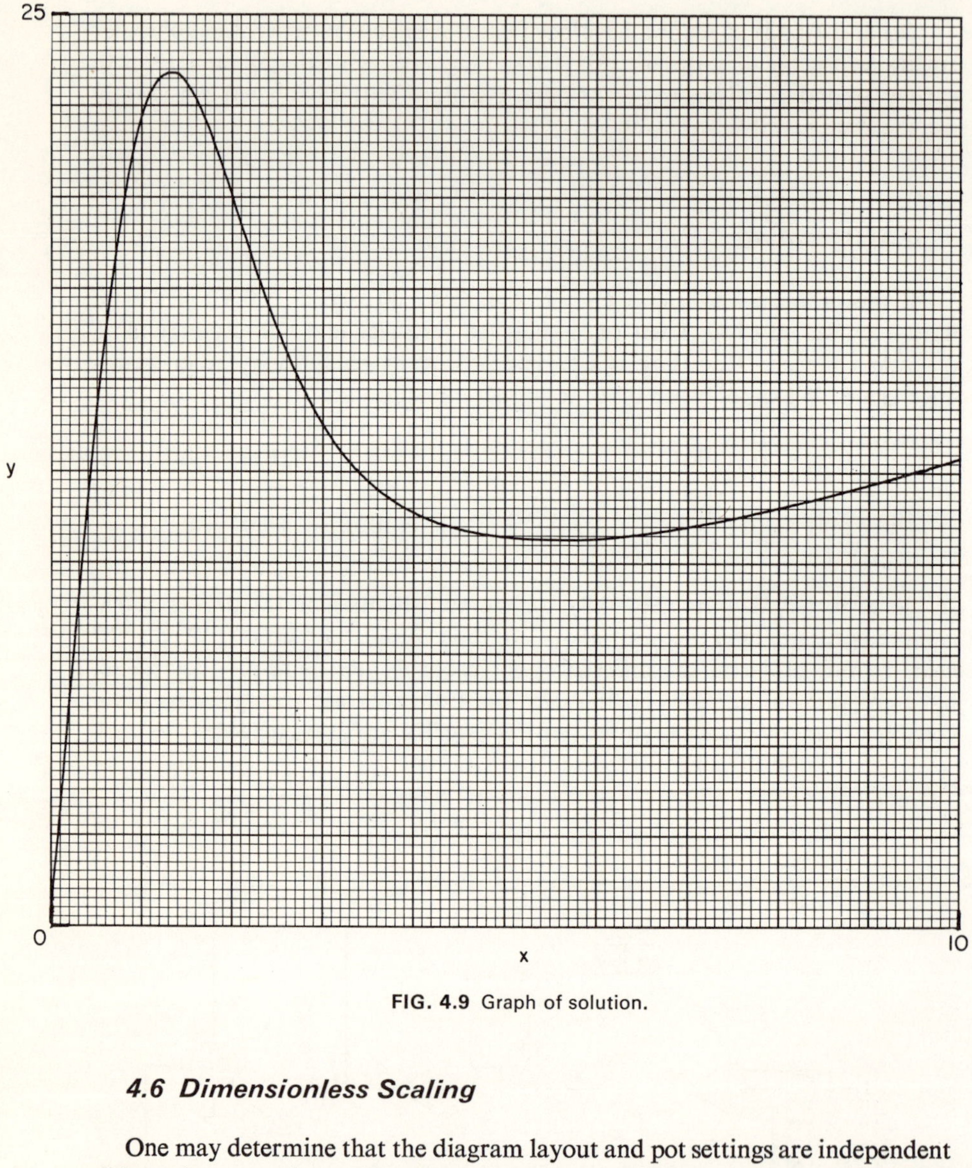

FIG. 4.9 Graph of solution.

4.6 Dimensionless Scaling

One may determine that the diagram layout and pot settings are independent of the reference voltage. To take advantage of this fact, a method of scaling referred to as the *dimensionless* or *normalized* method has been developed. Originally, computer systems with references that were not factors of ten used this system to minimize the arithmetic required in scaling. However, it is currently becoming more popular for several reasons. One reason, as it turns out, is that normalized scale factors for products and quotients need not be used. This is an important consideration since a programmer will often forget these factors, so that the final equations are incorrectly scaled, with serious consequences.

To see that the program is actually independent of the reference voltage, let us scale and solve Example 4.1 with a computing system that has an arbitrary reference voltage V_R, using the voltage scaling method. With

$$|x| \leq 10, \qquad |y| \leq 25$$

we must use the computer variables $(V_R x/10)$ and $(V_R y/25)$. Thus, from (4.1),

$$\left(\frac{V_R \dot{x}}{10}\right) = \frac{kV_R}{10}, \qquad \left(\frac{V_R x}{10}\right)_0 = 0$$

$$\left(\frac{V_R \dot{y}}{25}\right) = 10k\,\frac{(V_R x/10)[0.4(V_R x/10) - (V_R y/25)]}{V_R} + 1.2k(V_R), \qquad \left(\frac{V_R y}{25}\right)_0 = 0$$

so that, with $k = 1$,

$$\left(\frac{V_R \dot{x}}{10}\right) = 0.1(V_R), \qquad \left(\frac{V_R x}{10}\right)_0 = 0$$

$$\left(\frac{V_R \dot{y}}{25}\right) = 10\,\frac{(V_R x/10)[0.4(V_R x/10) - (V_R y/25)]}{V_R} + 1.2(V_R), \qquad \left(\frac{V_R y}{25}\right)_0 = 0 \tag{4.7}$$

The computer diagram appears as in Fig. 4.10 and is identical in every way to the one in Fig. 4.8 except for the interpretation of the computer output. In view of this, why not eliminate the reference from the mechanics of scaling?

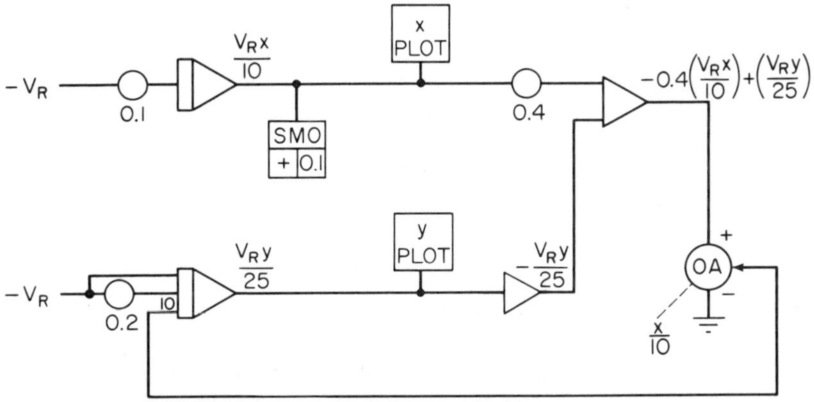

FIG. 4.10 Final diagram for arbitrary reference.

The most convenient value of V_R in (4.7) *is $V_R = 1$, which eliminates the normalizing factor for multiplication.* This procedure is enhanced if a ratiometer is available for reading voltage outputs, since the value $V_R x/10$ is read $x/10$ anyway. If $V_R = 1$, these values coincide. Dimensionless scaling is accomplished by considering the reference to be equal to one. Note that this is equivalent to using expressions such as $x/|x|_{\max}$ as computer variables, with maximum values of one. No confusion could result with a ratiometer; output equipment such as scopes, plotters, and recorders, however, rely strictly upon the actual voltage values, and must be adjusted on that basis.

In the remainder of this book, we shall use the dimensionless scaling method because of its simplicity and universal applicability. Thus (4.1) is scaled as

$$\left(\frac{\dot{x}}{10}\right) = 0.1, \qquad \left(\frac{x}{10}\right)_0 = 0$$

$$\left(\frac{\dot{y}}{25}\right) = 10\left(\frac{x}{10}\right)\left[0.4\left(\frac{x}{10}\right) - \left(\frac{y}{25}\right)\right] + 1.2, \qquad \left(\frac{y}{25}\right)_0 = 0$$

with the corresponding diagram shown in Fig. 4.11. There is no difference in the graphical result since the axes are labeled with problem variable notation in either case.

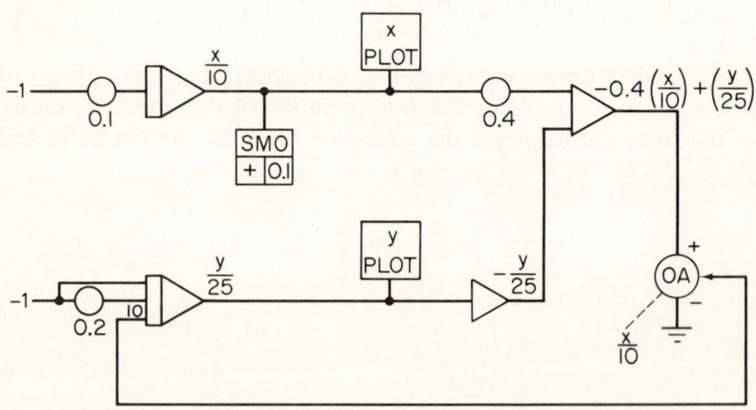

FIG. 4.11 Diagram for dimensionless scaling.

4.7 Scaling for Parameters

When parameters are involved in an investigation, as is virtually always the case, they are in a sense scaled, but not quite in the same way as amplifier outputs. For example, if a series of runs is to be made with the parameter

$0 \leq \alpha \leq 0.05$, then we arrange the pot setting to be no greater than 20α for convenience. It is almost always possible to do this, permitting the runs to be made conveniently. It should be remembered that errors in pot settings also affect accuracy, and that one should attempt to arrange for large settings. Thus, although it is quite annoying to interrupt a series of runs for rescaling, it is sometimes necessary to do so. The programmer should weigh the accuracy and convenience factors in this decision. The next example uses parameters to show how they enter into the scaled equations and computer diagram.

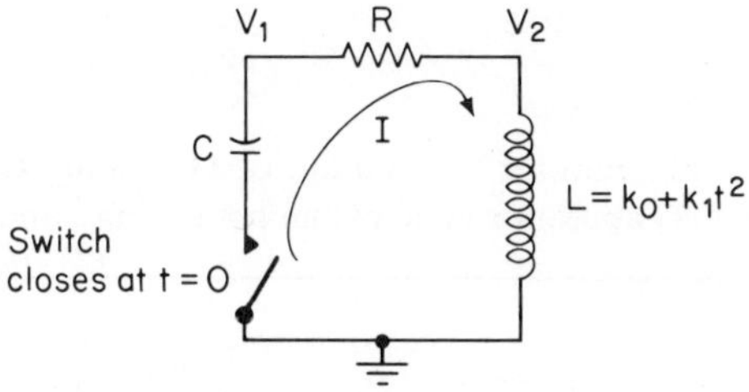

FIG. 4.12 Series *R-L-C* circuit with variable *L*.

Example 4.2. Investigate the series *R-L-C* circuit in Fig. 4.12, getting graphs of the voltage V_1 versus time t. This inductance is time-dependent:

$$L = k_0 + k_1 t^2$$

and energy is introduced into the circuit by means of an initial charge on the capacitor C, so that $(V_1)_0 \neq 0$. We will arrange for variable parameters about the nominal values

$$\begin{aligned} k_0 &= 0.054 \text{ H} \\ k_1 &= 200 \text{ H/s}^2 \\ C &= 200\ \mu\text{F} \\ R &= 48.5\ \Omega \\ (V_1)_0 &= 500 \text{ V} \end{aligned}$$

and also assume that only the first 0.02 s of the curve is of interest. Since

$$CV_1 = -\int_0^t I\,dt, \qquad V_1 - V_2 = IR, \qquad LI = \int_0^t V_2\,dt$$

we may substitute

$$I_1 = LI$$

and derive the equations

$$\frac{dI_1}{dt} = V_1 - \frac{I_1 R}{L}, \qquad (I_1)_0 = 0$$

$$\frac{dV_1}{dt} = -\frac{1}{C}\frac{I_1}{L}, \qquad (V_1)_0 = 500$$

As computer equations, we shall use

$$\dot{t} = kL = k(k_0 + k_1 t^2), \qquad t_0 = 0 \tag{4.8}$$

from which

$$\begin{aligned} \dot{I}_1 &= kV_1 L - kI_1 R, &\qquad (I_1)_0 = 0 \\ \dot{V}_1 &= -\frac{k}{C} I_1, &\qquad (V_1)_0 = 500 \end{aligned} \tag{4.9}$$

An unscaled computer diagram for the solution of (4.8) and (4.9) is shown in Fig. 4.13. The parameter k/C appears as one of the pot settings, and the pot controlling

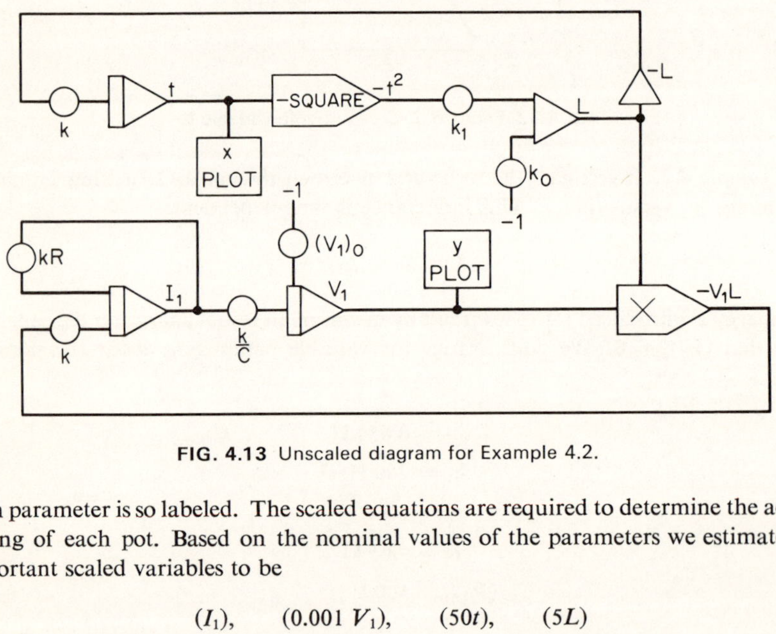

FIG. 4.13 Unscaled diagram for Example 4.2.

each parameter is so labeled. The scaled equations are required to determine the actual setting of each pot. Based on the nominal values of the parameters we estimate the important scaled variables to be

$$(I_1), \qquad (0.001\,V_1), \qquad (50t), \qquad (5L)$$

from which we may obtain the following scaled equations:

$$\begin{aligned} (5L) &= 5k_0 + \frac{k_1}{500}(50t)^2 & \\ (50\dot{t}) &= 10k(5L), & (50t)_0 = 0 \\ (\dot{I}_1) &= 200k(0.001V_1)(5L) - kR(I_1), & (I_1)_0 = 0 \\ (0.001\,\dot{V}_1) &= -\frac{0.001k}{C}(I_1), & (0.001V_1)_0 = 0.5 \end{aligned} \tag{4.10}$$

A good time scaling factor is $k = \frac{1}{50}$, which reduces (4.10) to

$$
\begin{aligned}
(50\dot{t}) &= 0.2(5L), & (50t)_0 &= 0 \\
(\dot{I}_1) &= 4(0.001V_1)(5L) - 0.02R(I_1), & (I_1)_0 &= 0 \\
(0.001\dot{V}_1) &= -\frac{2 \times 10^{-4}}{C}(I_1), & (0.001V_1)_0 &= 0.5
\end{aligned}
\tag{4.11}
$$

The pot settings are now clearly defined by (4.11) and are shown in the scaled diagram of Fig. 4.14. Nominal values of the five pot settings are therefore

$$
\begin{aligned}
5k_0 &= 0.2700 \\
\frac{k_1}{500} &= 0.4000 \\
\frac{2 \times 10^{-4}}{C} &= 1.0000 \\
0.02R &= 0.9700 \\
(0.001V_1)_0 &= 0.5000
\end{aligned}
$$

Each pot can be changed between runs to any value between 0 and 1. Note that C could not decrease below its nominal value, unless the pot output were fed to a 10

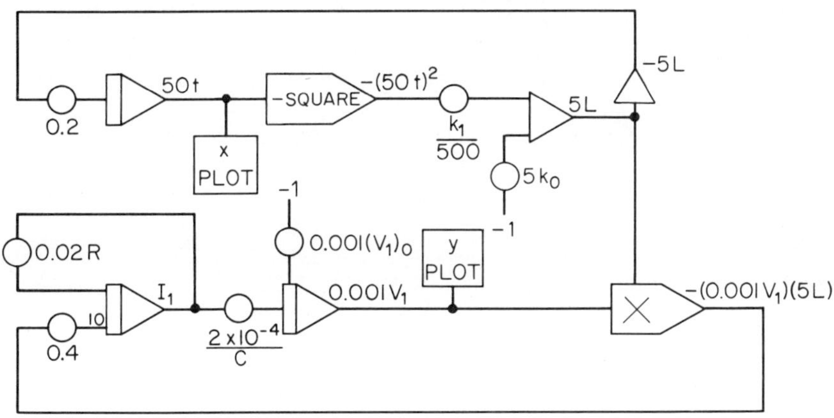

FIG. 4.14 Scaled diagram for Example 4.2.

gain and the pot setting labeled $2 \times 10^{-5}/C$. There will also be combinations of pot settings which will overload amplifiers, requiring rescaling. One must not think, for example, that R can take on any value between 0 and 50Ω without rescaling. The scale factor for I_1 was estimated on a value of $R \simeq 50$ and it would have to be rescaled for large deviations from the nominal.

Rescaling equations for variable parameters is the same process as rescaling with constants.

4.8 Scale Factors Containing Parameters

In Example 4.2, the magnitude of both V_1 and I_1 can be shown to be proportional to $(V_1)_0$ because the set of differential equations is linear. This being the case, it would be advantageous to include $1/(V_1)_0$ as a factor in the scale factors of both variables. Not only does this eliminate $(V_1)_0$ from every equation, but the number of parameters is reduced from five to four. By using parameters as scale factors, the effective number of parameters can often be reduced to a more manageable number. In Example 4.2, the number of parameters can actually be reduced to two by using scaled variables equal to

$$\left(\frac{t}{2RC}\right), \qquad \left[\frac{RI_1}{k_0(V_1)_0}\right], \qquad \left[\frac{V_1}{(V_1)_0}\right]$$

From (4.8) and (4.9), we obtain

$$\left(\frac{\dot{t}}{2RC}\right) = \frac{kk_0}{2RC}\left[\frac{k_0 + 4k_1(RC)^2(t/2RC)^2}{k_0}\right], \qquad \left(\frac{t}{2RC}\right)_0 = 0$$

$$\left[\frac{R\dot{I}_1}{k_0(V_1)_0}\right] = kR\left\{\left[\frac{V_1}{(V_1)_0}\right]\left[\frac{k_0+4k_1(RC)^2(t/2RC)^2}{k_0}\right] - \left[\frac{RI_1}{k_0(V_1)_0}\right]\right\}, \qquad \left[\frac{RI_1}{k_0(V_1)_0}\right]_0 = 0$$

$$\left[\frac{\dot{V}_1}{(V_1)_0}\right] = -\frac{kk_0}{RC}\left[\frac{RI_1}{(V_1)_0k_0}\right], \qquad \left[\frac{V_1}{(V_1)_0}\right]_0 = 1 \tag{4.12}$$

Now the time scale factor can be made proportional to RC/k_0, or

$$k = \frac{0.1RC}{k_0}$$

so that

$$\left(\frac{\dot{t}}{2RC}\right) = 0.05\left[1 + \frac{4k_1(RC)^2}{k_0}\left(\frac{t}{2RC}\right)^2\right], \qquad \left(\frac{t}{2RC}\right)_0 = 0$$

$$\left[\frac{R\dot{I}_1}{k_0(V_1)_0}\right] = \frac{0.1R^2C}{k_0}\left\{\left[\frac{V_1}{(V_1)_0}\right]\left[1 + \frac{4k_1(RC)^2}{k_0}\left(\frac{t}{2RC}\right)^2\right] - \left[\frac{RI_1}{k_0(V_1)_0}\right]\right\}, \qquad \left[\frac{RI_1}{k_0(V_1)_0}\right]_0 = 0$$

$$\left[\frac{\dot{V}_1}{(V_1)_0}\right] = -0.1\left[\frac{RI_1}{(V_1)_0k_0}\right], \qquad \left[\frac{V_1}{(V_1)_0}\right]_0 = 1$$

These equations can now accommodate wider ranges of parameters, of which there are only two, namely $4k_1R^2C^2/k_0$ and $0.1R^2C/k_0$. It may be necessary to provide additional constant factors to rescale for some combinations of parameters, say, for example, $t/(10RC)$, but the number of effective parameters remains unchanged. Since the nominal values are $4k_1R^2C^2/k_0 \simeq 1.6$ and $R^2C/k_0 \simeq 10$, we may provide for these values to reach about 8 and 100, respectively, by scaling as follows

$$\left(\frac{\dot{t}}{2RC}\right) = 0.4\left[0.125 + \frac{0.5k_1(RC)^2}{k_0}\left(\frac{t}{2RC}\right)^2\right], \qquad \left(\frac{t}{2RC}\right)_0 = 0$$

$$\left[\frac{R\dot{I}_1}{k_0(V_1)_0}\right] = \frac{0.8R^2C}{k_0}\left\{\left[\frac{V_1}{(V_1)_0}\right]\left[0.125 + \frac{0.5k_1(RC)^2}{k_0}\left(\frac{t}{2RC}\right)^2\right] - 0.125\left[\frac{RI_1}{k_0(V_1)_0}\right]\right\},$$

$$\left[\frac{RI_1}{k_0(V_1)_0}\right]_0 = 0$$

$$\left[\frac{\dot{V}_1}{(V_1)_0}\right] = -0.1\left[\frac{RI_1}{(V_1)_0k_0}\right], \qquad \left[\frac{V_1}{(V_1)_0}\right]_0 = 1$$

A scaled diagram is shown in Fig. 4.15 with representative results shown in Fig. 4.16. The graphical output contains normalized axes. If this is not satisfactory, i.e., more than the shape of the curve is wanted, then some compromise

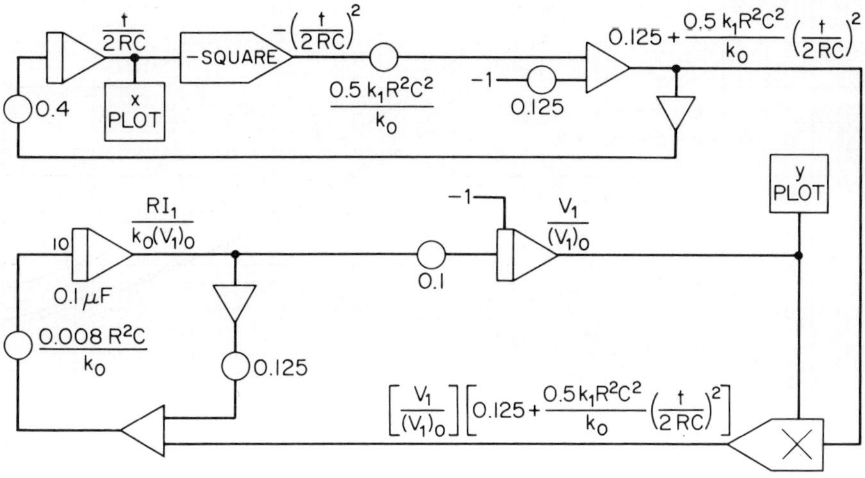

FIG. 4.15 Example 4.2 with variable scale factors.

can often be reached. For example, a mixture of variable and absolute scaling can reduce the number of parameters. Since there may be no interest in I_1, it can be scaled as $RI_1/[k_0(V_1)_0]$ to simplify matters somewhat in this problem even though absolute scaling is used for t and V_1. A programmer learns to be aware of this method to reduce the number of parameters in an investigation, particularly for simple problems.

The procedure used to determine the variable scale factors can best be determined from methods in dimensional analysis [4]. It often helps, however, to apply physical intuition once insight into the problem is gained. Insight often comes, in turn, just by examining and scaling the equations to be solved. In Example 4.2, all variables were made dimensionless, or normalized by applying a combination of experience and observation. From (4.9), $\dot{I}_1$ will tend

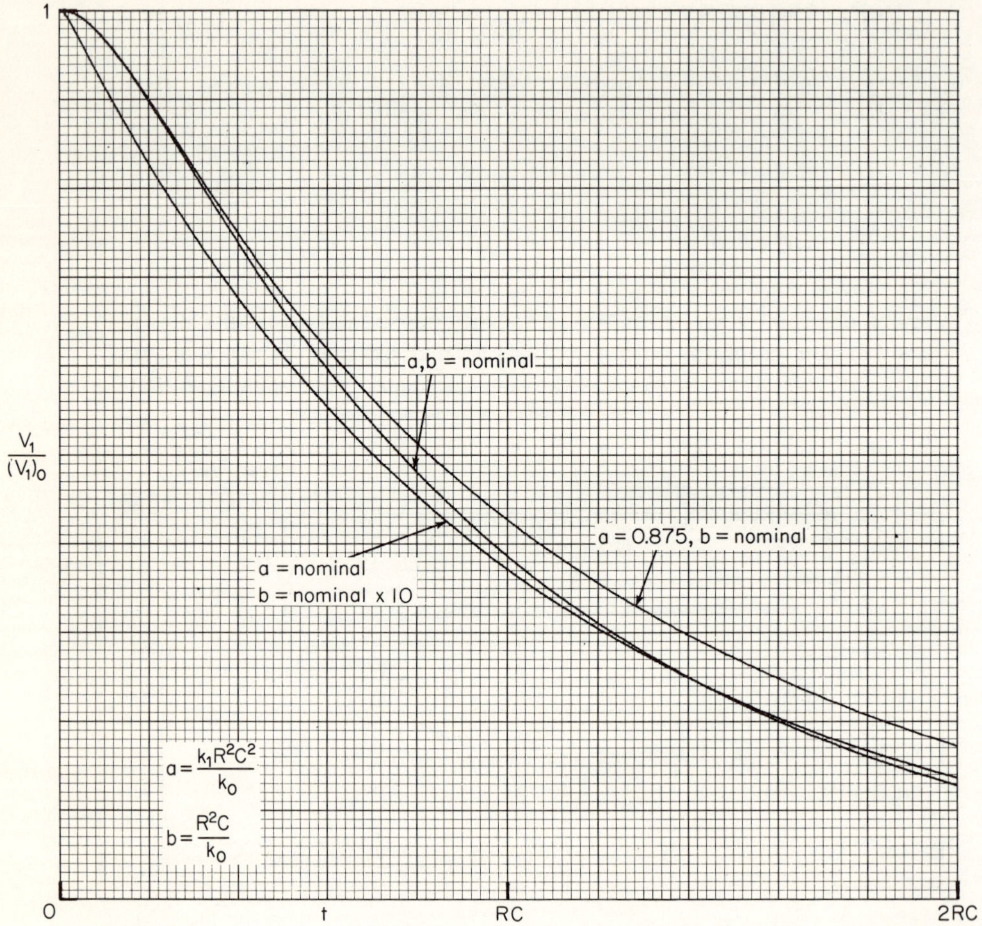

FIG. 4.16 Graphs of $V_1/(V_1)_0$ vs. t.

to remain close to zero because of "negative feedback," the term $-I_1R$. Hence under suitable conditions $I_1 \simeq V_1L/R$ which suggests the final scaling. Choosing k proportional to RC/k_0 eliminates the coefficient in the equation for $\dot{V}_1$. Finally, the scale factor for t was chosen proportional to $1/(RC)$ to reduce by one the number of remaining parameters. RC also has units of time, so that all variables finally used are dimensionless. This may not be true in other cases.

If by some foresight the transformations to obtain dimensionless variables are known, these can be effected first, and the resulting equations then mechanized and scaled. If we let

$$X_1 = \frac{t}{RC}, \qquad X_2 = \frac{RI_1}{k_0(V_1)_0}, \qquad X_3 = \frac{V_1}{(V_1)_0}$$

then we get

$$\dot{X}_1 = \frac{kk_0}{RC}\left[1 + \frac{k_1}{k_0}(RC)^2X_1^2\right], \qquad (X_1)_0 = 0$$

$$\dot{X}_2 = kR\left\{X_3\left[1 + \frac{k_1}{k_0}(RC)^2X_1^2\right] - X_2\right\}, \qquad (X_2)_0 = 0$$

$$\dot{X}_3 = -\frac{kk_0}{RC}X_2, \qquad (X_3)_0 = 1$$

which is just about equivalent to (4.12). The variable X_1 must be scaled as $(0.5X_1)$ to provide for the time range needed for this problem.

4.9 Summary; Other Examples

The general amplitude and time scaling procedures recommended can be summarized as follows:

A set of n computer equations are assumed to have been obtained in the form

$$\dot{y}_i = kf_i(y_1, y_2, \ldots, y_n), \qquad (y_i)_0 = c_i \tag{4.13}$$

for $i = 1, 2, \ldots, n$, where y_i are problem state variables, k and c_i are constants, and one of the y_i, say y_1, is the problem independent variable. Now

(a) Draw a computer diagram to indicate how various elements will be connected (or alternatively have some mental picture of the diagram).

(b) Select scale factors a_i so that the maximum M_i of $|a_iy_i|$ is such that $L_i \leq M_i \leq 1$, where L_i is large enough, say near $\frac{1}{2}$, to avoid large per cent errors due to nonideal computer elements. It is customary to use convenient a_i such as 1, 2, 2. 5, 4, or 5×10^j, j integral, in order to facilitate easy computation. The a_i can include factors which are constant parameters, if desired.

(c) Multiply (4.13) by a_i and substitute a_iy_i in the expression for f_i. In doing so, insert constant factors appropriately so the new expressions are equivalent to (4.13):

$$(a_i\dot{y}_i) = a_ikf_i\left[\frac{(a_1y_1)}{a_1}, \frac{(a_2y_2)}{a_2}, \ldots, \frac{(a_ny_n)}{a_n}\right], \qquad (a_iy_i)_0 = a_ic_i$$

(d) Redistribute the constants $a_1, a_2, \ldots, a_n$ so that the maximum absolute values of all amplifiers, multipliers, etc., are scaled to lie between $\frac{1}{2}$ and 1. (This may not always be possible with multipliers.)

(e) Rescale as necessary during computer runs, with the information obtained by them.

Although the process is straightforward, proficiency comes only with practice. We accordingly present three more problems here and suggest the reader attempt some of the exercises.

Example 4.3. Van der Pol's equation

$$\frac{d^2y}{dx^2} + \mu(y^2 - 1)\frac{dy}{dx} + y = 0 \tag{4.14}$$

for $0 \leq \mu \leq 2$ can be investigated in a straightforward manner. Let

$$\frac{dy}{dx} = y_1$$

Then

$$\frac{dy_1}{dx} = -y - \mu(y^2 - 1)y_1$$

and computer equations may be written

$$\begin{aligned} \dot{x} &= k, & x_0 &= 0 \\ \dot{y} &= ky_1, & |y_0| &< 4 \\ \dot{y}_1 &= -k[y + \mu(y^2 - 1)y_1], & (y_1)_0 &= 0 \end{aligned}$$

Good scaled variables turn out to be $(y_1/4)$ and $(y/4)$ with the scaling on x dependent on the range of interest. Since the phase-plane is most interesting, we will not mechanize the $\dot{x} = k$ equation, for x is not needed as feedback. Thus,

$$\left(\frac{\dot{y}}{4}\right) = k\left(\frac{y_1}{4}\right)$$

$$\left(\frac{\dot{y}_1}{4}\right) = -k\left(\frac{y}{4}\right) - 10k\mu\left[1.6\left(\frac{y}{4}\right)^2 - 0.1\right]\left(\frac{y_1}{4}\right)$$

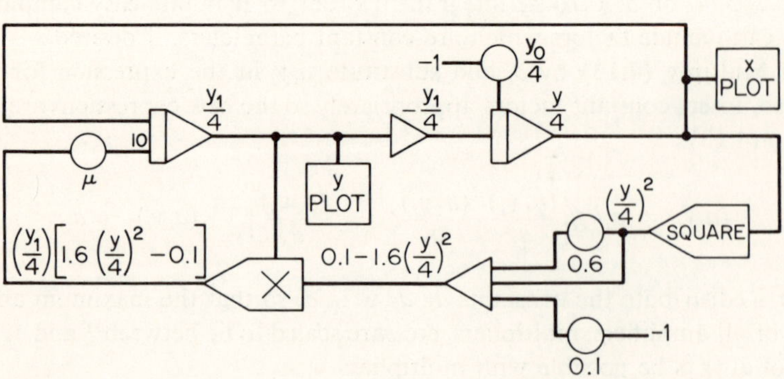

FIG. 4.17 Mechanization for Van der Pol's equation.

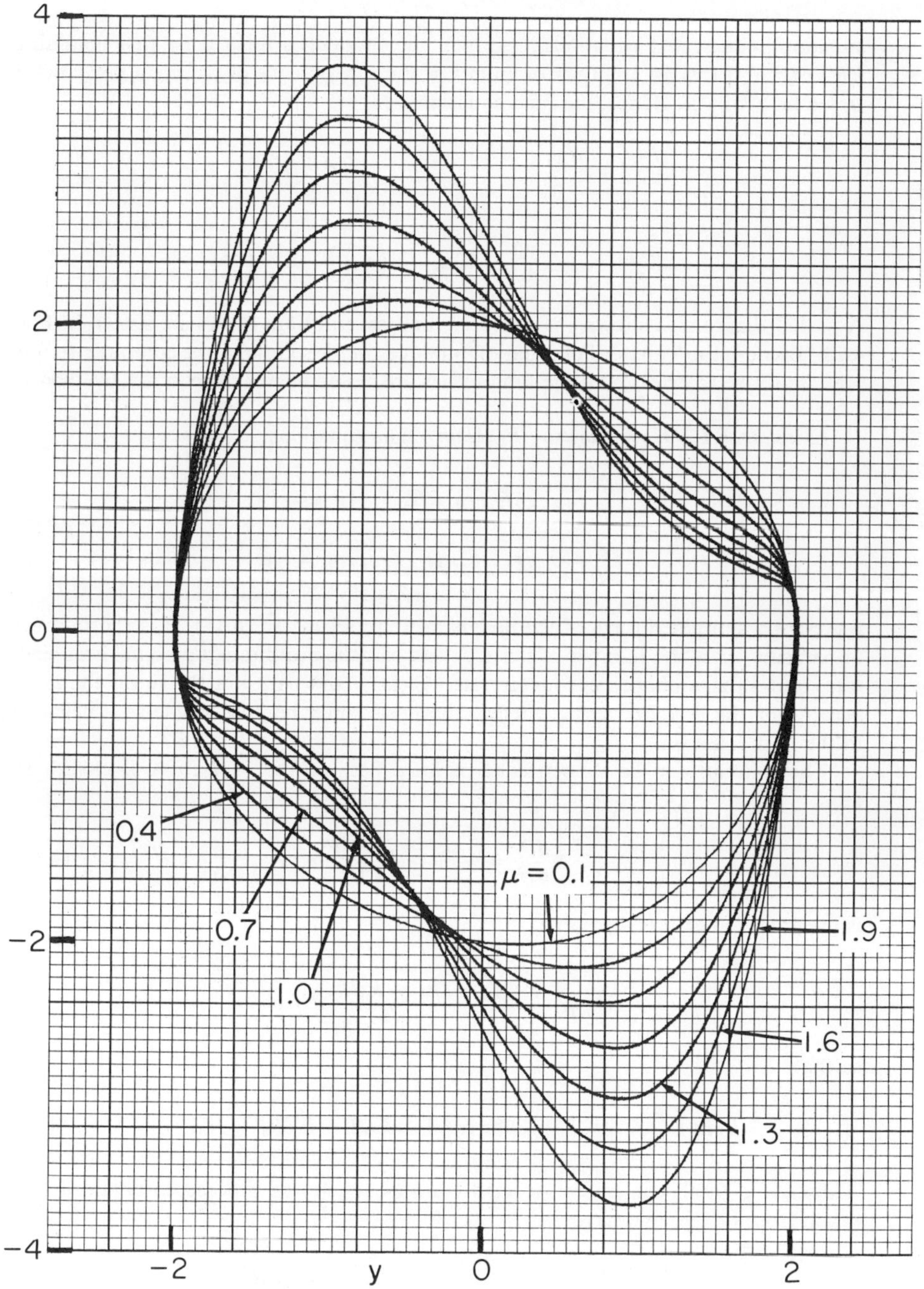

FIG. 4.18 Solutions to Van der Pol's equation.

are amplitude scaled equations, assuming the product is formed by first obtaining $1.6(y/4)^2 - 0.1$ as an amplifier output. For the convenience of having a pot labeled μ, we take $k = 1$. Hence, the final equations are

$$\left(\frac{\dot{y}}{4}\right) = \left(\frac{y_1}{4}\right)$$

$$\left(\frac{\dot{y}_1}{4}\right) = -\left(\frac{y}{4}\right) - 10\mu\left[1.6\left(\frac{y}{4}\right)^2 - 0.1\right]\left(\frac{y_1}{4}\right)$$

A scaled diagram is shown in Fig. 4.17, and several limit cycles are shown in Fig. 4.18. Because the Van der Pol equation produces *stable* solutions, the computer results are insensitive to errors. All the limit cycles are closed curves as required by theory.

***Example 4.4.* (*Host-Parasite Problem*).** The equations

$$\begin{aligned} \frac{dP}{dt} &= -k_1P + k_3PH, \qquad (P)_0 = P_0 \\ \frac{dH}{dt} &= k_2H - k_3PH, \qquad (H)_0 = H_0 \end{aligned} \tag{4.15}$$

represent a simplified form of the host-parasite problem. In (4.15), P is the number of parasites, H is the number of hosts, k_1 is a parasite death rate, k_2 is a net host birth rate, and k_3 is a chance encounter rate. Clearly, the parasites thrive on the hosts because of the $+k_3PH$ term, whereas the hosts suffer because this term is negative in the host equation.

The solution can be shown to be periodic because the P-H curve is closed. In fact,

$$\frac{dP}{dH} = \frac{(-k_1 + k_3H)P}{(k_2 - k_3P)H} \tag{4.16}$$

is integrable since it is separable:

$$\frac{(k_2 - k_3P)}{P}\,dP + \frac{(k_1 - k_3H)}{H}\,dH = 0$$

so that

$$k_2 \ln\left(\frac{P}{P_0}\right) - k_3P + k_1 \ln\left(\frac{H}{H_0}\right) - k_3H + k_3(P_0 + H_0) = 0 \tag{4.17}$$

Although a graph can be constructed from (4.17), let us proceed to obtain a P-H graph from (4.15). We have

$$\begin{aligned} \dot{t} &= k, & t_0 &= 0 \\ \dot{P} &= k(-k_1P + k_3PH), & (P)_0 &= P_0 \\ \dot{H} &= k(k_2H - k_3PH), & (H)_0 &= H_0 \end{aligned}$$

From (4.16), $dP/dH = 0$ when $H = k_1/k_3$ and $dH/dP = 0$ when $P = k_2/k_3$. Since $H_{\max}$ is at least equal to k_1/k_3 and $P_{\max} = k_2/k_3$, this suggests variable scaling proportional to these values. Let us obtain solutions when the constant of proportionality is four and take the scaled variables to be $(0.25k_3H/k_1)$ and $(0.25k_3P/k_2)$. Now, the main equations are scaled as

$$\left(0.25\frac{k_3}{k_2}\dot{P}\right) = k\left[-k_1\left(0.25\frac{k_3}{k_2}P\right) + 4k_1\left(0.25\frac{k_3}{k_2}P\right)\left(0.25\frac{k_3}{k_1}H\right)\right], \qquad \text{IC: } \frac{k_3P_0}{4k_2}$$

$$\left(0.25\frac{k_3}{k_1}\dot{H}\right) = k\left[k_2\left(0.25\frac{k_3}{k_1}H\right) - 4k_2\left(0.25\frac{k_3}{k_2}P\right)\left(0.25\frac{k_3}{k_1}H\right)\right], \qquad \text{IC: } \frac{k_3H_0}{4k_1}$$

Since k_1 is a factor in the $\dot{P}$ equation, and k_2, in the $\dot{H}$ equation, we may choose k proportional to either $1/k_1$ or $1/k_2$, obtaining k_1/k_2 as a critical parameter. We may also, however, choose k proportional to $1/\sqrt{k_1k_2}$ to achieve this property, and at the same time get better gain characteristics. Now we have

$$\left(0.25\frac{k_3}{k_2}\dot{P}\right) = 2\sqrt{\frac{k_1}{k_2}}\left[-0.5\left(0.25\frac{k_3}{k_2}P\right) + 2\left(0.25\frac{k_3}{k_2}P\right)\left(0.25\frac{k_3}{k_1}H\right)\right], \qquad \text{IC: } \frac{k_3P_0}{4k_2}$$

$$\left(0.25\frac{k_3}{k_1}\dot{H}\right) = 2\sqrt{\frac{k_2}{k_1}}\left[0.5\left(0.25\frac{k_3}{k_1}H\right) - 2\left(0.25\frac{k_3}{k_2}P\right)\left(0.25\frac{k_3}{k_1}H\right)\right], \qquad \text{IC: } \frac{k_3H_0}{4k_1} \tag{4.18}$$

It becomes clear that the shape of the *P-H* graph depends only upon three parameters k_1/k_2, k_3P_0/k_2, and k_3H_0/k_1, while the period requires the knowledge of k_1, k_2, k_3H_0, and k_3P_0. If the shape of the *P-H* curve is to be investigated, (4.18) is appropriate. For investigating the period, however, k can be taken proportional to some constant. In either case, there is a reduction in the original number of parameters (five) which can save considerable time in the investigation.

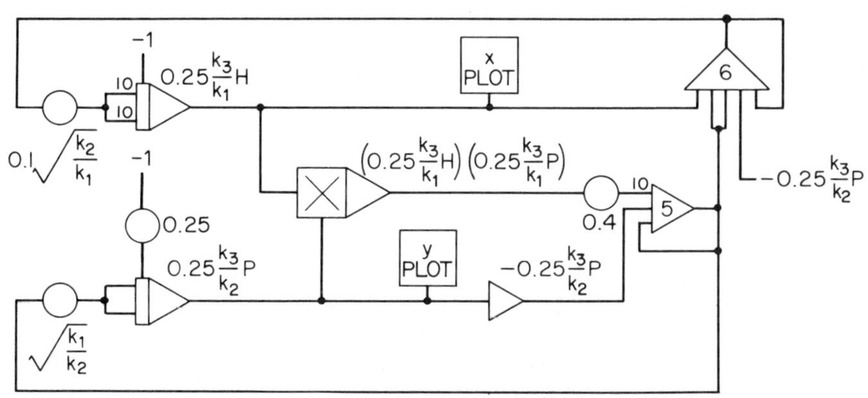

FIG. 4.19 Diagram for the host–parasite problem.

A diagram to solve (4.18) is shown in Fig. 4.19 and several *P-H* curves obtained are shown in Fig. 4.20, all for initial conditions $k_3H_0/(4k_1)=1$ and $k_3P_0/(4k_2)=0.25$ (*H* starts at its maximum value). Under these conditions, k_1 must be kept less than k_2; the diagram shows pot settings for $1 \leq \sqrt{k_2/k_1} \leq 10$. Additional 1-gain feedback on amplifiers 5 and 6 is used to multiply sums by 0.5 without pots. This will be, in general, more accurate, because input and feedback resistances are kept to within 0.01 per cent of nominal values.

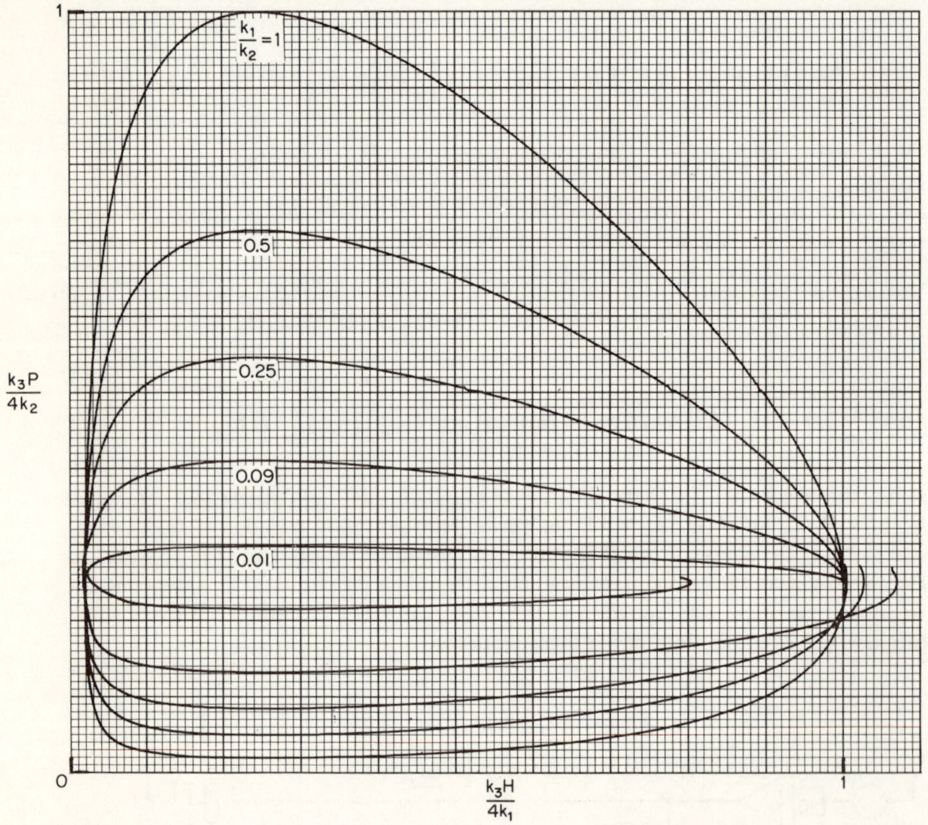

FIG. 4.20 Solutions to the host–parasite problem.

The curves in Fig. 4.20 do not close, a dramatic effect of analog computer error, probably due to errors in multiplication, when working with unstable equations. To combat this difficulty, we generally use automatic rescaling of the product when input voltages are too low (Chapter 6). We will return to this problem later.

Example 4.5. A simple frictionless pendulum as in Fig. 4.21 obeys the equation

$$\frac{d^2\theta}{dt^2} = -\frac{g}{\ell}\sin\theta \tag{4.19}$$

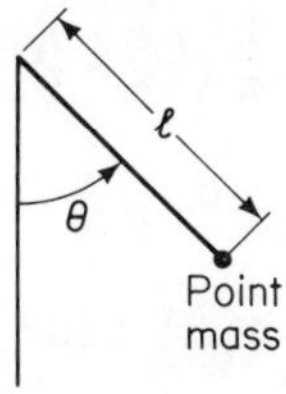

FIG. 4.21 A simple pendulum.

for gravitational acceleration g. Let us investigate the motion on the computer with the velocity 0 for some initial θ_0. With

$$\frac{d\theta}{dt} = \theta_1$$

we may take

$$\begin{aligned} \dot{t} &= k, & t_0 &= 0 \\ \dot{\theta} &= k\theta_1, & (\theta)_0 &= \theta_0 \\ \dot{\theta}_1 &= -k\frac{g}{\ell}\sin\theta, & (\theta_1)_0 &= 0 \end{aligned}$$

We are faced here with the problem of generating the transcendental function $\sin\theta$. Rather than use a resolver (Chapter 8), we shall generate this function with integrators. This will be discussed more fully in the next chapter. It suffices here to point out that $\sin\theta$ is the solution of a set of differential equations which must be found and solved on the computer. Let

$$S = \sin\theta, \qquad C = \cos\theta$$

Now

$$\dot{S} = C\dot{\theta} = kC\theta_1, \qquad \dot{C} = -S\dot{\theta} = -kS\theta_1 \tag{4.20}$$

which involves two multiplications. The set of equations now solved is

$$\begin{aligned} \dot{t} &= k, & t_0 &= 0 \\ \dot{\theta} &= k\theta_1, & (\theta)_0 &= \theta_0 \\ \dot{\theta}_1 &= -k\frac{g}{\ell}S, & (\theta_1)_0 &= 0 \\ \dot{S} &= kC\theta_1, & (S)_0 &= \sin\theta_0 \\ \dot{C} &= -kS\theta_1, & (C)_0 &= \sin\theta_0 \end{aligned} \tag{4.21}$$

Since $|\theta_1|_{\max} = 2\sqrt{g/\ell}$, $|\theta|_{\max} = \pi$, and $|C|_{\max} = |S|_{\max} = 1$, an appropriate set of scaled equations, except for t, is

$$\left(\frac{\dot{\theta}}{\pi}\right) = \frac{2}{\pi}\sqrt{\frac{g}{\ell}}\,k\left(0.5\sqrt{\frac{\ell}{g}}\,\theta_1\right)$$

$$\left(0.5\sqrt{\frac{\ell}{g}}\,\dot{\theta}_1\right) = -0.5\sqrt{\frac{g}{\ell}}\,k(S)$$

$$(\dot{S}) = 2\sqrt{\frac{g}{\ell}}\,k\left(0.5\sqrt{\frac{\ell}{g}}\,\theta_1\right)(C)$$

$$(\dot{C}) = -2\sqrt{\frac{g}{\ell}}\,k\left(0.5\sqrt{\frac{\ell}{g}}\,\theta_1\right)(S)$$

With $k = 0.5\sqrt{\ell/g}$, and t scaled proportional to $\sqrt{g/\ell}$, only one parameter remains, namely, θ_0 [the general case would involve a second parameter $(\theta_1)_0$]. For,

$$\begin{aligned}
\left(\sqrt{\frac{g}{\ell}}\,\dot{t}\right) &= 0.5, & \left(\sqrt{\frac{g}{\ell}}\,t\right)_0 &= 0 \\
\left(\frac{\dot{\theta}}{\pi}\right) &= \frac{1}{\pi}\left(0.5\sqrt{\frac{\ell}{g}}\,\theta_1\right), & \left(\frac{\theta}{\pi}\right)_0 &= \frac{\theta_0}{\pi} \\
\left(0.5\sqrt{\frac{\ell}{g}}\,\dot{\theta}_1\right) &= -0.25(S), & \left(0.5\sqrt{\frac{\ell}{g}}\,\theta_1\right)_0 &= 0 \\
(\dot{S}) &= \left(0.5\sqrt{\frac{\ell}{g}}\,\theta_1\right)(C), & (S)_0 &= \sin\theta_0 \\
(\dot{C}) &= -\left(0.5\sqrt{\frac{\ell}{g}}\,\theta_1\right)(S), & (C)_0 &= \cos\theta_0
\end{aligned} \tag{4.22}$$

Figure 4.22 shows a circuit diagram which solves (4.22) without t, for recording with a strip chart recorder. Figures 4.23 and 4.24 show various graphs of results. The proper initial conditions must be used on the integrators producing $\sin\theta$ and $\cos\theta$.

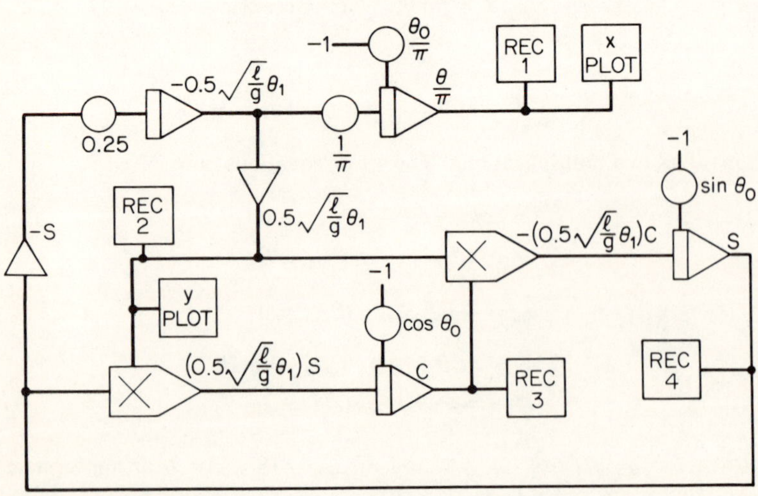

FIG. 4.22 Diagram for the simple pendulum problem.

The voltage for $\cos\theta_0$ is negative for $\pi/2 < \theta_0 < \pi$ and positive for $0 < \theta_0 < \pi/2$, so care must be taken to insure the proper polarity. The function $\sin\theta_0$ is always positive in these quadrants as indicated on the diagram.

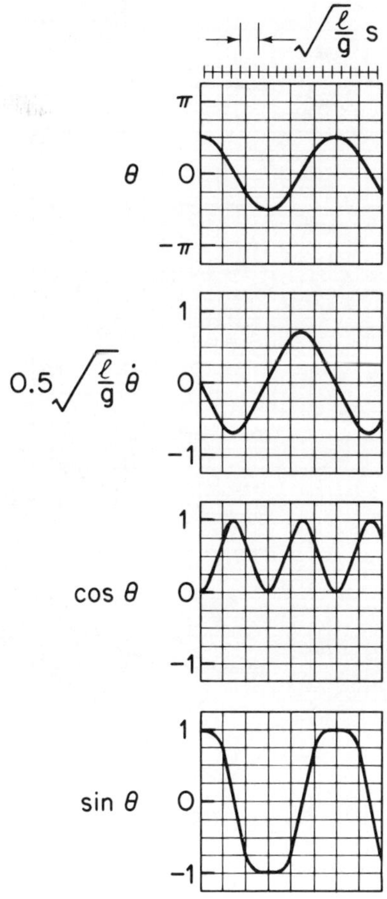

FIG. 4.23 Strip chart output for the pendulum problem.

Because of its scaling, $t\sqrt{g/\ell} = 1$ in 2 s. Hence this scaling is so indicated on the strip chart record, which contains timing marks every second. The phase-plane graph shows closed curves even for θ_0 as large as 0.99π, indicating less error here than in the host-parasite problem. It should be pointed out that the curves only appear to close. Errors actually prevent an exact solution because system (4.21) does not have a limit cycle. For $0.99\pi < \theta_0 < \pi$, more difficulty would be encountered.

The integrator producing θ is not involved in feedback for any equation. It need not have been used if the period of the pendulum were desired or the position were needed as a function of time, since θ could be obtained from the functions $\sin\theta$ and $\cos\theta$.

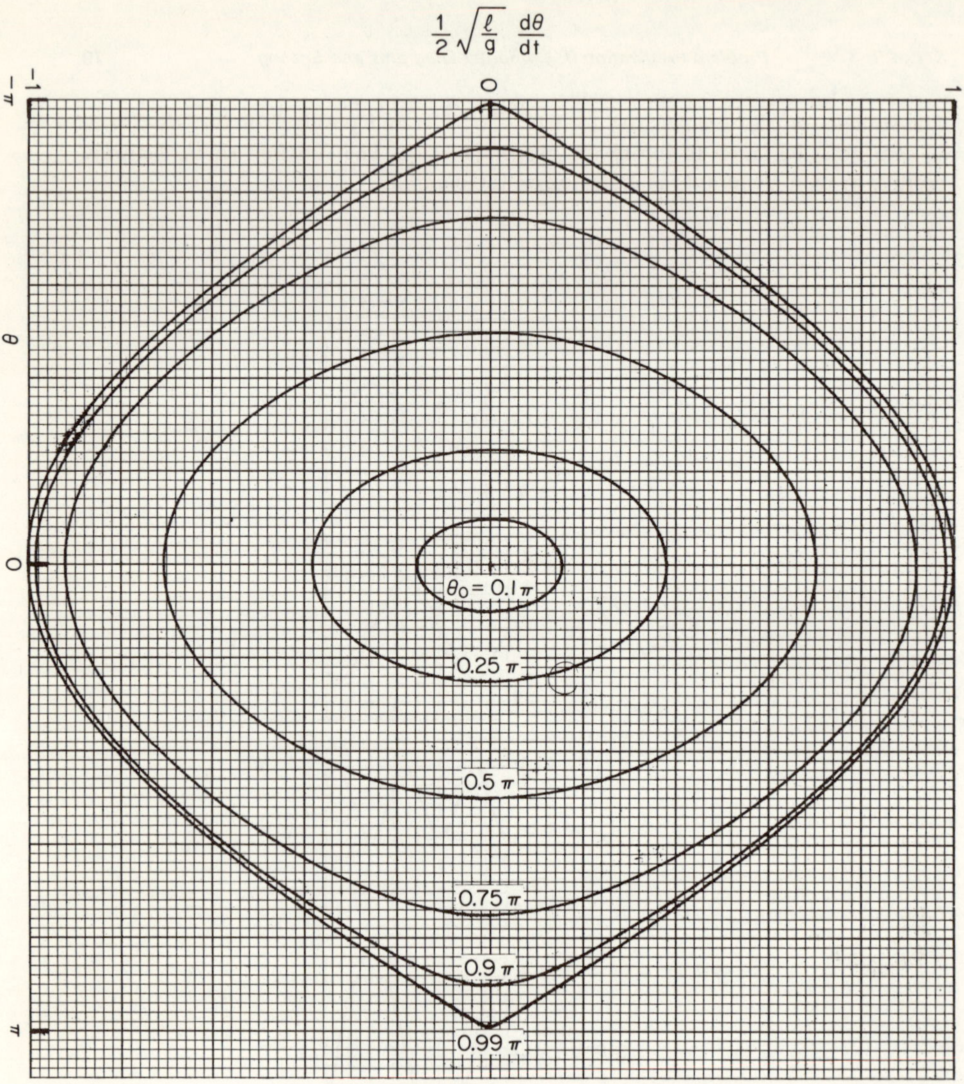

FIG. 4.24 Variplotter output for the pendulum problem.

PROBLEMS

4-1. Mechanize Example 4.2 using i = constant and dimensionless scaling about nominal values. Use a servo multiplier to mechanize the divison and do not use variable scaling.

4-2. Prepare a scaled diagram to obtain a family of solutions of

$$\frac{dy}{dx} = \frac{x + y + 1}{2x - y + 2}$$

for $-5 \leq x \leq 5$ and $-5 \leq y \leq 5$.

4-3. Prepare a scaled diagram for the solution of Legendre's equation

$$\frac{d^2y}{dx^2} = \frac{2x(dy/dx) - n(n+1)y}{1 - x^2}$$

for $0 \leq x \leq 1$ and all integral n, $0 \leq n \leq 6$.

By using the initial conditions of

$n =$	0	1	2	3	4	5	6
$y(0)$	1	0	$-\frac{1}{2}$	0	$\frac{3}{8}$	0	$-\frac{5}{16}$
$\frac{dy(0)}{dx}$	0	1	0	$-\frac{3}{2}$	0	$\frac{15}{8}$	0

the solution $y(x)$ yields, theoretically, the well known Legendre polynomials (Chapter 7). (Assume $|dy/dx| \leq 25$ in this interval and $|y| \leq 1$.)

4-4. Mechanize Problem 4-3 by generating the function $y_1 = (dy/dx)(1 - x^2)$ instead of dy/dx. There will be one less multiplication required [5].

4-5. Suppose we wished to obtain a graph of the function $y = \ln |1/x + 1|$ for $|x| < 2$ by solving a differential equation

$$\frac{dy}{dx} = \frac{y(1+x) - x}{x(1-x)}$$

The function y is unbounded at $x = -1$ so the ordinate of our graph must be arbitrarily limited. Let us assume we wish $|y| \leq 5$. There are now three distinct ranges which must be programmed: $0 < x \leq 2$, $-1 < x \leq 0$, and $-2 \leq x \leq -1$. Why is this so? Mechanize the equation and calculate the proper initial conditions for each run. If initial conditions are calculated for $x = \frac{1}{2}$, a positive $\dot{x}$ is needed for x to approach 0, and negative $\dot{x}$ for x to approach -1. Take this into account in your thinking. Manual function switches are available and one can be patched as shown in Fig. 4.25.

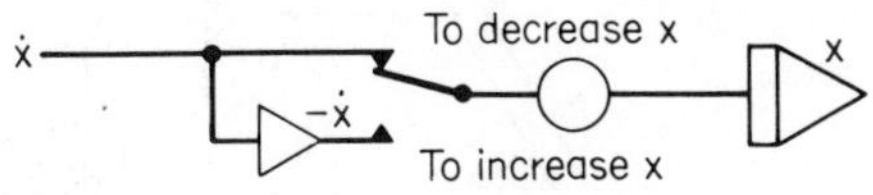

FIG. 4.25 Manual function switch application.

4-6. A simple chemical reaction is described by

$$\frac{dA}{dt} = -r_1 A, \qquad (A)_0 = A_0$$

$$\frac{dB}{dt} = r_1 A - r_2 B, \qquad B_0 = 0$$

$$\frac{dC}{dt} = r_2 B, \qquad C_0 = 0$$

where A, B, and C are distinct chemicals which interact, r_1 is a conversion rate of A to B, and r_2 is a conversion rate of B to C; both are constants. Assuming $r_2 < r_1$, prepare a scaled circuit diagram to solve the equations with variable scaling and r_2/r_1 as a parameter. (A, B, and C are all proportional to A_0.) Prepare a graph of A/A_0 versus t in units of $1/r_1$ s.

4-7. The equation describing the radial velocity V of a particle as a function of its radial distance r from the center of a disc used in a spray drier is given by

$$\frac{dV}{dr} = \frac{-AV^3 + Br}{V}$$

where $V = 0$ when $r = r_0$, and A and B are constants. Show that the shape of $V(r)$ is given by one parameter $r_0^2 A\sqrt{B}$ by scaling the variables proportional to $V/(\sqrt{B}r_0)$ and r/r_0. Set up a scaled diagram to plot these variables for up to 10 r/r_0 and obtain graphs for several values of $r_0^2 A\sqrt{B}$.

4-8. Set up the pendulum with scale factors for the variables such that every integrator output reaches the reference voltage at some time during the run, independent of θ_0.

4-9. In a simplified pure pursuit of a target by a missile, the missile velocity vector always points at the target as shown in Fig. 4.26. The equations of motion are then given by [6]

$$\frac{dr}{dt} = V_t \cos\phi - V_m, \qquad (r)_0 = r_0$$

$$\frac{d\phi}{dt} = \frac{-V_t \sin\phi}{r}, \qquad (\phi)_0 = \phi_0$$

Assuming V_m, V_t are constants and the motion is planar, show this may be programmed with scaled variables proportional to $(V_t t/r_0)$ and (r/r_0), resulting in the

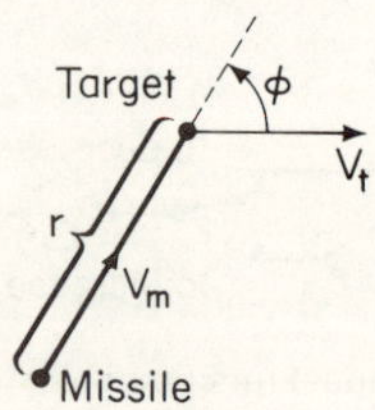

FIG. 4.26 Missile–target geometry.

critical parameters V_m/V_t and ϕ_0. By generating $\sin\phi$ and $\cos\phi$ in a manner similar to (4.20), draw a scaled computer diagram and obtain graphs of r/r_0 versus ϕ for various V_m/V_t and ϕ_0.

4-10. The deceleration of a reactor control rod is governed by the differential equation [7]

$$\frac{d^2x}{dt^2} = -\frac{ax(dx/dt)}{(b-x)^2}$$

where a and b are positive constants and initial conditions at $t = 0$ are $x = 0$ and $dx/dt = V_0$. It is clear that $x_{\max} = b$ and $(dx/dt)_{\max} = V_0$. Show that with variable scaling of (V_0t/b), (x/b), and $[(dx/dt)/V_0]$, there is only one critical parameter a/V_0. Prepare a scaled diagram and graphs of the solutions for several values of a/V_0.

4-11. In examining pressure variations in a packed bed reactor [7], the equation

$$\frac{dP}{dx} = \frac{-a(b+cx)P}{P^2 - cx}$$

applies, where $x_0 = 0$ and $(P)_0 = P_0$, and a, b, and c are constants. Show that this equation can be investigated with two critical parameters $c/(ab)$ and P_0^2/b using variable scaling proportional to (cx/b) and (P/P_0). Draw a scaled diagram to obtain graphs of the solution for some pairs of these parameters.

REFERENCES

[1] Emms, E. T., and K. H. Brinkmann, "The Minimization of the Effect of Drift in D. C. Analogue Computers," *Electronic Eng.*, vol. 32, Sept. 1960.

[2] Johnson, C. L., *Analog Computer Techniques*, 2nd ed. (New York: McGraw-Hill Book Co., 1963), p. 47.

[3] Korn, G. A., and T. M. Korn, *Electronic Analog and Hybrid Computers* (New York: McGraw-Hill Book Co., 1964), p. 59.

[4] Huntley, H. E., *Dimensional Analysis* (New York: Rinehart and Co., [1955?]).

[5] Hausner, A., "Parametric Techniques for Eliminating Division and Treating Singularities in Computer Solutions of Ordinary Differential Equations," *IRETEC*, vol. 11, Feb. 1962.

[6] Fifer, S., *Analogue Computation*, vol. 2 (New York: McGraw-Hill Book Co., 1961), p. 399.

[7] Bryant, L. T., *et al.*, "Engineering Applications of Analog Computers," *Argonne National Laboratory Report ANL-6319 Rev.*, Argonne, Ill., Oct. 1962.

[8] Lane, L. J., "A Method of Scaling and Checking Computer Circuits," *Trans. AIEE, Part II (Appl. and Ind.)*, vol. 77, May 1958.

[9] Rogers, A. E., and T. W. Connolly, *Analog Computation in Engineering Design* (New York: McGraw-Hill Book Co., 1960).

[10] Jackson, A. S., *Analog Computation* (New York: McGraw-Hill Book Co., 1960).

[11] Rubin, A. I., "Analog-Computer Programming, Scaling, and Problem Preparation," *Computer Handbook*, H. D. Huskey and G. A. Korn, eds. (New York: McGraw-Hill Book Co., 1962).

[12] Levy, S. M., "A Counter Example to an Assertion of Jackson," *IEEETEC*, vol. 12, Oct. 1963.

[13] Page, C., "Scaling for Analog Simulation," *Instr. and Control Systems*, vol. 37, Sept. 1964.

Chapter 5

GENERATING EXPLICIT FUNCTIONS WITH INTEGRATORS

As we have seen in Example 4.5, it is sometimes necessary to generate nonrational functions of problem variables. In such cases, the functions must be generated by means involving more than just the rational operations. Approximations, implicit functions, arbitrary or fixed function generators, etc., are methods to be discussed in subsequent chapters. In this chapter, we discuss functions generable on the computer by solving the differential equations they satisfy. Such a process, known as *generalized integration,* often provides better accuracy than the use of nonlinear equipment.

Generalized integration was developed quite early in the analog computer art [1], because the computer solves differential equations very well. To use this technique, we must find differential equations where right-hand expressions are rational functions or ones which can be easily generated on the computer. (Actually, right-hand expressions can be algebraic, such as a square root, because such functions are also easy to generate by implicit methods.) Not every analytic function, however, is the solution of a finite set of differential equations involving algebraic functions. The gamma function $\Gamma(x)$ is a classic example of a common function we could not generate on the computer by solving differential equations. It must be approximated by other functions generable on the computer.

5.1 Basic Functions Generable with Linear Equipment

Basic trigonometric, exponential, and other functions and their inverses can be generated with linear equipment with a proper choice of $\dot{x}$. These functions fall conveniently into groups, determined by their differential equations and the choice of $\dot{x}$. Three of the more common groups are those which require $\dot{x} = k =$ constant, $\dot{x} = kx$, and $\dot{x} = k(a^2 - x^2)^{1/2}$.

Group I ($\dot{x} = k =$ constant)

Let

$$y_1 = \sin ax$$
$$y_2 = \cos ax \qquad a, b \text{ constant}$$
$$y_3 = e^{bx}$$

Then

$$\begin{aligned} \dot{y}_1 &= ay_2\dot{x}, & (y_1)_0 &= \sin ax_0 \\ \dot{y}_2 &= -ay_1\dot{x}, & (y_2)_0 &= \cos ax_0 \\ \dot{y}_3 &= by_3\dot{x}, & (y_3)_0 &= e^{bx_0} \end{aligned} \tag{5.1}$$

It is clear that when $\dot{x} = k =$ constant, these functions are generable with linear equipment, since multiplication by the constants ak and bk can be performed with pots. If $\dot{x}$ is not constant, multipliers are required (as in the pendulum problem). We tacitly assume $\dot{x}$ is available in terms of some rational

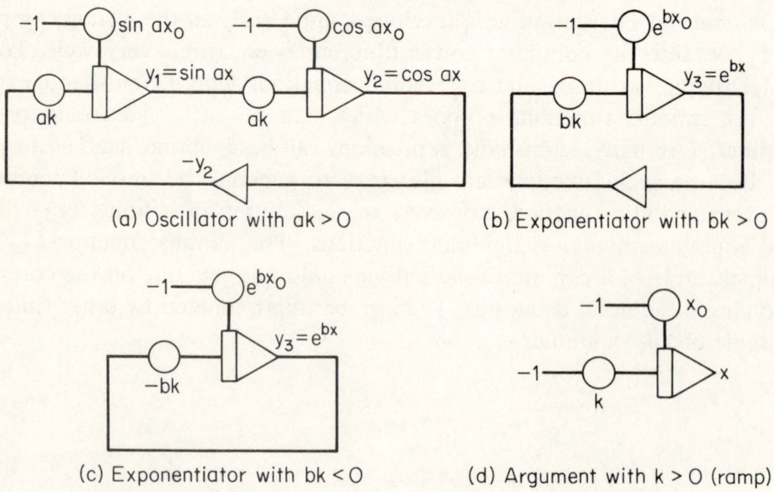

FIG. 5.1 Circuit diagrams for generating Group I functions ($\dot{x} = k$).

function. If this is not so, differentiating circuits are needed to obtain an $\dot{x}$ given an x (Chapter 10), but we then usually prefer to generate the function by some other means.

Figure 5.1 shows an unscaled circuit diagram solving (5.1) assuming $\dot{x} = k = \text{constant}$. The circuit generating sin ax and cos ax is called an *oscillator* (or *oscillating circuit*), and the circuits generating e^{bx} are called *exponentiators* (or *exponentiating circuits*). The circuit producing x is also called a *ramp.* The initial conditions must be correctly provided, and the actual mechanizations depend upon the signs of ak, bk, and k.

Group II ($\dot{x} = kx$)

Let

$$\begin{aligned} y_1 &= \ln x^a \\ y_2 &= x^b \\ y_3 &= \sin(\ln x^c) \\ y_4 &= \cos(\ln x^c) \end{aligned} \qquad a, b, c \text{ constants}$$

Then

$$\begin{aligned} \dot{y}_1 &= a\frac{\dot{x}}{x}, & (y_1)_0 &= \ln x_0^a \\ \dot{y}_2 &= by_2\frac{\dot{x}}{x}, & (y_2)_0 &= x_0^b \\ \dot{y}_3 &= cy_4\frac{\dot{x}}{x}, & (y_3)_0 &= \sin(\ln x_0^c) \\ \dot{y}_4 &= -cy_3\frac{\dot{x}}{x}, & (y_4)_0 &= \cos(\ln x_0^c) \end{aligned} \tag{5.2}$$

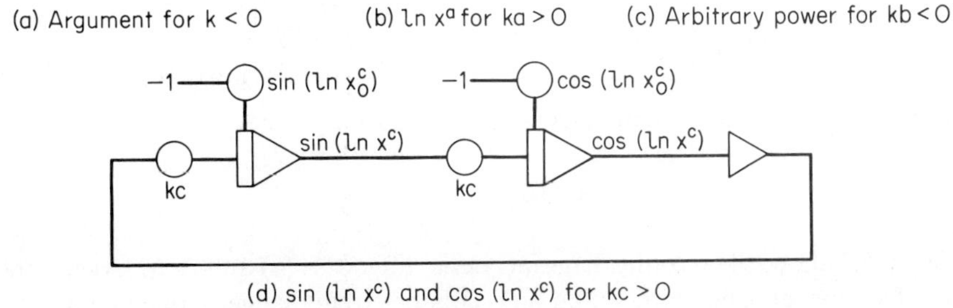

FIG. 5.2 Circuit diagrams for generating Group II functions ($\dot{x} = kx$).

These differential equations are generable with linear equipment when

$$\dot{x} = kx, \qquad (x)_0 = x_0 \tag{5.3}$$

with multipliers required otherwise. If $\dot{x}$ does not contain x as a factor, a division is also required. Figure 5.2 shows circuitry for generating the functions defined by (5.2) and (5.3) assuming the polarities of the constants indicated. The circuitry of Fig. 5.2 is similar to that of Fig. 5.1, namely in the use of a ramp, exponentiators, and an oscillator.

Group III ($\dot{x} = k\sqrt{a^2 - x^2}$)

Let

$$y_1 = \sqrt{a^2 - x^2}$$
$$y_2 = \exp\left[b \arccos\left(\frac{x}{a}\right)\right]$$
$$y_3 = \exp\left[c \arcsin\left(\frac{x}{a}\right)\right] \qquad a, b, c, d, f \text{ constants}$$
$$y_4 = d \arccos\left(\frac{x}{a}\right)$$
$$y_5 = f \arcsin\left(\frac{x}{a}\right)$$

Then

$$\begin{aligned}
\dot{y}_1 &= -x\frac{\dot{x}}{y_1}, & (y_1)_0 &= \sqrt{a^2 - x_0^2} \\
\dot{y}_2 &= -by_2\frac{\dot{x}}{y_1}, & (y_2)_0 &= \exp\left[b \arccos\left(\frac{x_0}{a}\right)\right] \\
\dot{y}_3 &= cy_3\frac{\dot{x}}{y_1}, & (y_3)_0 &= \exp\left[c \arcsin\left(\frac{x_0}{a}\right)\right] \\
\dot{y}_4 &= -d\frac{\dot{x}}{y_1}, & (y_4)_0 &= d \arccos\left(\frac{x_0}{a}\right) \\
\dot{y}_5 &= f\frac{\dot{x}}{y_1}, & (y_5)_0 &= f \arcsin\left(\frac{x_0}{a}\right)
\end{aligned} \tag{5.4}$$

A linear autonomous system again results, this time when

$$\dot{x} = ky_1 = k\sqrt{a^2 - x^2} \tag{5.5}$$

Figure 5.3 shows the circuit diagrams derived from (5.4) and (5.5). Again, the use of ramps, exponentiators, and an oscillator circuit suggests that these three

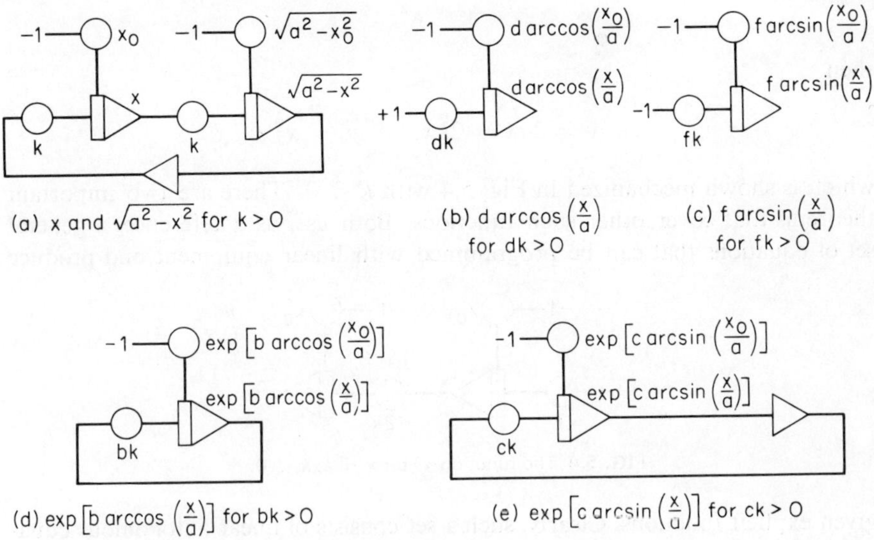

FIG. 5.3 Circuit diagrams for generating Group III functions ($\dot{x} = \sqrt{a^2 - x^2}$).

groups are derivable from one another by interchanging the role of x with any one of the dependent variables which is, in fact, the case.

In Fig. 5.3, some of the functions are multivalued. In fact, $(a^2 - x^2)^{1/2}$ can assume both positive and negative values, since the oscillator is used in its generation. Care must be taken when using multivalued functions to be sure the desired branch is generated.

Because of the simplicity of some of these circuits, our choice of $\dot{x}$ will be greatly influenced by the appearance of these functions in a problem. Of course, if functions of two different groups are needed, then both cannot be generated with linear equipment. Many other groups exist but are not frequently encountered. For example, functions of the form $(a + bx + cx^2)^{1/2}$ are generable with linear equipment when the proper choice of $\dot{x}$ is made.

These techniques are especially useful in generating functions with a more sophisticated computer containing point memory and mode control. In that case, $\dot{x}$ is not required, and the argument of the function can be any voltage. This is discussed further in Chapter 18.

5.2 Summation and Linear Equipment Theorems

The question arises about other functions generable with linear equipment within each group. Others exist, since it can be demonstrated that the function x^2 can be so generated using $\dot{x} = k$. For, if we let

$$y = x^2$$

then

$$\dot{y} = 2x\dot{x} = 2kx, \qquad y_0 = x_0^2$$

which is shown mechanized in Fig. 5.4 with $k < 0$. There are two important theorems that cover other such functions. Both use, as a reference, a general set of equations that can be programmed with linear equipment and produce

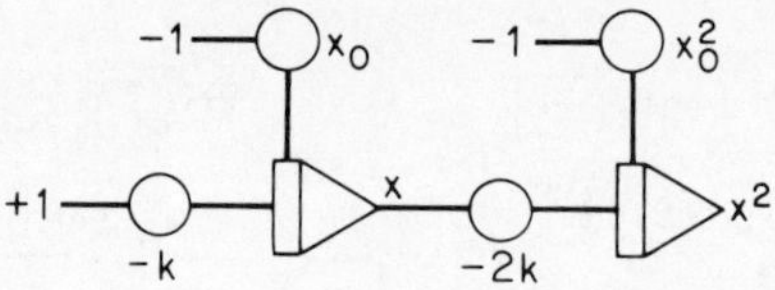

FIG. 5.4 The function x^2 for $x = k, k < 0$.

given explicit functions. Clearly, such a set consists of linear autonomous equations with constant coefficients. Let us consider, then, a set of n such equations

$$\dot{x}_i = \sum_{j=1}^{n} a_{i,j}x_j + b_i, \qquad i = 1, 2, \ldots, n, \qquad (x_i)_0 = c_i \tag{5.6}$$

where one of the x_i's, say x_1, is considered to be the problem independent variable. Our concern now lies in generating related functions on the computer.

***Theorem 5.1* (*Summation Theorem*).** If two specific functions x_t and x_r are generable at the output of integrators with linear equipment using (5.6), $x_t \pm x_r$ is generable at the output of an integrator with linear equipment. (t and r are integers in the closed interval $[1, n]$.)

Proof. By construction. Let

$$x_{t,r} = x_t \pm x_r \tag{5.7}$$

Then

$$\begin{aligned}\dot{x}_{t,r} &= \dot{x}_t \pm \dot{x}_r \\ &= \sum_{j=1}^{n} a_{t,j}x_j + b_t \pm \sum_{j=1}^{n} a_{r,j}x_j \pm b_r \\ &= \sum_{j=1}^{n} (a_{t,j} \pm a_{r,j})x_j + (b_t \pm b_r)\end{aligned} \tag{5.8}$$

where $(x_{t,r})_0 = (x_t)_0 \pm (x_r)_0$.

This seemingly trivial theorem has important uses, since feedback can sometimes be introduced in (5.8) by eliminating x_t or x_r from (5.7) thereby improving accuracy. Also, the application of the summation theorem frequently reduces equipment requirements. These benefits will be demonstrated later.

***Theorem 5.2* (*Linear Equipment Theorem*).** If two specific functions x_t and x_r are generable with linear equipment using (5.6), their product $x_t x_r$ is also generable with linear equipment. (t and r are integers in the closed interval $[1, n]$.)

Proof. By construction. Let $n(n + 1)/2$ new functions be defined as

$$x_{k,m} = x_k x_m \qquad \text{for all } k, m = 1, 2, \ldots, n \text{ such that } k \leq m.$$

Then, for all permissible k and m,

$$\begin{aligned} \dot{x}_{k,m} &= x_k \dot{x}_m + \dot{x}_k x_m \\ &= x_k\Big(\sum_{j=1}^{n} a_{m,j} x_j + b_m\Big) + x_m\Big(\sum_{j=1}^{n} a_{k,j} x_j + b_k\Big) \\ &= \sum_{j=1}^{n} (a_{m,j} x_k x_j + a_{k,j} x_m x_j) + b_m x_k + b_k x_m \\ &= \sum_{j=1}^{n} (a_{m,j} x_{k,j} + a_{k,j} x_{m,j}) + b_m x_k + b_k x_m \end{aligned} \tag{5.9}$$

(Since $x_{k,m}$ would equal $x_{m,k}$ if defined, permutation of these subscripts in (5.9) is necessary in the last line of the proof when $k > m$.) These equations together with (5.6) comprise a new linear system of order $n^2/2 + 3n/2$.

This theorem states, in effect, that positive integral powers and cross-products of functions generable with linear equipment are also generable with linear equipment. It is indeed a powerful theorem used frequently in lieu of multipliers. (Amplifiers are generally more plentiful than multipliers.) In terms of the proof, all cross-products and powers must be generated. In practice, many of the $a_{i,j}$ and b_i are zero, so that many functions are generable with less equipment than might seem to be required at first glance.

5.3 Examples of Functions Generable with Linear Equipment

In this section, we examine some of the more common functions generable with linear equipment and show the utility of the summation and linear equipment theorems.

5.3.1 When $\dot{x} = k$ (Group I functions)

To simplify our discussion when $\dot{x} = k$, we assume $k = 1$, since k is absorbed in time scaling anyway. Hence we refer to this case as $\dot{x} = 1$; if we wish to decrease the independent variable we use $\dot{x} = -1$.

Example 5.1. Generate powers of x. Let

$$y_i = x^i$$

Then

$$\dot{y}_i = ix^{i-1}\dot{x} = iy_{i-1}, \qquad (y_i)_0 = x_0^i \tag{5.10}$$

Figure 5.5 shows an unscaled diagram which solves (5.10). The even powers contain a sign inversion to save amplifiers. If needed, these can be inverted.

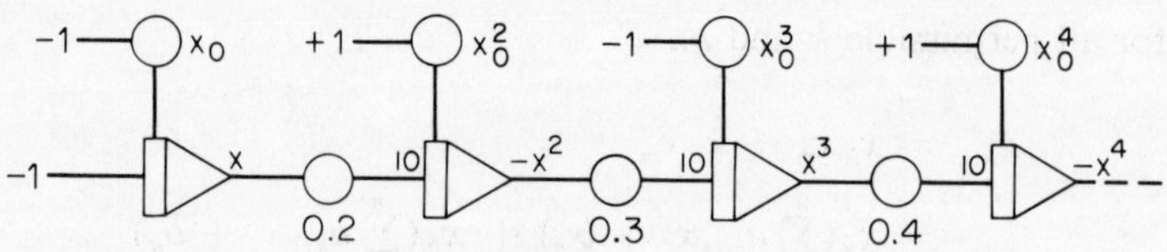

Shown for $x_0 \geq 0$ and $k=1$.

FIG. 5.5 Unscaled diagram for powers of x using $\dot{x} = 1$.

Example 5.2. Generate polynomials. Let

$$y = \sum_{i=0}^{n} a_i x^i \tag{5.11}$$

This function can be formed directly by summing the terms as shown in Fig. 5.6. If any a_i is negative, the input can come from the power $+x^i$. To save the summer,

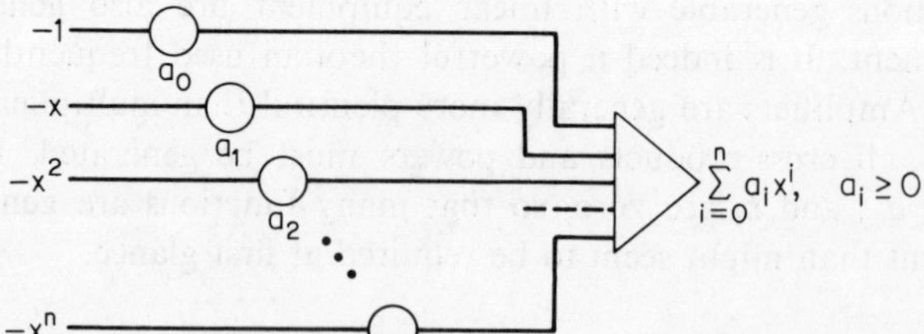

FIG. 5.6 Summing $a_i x^i$ for polynomial.

the summation theorem can be used by generating all derivatives of y as follows: Let

$$y_i = \frac{d^i y}{dx^i}, \qquad 0 \leq i \leq n-1$$

with $y_0 = y$. Then

$$\dot{y}_i = y_{i+1}\dot{x} = y_{i+1}, \qquad 0 \le i \le n-2, \qquad (y_i)_0 = \frac{d^i y(x_0)}{dx^i} \tag{5.12}$$

$$\dot{y}_{n-1} = \frac{d^n y}{dx^n} = a_n n!, \qquad (y_{n-1})_0 = \frac{d^{n-1}y(x_0)}{dx^{n-1}} \tag{5.13}$$

To minimize inverter requirements, the derivatives are generated with alternating signs as shown in Fig. 5.7. Amplitude and time scaling of the derivatives will generally require pots between integrators, which are not shown. The initial conditions

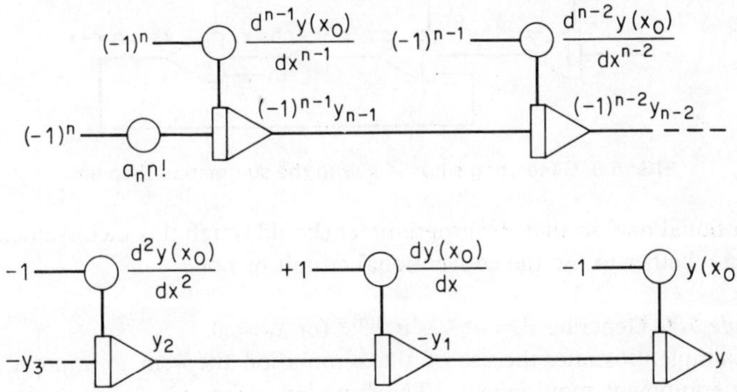

FIG. 5.7 Polynomial generation using summation theorem with $\dot{x} = 1$.

must be calculated from the formulas by actually differentiating (5.11) and substituting the initial value of x, namely x_0. The actual expressions for the derivatives are also useful for amplitude scaling.

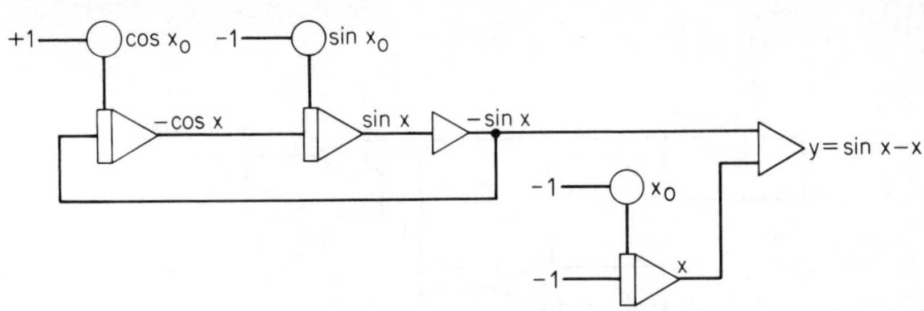

FIG. 5.8 Generating $\sin x - x$ conventionally.

Example 5.3. Generate $y = \sin x - x$.

This example again illustrates the use of the summation theorem to conserve equipment. The conventional method for generating y is simply by summing one of the outputs of an oscillator with $+x$ as shown in Fig. 5.8. (One integrator is labeled $-\cos x$ to make $-\sin x$ available.) To generate y at the output of an integrator, it is best to introduce feedback as follows: Let

$$C = +\cos x$$

Then

$$\dot{y} = C - 1, \qquad y_0 = \sin x_0 - x_0$$
$$\dot{C} = -\sin x = -y - x, \qquad C_0 = +\cos x_0$$

As shown in Fig. 5.9, one less amplifier is required than in the conventional circuit. Scaling pots have again been omitted. This circuit is more difficult to check out than

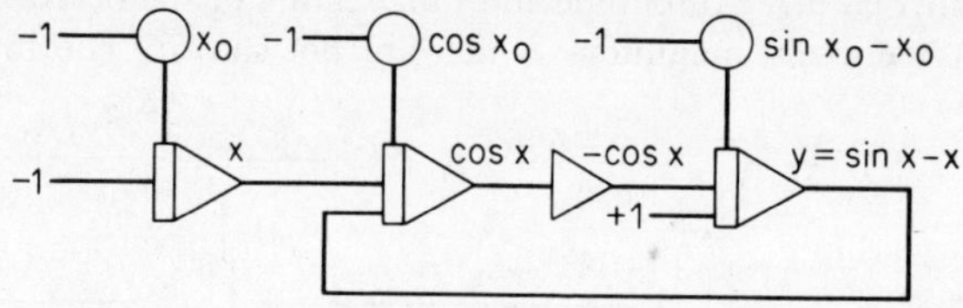

FIG. 5.9 Generating sin $x - x$ with the summation theorem.

the conventional one, so that the programmer should weigh this inconvenience factor in deciding whether to use the conventional circuit or not.

Example 5.4. Generate $y = e^{-x} - e^{-1.01x}$ for $x_0 = 0$.

This example illustrates the use of the summation theorem to improve accuracy and lessen equipment requirements. The function y for $x > 0$ has a maximum of about 0.01 and should be scaled as $(100y)$. If, however, the individual terms are generated as in Fig. 5.1c, these integrators should each be scaled for a maximum of 1. A gain of 100 is then required in the summer producing y, as shown in Fig. 5.10. This, in turn, magnifies noise and errors, resulting in an unsatisfactory mechanization.

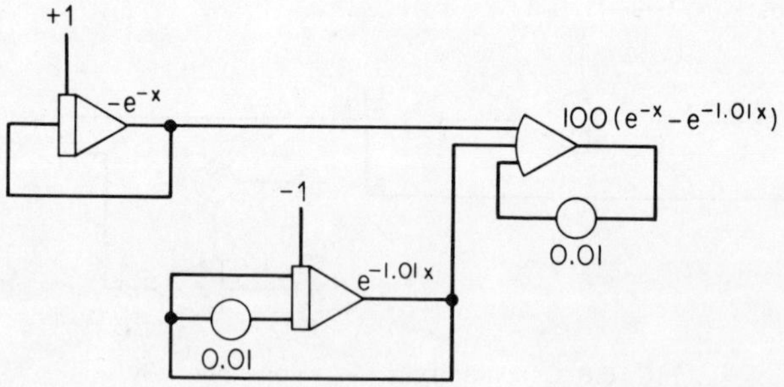

FIG. 5.10 Conventional generation of $100(e^{-x} - e^{-1.01x})$ with $\dot{x} = 1$, $x_0 = 0$.

If y is produced at the output of an integrator, much better results are obtained. For, we let

$$H = -e^{-x}$$

Then

$$\dot{H} = -H, \qquad H_0 = -1$$
$$\dot{y} = -e^{-x} + 1.01\, e^{-1.01x}, \qquad y_0 = 0$$

Now we may introduce feedback from the definition of y, since

$$e^{-1.01x} = -H - y$$

Hence

$$\begin{aligned} \dot{y} &= H + 1.01(-H - y) \\ &= -0.01H - 1.01y, \qquad y_0 = 0 \end{aligned}$$

The scaled equations are now compatible with each other:

$$(\dot{H}) = -(H), \qquad (H)_0 = -1$$
$$(100\dot{y}) = -(H) - 1.01(100y), \qquad (100y_0) = 0$$

The mechanization, shown in Fig. 5.11, produces excellent results, with minimal equipment.

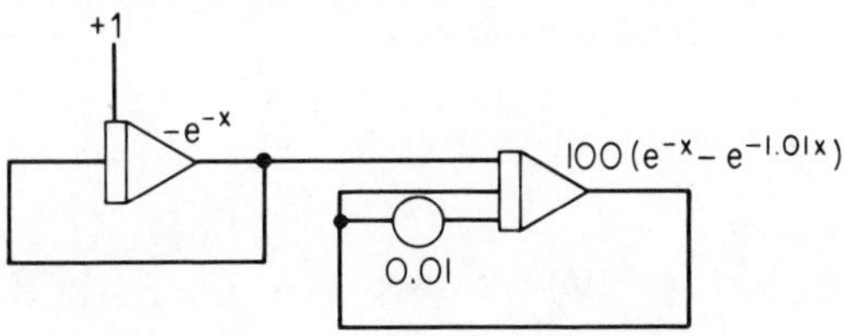

FIG. 5.11 Improved generation of $100(e^{-x} - e^{-1.01x})$ with $x = 1$, $x_0 = 0$.

Example 5.5. Generate $y = xe^{-10x}$ for $x_0 = 0$ and $0 \leq x \leq 1$.

This example illustrates two excellent uses of the linear equipment theorem: (1) improving the accuracy of products that are badly scaled, and (2) eliminating feedback from an increasing x when the function is bounded. If (x) and (e^{-10x}) are multiplied together, the product results in a maximum of $[1/(10e)]$ found by setting $dy/dx = 0$. The function y should therefore be scaled at about $(25y)$. This difficulty sometimes occurs with products because one factor is large when the other is small and vice versa. If we let

$$H = -e^{-10x}$$

then

$$\dot{H} = -10H, \qquad H_0 = -1$$

Also

$$\dot{y} = -10xe^{-10x} + e^{-10x}$$

which, by introducing feedback once again, becomes

$$\dot{y} = -10y - H, \qquad y_0 = 0$$

By amplitude scaling,

$$(\dot{H}) = -10(H), \qquad (H)_0 = -1$$
$$(25\dot{y}) = -10(25y) - 25(H), \qquad (25y)_0 = 0$$

which are mechanized in Fig. 5.12 with $k = 0.1$. The variable x is not required as

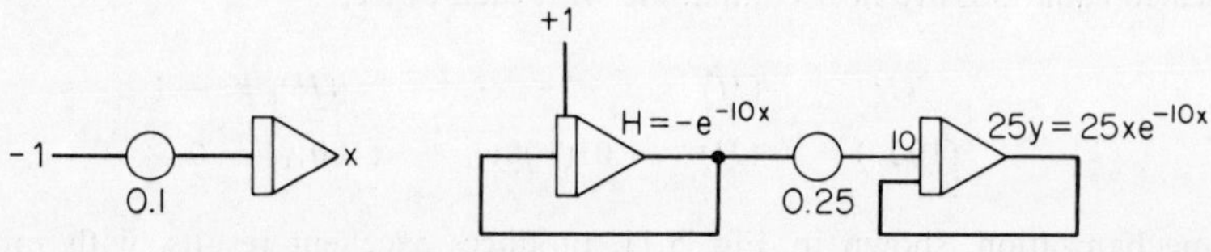

FIG. 5.12 Generation of $25xe^{-10x}$ with $\dot{x} = 0.1$ and $x_0 = 0$.

feedback, as would be required with a multiplier, so that the function could be used for $0 \leq x < \infty$ if x were not mechanized.

Example 5.6. Generate $y = \sin^n x$ with $x_0 = 0$.
Let

$$G = -\sin^{n-1} x \cos x$$

Then

$$\dot{y} = -nG, \qquad y_0 = 0$$
$$\dot{G} = -(n-1)\sin^{n-2} x \cos^2 x + \sin^n x$$

By substituting $1 - \sin^2 x$ for $\cos^2 x$, we obtain

$$\dot{G} = -(n-1)\sin^{n-2} x + ny, \qquad G_0 = 0 \text{ for } n > 1$$
$$= -1 \text{ for } n = 1$$

Thus, even or odd powers of $\sin x$ are generated by cascading oscillator circuits. A circuit for odd powers is shown in Fig. 5.13. Similar principles are involved in generating even powers of $\sin x$ and both odd and even powers of $\cos x$. The constraint $\sin^2 x + \cos^2 x = 1$ reduced the number of cross-products required in the general proof of the linear equipment theorem. This function could actually be mechanized with only two inverters for any power [2], but the mechanization is more tedious to set up, and the functions generated are not so simple.

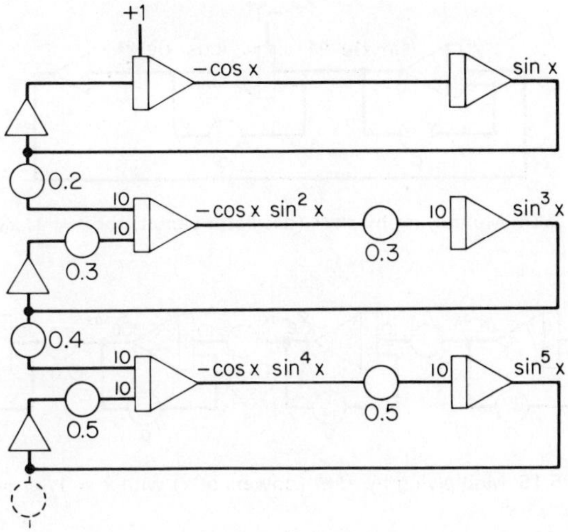

FIG. 5.13 Circuit for odd powers of sin x with $\dot{x} = 1$, $x_0 = 0$.

***Example** 5.7.* Generate $y = G(x)e^{-ax}$, $a > 0$, $\dot{x} = 1$, where $G(x)$ is generable with linear equipment when $\dot{x} = 1$.

Assume that $G(x)$ is part of a linear system of functions $G_i(x)$ with constant coefficients:

$$\dot{G}_i = \sum_{j=1}^{n} a_{i,j} G_j + b_i, \qquad i = 1, 2, \ldots, n \tag{5.14}$$

and let

$$y_i = e^{-ax} G_i$$

Then

$$\begin{aligned} \dot{y}_i &= -ae^{-ax} G_i + e^{-ax} \dot{G}_i \\ &= -ay_i + \sum_{j=1}^{n} a_{i,j} e^{-ax} G_j + b_i e^{-ax} \\ &= -ay_i + \sum_{j=1}^{n} a_{i,j} y_j + b_i e^{-ax} \end{aligned} \tag{5.15}$$

This result shows that identical circuitry is required in (5.14) and (5.15), except that (1) all integrators have an additional feedback path with gain a and (2) all constants b_i are replaced by decreasing exponential functions. Figures 5.14 and 5.15 show circuit diagrams for this variation on two previous circuits in Figs. 5.1 and 5.5.

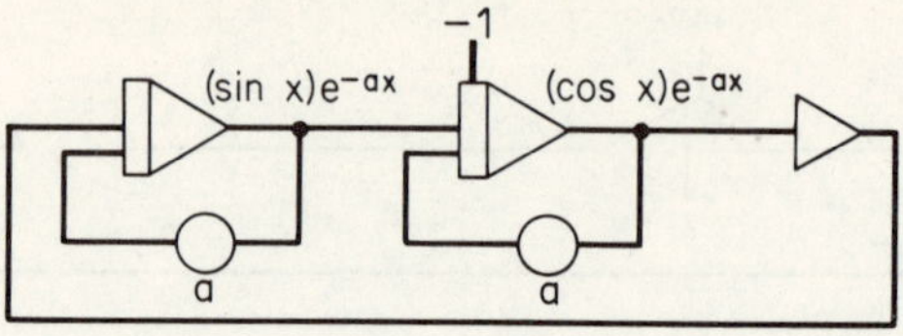

FIG. 5.14 Multiplying by e^{-ax} (oscillator circuit) for $\dot{x} = 1$, $x_0 = 0$.

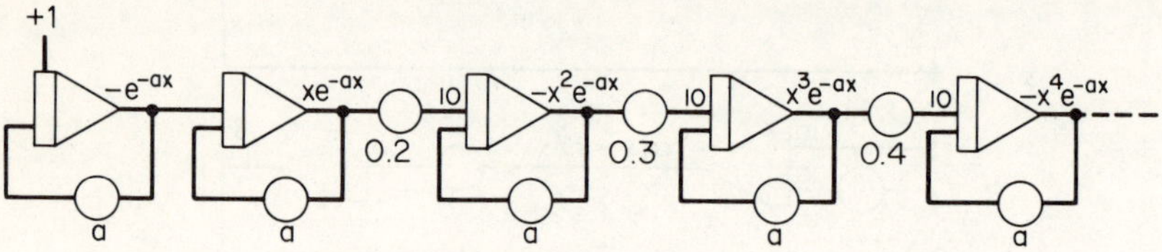

FIG. 5.15 Multiplying by e^{-ax} (powers of x) with $\dot{x} = 1$, $x_0 = 0$.

5.3.2 When $\dot{x} = kx$ (Group II functions)

This case has important stability properties when k is negative so that we attempt to arrange solving a problem containing group II functions with negative k. Since k again is absorbed in scaling, we refer to this case as

$$\dot{x} = -x, \qquad (x)_0 = x_0$$

Group II functions generable with linear equipment are now covered by the following single theorem.

Theorem 5.3. If $y(x)$ is generable with linear equipment with $\dot{x} = 1$, then $y(-\ln x)$ is generable with linear equipment using $\dot{x} = -x$ and identical circuitry.

Proof. Assume $y(x)$ is one of the y_i in the system

$$\dot{x} = 1, \qquad \dot{y}_i = f_i(x, y_1, y_2, \ldots, y_n) \tag{5.16}$$

with initial conditions specified. Then, we have

$$\frac{dy_i}{dx} = \frac{\dot{y}_i}{\dot{x}} = f_i[x, y_1(x), y_2(x), \ldots, y_n(x)]$$

To generate $y_i(-\ln x)$, we let

$$z_i = y_i(w), \qquad w = -\ln x \tag{5.17}$$

Now,

$$\frac{dz_i}{dx} = \frac{dy_i}{dw}\frac{dw}{dx} = \frac{dy_i}{dw}\left(-\frac{1}{x}\right)$$

But

$$\frac{dy_i}{dw} = f_i[w, y_1(w), y_2(w), \ldots, y_n(w)]$$

$$= f_i(w, z_1, z_2, \ldots, z_n)$$

Hence

$$\frac{dz_i}{dx} = -\frac{f_i(w, z_1, z_2, \ldots, z_n)}{x}$$

If we choose

$$\dot{x} = -x \tag{5.18}$$

then

$$\dot{z}_i = f_i(w, z_1, z_2, \ldots, z_n)$$

which results in the identical circuitry as in (5.16) because $\dot{w} = 1$ from (5.17), except that (5.18) is now included also. This is equivalent to simply replacing x by $-\ln x$ in any circuit diagram. Note carefully, however, that the initial conditions should be adjusted to correspond to the proper new x_0.

Example 5.8. Generate powers of $\ln x$ with $\dot{x} = -x$.

Let

$$z_i = (\ln x)^i$$

Then

$$\dot{z}_i = i(\ln x)^{i-1}\frac{\dot{x}}{x} = i(z_{i-1})\frac{\dot{x}}{x}$$

or, with

$$\dot{x} = -x$$

we have

$$\dot{z}_i = -iz_{i-1}, \qquad (z_i)_0 = (\ln x_0)^i \tag{5.19}$$

Figure 5.16 shows the mechanization of (5.19) with the negative powers generated to show how identical it is with the diagram of Fig. 5.5 when x in that diagram is replaced by $-\ln x$.

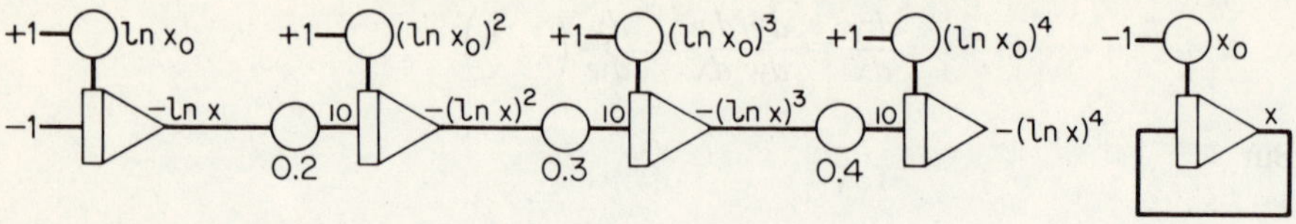

FIG. 5.16 Unscaled diagram for powers of ln x using $\dot{x} = -x$.

5.3.3 Other systems

Group III functions can be generated from the circuitry generating group I functions by replacing x by $\sin x$. Virtually any integrator output may be taken as the independent variable, thus providing the possibility of generating inverse functions of all group I functions with linear equipment. Thus, because $F = xe^{-x}$ is in group I, a function G defined by $x = Ge^{-G}$ is generable with linear equipment with the proper choice of $\dot{x}$. Most of the possibilities, however, do not arise in practice.

5.4 Functions Requiring Multipliers

All the functions generable with linear equipment require multipliers when $\dot{x}$ is chosen in a way other than that described in the three previous sections. In those cases, the differential equations derived are still applicable, but the actual $\dot{x}$ is used in each equation. Thus, system (5.4) can be solved with any $\dot{x}$ by replacing k by $\dot{x}/y_1$ in the circuits of Fig. 5.3. Furthermore, such functions of one or more dependent variables are nearly always treated in this way.

Example 5.9. Solve the falling ball problem (Example 3.1) using $A(y) =$ 0.001 exp $(-0.00004328y)$ per foot, y in feet. This term is approximately proportional to the standard atmospheric weight density of air.

We have, assuming no change of g with height,

$$\frac{dy}{dt} = v, \qquad (y)_0 = y_0$$

$$\frac{dv}{dt} = -32.17 + 0.001e^{-0.00004328y}v^2, \qquad (v)_0 = v_0$$

where the units of y, t, and v are feet, seconds, and ft/s. Let

$$D = e^{-0.00004328y}$$

Then

$$\frac{dD}{dt} = -0.00004328D\frac{dy}{dt} = -0.00004328vD,$$

so that

$$\dot{y} = v\dot{t}, \qquad (y)_0 = y_0$$
$$\dot{v} = (-32.17 + 0.001Dv^2)\dot{t}, \qquad (v)_0 = v_0$$
$$\dot{D} = -0.00004328vD\dot{t}, \qquad (D)_0 = e^{-0.00004328y_0}$$

There is no special advantage in choosing $\dot{t}$ to enable D to be generated with linear equipment ($\dot{t}$ inversely proportional to v) since the other equations become unnecessarily complicated, particularly if $v = 0$ sometime during the computation. Here, $\dot{t} = k$ is appropriate.

There are also a myriad of functions that could not be generated with linear equipment under any circumstances. For example, the function

$$y = e^{-ax^2}$$

is generated by solving the differential equation

$$\dot{y} = -2axy\dot{x}$$

and no choice of $\dot{x}$ will simplify the expressions of both $\dot{x}$ and $\dot{y}$ so as not to involve any multiplication or division of variables. Other elementary functions in this class are tan x, sec x, ctn x, csc x, x^x, etc. The former functions can be obtained by generating sin x and cos x and forming the appropriate quotients, although differential equations may also be found with these functions as solutions. The function x^x, on the other hand, must be generated from scratch.

Example 5.10. Obtain a graph of x^x for $0 \leq x \leq 1$ by solving its differential equation. With

$$y = x^x$$

we have

$$\frac{dy}{dx} = y(1 + \ln x) = \frac{y(x + x \ln x)}{x}$$

Let

$$z = x \ln x$$

Then

$$\frac{dz}{dx} = 1 + \ln x = 1 + \frac{z}{x}$$

Hence

$$\dot{y} = \frac{y(x+z)}{x}\dot{x}, \qquad \dot{z} = \frac{x+z}{x}\dot{x}$$

so that a good mechanization results from

$$\begin{aligned} \dot{x} &= -x, & x_0 &= 1 \\ \dot{z} &= -(x+z), & z_0 &= 0 \\ \dot{y} &= -y(x+z), & y_0 &= 1 \end{aligned} \tag{5.20}$$

Figure 5.17 is a scaled diagram solving (5.20) using a servo multiplier to obtain the indicated product.

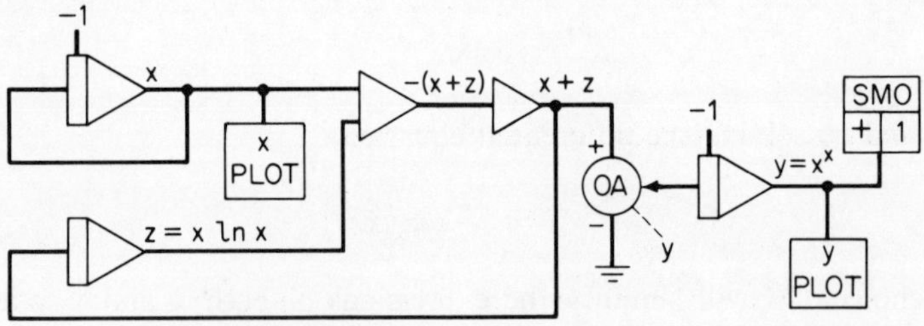

FIG. 5.17 Mechanization to plot x^x for $0 < x \leq 1$.

Many other often-needed functions are solutions of higher-order differential equations that can be solved on the computer. A partial list is shown in Table 5.1. The mechanization of some of these equations requires special techniques to overcome difficulties due to singularities or indeterminate expressions. An early technique introduced by Mengel and Melahn [1] approximates the differential equation near the point of indeterminacy. For example, in generating Bessel functions for integral $n \geq 1$, $\dot{x} = 1$, and $x_0 = 0$, the equation

$$\frac{d^2y}{dx^2} = \frac{-x\dfrac{dy}{dx} - (x^2 - n^2)y}{x^2}$$

contains an indeterminate expression initially, but the equation

$$\frac{d^2y}{dx^2} = \frac{\dfrac{x^{n-2}}{2^n(n-2)!}be^{-ax} - x\dfrac{dy}{dx} - (x^2 - n^2)y}{be^{-ax} + x^2}$$

not only is free of the indeterminacy but also approximates the true differential equation at the origin (it can be shown that $d^2y/dx^2 \simeq x^{n-2}/[2^n(n-2)!]$ near $x = 0$). By suitably choosing a and b, errors due to the approximation presumably can be made small. This method was also successfully used with the function

TABLE 5.1

Function name, symbols, and definitions	Differential equations (see [3] for initial conditions)
Error function $y = erf\, x = \frac{2}{\sqrt{\pi}} \int_0^x e^{-t^2}\, dt$	$\frac{d^2y}{dx^2} = -2x\frac{dy}{dx}$
Dawson's integral $y = e^{-x^2} \int_0^x e^{t^2}\, dt$	$\frac{dy}{dx} = 1 - 2xy$
Fresnel integrals $C(x) = \int_0^x \cos\left(\frac{\pi t^2}{2}\right) dt$ $S(x) = \int_0^x \sin\left(\frac{\pi t^2}{2}\right) dt$	$\frac{dC}{dx} = C_1, \quad \frac{dS}{dx} = S_1$ $\frac{dS_1}{dx} = \pi x C_1, \quad \frac{dC_1}{dx} = -\pi x S_1$
Legendre functions, y	$(1 - x^2)\frac{d^2y}{dx^2} - 2x\frac{dy}{dx} + n(n+1)y = 0$
Bessel functions, y	$x^2\frac{d^2y}{dx^2} + x\frac{dy}{dx} + (x^2 - n^2)y = 0$
Kummer functions, y	$x\frac{d^2y}{dx^2} + (b - x)\frac{dy}{dx} - ay = 0$
Whittaker functions, y	$\frac{d^2y}{dx^2} + \left(-\frac{1}{4} + \frac{K}{x} + \frac{0.25 - \mu^2}{x^2}\right)y = 0$
Jacobian elliptic integrals, ϕ $\left(\text{related to } u = \int_0^\phi \frac{d\theta}{(1 - m\sin^2\theta)^{1/2}}\right)$	$\frac{d^2\phi}{du^2} = -m \sin\phi \cos\phi$
Parabolic cylinder functions, y	$\frac{d^2y}{dx^2} - (0.25x^2 + a)y = 0$ or $\frac{d^2y}{dx^2} + (0.25x^2 - a)y = 0$
Mathieu functions, y	$\frac{d^2y}{dx^2} + (a - 2q\cos 2x)y = 0$ or $(1 - t^2)\frac{d^2y}{dt^2} - t\frac{dy}{dt} + (a + 2q - 4qt^2)y$ $= 0$ (for $\cos x = t$)
Hypergeometric function, $y(a, b, c, x)$	$x(1 - x)\frac{d^2y}{dx^2} + [c - (a + b + 1)x]\frac{dy}{dx}$ $- aby = 0$

$$y = \frac{\sin x}{x}$$

by generating

$$y = \frac{be^{-ax} + \sin x}{be^{-ax} + x}$$

Another method for handling some indeterminate expressions in differential equations was offered by Van Remoortere [4], who replaced the equation by one derived from an infinite series solution. For small x, the series can be truncated, and the mechanization can be based on solving an equation yielding approximations that are very accurate by analog computer standards. For larger x, the exact differential equation is solved, with the two solutions combined automatically by switching (Chapter 6).

Still another method [5] is to approximate the solution for small x by any means (preferably Chebyshev polynomials, Chapter 7) and then solve the differential equation exactly for larger x, again combining the two solutions by switching. All of these methods rely on solving the exact equation away from the point of indeterminacy, so that effective methods for such a solution are quite important.

5.5 Rational Functions

Rational functions, i.e., the quotient of two polynomials, can be generated by multipliers and summers. Why then should we consider the generation of rational functions with integrators? The answer is quite simple: There are many cases where accuracy and convenience can be improved by this method. In this section, we examine a general procedure to generate rational functions with integrators and illustrate several important applications.

Consider the function

$$y = \frac{N}{D}$$

where

$$N = \sum_{j=0}^{n} a_j x^j, \qquad D = \sum_{j=0}^{m} b_j x^j, \qquad m \geq n \tag{5.21}$$

We assume that D has no zeros for $x > x_0$ and are concerned with the case where $\dot{x} = 1$ and $(x)_0 = x_0$. Let

$$y_i = \frac{x^i}{D}, \qquad i = 0, 1, \ldots, m - 1 \tag{5.22}$$

so that

$$y = \sum_{j=0}^{n} a_j y_j$$

with the function $x^m/D = y_m$ definable in terms of the other y_i. Now

$$\dot{y}_i D + y_i \frac{dD}{dx}\dot{x} = i x^{i-1}\dot{x}$$

or, since $\dot{x} = 1$,

$$\dot{y}_i = \frac{i x^{i-1} - y_i \dfrac{dD}{dx}}{D}$$

Because

$$\frac{dD}{dx} = \sum_{j=0}^{m} j b_j x^{j-1}$$

we have

$$\dot{y}_i = i y_{i-1} - y_i \sum_{j=0}^{m} j b_j y_{j-1} \tag{5.23}$$

where initial conditions are defined by (5.21) and (5.22) at $x = x_0$. If each product in (5.23) is generated by servo multiplication, only one servo with n cups is needed, with the function

$$\frac{dD/dx}{D} = \sum_{j=0}^{m} j b_j y_{j-1}$$

as the driver. The chief advantage of using this method is that x is eliminated from the feedback equations. This factor not only can be a big convenience in any mechanization, but can permit accurate generation of functions for larger ranges.

Example 5.11. Do example 4.2, but plot a graph of the current $I(t)$ for large t on a recorder.

There is now no advantage in not choosing

$$i = k$$

so that instead of (4.9), we obtain

$$\dot{I}_1 = k\left(V_1 - \frac{RI_1}{L}\right), \qquad \dot{V}_1 = -\frac{k}{C}\frac{I_1}{L}$$

The function

$$\frac{1}{L} = \frac{1}{k_0 + k_1 t^2}$$

is a rational function satisfying the conditions for its generation by integration by using (5.23). We let

$$F_0 = \frac{1}{k_0 + k_1 t^2}, \qquad F_1 = \frac{t}{k_0 + k_1 t^2} \tag{5.24}$$

We now have

$$\begin{aligned} \dot{F}_0 &= -2kk_1 F_0 F_1, \qquad & (F_0)_0 &= \frac{1}{k_0} \\ \dot{F}_1 &= k(F_0 - 2k_1 F_1^2), \qquad & (F_1)_0 &= 0 \end{aligned} \tag{5.25}$$

as well as

$$\begin{aligned} \dot{t} &= k, & (t)_0 &= 0 \\ \dot{I}_1 &= k(V_1 - RI), & (I_1)_0 &= 0 \\ \dot{V}_1 &= -\frac{k}{C} I, & (V_1)_0 &= (V_1)_0 \\ I &= I_1 F_0 \end{aligned}$$

Using variable scale factors as before: $\{RI_1/[k_0(V_1)_0]\}$, and $[V_1/(V_1)_0]$, together with $(k_0 F_0)$, $(2\sqrt{k_1 k_0} F_1)$, $[t/(RC)]$, and $[RI/(V_1)_0]$ we obtain, with $k = RC$,

$$\begin{aligned} (k_0 \dot{F}_0) &= -RC\sqrt{\frac{k_1}{k_0}}\,(k_0 F_0)(2\sqrt{k_1 k_0} F_1), & (k_0 F_0)_0 &= 1 \\ (2\sqrt{k_1 k_0} \dot{F}_1) &= 2RC\sqrt{\frac{k_1}{k_0}}\,[(k_0 F_0) - 0.5(2\sqrt{k_1 k_0} F_1)^2], & (2\sqrt{k_0 k_1} F_1)_0 &= 0 \\ \left[\frac{R\dot{I}_1}{k_0(V_1)_0}\right] &= \frac{R^2 C}{k_0}\left\{\left[\frac{V_1}{(V_1)_0}\right] - \left[\frac{RI}{(V_1)_0}\right]\right\}, & \left[\frac{RI_1}{k_0(V_1)_0}\right]_0 &= 0 \\ \left[\frac{\dot{V}_1}{(V_1)_0}\right] &= -\left[\frac{RI}{(V_1)_0}\right], & \left[\frac{V_1}{(V_1)_0}\right]_0 &= 1 \\ \left(\frac{\dot{t}}{RC}\right) &= 1, & \left(\frac{t}{RC}\right)_0 &= 0 \end{aligned} \tag{5.26}$$

with

$$\left[\frac{RI}{(V_1)_0}\right] = \left[\frac{RI_1}{k_0(V_1)_0}\right](k_0 F_0) \tag{5.27}$$

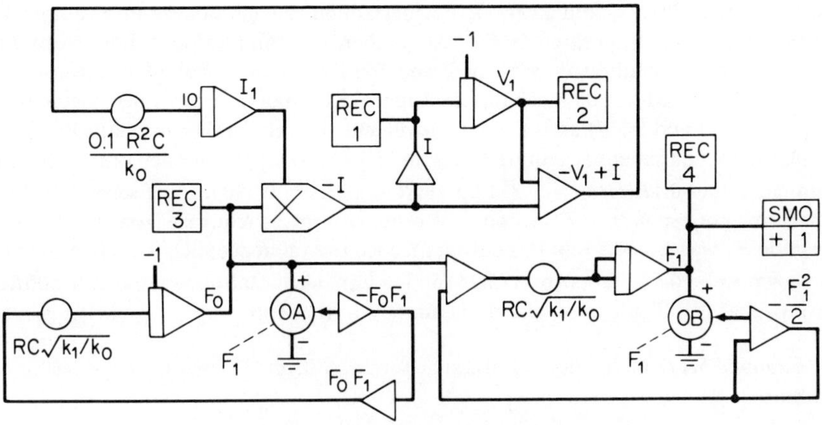

FIG. 5.18 Mechanization of (5.26) and (5.27).

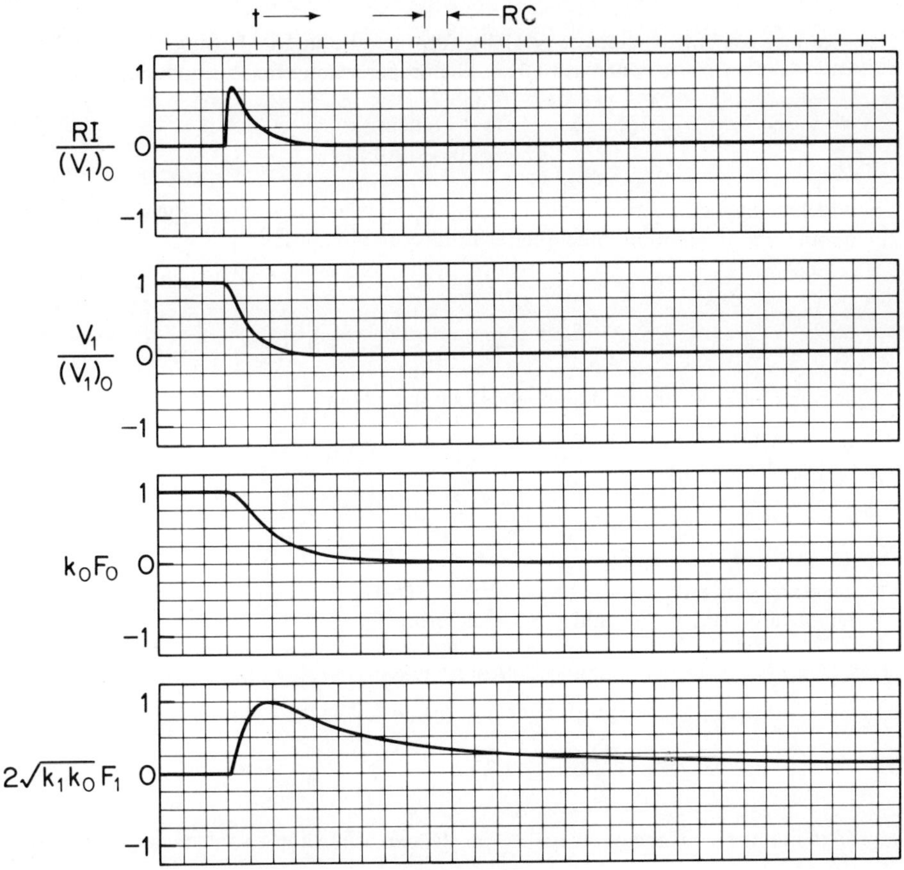

FIG. 5.19 Recorder output of the circuit in Fig. 5.18.

Two constants, R^2C/k_0 and $RC(k_1/k_0)^{1/2}$, determine the solution as in Example 4.2 (the second constant appeared as $R^2C^2k_1/k_0$ there), but in (5.26), t is not involved in feedback, so the solution may be observed for large t on a recorder because the equation for t need not be mechanized. Figure 5.18 shows an unscaled mechanization for (5.26) and (5.27) and is more convenient than the earlier mechanization, although more equipment is required. Figure 5.19 shows the recorder output for the nominal values $0.1R^2C/k_0 = 0.8712$ and $RC\sqrt{k_1/k_2} = 0.5903$. Each second of computer time corresponds to RC seconds of problem time as required from (5.26). The function F_1 was scaled so that its output reached the reference; i.e., its maximum can be shown to be $0.5/\sqrt{k_1k_0}$ from (5.24) and (5.25). The graph shows results for $30RC$, a range that is difficult to obtain if t is directly in the loop.

Example 5.12. Generate $y = \arctan x$ for $x_0 = 0$, $\dot{x} = 1$, and $0 \leq x \leq 100$. Since

$$\frac{dy}{dx} = \frac{1}{1 + x^2}$$

y can be generated (since $\dot{x} = 1$) by forming the quotient using $\dot{y} = 1/(1 + x^2)$. However, since x is scaled as $(0.01x)$, the proper scaling for this quotient is

$$\dot{y} = \frac{0.0001}{0.0001 + (0.01x)^2}$$

which cannot be obtained without large error when x is small. By generating this quotient by integration, much better results are obtained.

Let

$$F_0 = \frac{1}{1 + x^2}, \qquad F_1 = \frac{x}{1 + x^2}$$

as was similarly done in the previous example. Now

$$\begin{aligned} \dot{F}_0 &= -2F_0F_1, & (F_0)_0 &= 1 \\ \dot{F}_1 &= F_0 - 2F_1^2, & (F_1)_0 &= 0 \\ \dot{y} &= F_0, & y_0 &= 0 \end{aligned}$$

generates $y = \arctan x$. Since $y_{\max} = \pi/2$, $(F_0)_{\max} = 1$, and $(F_1)_{\max} = 0.5$, the scale factors are $(2y/\pi)$, (F_0), and $(2F_1)$. Therefore

$$\begin{aligned} (\dot{F}_0) &= -(F_0)(2F_1), & (F_0)_0 &= 1 \\ (2\dot{F}_1) &= 2(F_0) - (2F_1)^2, & (2F_1)_0 &= 0 \\ \left(\frac{2\dot{y}}{\pi}\right) &= \frac{2}{\pi}(F_0), & \left(\frac{2y}{\pi}\right)_0 &= 0 \end{aligned} \tag{5.28}$$

If it is more convenient to scale y in degrees using the variable y_D, we note that since

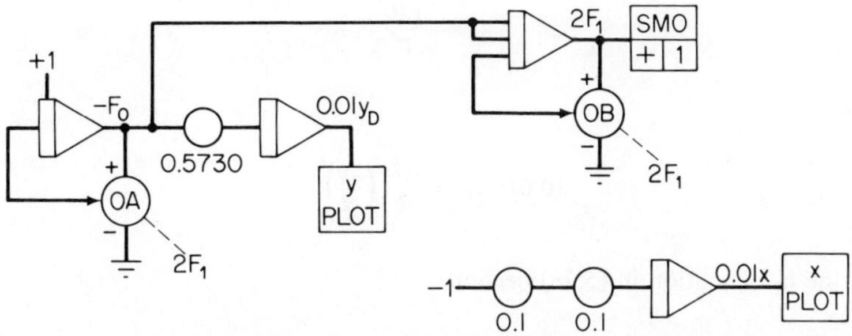

FIG. 5.20 Mechanization for $y_D = \arctan x$ in degrees.

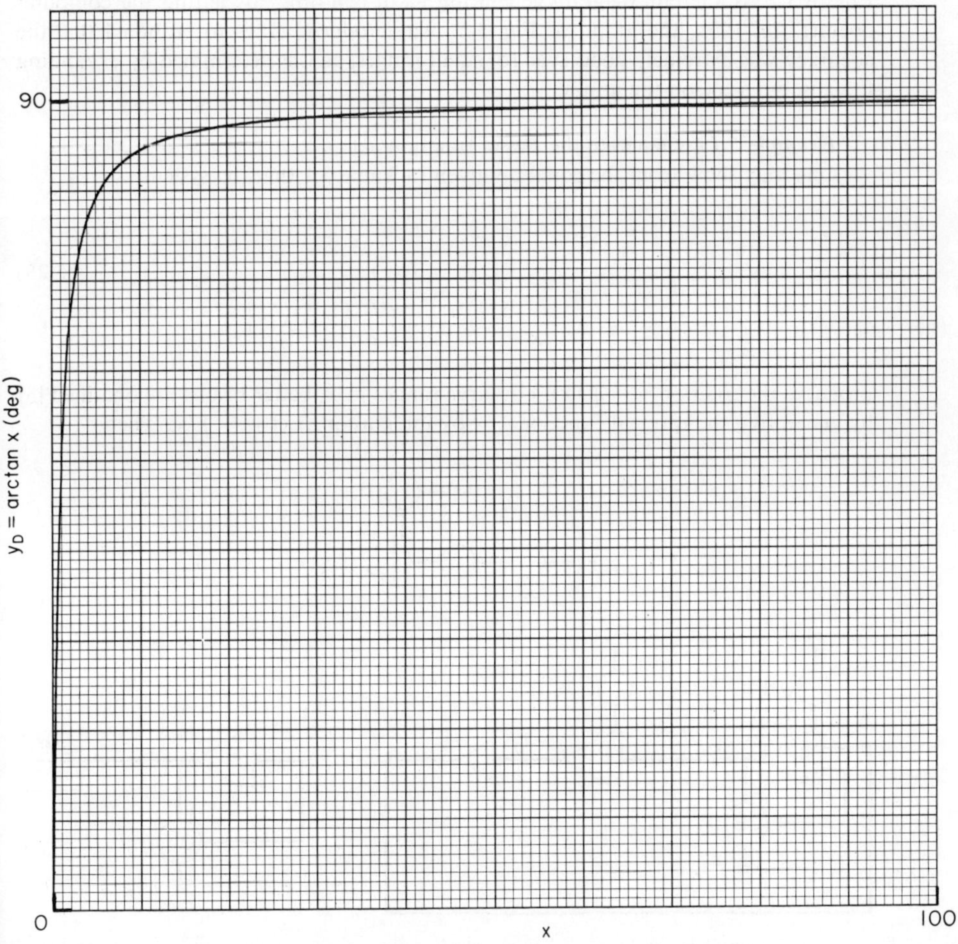

FIG. 5.21 Graph of arctan x.

$$y_D = \frac{180}{\pi} y$$

then

$$(0.01\dot{y}_D) = \frac{1.8}{2}\left(\frac{2\dot{y}}{\pi}\right)$$

so that the last equation in (5.28) becomes

$$(0.01\dot{y}_D) = \frac{1.8}{\pi}(F_0) = 0.5730(F_0)$$

Figure 5.20 is a circuit diagram generating y_D in real time. By letting the computer operate for 100 s, the graph of Fig. 5.21 was produced. It is to be noted that the accuracy was extremely good even for small ranges, as was determined by expanding the abscissa scale on the graph.

Example 5.13. Generate the Bessel function $J_0(x)$, $x \geq 1$, for $\dot{x} = 1$, $x_0 = 1$. One may show from the equation given in Table 5.1 that the system

$$\begin{aligned} \dot{x} &= 1, & x_0 &= 1 \\ \dot{y} &= -z, & y_0 &= J_0(1) \\ \dot{z} &= y - \frac{z}{x}, & z_0 &= -\frac{dJ_0(1)}{dx} = J_1(1) \end{aligned} \tag{5.29}$$

produces the solution $y = J_0(x)$. For large ranges in x, better results are obtained [5] by generating the rational function $1/x$ by integration. Let $w = 1/x$. Then

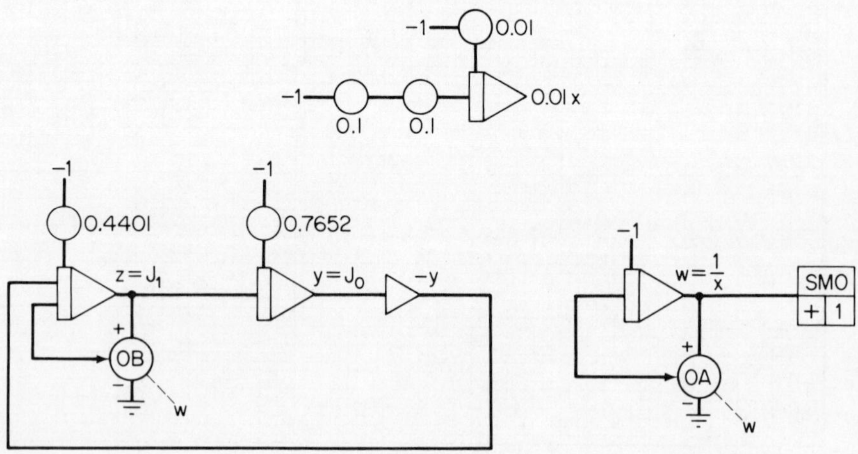

FIG. 5.22 Generation of the Bessel functions J_0 and J_1 with the rational function technique, $x_0 = 1$.

$$\dot{w} = -\frac{\dot{x}}{x^2} = -w^2, \qquad w_0 = \frac{1}{x_0} = 1 \tag{5.30}$$

System (5.29) is replaced by

$$\begin{aligned} \dot{x} &= 1, & x_0 &= 1 \\ \dot{y} &= -z, & y_0 &= 0.7652 \\ \dot{z} &= y - wz, & z_0 &= 0.4401 \\ \dot{w} &= -w^2, & w_0 &= 1 \end{aligned}$$

Figure 5.22 is a simple diagram that generates J_0 and J_1 for $1 \leq x \leq 100$.

5.6 Errors and Stability

There are many ways to generate a given function, and one of the most difficult jobs of a programmer is to determine the best way. This is part of the problem preparation procedure. For example, the function x^2 is generable with a squarer or multiplier, by integrating $2x\dot{x}$, and also by other methods not yet discussed. Generating x^2 with a multiplier automatically bounds the error; with an integrator the error may grow with time if the equation is not being solved exactly.

An analog computer never solves equations without error. In generalized integration, we must determine how any error will propagate with time. This depends primarily on the stability of the equations involved.

A good method to check stability is to introduce small experimental or theoretical errors when solving the equations and see how the errors propagate. It is convenient to consider just two sources of error: (1) errors in pot settings, (2) multiplier errors. Other errors might sometimes be considered. For example, amplifiers have offset; integrators are frequency-dependent and drift with time; noise is present; etc. (see Chapter 17). The fact that integrators do not integrate with exactly the time constant desired can be equivalently thought of as an error in the pot settings of the leads feeding it.

Suppose a system that produces a solution $f(\tau)$ is disturbed at time $\tau = \tau_0$. If the new solution $g(\tau)$ is such that $|f(\tau) - g(\tau)|$ remains "small" for all $\tau > \tau_0$, then the disturbance had very "little" effect, and the solution is said to be stable. If not, the solution is said to be unstable. It remains for mathematical preciseness to define the concepts of "small" and "little," but the reader can appreciate that unstable systems of equations are less likely to produce satisfactory results on the computer than do stable ones. We noted, for example, that the van der Pol equation had a stable limit cycle, which all solutions approached, irrespective of the initial conditions. Solutions to the host-parasite problem, however, were unstable and made us painfully aware that computer solutions are sometimes solved with very noticeable error. On the other hand, the pendulum problem showed us that solutions of unstable systems of equations can sometimes be produced with good results on the computer. This was only because we operated in a *finite time*, sufficiently small, during which errors did not have a chance to propagate and become noticeable.

5.6.1 Integration with negative and positive feedback

Consider the equation

$$\dot{x} = ax, \qquad (x)_0 = x_0 \tag{5.31}$$

the solution of which is $x = x_0 e^{a\tau}$. Suppose we disturb the solution at $\tau = 0$ by using a wrong initial condition, $x_0 + \epsilon_1$, and a poor pot setting, $a + \epsilon_2$. The disturbed solution $\bar{x}$ is clearly

$$\bar{x} = (x_0 + \epsilon_1)e^{(a+\epsilon_2)\tau} = (x_0 + \epsilon_1)e^{a\tau}e^{\epsilon_2\tau}$$

with the error E given by

$$\begin{aligned} E = x - \bar{x} &= x_0 e^{a\tau} - (x_0 + \epsilon_1)e^{a\tau}e^{\epsilon_2\tau} \\ &= e^{a\tau}[x_0 - (x_0 + \epsilon_1)e^{\epsilon_2\tau}] \end{aligned}$$

Since $|\epsilon_2| \ll |a|$, and in no case can $a + \epsilon_2$ reverse its sign, $|E|$ grows with time or decays with time depending upon whether a is positive or negative. If (5.31) is solved on the computer, we recognize that the circuit of Fig. 5.1(b) is used for the former case, while that of Fig. 5.1(c) is used for the latter.

When $a < 0$, the term ax is said to be a *negative feedback* term, and the associated integrator is said to contain negative feedback. When $a > 0$, the term *positive* is used instead of negative in the definition. These definitions are extended to include cases where there are multiple inputs or multiple amplifiers. The integrators in Fig. 5.14 both contain negative feedback. Also the circuits producing F_0 and F_1 in Fig. 5.18 contain negative feedback, since an even number of inverters are between the output and input. (An odd number of amplifiers are in the loop including the integrator.) We see this from (5.28) because both equations describing $\dot{F}_i$ contain terms of the form $-aF_i$ where a is positive, even though variable. A programmer learns to recognize negative feedback as a highly stabilizing influence. When positive feedback is present, results could not be good for a long time. If both terms are present, further analysis is required to determine the dominant influence.

5.6.2 Open-loop integration

In generating the polynomial with (5.12) and (5.13), errors ϵ_i and ϕ_i are introduced which actually solve the equations

$$\dot{y}_i = (1 + \epsilon_i)y_{i+1}, \qquad 0 \leq i \leq n-2, \qquad (y_i)_0 = (1 + \phi_i)\frac{d^i y(x_0)}{dx^i} \tag{5.32}$$

$$\dot{y}_{n-1} = (1 + \epsilon_{n-1})a_n n!, \qquad (y_{n-1})_0 = (1 + \phi_{n-1})\frac{d^{n-1}y(x_0)}{dx^{n-1}}$$

These equations require the mechanization of a set of cascaded integrators without any feedback, even from the integrators. Such cases are known as *open-loop integrations.* By integrating (5.32), the polynomial $p(\tau)$ produced can be shown to be

$$p(\tau) = \sum_{i=0}^{n} a_i(1 + \phi_i)(\tau - \tau_0)^i \left[\prod_{j=0}^{j=i-1}(1 + \epsilon_j)\right]$$

where the product term is equal to one when $i = 0$. The error is therefore a polynomial of degree n in τ, so that the solution is unstable for nonconstant polynomials. One can expect open-loop integration schemes to eventually produce intolerable error, although excellent results occur for small ranges. This case also includes the powers-of-x mechanization shown in Fig. 5.5. For accurate polynomial generation using negative feedback, see Chapter 11 (Example 11.1).

5.6.3 *Oscillator circuit*

Instead of solving (5.1), the circuit of Fig. 5.1(a) solves the equations

$$\begin{aligned} \dot{y}_2 &= -\omega(1 + \epsilon_1)y_1, \qquad (y_2)_0 = (1 + \phi_1)\cos ax_0 \\ \dot{y}_1 &= +\omega(1 + \epsilon_2)y_2, \qquad (y_1)_0 = (1 + \phi_2)\sin ax_0 \end{aligned} \tag{5.33}$$

where $\omega = ak$. By dividing one equation by another we obtain

$$\frac{dy_2}{dy_1} = -\frac{(1 + \epsilon_1)y_1}{(1 + \epsilon_2)y_2}$$

from which

$$\begin{aligned} (1 + \epsilon_2)y_2^2 + (1 + \epsilon_1)y_1^2 = (1 + \epsilon_2)&(1 + \phi_1)^2 \cos^2 ax_0 \\ &+ (1 + \epsilon_1)(1 + \phi_2)^2 \sin^2 ax_0 \end{aligned}$$

which is the equation of an ellipse instead of a circle (unless $\epsilon_1 = \epsilon_2$), although the error in amplitude is small if other errors are small. By differentiating one of the equations in (5.33) we may obtain

$$\ddot{y}_2 + \omega^2(1 + \epsilon_1)(1 + \epsilon_2)y_2 = 0$$

which produces a sine wave of frequency $\omega\sqrt{(1 + \epsilon_1)(1 + \epsilon_2)}$. Eventually, the mechanization of Fig. 5.1(a) becomes out of phase with the desired oscillator. This mechanization is therefore unstable. In actual practice, the amplitude will also decay or expand because the integrators are frequency-dependent and the inverter is not ideal (Chapter 17). When only several cycles of $\sin x$ or $\cos x$ are desired, good results can be obtained.

5.6.4 A nonlinear case

For most cases, it is much easier to attempt to mechanize a function and determine the error by trial rather than by analysis. This is true because the function desired can be calculated easily or is well tabulated, so that a spot check of a few points is all that is needed to see if the error is intolerable for the purposes required. This is certainly true for functions generated with nonlinear equations because a difficult, if not impossible, analytical investigation would be required otherwise.

Nonlinear equations arise when rational functions are generated by integration. We could suspect from (5.23), however, that the results would be quite accurate since negative feedback is involved [$(dD/dx)/D$ must become positive for large enough x when D is a nonconstant polynomial]. To show what calculations are involved in estimating an error, we attempt an analysis of the rational function $1/x$, generated by (5.30). Let us assume that the product w^2 is performed with a servo multiplier, which, because of a small lag, nonlinearities, and cup misalignment actually produces the product $w(w + \epsilon_1)$. The differential equation solved is then

$$\dot{w} = -(1 + \epsilon_2)w(w + \epsilon_1), \qquad w_0 = \frac{1}{x_0} \tag{5.34}$$

where $|\epsilon_1| \ll w_0$ and $|\epsilon_2| \ll 1$, and assuming no error in initial condition. (Initial condition error is of secondary importance in this case.) Since

$$\dot{x} = 1, \qquad (x)_0 = x_0$$

we know that

$$x = x_0 + \tau \tag{5.35}$$

again by assuming no error in initial condition, and ignoring any integrator rate error that can be included as part of ϵ_2. The solution of (5.34) is

$$w = \frac{\epsilon_1}{(1 + \epsilon_1 x_0)\exp[\epsilon_1(1 + \epsilon_2)\tau] - 1}$$

or, eliminating τ from (5.35),

$$w = \frac{\epsilon_1}{(1 + \epsilon_1 x_0)\exp[\epsilon_1(1 + \epsilon_2)(x - x_0)] - 1} \tag{5.36}$$

Assuming now that $\epsilon_1(1 + \epsilon_2)(x - x_0)$ is small enough in the range of interest to permit a linear expansion of the exponential, we may show that a good approximation to (5.36) is

$$w \simeq \frac{1}{x_0 + (1 + \epsilon_1 x_0 + \epsilon_2)(x - x_0)}$$

As a function of x, the error E is now given by

$$E = w - \frac{1}{x}$$

To determine the $x = \bar{x}$ for maximum error, we set $dE/dx = 0$:

$$\frac{dw}{dx} + \frac{1}{\bar{x}^2} = 0 \tag{5.37}$$

where dw/dx is evaluated at $\bar{x}$. If $\epsilon_1 x_0$ and ϵ_2 are much less than 1, the solution of (5.37) is approximately

$$\bar{x} \simeq 2x_0$$

which results in

$$|E_{\text{max}}| \simeq \frac{|\epsilon_1 x_0 + \epsilon_2|}{4x_0}$$

In terms of error on the computer, we first scale the function w as $(x_0 w)$ to take full advantage of multiplier accuracy. The error term then is scaled to become

$$|x_0 E_{\text{max}}| \simeq \frac{|\epsilon_1 x_0 + \epsilon_2|}{4}$$

Since (5.34) is now scaled at

$$(x_0 \dot{w}) = -\frac{(1 + \epsilon_2)}{x_0}(x_0 w)[(x_0 w) + \epsilon_1 x_0]$$

the term $\epsilon_1 x_0$ becomes the error in linearity, etc., a typical value of which is 0.001. By carefully selecting integrators to match *RC* times and avoiding the use of low pot settings, ϵ_2 can be made less than 0.001. The maximum error can therefore be kept below 0.0005 of reference, which has been verified by experience.

The analysis does not apply if electronic multipliers are used, since the equation solved becomes

$$\dot{w} = -(1 + \epsilon_2)(w^2 + \epsilon_1), \qquad w_0 = \frac{1}{x_0}$$

the solution of which is unstable if $\epsilon_1 > 0$, and approaches $\sqrt{-\epsilon_1}$ if $\epsilon_1 < 0$. Thus eventual errors depend mostly on the zero error of the product. Even if ϵ_1 is of the order of -0.0001, the terminal value of 0.01, or one per cent of reference, is too large for some applications.

5.7 Summary and Conclusions

We have seen that a large number of explicit functions are generable with integrators using the generalized integration technique, although error considerations will sometimes force us to use another means of function generation.

Possible advantages are as follows:

(1) Nonlinear equipment savings may result for a given choice of $\dot{x}$. This is the primary motive for generating polynomials and algebraic functions by integrations.

(2) Nonalgebraic functions that are solutions of differential equations are conveniently generable.

(3) The accuracy for short time runs is unsurpassed by any other method. Initial conditions are exact by analog computer standards since they are calculated and supplied by potentiometers. This is often the reason for generating rational functions with integrators.

About the only disadvantage of this method is that the error produced may increase with time and not become bounded as τ becomes large, or may be bounded but intolerable. This results from small computing errors that are inevitable in all analog computations. If the equations required are stable, the errors generally stay small. Negative feedback circuits usually exhibit this characteristic, a reason that their generation with integrators is quite accurate. To minimize errors, in any case, it is important to insure that equipment is carefully balanced, that components (e.g., *RC* time constants) are matched, that pot settings are not too small, and, in general, that all usual methods to improve computing accuracy are used.

Since the functions themselves are easy to check because they may be calculated or are well tabulated, actual trial is the best way to determine if errors are tolerable. If they are not, other methods for generating the functions must be tried. Occasionally a problem can be restated in a way that creates stable equations for the generated functions, but this is not always the case.

Corrective circuits (Chapter 13) when algebraic constraints are known to exist (e.g., $\sin^2 x + \cos^2 x = 1$) are also useful to reduce error, although more nonlinear equipment is required.

PROBLEMS

5-1. Set up a program to investigate the behavior of solutions to Mathieu's equation

$$\frac{d^2y}{dx^2} + (a - 2q \cos 2x)y = 0$$

for various a and q.

5-2. Obtain a graph of

$$y(x) = \int_1^x \ln x \, dx$$

for $0 < x \leq 1$, by obtaining and solving a differential equation on the computer.

5-3. Show that the function

$$y = \int x_1 \, dx_2$$

is generable with linear equipment when x_1 and x_2 are outputs of a linear system such as (5.6).

5-4. Show that the function

$$y = \sqrt{a + bx + cx^2}$$

is generable with linear equipment with a suitable choice of $\dot{x}$, and set up the circuit diagram.

5-5. Generate the functions sinh x and cosh x with $\dot{x} = 1$, $x_0 = 0$ with integrators for $0 \leq x \leq 5$. Are these functions solutions of stable equations? Make an estimate of errors based on the initial condition error only.

5-6. Generate the functions sec x and tan x for $\dot{x} = 1$, $x_0 = 0$, and $0 \leq x \leq 1.47$ with integrators and a servo multiplier.

5-7. How many integrators would be required to generate $x^3 \sin x$ with linear equipment? In view of this large number, how would you generate this function for $\dot{x} = 1$?

5-8. Generate the function

$$y = \arccos (1 - x) - (1 - x)\sqrt{2x - x^2}$$

with linear equipment for $x_0 = 0$ and $0 \leq x \leq 2$.

5-9. Prove the circuits in Figs. 5.14 and 5.15 generate the indicated functions.

5-10. Show that a three-phase oscillator generating the functions

$$y_1 = \sin ax, \qquad y_2 = \sin\left(ax + \frac{2\pi}{3}\right)$$

$$y_3 = \sin\left(ax - \frac{2\pi}{3}\right)$$

with $\dot{x} = 1$ can be generated by the equations

$$\dot{y}_1 = \frac{a}{\sqrt{3}}(-y_1 - 2y_3)$$

$$\dot{y}_2 = \frac{a}{\sqrt{3}}(-y_2 - 2y_1)$$

$$\dot{y}_3 = \frac{a}{\sqrt{3}}(-y_3 - 2y_2)$$

and as such requires three integrators and no inverters [6].

5-11. In trajectory studies, the dynamic pressure

$$q = \tfrac{1}{2}\rho(h)v^2$$

is required for the drag term and $\rho(h)$ is often approximated as

$$\rho(h) \simeq \rho_0 e^{-ah}$$

as was done in the falling ball problem (Example 5.9). It frequently happens that v is small when ρ is large, in much the same way as in Example 5.5, and vice versa, so that better results can be obtained by generating q at the output of an integrator. Derive the differential equation for q assuming $\dot{h}$ and $\dot{v}$ are available for use [7].

5-12. In antenna radiation studies, the function

$$E(x) = A_0 + \sum_{i=1}^{n} A_i \cos(a_i \sin x)$$

is generated for $0 \leq x < \pi/2$, $x_0 = 0$, and gives much information concerning the efficiency of the design. Show that much nonlinear equipment is saved if $\dot{x}$ is chosen proportional to $1/\cos x$ [8].

5-13. Obtain a graph of

$$y = \frac{w^2 - a}{w(w-1)^{1/2}}$$

for several values of a between 0.9 and 1.0, $1.09 \leq w \leq 2$, by first transforming the independent variable $w - 1 = x^2$ and then developing the resultant expression by the rational function method. (Plot y versus x^2 and label the abscissa values in terms of w.)

5-14. The generalized integration technique is extendable to functions of more than one variable. Draw a diagram to solve the equation

$$\frac{dy}{dx} = \sqrt{1 + xy}$$

for various initial conditions, $-5 \leq x, y \leq 5$, by generating the radical by integration.

5-15. The normal distribution function

$$\Phi\left(\frac{x - \mu_1}{\sigma_1}\right) = \frac{1}{\sqrt{2\pi}} \int_{-\infty}^{\frac{x-\mu_1}{\sigma_1}} e^{-t^2/2}\, dt$$

can be generated by integration since

$$\frac{d\Phi}{dx} = \frac{1}{\sqrt{2\pi}\sigma_1} e^{-\left(\frac{x-\mu_1}{\sigma_1}\right)^2/2}$$

or, with

$$y = \frac{1}{\sqrt{2\pi}\sigma_1} e^{-\left(\frac{x-\mu_1}{\sigma_1}\right)^2/2}$$

$$\frac{dy}{dx} = -\frac{y(x - \mu_1)}{\sigma_1^2}$$

$$\frac{d\Phi}{dx} = y$$

Because the equation for y involves positive feedback when $x < \mu_1$, the generation of y cannot be accurate for x_0 much less than μ_1. To graph $\Phi(x)$, it is best to start at $x = \mu_1$, and use $\dot{x} = 1$ to graph $\Phi(x)$ for $x > \mu_1$ and $\dot{x} = -1$ to graph $\Phi(x)$ for $x < \mu_1$. The initial condition for Φ_0 is 0.5, because $\Phi(0) = 0.5$, as can be determined from tables [3]. Set up the circuit to graph $\Phi(x)$ for $-5\sigma_1 \leq x \leq 5\sigma_1$.

5-16. Suppose we now wish to graph the distribution function based on the interaction of two normal distributions:

$$G = \int_{-\infty}^{x} \{(1 - R_1)y_2(t) + R_1[y_2(t)\Phi_1(t) + y_1(t)\Phi_2(t)]\}\, dt$$

where y_1 and y_2 are normal density functions:

$$y_1(x) = \frac{1}{\sqrt{2\pi}\sigma_1} e^{-\left(\frac{x-\mu_1}{\sigma_1}\right)^2/2}$$

$$y_2(x) = \frac{1}{\sqrt{2\pi}\sigma_2} e^{-\left(\frac{x-\mu_2}{\sigma_2}\right)^2/2}$$

and Φ_1 and Φ_2 are normal distribution functions:

$$\Phi_1(x) = \frac{1}{\sqrt{2\pi}} \int_{-\infty}^{\frac{x-\mu_1}{\sigma_1}} e^{-t^2/2}\, dt$$

$$\Phi_2(x) = \frac{1}{\sqrt{2\pi}} \int_{-\infty}^{\frac{x-\mu_2}{\sigma_2}} e^{-t^2/2}\, dt$$

and R_1 is constant. To minimize the amount of computation done with positive feedback, we may take $x_0 \simeq (\mu_1 + \mu_2)/2$ and determine the initial conditions used for y_1, y_2, Φ_1, and Φ_2 from tables. How shall we determine G_0? Set up an unscaled circuit diagram to generate G.

REFERENCES

[1] Mengel, A. S., and W. S. Melahn, "*RAND REAC Manual*," The RAND Corp. Research Memo. RM-525 (Revised), Santa Monica, Calif., Dec. 1, 1950 (ASTIA No. 210675).

[2] Hausner, A., "Selected Generalized Integration Techniques for Analog Computation," *Simulation*, vol. 9, July 1967.

[3] *Handbook of Mathematical Functions*, M. Abramowitz and I. A. Stegun, eds., NBS Applied Math. Series 55 (Washington, D. C.: National Bureau of Standards, June 1964).

[4] Remoortere, P. Van, "Étude Analogique des Equations Différentielles Linéaires à Coefficients Variables, Singulières à l'Origine," *AICA*, vol. 7, April 1965, pp. 124-128.

[5] Hausner, A. "Accurate Analog Computer Generation of Bessel Functions for Large Ranges," *Simulation*, vol. 9, Nov. 1967.

[6] Witsenhausen, H., "Three Phase Oscillator," *AICA*, vol. 4, April 1962.

[7] Kasischke, F. W., "Analog-Computer Technique for Simulation of Dynamic Pressure in Re-Entry Studies," *IRETEC*, vol. 11, Feb. 1962.

[8] Butler, R. F., "Some Examples of the Parametric Technique in Programming an Analog Computer," *Simulation*, vol. 6, May 1966.

See also:

[9] Bekey, G. A., "Generalized Integration on the Analog Computer," *IRETEC*, vol. 8, June 1959.

[10] Bingulac, S. P., and E. A. Humo, "Analog Computer Generation of Bessel Functions of Arbitrary Order," *IEEETEC*, vol. 14, Dec. 1965.

Chapter 6

GENERATING FUNCTIONS WITH DISCONTINUITIES

Many practical computer problems involve functions that are either discontinuous or contain discontinuous derivatives. Examples are functions with a finite number of discontinuities, such as $|x|$, as well as more complicated functions describing limit stops, saturation and hysteresis phenomena, backlash (in gears), etc. Early programmers quickly devised electromechanical or electronic means of generating these functions. Bennett [1] and Warshawsky and Braun [2] reported their experiences using electromechanical differential relays, while Morrill and Baum [3], [4] discussed the use of diodes. Electronic switches, also called digital-analog switches (DAS), were introduced later, when logic boards were incorporated in computers, to permit rapid electronic switching controlled by patchable digital logic signals. All methods used have the characteristic of permitting an abrupt (or nearly abrupt) change in voltage output or its derivative, and as such are very useful in generating approximations of functions having such properties.

Discontinuous functions considered here can be described in terms of the *unit step function* $U(x)$:

$$\begin{aligned} U(x) &= 0 \qquad \text{for } x < 0 \\ &= 1 \qquad \text{for } x > 0 \end{aligned}$$

and its complement $\bar{U}(x)$:

$$\bar{U}(x) \equiv 1 - U(x) = U(-x)$$

We dispense with any definitions at $x = 0$, even though mathematicians, to compel agreement in certain series expansions, customarily define discontinuities to be the arithmetic mean of the two limits when the point of discontinuity is approached from both right and left. In this case, the functions defined would be $\frac{1}{2}$ at $x = 0$. However, such a definition is pointless for our purposes, for we shall only be able to generate approximations of $U(x)$ and $\bar{U}(x)$ anyway, with a small interval about $x = 0$ subject to error due to various equipment limitations.

Another discontinuous function seen in the literature and used frequently in this book is the signum function, sign (x)

$$\begin{aligned} \text{sign}(x) &= -1 \qquad \text{for } x < 0 \\ &= +1 \qquad \text{for } x > 0 \end{aligned}$$

Clearly,

$$\begin{aligned} \text{sign}(x) &= -U(-x) + U(x) \\ &= -1 + 2U(x) = 1 - 2U(-x) \end{aligned} \tag{6.1}$$

or

$$U(x) = \tfrac{1}{2}[1 + \text{sign}(x)]$$

so that either can be taken as the more basic function. Because of the nature of the analog equipment, however, the step function $U(x)$ is more convenient and more direct to use. Whenever sign (x) is implemented on the computer, it

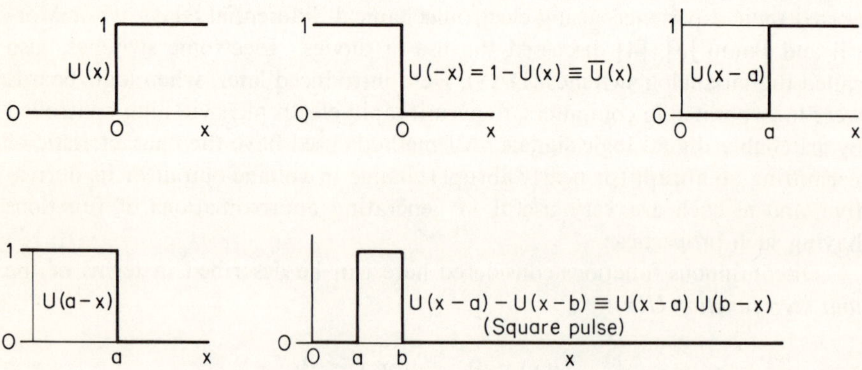

FIG. 6.1 Graphs of various functions involving the unit step.

is often better to think of it in terms of $U(x)$ as in (6.1). Figure 6.1 shows the function $U(x)$, as well as other related functions.

6.1 Digital Signals for Analog Control

The step function $U(x)$ is a two-valued (binary) function generable by two-state devices, sometimes using complicated electronic circuitry. The variable x is usually the difference of two other variables, say x_1 and x_2. Such devices are therefore called *comparators*, because they essentially compare the magnitudes of x_1 and x_2. Almost all analog comparators compare the sum of the two voltages with 0 to produce some linear combination of $U(x_1 + x_2)$ and $\bar{U}(x_1 + x_2)$.

To obtain an analog comparator, early relay users patched a "relay amplifier," simply a high-gain amplifier as shown in Fig. 6.2. This amplifier was

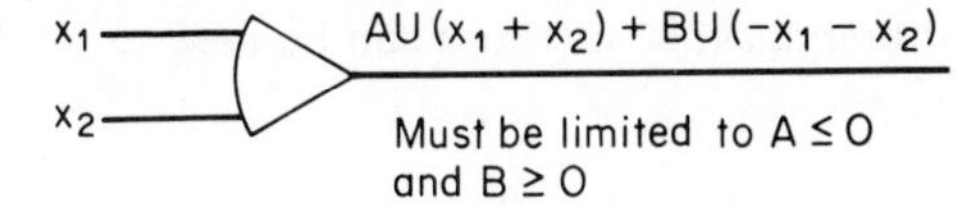

FIG. 6.2 Simple analog comparator.

somehow limited to the values A and B, as indicated, to prevent saturation, and then was used to drive a relay coil for automatic switching purposes. Later computer models provided a permanent limited amplifier for each relay; it was necessary only to patch x_1 and x_2 to use the relay. More modern computers have more flexibility. Since comparator outputs can be used for a variety of purposes—controlling electronic switches, relays, etc.—it is most convenient to isolate the output and patch the desired control. A separate digital patchboard, called the *logic board*, or *MLG board* (for *M*emory and *L*ogic *G*roup) is provided for that purpose, because of the convenience of grouping all digital-type signals together.

The comparator output is essentially a two-state digital signal rather than a continuously variable analog signal. On the MLG board, such binary digital signals can be patched into AND gates, digital inverters, etc, to produce other binary signals, any of which can be used to close electronic switches or operate relays and thereby control analog voltages. When electronic mode control is available (Chapter 18), these signals can also control the operating mode of individual integrators.

It is very convenient to take 0 and 1 as the binary digital signals even though the actual voltage outputs used in the designed circuitry are usually different. Binary signals may then be manipulated both algebraically and logically. In general, 0 is called the *low* state (or level), and 1 is called the *high* state.

6.1.1 Comparator

Figure 6.3 shows the working computer diagram for a comparator. Digital signals always appear within the digital patchboard (they are available, if needed, on the analog board via trunks). They may be distinguishable in a diagram by

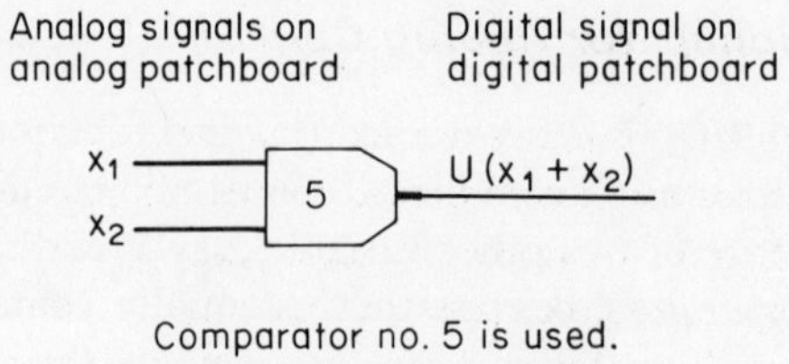

FIG. 6.3 Comparator symbol.

heavy accentuation at the beginning and end of each line whenever confusion of the two types of signals could result. We will assume throughout that a logic board is available, although other methods can be used to implement all of the circuits to be shown.

6.1.2 Digital inverter

All digital signals U usually have permanent digital inverters $\bar{U}$ associated with them or may be patched into an inverter. Digital inversion produces the binary complement

$$\bar{U} = 1 - U$$

Thus, a low state becomes high when inverted, and a high state becomes low. The diagram for inversion is shown in Fig. 6.4 together with a frequently used alternate symbol.

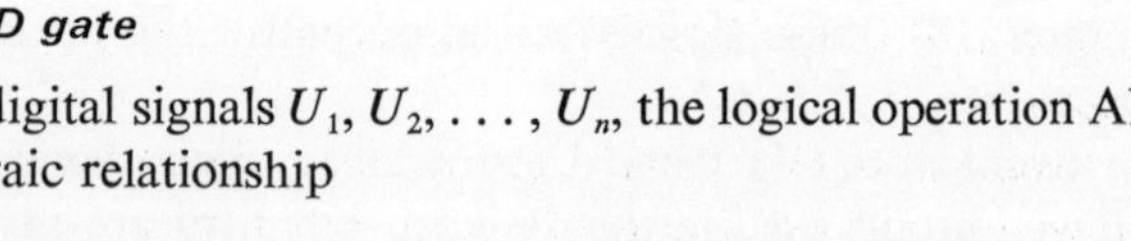

FIG. 6.4 Digital inversion (binary complement) symbols.

6.1.3 AND gate

Given n digital signals $U_1, U_2, \ldots, U_n$, the logical operation AND is defined by the algebraic relationship

$$U_{\text{AND}} = \prod_{i=1}^{n} U_i \tag{6.2}$$

U_{AND} is in the high state only when each U_i is high. AND gates are designed for typically two or four inputs with unused inputs not counting in determining the output state. They can be cascaded, however, when more inputs are needed. The symbols are shown in Fig. 6.5. We shall use both notations $U_1 U_2$ and U_1 AND U_2 to indicate this logical operation.

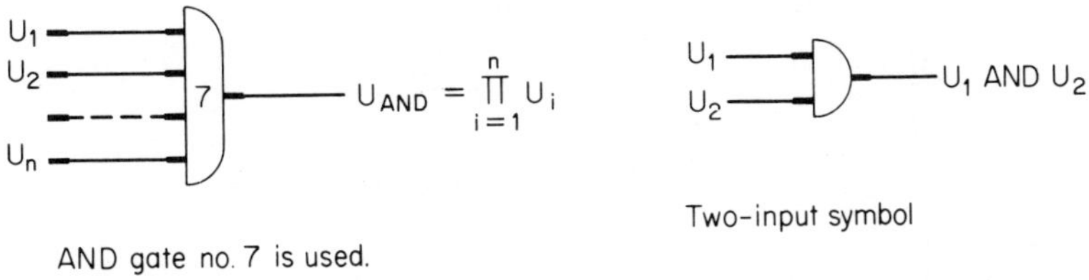

FIG. 6.5 AND gate symbols.

6.1.4 OR gate and others

An OR gate implements the logical OR operation defined algebraically by

$$U_{OR} = \overline{\left(\prod_{i=1}^{n} \bar{U}_i\right)} = 1 - \prod_{i=1}^{n}(1 - U_i) \tag{6.3}$$

U_{OR} is in the high state when either U_1 OR U_2 OR . . . OR U_n is high. Although an accepted OR gate symbol is given in Fig. 6.6, most computers do not contain OR gates directly. An electronic latching (EL) provision on comparators is

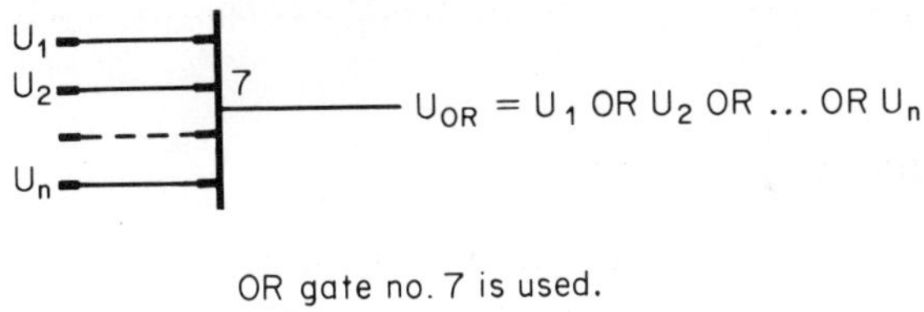

FIG. 6.6 OR gate symbol.

often provided to accomplish OR logic, as shown for two logic variables in Fig. 6.7. The output of any comparator is forced high whenever a digital latching signal U_L is high and has no effect otherwise. A separate EL input on the logic board is provided for this purpose.

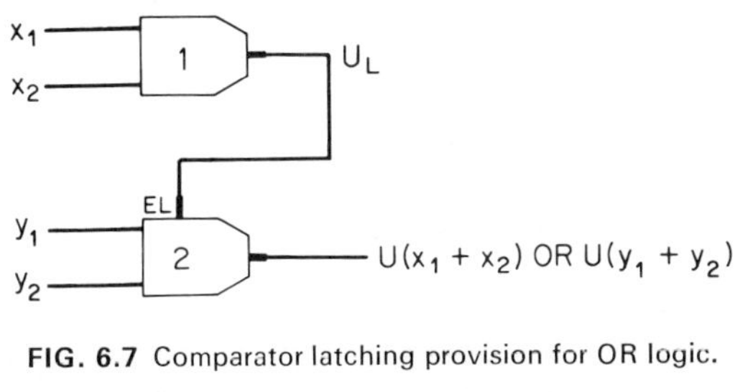

FIG. 6.7 Comparator latching provision for OR logic.

OR logic can also be mechanized with AND gates and inverters directly from (6.3), as is illustrated in Fig. 6.8 with two logic variables U_1 and U_2. In

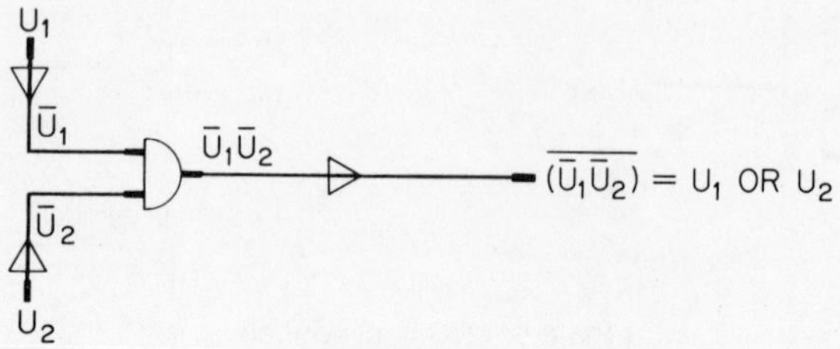

FIG. 6.8 OR logic from AND gate and inverter.

a similar manner, other Boolean logic functions can be patched as desired. In logic notation, the logic symbol + is often used instead of OR. Since we wish to retain the algebraic definition (6.3) and the logic + is indistinguishable from the algebraic +, much confusion can result. Accordingly, the symbol + will not be used in this book for logic OR.

All the equipment described for switching has frequency and sensitivity limitations that should be checked in critical situations; a logic output can drive only a limited number of logic inputs, relays, switches, etc. These are generally specified in the manuals supplied by the computer manufacturer.

6.2 Useful Relay Circuits

Most relays provided in commercial machines for switching analog voltages are double-pole double-throw (DPDT) units which switch typically in 0.1 to 1 ms. Often the contacts "make before break" to obtain high switching speeds. If constant voltages are to be switched, it is better in such cases to obtain a

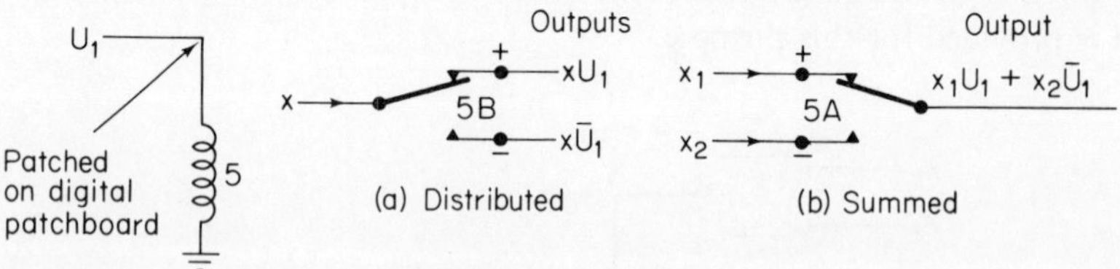

Relay no. 5 is used.

FIG. 6.9 Relay symbols.

source from an amplifier rather than the reference supply to prevent damage to the latter. In Fig. 6.9b, for example, x_1 or x_2 should not be supplied from pots whose inputs are the reference or from the reference itself.

The accepted method of indicating relay usage from a digital signal is shown in Fig. 6.9. Because relays are DPDT units, the A and B are used to indicate the channel. It is clear that the function produced when the logic output U_1 drives the relay coil with x_1 patched into the positive relay terminal and x_2 patched into the negative relay terminal is $x_1U_1 + x_2\bar{U}_1$. Distributed outputs are also shown in Fig. 6.9a. One of the outputs is 0. From these fundamental concepts, virtually all circuits involving relays can be devised. Although U_1 is a logic function, it is used algebraically in the final output. Thus, relay circuits suggest themselves when the desired result is expressed in terms of the unit step function, since we may interpret $U(-x)$ as $\bar{U}_1$ and $U(x)$ as U_1 when x is patched into the comparator.

Boolean logic can be directly implemented without digital logic by recognizing that the AND and OR relations are equivalent to two switches in series and in parallel, as shown in Fig. 6.10. The use of AND or OR logic on the digital patchboard permits the mechanization with only one relay. If patchable logic is available, the choice is up to the programmer.

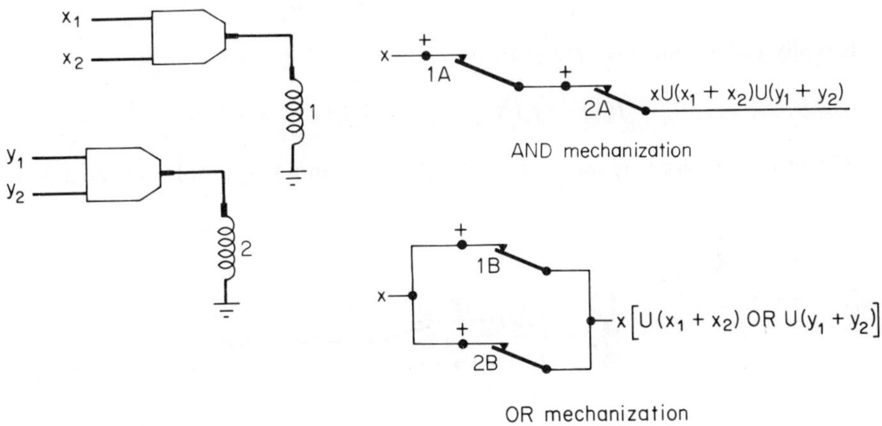

FIG. 6.10 The AND and OR relationships with relays.

In either case, a programmer should keep in mind the fundamental equipment reduction theorems of Boolean algebra when more complicated Boolean expressions are needed. If U_i are digital signals, then [20]

$$U_i^2 = U_i$$

$$U_i\bar{U}_i = 0$$

$$U_i \text{ OR } U_i = U_i$$

$$U_i \text{ OR } \bar{U}_i = 1$$

$$(U_1U_2) \text{ OR } (U_1U_3) = U_1(U_2 \text{ OR } U_3)$$

$$(U_1 \text{ OR } U_2)(U_1 \text{ OR } U_3) = U_1 \text{ OR } (U_2U_3)$$

$$U_1 \text{ OR } U_1 U_2 = U_1$$
$$U_1 \text{ OR } \bar{U}_1 U_2 = U_1 U_2$$
$$U_1(U_1 \text{ OR } U_2) = U_1$$
$$\overline{(\bar{U}_1)} = U_1$$
$$\overline{U_1 U_2} = \bar{U}_1 \text{ OR } \bar{U}_2$$
$$\overline{U_1 \text{ OR } U_2} = \bar{U}_1 \bar{U}_2$$

Most of these are provable algebraically from (6.3) and (6.2).

6.2.1 Simple control—one switching point

A function with one discontinuity

$$\begin{aligned} f(x) &= f_1(x), \qquad x < x_1 \\ &= f_2(x), \qquad x > x_1 \end{aligned}$$

is easily generated, since $f(x)$ can be written

$$f(x) = f_1 U(x_1 - x) + f_2 U(x - x_1)$$

The circuit diagram is shown in Fig. 6.11. A common example of such a function is $f(x) = |x|$ with $f_1(x) = -x, f_2(x) = x$, and $x_1 = 0$.

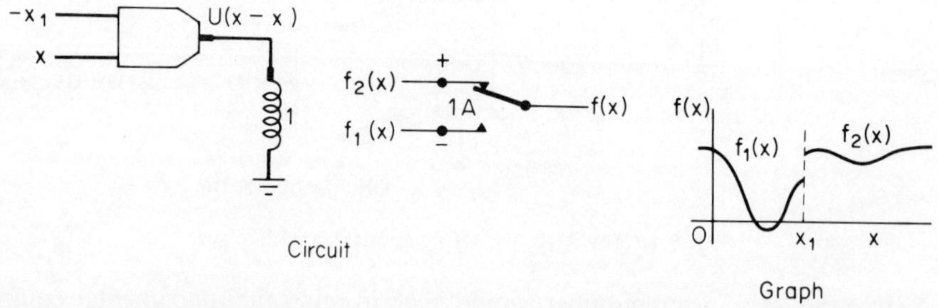

FIG. 6.11 Relay generation of a function with one point of discontinuity

6.2.2 Simple control—two or more switching points

When two switching points are required, i.e.,

$$\begin{aligned} f(x) &= f_1(x), \qquad x < x_1 \\ &= f_2(x), \qquad x_1 < x < x_2 \\ &= f_3(x), \qquad x_2 < x \end{aligned}$$

several variations are possible. We may write

$$f(x) = (f_3 - f_2)U(x - x_2) + f_2U(x - x_1) + f_1U(x_1 - x)$$

or

$$f(x) = f_3U(x - x_2) + f_2U(x_2 - x) + (f_1 - f_2)U(x_1 - x)$$

or utilizing the AND relationship

$$f(x) = f_1U(x_1 - x) + f_2U(x - x_1)U(x_2 - x) + f_3U(x - x_2) \quad (6.4)$$

Each expression gives rise to a different mechanization, the simplest being derivable from (6.4). Since $U(x_2 - x) = 1$ if $U(x_1 - x) = 1$, we may also write

$$f(x) = [f_1U(x_1 - x) + f_2U(x - x_1)]U(x_2 - x) + f_3U(x - x_2) \quad (6.5)$$

and mechanize (6.5) as shown in Fig. 6.12. Two common uses of the circuit in Fig. 6.12 are the limiter and the dead-space circuits shown in Figs. 6.13 and 6.14. Some variation is also possible in mechanizing these circuits.

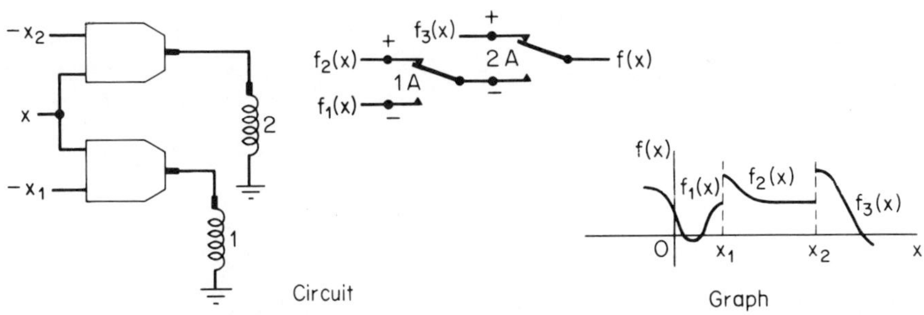

FIG. 6.12 A circuit for two switching points.

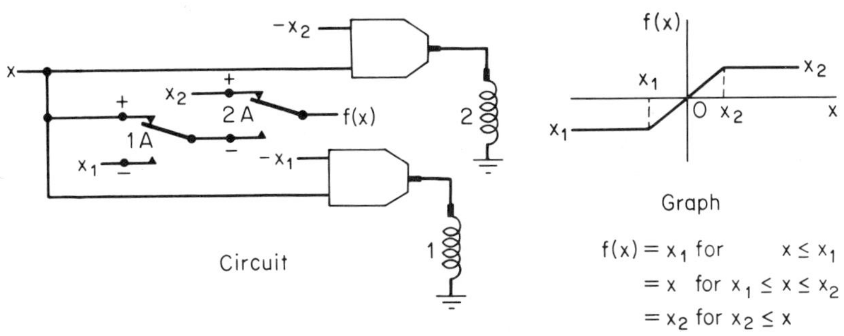

FIG. 6.13 Relay limiter circuit.

The extension to n switching points ($n > 2$) follows the same logic as in deriving (6.5). If

$$\begin{aligned} f(x) &= f_1(x), && x < x_1 \\ &= f_i(x), && x_{i-1} < x < x_i, \quad 1 \leq i \leq n \\ &= f_{n+1}(x), && x_n < x \end{aligned}$$

then $f(x)$ can be written

$$f(x) = f_1 U(x_1 - x) + \sum_{i=2}^{n} f_i U(x - x_{i-1}) U(x_i - x) + f_{n+1} U(x - x_n) \quad (6.6)$$

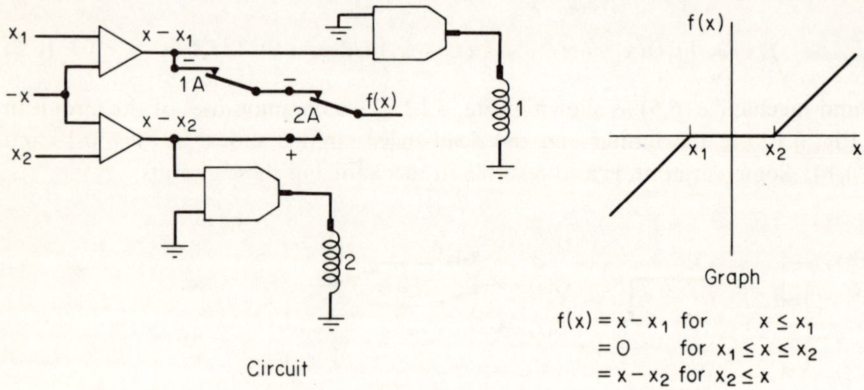

FIG. 6.14 Relay dead-space circuit.

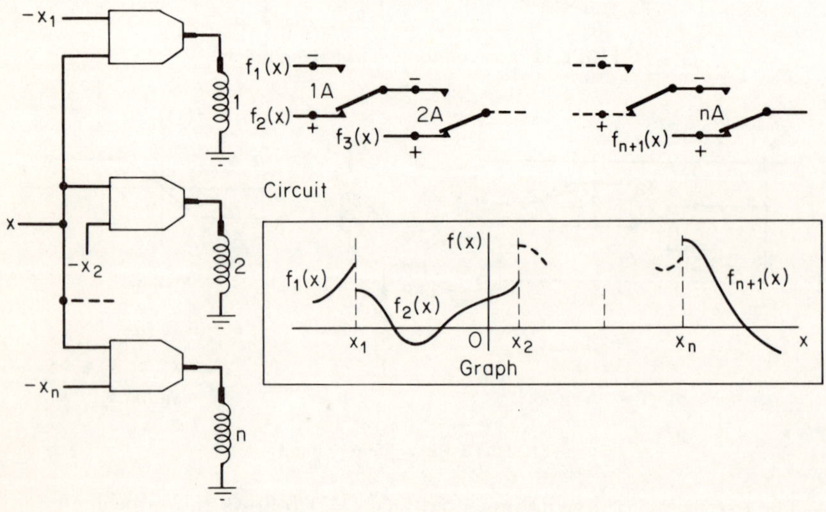

FIG. 6.15 Relay control of *n* switching points.

which may be mechanized as in Fig. 6.15. If the f_i are polynomials and $\dot{x} = 1$, a special technique can be used which requires switching at inputs to integrators [21].

6.2.3 Feedback circuits for multivalued functions

To handle certain types of multivalued functions, it is necessary to arrange to feed the relay output back into a comparator to sense its value. Thus, in the simulation of backlash in gears or hysteresis phenomenon, an entire region in the x-y plane can be obtained as shown in Fig. 6.16. The function desired can best be described as follows:

(1) The function $y(x)$ may take on values in a region bounded by $f_1(x)$ and $f_2(x)$ (the shaded area in Fig. 6.16).

(2) If $y = f_1(x_1)$ at any time, then $y = f_1(x_1) =$ constant if $\dot{x} > 0$ and $y = f_1(x)$ if $\dot{x} < 0$.

(3) If $y = f_2(x_2)$ at any time, then $y = f_2(x_2) =$ constant if $\dot{x} < 0$ and $y = f_2(x)$ if $\dot{x} > 0$.

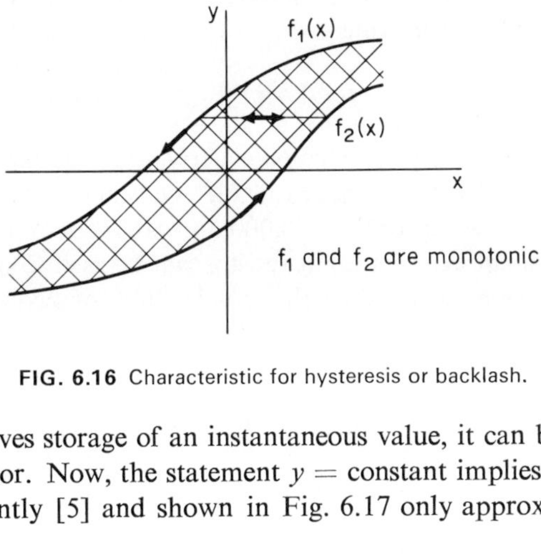

FIG. 6.16 Characteristic for hysteresis or backlash.

Since y involves storage of an instantaneous value, it can best be generated using an integrator. Now, the statement $y =$ constant implies $\dot{y} = 0$. The circuit used frequently [5] and shown in Fig. 6.17 only approximates $y(x)$. We see that

$$\dot{y} = k[(f_1 - y)U(y - f_1) + (f_2 - y)U(f_2 - y)]$$

If y lies between f_1 and f_2, then $\dot{y} = 0$. Otherwise, $\dot{y} = k(f_1 - y)$ [or $\dot{y} = k(f_2 - y)$]. The negative feedback produces a stable circuit; a large value of k forces y to remain close to f_1 (or f_2). But $y - f_1$ remains slightly positive (or $y - f_2$ remains slightly negative) if $\dot{x} < 0$ ($\dot{x} > 0$), so the relay remains closed. Ideally, k should be infinite to force the correct value, but too large a value keeps $y - f_1$ (or $y - f_2$) too close to zero to change the state of the comparator when it should change. Such a phenomenon is called *locking* or *sticking*, and

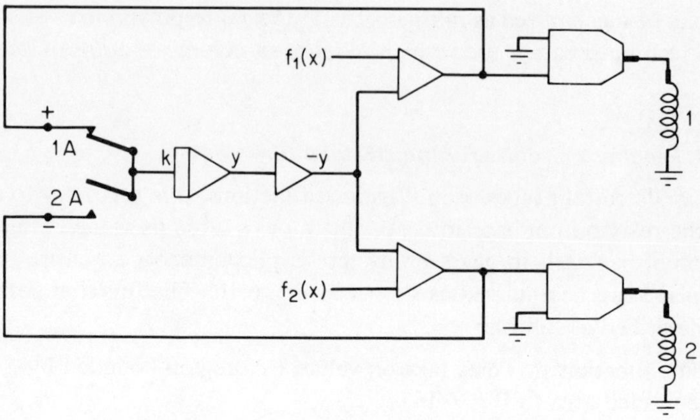

FIG. 6.17 Circuit for hysteresis or backlash.

can severely limit the usefulness of this circuit. Locking also occurs when x does not change rapidly enough or when the comparator zero point has not been carefully adjusted. These limitations can be removed by using a third relay controlled by $\dot{x}$, and mechanizing

$$\dot{y} = k[(f_1 - y)U(y - f_1)U(-\dot{x}) + (f_2 - y)U(f_2 - y)U(\dot{x})]$$

as shown in Fig. 6.18. An approximate derivative can be obtained by the methods in Chapter 10 if $\dot{x}$ is not available. Only the sign of $\dot{x}$ is important. This circuit will now work for $k \simeq 10{,}000$ and produces excellent results even for $\dot{x}$ near 0. A small external input resistor can be used to obtain large k. It need not be a precision resistor since the actual value is of secondary importance.

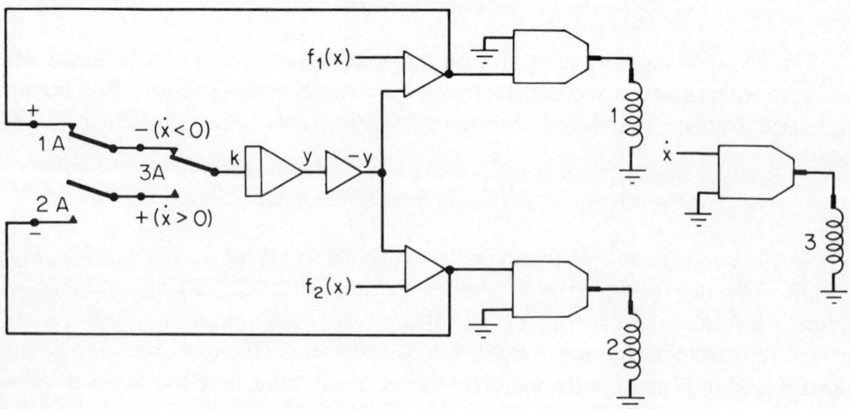

FIG. 6.18 Improved circuit for hysteresis or backlash.

For backlash simulation, the functions $f_1(x)$ and $f_2(x)$ are usually taken to be linear but need not be.

6.2.4 Limiting an integrator

To limit an integrator output x between x_1 and x_2, i.e., $x_1 \leq x \leq x_2$, we modify the input $\dot{x}^*$ as follows:

$$\dot{x} = \dot{x}^*\{[U(x_2 - x) \text{ OR } U(-\dot{x}^*)] \text{ AND } [U(x - x_1) \text{ OR } U(\dot{x}^*)]\} \quad (6.7)$$

so that $\dot{x} = 0$ only if the bounds are exceeded and $\dot{x}^*$ tends to make the error grow. Again, the use of $\dot{x}^*$ control is needed to prevent locking at a limit. This circuit is shown in Fig. 6.19 and should not be confused with other circuits simulating mechanical impact phenomena. When such a limit stop is hit, the value of $\dot{x}^*$ must also abruptly change an amount depending on the nature of the collision (elastic or inelastic). This will be discussed more fully in Chapter 12. In Fig. 6.19, series-parallel switches are used to implement the AND-OR logic in (6.7).

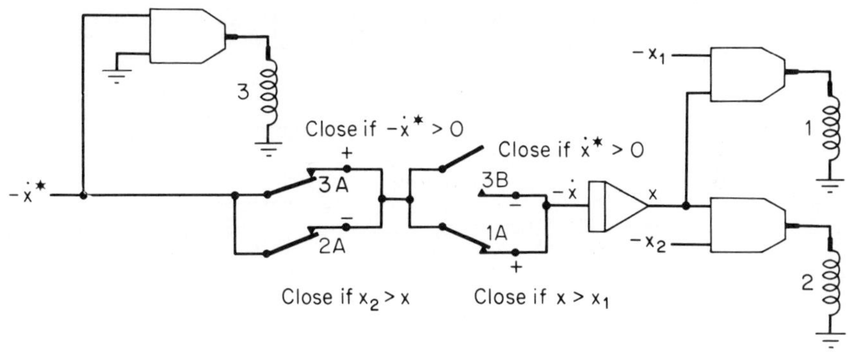

FIG. 6.19 Relay limiting of an integrator.

6.2.5 Maximum and minimum values

A relay circuit to store the maximum and minimum values of $x(\tau)$ is shown in Fig. 6.20. Again $\dot{x}$ control prevents locking so that using k in the order of 10,000 assures accurate results by analog computer standards.

6.2.6 Bistable multivibrator and sweep generator

A useful circuit providing a variety of waveforms is shown in Fig. 6.21, with the waveform outputs given in Fig. 6.22. There is no restriction on the y_2 or y_1 other than $y_2 > y_1$. No locking can occur because, as soon as C changes sign, it abruptly jumps to a value that maintains the proper sign and then approaches zero again. We see that

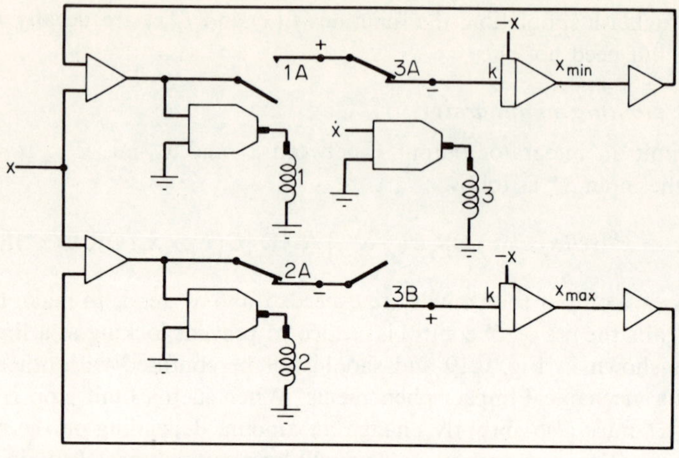

FIG. 6.20 Maximum and minimum circuits.

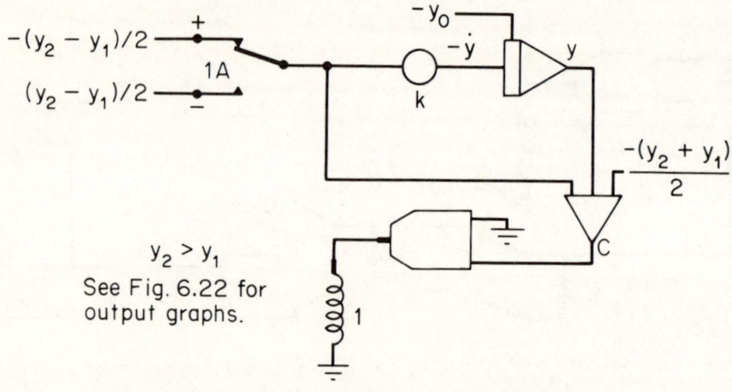

FIG. 6.21 Bistable multivibrator and sweep generator.

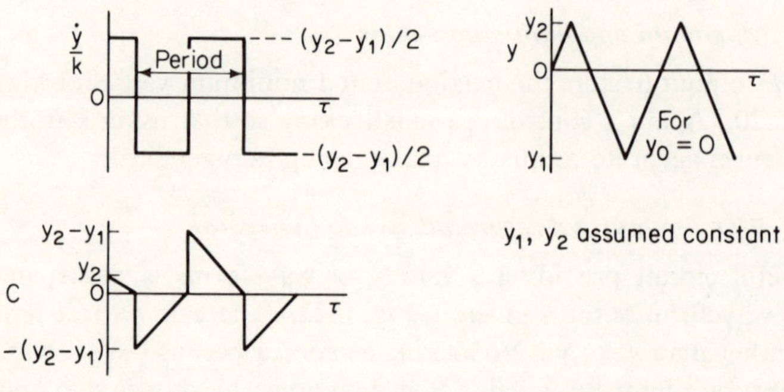

FIG. 6.22 Multivibrator and sweep generator outputs.

$$\dot{y} = \frac{k}{2}\left[(y_2 - y_1)\,\text{sign}\left(+\frac{\dot{y}}{k} - y + \frac{y_2}{2} + \frac{y_1}{2}\right)\right]$$

Both $\dot{y} = k(y_2 - y_1)/2$ and $\dot{y} = -k(y_2 - y_1)/2$ are stable states as long as $y_1 \leq y \leq y_2$. But as soon as y lies outside these limits, the only stable state is the value that forces y to return within the bounds of y_2 and y_1. The values of y_2 or y_1 or both need not be constant for stable operation, a fact which is quite important when computing multiple integrals (Example 6.3).

The relay has many applications in situations requiring switching, but it can only be used in slow-speed operation. For high-speed applications, the DAS or diode circuitry is mandatory.

6.3 Circuits with Digital-Analog Switches (DAS)

A digital-analog switch (DAS), also called an electronic switch, is designed to produce the function xU, where x is an analog signal and U is a digital signal. Its design is often diode oriented, but it is separately discussed here because it can be available for use as a general-purpose switch. The diagram symbol of the switch is shown in Fig. 6.23, although it may not be a working symbol

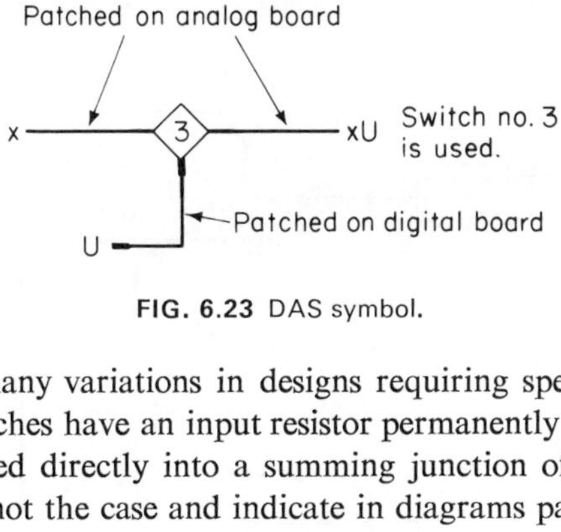

FIG. 6.23 DAS symbol.

because there are many variations in designs requiring special patching. For example, some switches have an input resistor permanently attached so that the output xU is patched directly into a summing junction of an amplifier. We will assume this is not the case and indicate in diagrams patching directly into input resistors.

A DAS is essentially equivalent to a single-pole single-throw switch, although by patching with amplifiers, a SPDT switch can be simulated (Fig.

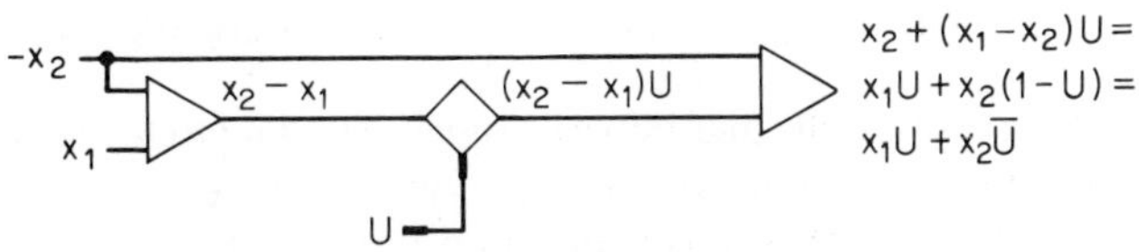

FIG. 6.24 Simulation of a relay with one DAS.

6.24). For complicated Boolean logic, however, such a procedure requires too many amplifiers. The use of the logic board for implementing Boolean functions is therefore much more common practice. Two DAS's can be considered to be an electronic relay when used in parallel and driven by complementary digital signals, as shown in Fig. 6.25. All relay circuits can be simulated by

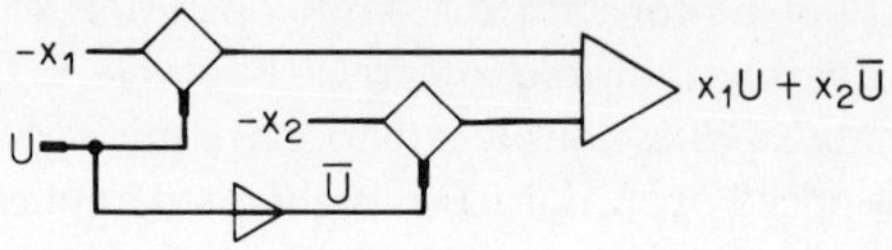

FIG. 6.25 Simulation of a relay with two DAS's.

combinations of circuits where AND and OR logic is patched on the logic board. It should be mentioned, however, that an electronic switch has a small leakage which introduces magnitude errors, unlike the relay. Thus, circuits depending upon an open relay contact will be less ideal when using a DAS (or diodes). Because of this, relay circuits are preferred when applicable.

To illustrate the conversion of a relay circuit to a DAS circuit, switching $n + 1$ functions corresponding to Fig. 6.15 is shown in Fig. 6.26. Equation (6.6) is mechanized directly, with the AND logic performed by AND gates instead of the series arrangement for relays.

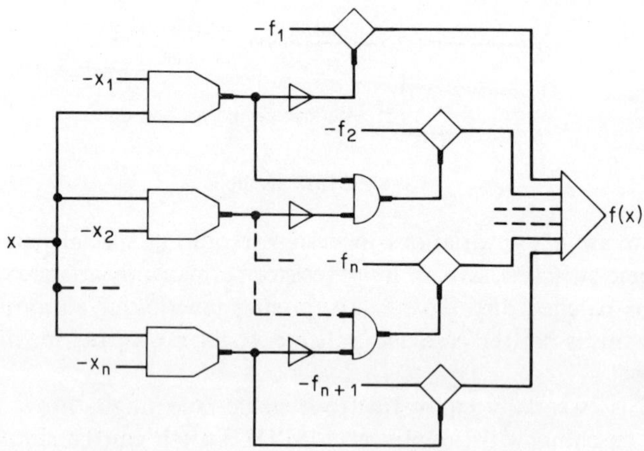

FIG. 6.26 Mechanization of (6.6) with DAS's.

To illustrate more complicated Boolean functions, the circuit to limit an integrator is shown in Fig. 6.27, where the logic function $[U(x_2 - x)$ OR $U(-\dot{x}^*)]$ AND $[U(x - x_1)$ OR $U(\dot{x}^*)]$ is mechanized on the logic board using the electronic latch inputs of the comparators for the OR logic.

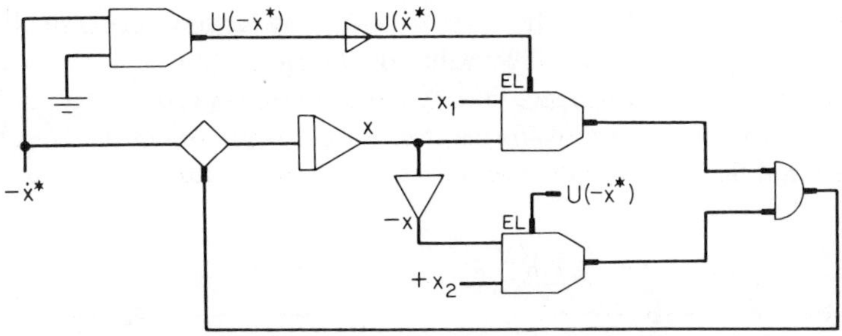

FIG. 6.27 Limiting an integrator with DAS.

6.4 Some Simple Diode Circuits

All computer manufacturers supply passive elements such as resistors, capacitors, and solid-state diodes, permanently wired in patch cords to enable rapid patching of circuits requiring these external components. Diode circuits have proved more useful for higher speed computation than relay circuits, particularly before the introduction of DAS capability. This section presents circuits that can be patched using diodes and ungrounded pots where accuracy is not of the utmost importance.

The standard symbol for the solid-state diode in shown in Fig. 6.28 together with its ideal voltage characteristic. Simply stated, the output is approximately $xU(x)$ or $xU(-x)$ provided it is connected through some load R_L to ground. The diode never allows its plate voltage to be larger than the cathode voltage. If these voltages are equal, the diode is conducting, otherwise it is not.

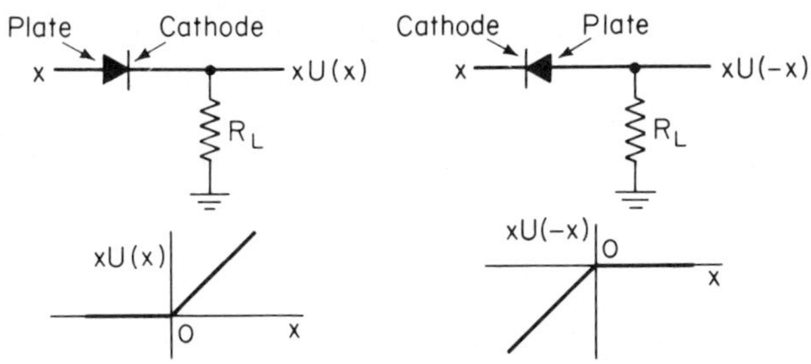

FIG. 6.28 Solid-state diode symbol and ideal voltage output.

In practice, diodes exhibit a small resistance (forward resistance) when conducting and a very large but finite resistance (back resistance) when not conducting. The current-voltage graph is continuous and passes through (0, 0).

With an ungrounded pot, the diode can be biased as shown in Fig. 6.29.

If the pot is set at $a/(1 + a)$ [or $-b/(1 - b)$] conduction occurs at $x = a$, if $a \geq 0$ (or $x = b$, if $b \leq 0$). The value of the slope must be calculated from (2.6), but if the load resistance is large, the slope is approximately $1/(1 + a)$ [or $1/(1 - b)$]. (The nonzero forward resistance of the diode makes an exact calculation difficult, so conduction points are usually found by trial.)

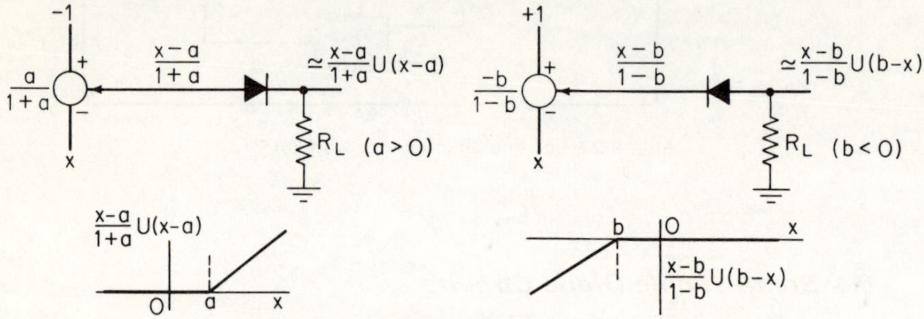

FIG. 6.29 Biasing the diode and shifting the conduction point.

In most of the circuits that follow, R_L is the value of an input resistor, or the diode is connected directly to the grid of an amplifier. In the latter case, the amplifier must be in a stable feedback loop that automatically adjusts the grid potential to zero.

6.4.1 Bang-bang circuit

The bang-bang circuit is a limited comparator circuit, as shown in Fig. 6.30, utilizing two ungrounded pots and producing an approximation of the function $aU(x_1 - x) + bU(x - x_1)$ where $a \geq 0$ and $b \leq 0$ are constants. Because there is no feedback resistor, the amplifier output moves as far away from zero as it can—to the conducting point of either diode, a or b. If $x - x_1 > 0$, the output is negative (b), and if $x_1 - x > 0$, the output is positive (a). The output is not flat but has a slight negative slope as indicated in the graph by the dashed lines. This is due to the finite forward resistance of the diode which acts as a feedback resistor. Because of this characteristic, the circuit is said to have a *soft*

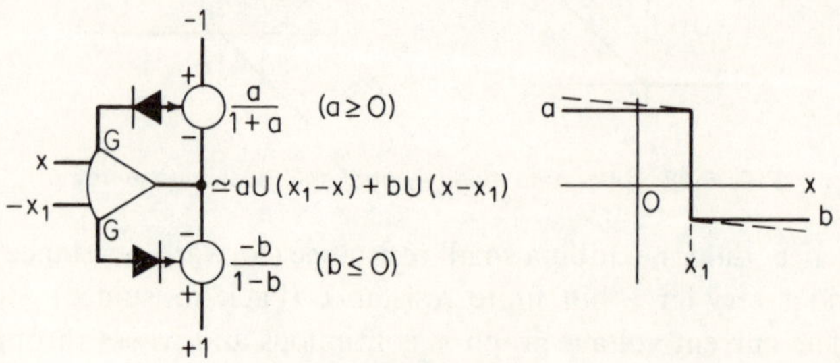

FIG. 6.30 Bang-bang circuit.

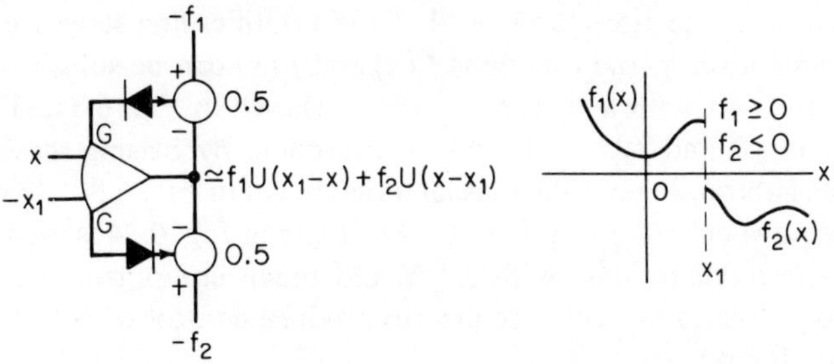

FIG. 6.31 Relay simulation for $f_1(x) \geq 0, f_2(x) \leq 0$.

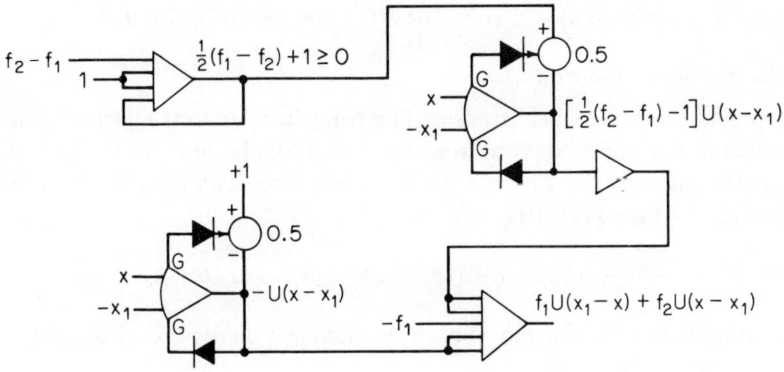

FIG. 6.32 DPST relay simulation with no restriction of f_1 and f_2.

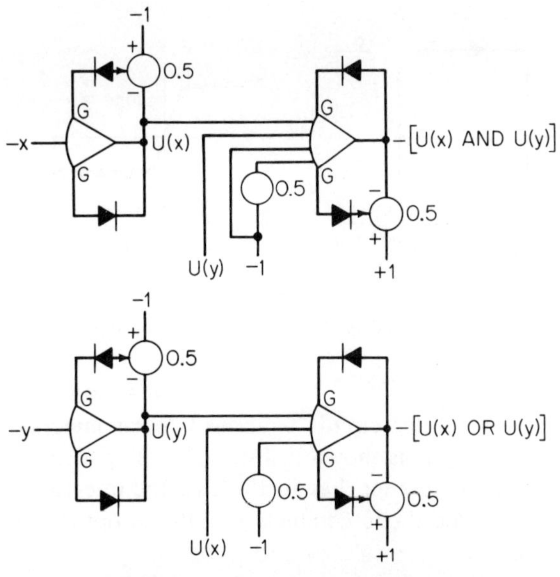

FIG. 6.33 Boolean logic function generation with bang-bang circuits.

limit. Note how connections to the grid of the amplifier first stage are indicated.

To simulate a relay, the functions $f_1(x)$ and $f_2(x)$ can be substituted for the reference 1 and -1 with $a = 1, b = -1$, as shown in Fig. 6.31. The restrictions of $f_1(x) \geq 0$ and $f_2(x) \leq 0$ can be overcome by biasing these functions first and then subtracting out the bias later, as shown in Fig. 6.32. (The absence of a pot is equivalent to setting it to 0. The logic in Fig. 6.24 is used.) Because the circuit approximates only within 1 V and much equipment is required, it is rarely used. Cascading such circuits can produce control of $n + 1$ functions, as in the case for relays.

Bang-bang circuits are also useful for constructing Boolean functions for driving other circuits. For example, the circuits for producing voltages proportional to $U(x)$ AND $U(y)$ and $U(x)$ OR $U(y)$ are shown in Fig. 6.33.

6.4.2 Feedback limiting

By including the feedback resistor in a bang-bang circuit, a normal summer is obtained in the nonconducting regions of each diode, but the limiting principle applies just as well. Figure 6.34 shows the circuit and graph of the function produced which is clearly

$$V = -kx + (kx + b)U(kx + b) + (kx + a)U(-kx - a)$$

Soft limiting is produced again, due to the nonideal diode; the actual output is shown dashed in the graph.

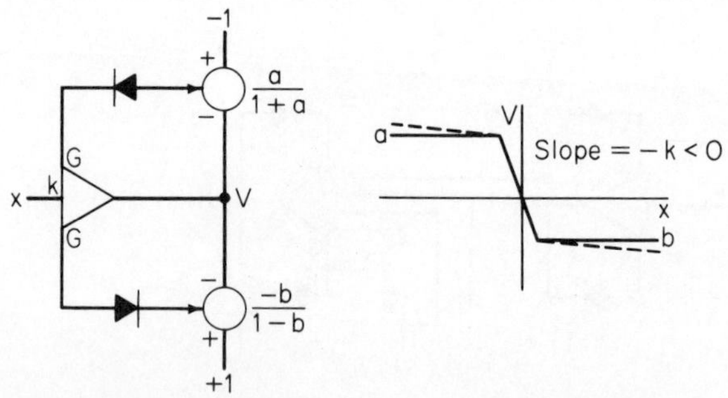

FIG. 6.34 Feedback limiter circuit and output graph.

6.4.3 Limiting an integrator

When a feedback limiter circuit is applied to an integrator output, x is limited so that $b \leq x \leq a$, as shown in Fig. 6.35. The identical reasoning as in the other cases applies except that in this case the integrator integrates at a very large rate when the diode conducts to balance out the other input $-\dot{x}$. No locking can occur.

Because a feedback limiter circuit can be applied in several ways, special feedback limiting units are often available in the patchboard, where it is neces-

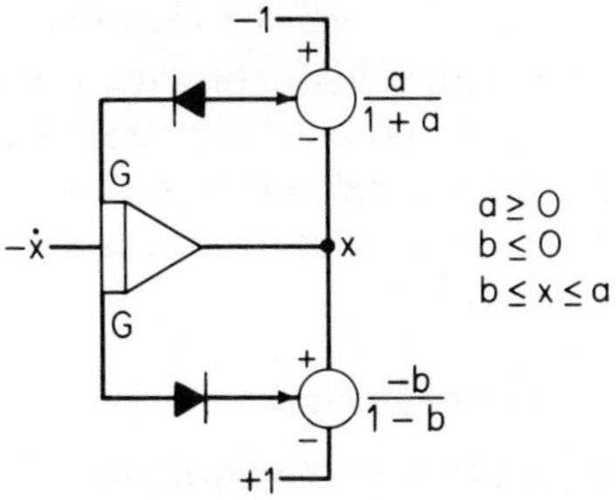

FIG. 6.35 Limiting an integrator with a feedback limiter.

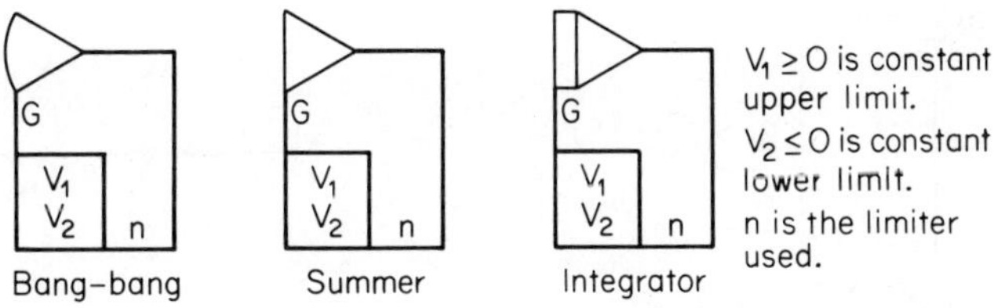

FIG. 6.36 Special feedback limiter symbols.

sary only to patch to the grid and output of the amplifier and adjust pots to the proper limits. Figure 6.36 shows the special limiter symbols used in these cases.

6.4.4 Soft limiting

By including a resistance prior to any grid connection, the slope of the bang-bang or feedback limiter circuit outputs may be forced to take on a wide range

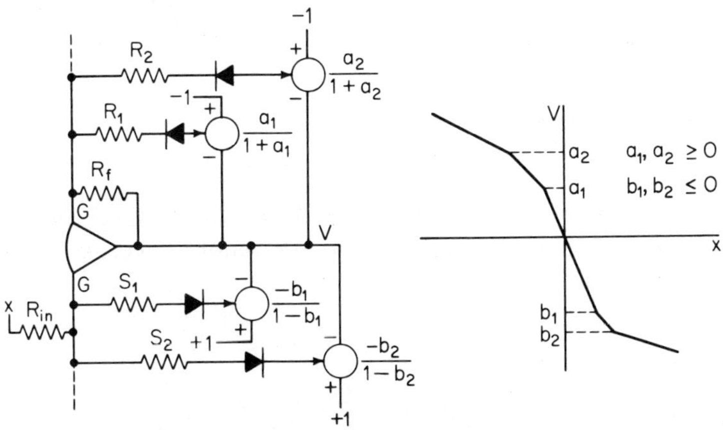

FIG. 6.37 Soft limiting.

of shapes. Several such connections can be used to decrease the magnitude of the slope at several points (Fig. 6.37). This circuit requires special precise resistors $R_1, R_2, \ldots, S_1, S_2 \ldots$, to obtain the proper slopes and is therefore not often used in general-purpose computers. Special diode function generators (Chapter 8), based on similar principles, are available to produce general piecewise linear functions when needed.

6.4.5 Dead-space and backlash circuits

A dead-space circuit can be constructed from Fig. 6.29 as shown in Fig. 6.38. This circuit has application in one frequently used for backlash and a

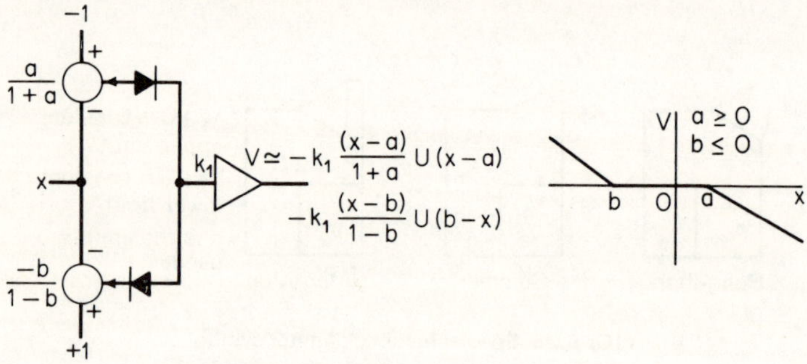

FIG. 6.38 Simple dead-space circuit.

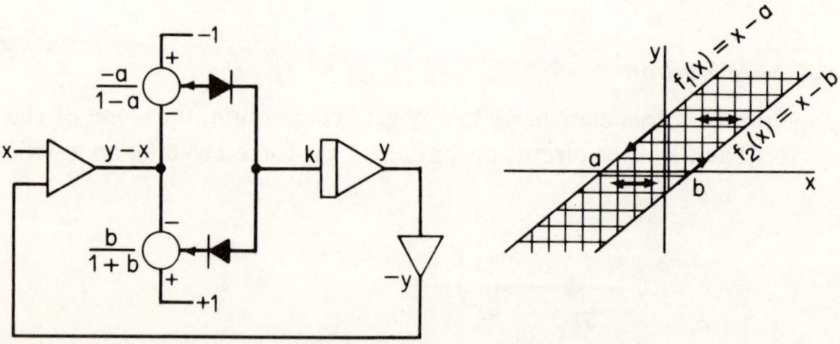

FIG. 6.39 Circuit for backlash.

simple form of hysteresis as shown in Fig. 6.39. One diode conducts outside the shaded region, and the integrator integrates until the conduction point is reached. Hence, either

$$y - x \simeq a$$

or

$$y - x \simeq b$$

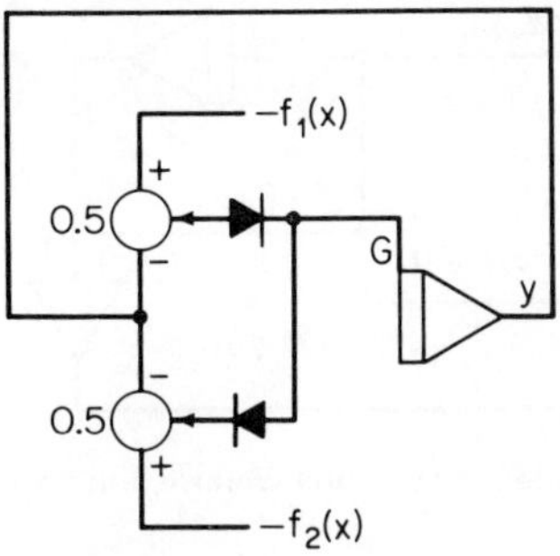

FIG. 6.40 Simple diode circuit for hysteresis or backlash.

A large value of k minimizes any time delays, and no locking can occur. Because the diodes are not ideal, y drifts within the shaded area, unlike the relay circuit equivalents. To generate a characteristic such as in Fig. 6.16, a simpler circuit, shown in Fig. 6.40, requires variable voltages $-f_1$ and $-f_2$ to substitute for the reference. A diode conducts when $y \geq f_1(x)$ or $y \leq f_2(x)$ and forces integration to either of these limits. If $f_2(x) < y < f_1(x)$, neither diode conducts. Because the feedback is direct when the diode conducts, the input may be made directly to the grid.

6.4.6 Maximum and minimum circuits

Using virtually the same ideas as in the previous circuit, a simple diode circuit for a maximum and minimum is shown in Fig. 6.41. These outputs drift sufficiently to prevent long-term operation.

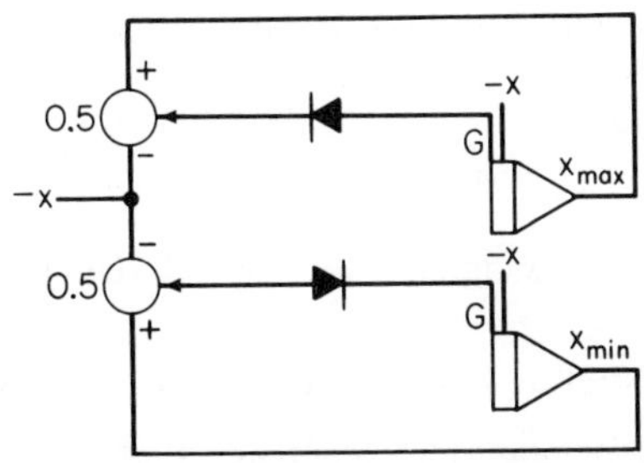

FIG. 6.41 Maximum and minimum circuits.

6.4.7 Bistable multivibrator and sweep generator

The circuit of Fig. 6.42 is a diode equivalent of that in Fig. 6.21, producing the outputs as indicated in Fig. 6.22.

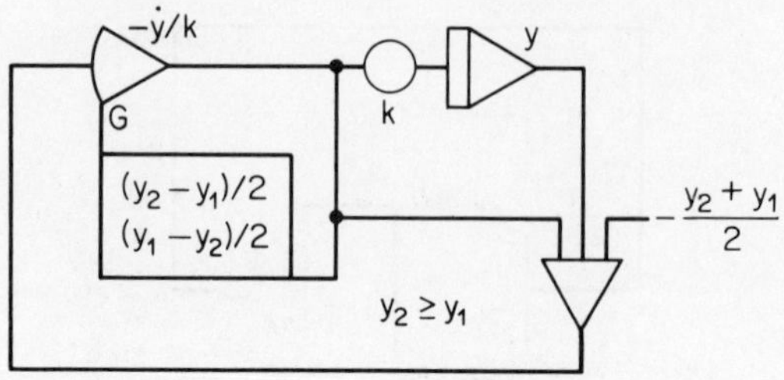

FIG. 6.42 Diode bistable multivibrator and sweep generator.

6.5 Idealized Diode(Precision) Circuits

For accurate function generation using diodes, circuits employing the so-called idealized diode circuits can be used (Fig. 6.43).

When the diodes D_1 do not conduct, the outputs provide a near unity gain. When D_1 conducts, each amplifier output is near zero, but the voltage is then further attenuated by D_2 to be much less than 0.001 V in absolute value. The net result is a voltage equal to $-(x - x_1)U(x - x_1)$ [or $-(x - x_1)U(x_1 - x)$] to within 0.01 V, which is as precise as is measurable on 100-V machines.

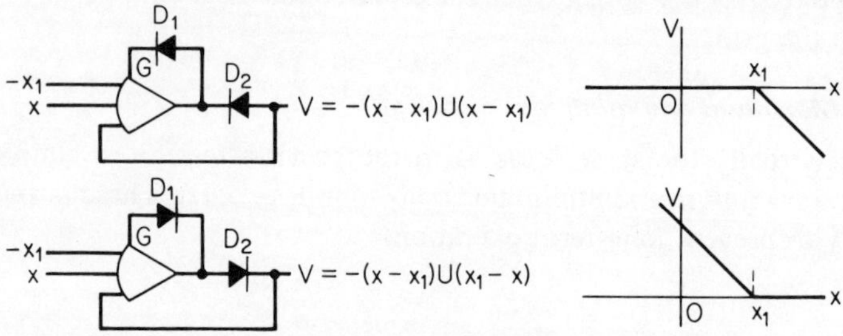

FIG. 6.43 Idealized diode circuits.

6.5.1 Absolute value

The function

$$y = |x| = x\,\text{sign}\,(x) = -x + 2xU(x)$$

or

$$y = -|x| = -x\,\text{sign}\,(x) = -x + 2xU(-x)$$

from (6.1) is accurately formed with idealized diodes as shown in Fig. 6.44.

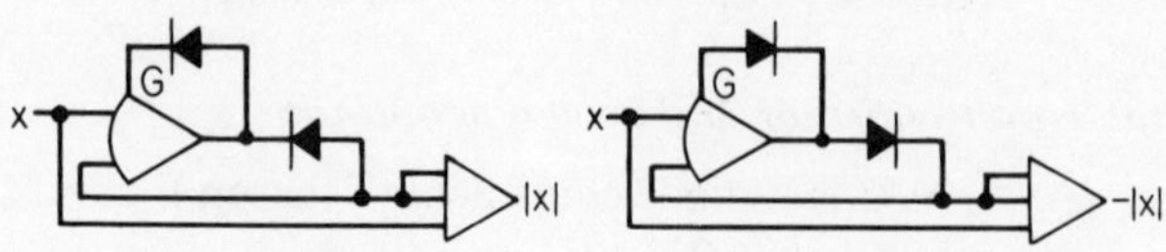

FIG. 6.44 Accurate absolute-value circuits.

6.5.2 Dead space

A dead-space circuit as described in Fig. 6.14 is written as

$$y = (x - x_1)U(x_1 - x) + (x - x_2)U(x - x_2)$$

The idealized diode circuit simply sums these two components, as shown in Fig. 6.45.

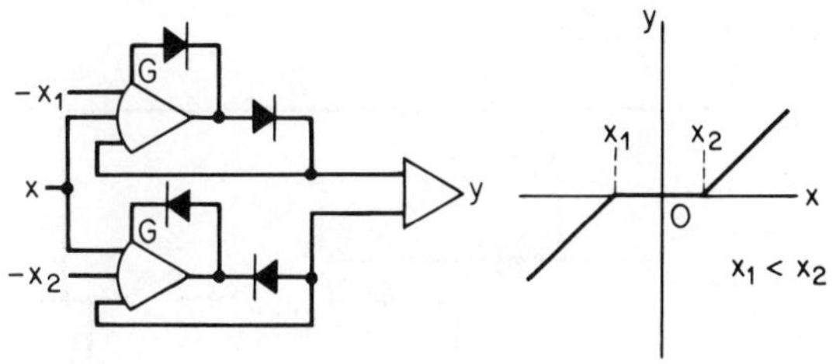

FIG. 6.45 Ideal diode dead-space circuit.

6.5.3 Limiter circuit

A limiter circuit can be characterized by

$$y = -x + (x - x_1)U(x_1 - x) + (x - x_2)U(x - x_2)$$

which may be generated by summing each term as shown in Fig. 6.46.

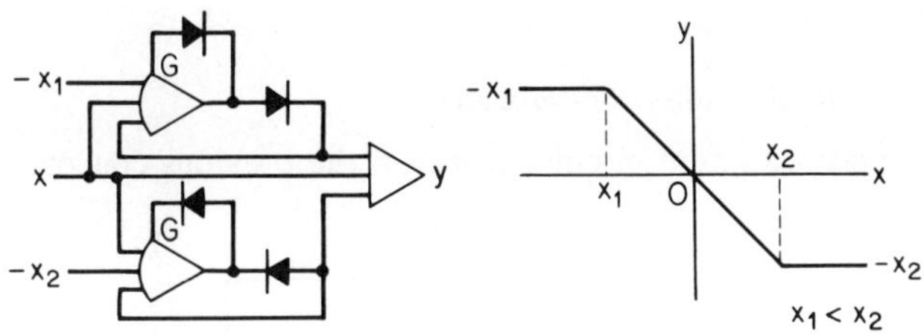

FIG. 6.46 Ideal diode limiter circuit.

6.5.4 Limiting an integrator—hysteresis and backlash

For use in integrator feedback applications, the 1-gain feedback from the output diode can be removed making the amplifier high gain when the diode connected to the grid does not conduct, as shown in Fig. 6.47. The circuits are

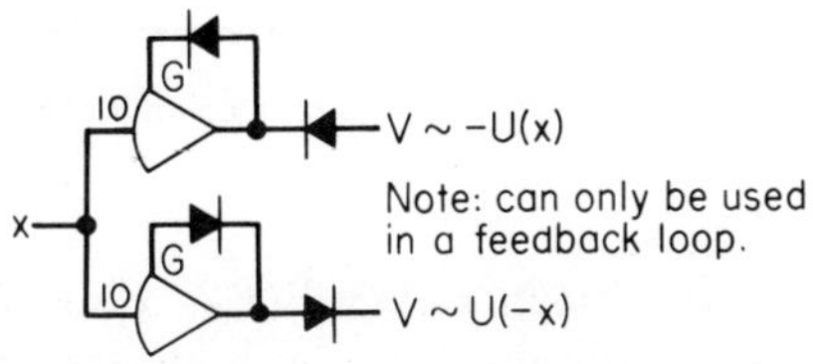

FIG. 6.47 Removing feedback for integrator feedback application.

characterized by the outputs being very close to zero when they should be, permitting virtually no drift when fed to integrators, even with high gain.

A circuit limiting an integrator by this method is shown in Fig. 6.48. Even at gains of 1000 (obtained by using an external 1-kΩ resistor connected to the summing junction) there is virtually no drift when $\dot{y} = 0$. The output of the high-gain amplifiers provides just enough voltage to buck out any voltage $-\dot{y}$ tending to cause y to move outside its limits. Inputs are connected to 10 gains to improve comparator operation.

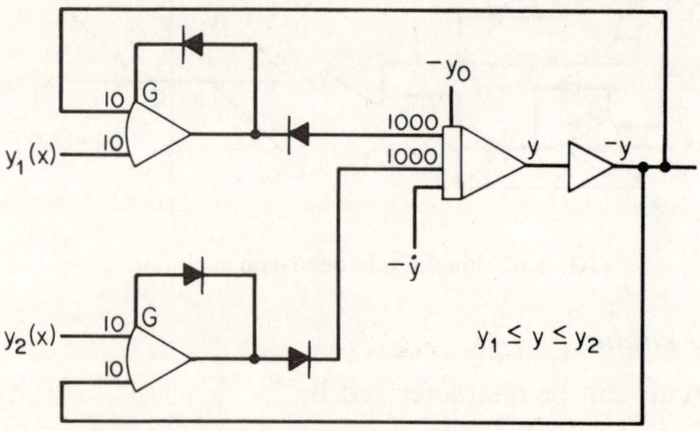

FIG. 6.48 Limiting an integrator with ideal diode circuits.

By removing the lead $-\dot{y}$, the circuit becomes a hysteresis or backlash circuit, depending on the supplied values of $y_1(x)$ and $y_2(x)$.

6.5.5 Maximum and minimum circuits

Maximum and minimum circuits based on the previous concept are shown in Fig. 6.49.

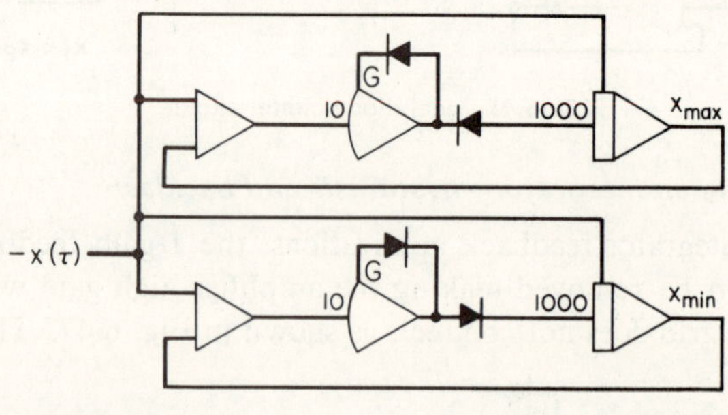

FIG. 6.49 Maximum and minimum circuits using idealized diodes.

6.5.6 Relay simulation

A circuit to generate the function $f_1(x)U(x - x_1)$ is shown in Fig. 6.50. Because each idealized diode circuit is biased with the reference voltage, it either transmits the voltages or is zero. Thus the output simulates SPST relay operation quite exactly. To include an $f_2(x)U(x_1 - x)$, we use the logic of Fig. 6.24 to minimize equipment (Fig. 6.51). Simpler circuits are sometimes possible if f_1 and f_2 are monotonic functions. These circuits are never used if DAS's are available.

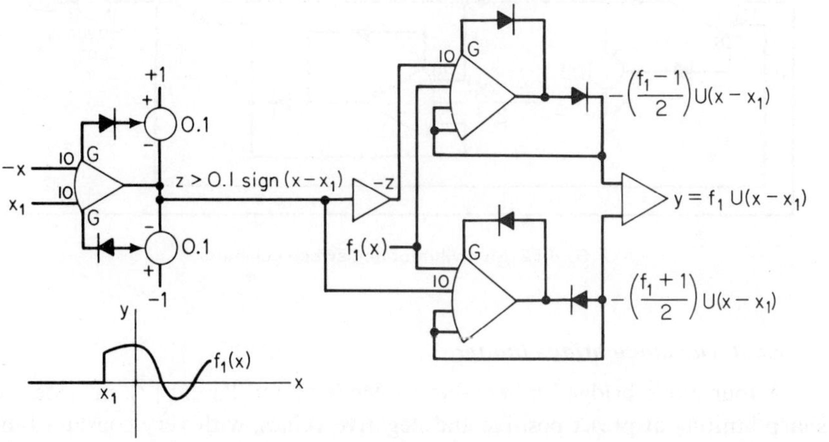

FIG. 6.50 SPST relay simulation.

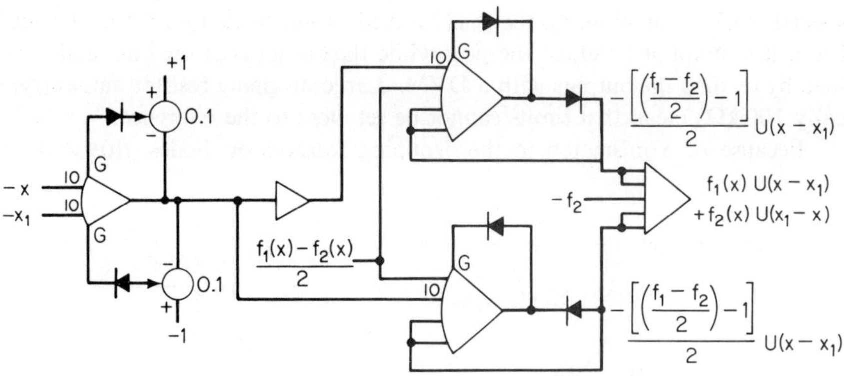

FIG. 6.51 SPDT relay simulation with idealized diodes.

6.5.7 Bistable multivibrator and sweep generator

The idealized diode version of this circuit is shown in Fig. 6.52. Each diode is biased alternately by the reference (1 and -1) with switching occurring at the limits. The outputs of both idealized diodes alternate between zero and a nonzero voltage and are out of phase with each other. The output y is slope-symmetric as in Fig. 6.22. To obtain asymmetry in $\dot{y}$, an additional constant voltage can be fed into the integrator. This circuit requires one less amplifier than a circuit proposed by Langill [6] and can accommodate variable y_1 and y_2.

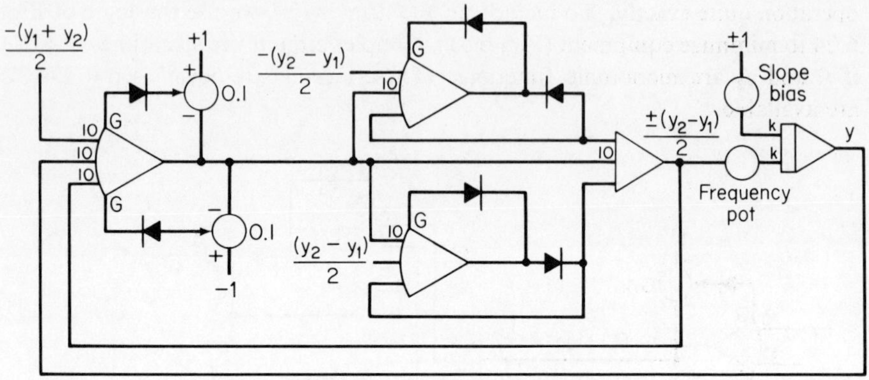

FIG. 6.52 Multivibrator and sweep generator.

6.5.8 The diode bridge limiter

A four diode bridge limiter, due to Medkeff and Parent [7], provides very sharp limiting at preset positive and negative values, with very constant limits. Its operation requires a low-impedance source (an amplifier), and it must drive a constant load R_L as shown in Fig. 6.53. Since some computers provided specially wired units where the input x and references need only be patched, a special working diagram symbol has evolved as shown in Fig. 6.53. To set a limit, it is simplest to adjust the pots while the output is at the limit and under load by reading the outputs with a DVM. Large dropping resistor values (typically 100 kΩ) mean that limits cannot be set close to the reference.

Because of a mismatch in the dropping resistors or diodes, $f(0) \neq 0$. In

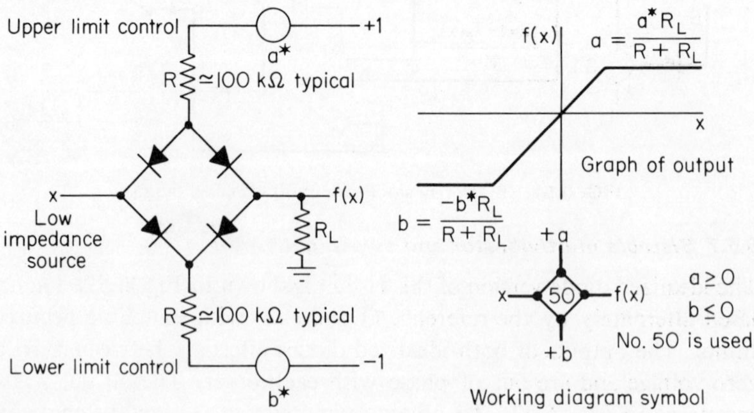

FIG. 6.53 Diode bridge limiter.

applications requiring a precise function, this offset should be nulled by introducing a constant voltage. Variable voltages can be patched in lieu of the references, but then the offset is not constant. Two other uses of the bridge limiter are shown in Fig. 6.54.

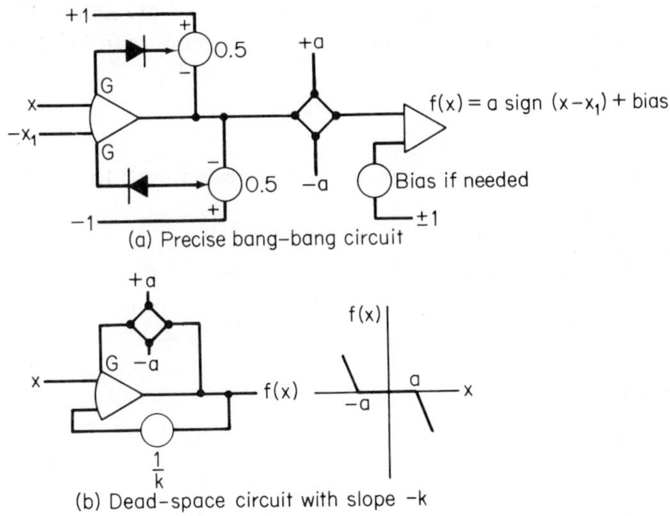

FIG. 6.54 Other uses of a diode bridge limiter.

6.6 Summary and Examples

We have seen the many circuits possible for generating functions with discontinuities, using relays, diodes, and digital-analog switches. The list is by no means complete since the literature abounds with special circuits and techniques (partially listed in [4]–[19]). In general, relay circuits are used for slow-speed computation. Because of precious metal contacts, amplitude errors are negligible, and an open switch has virtually infinite resistance. Diode circuits have less of a frequency-response problem but are not so accurate as relay circuits. If more diodes and equipment are available, these can be made quite accurate with idealized diode circuits. Finally, DAS's are intended for use in iterative computers at fairly high speeds and are easy to program when a logic board is available. Usually there is a clear-cut choice of which type switching should be used, depending upon the application and equipment available.

Example 6.1. Scattering Integral. A scattering integral arises in optics:

$$I = -C_1 \int_1^{-k} (\arccos|x| - |x|\sqrt{1-x^2}) \exp\left(\frac{-C_2}{x+k}\right) dx$$

As with all definite integrals, it is convenient to convert it to differential equation form:

$$\frac{dI}{dx} = -C_1(\arccos|x| - |x|\sqrt{1-x^2})\exp\left(\frac{-C_2}{x+k}\right)$$

with initial conditions of $x_0 = 1$ and $I_0 = 0$. I is found when $x = -k, 0 \leq k \leq 1$, and C_1 and C_2 are constants. Since $\arccos x$ and $\sqrt{1-x^2}$ are group III functions (Chapter 5), the function $\arccos x - x\sqrt{1-x^2}$ is generable with linear equipment with

$$\dot{x} = -\sqrt{1-x^2}$$

where the negative sign is taken because x must decrease from 1 to $-k$. If we let

$$F_1 = \sqrt{1-x^2}, \qquad F_4 = x^3$$

$$F_2 = \sqrt{1-x^2}\,(\arccos|x|), \qquad F_5 = x^2\sqrt{1-x^2}$$

$$F_3 = x\arccos|x|, \qquad F_6 = \exp\left(\frac{-C_2}{x+k}\right)$$

we may show that

$$\begin{aligned}
\dot{x} &= -F_1, & x_0 &= 1\\
\dot{F}_1 &= x, & (F_1)_0 &= 0\\
\dot{F}_2 &= F_1 \operatorname{sign} x + F_3, & (F_2)_0 &= 0\\
\dot{F}_3 &= |x| - F_2, & (F_3)_0 &= 0\\
\dot{F}_4 &= -3F_5, & (F_4)_0 &= 1\\
\dot{F}_5 &= -2x + 3F_4, & (F_5)_0 &= 0\\
\dot{F}_6 &= -\frac{C_2F_1F_6}{(x+k)^2}, & (F_6)_0 &= \exp\left(\frac{-C_2}{1+k}\right)\\
\dot{I} &= -\frac{dI}{dx}F_1 = C_1F_6(F_2 - |x - F_4|), & I_0 &= 0
\end{aligned} \tag{6.8}$$

An unscaled diagram to solve system (6.8) is shown in Fig. 6.55 using servo multipliers and relays for slow-speed operation. Relay 3 is used in lieu of a hold circuit (Chapter 18) to enable I to be read with a DVM. Both quotients $F_1F_6/(x+k)$ and $F_1F_6/(x+k)^2$ approach 0 as $x \to -k$. If these amplifiers overload because of other inaccuracies, it happens in the last small fraction of the total computing time. Because F_6 is very near zero at this point and most of the contribution to the integral has already been obtained, very little error is introduced. In fact, the computation can be stopped before $x = -k$ with virtually no error.

Note the use of the parametric technique and the linear equipment theorem for this problem.

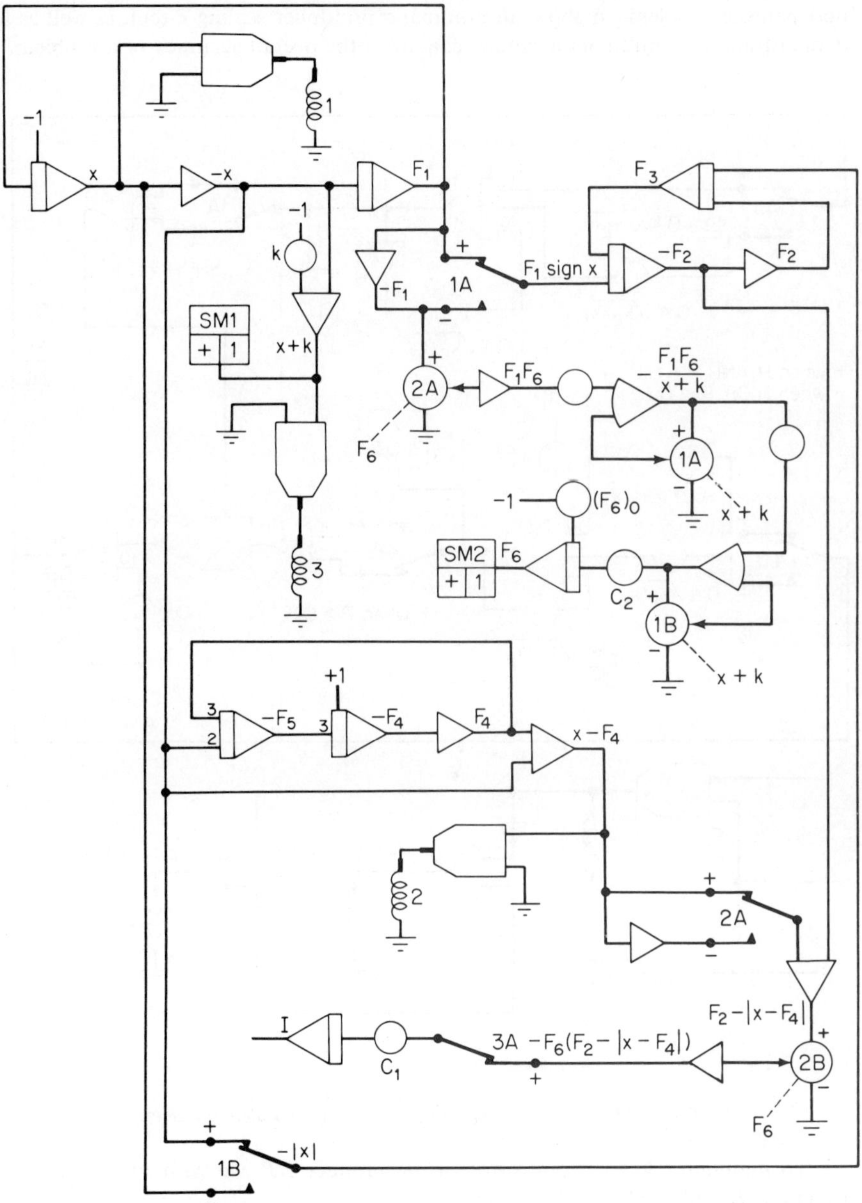

FIG. 6.55 Scattering integral mechanization.

Example 6.2. Automatic Multiplier Rescaling. We return to Example 4.4, the host-parasite problem, to show an automatic multiplier scaling circuit as well as to demonstrate how multiplier accuracy can affect the overall accuracy of a problem.

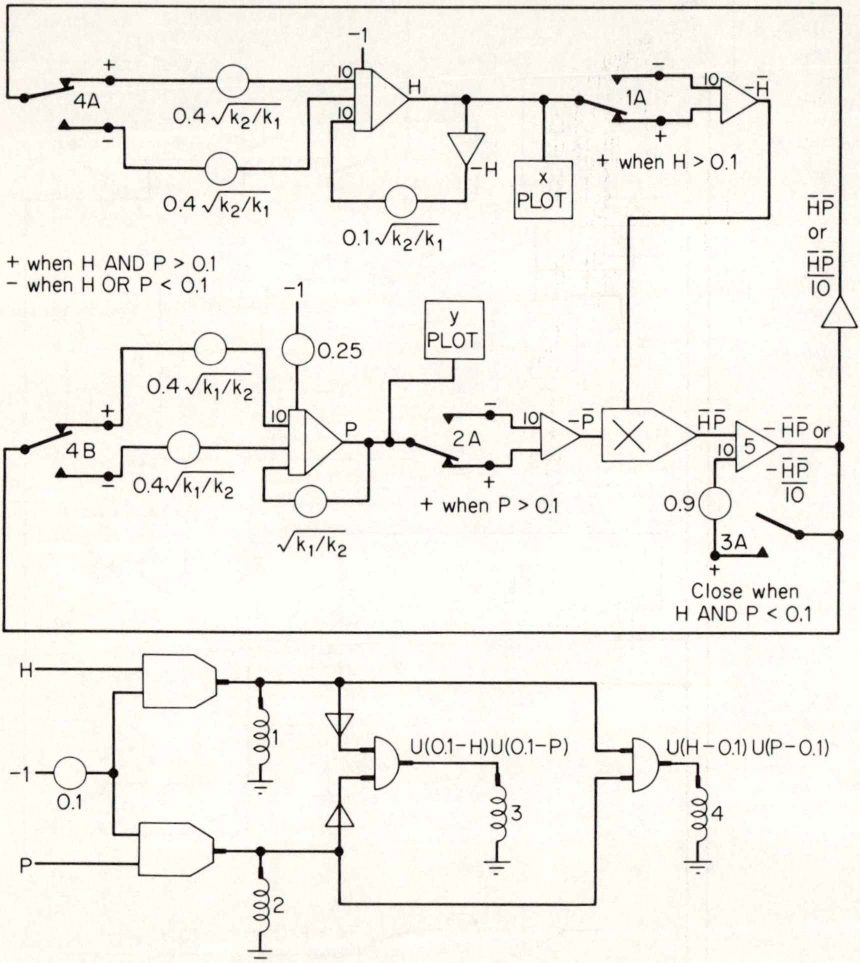

FIG. 6.56 Automatic rescaling for the host–parasite problem.

To maintain a low per cent error in the product *HP*, we wish to multiply the functions $\bar{H}$ and $\bar{P}$ where

$$\begin{aligned} \bar{H} &= H, & H &> 0.1 \\ &= 10H, & H &< 0.1 \end{aligned}$$

$$\text{i.e., } \bar{H} = HU(H - 0.1) + 10HU(0.1 - H)$$

and

$$\begin{aligned} \bar{P} &= P, \qquad P > 0.1 \\ &= 10P, \qquad P < 0.1 \\ \text{i.e., } \bar{P} &= PU(P - 0.1) + 10PU(0.1 - P) \end{aligned}$$

and account for any increase in the product $\bar{H}\bar{P}$ over HP by reducing pot settings as needed. A circuit to accomplish this using relays and the logic board is shown in Fig. 6.56. Scale factors have been omitted for clarity, but the circuit is nevertheless comparable to that of Fig. 4.19. When H and P are both less than 0.1, the product $-\bar{H}\bar{P}$ is attenuated at amplifier 5 by a factor of 10 by introducing an additional 9 gain feedback (a total of 10). An alternate scheme requires an additional relay at the H and P integrator inputs, as shown in Fig. 6.57, which simply attenuates the inputs by an additional factor of 10.

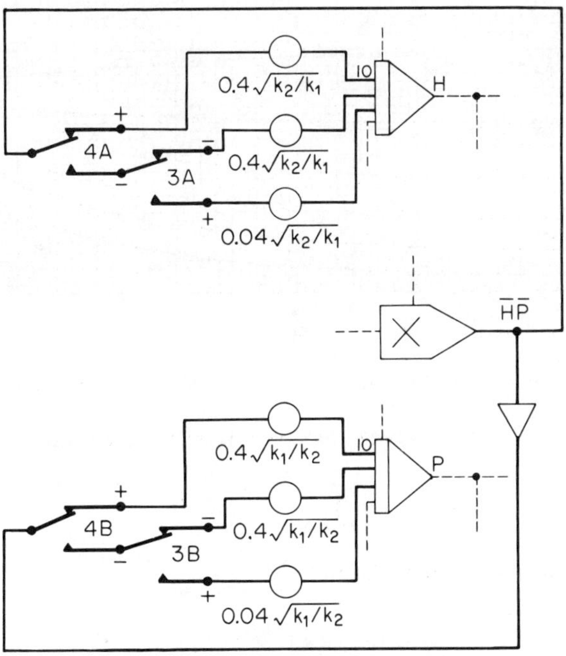

FIG. 6.57 Alternate switching scheme when H AND $P < 0.1$.

Figure 6.58 is a set of graphs made from the circuit of Fig. 6.56, for comparison with those in Fig. 4.20. Although the curves still do not close in some cases because some error is unavoidable, it is clear that automatic rescaling improves the results considerably.

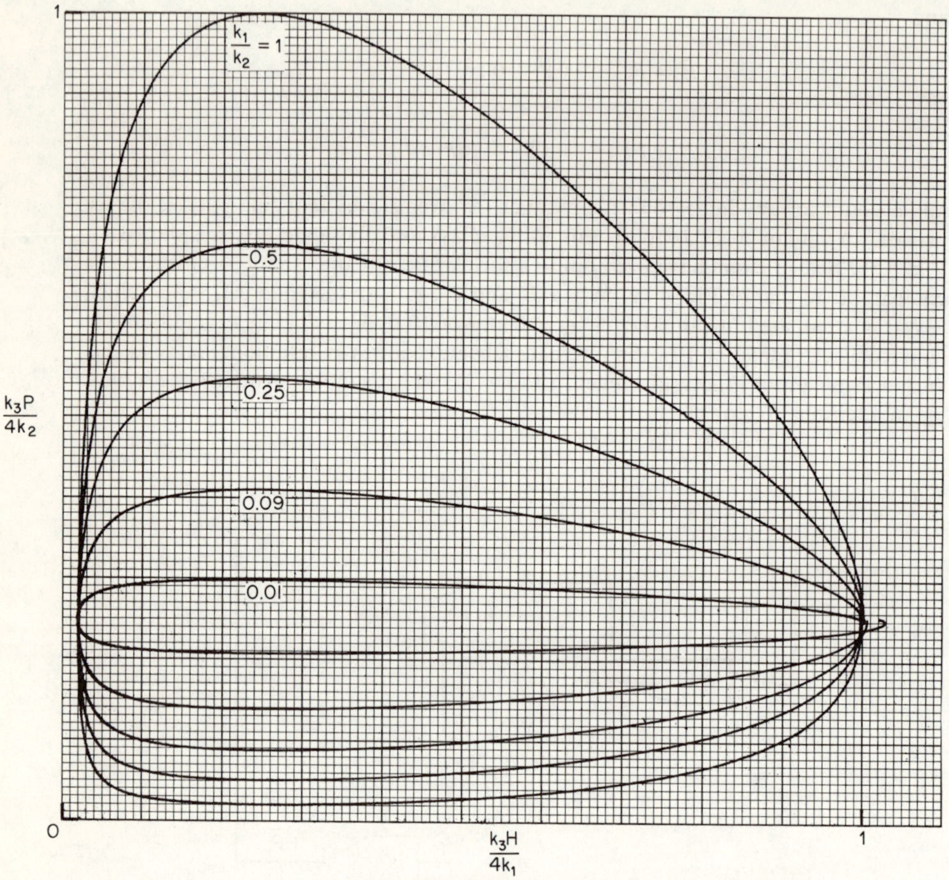

FIG. 6.58 Graphs of host–parasite solutions with automatic rescaling.

Example 6.3. Multiple Integrals. Double integrals of the form

$$I = \int_a^b \int_{y_1(x)}^{y_2(x)} f(x, y)\, dy\, dx$$

can be approximated by a single integral

$$I \simeq \int_a^b [y_2(x) - y_1(x)] f[x, y(x)]\, dx \tag{6.9}$$

provided $y(x)$ is a high frequency triangular wave that oscillates between $y_2(x)$ and $y_1(x)$ [13], [22]. The frequency need not be constant; as long as it is large compared with the frequencies of the other functions y_2, y_1, and f, good results can be obtained. We may use, therefore, the output of a bistable multivibrator and sweep generator circuit for $y(x)$. To demonstrate this technique, let us devise the circuit to find the volume of sphere of radius 1:

$$I = 4\pi \int_0^1 \int_0^{\sqrt{1-x^2}} y \, dy \, dx$$

Although a trivial analytical problem, it serves well to demonstrate the method. Using (6.9), we replace I by

$$I = \int_0^1 y\sqrt{1 - x^2} \, dx$$

and use

$$\frac{dI}{dx} = y\sqrt{1 - x^2}$$

with initial conditions of $x = 0$ and $I = 0$, and stop the computation when $x = 1$. Although the reference [13] generates $\sqrt{1 - x^2}$ with a standard square-root circuit (Chapter 9), we shall use the linear equipment theorem for its generation after noting that it is a group III function. Thus if we let

$$F_1 = \sqrt{1 - x^2}, \qquad F_2 = x\sqrt{1 - x^2}, \qquad F_3 = 1 - x^2 = F_1^2$$

we obtain

$$\frac{dF_1}{dx} = \frac{-x}{F_1}, \qquad (F_1)_0 = 1$$

$$\frac{dF_2}{dx} = F_1 - \frac{x^2}{F_1}, \qquad (F_2)_0 = 0$$

$$\frac{dF_3}{dx} = -2x, \qquad (F_3)_0 = 1$$

With

$$\dot{x} = F_1, \qquad x_0 = 0 \tag{6.10}$$

we also get

$$\begin{aligned} \dot{F}_1 &= -x, & (F_1)_0 &= 1 \\ \dot{F}_2 &= F_1^2 - x^2 = 1 - 2x^2 = 2F_3 - 1, & (F_2)_0 &= 0 \\ \dot{F}_3 &= -2F_2, & (F_3)_0 &= 1 \\ \dot{I} &= 4\pi y F_3, & I_0 &= 0 \end{aligned} \tag{6.11}$$

A scaled diagram for solving (6.10) and (6.11) is shown in Fig. 6.59 using the multivibrator circuit of Fig. 6.52 with $y_2 = \sqrt{1 - x^2}$, $y_1 = 0$, and $k = 20$. The out-

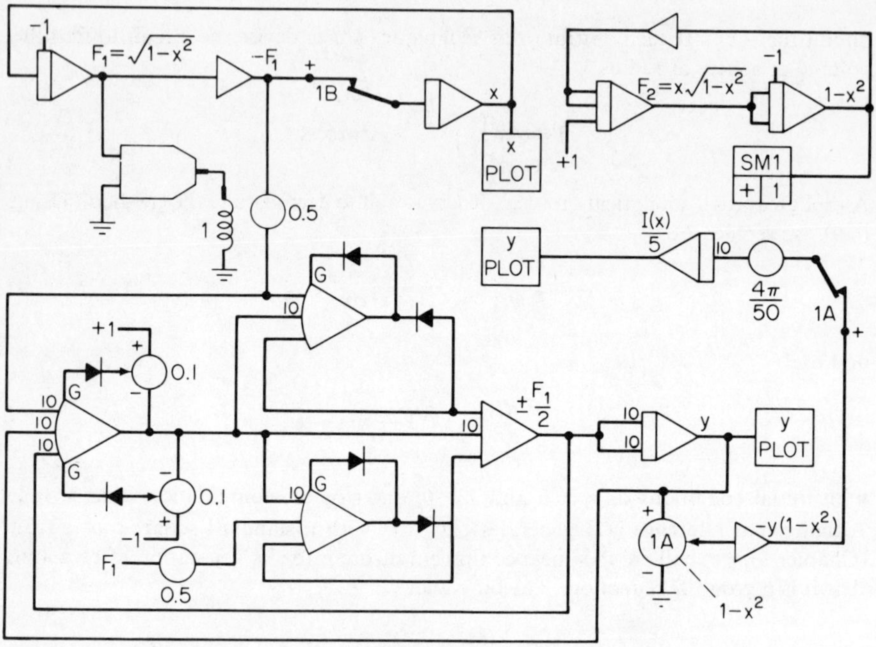

FIG. 6.59 Scaled diagram for solving double integral.

put of $y(x)$ is shown in Fig. 6.60 (10-μF capacitors were used to slow the problem down sufficiently for plotting). The slopes are constant only because $\dot{y}$ and $\dot{x}$ are proportional to the same function of x so that $dy/dx = \dot{y}/\dot{x} = \pm$ constant. In general, this would not be so. Figure 6.61 is a graph of the function $I(x)$ which should approximate

$$I(x) = 4\pi\int_0^x \left(\int_0^{\sqrt{1-z^2}} y\,dy\right) dz = 2\pi\left(x - \frac{x^3}{3}\right)$$

By increasing k to 1000 (using a 0.1-μF capacitor), the oscillations of y are so frequent that $I(x)$ does indeed approach the correct function without noticeable oscillation. Although extendable to higher-order integrals, the difficulty in generating $f[x, y(x)]$ makes this method less practical than those involving memory (Chapter 18). If a repetitive computer with memory is *not* available, however, this technique can be used.

Example 6.4. Counting Circuits. Sometimes it is convenient to count on the computer—a common situation is to obtain a voltage proportional to the number of times an event occurs, the event being a function $f(\tau)$ crossing a given level y_1. By means of other comparators, other analog voltages can be controlled after n events occur. A simple relay circuit accomplishing this is shown in Fig. 6.62. In lieu of using relays at the integrator inputs, mode relay drivers (Chapter 18) can control the HOLD relays. Each time $f(\tau)$ crosses y_1, one pair of relays opens while the other closes. If $f(\tau) - y_1$ becomes negative, y increases by the amount a, while x increases by this amount each time $f(\tau) - y_1$ becomes positive. The sum of x and y increments by the

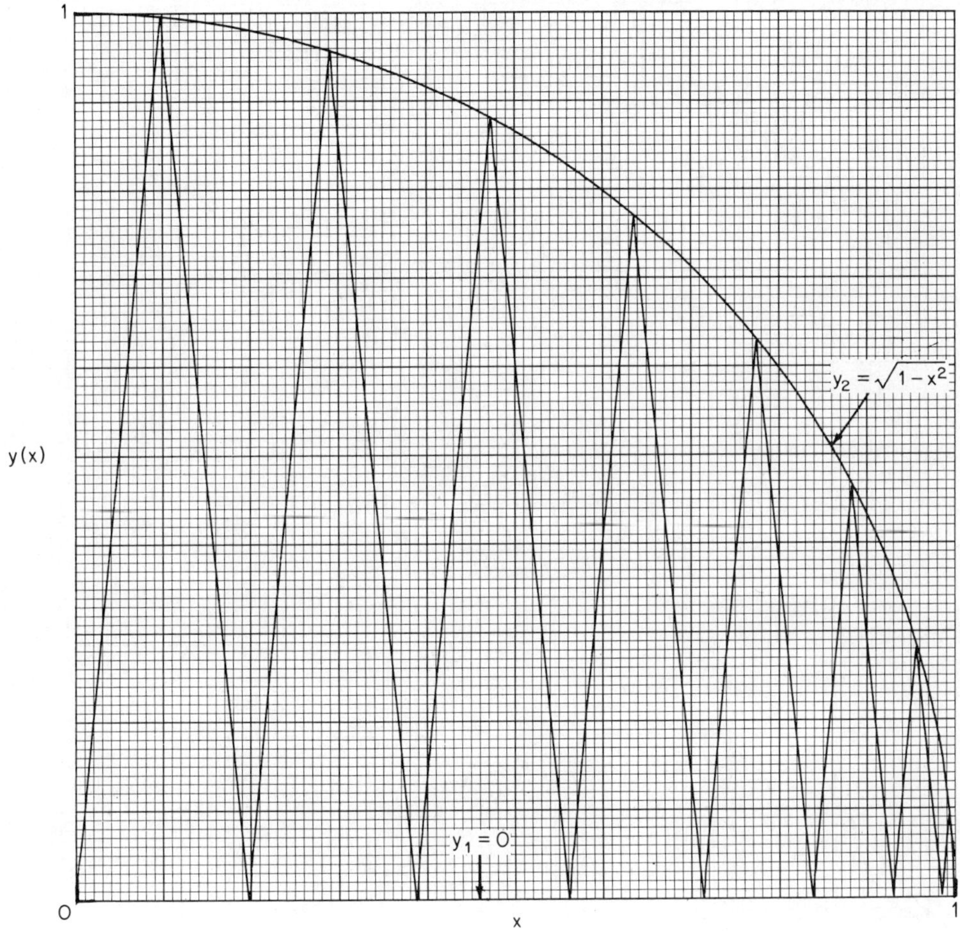

FIG. 6.60 The output $y(x)$.

amount a every time $f(\tau) - y_1$ passes through zero. Because of timing (one switch should completely open before the other closes), it may be necessary to introduce a small dead zone where both sets of switches are open. By making a small, counts as high as 1000 can be made.

For higher frequency signals, diode circuits may be substituted, The circuit in Fig. 6.63 has poor resolution since only a simple (and inaccurate) diode simulation of SPST switches is used, but it is adequate for a small number of counts. The diode in the output path reduces the voltage to zero as in an idealized diode circuit, permitting the large gain at the integrator input with negligible drift. Also, the two pots labeled ϵ can create a dead zone where both switches are open, in case there are timing problems. For precision counting with diodes, the circuit of Fig. 6.51 is adaptable by its conversion to a two-SPST simulation.

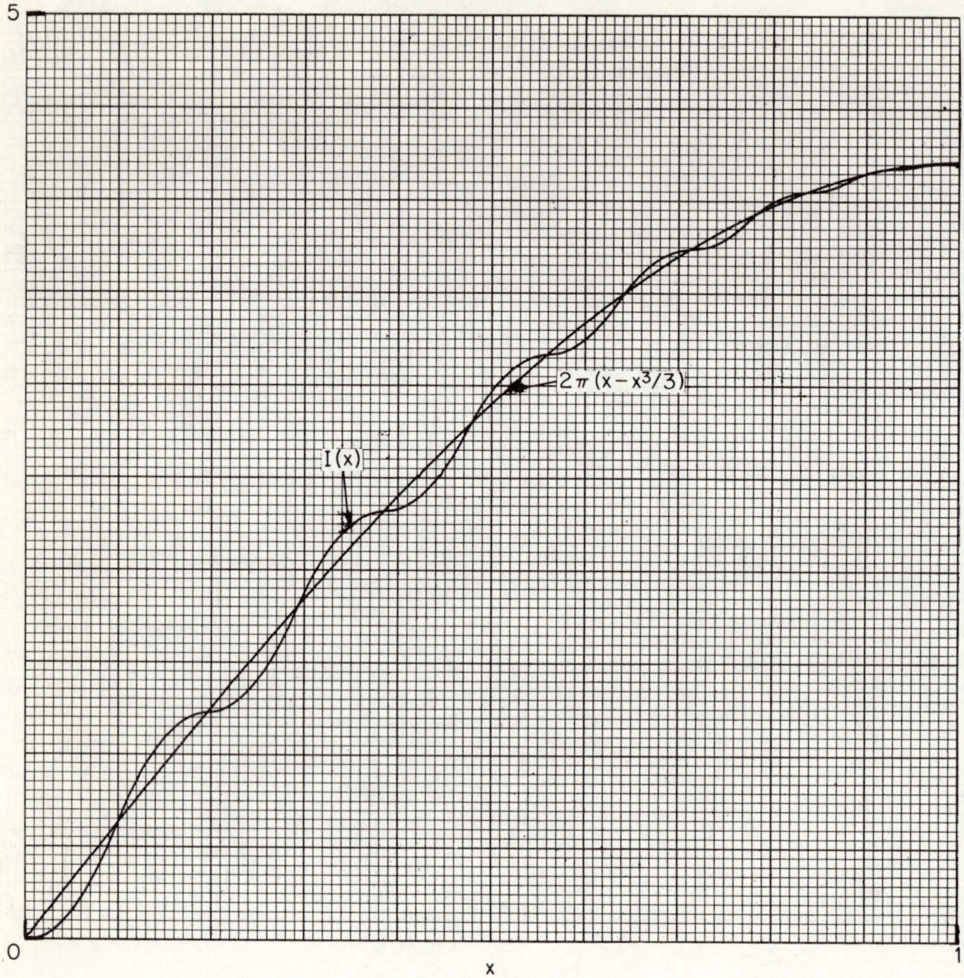

FIG. 6.61 The output $I(x)$.

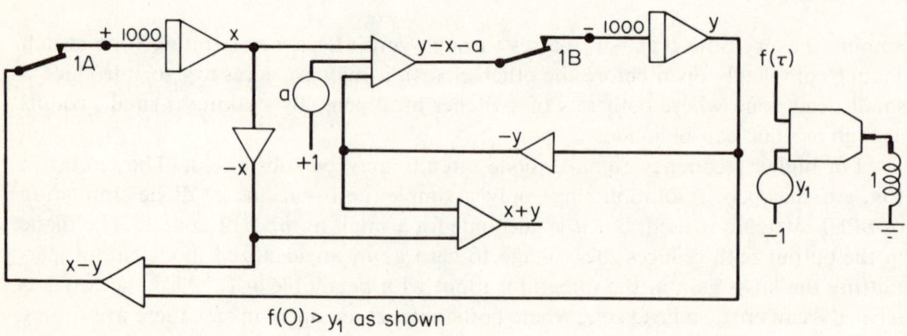

FIG. 6.62 A counting circuit.

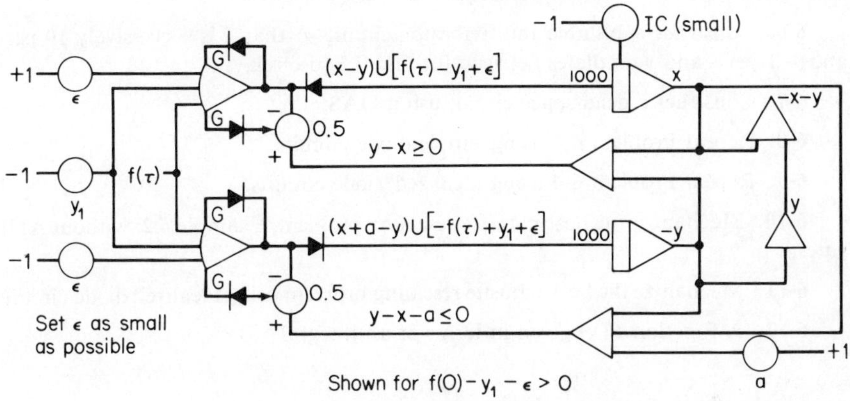

FIG. 6.63 A counting circuit using simple diodes.

PROBLEMS

6-1. Given two logic variables U_1 and U_2, show that the logical OR relation is given algebraically by

$$U_1 \text{ OR } U_2 = U_1 + U_2 - U_1 U_2$$

6-2. Given the digital signals U_1, U_2, and U_3 and the analog voltage x, construct the relay circuit producing

$$x[(U_1 \text{ OR } U_2) \text{ AND } (U_2 \text{ OR } U_3) \text{ AND } (U_1 \text{ OR } U_3)]$$

and show it is algebraically equal to

$$x(U_1U_2 + U_1U_3 + U_2U_3 - 2U_1U_2U_3)$$

6-3. Construct a bang-bang circuit using relays:

$$\begin{aligned} f(x) &= a, & y &< y_1 \\ &= -a, & y &> y_1 \end{aligned}$$

6-4. Construct a relay circuit generating the function

$$\begin{aligned} y &= 10x, & 0 &\le x \le 0.1 \\ &= 1 - 2.5(x - 0.1), & 0.1 &\le x \le 0.9 \\ &= -1 + 10(x - 0.9), & 0.9 &\le x \le 1 \end{aligned}$$

Be sure to prevent every amplifier from overloading.

6-5. Prepare a backlash circuit (with relays) with $f_1(x) = x - 0.01$ and $f_2(x) = x + 0.01$.

6-6. Construct a bistable multivibrator circuit so that $\dot{y}$ is successively 10 per s and -1 per s and y oscillates between 0.1 and 0.5 (use relays).

6-7. Construct a dead-space circuit using DAS.

6-8. Repeat Problem 6-4 using simple diode circuits.

6-9. Repeat Problem 6-4 using idealized diode circuits.

6-10. Mechanize the host-parasite rescaling problem, Example 6.2, without AND gates.

6-11. Mechanize the host-parasite rescaling problem using idealized diode circuits.

6-12. A function $F(x)$ given in terms of an integral

$$F(x) = \int_{a(x)}^{b(x)} f(x, y)\, dy$$

can be reduced to the solution of a double integral when $F(x_0)$ is known. For, by differentiating using the Leibnitz rule,

$$\frac{dF}{dx} = \int_{a(x)}^{b(x)} \frac{\partial f(x, y)}{\partial x}\, dy + f(x, b)\frac{db}{dx} - f(x, a)\frac{da}{dx}$$

from which

$$F = \int_{x_0}^{x} \left[\int_{a(x)}^{b(x)} \frac{\partial f(x, y)}{\partial x}\, dy\right] dx + \int_{x_0}^{x} \left[f(x, b)\frac{db}{dx} - f(x,a)\frac{da}{dx}\right] dx + F(x_0)$$

Generate the function

$$F(x) = \int_0^x e^{-xy^2}\, dy$$

in this way, assuming $x_0 = 0$, $0 \leq x \leq 2$, and $\dot{x} =$ constant.

6-13. Given the signal $f(\tau)$, construct a circuit to determine the per cent time that $|f(\tau)|$ is above the level f_1, assuming $0 \leq \tau \leq 100$.

6-14. Devise a counting circuit with DAS.

6-15. Devise a counting circuit with idealized diodes.

6-16. Using circuits as in Problem 6-15, construct a diagram to determine the percentage of peaks of a signal $|f(\tau)|$ above the level f_1. A peak is defined to be a point where the derivative of $|f(\tau)|$ passes from positive to negative. Assume the signal contains 1000 peaks, and that $\dot{f}$ is available.

6-17. Draw an idealized diode circuit which produces max $(x_1, x_2, \ldots, x_n)$ using $n + 1$ summers and $2n$ diodes. [*Hint:* $x_1 U(x_1 - x_2) + x_2 U(x_2 - x_1) = x_1 + (x_2 - x_1)U(x_2 - x_1)$.]

REFERENCES

[1] Bennett, R. R, "The Generation of Straight Line Transfer Relationships," *Symposium I on REAC Techniques*, Reeves Instr. Corp., New York, March 15–16, 1951 (ASTIA No. 108488).

[2] Warshawsky, L. M., and W. Braun, "Applications of Differential Relays to Solution of REAC Problems," *Symposium II on Simulation and Computing Techniques, part 2*, Reeves Instr. Corp., New York, April 28–May 2, 1952 (ASTIA No. 170848).

[3] Morrill, C. D., and R. V. Baum, "The Role of Diodes in an Electronic Differential Analyzer," *Symposium II on Simulation and Computing Techniques, part 2*, Reeves Instr. Corp., New York, April 28–May 2, 1952 (ASTIA No. 170848).

[4] Morrill, C. D., and R. V. Baum, "Diode Limiters Simulate Mechanical Phenomena," *Electronics*, vol. 25, Nov. 1952.

[5] Fifer, S., *Analogue Computation*, Vol. 1 (New York: McGraw-Hill Book Co., 1961), p. 275.

[6] Langill, A. W., Jr., "Accurate Simulation of Nine Common Nonlinearities," *Electronic Design*, vol. 9, June 21, 1961.

[7] Medkeff, R. J., and R. J. Parent, "A Diode Bridge Limiter for Use with Electronic Analogue Computers," *Trans. AIEE, part I*, vol. 70, 1951, pp. 913–916.

[8] Alliston, W. H., "Quadrature Triangular Wave Oscillator," *Instr. and Control Systems*, vol. 36, Oct. 1963, p. 125. This circuit generates two triangular waves 90° apart using diodes.

[9] Brown, T. S., and K. F. Nelson, "The Logical Simulation of Discontinuous Nonlinearities," *Simulation*, vol. 6, Feb. 1966. This paper discusses uses of DAS and the logic board.

[10] Earls, J. C., "Backlash Simulation in d.c. Electronic Analogue Computers," *AICA*, vol. 3, July 1961. Analyzes conditions under which two backlash simulators operate well.

[11] Foote, D.D., "Pulse-Width Modulator," *Instr. and Control Systems*, vol. 36, May 1963.

[12] Kasischke, F. W., "Analog-Computer Technique for Simulation of Dynamic Pressure in Re-Entry Studies," *IRETEC*, vol. 11, Feb. 1962. Discusses an automatic rescaling circuit for very large ranges.

[13] Hausner, A., "Multiple Integrals on a Non-Repetitive Analog Computer," *Proc. Spring Joint Comp. Conference*, 1963.

[14] Blache, Y. J. M., "Automatic Time-Mark Generation for Analog Computer Phase-Plane Trajectories," *IRETEC*, vol. 11, Aug. 1962.

[15] Dannback, T. M., "Squaring Circuits," *Simulation*, vol. 3, July 1964. Generates a voltage which averages x^2 using diodes.

[16] Alliston, W. H., "A Four-Quadrant Multiplier," *Simulation,* vol. 4, Jan. 1965. Produces a quarter-square multiplier using Dannback's squaring circuits [15].

[17] Eubanks, R. A., "Simple Analog-Computer Sawtooth Generator," *Instr. and Control Systems,* vol. 33, Feb. 1960. This circuit uses a relay and feedback to generate a sawtooth—a repetitive function of time of the form

$$f(\tau) = a\tau - a\tau_p \sum_{i=1}^{\infty} U(\tau - i\tau_p)$$

where $a\tau_p$ is the amplitude and $1/\tau_p$ is the frequency of the wave.

[18] Howe, R. M., "The Use of Silicon-Junction Diodes in Analog Simulation," *Instr. and Control Systems* (In Simulation Council newsletter), vol. 34, Aug. 1961.

[19] Witsenhausen, H., "Circuit for Discrete Summation," *AICA,* vol. 3, July 1962. Produces a circuit to generate

$$f(\tau) = a \sum_{i=1}^{\infty} U(\tau - i\Delta\tau)$$

by subtracting a sawtooth wave from a ramp.

[20] Federal Electric Corporation, *Boolean Algebra* (Englewood Cliffs, N. J.: Prentice-Hall, Inc., 1966).

[21] Cartianu, G., and V. I. Vlad, "The Synthesis and Processing of the Signals with Discontinuities in the Time Domain," *Proceedings of the IFIP Congress,* 1968.

[22] Bayly, J. G., and A. C. Soudack, "Use of the Berkeley EASE Computer in Evaluating Double Integrals," *AICA,* vol. 5, April 1963.

Chapter 7

GENERATING APPROXIMATIONS OF FUNCTIONS

Generalized integration fails when unstable equations must be solved for an extended period of time. In such a case, we seek to generate the required function by one of several other ways, all of which can be characterized by a final generation which has a *bounded error*.

In this chapter, we consider the approximation of functions by other easily generated functions. Because of the availability of multipliers, a polynomial in x has priority over all other types of such approximations. Accordingly, we emphasize polynomial approximations.

Readers familiar with an approximation using a truncated Taylor series expanded about $x = a$

$$f(x + a) \simeq f(a) + x\frac{df(a)}{dx} + \ldots + \frac{x^n}{n!}\frac{d^n f(a)}{dx^n} \tag{7.1}$$

may feel that this is a sufficient method for all continuously differentiable cases. Indeed, application of (7.1) is quite effective in many cases, particularly when relatively few terms are needed for the required accuracy. As an approximation to $f(x)$, however, the Taylor series is inefficient. Many more terms are needed for a given accuracy than for other approximations to be described. This is so because the $n + 1$ coefficients of the polynomial in a Taylor series are chosen so that $f(x)$ and its first n derivatives are exact at the *single* point a. In most instances, we do not care about the derivatives at a, and this permits error characteristics to be improved at other points in an interval of interest.

To use a Taylor series, it is well to remember that fewer terms are generally needed for the accuracy required when $f(x)$ is expanded about the midpoint of the interval needed than about any other point. For the function $f = (\sin x)/x$ in the interval $0 \leq x \leq \pi$, for example, we have

$$xf = \sin x, \quad x\frac{df}{dx} + f = \cos x$$

$$x\frac{d^2f}{dx^2} + 2\frac{df}{dx} = -\sin x, \quad x\frac{d^3f}{dx^3} + 3\frac{d^2f}{dx^2} = -\cos x, \quad \text{etc.}$$

from which we may determine

$$f(x) = 0.6366 - 0.4053\left(x - \frac{\pi}{2}\right) - 0.06030\left(x - \frac{\pi}{2}\right)^2 + 0.03889\left(x - \frac{\pi}{2}\right)^3 \cdots$$

to any accuracy desired. In this case, the fifth-order polynomial produces $f(x)$ to within 0.002, with high accuracy near $\pi/2$, and with the worst error occurring at $x = 0$.

All such series have an interval of convergence that may or may not include the interval needed, so care should be taken in this regard.

Other series expansions, such as Laurent's:

$$f(x) = \sum_{i=0}^{\infty} \frac{a_i}{(x - x_0)^i}$$

are not usually convenient for analog computation and are rarely used.

7.1 Basic Approximation Theory

The concept of approximating one function by others has been used by virtually all the sciences since their inception. We have, therefore, a large body of knowledge in this area, most of which can be used to great advantage in analog computation. In this section we examine some fundamental ideas on this subject.

To begin with, we recognize that two important distinct situations arise in practice:

(1) A given continuous function $f(x)$ must be generated on the computer for a given known interval, $a \leq x \leq b$. If no equipment can generate $f(x)$ directly, how may we obtain a good approximation to $f(x)$ with available equipment?

(2) A point set of m number pairs (x_j, y_j), $j = 1, 2, \ldots, m$, is available from experimental data. What function $y = f(x)$ should be used to pass close to these points; i. e., what is a good approximation to the point set? It should be noted that the points (x_j, y_j) usually contain experimental error.

7.1.1 Terminology

In general, we look at specific functional forms, $F(A, x)$, as approximation functions to $f(x)$ in an interval $a \leq x \leq b$, where A is a set of n parameters, a_1, a_2, $\ldots$, a_n. The notation $F(A_1, x)$ and $F(A_2, x)$ denotes two different approximating functions (provided the set A_1 is not equal to the set A_2) of the same functional form. A simple example is the function

$$F(A, x) = a_1 + a_2 x^2$$

which is a linear† approximation containing two parameters. The function $f(x)$ could be just defined at a finite set of points x_i.

We wish to compare two functions to determine which is a better approximation to $f(x)$, so we need some distance function, $\rho[F(A, x), f(x)]$. This is a number that determines how close $F(A, x)$ is to $f(x)$. A distance function should have the property that it is always nonnegative. If

$$\rho[F(A_1, x), f(x)] < \rho[F(A_2, x), f(x)] \tag{7.2}$$

we say that $F(A_1, x)$ is a better approximation to $f(x)$ than is $F(A_2, x)$. We are now in the position to look for a best approximation for a specific distance function. This is the classical approximation problem:

For a given distance function ρ, determine the set of parameters A^* so that

$$\rho[F(A^*, x), f(x)] \leq \rho[F(A, x), f(x)]$$

for all other permissible sets of parameters. In other words, we wish to find the parameters that minimize the distance function. A best approximation is not necessarily unique; neither is existence guaranteed. Methods to find the A^* depend heavily upon ρ, F, and whether or not $f(x)$ is a point set or depends continuously on x. (F, of course, depends continuously on x.) Rice [1] gives an excellent discussion of best approximations.

7.1.2 Common distance functions

Distance functions are ordinarily norms that satisfy properties additional to (7.2). For this reason, one usually speaks of minimizing a norm or using a norm in a given approximation problem, without confusion. Table 7.1 shows the three most commonly used norms.

A weighting function $w(x) \geq 0$ is included in each definition because some values of x may be more important than are other values. For example, if $f(x)$ is positive, we often take $w(x)$ inversely proportional to $f(x)$ {$[f(x)]^2$ in the case

†The term linear applies to the approximation, not the function. An approximation is linear if it is of the form $\sum_{i=1}^{n} a_i \phi_i(x)$.

TABLE 7.1

Commonly Used Norms

Name of norm	Norm for continuous x	Norm for discrete x_j
L_1 or absolute error	$\rho = \int_a^b w(x)\lvert F(A, x) - f(x)\rvert\, dx$	$\rho = \sum_{\text{all } j} w(x_j)\lvert F(A, x_j) - f(x_j)\rvert$
L_2 or least squares	$\rho = \int_a^b w(x)[F(A, x) - f(x)]^2\, dx$	$\rho = \sum_{\text{all } j} w(x_j)[F(A, x_j) - f(x_j)]^2$
L_∞ or Chebyshev	$\rho = \max_{a \leq x \leq b} w(x)\lvert F(A, x) - f(x)\rvert$	$\rho = \max_{\text{all } j} w(x_j)\lvert F(A, x_j) - f(x_j)\rvert$

of least squares} to optimize the relative error. With a Chebyshev norm, we can then state that the maximum per cent error (equal to 100 times the relative error) is the smallest possible. The weighting function is most often taken to be equal to 1 otherwise and is then called a uniform weighting function. If not specifically stated, $w(x) = 1$ will be understood.

The Chebyshev norm is also called an L_∞ norm because of its relationship to a general L_k norm defined in the continuous case by

$$\rho = \int_a^b w(x)|F(A, x) - f(x)|^k\, dx$$

An important approximation theorem states that the sequence of approximations obtained using an L_k norm approaches that of the Chebyshev norm as $k \to \infty$ provided the Chebyshev approximation is unique.

The absolute error norm is used frequently in analog computations because of its simplicity. As we shall see in subsequent chapters, its use in optimization problems also keeps voltages large and allows for more accurate computations. For a uniform weighting function, this norm minimizes the area of the absolute error $|F(A, x) - f(x)|$.

The least squares norm is generally the easiest to calculate analytically and is often used for that reason. There is also much justification for using a least squares norm when fitting normally distributed discrete data to a functional form that is known to satisfy the physics of a situation [2]. In that case, we determine the appropriate continuous function first from the discrete data and then generate the continuous function on the analog computer.

The Chebyshev norm is that most frequently desired (although difficult to calculate and therefore not often used) because of several reasons:

(a) By minimizing the maximum absolute error, we may make the statement, "The maximum absolute error is __ per cent of full scale and cannot be reduced without changing the functional form." This is a desirable statement from the point of view of error analysis.

(b) The Chebyshev norm has a simple geometric interpretation that allows graphical methods of approximations for some functional forms, among which are piecewise linear functions (used in Chapter 8).

(c) The characterization of a best Chebyshev approximation is usually that the error function contains $n + 1$ points of extreme error (*extremal* points) all equal in absolute value (called the deviation) but alternating in sign as x increases from a to b (see Hastings, [3]). This property is called *alternation*, and the curve is said to alternate n times. It is a property that is simple to check.

Which norm should be used? It generally makes very little difference, but a great deal depends upon the purpose for which the function $f(x)$ is needed. If it is to be integrated, and we wish $\int_a^b f(x)\,dx$ to have little error, the absolute error norm is an appropriate choice [1]. Probably the best norm to use, if any, is one that can be minimized quickly. The actual methods of minimizing each norm vary and will be discussed in subsequent examples.

7.2 Approximations for Discrete Data

Invariably, a polynomial is fitted to discrete data by the method of least squares when no other specific functional form is required by the physics of the problem. This is so for three reasons: (1) When the data are normally distributed, there is strong theoretical justification for using a least squares norm for a given functional form; (2) least squares approximations have been intensively investigated with much theoretical material available on the subject; and (3) the determination of the approximation parameters involves simple calculations that can often be performed easily with a desk calculator.

If there are m data points (x_j, y_j), and we wish

$$F(x) = \sum_{i=0}^{n} a_i x^i$$

so that the distance function

$$\rho = \sum_{j=1}^{m} [y_j - F(x_j)]^2 = \sum_{j=1}^{m} \left(y_j - \sum_{i=0}^{n} a_i x_j^i\right)^2 \tag{7.3}$$

is minimized; then a necessary and sufficient condition for the unique best approximation is that the a_i satisfy the $n + 1$ equations

$$\frac{\partial \rho}{\partial a_i} = 0, \qquad i = 0, 1, \ldots, n \tag{7.4}$$

In fact, (7.4) is necessary (but not always sufficient) for all continuous and differentiable distance functions and will be referred to often.

Application of (7.4) to (7.3) results in

$$
\begin{aligned}
\sum_{j=1}^{m} y_j &= a_0 m + a_1 \sum_{j=1}^{m} x_j + \ldots + a_n \sum_{j=1}^{m} x_j^n \\
\sum_{j=1}^{m} x_j y_j &= a_0 \sum_{j=1}^{m} x_j + a_1 \sum_{j=1}^{m} x_j^2 + \ldots + a_n \sum_{j=1}^{m} x_j^{n+1} \qquad (7.5) \\
&\vdots \\
\sum_{j=1}^{m} x_j^n y_j &= a_0 \sum_{j=1}^{m} x_j^n + a_1 \sum_{j=1}^{m} x_j^{n+1} + \ldots + a_n \sum_{j=1}^{m} x_j^{2n}
\end{aligned}
$$

The parameters are therefore found by solving system (7.5), which involves $n + 1$ linear equations in $n + 1$ unknowns. When n is no greater than 6, little difficulty is encountered in its solution. If a digital computer is available, there are undoubtedly canned programs to enable not only a rapid error-free computation but also an examination of the final approximation to determine the degree of the polynomial that best suits the data.

For functional forms other than polynomials without any missing powers, the situation is not nearly so simple. When the approximating function is linear in the parameters, application of (7.4) results in linear algebraic equations; otherwise nonlinear algebraic equations must be solved. When a least squares norm is not used, more complicated equations often result. It should be remembered, however, that the choice of the form often determines just how good an approximation is. Some data are clearly not suitable for a polynomial approximation—those involving sharp corners, oscillatory shapes, a function asymptotic to a constant as x gets larger, etc. We often split the range into more than one subrange, find approximations in each subrange, and connect them by switching during the problem. This procedure is preferred over choosing some form inconvenient to use on an analog computer—rational functions, continued fractions, etc.

To determine a best polynomial fit for discrete data using a Chebyshev norm, a digital computer is a virtual necessity. The characterization of the best fit is the usual one; i. e., the error curve alternates $n + 1$ times (for a polynomial of degree n, there are $n + 1$ parameters). Thus, if we could determine the $n + 2$ points (x_{j_k}, y_{j_k}) for $k = 1, 2, \ldots, n, n + 1, n + 2$, where j_k is integral and $1 \leq j_k \leq m$, which control the best approximation, the $n + 2$ equations

$$
\begin{aligned}
\sum_{i=0}^{n} a_i x_{j_1}^i - y_{j_1} &= d \\
\sum_{i=0}^{n} a_i x_{j_2}^i - y_{j_2} &= -d \qquad (7.6) \\
&\vdots \\
\sum_{i=0}^{n} a_i x_{j_{n+2}}^i - y_{j_{n+2}} &= (-1)^{n+1} d
\end{aligned}
$$

can be solved for the $n + 2$ unknowns, $a_0, a_1, \ldots, a_n$, and d. (The solution always exists provided the x_{j_k} are distinct.) Unfortunately, finding the proper point subset could be a tedious task on a desk calculator, so that such a procedure is not recommended unless a digital computer can be used. (See, however, Problem 7-4.) In that case, several canned programs usually exist to obtain the best fit.

An absolute error norm is almost never used because the actual computation of the best fit to discrete data is considerably more difficult than both previously discussed procedures.

7.3 Absolute Error Norm on a Continuous Interval

A computational procedure for computing the unique, best polynomial approximation to $f(x)$ defined on $a \leq x \leq b$ using the absolute error norm is based on the following remarkable theorem [1]:

Theorem 7.1 $F(x) = \sum_{i=0}^{n} a_i x^i$ is the unique, best absolute error fit to $f(x)$ if the error $f(x) - F(x)$ changes sign at the points

$$x_k = \frac{a+b}{2} + \frac{a-b}{2} \cos\left(\frac{k\pi}{n+2}\right)$$

for $k = 1, 2, \ldots, n+1$, and at no other points in the interval.

We therefore set up the $n + 1$ equations specifying equality at these points:

$$f(x_k) = \sum_{i=0}^{n} a_i x_k^i, \qquad k = 1, 2, \ldots, n+1$$

and solve for the a_i. These equations are linear so that standard techniques for solution can be used. If the x_k are the only zeros of the error function in the interval, the best polynomial is found. If not, the theorem does not apply since it provides only a sufficient condition. We may then increase the degree of the polynomial or use some other norm.

Example 7.1 Determine the best absolute error fit of the parabola

$$F(x) = a_0 + a_1 x + a_2 x^2$$

to the function $f(x) = (\sin x)/x$ in the interval $0 \leq x \leq \pi$. Applying Theorem 7.1, we find

$$x_k = \frac{\pi}{2} - \frac{\pi}{2} \cos\left(\frac{k\pi}{4}\right), \qquad k = 1, 2, 3$$

or

$$x_1 = \frac{\pi}{2}\left(1 - \sqrt{\frac{1}{2}}\right) = 0.46007$$

$$x_2 = \frac{\pi}{2} = 1.57080$$

$$x_3 = \frac{\pi}{2}\left(1 + \sqrt{\frac{1}{2}}\right) = 2.68152$$

We calculate

$$f(x_1) = 0.96509, \qquad f(x_2) = 0.63662, \qquad f(x_3) = 0.16558$$

and solve the equations

$$0.96509 = a_0 + a_1(0.46007) + a_2(0.46007)^2$$

$$0.63662 = a_0 + a_1(1.57080) + a_2(1.57080)^2$$

$$0.16558 = a_0 + a_1(2.68152) + a_2(2.68152)^2$$

to find

$$a_0 = 1.0594, \qquad a_1 = -0.1784, \qquad a_2 = -0.05778.$$

A spot check of other points $F(x)$ shows these x_k are the only zeros of the error function so that it is indeed the unique best quadratic absolute error approximation.

7.4 Chebyshev Norm on a Continuous Interval

To compute a best approximation using a Chebyshev norm, we make use of the alternation property of the error curve as described in Section 7.1. For *polynomial* approximating functions we have the following important theorem [1]:

Theorem 7.2 $F(x) = \sum_{i=0}^{n} a_i x^i$ is the unique best Chebyshev fit to $f(x)$ if and only if the error $E(x) = f(x) - F(x)$ alternates at least $n + 1$ times in the interval $a \leq x \leq b$; i. e., there are $n + 2$ extremal points x_k where $E(x_k) = +d$ or $-d$. If $x_1 < x_2 < \cdots < x_{n+2}$ and $E(x_k) = d$, then $E(x_{k-1}) = E(x_{k+1}) = -d$. $|d|$ is called the deviation and all $|E(x)| \leq |d|$.

Theorem 7.2 actually applies to some (but not all) other approximating functions besides polynomials. This characterization, although quite elegant, does not lead directly to the determination of a best Chebyshev fit. Iterative computations are required in all but the simplest cases. To see this, we consider one method to obtain a best Chebyshev approximation (there are several methods—see Rice [1]) in the following example:

Example 7.2 Determine the best Chebyshev fit of the parabola

$$F(x) = a_0 + a_1 x + a_2 x^2$$

to the function $f(x) = (\sin x)/x$ in the interval $0 \leq x \leq \pi$. There must be *at least* four extremal points. We assume exactly four initially. We also assume that two of the extremal points are the endpoints 0 and π. The other two we call x_1 and x_2 with $x_1 < x_2$.

Since x_1 and x_2 are interior extremal points, the error function must have zero derivatives. Thus

$$\frac{dE(x_1)}{dx} = \frac{x_1 \cos x_1 - \sin x_1}{x_1^2} - (a_1 + 2a_2 x_1) = 0$$

$$\frac{dE(x_2)}{dx} = \frac{x_2 \cos x_2 - \sin x_2}{x_2^2} - (a_1 + 2a_2 x_2) = 0$$

We also have four equations because of the alternation requirement:

$$\text{at } x = 0, \qquad 1 - (a_0) = -d$$

$$\text{at } x = x_1, \qquad \frac{\sin x_1}{x_1} - (a_0 + a_1 x_1 + a_2 x_1^2) = +d$$

$$\text{at } x = x_2, \qquad \frac{\sin x_2}{x_2} - (a_0 + a_1 x_2 + a_2 x_2^2) = -d$$

$$\text{at } x = \pi, \qquad -(a_0 + a_1 \pi + a_2 \pi^2) = +d$$

The six nonlinear equations are solved for the six unknowns, a_0, a_1, a_2, x_1, x_2, and d. If a solution is found, it is the unique best Chebyshev fit, only if all $|E(x)| \leq |d|$. If this is not so, there are more than two internal extremal points. An x_3 must be introduced, requiring seven equations for solution [$dE(x_3)/dx = 0$ is the other one]. The worst that can happen is that an x_4 will have to be introduced requiring eight equations for solutions.

This procedure requires at least $2(n + 1)$ nonlinear equations to be solved, a reason why it is almost never attempted without computer aid (see Chapter 11 for analog computer aid). A digital computer computation showed that

$$a_0 = 1.03348, \qquad a_1 = -0.169334, \qquad a_2 = -0.0542058$$

$$x_1 = 0.837873, \qquad x_2 = 2.37182, \qquad d = 0.0334842$$

did indeed give a best Chebyshev fit.

Despite our preference for approximations using a Chebyshev norm, there is certainly a computational disadvantage that tends to discourage its use among analog computer programmers. For functions commonly used, however, the effort required may be worthwhile. Hastings [3] has tabulated many useful Chebyshev polynomial approximations.

7.5 Least Squares Norm on a Continuous Interval–Orthogonal Functions

Just as in the case of discrete data, the least squares norm provides the simplest computational methods for polynomial approximations to the function $f(x)$ defined on a continuous interval $a \leq x \leq b$. Many procedures are also available. Let us, for example, derive the $n + 1$ equations directly from (7.4): Since

$$\rho = \int_a^b \left(f(x) - \sum_{i=0}^{n} a_i x^i \right)^2 dx$$

then

$$\frac{\partial \rho}{\partial a_j} = -2 \int_a^b \left(f(x) - \sum_{i=0}^{n} a_i x^i \right) x^j \, dx = 0$$

for $j = 0, 1, \ldots, n$. Thus the system of $n + 1$ linear equations:

$$\begin{aligned} \int_a^b f(x) x^j \, dx &= \sum_{i=0}^{n} a_i \int_a^b x^{i+j} \, dx \\ &= \sum_{i=0}^{n} a_i \frac{b^{i+j+1} - a^{i+j+1}}{i + j + 1} \end{aligned} \tag{7.7}$$

must be solved after the integrals on the left and the coefficients on the right are evaluated. Before proceeding along these lines, we observe that the equations are linear when linear approximation functions of the form

$$F(x) = \sum_{j=0}^{n} a_j \phi_j(x)$$

are used for any set of functions $\phi_j(x)$. The final equations become

$$\int_b^b f(x)\phi_j(x) \, dx = \sum_{i=0}^{n} a_j \int_a^b \phi_i(x)\phi_j(x) \, dx, \qquad j = 0, 1, \ldots, n \tag{7.8}$$

This is precisely the reason least squares approximations are so amenable to calculation. Were the final equations nonlinear in the a_j, there could be little hope of a simple calculation without computer aid.

When the $\phi_j(x)$ are linearly independent (as are the powers of x), we have [1]:

Theorem 7.3. The function $F(x)$ is the unique, best least squares approximation to $f(x)$ if and only if each a_j satisfies (7.8).

Although Theorem 7.3 is the basis for obtaining best least squares approximations, much simplification is obtained by the use of orthogonal functions, and in particular orthogonal polynomials.

Orthogonal functions are sets of functions $\phi_j(x)$ such that

$$\begin{aligned} \int_a^b \phi_i(x)\phi_j(x) \, dx &= 0, \qquad i \neq j \\ &= h_j, \qquad i = j \end{aligned}$$

For then we instantly have from (7.8)

$$a_j = \frac{1}{h_j}\int_a^b f(x)\phi_j(x)\,dx \tag{7.9}$$

The a_j, called *Fourier coefficients*, are independent of each other—a great convenience in their calculation. If $h_j = 1$, the set of functions is called *orthonormal*.

In practice many sets of orthogonal functions are tabulated with fixed limits, so that their use requires transforming $f(x)$ to correspond to these limits. Most are defined orthogonal with respect to a weight function $w(x) \geq 0$, i. e.,

$$\begin{aligned}\int_a^b w(x)\phi_i(x)\phi_j(x)\,dx &= 0, \qquad i \neq j\\ &= h_j, \qquad i = j\end{aligned} \tag{7.10}$$

to permit using a weighted least squares norm.

The system of trigonometric functions $\sin jx$ and $\cos jx$, $j = 0, 1, \ldots$ is the best known example of orthogonal functions—they are orthogonal in the interval $-\pi \leq x \leq \pi$ with $w(x) = 1$. From these, the familiar Fourier series may be evolved, but because they are rarely used in analog computation as approximations, they will not be discussed here. However, Fourier analysis methods are often applied to periodic signals (Problem 7-20).

Systems of orthogonal polynomials, however, provide the most convenient means for obtaining a best weighted least squares polynomial fit to $f(x)$. An extensive list of orthogonal polynomials, their properties, and a good bibliography are provided in [4]. We shall concern ourselves with three of the sets as shown in Table 7.2. Each is governed by a three-term recurrence relation from which all the polynomials can be calculated given the first two. Tables 7.3 through

TABLE 7.2

Three Common Orthogonal Polynomials

Name	Symbol	Interval $a < x < b$	$w(x)$	h_j	*Recurrence Relation*
Chebyshev of the first kind	$T_j(x)$	$-1 \leq x \leq 1$	$\frac{1}{\sqrt{1-x^2}}$	π for $j = 0$; $\frac{\pi}{2}$ for $j \neq 0$	$T_{j+1} = 2xT_j - T_{j-1}$
Legendre	$P_j(x)$	$-1 \leq x \leq 1$	1	$\frac{2}{2j+1}$	$(j+1)P_{j+1} = (2j+1)xP_j - jP_{j-1}$
Laguerre	$L_j(x)$	$0 \leq x \leq \infty$	e^{-x}	1	$(j+1)L_{j+1} = [(2j+1) - x]L_j - jL_{j-1}$

TABLE 7.3

Chebyshev Polynomials to Order 8

$$
\begin{aligned}
T_0 &= 1\\
T_1 &= x\\
T_2 &= -1 + 2x^2\\
T_3 &= -3x + 4x^3\\
T_4 &= 1 - 8x^2 + 8x^4\\
T_5 &= 5x - 20x^3 + 16x^5\\
T_6 &= -1 + 18x^2 - 48x^4 + 32x^6\\
T_7 &= -7x + 56x^3 - 112x^5 + 64x^7\\
T_8 &= 1 - 32x^2 + 160x^4 - 256x^6 + 128x^8
\end{aligned}
$$

$$
\begin{aligned}
x^0 &= T_0\\
x^1 &= T_1\\
x^2 &= (1/2)(T_0 + T_2)\\
x^3 &= (1/4)(3T_1 + T_3)\\
x^4 &= (1/8)(3T_0 + 4T_2 + T_4)\\
x^5 &= (1/16)(10T_1 + 5T_3 + T_5)\\
x^6 &= (1/32)(10T_0 + 15T_2 + 6T_4 + T_6)\\
x^7 &= (1/64)(35T_1 + 21T_3 + 7T_5 + T_7)\\
x^8 &= (1/128)(35T_0 + 56T_2 + 28T_4 + 8T_6 + T_8)
\end{aligned}
$$

TABLE 7.4

Legendre Polynomials to Order 8

$$
\begin{aligned}
P_0 &= 1\\
P_1 &= x\\
P_2 &= (1/2)(-1 + 3x^2)\\
P_3 &= (1/2)(-3x + 5x^3)\\
P_4 &= (1/8)(3 - 30x^2 + 35x^4)\\
P_5 &= (1/8)(15x - 70x^3 + 63x^5)\\
P_6 &= (1/16)(-5 + 105x^2 - 315x^4 + 231x^6)\\
P_7 &= (1/16)(-35x + 315x^3 - 693x^5 + 429x^7)\\
P_8 &= (1/128)(35 - 1260x^2 + 6930x^4 - 12{,}012x^6 + 6435x^8)
\end{aligned}
$$

$$
\begin{aligned}
x^0 &= P_0\\
x^1 &= P_1\\
x^2 &= (1/3)(P_0 + 2P_2)\\
x^3 &= (1/5)(3P_1 + 2P_3)\\
x^4 &= (1/35)(7P_0 + 20P_2 + 8P_4)\\
x^5 &= (1/63)(27P_1 + 28P_3 + 8P_5)\\
x^6 &= (1/231)(33P_0 + 110P_2 + 72P_4 + 16P_6)\\
x^7 &= (1/429)(143P_1 + 182P_3 + 88P_5 + 16P_7)\\
x^8 &= (1/6435)(715P_0 + 2600P_2 + 2160P_4 + 832P_6 + 128P_8)
\end{aligned}
$$

TABLE 7.5

Laguerre Polynomials to Order 8

$L_0 = 1$
$L_1 = 1 - x$
$L_2 = (1/2)(2 - 4x + x^2)$
$L_3 = (1/6)(6 - 18x + 9x^2 - x^3)$
$L_4 = (1/24)(24 - 96x + 72x^2 - 16x^3 + x^4)$
$L_5 = (1/120)(120 - 600x + 600x^2 - 200x^3 + 25x^4 - x^5)$
$L_6 = (1/720)(720 - 4320x + 5400x^2 - 2400x^3 + 450x^4 - 36x^5 + x^6)$
$L_7 = (1/5040)(5040 - 35{,}280x + 52{,}920x^2 - 29{,}400x^3 + 7350x^4 - 882x^5 + 49x^6 - x^7)$
$L_8 = (1/40{,}320)(40{,}320 - 322{,}560x + 564{,}480x^2 - 376{,}320x^3 + 117{,}600x^4 - 18{,}816x^5 + 1586x^6 - 64x^7 + x^8)$

$x^0 = L_0$
$x^1 = L_0 - L_1$
$x^2 = 2L_0 - 4L_1 + 2L_2$
$x^3 = 6L_0 - 18L_1 + 18L_2 - 6L_3$
$x^4 = 24L_0 - 96L_1 + 144L_2 - 96L_3 + 24L_4$
$x^5 = 120L_0 - 600L_1 + 1200L_2 - 1200L_3 + 600L_4 - 120L_5$
$x^6 = 720L_0 - 4320L_1 + 10{,}800L_2 - 14{,}400L_3 + 10{,}800L_4 - 4320L_5 + 720L_6$
$x^7 = 5040L_0 - 35{,}280L_1 + 105{,}840L_2 - 176{,}400L_3 + 176{,}400L_4 - 105{,}840L_5 + 35{,}280L_6 - 5040L_7$
$x^8 = 40{,}320L_0 - 322{,}560L_1 + 1{,}128{,}960L_2 - 2{,}257{,}920L_3 + 2{,}822{,}400L_4 - 2{,}257{,}920L_5 + 1{,}128{,}960L_6 - 322{,}560L_7 + 40{,}320L_8$

7.5 contain these polynomials to order 8 for convenience, since it is sometimes necessary to calculate points in the interval. Also, since each jth-order polynomial is one of degree j, the function x^j is expressible in terms of the polynomials of order j and less. These are given in each table, again for convenience. We now examine methods of using these polynomials for least squares fitting.

7.5.1 Chebyshev polynomials

When Chebyshev polynomials are used for a weighted least squares polynomial approximation, the weight emphasizes the endpoints more than the center section. If $f(x)$ has the range $c \le x \le d$, and a weighting function $w(x) = 1/[(x - c)(d - x)]^{1/2}$, we transform to the interval $-1 \le y \le 1$ by

$$x = \left(\frac{d-c}{2}\right)y + \left(\frac{d+c}{2}\right) \tag{7.11}$$

and the corresponding weight function becomes $w(y) = 2/[(d - c)\sqrt{1 - y^2}]$. The constant factor $2/(d - c)$ in the new weight function can be ignored, however. For if we compute

$$\frac{1}{h_j}\int_{-1}^{1}\left(\frac{2}{d-c}\right)\frac{f[\frac{1}{2}(d-c)y+\frac{1}{2}(d+c)]T_j(y)\,dy}{\sqrt{1-y^2}}$$

then the approximation coefficients for the function

$$\frac{2f[\frac{1}{2}(d-c)y+\frac{1}{2}(d+c)]}{d-c}$$

are obtained, so that the final answer must be multiplied by $(d-c)/2$ anyway. Thus we actually have

$$a_j = \frac{1}{h_j}\int_{-1}^{1}\frac{f[\frac{1}{2}(d-c)y+\frac{1}{2}(d+c)]T_j(y)\,dy}{\sqrt{1-y^2}} \tag{7.12}$$

To use an analog computer to evaluate these integrals, we convert (7.12) to differential equation form:

$$\frac{da_j}{dy} = \frac{f[\frac{1}{2}(d-c)y+\frac{1}{2}(d+c)]T_j(y)}{h_j\sqrt{1-y^2}}$$

If we let $F = \sqrt{1-y^2}$, then

$$\begin{aligned} \dot{y} &= F, & y_0 &= -1 \\ \dot{F} &= -y, & F_0 &= 0 \end{aligned} \tag{7.13}$$

$$\dot{a}_j = \frac{1}{h_j}f[\tfrac{1}{2}(d-c)y+\tfrac{1}{2}(d+c)]T_j(y), \qquad (a_j)_0 = 0 \tag{7.14}$$

Now, $T_0 = 1$, and $T_1(y) = y$, so no difficulty results in evaluating a_0 and a_1 (assuming f can be generated). For the higher orders, very simple circuitry is involved if we recognize that $T_j(y)$ is an expansion of $\cos jz$ in disguise [4]. With

$$z = \arccos y \qquad \text{or} \qquad y = \cos z$$

then $T_j(\cos z) = \cos jz$. We see that y and F are in an oscillator circuit from (7.13); let

$$F_j = \sin jz$$

Now

$$\dot{F}_j = jT_j\dot{z}, \qquad \dot{T}_j = -jF_j\dot{z}$$

But since

$$\dot{y} = -(\sin z)\,\dot{z} = \sqrt{1 - y^2}\,\dot{z}$$

we must have $\dot{z} = 1$ from (7.13). Hence, we generate T_j from an oscillator circuit

$$\begin{aligned} \dot{F}_j &= jT_j, \qquad (F_j)_0 = 0 \\ \dot{T}_j &= -jF_j, \qquad (T_j)_0 = (-1)^j \end{aligned} \tag{7.15}$$

We may alternatively start at $y = 1$ and decrease y with the proper choice of $\dot{y}$.

Now we show that the a_j in (7.14) can actually be generated without multipliers. Consider the output of the system (7.15) together with

$$\begin{aligned} \dot{Q}_j &= jR_j + \frac{1}{h_j} f[\tfrac{1}{2}(d - c)y + \tfrac{1}{2}(d + c)], \qquad (Q_j)_0 = 0 \\ \dot{R}_j &= -jQ_j, \qquad (R_j)_0 = 0 \end{aligned}$$

Form

$$T_j\dot{Q}_j + Q_j\dot{T}_j = jT_jR_j + \frac{1}{h_j} fT_j + jQ_jF_j$$

and integrate with respect to τ:

$$T_jQ_j = \frac{1}{h_j}\int_0^\tau fT_j\,d\tau + j\int_0^\tau (T_jR_j + Q_jF_j)\,d\tau$$

When $\tau = \pi$, corresponding to $y = 1$, we also have $T_j = 1$ and $F_j = 0$. Thus, at $\tau = \pi$,

$$Q_j = \frac{1}{h_j}\int_0^\pi fT_j\,d\tau + j\int_0^\pi (T_jR_j + Q_jF_j)\,d\tau \tag{7.16}$$

We may now evaluate the latter integral by parts and show it is equal to 0, to get

$$\begin{aligned} Q_j &= \frac{1}{h_j}\int_0^\tau fT_j\,d\tau = \frac{1}{h_j}\int_{-1}^1 fT_j\left(\frac{d\tau}{dy}\right)dy \\ &= \frac{1}{h_j}\int_{-1}^1 \frac{fT_j\,dy}{\sqrt{1 - y^2}} \\ &= a_j \qquad \text{when } y = 1 \end{aligned}$$

Thus, the output Q_j when $y = 1$ is the proper coefficient. We recognize this circuit to be the response of an undamped harmonic oscillator to the forcing function f/h_j.

Example 7.3. To illustrate the procedure, we approximate $f(x) = e^{-x}$ for $0 \leq x \leq 3$ using a least squares norm and weighting function $w(x) = 1/[x(3-x)]^{1/2}$. Transforming the interval to $-1 \leq y \leq 1$, we have

$$x = \tfrac{3}{2}y + \tfrac{3}{2}$$

which results in the weighting function $w(y) = 2/[3(1-y^2)^{1/2}]$. Thus, from (7.12),

$$a_j = \frac{1}{h_j} \int_{-1}^{1} \frac{e^{-(3y/2+3/2)}\, T_j(y)\, dy}{\sqrt{1-y^2}}$$

Now, we let

$$G = e^{-(3y/2+3/2)}$$

so that

$$\dot{G} = -\tfrac{3}{2}G\dot{y} = -\tfrac{3}{2}GF, \qquad (G)_0 = 1$$

with F defined as in (7.13). Figure 7.1 shows the unscaled circuit diagram, which can be used to determine each a_j one at a time. Since $h_0 = \pi$, and $h_j = \pi/2$ for $j \neq 0$ (see Table 7.3), Fig. 7.1 shows both situations. As many a_j could be computed simultaneously as desired, with many oscillator circuits. The circuit produced the values

$$\begin{aligned} a_0 &= 0.3668 \\ a_1 &= -0.4384 \\ a_2 &= 0.1513 \\ a_3 &= -0.0365 \\ a_4 &= 0.0068 \\ a_5 &= 0.0015 \end{aligned} \tag{7.17}$$

so that a fourth-degree approximation is

$$\begin{aligned} e^{-x} \simeq 0.3668 &- 0.4384T_1\left(\frac{2x}{3}-1\right) + 0.1513T_2\left(\frac{2x}{3}-1\right) \\ &-0.0365T_3\left(\frac{2x}{3}-1\right) + 0.0068T_4\left(\frac{2x}{3}-1\right) \end{aligned}$$

Because Chebyshev polynomials oscillate with a maximum absolute value of 1, we can judge the accuracy of the approximation by looking at the terms not used. However, we must also remember that the computer values of the a_j are in error.

To convert the estimate back to a function of x we use Table 7.3. Thus, the final fourth-degree approximation, as obtained from the analog computer, is

$$e^{-x} \simeq 0.9998 - 0.98727x + 0.45004x^2$$
$$-0.10773x^3 + 0.010746x^4, \qquad 0 \le x \le 3$$

It sometimes is convenient to generate Chebyshev polynomials with $\dot{x} = 1$. Since these are very oscillatory, such a procedure is subject to large error for the high orders, so care should be taken to check the accuracy. In lieu of generating powers of x, the

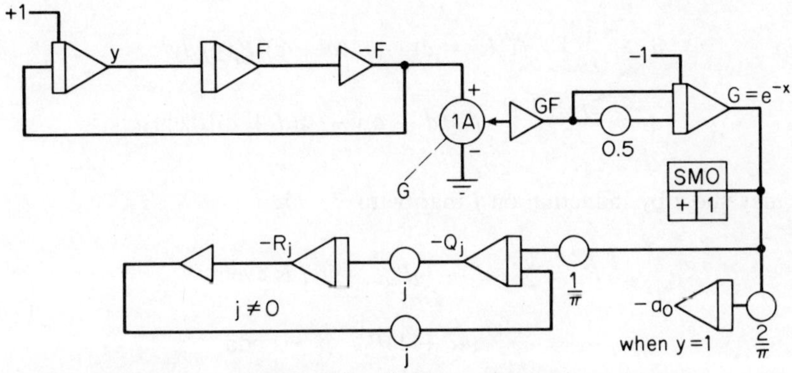

FIG. 7.1 Least squares fitting of e^{-x}, $0 \le x \le 3$, with Chebyshev polynomials.

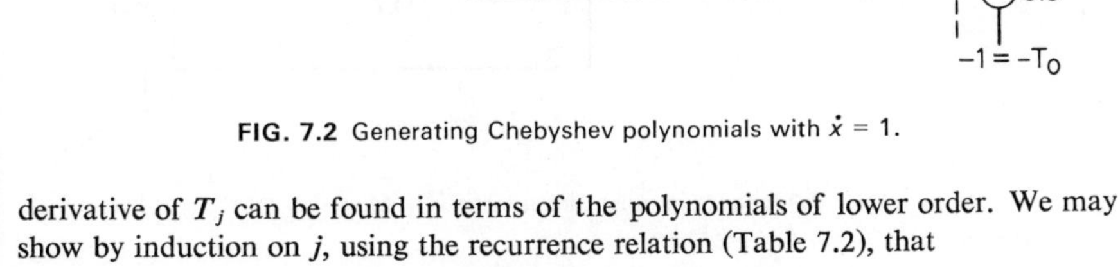

FIG. 7.2 Generating Chebyshev polynomials with $\dot{x} = 1$.

derivative of T_j can be found in terms of the polynomials of lower order. We may show by induction on j, using the recurrence relation (Table 7.2), that

$$\begin{aligned} \dot{T}_j &= 2j \sum_{i=1}^{j/2} T_{2i-1} && \text{if } j \text{ is even} \\ &= 2j \sum_{i=1}^{(j-1)/2} T_{2i} + jT_0 && \text{if } j \text{ is odd} \end{aligned} \tag{7.18}$$

so that the circuit of Fig. 7.2 can be used.

7.5.2 Legendre polynomials

Legendre polynomials are used for least squares fitting when $w(x) = 1$. This is the more usual case and is to be used whenever no concern exists about weighting any points more than any other. Again the linear transformation (7.11) transforms any interval of interest for $f(x)$, $c \leq x \leq d$, to $-1 \leq y \leq 1$, so that the Fourier coefficients are determined by

$$\begin{aligned} a_j &= \frac{1}{h_j}\int_{-1}^{1} f[\tfrac{1}{2}(d-c)y + \tfrac{1}{2}(d+c)]P_j(y)\,dy \\ &= \frac{2j+1}{2}\int_{-1}^{1} f[\tfrac{1}{2}(d-c)y + \tfrac{1}{2}(d+c)]P_j(y)\,dy \end{aligned}$$

We may show by induction on j that, with $\dot{x} = 1$,

$$\begin{aligned} \dot{P}_j &= \sum_{i=1}^{j/2} (4i-1)P_{2i-1} \quad \text{if } j \text{ is even} \\ &= \sum_{i=0}^{(j-1)/2} (4i+1)P_{2i} \quad \text{if } j \text{ is odd} \end{aligned} \tag{7.19}$$

as shown in Fig. 7.3. Again, this open-loop integration is prone to error and so should be carefully checked before proceeding. It may be necessary, depending upon how carefully one's computer is balanced, to generate the Legendre polynomials from the differential equation for the purpose of finding each a_j (see Problem 4-3).

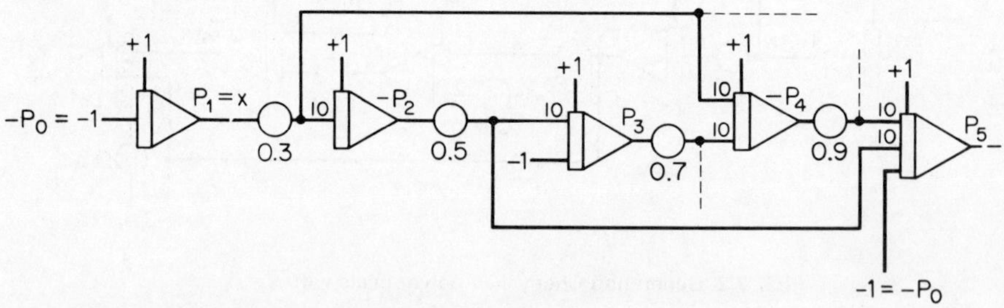

FIG. 7.3 Legendre polynomials with $\dot{x} = 1$, $x_0 = -1$.

It is also possible to set up coupled oscillator circuits to improve accuracy, since powers of x are generable in any linear system (see Problem 7-13).

7.5.3 Laguerre polynomials and functions

Because of the infinite limit in the defining integral, Laguerre polynomials are useful for approximating functions [5], [6] for $0 \leq x < \infty$ (x is usually time) by Laguerre functions

$$\lambda_j = e^{-x/2}L_j(x)$$

If any function $f(x)$ is to be represented by

$$F(x) = e^{-x/2}\sum_{j=0}^{n} a_j L_j(x) = \sum_{j=0}^{n} a_j \lambda_j$$

such that

$$\rho = \int_0^\infty [f(x) - F(x)]^2\, dx \tag{7.20}$$

is a minimum, we must have

$$\begin{aligned}\frac{\partial \rho}{\partial a_j} &= 2\int_0^\infty [f(x) - F(x)]\frac{\partial F}{\partial a_j}\, dx \\ &= 2\int_0^\infty [f(x) - \sum_{i=0}^{n} a_i\lambda_i]\lambda_j\, dx = 0\end{aligned}$$

Hence, because of the orthonormal condition,

$$a_j = \int_0^\infty f(x)e^{-x/2}L_j(x)\, dx$$

In practice, $f(t)$ is given, but the transformation

$$x = \alpha t \tag{7.21}$$

retains the identical limits for all $\alpha > 0$. Thus it is necessary to obtain α by trial and error by evaluating $\rho = \rho(\alpha)$ in (7.20) and by choosing an α and set of a_j that gives the smallest ρ. This, in turn, will depend upon the number of terms desired. Clearly, the weighting also suggests that good approximations will be obtained only for functions $f(x)$ that approach zero as $x \to \infty$.

To generate the Laguerre functions, we first can show by induction on j using the recurrence relation that

$$\frac{dL_j}{dx} = -\sum_{i=0}^{j-1} L_i, \qquad (L_j)_0 = 1 \tag{7.22}$$

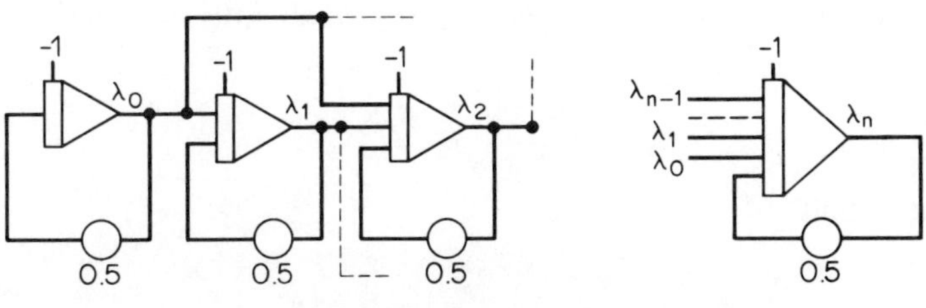

FIG. 7.4 Generation of Laguerre functions with $\dot{x} = 1$.

We then may generate λ_j with $\dot{x} = 1$ by applying the results of Example 5.7—adding a feedback pot set to 0.5 for each integrator, as shown in Fig. 7.4.

7.5.4 Avoiding multiplication in the Fourier coefficient determination

If the orthogonal functions can be generated by a linear system with constant coefficients and initial conditions equal to one, the Fourier coefficients can be determined without multipliers. Before showing this, we note first the controlling theorem of such systems [7], here written in a nonmatrix notation:

Theorem 7.4. If $a_{i,j}$ are constant, the solution of

$$\frac{dz_i}{dt} = \sum_{j=1}^{n} a_{i,j} z_j + w_i(t), \qquad (z_i)_0 = 0$$

is

$$z_i = \int_0^t \sum_{j=1}^{n} Q_{i,j}(t - y) w_j(y)\, dy$$

for $i = 1, 2, \ldots, n$, where the n^2 functions $Q_{i,j}$ are solutions of

$$\frac{dQ_{i,j}}{dt} = \sum_{j=1}^{n} a_{i,j} Q_{j,i}, \qquad \begin{aligned} (Q_{i,j})_0 &= 1 \text{ if } i = j \\ &= 0 \text{ if } i \neq j \end{aligned} \tag{7.23}$$

and the $w_i(t)$ are arbitrary functions of t. Further, the solution of

$$\frac{d\phi_i}{dt} = \sum_{j=1}^{n} a_{i,j} \phi_j, \qquad (\phi_i)_0 = 1$$

is given by

$$\phi_i(t) = \sum_{j=1}^{n} Q_{i,j}(t) \tag{7.24}$$

To apply this powerful theorem to approximate a function $f(x)$ with any set of orthogonal functions $\phi_i(x)$ for $0 \leq x \leq b$, all $\phi_i(0) = 1$, and

$$f(x) = \sum_{j=1}^{n} a_j \phi_j$$

we need to find

$$a_j = \frac{1}{h_j} \int_0^b f(x) \phi_j(x)\, dx \tag{7.25}$$

Since we may assume that the ϕ_i are to be generated from

$$\dot{x} = 1, \qquad x_0 = 0$$

$$\dot{\phi}_i = \sum_{j=1}^{n} a_{i,j}\phi_j, \qquad (\phi_i)_0 = 1$$

then in terms of time, (7.25) transforms to

$$a_j = \frac{1}{h_j}\int_0^b f(\tau)\phi_j(\tau)\,d\tau \tag{7.26}$$

Now let us construct the system to find the a_j in (7.26). If we generate $f(x)$ with decreasing x and solve the equations

$$\dot{x} = -1, \qquad x_0 = b$$

$$\dot{z}_i = \sum_{j=1}^{n} a_{i,j}z_j + f(b - \tau)$$

then from Theorem 7.4

$$z_i = \int_0^\tau \sum_{j=1}^{n} Q_{i,j}(\tau - y)f(b - y)\,dy$$

where the $Q_{i,j}$ are found from (7.23). From (7.24), we also have

$$z_i = \int_0^\tau \phi_i(\tau - y)f(b - y)\,dy$$

Now we let

$$w = \tau - y$$

so that

$$z_i(\tau) = -\int_\tau^0 f(b + w - \tau)\phi_i(w)\,dw = \int_0^\tau f(b + w - \tau)\phi_i(w)\,dw$$

When $\tau = b$, then we have

$$z_i(b) = \int_0^b f(w)\phi_i(w)\,dw = h_i a_i$$

as needed.

Thus we must

(a) generate $f(x)$ with x decreasing,

(b) supply the result to all integrators in the system,
(c) hold the result when $x = 0$, and
(d) read the answers $h_i a_i$.

The circuit may then be supplied with initial conditions to generate $f(x)$. Figure 7.5 shows the determination of the a_i using the Laguerre functions λ_i. The infinite limit is actually replaced by a finite one, since $f(x)$ is assumed to be close

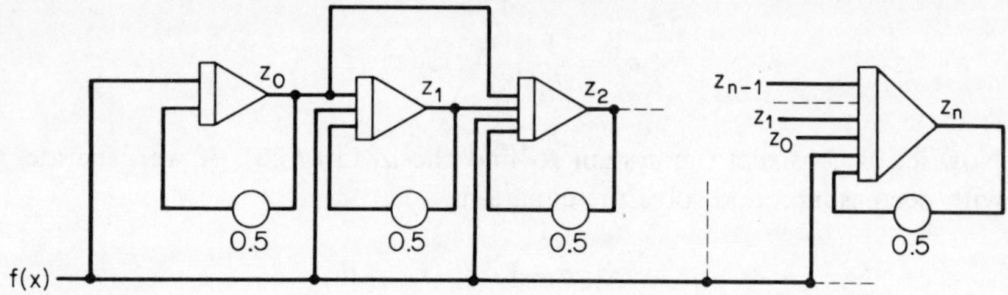

Notes: (1) $\dot{x} = -1$.
(2) x_0 is such that $f(x_0) \simeq 0$ and $f(x_1) \simeq 0$ when $x_1 > x_0$.
(3) Hold when $x = 0$. (All $z_i = a_i$ when $x = 0$.)

FIG. 7.5 Determination of a_i by driving all integrators with $f(x)$ using $\dot{x} = -1$.

enough to zero for $x > x_0$ to produce good results. If it is inconvenient to generate $f(x)$ with $\dot{x} = -1$, the a_j can be found directly with multipliers from (7.25) with $\dot{x} = 1$ but the error can be expected to be somewhat larger than by using linear equipment alone.

7.5.5 General linear system orthogonal functions

Gilbert [8] devised a method of generating orthogonal functions with linear equipment that is much more efficient (i. e., requires less integrators) for generating $f(x)$, $0 \leq x \leq \infty$, $\dot{x} = 1$, than by using Laguerre functions. Gilbert showed that the functions $\phi_j(x)$ produced by cascading circuits as shown in Fig. 7.6 are indeed orthogonal. In practice, the α_i and β_j are adjusted by trial so that when Theorem 7.4 is applied, the largest possible $\sum_{i=0}^{n} a_i^2 h_i$ are obtained. We attempt this because the error ρ can easily be shown to be

$$\rho = \int_0^\infty [f(x)]^2 \, dx - \sum_{i=0}^{n} a_i^2 h_i \tag{7.27}$$

The amplifiers labeled F_i and F_{j+1} are not really needed, and all amplifiers may sometimes need rescaling. Again, if it is inconvenient to obtain an $f(x)$ with $\dot{x} = -1$ (such as when we are trying to obtain a linear system approximation to an actual output), the a_j are found by multipliers.

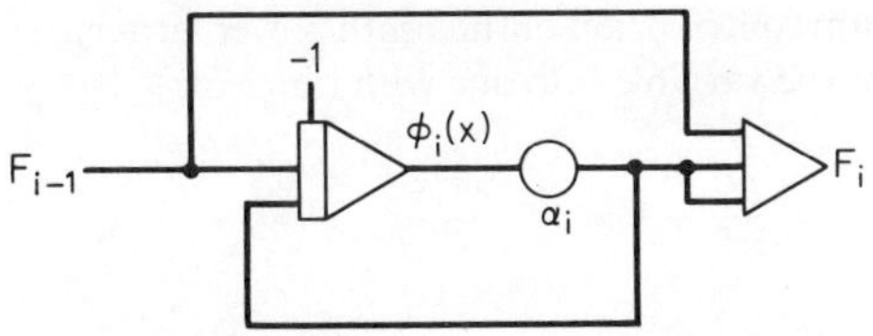

(a) Decreasing exponential: $h_i = \frac{1}{\sqrt{2\alpha_i}}$

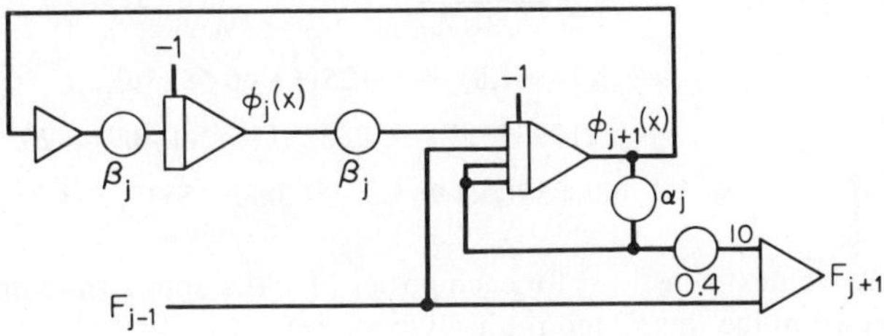

(b) Exponentially decreasing oscillator: $h_j = h_{j+1} = \frac{1}{2\sqrt{\alpha_j}}$

Fig. 7.6 Sections of orthogonal functions (Gilbert [8]).

Another linear system approximation method based on Padé approximants, has been discussed by Teasdale [13].

7.6 Economization

Whenever $f(x)$ can be accurately expressed as a powers series in x (by a Taylor series), Table 7.3 or 7.4 can be used to quickly express $f(x)$ in terms of the orthogonal functions needed, without the need of calculating any integrals. As few terms as are desired can then be used. In the case of using Chebyshev polynomials, the maximum error is easily estimated because these are equiripple functions.

Example 7.4. We will compute $f(x) = e^{-x}$, $0 \leq x \leq 3$, in terms of Chebyshev polynomials to illustrate this method and to make possible the comparison of the computer results obtained earlier in Example 7.3.

First, expand e^{-x} at the midpoint of the interval $x = 1.5$ in a Taylor series:

$$f(x) = e^{-x} = e^{-1.5} \sum_{i=0}^{\infty} (-1)^i \frac{(x - 1.5)^i}{i!} \tag{7.28}$$

Next, determine the number of terms required for the most accurate analog computer result. Nine terms in this case produce e^{-x} to within 0.00002, since the series alternates in sign and the maximum contribution of the ninth power term is $e^{-1.5}(1.5)^9/9! \simeq 0.00002$. We now transform the variable into one with range of $-1 \leq y \leq +1$ by

$$y = \frac{x}{1.5} - 1 = \frac{x - 1.5}{1.5}$$

Thus, $f(1.5y + 1.5)$ from (7.28) is found to be

$$\begin{aligned} f(1.5y + 1.5) &\simeq e^{-1.5} \sum_{i=0}^{8} \frac{(-1.5y)^i}{i!} \\ &= e^{-1.5}(1 - 1.5y + 1.125y^2 - 0.5625y^3 \\ &\quad + 0.2109375y^4 - 0.06328125y^5 + 0.0158203125y^6 \\ &\quad - 0.003390066964y^7 + 0.0006356375558y^8) \end{aligned}$$

By Table 7.3, we next substitute for each power of y the appropriate linear combination of $T_i(y)$, combine them, and multiply by $e^{-1.5}$:

$$\begin{aligned} f(1.5y + 1.5) \simeq{}& 0.3674327225T_0 - 0.4380669191T_1 \\ &+ 0.150760707T_2 - 0.03603836717T_3 + 0.00657621293T_4 \\ &- 0.000965T_5 + 0.000119T_6 - 0.0000118T_7 \\ &+ 0.0000011T_8 \end{aligned}$$

Thus the computer-found coefficients in (7.17) were in error by at most 0.0006.

If a fourth-degree approximation is to be used, the error E can be at most the sum of the absolute values of the coefficient of the terms not used:

$$E \leq 0.000965 + 0.000119 + 0.000012 + 0.000001 \simeq 0.0011$$

The remaining expression must now be transformed back into powers of y and then to powers of x:

$$\begin{aligned} f(1.5y + 1.5) \simeq{}& 0.2232482285 - 0.3299518176y \\ &+ 0.2489117102y^2 - 0.1441534687y^3 \\ &+ 0.0526097038y^4 \end{aligned} \tag{7.29}$$

and

$$\begin{aligned} f(x) \simeq{}& 0.99887 - 0.98045x + 0.44312x^2 - 0.10506x^3 \\ &+ 0.010392x^4 \end{aligned} \tag{7.30}$$

Either (7.29) or (7.30) will finally be used [probably (7.29) because of scaling on x].

The process of economization requires a polynomial expression but thereafter saves the evaluation of complicated integrals. If it is too tedious to obtain the accurate polynomial expansion, the required integrals must be evaluated.

PROBLEMS

7-1. Using three electronic multipliers, construct circuits to generate the truncated Taylor series approximations

$$\sin x \simeq x - \frac{x^3}{3!} + \frac{x^5}{5!}, \qquad \cos x \simeq 1 - \frac{x^2}{2!} + \frac{x^4}{4!}$$

What range in x theoretically permits these functions to be used so that the error does not exceed 0.01? Assuming no summer errors, and a maximum scaled multiplier error of 0.002, what actual error can result?

7-2. Construct the best least squares fit of the form $F = a_0 + a_1x + a_2x^2$ to the following data:

$$\begin{aligned}(x_i, y_i) = {} & (0, 1.000), (0.314, 0.984), (0.628, 0.935),\\ & (0.942, 0.858), (1.257, 0.757),\\ & (1.571, 0.637), (1.885, 0.505),\\ & (2.199, 0.368), (2.513, 0.234),\\ & (2.827, 0.109), \text{ and } (3.142, 0).\end{aligned}$$

using system (7.5).

7-3. The data in Problem 7-2 are known to come from an even function, and the approximation is needed for negative values of x. How may system (7.5) be modified for obtaining a least squares fit with the approximating function $F = a_0 + a_2x^2$?

7-4. Information obtained from least squares fits is often useful in getting Chebyshev fits. From the result in Problem 7-2, estimate the four extremal points for a best Chebyshev fit using the same functional form, and determine the coefficients from system (7.6).

7-5. Find the best absolute error fit of $F = a_0 + a_1x + a_2x^2$ to the function e^{-x}, $0 \leq x \leq 3$, using Theorem 7.1.

7-6. For the approximation problem of Problem 7-5, set up the equations using a Chebyshev norm with Theorem 7.2, assuming the endpoints are extremal points.

7-7. For the approximation problem of Problem 7-5, set up and solve the equations using a least squares norm with equation (7.7). Assume uniform weighting.

7-8. Show that $\int_0^{\pi} (T_jR_j + Q_jF_j)\,d\tau$ in (7.16) is equal to zero by its evaluation by parts.

7-9. Set up a circuit to determine the Fourier coefficients for the function $\sin x$, $-\pi \leq x \leq \pi$, using Chebyshev polynomials, as in Example 7.3.

7-10. Prove Eq. (7.18) by induction on j.

7-11. Prove Eq. (7.19) by induction on j.

7-12. Repeat Problem 7-7 using Legendre polynomials. Set up a computer circuit to determine the Fourier coefficients. How can the results of Theorem 7.4 and the discussion following be applied in this case to eliminate the need for multipliers?

7-13. Let

$$F_1 = \sqrt{1-x^2}, \qquad F_2 = xF_1, \qquad F_3 = x^2F_1, \qquad F_4 = x^3F_1$$

and determine the differential equations for the first five Legendre polynomials, P_0, P_1, P_2, P_3, and P_4 in terms of each P_i and F_j. This results in coupled oscillator circuits that tend to improve accuracy over the open-loop circuits. For the utmost in accuracy using negative feedback, see Chapter 11.

7-14. Set up a circuit to determine the Fourier coefficients for the function

$$\begin{aligned} f(t) &= 1, & 0 \le t \le 1 \\ &= 1-(t-1), & 1 \le t \le 2 \\ &= 0, & 2 \le t \end{aligned}$$

using Laguerre functions. Determine the optimum α [Eq. (7.21)] by trial for using the first six terms.

7-15. Prove Eq. (7.22) by induction on j.

7-16. Attempt the approximation required in Problem 7-14 using cascaded sections of orthogonal functions as shown in Fig. 7.6a.

7-17. Prove Eq. (7.27).

7-18. Use the method of economization to obtain a polynomial approximation for $f(x) = e^{\sin x}$, $-1 \le x \le 1$, so that the error does not exceed 0.002.

7-19. Use the method of economization to obtain a third-degree polynomial approximation of the Bessel function $J_0(x)$ for $0 \le x \le 1$ [9].

7-20. The process of evaluating the coefficients a_k and b_k of a Fourier series,

$$f(t) = \frac{a_0}{2} + \sum_{k=1}^{\infty} a_k \sin kt + \sum_{k=1}^{\infty} b_k \cos kt$$

where $f(t)$ is periodic with period 2π is called *harmonic analysis* of the signal $f(t)$. Because the functions are orthogonal in the interval $-\pi \le t \le \pi$, with $h_k = \pi$,

$$a_k = \frac{1}{\pi}\int_{-\pi}^{\pi} f(t)\sin kt\,dt$$

$$b_k = \frac{1}{\pi}\int_{-\pi}^{\pi} f(t)\cos kt\,dt, \qquad k = 0, 1, 2, \cdots$$

Show that these can be obtained without multipliers [10]-[12] in a manner similar to

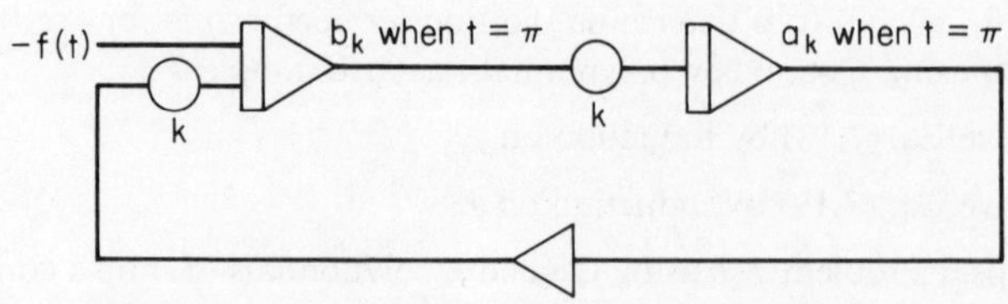

FIG. 7.7 Determination of Fourier series coefficients.

Chebyshev polynomial coefficient evaluation, by the circuit in Fig. 7.7. *Note:* Let t vary from $-\pi$ to $+\pi$ and hold when $t = \pi$.

REFERENCES

[1] Rice, J. R., *The Approximation of Functions* (Reading, Mass.: Addison-Wesley, 1964).

[2] Hamming, R. W., *Numerical Methods for Scientists and Engineers* (New York: McGraw-Hill Book Co., 1962).

[3] Hastings, C., Jr., *Approximations for Digital Computers* (Princeton, N. J.: Princeton University Press, 1955).

[4] Abramowitz, M., and I. A. Stegun, ed., *Handbook of Mathematical Functions*, NBS Applied Math. Series 55 (Washington, D.C., National Bureau of Standards, June 1964).

[5] Burgin, G. H., "Function Generation on an Analog Computer Using Generalized Laguerre Functions," *Simulation*, vol. 4, March 1965.

[6] Otterman, J., "Time Domain Network Synthesis for an Analog Computer Setup," *Proc. National Simulation Conference*, Jan. 19-21, 1956, Dallas, Texas.

[7] Bellman, R., *Stability Theory of Differential Equations* (New York: McGraw-Hill Book Co., 1953), pp. 12-14.

[8] Gilbert, E. G., "Linear System Approximation by Differential Analyzer Simulation of Orthonormal Approximation Functions," *IRETEC*, vol. 8, June 1959.

[9] Hausner, A., "Accurate Analog Computer Generation of Bessel Functions for Large Ranges," *Simulation*, vol. 9, Nov. 1967.

[10] Pode, L., "Some REAC Techniques Employed at the David Taylor Model Basin," *Symposium II on Simulation and Computing Techniques, part 2*, Reeves Instr. Corp., New York, April 28-May 2, 1952 (ASTIA No. 170848).

[11] Lauber, R., "A Comparison of Fourier Analysis Methods Using an Analogue Computer," *Proceedings of the 4th Inter. Analogue Comp. Meetings* (Brussels: Presses Académiques Européennes, 1966).

[12] Harbert, F. C., "The Generation of Fourier Transforms and Coefficients on an Analogue Computer," *Electronic Eng.*, vol. 32, Aug. 1960.

[13] Teasdale, R. D., "Time Domain Approximation by Use of Padé Approximants," *IRE National Convention Record, part 5*, 1953.

See also:

[14] Petric, J., "The Application of Tchebysheff Polynomials to the Solution of Some Mathematical and Technical Problems by Analog Computers," *Proceedings of the 4th Inter. Analogue Comp. Meetings* (Brussels: Presses Académiques Européennes, 1966).

[15] McCool, W. A., "Frequency Analysis by Electronic Analog Methods," *Symposium I on REAC Techniques*, Reeves Instr. Corp., New York, March 15-16, 1951 (ASTIA No. 108488).

Chapter 8

GENERATING FUNCTIONS WITH SPECIALIZED EQUIPMENT

Another method of generating functions with bounded error is to use units especially designed for that purpose and available in most analog computers. These include the so-called fixed function generators such as resolvers for generating sin ax and cos ax, and arbitrary function generators such as curve followers, pot padders (tapped potentiometers), and diode function generators (DFG).

8.1 Fixed Function Generators

A unit designed to generate a specific function $f(x_1, x_2, \ldots, x_n)$ for the n input variables $x_1, x_2, \ldots, x_n$, with bounded error independent of time, is called a *fixed function generator*. Even some functions generable with integrators should be generated with such equipment. The reason is apparent from the discussions in Sec. 5.6, where it was shown that the errors often become intolerable after long periods of time. If this is the case, the use of fixed function generators, if available, is appropriate to minimize the final error.

In some applications, failure of the generalized integration method can be determined only by trial. To avoid the possibility of solution delays, many always prefer to use fixed function generators, if available. The choice is up to the programmer.

8.1.1 *Multiplier tricks*

Multipliers and summers are fixed function generators that were discussed earlier because they are so basic to all computations. Early users of analog computers discoverd interesting and useful techniques to simplify the generation of certain functions with servo multipliers. These might be called multiplier tricks and can be best understood by considering Fig. 8.1, which is comparable

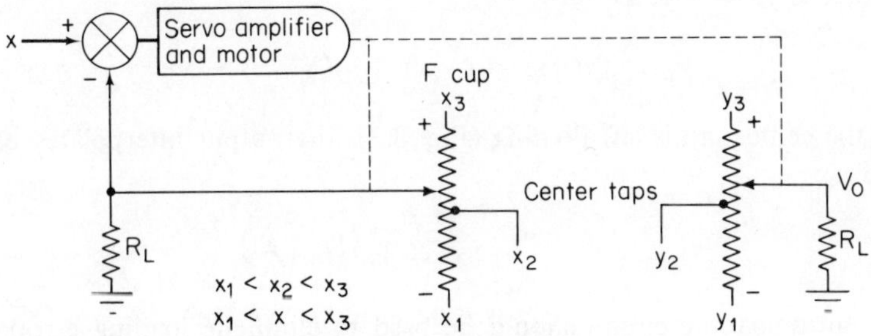

FIG. 8.1 General servo multiplier patching.

to the servomultiplier schematic shown in Fig. 2.14. The difference is simply in how the various holes are patched. As shown, the output is given by

$$V_0 = y_2 + \left(\frac{x - x_2}{x_3 - x_2}\right)(y_3 - y_2)U(x - x_2) + \left(\frac{x_2 - x}{x_2 - x_1}\right)(y_1 - y_2)U(x_2 - x) \tag{8.1}$$

because this circuit interpolates linearly. The load resistance R_L does not quite compensate for loading errors unless $x_2 = y_2 = 0$. We see from (8.1), however, that no loading error occurs at $x = x_1$, x_2, or x_3.

An unloading circuit commonly used if the errors are intolerable is shown

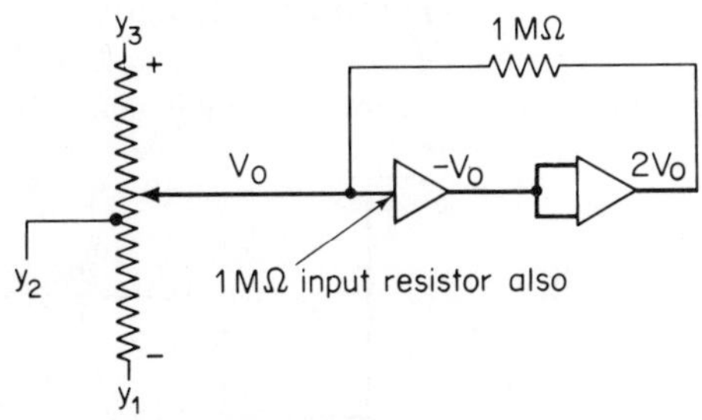

FIG. 8.2 Common unloading circuit.

in Fig. 8.2. No load resistor is used for the follow-up cup, but the voltage $2V_0$ is connected to the multiplier output through a 1-MΩ resistor (assuming the input resistor is 1 MΩ also.) It may be shown [1] that no current is drawn at the output so that we effectively have $R_L = \infty$. The amplifier output of $2V_0$ limits $|V_0|$ to less than $\frac{1}{2}$. In addition, servo design limitations usually require $x_1 < x_2 < x_3$, $x_2 - x_1 \gtrsim 0.05$ and $x_3 - x_2 \gtrsim 0.05$ for satisfactory operation. The input value x should lie between x_1 and x_3 so that the servo system can null properly. The normal situation has $x_3 = 1$, $x_2 = 0$, $x_1 = -1$, $y_3 = y$, $y_2 = 0$, and $y_1 = -y$, so that

$$V_0 = xyU(x) + (-x)(-y)U(-x) = xy$$

If the center tap is left floating (Fig. 8.3), the output interpolates between x_1 and x_2:

$$V_0 = y_1 + \frac{x - x_1}{x_2 - x_1}(y_2 - y_1) \tag{8.2}$$

Again, an unloading circuit should be used to eliminate loading errors if the pairs x_1 and y_1 or x_2 and y_2 are not zero. If one center tap is used and the

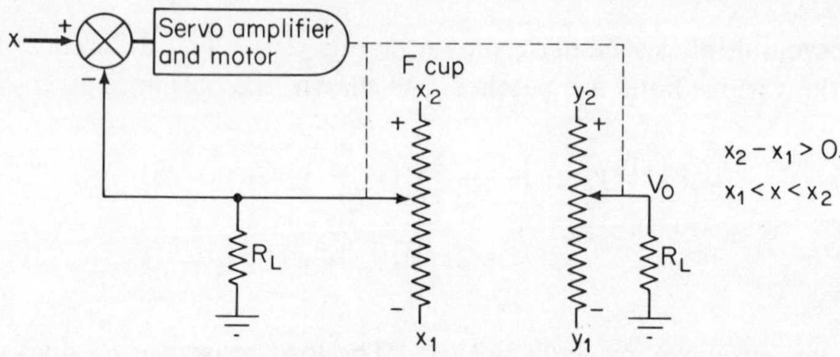

FIG. 8.3 Servo multiplier patching with floating center tap.

other is not, the average voltage can be assumed at the center tap and (8.1) used, but in such a case an unloading circuit will give better results.

The most common use of (8.1) occurs in generating the function $-|x|y$ as shown in Fig. 8.4, with $x_2 = y_2 = 0$, $x_1 = -1$, $x_3 = 1$, and $y_1 = y_3 = y$. Thus, despite the normal connection, the function x^2y is generable without requiring

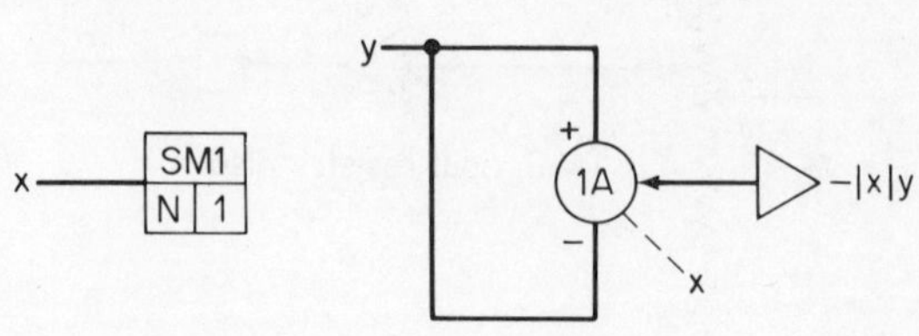

FIG. 8.4 Generation of $-|x|y$ with a servo multiplier.

two amplifiers at each multiplication stage, as shown in Fig. 8.5. If $y = x$, odd powers may be generated with one amplifier per power.

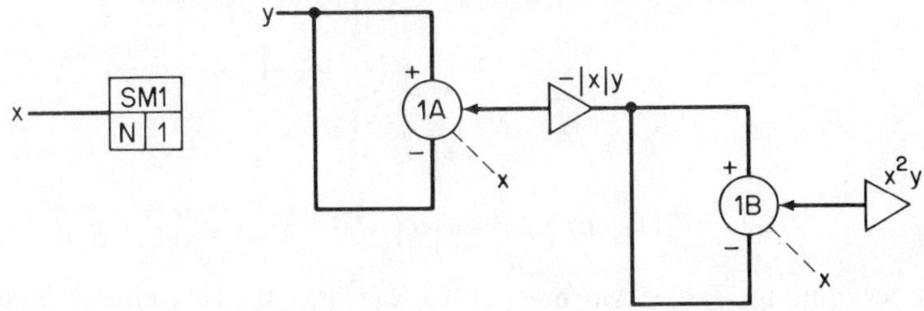

FIG. 8.5 Optimum generation of x^2y using a normal connection.

Two other interesting circuits are described by Witsenhausen [2] and are shown in Figs. 8.6 and 8.7.

The weighted mean, given by

$$y = \frac{y_1x_1 + y_2x_2}{x_1 + x_2}$$

is shown produced in Fig. 8.6 for $x_1 + x_2 > 0$. It follows directly from substitution into (8.2) with $x = 0$ and x_1 inverted.

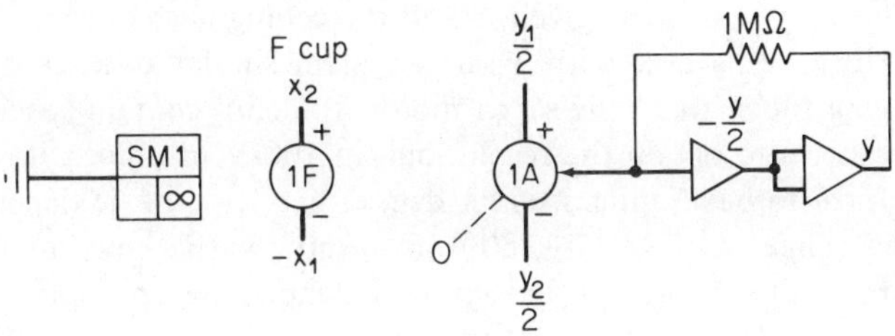

FIG. 8.6 Circuit for weighted mean.

The function

$$z = (x^2 - y^2)^{1/2}$$

can be generated by the circuit in Fig. 8.7. From (8.2), with $x_1 = 0$, $y_1 = 0$,

$$-z = \frac{x - y}{z}(-x - y)$$

from which

$$z^2 = x^2 - y^2$$

provided the circuit is stable. Stability is assured if $x - y \geq 0$ and $x + y \geq 0$ (forcing $z \geq 0$ in Fig. 8.7). However, $x - y$ must also be less than $(x^2 - y^2)^{1/2}$

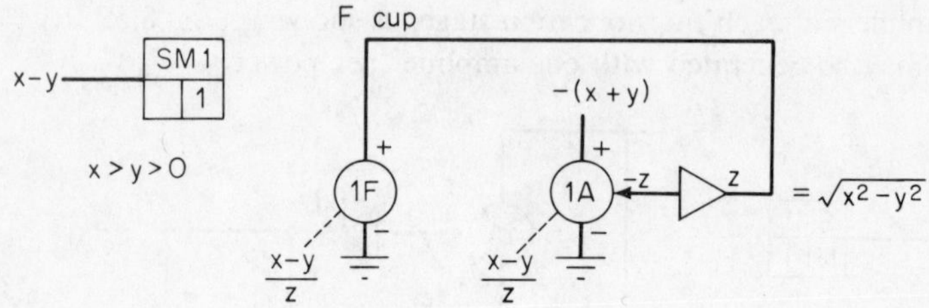

FIG. 8.7 Generation of $\sqrt{x^2 - y^2}$.

for proper nulling. Hence we must have $x > y > 0$. This circuit is similar to generating functions defined implicitly (Chapter 9), but the feedback to the follow-up cup is slightly different in principle.

Many variations are possible with nonconstant follow-up cup voltages, but care should be taken to insure the computer stability of the final arrangement.

It is sometimes possible to arrange circuitry to obtain $x|x|$ with electronic multipliers; e.g., see Brown [3].

8.1.2 Servo resolver

Early users of analog computers were heavily oriented in the aerospace industries. Problems involving six-degree-of-freedom trajectories require many trigonometric functions for coordinate transformations. A resolver determines $x = R\cos\theta$ and $y = R\sin\theta$ given R and θ (rectangular conversion). A servo resolver performs this task with a servo system similar to a servo multiplier. With θ driving the shaft of the servo motor, the cups contain specially wound nonlinear resistors to obtain the resolution. In 100-V machines, it is customary to arrange for θ to have units of volts/deg, either 1, $\frac{1}{2}$, or $\frac{1}{4}$, depending upon the designed range of $\pm 90°$, $\pm 180°$ or $\pm 360°$, with a maximum limitation of 90-V input. The range can always be extended by appropriate switching circuits. Some models permit multiturn operation for $-1800° \leq \theta \leq 1800°$.

A resolver symbol together with a working diagram symbol for a servo resolver is shown in Fig. 8.8. A load resistor is not connected to the wiper arm of the linear follow-up cup because the resistors are wound assuming that the outputs will be patched directly into unity gains. The symbol R in the space

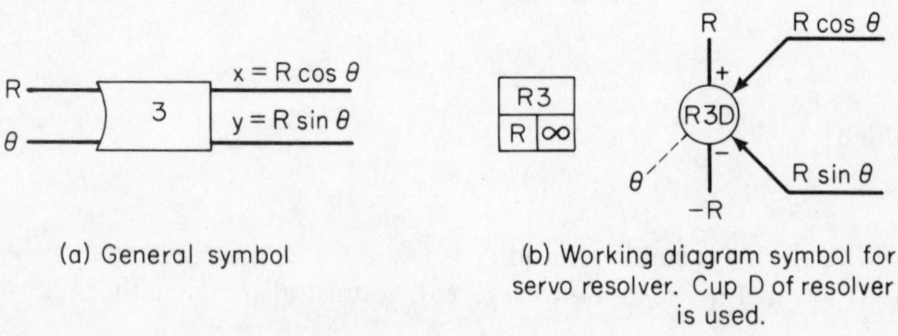

(a) General symbol

(b) Working diagram symbol for servo resolver. Cup D of resolver 3 is used.

FIG. 8.8 Resolver symbols.

usually allotted for type of connection stands for rectangular conversion mode, since it is also possible to patch the resolver in the polar conversion mode (in which case the symbol P is used). In the polar conversion mode, the inverse problem is solved. Given x and y, find R and θ.

Some servo resolver models provide for continuous rotation of the shaft and can be used by supplying a voltage equal to $\dot{\theta}$. In that case, the feedback into the servo amplifier comes from a tachometer driven by the servo motor. Called a *rate* resolver, it is convenient when $\dot{\theta}$ is available and when many cycles may be required.

8.1.3 Electronic resolver

An electronic resolver is to a servo resolver as a quarter-square multiplier is to a servo multiplier. Such a unit usually combines good diode shaping networks that approximate $\sin\theta$ for $-90° \leq \theta \leq 90°$ by ten or more straight line segments in a manner similar to a squaring network (see Sec. 8.1.4) of an electronic multiplier. By automatic switching circuits, larger ranges can be obtained. There are often means to operate the unit in the polar conversion mode or as a rate resolver.

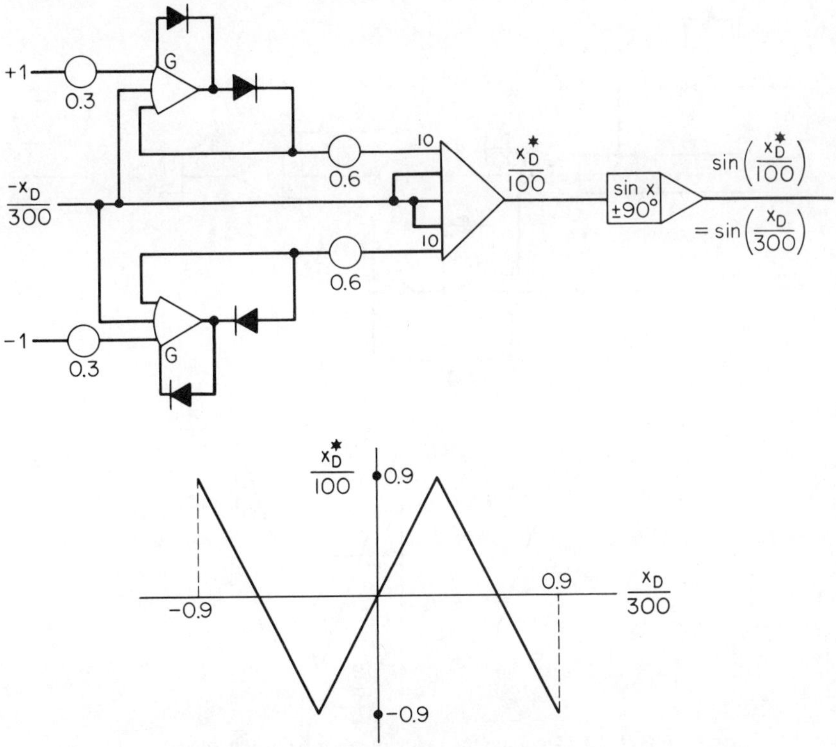

FIG. 8.9 Extending the range of sin x with switching.

8.1.4 Fixed function cards

By means of diode and resistor shaping networks, commonly used functions are approximated with piecewise linear functions; ten straight lines generally provide 0.1-per cent accuracy for the simplest functions. Functions such as sin ax, cos ax (for $-90° \leq x \leq 90°$ or $-180° \leq x \leq 180°$), x^2, x^4, and $a \log_{10} bx$ where a and b depend on the reference, have been so approximated. With implicit function techniques (Chapter 9), inverse functions may also be generated (arcsin x, e^{bx}, $x^{1/2}$, $x^{1/4}$, etc.). A unit generating sin ax and cos ax is often called a *sinusoid* generator.

To extend the range of a sinusoid generator, switching circuits of the type discussed in Chapter 6 can be used. For example, if the function sin x is generable in the range $-90° \leq x_D \leq 90°$ with an input variable scaled at 0.01/deg, the circuit in Fig. 8.9 accurately extends the range to $-270° \leq x_D \leq 270°$. An idealized diode dead-space circuit converts the variable $x_D/300$ to another variable x_D^* so that $-90° \leq x_D^* \leq 90°$. This technique makes use of the reflection symmetry of sin x at $x = -90°$ and $+90°$. Because cos x has reflection symmetry at 0° and $\pm 180°$, the range of cos x can be extended without other switching circuits only if units operating in the range of $-180° \leq x$

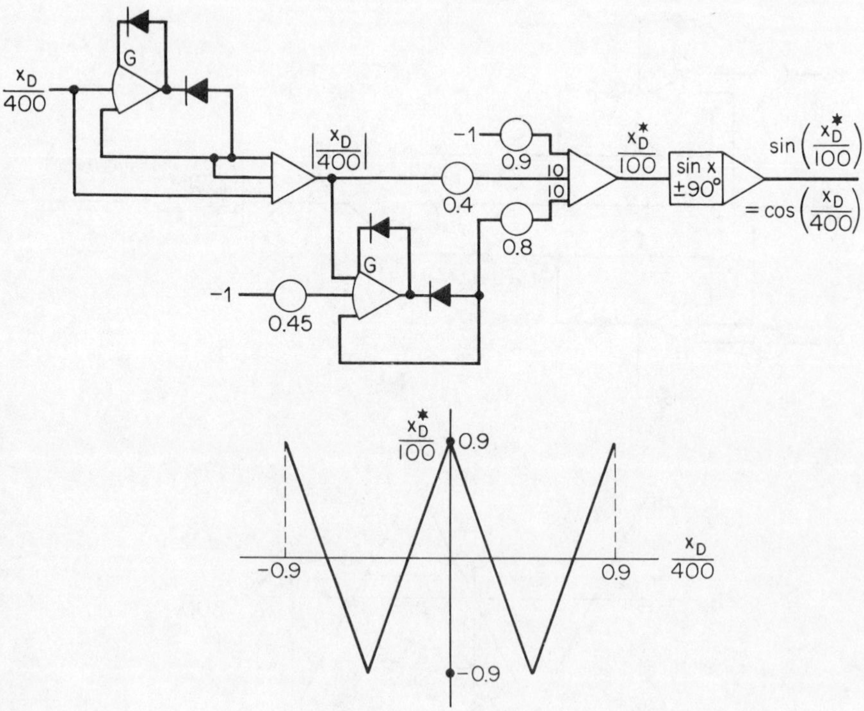

FIG. 8.10 Extending the range of cos x with switching.

$\leq 180°$ are available. It is more common to offset x_D by 90° and use a sin x generator, since sin $x_D = \cos(90° - x_D)$. The circuit of Fig. 8.10 takes advantage of the reflection symmetry of x_D^* about 90° by first using an absolute value circuit. The ranges can be extended even further by additional switching.

To operate a sinusoid generator as a rate resolver, much logic is required to allow for both positive and negative rates. In Fig. 8.11, the digital signal U goes high whenever $x_D/100$ exceeds its bounds (-0.9 and 0.9 in this case) AND the sign of $-\dot{x}_D^*$ causes $x_D/100$ to further increase in magnitude. It is then necessary to change the state of W, which reverses the sign of $-\dot{x}_D^*$. (Because of reflection symmetry of sin x_D at $-90°$ and $+90°$, the function is then

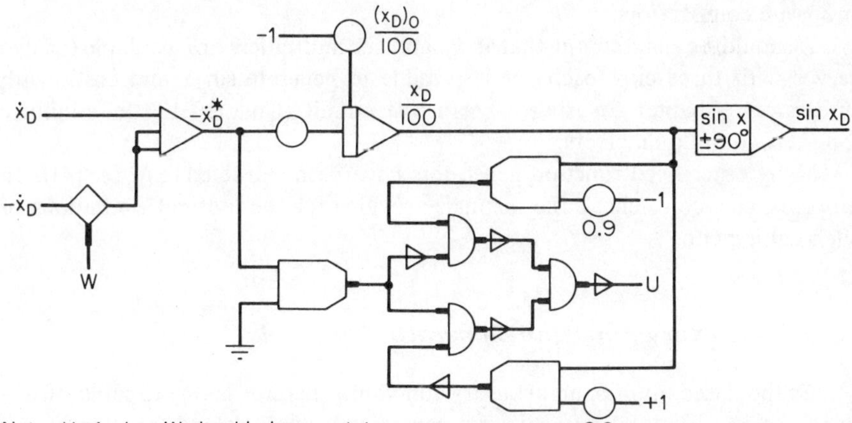

Note: U=1 when W should change state

FIG. 8.11 Operating a sinusoid generator as a rate resolver.

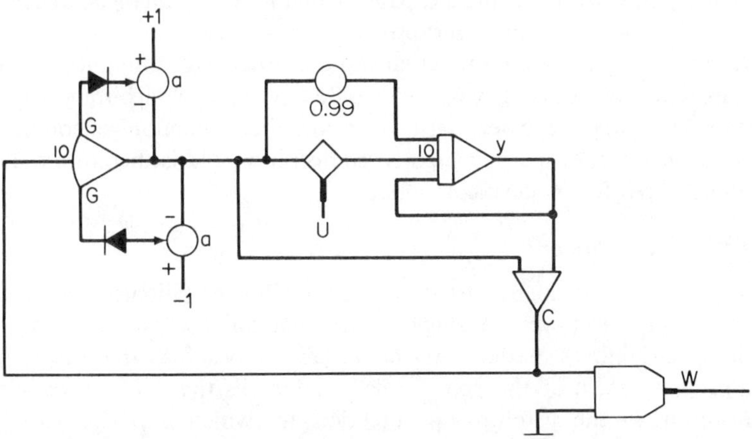

FIG. 8.12 Generating *W* for Fig. 8.11.

properly generated.) This is a particularly easy task if flip-flop digital elements are available (Chapter 20), but a modified multivibrator and sweep generator circuit (Fig. 6.42) can do the job otherwise. The modifications are shown in Fig. 8.12. If the signal $U = 1$, y integrates until it reaches one of its limits. The waveform C (Fig. 6.22) reverses its sign and changes the state of W which in turn makes U low and stops y from oscillating. We must judiciously adjust the values of a and the feedback capacitor of y to insure that this happens quickly. If it is too quick, however, the delay in transmission of $-\dot{x}_D^*$ through the comparator, AND gates, and digital inverters will not make U low soon enough. The delay, however, is a few microseconds or less, so that little trouble can be expected. We must also insure that the state of W yields the initial correct value of $-\dot{x}_D^*$, by either adjusting the initial value of y or latching appropriate comparators.

It should be pointed out that if $\dot{x}_D$ and six multipliers are available (or two servos with three cups each), it is possible to generate sin x and cos x with integrators (Chapter 5) using a corrective circuit (Chapter 13) to minimize long-term amplitude errors.

Many other fixed function generators have been marketed (e.g., see [4]). In all cases, be sure to check the manuals provided for the use and limitations of such equipment.

8.2 Arbitrary Function Generators

As the name implies, an arbitrary function generator is one capable of approximating virtually any arbitrary, continuous, single-valued function of one variable with a small bounded error, independent of time. It is used to generate an explicit continuous function under the same circumstances as any fixed function generator, with the added capability that it may be changed to generate some other function for some other problem.

The means of generating a function and the associated techniques vary with the equipment. We discuss here three different types of arbitrary function generators—the curve follower, pot padder, and diode function generator. The diode function generator permits high-speed operation and is therefore the only one manufactured for modern computers.

8.2.1 Curve follower

Most x-y plotters can be fitted with an attachment (adaptor kit) to convert it to a curve follower. A single-valued function $f(x)$ is drawn on graph paper in special ink as needed. As the variable x positions the arm servo, a sensing probe in place of the pen "reads" and tracks the curve, producing a voltage output in the y follow-up potentiometer which is linear with $f(x)$. Provision is usually made for a varying voltage z to be placed across the follow-up potentiometer, thereby multiplying simultaneously to produce $zf(x)$. Any convenient working diagram symbol can be used depending upon the patching needed.

8.2.2 Pot-padding unit

Some servo multiplier potentiometers are equipped with many taps each of which may be energized at a constant voltage level. Since the wiper arm output interpolates (approximately) linearly between voltages, this method can generate a piecewise linear (polygonal) approximation to any function. A typical arrangement is shown in Fig. 8.13. The taps are equispaced; i.e., the resistance

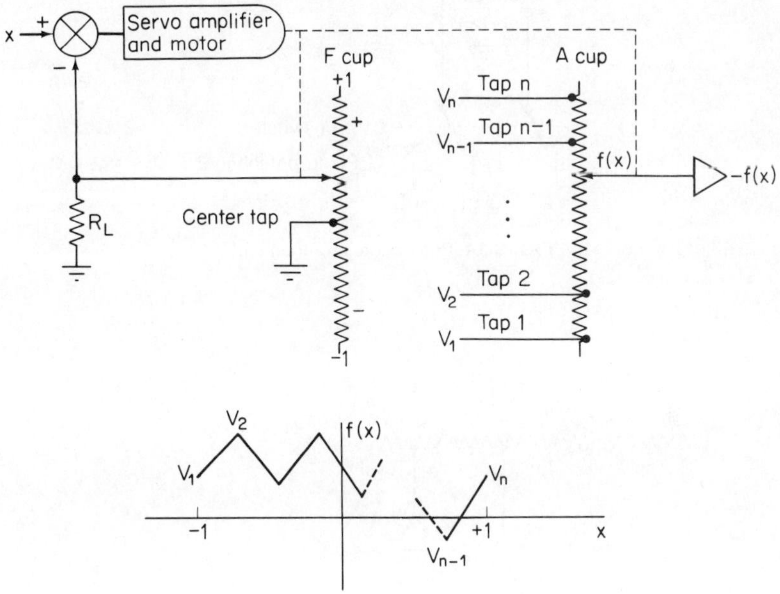

FIG. 8.13 Pot-padder principles.

between each tap is constant, so that the V_i must be chosen for fixed, equispaced values of x.

The typical pot-padding unit contains an adjustable pot and a selector switch for each tap, and one control switch, arranged as shown in Fig. 8.14. They are accessible for adjustment, while the leads to each are transmitted via cable to a plug fitting into the multiplier chassis.

To set pot i, an arrangement such as in Fig. 8.15 is needed to avoid interaction of one setting with another. The control switch is set to supply the reference. Voltages V_{i-1} and V_{i+1} are placed on taps $i - 1$ and $i + 1$ and the pot output (equal to the output at tap i) must be connected to the proper load R_L. The selector switch is then adjusted as follows: For $i \neq 1$ or n,

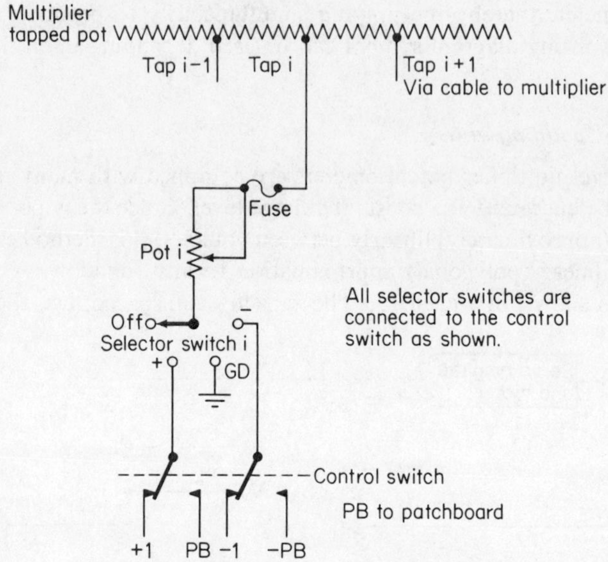

FIG. 8.14 Pot-padder arrangement.

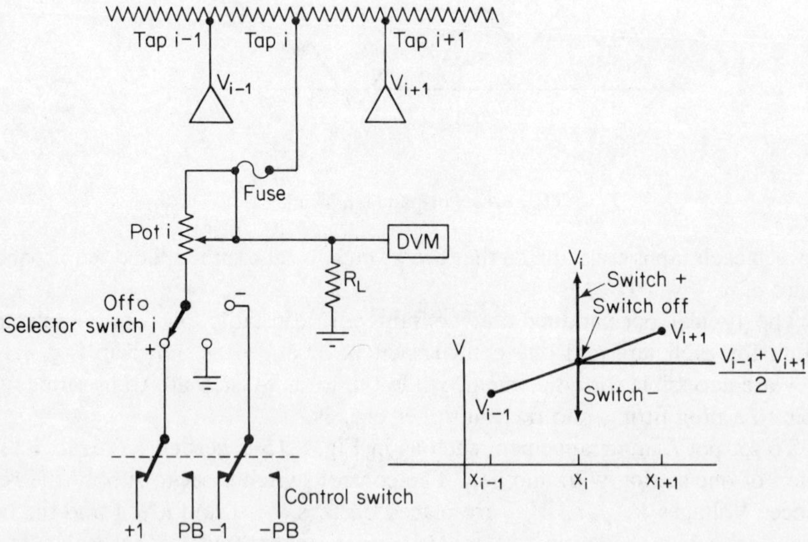

FIG. 8.15 Connections to set pot *i*.

(a) If $(V_{i-1} + V_{i+1})/2 > V_i$, put selector switch on +
(b) If $(V_{i-1} + V_{i+1})/2 < V_i$, put selector switch on −
(c) If $(V_{i-1} + V_{i+1})/2 = V_i$, keep selector switch on OFF.

Occasionally, the selector switch is placed on ground when an exact zero is needed. Thus, for any intermediate point, the position of the selector switch depends on the second difference of $f(x)$, the reason being geometrically shown in Fig. 8.15. For $i = 1$ or n, the selector switch is set to the polarity of V_i. Pot i is then adjusted so that the DVM reads V_i.

A control panel is generally available for manual adjustment of the pot padder. All V_i are first set with precision pots, and a pot selector switch is set at number i. The load R_L is then automatically connected to the pot wiper arm and V_{i-1} and V_{i+1} are automatically connected to the proper taps. One simply adjusts the selector switch and pot i by either directly reading a DVM or by some nulling procedure.

During operation, the control switch can be placed in the PB (patch board) position enabling the programmer to place any voltage through the padders in place of the reference, in the normal place one multiplics with the potentiometer. Figurc 8.16 shows a working diagram symbol in that case.

The process of multiplying $f(x)$ by z destroys it for any other use. If needed more than once, the reference is connected at the pot to generate $f(x)$ as an amplifier output; it then can be used as needed.

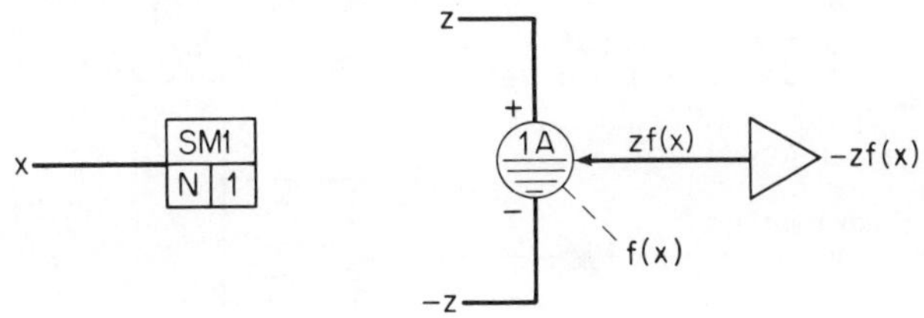

FIG. 8.16 Working symbol for pot-padder units.

There is an interpolation error between taps, although it is usually small. If V is the larger of two voltages at adjacent taps, and R_T and R_L are the resistances between the taps and the load resistance, the maximum error E is given by [5]

$$E \leq \frac{2R_T|V|}{R_L}$$

For a 1-MΩ load, a 50-kΩ potentiometer, and 20 segments, this error does not exceed 0.005 of the reference. If too great, of course, an unloading circuit can be used, as in Fig. 8.2.

Pot padders are available on older equipment and must always be set manually.

8.2.3 Diode function generator

A diode function generator consists of adjustable, parallel, biased diode and resistor networks usually used as input impedances to an operational amplifier, such as is shown in Fig. 8.17. Each of n (typically 20) pairs of breakpoint and slope potentiometers may be arranged by a quadrant selector switch in any of the four ways shown, to obtain four-quadrant operation. When any diode starts to conduct, the output changes linearly with that input. Thus, the output is

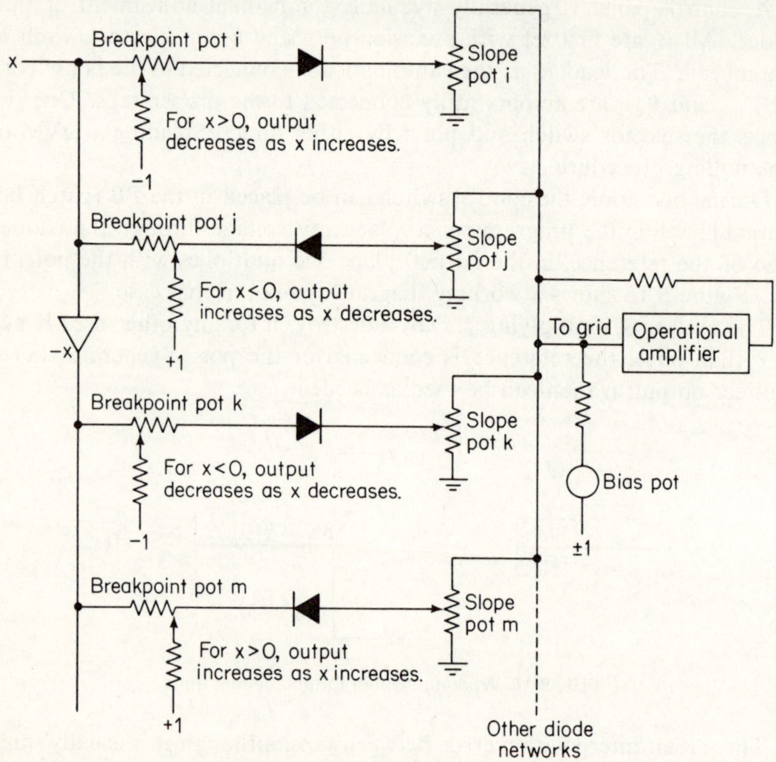

FIG. 8.17 Four ways to arrange input diodes in a DFG.

once again a piecewise linear function, as it was with the pot-padder unit. Here, however, the breakpoints x_i (where the slope changes) are adjustable, so that a given function $f(x)$ can be approximated with less error. A bias pot can shift the output so that $f(0)$ need not be zero.

A control panel is often available to simplify the setup of the function. Even without a control panel, however, the DFG is easily set up because the diodes hardly interact with the pot settings. We first rotate all breakpoint pots so all diodes do not conduct. Starting from zero, the pots then are adjusted, one by one, to conduct for positive x_i. At each x_i, the slope pot $i - 1$ is adjusted to obtain $f(x_i)$. By reading the DVM during the adjustment period, the coordinates of each x_i and $f(x_i)$ may be set with little error. The entire procedure is then repeated for negative x_i. Many variations are possible to the description of both the DFG and its setup.

More modern computers have automatic setup features where it is only necessary to specify the breakpoints and slopes, or sometimes just the coordinates (x_i, y_i) of each breakpoint. The information is prepared either on paper tape or digital cards, and the DFG can then be automatically set on command. Such conveniences are extremely worthwhile if functions must be set up frequently.

8.3 Approximations for Pot-Padding Units

In this and the following section, approximations for both the pot-padding units and diode function generators are discussed. We do not concern ourselves with using the curve follower. Approximation methods there apply only to discrete data, and the concept of experimental error forces us to smooth the data rather than to set up the function generator to pass through every point. Smoothing techniques are beyond the scope of this book.

In the case of a pot-padding unit, the approximation concepts of Chapter 7 apply, except that the functional form is fixed. For a unit producing n segments ($n + 1$ equispaced taps), the functional form can be best expressed as the sum of ramp functions:

$$F(M, x) = \sum_{i=1}^{n} m_i(x - x_i)U(x - x_i) + m_0 \tag{8.3}$$

where the m_i are the $n + 1$ parameters to be determined and $\Delta x = x_{i+1} - x_i$ is constant. A typical approximation is shown in Fig. 8.18. It is clear that

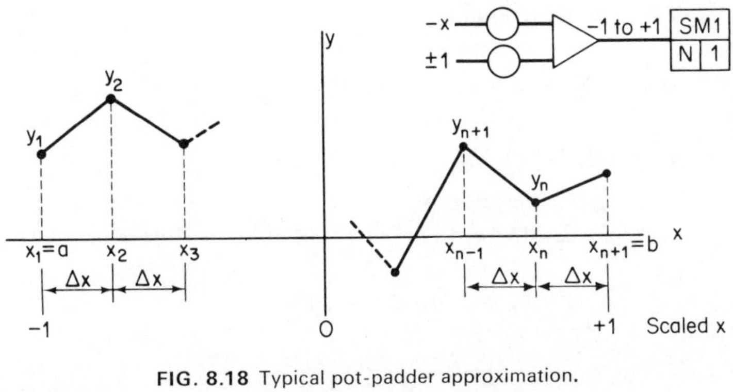

FIG. 8.18 Typical pot-padder approximation.

once the m_i in (8.3) are determined, the ordinates $y_i, i = 1, 2, \ldots, n + 1$, are found by

$$y_i = m_0 + \Delta x \sum_{k=1}^{i} (i - k)m_k \tag{8.4}$$

Also the independent variable x in Fig. 8.18 is transformed to range from its interval $a \leq x \leq b$ to $-1 \leq x \leq +1$ and a normal connection is made to the servo multiplier. In this way, the maximum number of segments is available for the approximation.

In most cases, it is possible to obtain adequate approximations by eye, using an accurate graph (or point plot in the case of discrete data) of the function needed. However, there are analytical methods that sometimes allow for optimum fitting. These should be used in case the final result is somewhat sensitive to the needed function.

8.3.1 Discrete data

For discrete data, the least squares norm again provides for the simplest analytical method. If there are m data points $(X_k, Y_k), k = 1, 2, \ldots, m$, then minimization of

$$\rho = \sum_{k=1}^{m} \left[\sum_{j=1}^{n} m_j(X_k - x_j)U(X_k - x_j) + m_0 - Y_k \right]^2$$

requires the solution of (7.4):

$$\frac{\partial \rho}{\partial m_0} = 2 \sum_{k=1}^{m} \left[\sum_{j=1}^{n} m_j(X_k - x_j)U(X_k - x_j) + m_0 - Y_k \right] = 0$$

$$\frac{\partial \rho}{\partial m_i} = 2 \sum_{k=1}^{m} \left[\sum_{j=1}^{n} m_j(X_k - x_j)U(X_k - x_j) + m_0 - Y_k \right](X_k - x_i) \times U(X_k - x_i) = 0$$

for $i = 1, 2, \ldots, n$. Rearranging,

$$\begin{aligned} m m_0 + \sum_{k=1}^{m} \sum_{j=1}^{n} m_j(X_k - x_j)U(X_k - x_j) &= \sum_{k=1}^{m} Y_k \\ m_0 \sum_{k=1}^{m} (X_k - x_i)U(X_k - x_i) & \\ + \sum_{k=1}^{m} \sum_{j=1}^{n} m_j(X_k - x_j)(X_k - x_i)U(X_k - x_j)U(X_k - x_i) & \\ = \sum_{k=1}^{m} Y_k(X_k - x_i)U(X_k - x_i) & \end{aligned} \tag{8.5}$$

for $i = 1, 2, \ldots, n$.

These equations are linear in the m_i, which makes their use attractive. Since n, however, is usually at least 10, a hand calculation is not feasible, and the solution of (8.5) should be attempted only when a digital computer is available. A unique solution requires that data points exist within two consecutive segments.

8.3.2 Chebyshev norm for functions on an interval

For this functional form, a Chebyshev approximation does not have its usual characterization of the error alternating $n + 1$ times in sign at a maximum absolute amplitude. Neither is a best approximation unique. The reason is obvious from Fig. 8.19, where an unconnected set of straight lines is shown as a best Chebyshev approximation to the function $f(x)$. Each segment crosses $f(x)$ at least twice, and there are at least three points in each segment for which the deviation assumes a maximum. (The latter also applies to a point set.) The maximum deviation, say, is E_i, and clearly cannot be lessened. Segment i is called a controlling segment. If we now rotate the other segments so that all

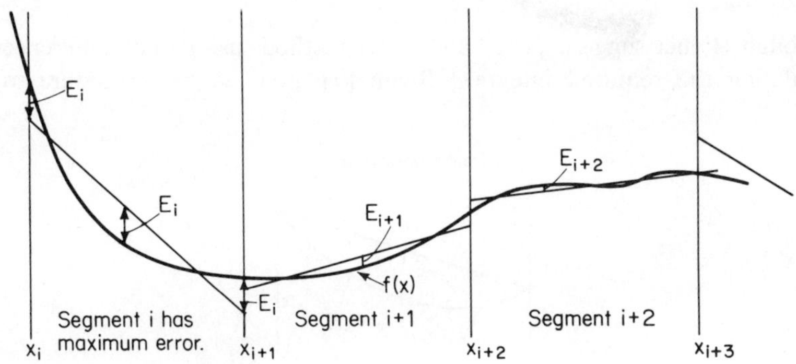

FIG. 8.19 Unconnected best approximation.

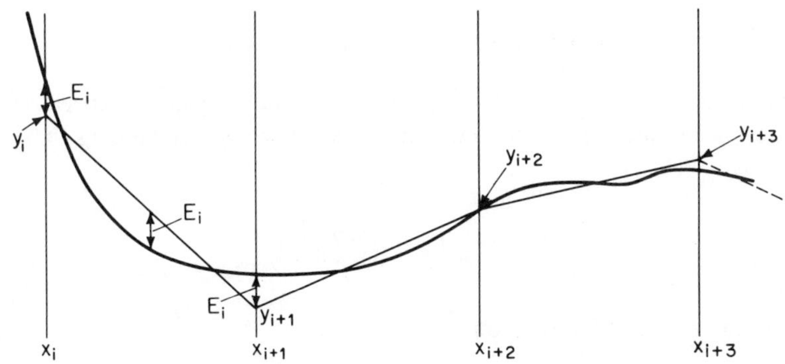

FIG. 8.20 Rotating and translating other segments.

lines are connected (Fig. 8.20), the maximum error remains E_i, as shown, more often than not. The result is a best Chebyshev approximation of the required form that is not unique because there will be considerable leeway in positioning many of the other y_i.

An excellent approximation is obtained by rotating each uncontrolling segment about its midpoint until all segments are connected.

If the process creates errors greater than E_i, the errors are balanced by rotating and translating the noncontrolling segments. One can usually adjust the y_i to obtain a best Chebyshev approximation sufficiently good for any investigation.

Hamer [6] derived an approximate relationship that helps fit linear segments to given functions when these are available as mathematical expressions or as accurate tables. By a Taylor series expansion about the midpoint $(x_i + x_{i+1})/2$, we may show that (Fig. 8.21)

$$(\Delta x)^2 \simeq \frac{8E}{\left|\dfrac{d^2f}{dx^2}\right|} \tag{8.6}$$

Although Hamer suggests pencil and paper methods using finite differences for calculating the required integrals, Ream [7] modified his procedure to use

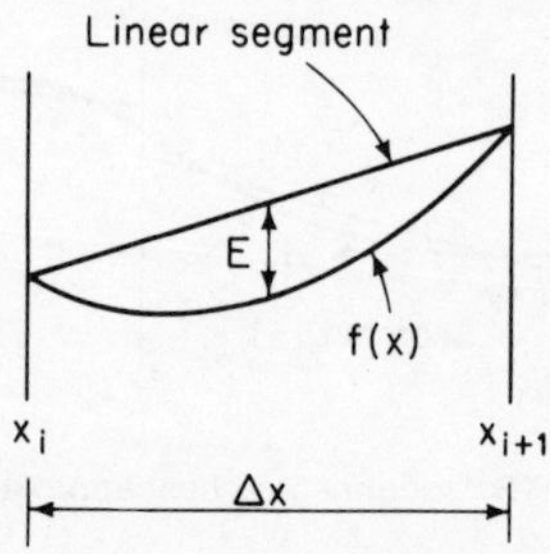

FIG. 8.21 Diagram for approximate deviation.

continuous integrals instead. This method will be developed here because the computer sometimes can aid in determining best fits. By rearranging (8.6) and integrating,

$$E \simeq \frac{1}{8}\left(\int_{x_i}^{x_{i+1}} \sqrt{\left|\frac{d^2f}{dx^2}\right|}\, dx\right)^2 \tag{8.7}$$

If we could generate the function

$$z(x) = \int_a^x \sqrt{\left|\frac{d^2f(u)}{du^2}\right|}\, du$$

a graphical and/or analytical procedure could be used to determine the unconnected E_i [the E_i are half the E in (8.7)], as shown in Fig. 8.19, when there are no inflection points in the interval. When inflection points exist, an excellent fit is usually obtained in that segment anyway, so that it will not be a controlling one and is therefore not particularly important.

Example 8.1. To illustrate the above semianalytical and graphical method of obtaining a best Chebyshev fit for a pot padder, we approximate the function

$$f(x) = \frac{\sin x}{x}, \qquad 0 \leq x \leq \pi$$

assuming a multiplying pot with 11 taps (10 segments).

Step 1. Obtain an accurate graph of $f(x)$ on graph paper where major coordinate lines occur where the potentiometer taps are. A visual aid can supplement other

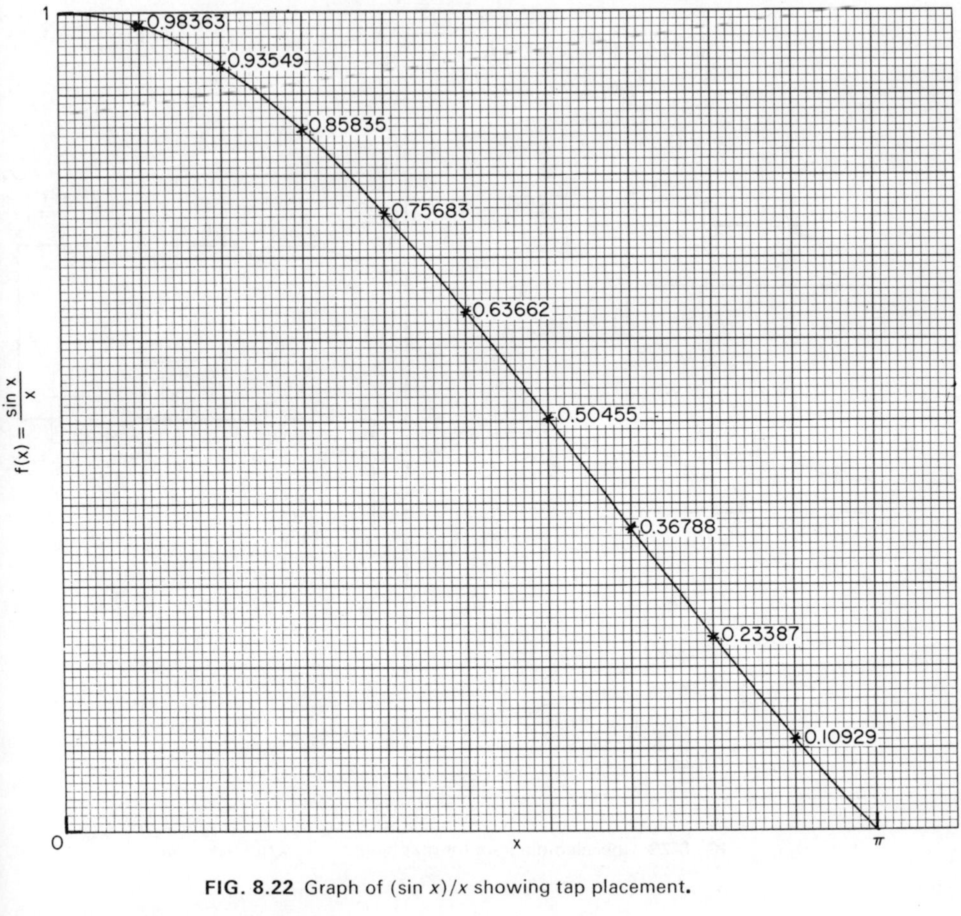

FIG. 8.22 Graph of (sin x)/x showing tap placement.

knowledge of the function. Figure 8.22 is such a graph showing accurate functional values at each tap. We may note from the graph whether or not $d^2f(x)/dx^2$ is positive or negative in the interval and if an inflection point exists.

Step 2. We next need a graph of $z(x)$

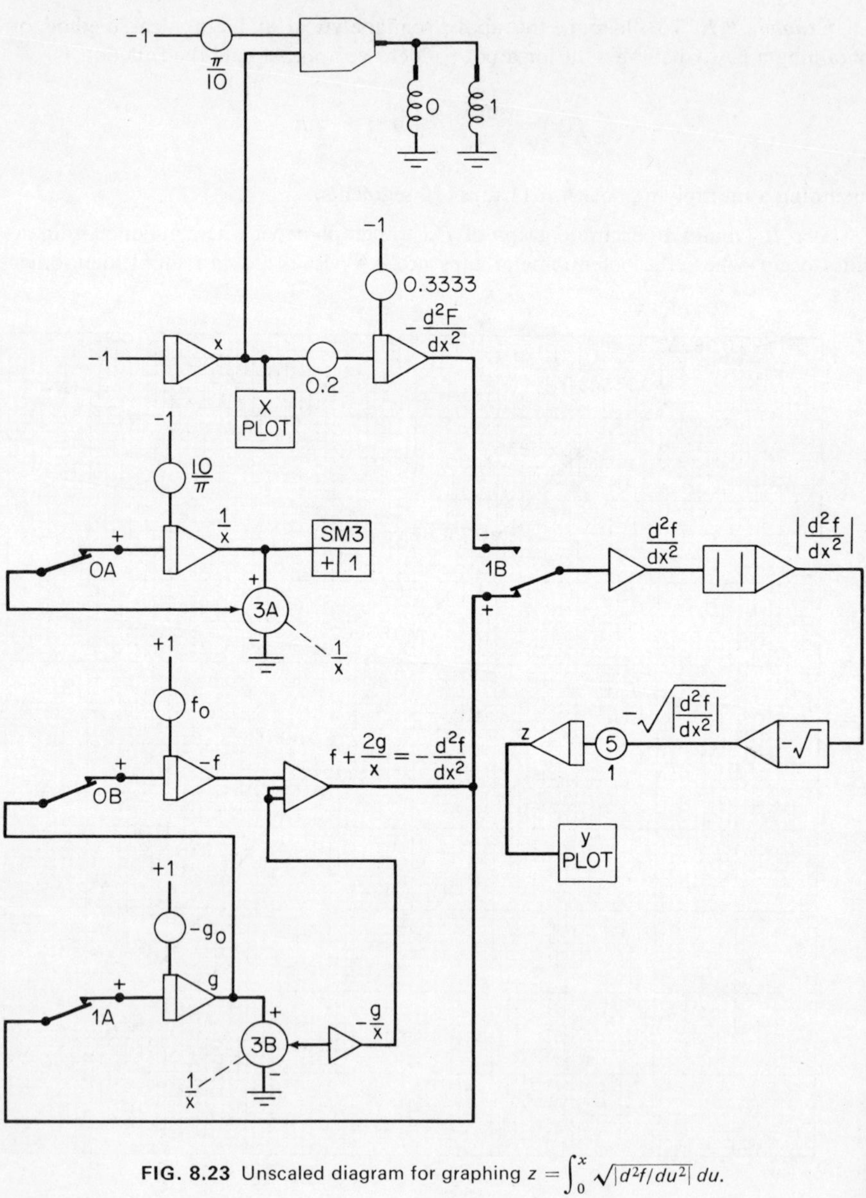

FIG. 8.23 Unscaled diagram for graphing $z = \int_0^x \sqrt{|d^2f/du^2|}\, du$.

$$z(x) = \int_0^x \sqrt{\left|\frac{d^2 f(u)}{du^2}\right|}\, du$$

for $0 \leq x \leq \pi$, from which we may graphically determine E for each segment from (8.7). This may be obtained in many ways, but the analog computer is used for that purpose since $f(x)$ and its second derivative can be obtained quite easily.

We generate $f = (\sin x)/x$ by integration, except that the initial indeterminacy forces us to use an approximation for small x and later switch to the differential equation that generates f (see Sec. 5.5). For small x, say $0 \leq x \leq \pi/10$, we use a truncated Taylor series and call the function F:

$$F = 1 - \frac{x^2}{6}, \qquad \frac{d^2F}{dx^2} = -\frac{1}{3} + \frac{x^2}{10}$$

In the range indicated,

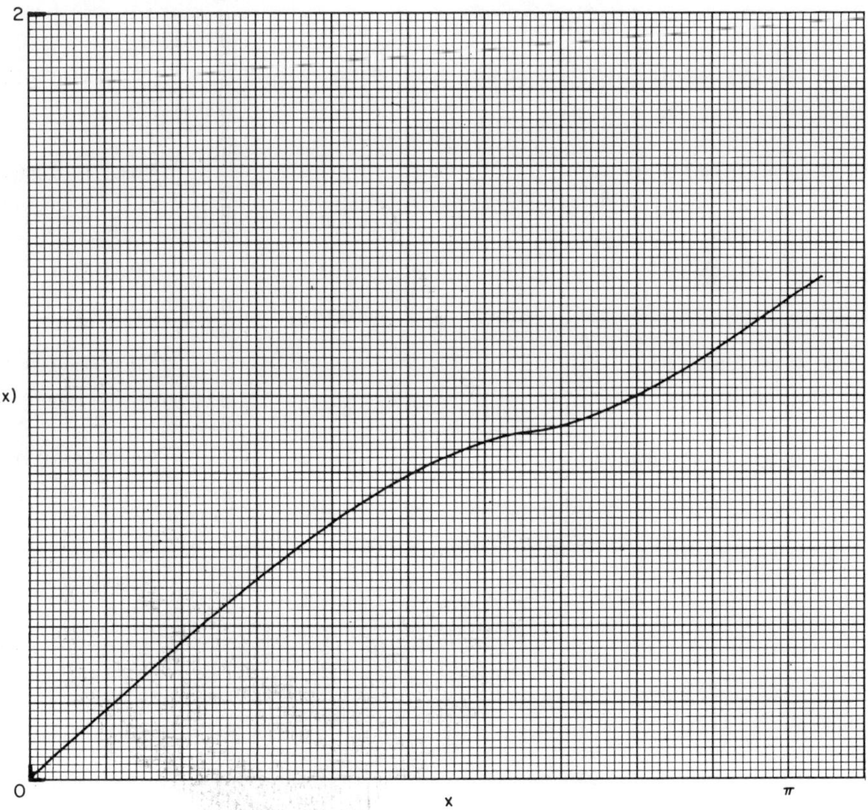

FIG. 8.24 Graph of $z = \int_0^x \sqrt{|d^2f/du^2|}\, du$.

TABLE 8.1

Calculations for Chebyshev Norm Using the Pot Padder

i	x_i	f_i	z_i *from graph*	Δz_i	$E=\frac{1}{8}(\Delta z_i)^2$	$C=-\frac{E}{2}\operatorname{sign}\frac{d^2f_i}{dx^2}$	$U=f_i+C$	*final adjustment*	$\lvert Error\rvert$
1	0	1.0000	0					1.00207	0.00207
				0.182	0.00414	+0.00207	1.00207 0.98570		
2	0.1π	0.98363	0.182				+11	0.98570	0.00207
				0.177	0.00392	+0.00196	0.98559 0.93745		
3	0.2π	0.93549	0.359				−11 +19	0.93734	0.00185
				0.163	0.00332	+0.00166	0.93715 0.86005		
4	0.3π	0.85839	0.522				−19 +10	0.85986	0.00147
				0.148	0.00274	+0.00137	0.85976 0.75820		
5	0.4π	0.75683	0.670				−10 +35	0.75810	0.00127
				0.121	0.00184	+0.00092	0.75775 0.63754		
6	0.5π	0.63662	0.791				−35 + 7	0.63719	0.00057
				0.089	0.00100	+0.00050	0.63712 0.50505		
7	0.6π	0.50455	0.880				− 7 +32	0.50498	0.00043
				0.042	0.00022	±0.00011	0.50466 0.36777		
8	0.7π	0.36788	0.922				−32 − 5	0.36745	0.00043
				0.078	0.00076	−0.00038	0.36750 0.23349		
9	0.8π	0.23387	1.000				+ 5 +49	0.23354	0.00033
				0.115	0.00166	−0.00083	0.23305 0.10846		
10	0.9π	0.10929	1.115				−49 −18	0.10797	0.00132
				0.135	0.00228	−0.00114	0.10815 −0.00114		
11	π	0	1.250				+18	−0.00096	0.00096

$$|F - f| < 0.0001, \qquad \left|\frac{d^2F}{dx^2} - \frac{d^2f}{dx^2}\right| < 0.0001$$

and are therefore quite satisfactory. Now, we let

$$g = \frac{\cos x - f}{x}$$

from which the desired differential equations are derived:

$$\frac{df}{dx} = g, \qquad f_0 = \frac{\sin 18^\circ}{(\pi/10)}$$

$$\frac{dg}{dx} = -f - \frac{2g}{x}, \qquad g_0 = \frac{\cos 18^\circ - f_0}{(\pi/10)}$$

It is clear that $d^2f/dx^2 = dg/dx$ so that $(|d^2f/dx^2|)^{1/2}$ can be generated with an absolute-value circuit (Chapter 6) and square-root circuit (Chapter 9). An unscaled diagram is shown in Fig. 8.23 using $\dot{x} = 1$, relays to hold the integrators until $x = \pi/10$, and by generating $1/x$ by integration for accuracy. Figure 8.24 shows the graph of $z(x)$. We may read the value of z at each x_i directly from the graph.

Step 3. Calling these values z_i, we now prepare a table such as Table 8.1, showing the various quantities indicated. The quantity $-E[\text{sign}\,(d^2f_i/dx^2)]/2$ is the amount added to f_i to determine the ordinates y_i of the endpoints of each unconnected straight line segment that best fits the function in the interval.

Step 4. The final adjustment is now made by rotating all segments about their center so they are connected, except the one we wish to disturb least, i.e., the segment with the largest E value. Rotation about the midpoint involves adding an ϵ to one end and subtracting it from the other. This column now contains the endpoints of all segments and may be set up on the pot padding unit. The absolute error in the last column is calculated to insure that no further adjustment is necessary.

This result is a best Chebyshev approximation [within the error inherent in (8.7)], but is not a unique best approximation using the distance function

$$\rho = \int_a^b |F(M, x) - f(x)|^k \, dx \tag{8.8}$$

where we let $k \to \infty$. Such a limiting case is normally the best Chebyshev approximation when the latter is unique. Much more manipulation (rotating other segments about their endpoints) is required using (8.8) as a norm, but it is not usually worth the effort.

8.3.3 Least squares norm for a function on an interval

It is also possible to set up linear equations for obtaining a best least squares fit when using a pot padder for a function defined on an interval $a \leq x \leq b$. From the representation in (8.3) and application of (7.4), the equations to be solved are

$$(b - a)m_0 + \sum_{j=1}^{n} \frac{(b - x_j)^2}{2} m_j = \int_a^b f(x)\,dx$$

$$\frac{(b - x_i)^2}{2} m_0 + \sum_{j=1}^{n} m_j \int_{x_k}^{b} (x - x_i)(x - x_j)\,dx$$

$$= \int_{x_i}^{b} f(x)(x - x_i)\,dx, \qquad i = 1, 2, \ldots, n$$

where $x_k = \max(x_i, x_j)$. As in the case involving discrete data, the fact that n is at least 10 precludes using this method unless a digital computer is available.

In all cases discussed, only slight modification is required for including a weighting function in the equations.

8.4 Approximations for Diode Function Generators

A typical approximation for use with a diode function generator is shown in Fig. 8.25. Again, a transformation is made so that the input to the DFG ranges from -1 to $+1$, primarily to aid the repeatability of the function set

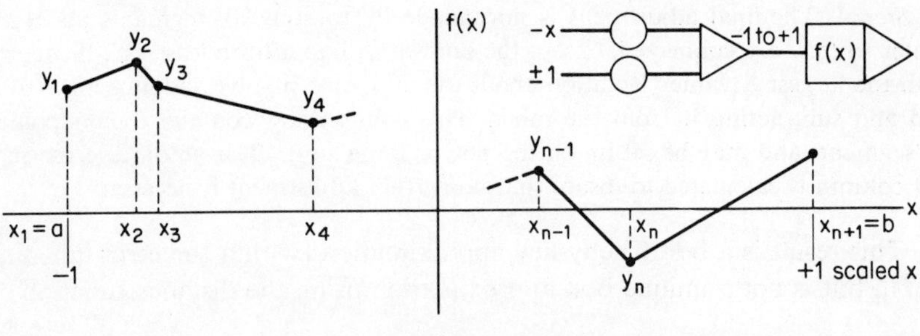

FIG. 8.25 Typical DFG approximation.

up. Also, there are slope limitations minimized in this way. For n segments there are not only $n + 1$ ordinates to be determined, but also $n - 1$ breakpoints ($x_1 = a$ and $x_{n+1} = b$ are given). The approximation therefore requires the determination of $2n$ parameters.

Using (8.3) for a general representation of the approximation function then requires that the y_i be constructed from

$$y_i = m_0 + \sum_{k=1}^{i} (x_i - x_k)m_k \tag{8.9}$$

because the Δx are not equal.

It should be pointed out that all equations derived from the application of (7.4) are nonlinear. As such, best fits can be determined only by iterative methods used in solving nonlinear algebraic equations. In the case of using discrete data, graphical methods have universally been used by programmers because of the difficulty involved in solving the proper equations.

For fitting a continuous function with continuous derivatives, methods were developed, notably by Hamer [6] and Ream [7], which in many cases lead to an accurate best fit. Inflection points create certain difficulties, and more often than not are simply ignored in the interest of expediency. In that case, we compromise to get a very good fit instead of a best one.

The following two theorems are due to Ream:

Theorem 8.1. Assuming $d^2y/dx^2 \neq 0$ in the interval covered, a first approximation to a best piecewise linear Chebyshev fit to $f(x)$ occurs when the breakpoints x_i are determined so that

$$C_1 = \int_{x_i}^{x_{i+1}} \sqrt{\left|\frac{d^2f(x)}{dx^2}\right|}\, dx \tag{8.10}$$

is constant in each interval $x_i \leq x \leq x_{i+1}$. Each breakpoint ordinate y_i is then given by

$$y_i = f(x_i) - \frac{C_1^2}{16} \operatorname{sign}\left[\frac{d^2f(x_i)}{dx^2}\right] \tag{8.11}$$

Theorem 8.2. Assuming $d^2y/dx^2 \neq 0$ in the interval covered, a first approximation to a best piecewise linear weighted least squares fit to $f(x)$ occurs when the breakpoints x_i are determined so that

$$C_2 = \int_{x_i}^{x_{i+1}} \left\{w(x)\left[\frac{d^2f(x)}{dx^2}\right]^2\right\}^{1/5} dx \tag{8.12}$$

is constant in each interval $x_i \leq x \leq x_{i+1}$. Each breakpoint ordinate y_i is then given by

$$y_i = f(x_i) - \frac{C_2^2}{12}\left[\frac{|d^2f(x_i)/dx^2|}{w(x_i)^2}\right]^{1/5} \operatorname{sign}\left[\frac{d^2f(x_i)}{dx^2}\right] \tag{8.13}$$

In a manner identical with that described by Ream, we may also prove the following two important theorems:

Theorem 8.3. Assuming $d^2y/dx^2 \neq 0$ in the interval covered, a first approximation to a best piecewise linear weighted absolute error fit to $f(x)$ occurs when the breakpoints x_i are determined so that

$$C_3 = \int_{x_i}^{x_{i+1}} \left[w(x) \left| \frac{d^2 f(x)}{dx^2} \right| \right]^{1/3} dx \tag{8.14}$$

is constant in each interval $x_i \leq x \leq x_{i+1}$. Each breakpoint ordinate y_i is then given by

$$y_i = f(x_i) - \frac{3C_3^2}{32} \left[\frac{|d^2 f(x_i)/dx^2|}{w(x_i)^2} \right]^{1/3} \operatorname{sign} \left[\frac{d^2 f(x_i)}{dx^2} \right] \tag{8.15}$$

Theorem 8.4. Assuming $d^2y/dx^2 \neq 0$ in the interval covered, a first approximation to a best piecewise linear weighted Chebyshev fit to $f(x)$ occurs when the breakpoints x_i are determined so that

$$C_4 = \int_{x_i}^{x_{i+1}} \sqrt{w(x) \left| \frac{d^2 f(x)}{dx^2} \right|} \, dx \tag{8.16}$$

is constant in each interval $x_i \leq x \leq x_{i+1}$. Each breakpoint ordinate y_i is then given by

$$y_i = f(x_i) - \frac{C_4^2}{16 w(x_i)} \operatorname{sign} \left[\frac{d^2 f(x_i)}{dx^2} \right] \tag{8.17}$$

This last theorem includes the statement of Ream's first theorem when $w(x) \equiv 1$.

These fitting theorems are easy to apply and often produce accurate results, sometimes even when relatively few segments are used. They are, however, first approximations and should not be considered as actual best fits. The errors introduced depend upon the third and higher derivatives of $f(x)$. To apply the theorems, we first obtain a graph of the required integral for $x_1 \leq x \leq x_n$ as a function of x. (See Chapter 9 for generating square, cube, and fifth roots.) The constant C_k may then be determined, and each x_i and y_i calculated from the formulas. In some cases, the integrals may be evaluated exactly by formula, making the application of the fitting theorems easier.

When d^2f/dx^2 is equal to zero in an interval, all the arguments in the derivation of the formulas break down. However, we may show that the contribution to the distance function is less than that of other segments; the segment can be larger and the error smaller, but it is nevertheless a very good fit, a term used here whenever $d^2f/dx^2 = 0$ somewhere in the range of interest. Several schemes to improve the fit then could be tried. These generally involve trial and error (or digital computer aid) and are not worthwhile unless an optimum fit is really necessary.

Example 8.2. We shall fit the function $f(x) = (\sin x)/x$ as in Example 8.1 using a Chebyshev norm and Theorem 8.1, assuming 10 segments will be used. It is convenient to normalize the integral

$$z(x) = \int_a^x \sqrt{\left|\frac{d^2 f(u)}{du^2}\right|}\, du$$

so that $z(x_i)$ is recorded on a major grid line. By adjusting pot 5 in Fig. 8.23, we produce the graph in Fig. 8.26 where each x_i is now clearly marked, since each Δz_i between major grid lines is constant. The constant

$$C_1 = \frac{z(x_n)}{n-1} = \frac{1.25}{10} = 0.125$$

is determined with sufficient accuracy from the graph. After the x_i are tabulated and $f(x_i)$ is computed, the y_i are found from (8.11) (see Table 8.2).

The segment containing the inflection point is not optimum but in no case can it

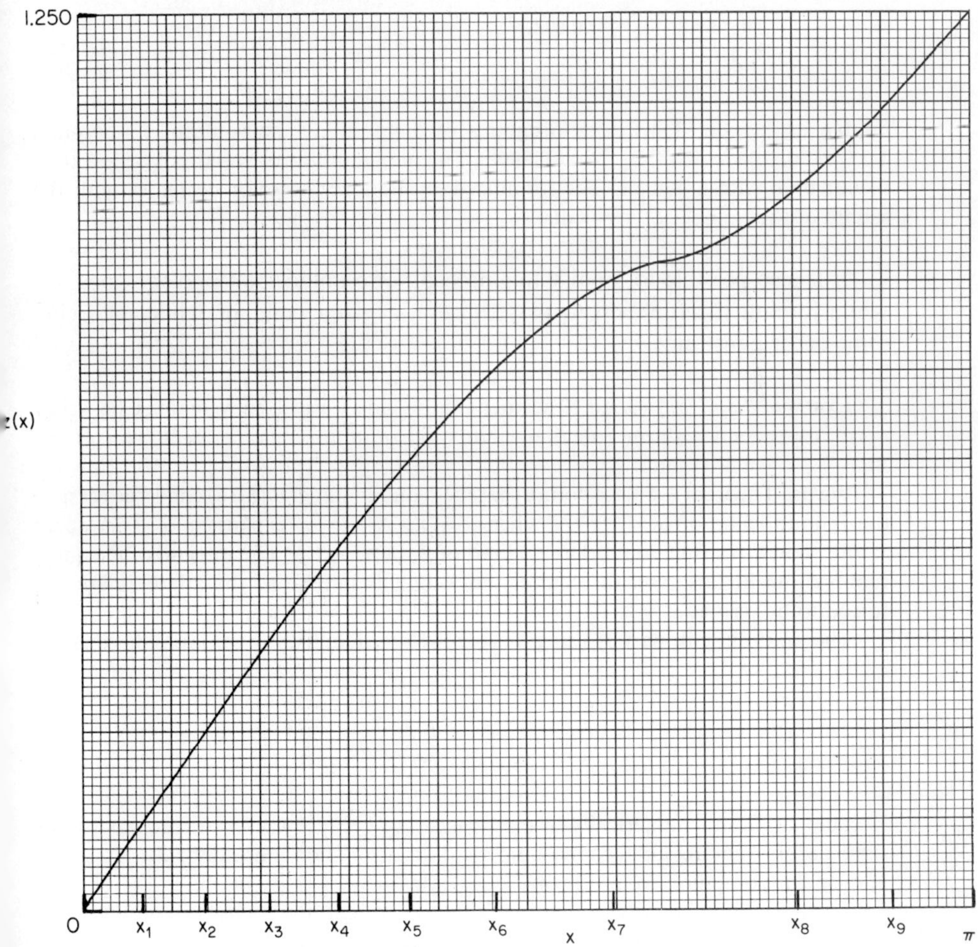

FIG. 8.26 Normalized graph of $z(x)$.

TABLE 8.2

Approximate Chebyshev Fit for (sin x)/x (10 Segments, $C_1^2/16 = 0.00098$)

x_i	$\frac{\sin x_i}{x_i} = f(x_i)$	$-\text{sign}\,\frac{d^2f(x_i)}{dx^2}$	y_i
0	1.00000	+1	1.00098
0.2136	0.99241	+1	0.99339
0.4367	0.96851	+1	0.96949
0.6629	0.92835	+1	0.92933
0.9016	0.86992	+1	0.87090
1.1592	0.79062	+1	0.79160
1.4577	0.68163	+1	0.68261
1.8692	0.51135	+1	0.51233
2.5196	0.23125	−1	0.23027
2.8557	0.09875	−1	0.09777
3.1416	0	−1	−0.00098

double its length. Thus, weighting this segment twice, the value of C_1 becomes $1.25/11 = 0.1135$ with $C_1^2/16 = 0.0008$. The maximum error could not be less than 0.0008 even with a best fit. The formulas indicate that the error decreases with the square of the number of segments.

It is also possible to modify this procedure to obtain best or very good fits while forcing the function to pass through a given point or while enforcing some other constraint.

Example 8.3. Find a good 10-segment fit for $(\sin x)/x$ using a weighted Chebyshev norm, using the weighting function $x/\sin x$, i.e., minimize the per cent error of the fit.

We use Theorem 8.4 after noting that the required integral (8.16) converges despite the infinite value of $|d^2f/dx^2|/f$ when $x = \pi$. A Taylor expansion of $f(x)$ about $x = \pi$ shows this function to have a limit of $2/[\pi(\pi - x)]$. The integral of its square root is finite in this case. Of course, the weight is infinite so the approximation is forced to be zero at $x = \pi$.

The most difficult part of using Theorem 8.4 is in obtaining the normalized graph of

$$z(x) = \int_0^x \sqrt{\left|\frac{d^2f}{dx^2}\right|/f}\, dx$$

It could be found numerically in most cases, but because $f(\pi) = 0$ here, an infinite slope occurs, making an accurate numerical determination more difficult. To graph $z(x)$ on the computer, we must solve

$$\frac{dz}{dx} = \sqrt{\left|\frac{x^2 \sin x + 2x \cos x - 2 \sin x}{x^2 \sin x}\right|}$$

The parametric technique is suitable if $\dot{x} \neq 0$. Accordingly, we integrate the first three terms of a Taylor expansion of dz/dx and find

$$z(x) \simeq \frac{x}{\sqrt{3}}\left(1 - \frac{x^2}{45} - \frac{2x^4}{525}\right)$$

is quite accurate for $0 \leq x \leq 0.1\pi$, and that $z(0.1\pi) \simeq 0.181$. With the definitions

$$S_2 = \sqrt{\sin x}, \qquad C = \cos x$$
$$R = x^2 S_2^2 + 2xC - 2S_2^2 \tag{8.18}$$
$$F = \frac{1}{x}\sqrt{R} \text{ for } R > 0, \qquad F = -\frac{1}{x}\sqrt{-R} \text{ for } R < 0$$

we may show that the system

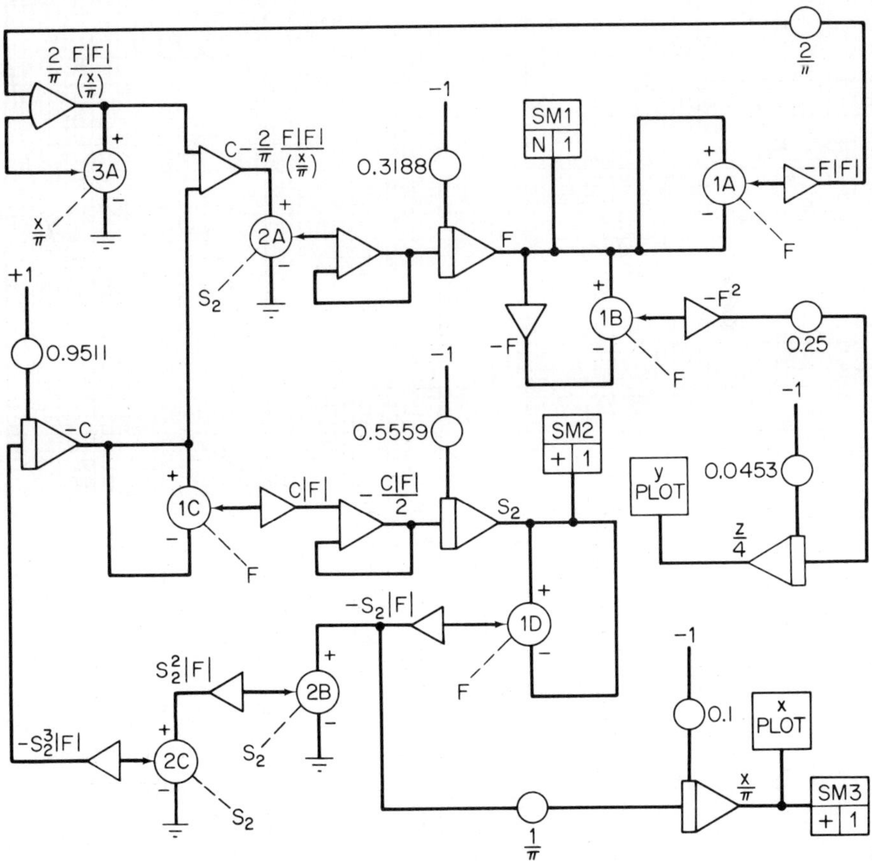

FIG. 8.27 Mechanization for graphing $z(x)$.

$$\begin{aligned}
\dot{z} &= F^2, & z_0 &= 0.181 \\
\dot{x} &= S_2|F|, & x_0 &= 0.1\pi \\
\dot{S}_2 &= \frac{C|F|}{2}, & (S_2)_0 &= 0.5559 \\
\dot{C} &= -S_2^3|F|, & C_0 &= 0.9511 \\
\dot{F} &= \left(-\frac{F|F|}{x} + \frac{C}{2}\right)S_2, & F_0 &= 0.3188
\end{aligned} \tag{8.19}$$

graphs the function $z(x)$ for $0.1\pi \leq x \leq \pi$. Initial conditions for system (8.19) are found by definitions (8.18) at $x = 0.1\pi$. Mechanization of system (8.19) is straightfoward; a scaled diagram using servo multipliers is shown in Fig. 8.27, while Fig. 8.28 shows the graph of $z(x)$ obtained, normalized so that $z(\pi)$ is on a major grid line. We easily read the x_i for equal Δz_i and construct Table 8.3 to find y_i from (8.17).

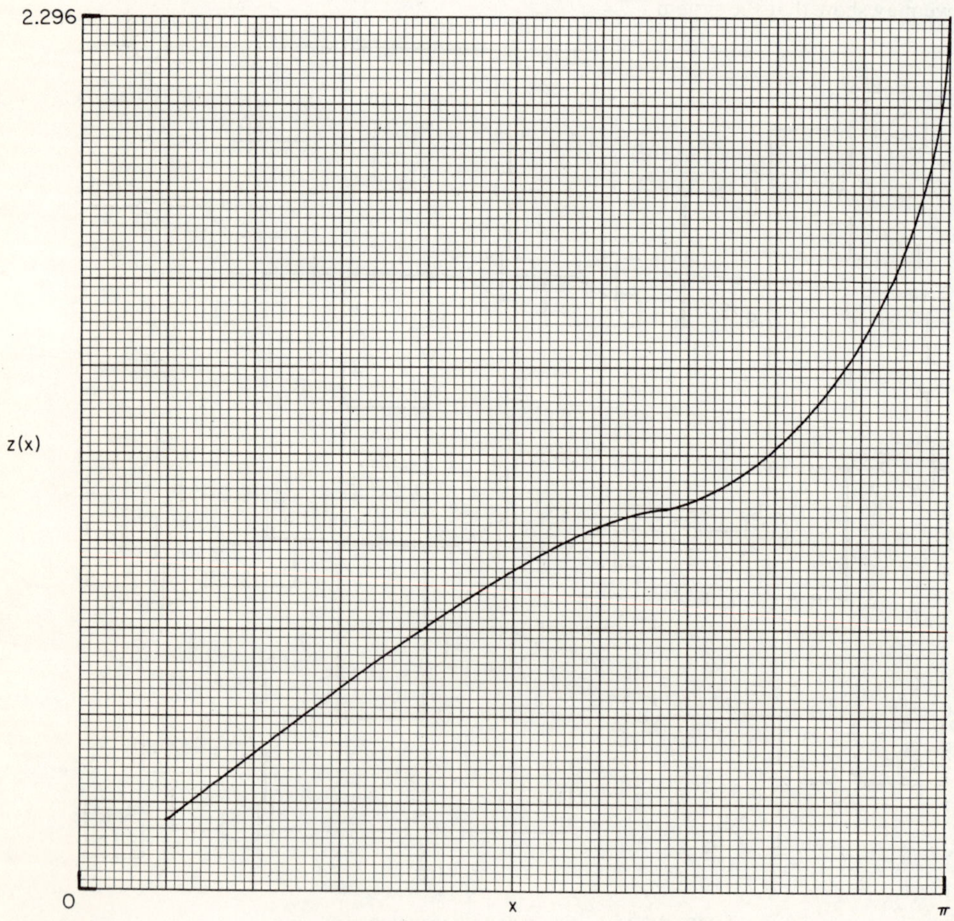

FIG. 8.28 Graph of $z(x) = \int_0^x \sqrt{\frac{1}{f}\left|\frac{d^2f}{dx^2}\right|}\,dx.$

Thus, the values of y_i deviate at most $\frac{1}{3}$ of one per cent, not of full scale, but of the value of the function.

TABLE 8.3

Approximate Weighted Chebyshev Fit for (sin x)/x Using Ten Segments of a DFG ($C_4^2 = 0.003295$)

x_i	$\dfrac{\sin x_i}{x_i} = f_i = \dfrac{1}{w^i}$	$-f_i \dfrac{C_4^2}{16} \operatorname{sign} \dfrac{d^2 f(x_i)}{dx^2}$	y_i
0	1.00000	+0.00330	1.00330
0.3958	0.97409	+0.00321	0.97730
0.8074	0.89484	+0.00295	0.89779
1.2409	0.76241	+0.00251	0.76492
1.7561	0.55969	+0.00184	0.56153
2.4850	0.24564	−0.00081	0.24483
2.7677	0.13197	−0.00043	0.13154
2.9405	0.06793	−0.00022	0.06771
3.0505	0.02982	−0.00010	0.02972
3.1165	0.00805	−0.00003	0.00802
3.1416	0	0	0

8.5 Transformations

Before placing any function on arbitrary function generation equipment, it is well to consider whether or not a simple transformation will help improve accuracy. Steep slopes, for example, are generally a source of difficulty. In such a case, it may pay to generate $h(g)$ instead of $f(x)$, assuming that $g(x)$ is generable accurately and $h[g(x)] = f(x)$. Figure 8.29 shows some examples of how to "straighten out" the function. This is desirable since the smaller the second derivative, the more accurate a function can be set up. It remains to choose a convenient $g(x)$ that offers a significant reduction in the total required integral, a task which is by no means simple.

Example 8.4. For the approximation problem we have been considering, $f = (\sin x)/x$ for $0 \leq x \leq \pi$, determine the breakpoints and ordinates using a Chebyshev norm for a function $h(g)$ where $g = x^2$. Our motivation is that $f(x)$ is a power series in x^2, but $h(g)$ is a power series in g and is "straighter." We have

$$f(x) = \frac{\sin \sqrt{g}}{\sqrt{g}} = h(g)$$

A six-term Taylor expansion is sufficient for 0.01-per cent accuracy in d^2h/dg^2.

$$\frac{d^2h}{dg^2} \simeq \frac{2}{5!}(1 - 0.0714g + 0.198 \times 10^{-2}g^2 - 0.301 \times 10^{-4}g^3 + 0.289 \times 10^{-6}g^4 - 0.193 \times 10^{-8}g^5)$$

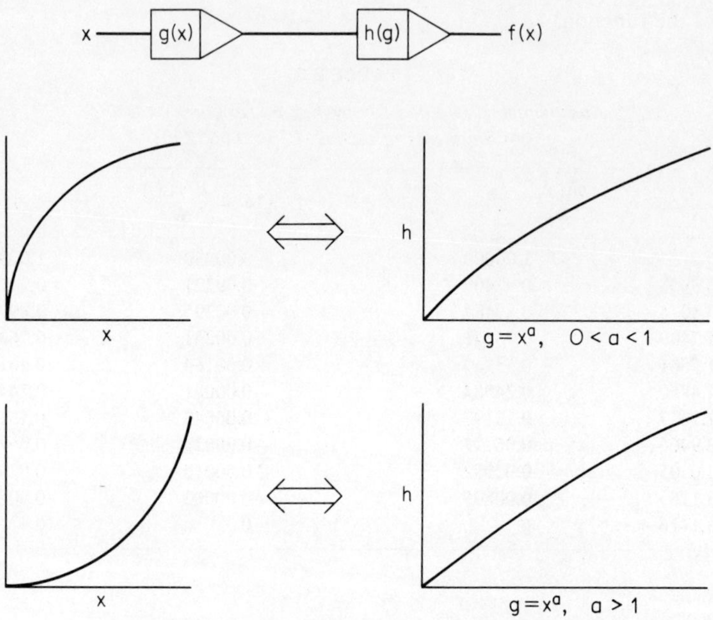

FIG. 8.29 Transforming the independent variable.

There is no inflection point, so we may expect a close approximation to a best fit using (8.10). Generating d^2h/dg^2 with $\dot{g} = 1$ as in Example 5.2, we find its square root, integrate, and hold when $g = \pi^2$. The value of C_i is read directly from the DVM, the normalized graph of $z(g)$ is drawn (Fig. 8.30), each x_i is read from the graph,

TABLE 8.4

Approximate Chebyshev Fit for $(\sin \sqrt{g})/\sqrt{g}$ Using Ten Segments of a DFG, $g = x^2$, $C_1^2/16 = 0.00071$

g_i	$\dfrac{\sin \sqrt{g_i}}{\sqrt{g_i}}$	y_i
0	1.00000	0.99929
0.83	0.86719	0.86648
1.70	0.73981	0.73910
2.59	0.62093	0.62022
3.51	0.50949	0.50878
4.46	0.40591	0.40520
5.46	0.30849	0.30778
6.50	0.21886	0.21815
7.57	0.13824	0.13753
8.69	0.06533	0.06462
π^2	0	−0.00071

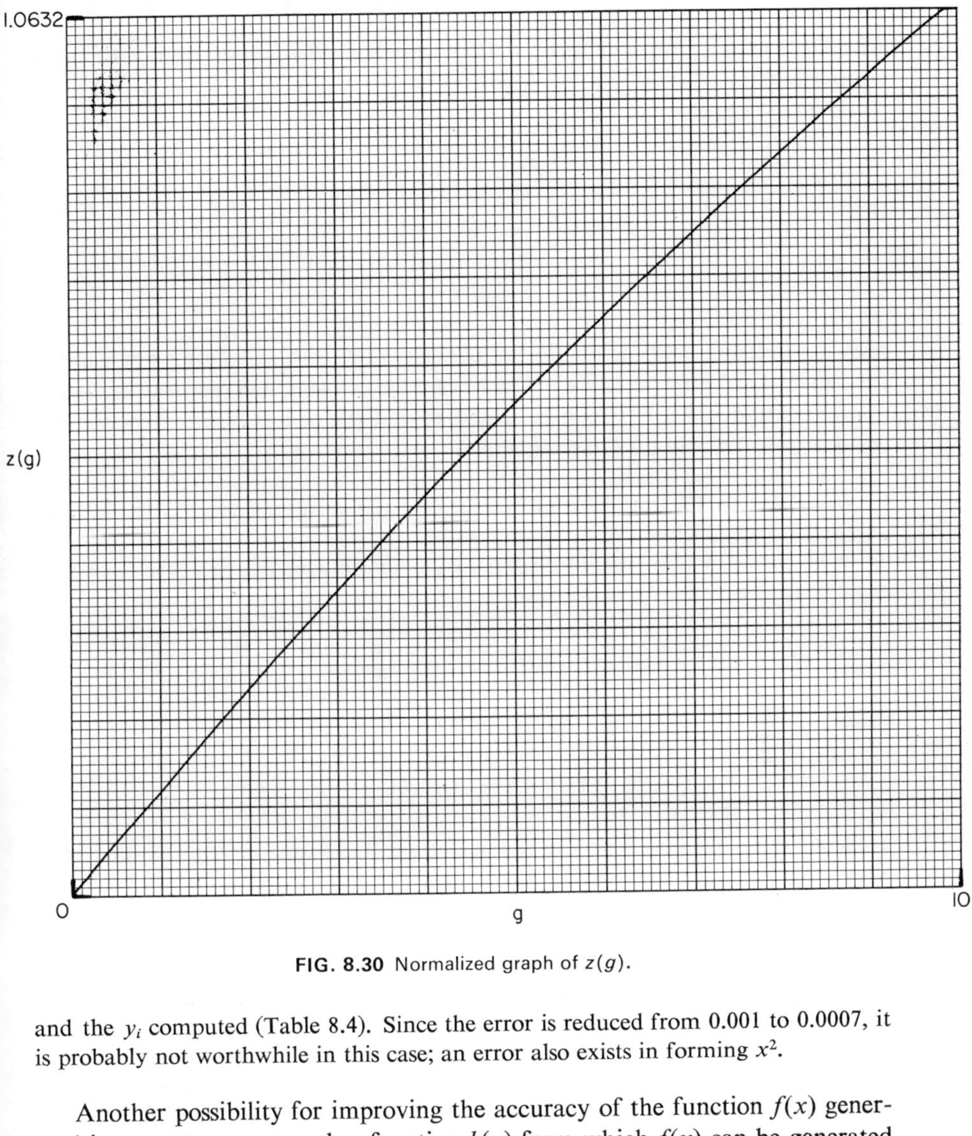

FIG. 8.30 Normalized graph of $z(g)$.

and the y_i computed (Table 8.4). Since the error is reduced from 0.001 to 0.0007, it is probably not worthwhile in this case; an error also exists in forming x^2.

Another possibility for improving the accuracy of the function $f(x)$ generated is to generate some other function $h(x)$ from which $f(x)$ can be generated with other analog equipment. Rubin [8], for example, had considerable success generating the function $f = \exp(b/x)$, $0 < a \leq x < b$ for large ranges by setting up the function $h(x) = f(x)(x + a)^2$. Division by $(x + a)^2$ is then required to generate $f(x)$, but the accuracy improvement for large x was considerable, despite the small error introduced by the division.

The most common situation is generating a correction function $C(x)$:

$$C(x) = f(x) - F(x)$$

where $F(x)$ is an approximation to $f(x)$, which is easily generated with other components.

If $F(x)$ is linear:

$$F(x) = dx + c$$

then

$$\frac{d^2C}{dx^2} = \frac{d^2f}{dx^2}$$

showing that the breakpoints do not change when using all norms nor can the final theoretical error decrease. The advantage of using a correction to a linear approximation is that $C(x)$ can be scaled so that the effects of incorrect settings (or drift, etc.) of the DFG are smaller.

Example 8.5. Fit a straight line to $(\sin x)/x$ and set up the error as found from an approximate Chebyshev fit in Example 7.2.

We see a reasonable straight line approximation (it need not be a "best" one) to be

$$F = 1.10098 - \frac{x}{\pi}$$

so that

$$C(x_i) = y_i - 1.10098 + \frac{x_i}{\pi}$$

Table 8.5 shows the new ordinates C_i for each x_i. Reconstructing $f(x)$ requires that

TABLE 8.5

Correction to a Linear Approximation

x_i	y_i	$C_i = y_i + \frac{x_i}{\pi} - 1.10098$	*Scaled as* $10C_i$
0	1.00098	−0.10000	−1.0000
0.2136	0.99339	−0.03959	−0.3959
0.4367	0.96949	+0.00751	+0.0751
0.6629	0.92933	+0.03935	+0.3935
0.9016	0.87090	+0.05692	+0.5692
1.1592	0.79160	+0.05962	+0.5962
1.4577	0.68261	+0.04563	+0.4563
1.8692	0.51233	+0.00635	+0.0635
2.5196	0.23027	−0.06871	−0.6871
2.8557	0.09777	−0.09421	−0.9421
3.1416	−0.00098	−0.10000	−1.0000

$$f(x) = C(x) + F(x)$$

as shown in Fig. 8.31. The value $10C$ is set up on the generator so that any error is reduced by a factor of 10.

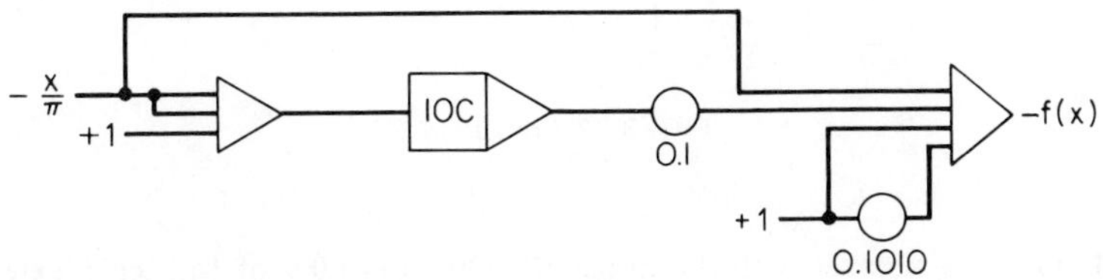

IOC is set up as a function of $-1+\frac{2x}{\pi}$

FIG. 8.31 Correction to a linear approximation.

An approximating function requiring nonlinear equipment changes the second derivative of the function set up on the generator. Not only can we expect to obtain a very small dependence of the generator on the final function, but we may also expect to improve the total theoretical error considerably. The simplest such function is

$$F = p + qx + rx^2$$

where the value of r determines the final theoretical accuracy of function $f(x)$, while p and q are adjusted for a close fit to minimize the dependence of $f(x)$ on the DFG settings. Finding the value r is an optimization problem (Chapter 19) requiring the minimization of one of the integrals depending upon the norm used. A difficulty with this procedure is that the multiplier itself produces error. Thus, the procedure can be used only when accurate multipliers are available.

PROBLEMS

8-1. Show that the unloading circuit in Fig. 8.2 draws no current and effectively presents no load.

8-2. Draw an unscaled diagram to investigate the solution of the forced nonlinear oscillator

$$\frac{d^2y}{dt^2} + ay^3 = b \sin \omega t$$

with initial conditions at $t = 0$ of $\dot{y} = y = 0$, using a servo multiplier to generate y^3.

8-3. Show that Witsenhausen's circuit in Fig. 8.7 can be modified to produce the circuit in Fig. 8.32, so that with $a = 2\sqrt{2}\,y$, the radical term of $z + a$ becomes $(x^2 + y^2)^{1/2}$. Draw the addition to obtain the function $(x^2 + y^2)^{1/2}$. What conditions are required on x and y for stability?

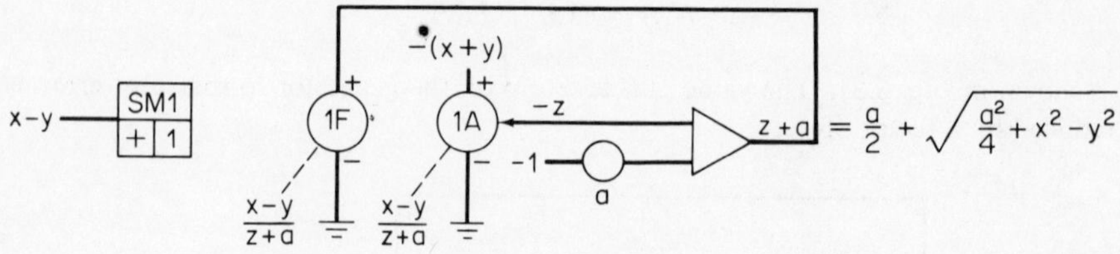

FIG. 8.32 Problem 8-3.

8-4. Given a resolver with the range of $\pm 90°$ for ± 0.9 of full scale, extend the range to $\pm 360°$ by means of switching circuits.

8-5. To the data given in Problem 7-2, find the best least squares fit for a pot padder having three segments. What are the four ordinates to be set up?

8-6. Determine analytically the best Chebyshev fit for a three-segment pot padder for the function $f(x) = e^{-x}$, $0 \leq x \leq 3$.

8-7. Do Problem 8-6 for the best least squares fit.

8-8. By applying Eq. (7.4), determine the proper equations for least squares fitting a set of m data points for n segments of a DFG, $m > n$.

8-9. Using Theorems 8.1 through 8.4, analytically determine approximate best fits for the function $f(x) = e^{-x}$, $0 \leq x \leq 3$, for a six-segment DFG. Assume the weight function e^x where appropriate.

8-10. It is occasionally necessary to generate arbitrary periodic functions of time. Witsenhausen [9] indicates that two DFG's are necessary to minimize transients and that switching is also required. Draw a circuit for such a generation. *Hint:* Use a slope-symmetric triangular wave to drive the DFG's and switch from no. 1 to no. 2 when the slope changes.

8-11. Make a linear approximation to any one of the fitting problems in Problem 8-9 find a correction function for placement on a DFG.

8-12. Using the results of Problem 7-5, construct a circuit to determine the optimum Chebyshev fit to the correction function

$$C(x) = e^{-x} - a_0 - a_1 x - a_2 x^2$$

for use with a six-segment DFG. How does the final error compare with the similar case in Problem 8-9? Assuming a multiplier error of 0.002 and a DFG error of 0.002, what maximum absolute final error can be expected?

REFERENCES

[1] Fifer, S., *Analogue Computation*, vol. 1 (New York: McGraw-Hill Book Co., 1961), p. 137.

[2] Witsenhausen, H., "Utilization Optimum des Multiplieurs Électro-Mécaniques," *Proceedings of the 2nd Inter. Analogue Comp. Meetings* (Brussels: Presses Académiques Européennes, 1959).

[3] Brown, D. R., "Squaring Technique That Maintains Sign of Input Variable," *Simulation*, vol. 6, Jan. 1966.

[4] Kovach, L. D., and W. Comley, "A New Solid-State Nonlinear Analog Component," *IRETEC*, vol. 9, Dec. 1960.

[5] Korn, G. A., and T. M. Korn, *Electronic Analog and Hybrid Computers* (New York: McGraw-Hill Book Co., 1964).

[6] Hamer, H., "Optimum Linear-Segment Function Generation," *Trans. AIEE, part I* (*Comm. and Elect.*), vol. 75, Nov. 1956.

[7] Ream, N., "Approximation Errors in Diode Function-Generators," *J. Electronics and Control*, vol. 7, July 1959.

[8] Rubin, A., "Function Generators," *Princeton Computation Center Report No. 116*, Electronic Associates, Inc., Princeton, N. J.

[9] Witsenhausen, H., "Generation of Arbitrary Functions of Time," *AICA*, vol. 4, April 1962.

See also

[10] Kislitsyn, S. G., and F. L. Litvin, "On Calculating Function-Generating Shunted Potentiometers for Chebyshev Approximations," *Automatika i Telemekhanika*, vol. 20, Nov. 1959.

[11] Deiters, R. M., "Optimum Design of a Squarer by Applying the Criterion of Square Root of the Integral of the Per Cent Error Squared," *IEEETEC*, vol. 14, June 1965.

Chapter 9

GENERATING IMPLICIT FUNCTIONS AND FUNCTIONS OF TWO VARIABLES

The last remaining formal function generating technique for analog computers is that of generating a function implicitly by solving one algebraic equation. This powerful technique was discovered early in the art of computation [1], [2]. It uses a high-gain amplifier to solve the algebraic equation, which defines the function to be generated implicitly, in terms of other easily generated functions.

9.1 Implicit Function Techniques—Stability and Dynamic Errors

The implicit function generation technique uses a high-gain amplifier as shown in Fig. 9.1. The function $F(x, \tau)$ must be generated by some fixed function method. The τ is not used to suggest that time explicitly appears in F; F

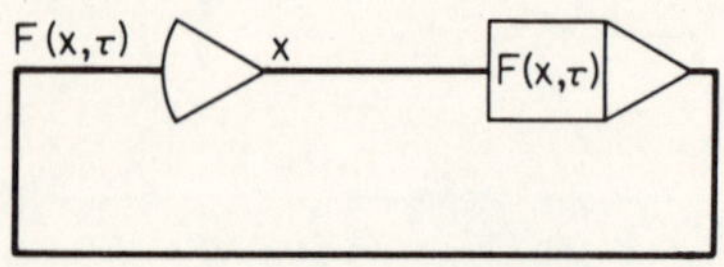

F(x,τ) = 0 because of high gain connection.
Note: F(x,τ) must be generated by a fixed function generator.

FIG. 9.1 Forcing $F(x, \tau) = 0$.

contains parameters changing with time. As described in Sec. 2.2.3, the amplifier does not overload because F is forced very close to zero, so close that we use the equality sign. The general rule is simply stated:

When $F(x, \tau)$ is at the input of a high-gain amplifier, x adjusts itself so that $F(x, \tau) = 0$ if the circuit is stable.

9.1.1 Circuit stability

A high-gain amplifier behaves very much as an integrator with an extremely small time constant (or large input gain), and Fig. 9.2 is a simple model of the circuit. The equation of the model is

$$\dot{x} = -\mu F(x, \tau) \tag{9.1}$$

where μ is very large but can be made smaller. Thus the circuit stability can be investigated by the mathematical theory of stability of differential equations.

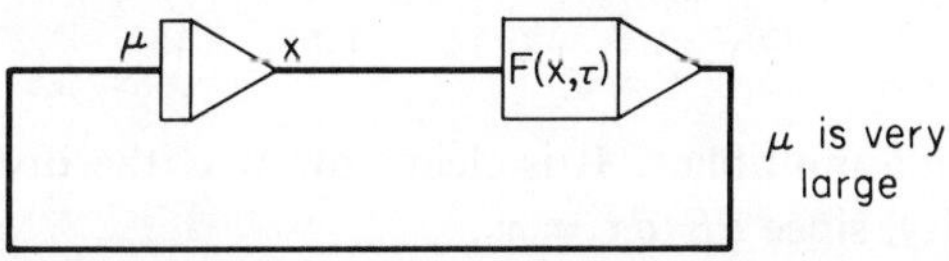

FIG. 9.2 Equivalent circuit of implicit function generation.

This situation is discussed at length by El'sgol'ts [3], but we may intuitively determine the stability condition by applying the principle of negative feedback:

(1) If $F > 0$, we wish F to decrease when x decreases because the output x will decrease.

(2) If $F < 0$, we wish F to increase when x increases because the output x will increase.

These conditions are summarized by the simple single condition:

Theorem 9.1. A necessary condition for the stability of (9.1) in a neighborhood of $F(x, \tau) = 0$ is that $\partial F/\partial x > 0$ whenever $F \neq 0$.

We note that this condition must hold for the circuit in Fig. 9.1, but because of "delays" in the loop [depending on how $F(x, \tau)$ is generated], it is

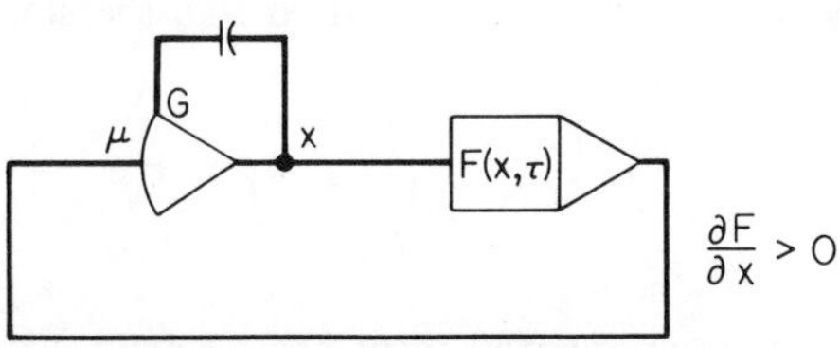

FIG. 9.3 Feedback capacitor sometimes needed for stability.

sometimes necessary to lower the gain μ by introducing a small feedback capacitor as shown in Fig. 9.3. In effect, we are forcing the time constants of other amplifiers used in the loop to be near zero by comparison, so that (9.1) applies more closely. Presumably, no other high-gain amplifiers are in the loop, for we would then be solving more than one algebraic equation in as many unknowns. There are similar conditions for stability in that case, but discussion of such conditions is deferred until we specifically discuss the solution of algebraic equations (Chapter 11).

Any required small feedback capacitor will always be omitted from the diagrams, it being understood that it may be necessary. Also, an actual integrator with a small capacitor could be used in all cases and should be used if an initial condition is important.

We have seen two cases where high-gain amplifiers have already been used. In Sec. 2.2.3, a pot in a feedback loop was used to obtain large voltage gains, and in Sec. 2.4, division of y_2 by y_1 was performed by solving the equation

$$F(x, \tau) \equiv y_1(\tau)x - y_2(\tau) = 0$$

implicitly by using a multiplier. It is clear now why the divisor $y_1(\tau)$ must be positive for stability, since $\partial F/\partial x = y_1$.

9.1.2 Dynamic errors

Although we wish to solve $F(x, \tau) = 0$, we actually produce a computer solution $x_c(\tau)$, which is slightly different, because μ is finite and F is changing with time. A dynamic error Δx is introduced so that $\Delta x = x - x_c$. Let us examine this possible error by expanding $F(x, \tau)$ in a Taylor series about x_0 and τ_0, retaining only the linear terms. From (9.1),

$$\dot{x} \simeq -\mu[F_0 + F_x(x - x_0) + F_\tau(\tau - \tau_0)] \tag{9.2}$$

where the symbols are defined as $F_0 = F(x_0, \tau_0)$, $F_x = \partial F/\partial x$ at x_0 and τ_0, and $F_\tau = \partial F/\partial \tau$ at x_0 and τ_0. Assuming an equality sign and constant partial derivatives, the solution of (9.2) can be shown to be

$$x_c = Ce^{-\mu F_x \tau} - \frac{F_\tau}{F_x}\tau + \frac{F_\tau}{\mu F_x^2} + \frac{F_x x_0 + F_\tau \tau_0 - F_0}{F_x} \tag{9.3}$$

where C is such that $x_c = x_0$ when $\tau = \tau_0$. Because μ is large, the exponential term quickly disappears. From (9.2), we would like the small-value solution to be $\dot{x} = 0$, or

$$x = \frac{F_x x_0 + F_\tau(\tau_0 - \tau) - F_0}{F_x}$$

From (9.3), assuming the exponential term has decayed, we have

Theorem 9.2. The first-order dynamic error Δx introduced in the solution of (9.1) due to finite μ is given by

$$\Delta x = \frac{-F_\tau}{\mu F_x^2} \tag{9.4}$$

if the equation is stable.

Equation (9.4) is useful for estimating the approximate error at any instant in high-gain amplifier connections, particularly if a feedback capacitor is included to reduce μ.

In the case of division, the dynamic error Δx is given by

$$\Delta x = -\frac{(\dot{y}_1 x_0 - \dot{y}_2)}{\mu y_1^2}$$

and since $x_0 \simeq y_2/y_1$, then for a division circuit,

$$\Delta x = -\frac{(\dot{y}_1 y_2 - \dot{y}_2 y_1)}{\mu y_1^3} = \frac{1}{\mu y_1}\frac{d}{d\tau}\left(\frac{y_2}{y_1}\right)$$

For large enough μ or small enough time scale factor (which reduces $\dot{y}_1$ and $\dot{y}_2$), the error can be made quite small, as is true in general from (9.4). This dynamic error is usually much smaller than are multiplication errors.

It should be emphasized that any set of summing or high-gain amplifiers in a closed loop containing no integrators constitutes an implicit computation, since a summer is a high-gain amplifier with a unity feedback. Therefore, it may sometimes be necessary to reduce μ in such cases by introducing a small feedback capacitor (Fig. 9.4).

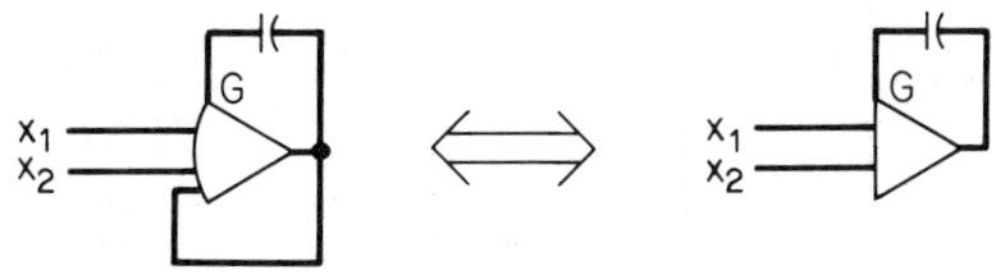

FIG. 9.4 Summer in an implicit computation.

This technique does not complete the subject of function generation, for we shall continually allude to techniques that permit more accurate and/or inclusive methods. Thus, the transfer function mechanization (Chapter 10), polynomial and algebraic equation solutions (Chapter 11), corrective systems (Chapter 13), etc., all are very much related to generating functions. In fact, analog computation itself pertains to generating functions, although the final solution cannot be so readily checked for errors as the functions discussed.

9.2 Typical Circuits Using Implicit Function Techniques

Implicit function techniques are most often used to generate inverses of fixed functions and roots of simple algebraic equations. We first discuss some applications using servo multipliers [4].

9.2.1 Servo multiplier circuits

When a divisor y is negative, x/y can be generated with y driving the shaft of a servo patched in the "$-$" connection as in Fig. 9.5. We have

$$F(Q, \tau) = x(\tau) - Q\,y(\tau)$$

with $\partial F/\partial Q = -y > 0$.

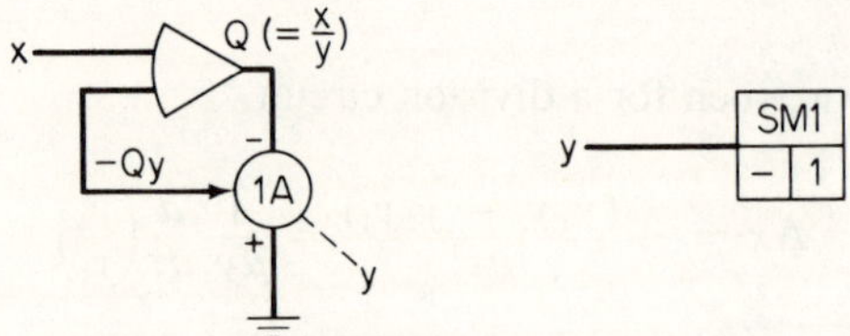

FIG. 9.5 Servo division for negative divisor.

By using absolute-value circuits in a normal connection, a four-quadrant servo division circuit is possible (Fig. 9.6). In that case, the circuits are stable except when $y = 0$.

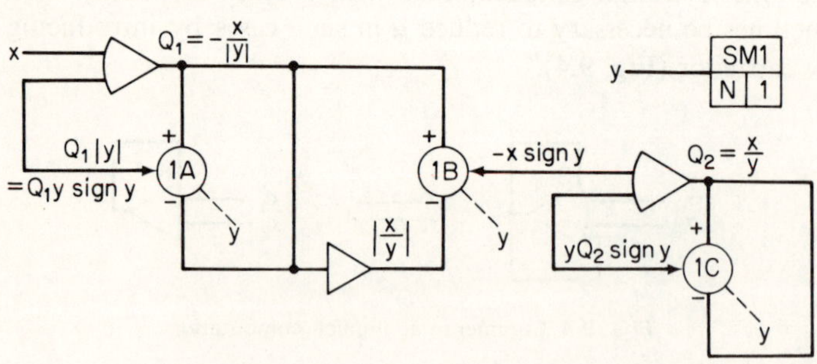

FIG. 9.6 Four-quadrant servo division circuit.

Division of x by $a + y$ requires the solution of the implicit equation

$$Q(a + y) - x = 0$$

When $a + y > 0$, the circuit of Fig. 9.7 allows for $a > 0$ and slightly negative y. By feeding $+x$ into the high-gain amplifier, $-x/(a + y)$ is produced

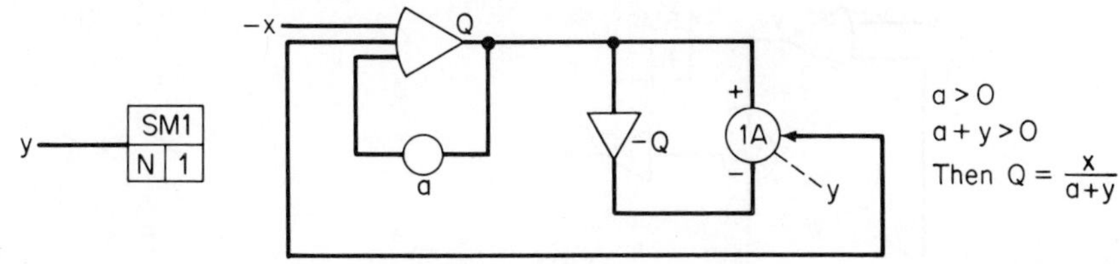

FIG. 9.7 Generating $x/(a+y)$.

at its output. In either case both polarities are generated because of the inverter.

To generate the square root of x, we solve one of the implicit equations

$$F_1 \equiv y^2 - x(\tau) = 0, \qquad F_2 \equiv -y^2 + x(\tau) = 0$$

The final circuit depends on whether we wish $+\sqrt{x}$ or $-\sqrt{x}$. Since

$$\frac{\partial F_1}{\partial y} = 2y, \qquad \frac{\partial F_2}{\partial y} = -2y$$

the circuits shown in Fig. 9.8 are both common.

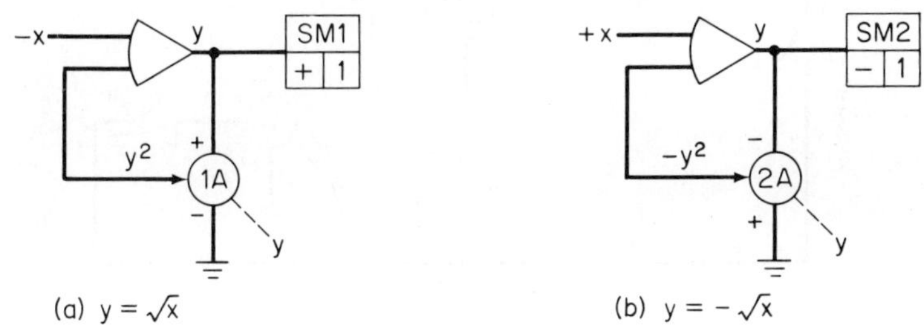

FIG. 9.8 Square-root circuits.

The nth root of x is found by solving the equation

$$F \equiv y^n - x = 0$$

When n is even, we have

$$\frac{\partial F}{\partial y} = ny^{n-1} \qquad \begin{array}{l} > 0 \text{ for positive root} \\ < 0 \text{ for negative root} \end{array}$$

Hence we must decide beforehand whether the positive or negative root is required and feed $+F$ or $-F$ into the high-gain amplifier input (Fig. 9.9). The fractional powers $x^{2/n}, x^{3/n}, \ldots, x^{(n-1)/n}$, are generated simultaneously.

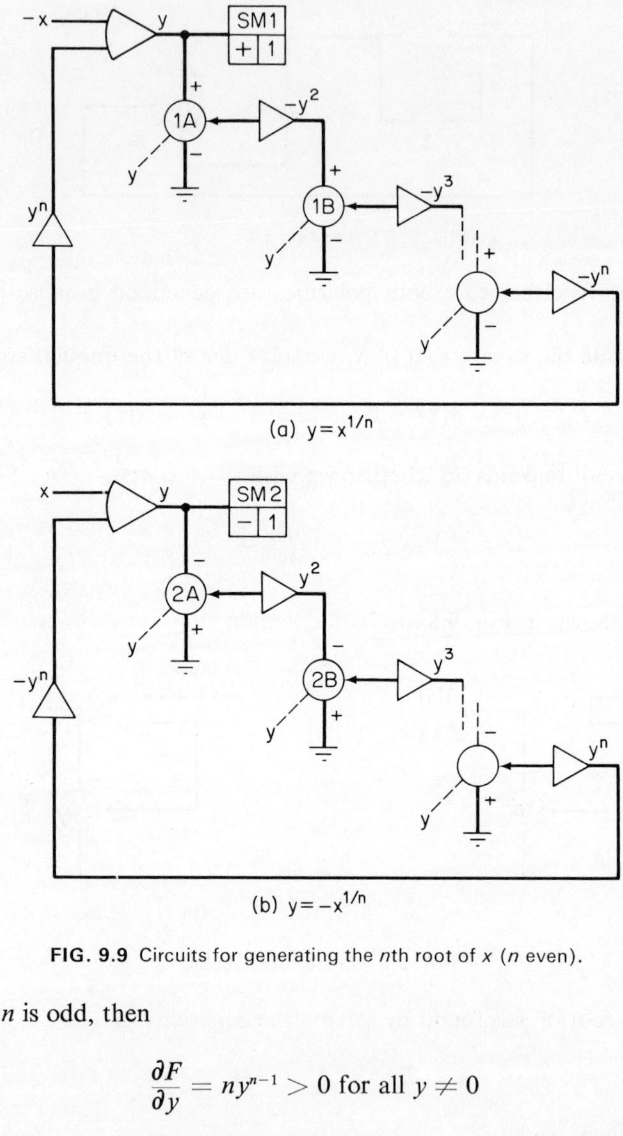

FIG. 9.9 Circuits for generating the *n*th root of x (n even).

When n is odd, then

$$\frac{\partial F}{\partial y} = ny^{n-1} > 0 \text{ for all } y \neq 0$$

Solutions exist here for positive and negative x, so the optimum circuit depends upon what is needed. If only the nth root of x is needed, the circuit in Fig. 9.10 eliminates the many inverters usually associated with normal connections. The functions $x^{m/n}$ with m odd and $m \leq n$ are also generated with the proper polarity, but the other powers are proper only when $y > 0$. If y^m, m even, is needed, extra inverters will be required. For example, to generate $(d^2y/dx^2)^{2/5}$ as is required when finding an unweighted least squares piecewise linear fit to

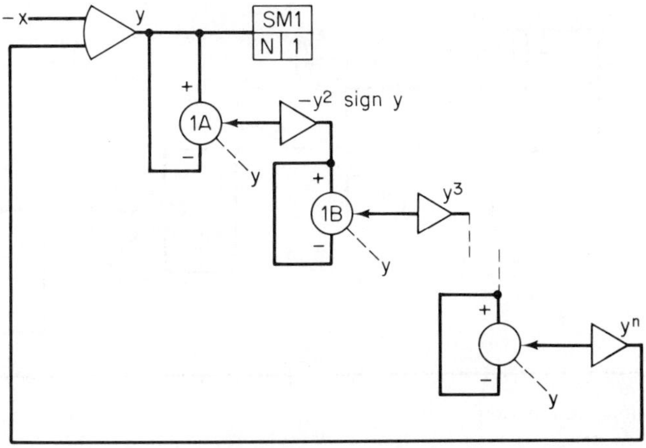

FIG. 9.10 Generating the nth root of x (n odd).

$f(x)$ from Theorem 8.2, the circuits of Fig. 9.11 are both applicable. We see also that any polynomial may be used as an input to a high-gain amplifier. The output is a root of the polynomial provided the circuit is stable. For example, if

$$F \equiv x^2 + y_1 x + y_2 = 0$$

the positive root

$$x = \frac{-y_1 + \sqrt{y_1^2 - 4y_2}}{2} \tag{9.5}$$

is found with the circuit of Fig. 9.12. The stability requirement that $2x + y_1 > 0$ is automatically satisfied, from (9.5), when $y_2 < 0$.

9.2.2 Fixed and arbitrary function generator circuits

With the fixed function cards producing x^2 and x^4, the positive inverses are easily generated as in Fig. 9.13. In fact, this principle applies to any monotonic function set up on arbitrary function-generating equipment. If, for example, we set up $(\sin x)/x$ on a DFG, the inverse is easily found by the circuit in Fig. 9.14. An inverter is required because the derivative of $(\sin x)/x$ is ≤ 0 in the range considered. Poor results are obtained near $x = 0$ (if y is near 1) because the slope is zero at that point. This is true also for the circuits in Fig. 9.13 at $x = 0$, since in all cases $dx/dy \rightarrow \infty$ as $x \rightarrow 0$. *Implicit methods invariably are poor when $\partial F/\partial x = 0$ at any x*, not only because the dynamic response is poor, but also because the slope is never quite set to zero at these points. Since the segments are straight lines, any one with zero slope could not null at the input of the amplifier and is therefore unstable in this region.

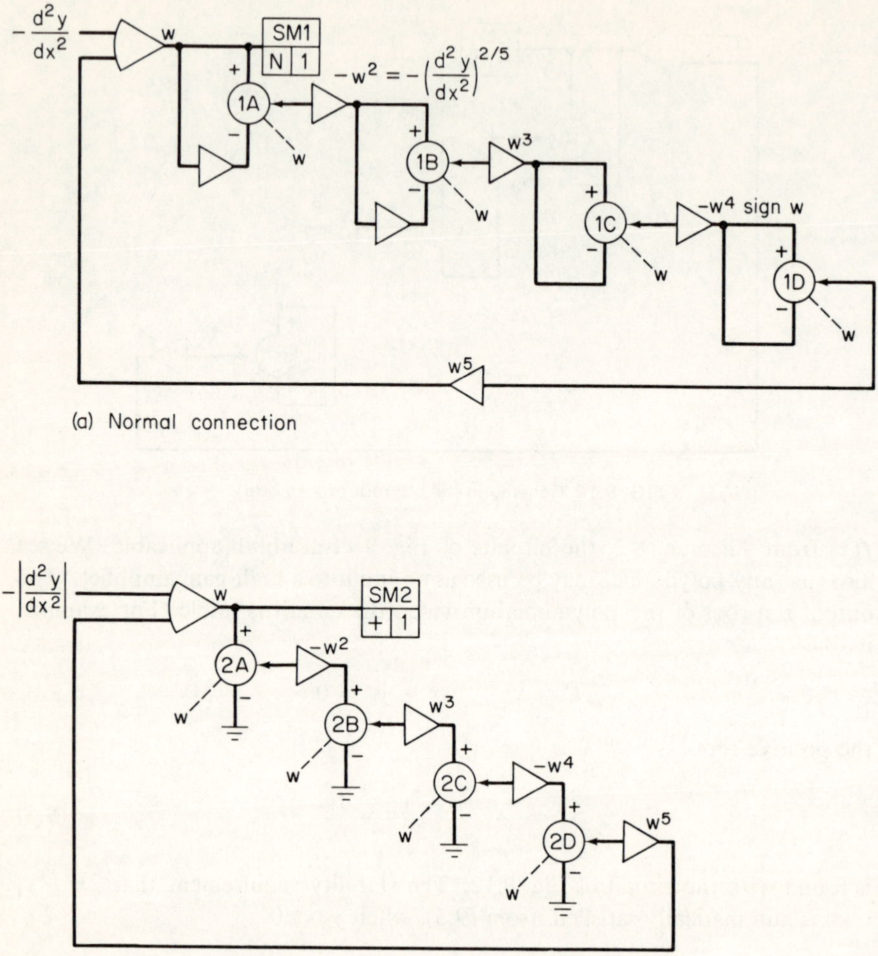

(a) Normal connection

(b) Plus connection

FIG. 9.11 Generating $(d^2y/dx^2)^{2/5}$.

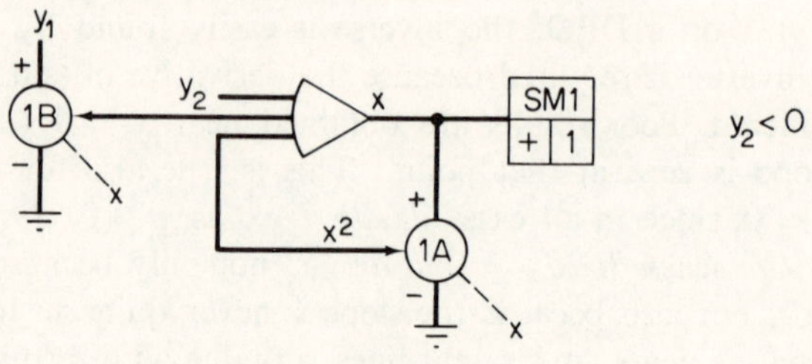

FIG. 9.12 Solving for $x^2 + y_1x + y_2 = 0$.

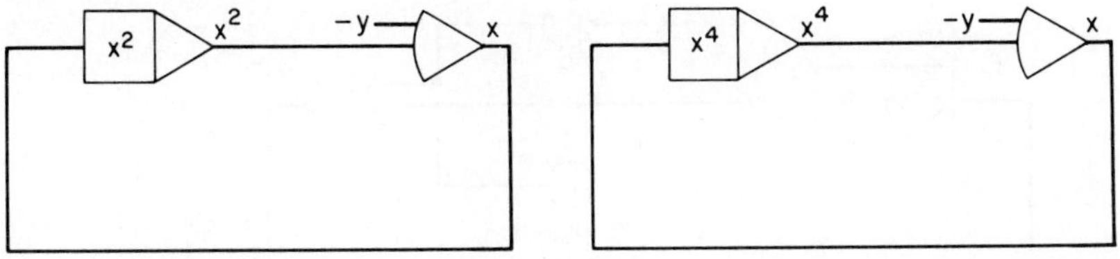

(a) $x = y^{\frac{1}{2}}$ (b) $x = y^{\frac{1}{4}}$

FIG. 9.13 Common inverses with fixed function cards.

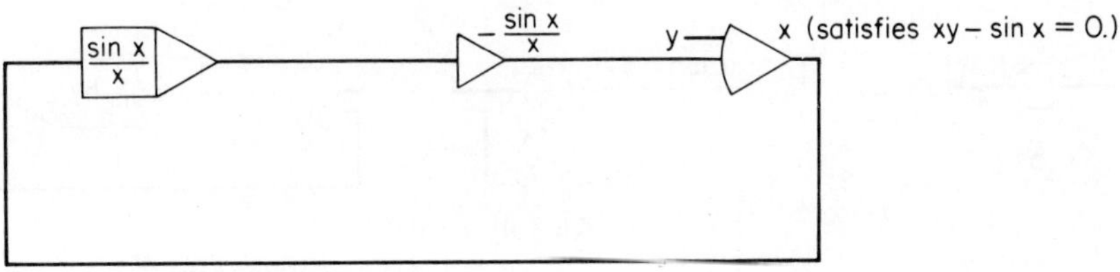

FIG. 9.14 Solution for $(\sin x)/x - y = 0$ when $(\sin x)/x$ is on a DFG.

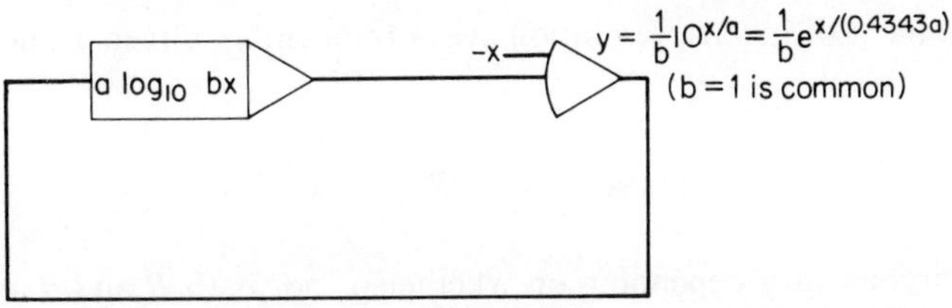

FIG. 9.15 Generation of $y = (1/b)[10]^{x/a} = (1/b)e^{x/(0.4343a)}$.

Exponentials can be generated with a log card producing $a \log_{10} bx$ as shown in Fig. 9.15. With two additional pots, the exponent and amplitude may be controlled, since

$$a \log_{10} bx = a \log_{10} b + a \log_{10} x$$

Figure 9.16 shows the diagram for producing $y = de^{cx/(0.4343a)}$. Since the derivative of $a \log_{10} bx$ is positive for all $x > 0$, these circuits are quite stable.

An important use of log cards is that of raising x to some arbitrary power p and obtaining x^p. If $y = x^p$

$$a \log_{10} by = ap \log_{10} b\left(\frac{x}{b}\right) + a \log_{10} b$$

so that y can be generated from the circuit of Fig. 9.15. Figure 9.17 shows the combined circuits. Since log cards are applicable only for $0 < x_{\min} \leq x \leq 1$, these circuits are never adequate for inputs too close to zero.

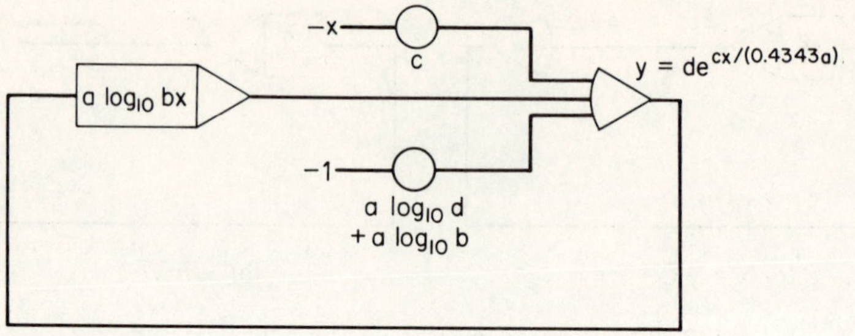

FIG. 9.16 Generating $y = de^{cx/(0.4343a)}$

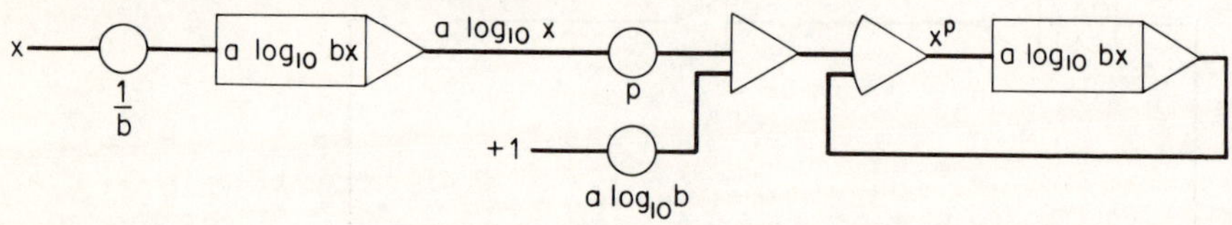

FIG. 9.17 Generating $y = x^p$ with 2 log cards.

9.2.3 Inverse resolver circuits

The problem of inverse resolution arises frequently: Given x and y, determine

$$\theta = \arctan \frac{y}{x}, \qquad R = \sqrt{x^2 + y^2} \tag{9.6}$$

Proposed circuits vary depending on whether or not both R and θ are needed, and if θ is needed, will $\sin \theta$ or $\cos \theta$ do. We first show that

$$R = y \sin \theta + x \cos \theta \tag{9.7}$$

From (9.6)

$$x \sin \theta - y \cos \theta = 0, \qquad R^2 = x^2 + y^2$$

By squaring

$$x^2 \sin^2 \theta - 2xy \sin \theta \cos \theta + y^2 \cos^2 \theta = 0$$

or

$$(R^2 - y^2) \sin^2 \theta - 2xy \sin \theta \cos \theta + (R^2 - x^2) \cos^2 \theta = 0$$

which leads to

$$\begin{aligned} R^2 &= y^2 \sin^2 \theta + 2xy \sin \theta \cos \theta + x^2 \cos^2 \theta \\ &= (y \sin \theta + x \cos \theta)^2 \end{aligned}$$

A check into the polarity of R as defined by (9.7) shows it is always non-negative in all four quadrants. It now follows quickly that

$$\sin\theta = \frac{y}{R}, \qquad \cos\theta = \frac{x}{R} \tag{9.8}$$

which are the usual polar coordinate relationships.

Method 1. θ not needed. Compute R directly from (9.6) and if $\sin\theta$, $\cos\theta$ or both are needed, compute these from (9.8) (Fig. 9.18). The divisions and

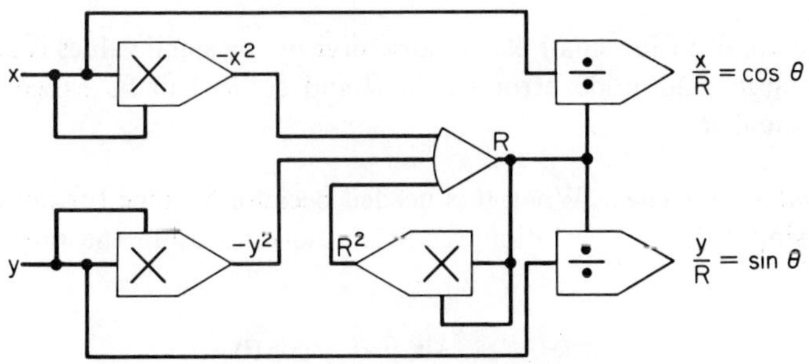

FIG. 9.18 Direct algebraic method for inverse resolution.

square roots involve direct feedback permitting very large μ so that dynamic response is excellent. Because the error of a square root circuit is large for small R, however, large errors may occur in this region.

Method 2. θ not needed. We use an implicit method [5]:

$$\dot{R} = -\mu\left(R - \frac{x^2 + y^2}{R}\right)$$

as shown in Fig. 9.19. Since

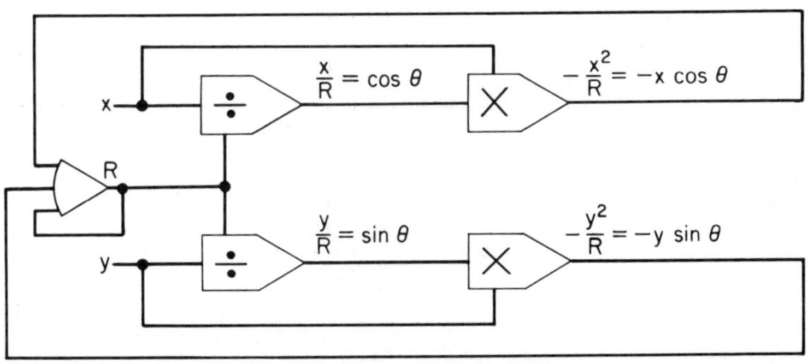

FIG. 9.19 Circuit for method 2.

$$F_R = 1 + \frac{x^2 + y^2}{R^2} \simeq 2$$

$$F_\tau = -\frac{2(x\dot{x} + y\dot{y})}{R}$$

the approximate dynamic error ΔR is

$$\Delta R \simeq \frac{(x\dot{x} + y\dot{y})}{2\mu R} = \frac{\dot{x}\cos\theta + \dot{y}\sin\theta}{2\mu}$$

which is small even for small R. Because division by small values is inaccurate, however, large magnitude error (in $\sin\theta$ and $\cos\theta$) can be expected in the region of small R.

Method 3. θ needed. When θ is needed because of some further computation [say $\sin(a\theta)$], the resolution requires resolvers, using the implicit equation [6]

$$\dot{\theta} = -\mu(x\sin\theta - y\cos\theta)$$

as shown in Fig. 9.20. The resolvers have limited range causing discontinuities at the limits. It is possible for θ to jump from $+180°$ to $-180°$ or vice versa. The first-order dynamic error is

$$\Delta\theta = -\frac{F_\tau}{\mu(F_\theta)^2} = -\frac{(\dot{x}\sin\theta - \dot{y}\cos\theta)}{\mu R^2}$$

which can have poor characteristics near $R = 0$. The value of R can now be found also with two multipliers and a summer.

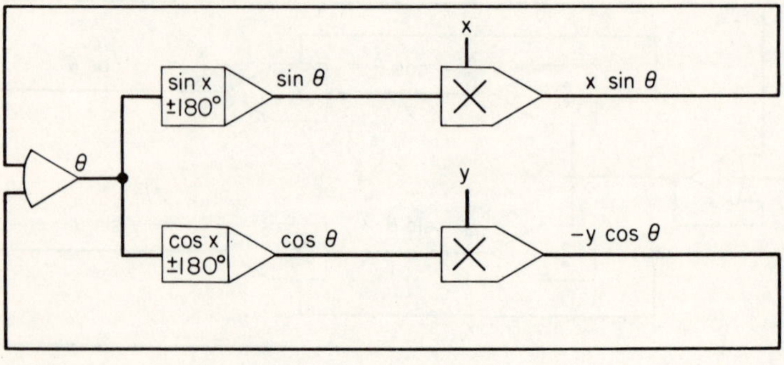

FIG. 9.20 Obtaining θ when needed.

9.3 Multiple Solutions in Implicit Computations

If $\partial F/\partial x < 0$ near the root of $F(x, \tau)$ we solve

$$\dot{x} = +\mu F(x, \tau)$$

But if both polarities of $\partial F/\partial x$ occur during a computation, special consideration should be given to the circuit, since an instability may exist when $\partial F/\partial x = 0$. This occurs if both polarities are possible and is also an indication that multiple solutions may exist in the range of computation. For example, if we wish to generate $y = \arccos x$ by implicitly solving

$$\dot{y} = -\mu(x - \cos y) \tag{9.9}$$

using a resolver with $\pm 180°$ range, two solutions are possible for $-1 \leq x \leq 1$. Furthermore, (9.9) is stable only if $y > 0$, so that no negative roots are obtained, even if needed, unless the polarity of the input is reversed. Occasionally, it is convenient to obtain one circuit for all the possible roots, with the one obtained being the closest to y. We have the following simple built-in sign changer.

Theorem 9.3. The equation

$$\dot{x} = -\mu F(x, \tau)\frac{\partial F(x, \tau)}{\partial x} \tag{9.10}$$

is always stable near a root of $F(x, \tau)$. Since

$$\begin{aligned}\frac{\partial}{\partial x}[F(x, \tau)F_x(x, \tau)] &= F_x^2(x, \tau) + F(x, \tau)F_{xx}(x, \tau)\\ &= F_x^2(x, \tau) \text{ when } F = 0\\ &\geq 0\end{aligned}$$

Actually, this is a special case of the method of steepest descent (Chapter 11) used often for finding solutions to algebraic equations. However, the first-order dynamic error now increases to $\Delta x = -F_\tau/(\mu F_x^3)$, which can be a problem when F_x is near zero. Because of this, many prefer to use the equation

$$\dot{x} = -\mu F(x, \tau) \operatorname{sign}\left[\frac{\partial F(x, \tau)}{\partial x}\right]$$

as a sign changer. Stability characteristics are preserved as before, but the first-order dynamic error reduces to (9.4).

9.4 Functions of More Than One Variable

It is often necessary to generate functions of two or more variables that agree with experimental data. For example, the drag coefficient of an aircraft is a function of Mach number, angles of attack and slideslip, as well as elevator, rudder and aileron positions, all obtained in point set form by wind tunnel tests. There is an inherent difficulty in generating functions of two or more variables. Without a functional form to approximate the data, arbitrary function generators of several variables are needed. Although Wilkinson [7] showed that diode-resistor networks can be used to generate a general n-dimensional polyhedron for functions of n variables, no commercial unit to do so conveniently has ever been produced. In general, arbitrary functions of more than two variables are approximated by functions of one or two variables and will not be considered here.

Early methods generating arbitrary functions of two variables included a three-dimensional servo controlled cam in a mechanical arrangement and a wire pickoff device [8], both of which were somewhat inflexible and tedious to use. About the only commercial unit supplied by some manufacturers uses resistive sheets (conductive paper) [9], where an x-y plotter is modified to accept a probe in place of a pen. On the resistive sheet, equipotential lines are drawn in silver (or other conducting) ink. When the arm and pen servos position the probe according to the values of x and y, a functional output $z(x, y)$ is obtained that is an approximate linear interpolation between the lines (see Fig. 9.21).

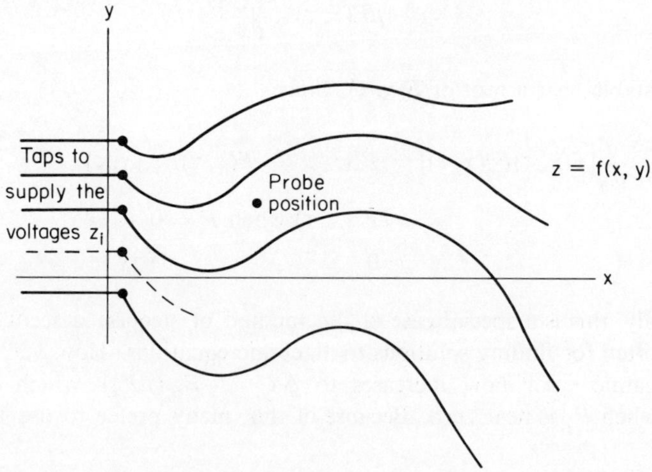

FIG. 9.21 Resistive sheet generation of an arbitrary function of two variables.

In an effort to generate high-speed functions of two variables, many designs were devised [10]–[14]. These require modifications of diode function generators or some specialized equipment and will not be discussed here.

The principal method used to generate a function $z = f(x, y)$ of two variables is by approximation using functions of single variables in combination. The functions are generated by any one of the methods previously described. It is the determination of these functions and how to combine them that comprise the difficult task.

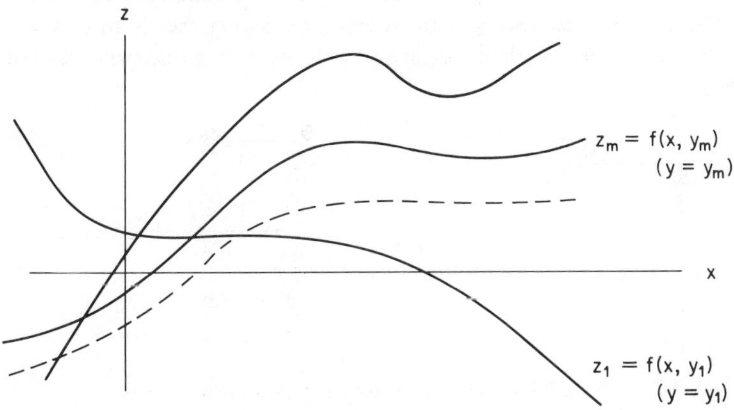

FIG. 9.22 Arrangement of data.

In what follows, it is more convenient to arrange the data as a family of curves with a set of fixed y, say $y_1, y_2, \ldots, y_m$, from which the set of m functions

$$z_i(x) = f(x, y_i), \qquad i = 1, 2, \ldots, m$$

are available as in Fig. 9.22. We also assume

$$y_1 < y_2 < \ldots < y_m$$

9.4.1 Linear interpolation

Since there is a very smooth transition between $z_i(x)$ and $z_{i+1}(x)$ for most practical cases, our first inclination is to interpolate linearly between these curves:

$$f(x, y) \simeq \sum_{i=1}^{m-1} \left\{ z_i(x) + \frac{(y - y_i)}{(y_{i+1} - y_i)} [z_{i+1}(x) - z_i(x)] \right\} \times U(y - y_i)U(y_{i+1} - y) \tag{9.11}$$

If a tapped servo potentiometer is available and servo speeds suffice in the problem, natural interpolation [15] is achieved by prearranging $y_{i+1} - y_i$ to be constant and supplying the voltages $z_i(x)$ at each tap (Fig. 9.23).

When higher speeds are needed, we may mechanize (9.11) directly by recognizing that the summation contains only one nonzero term. If $y_j \leq y \leq y_{j+1}$, then

$$f(x, y) \simeq z_j(x) + \frac{(y - y_j)}{(y_{j+1} - y_j)} [z_{j+1}(x) - z_j(x)]$$

Since there are $m - 2$ switching points, and three functions are needed, Fig. 9.24 shows a relay switching arrangement based on the circuit in Fig. 6.15. Only one multiplier is then required with $m - 2$ triple-pole double-throw switches.

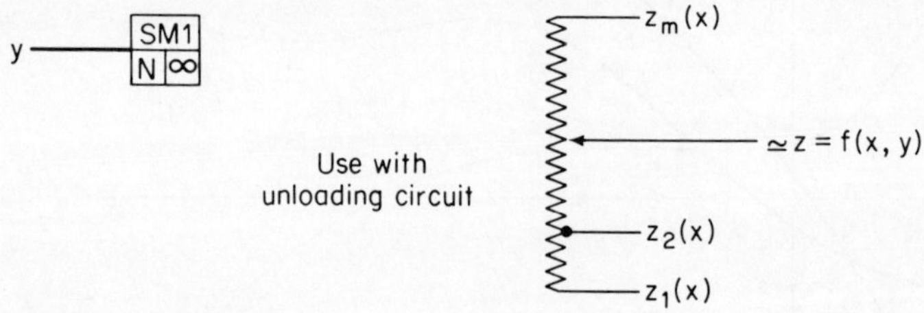

FIG. 9.23 Interpolation with a tapped potentiometer.

To use electronic switching, three sets of circuits as in Fig. 6.26 are required, transmitting $(y - y_j)/(y_{j+1} - y)$, z_j, and z_{j+1} as in the relay case. One multiplier is again required. It is also possible, however, to arrange a mechanization requiring m multipliers and m idealized diode circuits [16] but no switching. It is clear that a high-speed generation of $f(x, y)$, in any case, requires an enormous amount of equipment for m sufficiently large.

9.4.2 Quadratic interpolation

To reduce the number of functions of a single variable required, as well as the amount of switching or multipliers, Levitt proposed using quadratic instead of linear interpolation [17]. This process involves passing a quadratic polynomial through three successive curves, $z_j(x)$, $z_{j+1}(x)$, and $z_{j+2}(x)$. If $y_j \leq y \leq y_{j+2}$, we find that

$$f(x, y) \simeq z_j(x) + \frac{(y - y_j)}{(y_{j+1} - y_j)} f_{1,j}(x) + \frac{(y - y_j)}{(y_{j+1} - y_j)} \frac{(y - y_{j+1})}{(y_{j+2} - y_{j+1})} f_{2,j}(x) \quad (9.12)$$

where $f_{1,j}(x)$ is chosen so that $z = z_{j+1}$ when $y = y_{j+1}$, and $f_{2,j}(x)$ is chosen so that $z = z_{j+2}$ when $y = y_{j+2}$. In terms of z_j, these calculate to be

$$f_{1,j}(x) = z_{j+1}(x) - z_j(x) \quad (9.13)$$

$$f_{2,j}(x) = \frac{\{[z_{j+2}(x) - z_j(x)](y_{j+1} - y_j) - [z_{j+1}(x) - z_j(x)](y_{j+2} - y_j)\}}{(y_{j+2} - y_j)} \quad (9.14)$$

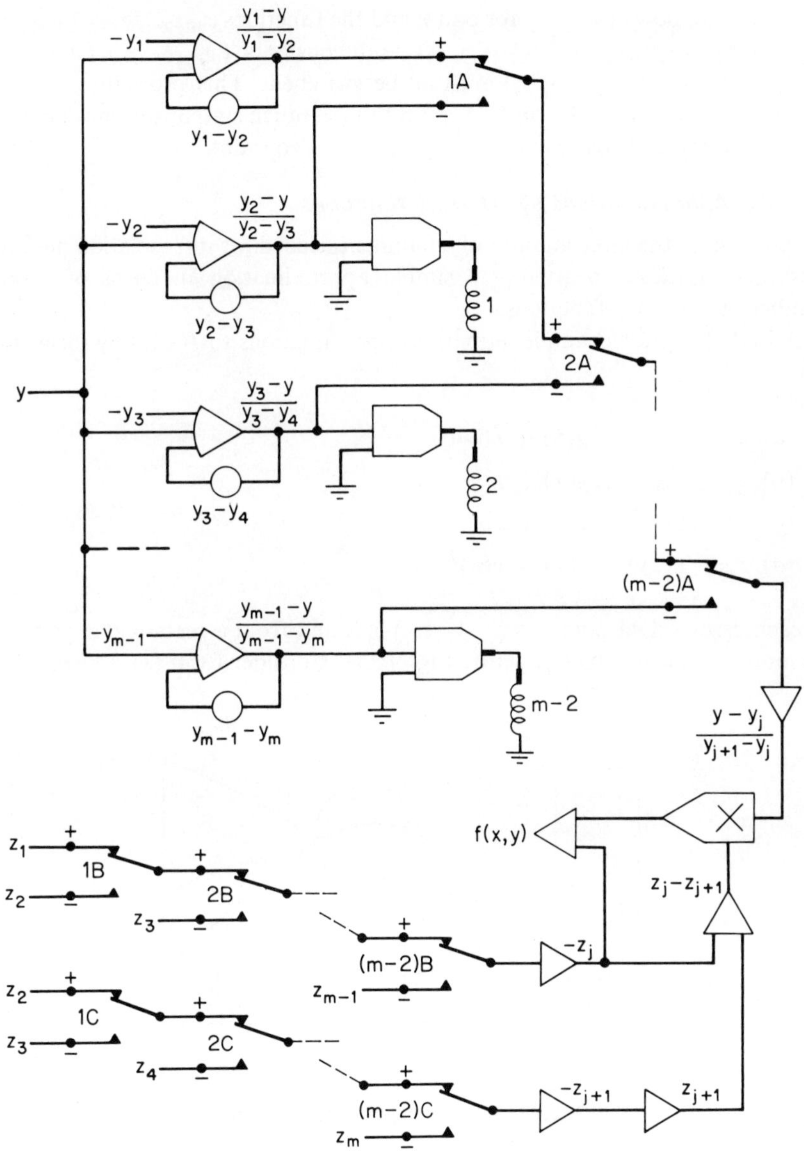

FIG. 9.24 Interpolation by switching using one multiplier.

For $m = 3$, the functions f_1 and f_2 can be supplied directly to (9.12) in terms of z_1, z_2, and z_3. Levitt also suggests a modification of (9.12) for a simple switching system when $m = 5$. For larger m, (9.13) and (9.14) must be formed after the proper z_j are switched in order to require only two multiplications. The switching points are now at $y = y_j$, for odd j, and the functions z_j, z_{j+1}, z_{j+2}, $(y - y_j)/(y_{j+1} - y_j)$, $(y - y_{j+1})/(y_{j+2} - y_{j+1})$ and constants $(y_{j+1} - y_j)/(y_{j+2} - y_j)$ and $(y_{j+2} - y_{j+1})/(y_{j+2} - y_j)$ must all be switched. This procedure is therefore not recommended for $m > 5$, even though much simplification is achieved if it is prearranged that each $y_{i+1} - y_i = \Delta y =$ constant.

9.4.3 Approximations by arbitrary functions

Because of the large amount of equipment needed for interpolation methods, it is very desirable to attempt a simple approximation in terms of a small number of arbitrary functions.

Kindle [18], for example, attempts approximations to $f(x, y)$ by forms such as

$$\begin{aligned}
&\text{(a)}\quad z = f(x, y) \simeq g(x) + h(y) \\
&\text{(b)}\quad z = f(x, y) \simeq g(x)h(y) \\
&\text{(c)}\quad z = f(x, y) \simeq A + g(x)h(y) \\
&\text{(d)}\quad z = f(x, y) \simeq u(x) + g(x)h(y)
\end{aligned} \tag{9.15}$$

In each case, a data point set $(x_i, y_j, z_{i,j})$ is found from a rectangular grid of n values of x and m values of y (see Fig. 9.25). Consider form (a) above, which

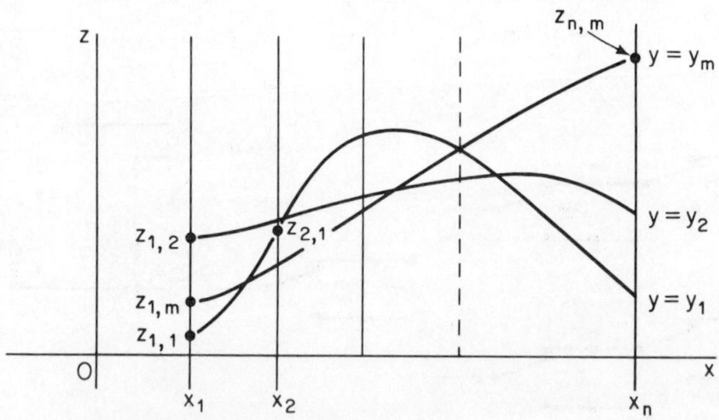

FIG. 9.25 Obtaining a point set $(x_i, y_j, z_{i,j})$.

is suitable when the curves are nearly parallel. We wish to find n values of g and m values of h so that

$$z_{i,j} \simeq g(x_i) + h(y_j)$$

or, calling g_i the value of g when $x = x_i$ and h_j the value of h when $y = y_j$,

$$z_{i,j} \simeq g_i + h_j$$

However, there are mn equations, but only $m + n$ unknowns, so that some norm must be used to approximate the g_i and h_j. The least squares norm is the most amenable to calculation. Thus we attempt to minimize

$$\rho = \sum_{i=1}^{n} \sum_{j=1}^{m} (z_{i,j} - g_i - h_j)^2$$

and by applying (7.4), the equations

$$\begin{aligned} &\sum_{i=1}^{n} (z_{i,j} - g_i - h_j) = 0 \text{ for each } j \\ &\sum_{j=1}^{m} (z_{i,j} - g_i - h_j) = 0 \text{ for each } i \end{aligned} \tag{9.16}$$

must be solved. We should actually fix one of the unknowns, say $g_1 = 0$, since from (9.15a) any constant could be added to $g(x)$ and subtracted from $h(y)$. There are therefore $m + n - 1$ linear equations in as many unknowns. Rearranging (9.16), we get

$$\begin{aligned} &\text{(a)}\ h_j = \frac{\sum_{i=1}^{n} z_{i,j} - \sum_{i=1}^{n} g_i}{n}, \qquad j = 1, 2, \ldots, m \\ &\text{(b)}\ g_i = \frac{\sum_{j=1}^{m} z_{i,j} - \sum_{j=1}^{m} h_j}{m}, \qquad i = 2, 3, \ldots, n \end{aligned} \tag{9.17}$$

The matrix of these equations is *positive definite* (Chapter 11), so that they are solvable by a *Gauss–Seidel* iteration scheme:

(1) Assume all $g_i = 0$
(2) Calculate all h_j from (9.17a)
(3) Calculate all g_i from (9.17b) using the new values of the h_j.
(4) Go to step 2. Always calculate new h_j using the most recently calculated g_i, and vice versa.

This process converges rather rapidly, i.e., one notices that successive values of h_j and g_i do not differ by much after several iterations. As soon as the proper number of significant figures is achieved, the iteration is stopped, and the most current values are used. The entire process is easily done with a desk calculator. The functions can then be set up on a DFG, or, if equal intervals are chosen, on a pot padding unit.

The functional form in (9.15b) will yield a good fit whenever the set of curves $z_i(x)$ with fixed y can be rotated about the origin into one another. Again, using a least squares norm

$$\rho = \sum_{i=1}^{n} \sum_{j=1}^{m} (z_{i,j} - g_i h_j)^2 \tag{9.18}$$

requires the solution of

$$\sum_{j=1}^{m} (z_{i,j} - g_i h_j) h_j = 0, \qquad i = 1, 2, \ldots, n$$

$$\sum_{i=1}^{n} (z_{i,j} - g_i h_j) g_i = 0, \qquad j = 1, 2, \ldots, m$$

Because any constant that multiplies all g_i can be used to divide all h_j, we may fix one of the unknowns, say $g_1 = 1$. By rearranging we get

$$\begin{aligned} g_i &= \frac{\sum_{j=1}^{m} z_{i,j} h_j}{\sum_{j=1}^{m} h_j^2}, \qquad i = 2, 3, \ldots, n \\ h_j &= \frac{\sum_{i=1}^{n} z_{i,j} g_i}{\sum_{i=1}^{n} g_i^2}, \qquad j = 1, 2, \ldots, m \end{aligned} \tag{9.19}$$

These equations are nonlinear, but Kindle suggests that the Gauss–Seidel iteration scheme converges anyway.

The other forms, (9.15c) and (9.15d), require the solution of more nonlinear equations, although yielding better fits. Again, Kindle claims that the Gauss–Seidel iteration scheme converges.

Fogarty [16] modified Kindle's procedure to include a weight function $w_{i,j}$ and a more complicated functional form

$$z = f(x, y) \simeq g_1(x) + h_1(y) + u_1(x)v_1(y) + u_2(x)v_2(y) + \cdots$$

where product terms are successively added until a desired accuracy is reached. He first considers all $u_i v_i = 0$ and optimizes the g_i and h_j as in (9.18) (with the weight function added). New data points

$$z'_{i,j} = z_{i,j} - g_i - h_j$$

are formed and are used in the optimization of the functions $u_{1,i}$ and $v_{1,j}$ using (9.19) and a weight function. This is continued with all the u and v functions. Fogarty also states that he had no difficulty with the convergence of the iteration scheme in either case. Unfortunately, his method does not yield optimum results because the sum terms are not allowed to interact with the product terms, and each product term is not allowed to interact with subsequent product terms. Since the functions are not orthogonal, all unknowns must be solved simultaneously for optimum results. If a digital computer is used, be sure to use one of the many optimization programs available. For desk calculator computation, some compromise will often be required to minimize the tedium involved.

The functional form one assumes is quite important and often depends upon the characteristics of the function needed. Suppose for example $f(0, y) = 0$ and it was important that this relationship be maintained. A functional form of

$$z(x, y) \simeq x[f(x) + g(y) + u(x)v(y)]$$

may be more appropriate. In each case the application of (7.4) to obtain the necessary equations for solution is straightforward.

9.4.4 Graphical methods

When neither desk calculator nor digital computer is available, it may be possible to obtain a reasonable approximation graphically. Krasil'shchikov and Kulakovskii [19] describe a graphical procedure that can produce good results if the approximating function $g(x)h(y)$ is appropriate. In a more elaborate procedure, Elliott and Caldwell [20] describe methods for determining the three functions f_1, f_2, and f_3 and constant a for the approximating form

$$f(x, y) \simeq f_1[ax + f_2(y)] + ax + axf_3(y) + f_3(y)$$

Such a form is particularly appropriate if the family of curves $z_i(x)$ appear similar to each other except for translations and rotations. As in all graphical procedures, the results are not optimum in any sense, with different programmers obtaining dissimilar results. Results, however, can be not only adequate but often excellent.

Generating arbitrary functions of two or more variables has always been a source of difficulty in analog computers—one reason why hybrid computers were embraced so rapidly by the aerospace industries. Automatic interpolation schemes can generate $f(x_1, x_2, \ldots, x_n)$ from a grid of values stored in the digital computer memory and supplied as a voltage to the analog program by digital-to-analog converters. Such schemes are beyond the scope of this book.

PROBLEMS

9-1. The equation of a simple pendulum rotating on a bearing of radius r can be shown to be

$$\frac{d^2\theta}{dt^2} + \frac{Cr}{L}\left(\text{sign}\,\frac{d\theta}{dt}\right)\sqrt{\left(\frac{g}{L}\right)^2 + \left(\frac{d\theta}{dt}\right)^4 + \left(\frac{d^2\theta}{dt^2}\right)^2 + \frac{2g}{L}\left[\left(\frac{d\theta}{dt}\right)^2 \cos\theta + \frac{d^2\theta}{dt^2}\sin\theta\right]}$$
$$+ \frac{g}{L}\sin\theta = 0$$

where

$g =$ acceleration of gravity
$L =$ length
$mL^2 =$ assumed moment of inertia
$C =$ coefficient of friction
$\theta =$ coordinate of rotation
$t =$ time

This model assumes a rubbing friction force uniformly distributed around the bearing. Determine the conditions on Cr/L so that an implicit mechanization of the equation is stable under all conditions. *Hint:* Use Theorem 9.1 and remember that $d\theta/dt$ and θ are, effectively, functions of τ when mechanized on the computer since they are outputs of low-gain integrators. They therefore cannot adjust significantly during the nulling period of the high-gain amplifier.

9-2. Compute the worst dynamic error, from Theorem 9.2, in the implicit solution of the pendulum equation of Problem 9-1. Ignore the switching effect when $d\theta/dt$ changes sign.

9-3. Mechanize the pendulum equation of Problem 9-1. What can be expected in the mechanization when the pendulum stops?

9-4. As we have seen in Chapter 5, square-root circuits are generable by integration. Modify Example 8.1 so that

$$\int_0^x \sqrt{\left|\frac{d^2 f(u)}{du^2}\right|}\, du$$

is generated by integration.

9-5. Verify Eq. (9.3).

9-6. Mechanize Example 8.1 for least squares fitting using Theorem 8.2 and $w(x) = 1$.

9-7. Mechanize Example 8.1 for absolute error fitting using Theorem 8.3 and $w(x) = 1$.

9-8. Abrahanian [21] devised a square-root circuit for generating $y = x^{1/2}$ which improves zero input characteristics. He noted that

$$(y+1)(y-1) = x - 1$$

so that

$$y = 1 + \frac{x-1}{y+1}$$

Mechanize this equation.

9-9. The trajectory of a smooth sphere in the atmosphere under the influence of a head or tail wind is found from the equations

$$\dot{V}_x = -\tfrac{1}{2}\frac{\rho C_D A}{M}(V_x + V_A)\sqrt{(V_x + V_A)^2 + V_y^2}, \qquad (V_x)_0 = V_{x0}$$

$$\dot{V}_y = -\tfrac{1}{2}\frac{\rho C_D A}{M}(V_y)\sqrt{(V_x + V_A)^2 + V_y^2} - g, \qquad (V_y)_0 = V_{y0}$$

$$\dot{x} = V_x, \qquad x_0 = 0$$

$$\dot{y} = V_y, \qquad y_0 = 0$$

where V_x = horizontal velocity component
V_y = vertical velocity component
V_A = wind velocity
g = gravity
ρ = mass density of air (as a function of y)
C_D = drag coefficient, as a function of net velocity $\sqrt{(V_x + V_A)^2 + V_y^2}$
A = cross-sectional area of sphere
M = mass of sphere
x = range
y = height

Draw an unscaled diagram for the solution of these equations using the circuit of Method 2 (Fig. 9.19) for generating the net air velocity. If the net air velocity is never near zero, Method 1 (Fig. 9.18) will suffice.

9-10. Devise implicit circuits to track both roots

$$x^2 + a(\tau)x - b(\tau) = 0$$

where $a, b \geq 0$.

9-11. Set up an implicit circuit to solve for the function $x(\tau)$ given by

$$x \sin x + a(\tau) \cos x = 1$$

Allow for the range $-180° \leq x \leq 180°$. (Use Theorem 9.3.)

9-12. We wish to generate the function

$$z = x^y, \qquad 1 \leq x \leq 2, \qquad 0 \leq y \leq 2$$

with fixed bounded error by interpolating linearly the following three functions obtained with $y = 0$, 1, and 2:

$$z_1(x) = 1, \qquad z_2(x) = x, \qquad z_3(x) = x^2$$

Verify that relative errors can exceed 5 per cent when $x = 2$.

9-13. Not wishing to obtain more that a 2 per cent relative error, we try a quadratic interpolation scheme using (9.12) to (9.14). Show that this is equivalent to using the approximation function

$$z_a = 1 + y(x - 1) + y(y - 1)\left(\frac{x^2}{2} + \frac{1}{2} - x\right)$$

and that the error still exceeds 2 per cent.

9-14. We let $m = 5$ and use

$$\begin{aligned} z_1 &= 1 && \text{for } y = 0 \\ z_2 &= x^{1/2} && \text{for } y = 0.5 \\ z_3 &= x && \text{for } y = 1 \\ z_4 &= x^{3/2} && \text{for } y = 1.5 \\ z_5 &= x^2 && \text{for } y = 2 \end{aligned}$$

Linear and quadratic interpolations yield better than 2 per cent relative accuracy, with the quadratic interpolation being much better than 1 per cent. Draw circuit diagrams for the linear and quadratic interpolation case using electronic switching.

9-15. Set up the equations for least squares fitting

$$z = x^y$$

using the functional form

$$z \simeq 1 + (x - 1)yf(x)g(y)$$

using Kindle's method. Assume the data point set comes from an 11×11 grid so that the functions $f(x)$ and $g(y)$ finally obtained can be set up on 10-segment DFGs. (The final error is about 2 per cent).

9-16. Functions of two or more variables can be generated by integration when applicable. Generate the function

$$z = x^y$$

this way, assuming $\dot{x}$ and $\dot{y}$ are available.

REFERENCES

[1] Mengel, A.S., and W. S. Melahn, *RAND REAC Manual*, The RAND Corporation. Res. Memo. RM-525 (revised), Santa Monica, Calif., Dec. 1, 1950 (ASTIA No. 210675).

[2] Richmond, W.F., Jr., "REAC Techniques," *Symposium I on REAC Techniques*, Reeves Instr. Corp., New York, March 15-16, 1951 (ASTIA No. 108488).

[3] El'sgol'ts, L. E., *Differential Equations* (Delhi, India: Hindustan Publishing Corp., 1961), p. 264 ff.

[4] Witsenhausen, H., "Utilization Optimum des Multiplieurs Electro-Mécaniques," *Proceedings of the 2nd Inter. Analogue Comp. Meetings* (Brussels: Presses Académiques Européennes, 1959).

[5] Anderson, T., and C. J. Sleehan, "Continuous Electronic Polar Resolver," *Instr. and Control Systems*, vol. 36, Feb. 1963.

[6] Rubin, A. I., "Inverse Resolution," *The Martin Co. Analog Report No. 50*, Baltimore, Md., Nov. 15, 1965.

[7] Wilkinson, R. H., "A Method of Generating Functions of Several Variables Using Analog Diode Logic," *IEEETEC*, vol. 12, April 1963.

[8] Meissinger, H. F., "Representation of Functions of Several Variables on REAC Equipment," *Symposium I on REAC Techniques*, Reeves Instr. Corp., New York, March 15–16, 1951 (ASTIA No. 108488).

[9] Larrowe, V. L., and M. M. Spencer, "Use of Semi-Conducting Surfaces in Analog Function Generation," *Proc. National Simulation Conference*, Jan. 19–21, 1956, Dallas, Texas.

[10] Meissinger, H. F., "An Electronic Circuit for the Generation of Functions of Several Variables," *IRE National Convention Record, part 4*, 1955.

[11] Polimerou, L. G., "A New Method for Generating a Function of Two Independent Variables," *IRETEC*, vol. 6, Sept. 1957.

[12] Tomovich, R., "New Applications of an Electronic Function Generator," *IRE-TEC*, vol. 7, March 1958.

[13] Visman, A. C. P., "A Diode Function Generator Made Suitable for Two Variables," *AICA*, vol. 7, Jan. 1965.

[14] Wallman, H., "An Electronic Integral-Transform Computer and the Practical Solution of Integral Equations," *J. Franklin Inst.*, vol. 250, July 1950.

[15] Langill, Jr., A. W., "Multivariant Function Generation on an Analog Computer," *Electro-Technology*, vol. 71, June 1963.

[16] Fogarty, L. E., "Computer Generation of Arbitrary Functions," *Simulation*, vol. 7, Aug. 1966.

[17] Levitt, J. R., "An Application of Numerical Mathematical Analysis to Analog Computation," *Proceedings of the 2nd Inter. Analogue Comp. Meetings* (Brussels: Presses Académiques Européennes, 1959).

[18] Kindle, W. K., "Methods of Approximating Functions of Two Variables with Functions of Single Variables," *Princeton Computation Center Report No. 53*, Electronic Associates, Inc., Princeton, N. J., July 2, 1956.

[19] Krasil'shchikov, M. N., and A. I. Kulakovskii, "A Method of Generating Functions of Several Variables and Schemes for its Instrumentation," *Automatika i Telemekhanika*, vol. 24, Aug. 1963.

[20] Elliott, D. A., and N. J. Caldwell, "Representation of Nonlinear Functions of Two Input Variables on Analog Equipment," *Trans. ASME*, vol. 79, April 1957.

[21] Abrahanian, V.V., "An Improved Squaring Circuit," *Instr. and Control Systems*, vol. 36, April 1963.

Chapter 10

THE MECHANIZATION OF TRANSFER FUNCTIONS

10.1 Transfer Functions

The concept of a transfer function arises in systems engineering wherever linear differential equations are encountered. Although a strict justification of many of the results obtained by using transfer functions is intimately associated with Laplace transform theory, for our purposes we consider the equation

$$x = \frac{\sum_{j=0}^{m} a_j p^j}{\sum_{j=0}^{n} b_j p^j} f(t), \qquad \begin{matrix} a_m \neq 0 \\ b_n \neq 0 \end{matrix} \tag{10.1}$$

to merely be a convenient notation for the differential equation

$$\sum_{j=0}^{n} b_j \frac{d^{(j)}x}{dt^{(j)}} = \sum_{j=0}^{m} a_j \frac{d^{(j)}f(t)}{dt^{(j)}} \tag{10.2}$$

where $f(t)$ is a given input function, $x(t)$ is the output, and $m \leq n$. In (10.1), we recognize the Heaviside operator p (Chapter 2) and use the notation

$$G_{m,n}(p) = \frac{\sum_{j=0}^{m} a_j p^j}{\sum_{j=0}^{n} b_j p^j}$$

so that

$$x = G_{m,n}(p)f(t)$$

represents the linear differential equation and is called the nth-order transfer function of the system.

10.1.1 Factoring transfer functions

Transfer functions have very interesting properties, most of which can be proved by substitution directly into (10.2). For example, in Chapter 2 we alluded to the fact that they may be manipulated algebraically. We show now as a theorem that they can be factored and cascaded if the coefficients are constant.

Theorem 10.1. If

$$y = G_{r,n}(p)f(t), \qquad z = H_{s,m}(p)y$$

then

$$z = F_{(r+s),(n+m)}(p)f(t) \tag{10.3}$$

where

$$F_{(r+s),(n+m)}(p) = G_{r,n}(p)H_{s,m}(p) \tag{10.4}$$

Proof. Let

$$G_{r,n}(p) = \frac{\sum_{i=0}^{r} a_i p^i}{\sum_{i=0}^{n} b_i p^i}, \qquad H_{s,m}(p) = \frac{\sum_{j=0}^{s} \alpha_j p^j}{\sum_{j=0}^{m} \beta_j p^j}$$

so that, from (10.2),

$$\sum_{i=0}^{n} b_i \frac{d^{(i)}y}{dt^{(i)}} = \sum_{i=0}^{r} a_i \frac{d^{(i)}f(t)}{dt^{(i)}} \tag{10.5}$$

$$\sum_{j=0}^{m} \beta_j \frac{d^{(j)}z}{dt^{(j)}} = \sum_{j=0}^{s} \alpha_j \frac{d^{(j)}y}{dt^{(j)}} \tag{10.6}$$

Now, differentiate (10.5) j times with respect to time, multiply by α_j, and sum s times over j. We have then

$$\sum_{j=0}^{s} \alpha_j \sum_{i=0}^{n} b_i \frac{d^{(i+j)}y}{dt^{(i+j)}} = \sum_{j=0}^{s} \alpha_j \sum_{i=0}^{r} a_i \frac{d^{(i+j)}f(t)}{dt^{(i+j)}}$$

provided a_i and b_i are constants. Similarly, we differentiate (10.6) i times, multiply by b_i, and sum n times over i.

$$\sum_{i=0}^{n} b_i \sum_{j=0}^{m} \beta_j \frac{d^{(j+i)}z}{dt^{(j+i)}} = \sum_{i=0}^{n} b_i \sum_{j=0}^{s} \alpha_j \frac{d^{(j+i)}y}{dt^{(j+i)}}$$

provided α_j and β_j are constants. Because α_j and b_i have the proper indices, these factors may be placed within the double summation signs, and the order of the summation can be reversed. Thus we are led to the equations

$$\sum_{j=0}^{s} \sum_{i=0}^{n} \alpha_j b_i \frac{d^{(i+j)}y}{dt^{(i+j)}} = \sum_{j=0}^{s} \sum_{i=0}^{r} \alpha_j a_i \frac{d^{(i+j)}f(t)}{dt^{(i+j)}}$$

$$\sum_{i=0}^{n} \sum_{j=0}^{m} b_i \beta_j \frac{d^{(i+j)}z}{dt^{(i+j)}} = \sum_{i=0}^{n} \sum_{j=0}^{s} b_i \alpha_j \frac{d^{(i+j)}y}{dt^{(i+j)}}$$

from which it follows that

$$\sum_{i=0}^{n} \sum_{j=0}^{m} b_i \beta_j \frac{d^{(i+j)}z}{dt^{(i+j)}} = \sum_{i=0}^{r} \sum_{j=0}^{s} a_i \alpha_j \frac{d^{(i+j)}f(t)}{dt^{(i+j)}}$$

We now define

$$F_{(r+s),(n+m)}(p) = \frac{\sum_{i=0}^{r} \sum_{j=0}^{s} a_i \alpha_j p^{i+j}}{\sum_{i=0}^{n} \sum_{j=0}^{m} b_i \beta_j p^{i+j}}$$

so that (10.3) is true. Equation (10.4) follows instantly because the double summation also defines polynomial products.

Theorem 10.1 permits writing a transfer function in factored form without confusion, and allows, in any block diagram, for treating the functions as black boxes that operate on signals to produce other signals. It can be shown that the order of the blocks is unimportant when the coefficients are constant, so

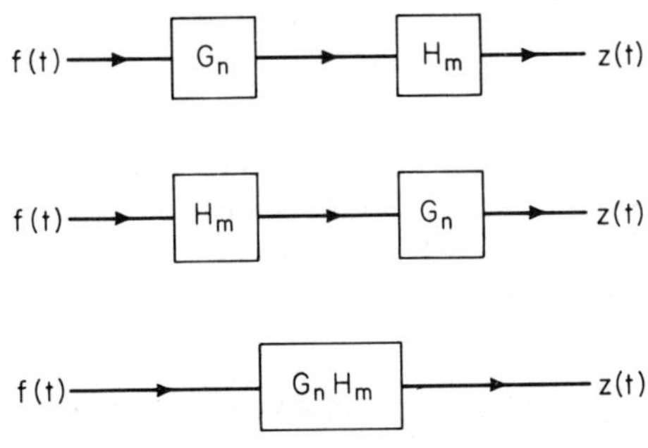

FIG. 10.1 Transfer functions may be factored and the order of factors interchanged.

that Fig. 10.1 shows three representations of the signal $z(t)$. It can also be shown by induction that a transfer function containing more than two factors can be similarly treated.

10.1.2 Frequency response evaluation

An important consideration in engineering design is the stability and steady-state frequency response of a given transfer function. The stability is completely defined by the *homogeneous* equation; i.e., the equation obtained from (10.2) when $f(t) = 0$. In that case, the stability criteria that force $x \to 0$ with time, despite any set of initial conditions, are well known. Since unstable transfer functions are not particularly useful, we assume that all transfer functions needed are stable. Steady-state frequency response analysis often is needed then to obtain proper designs. We also need to evaluate the frequency response in connection with other work in this chapter.

The question is this: If $f(t) = \sin \omega t$, what is the output $x(t)$ after the homogeneous solution decays to zero? The answer is provided in a simple theorem:

Theorem 10.2. Given a stable transfer function with constant coefficients

$$x = \frac{\sum_{j=0}^{m} a_j p^j}{\sum_{j=0}^{n} b_j p^j} f(t) = G_{m,n}(p) f(t)$$

the nonhomogeneous steady-state solution when $f(t) = \sin \omega t$ (i.e., the frequency response) is given by

$$x = A \sin \omega t + B \cos \omega t \tag{10.7}$$

where A and B are the real and imaginary parts of the complex function $G_{m,n}(i\omega)$, $i^2 = -1$, or

$$A + iB = G_{m,n}(i\omega)$$

Proof. By the *variation of constants* method [1], it is easy to show that form (10.7) is required, since $f(t)$ and its derivatives contain only $\sin \omega t$ and $\cos \omega t$ terms. We find A and B explicitly by substitution. Explicit representations of the required derivatives can be shown by induction on m to be

$$\sum_{j=0}^{m} a_j p^j \sin \omega t = c_1 \sin \omega t + c_2 \cos \omega t$$

where

$$c_1 = \sum_{j=0}^{[m/2]} (-1)^j \omega^{2j} a_{2j}$$

$$c_2 = \sum_{j=0}^{m-[m/2]-1} (-1)^j \omega^{2j+1} a_{2j+1}$$

and the symbol $[m/2]$ is the greatest integer $\leq m/2$. Similarly,

$$\sum_{j=0}^{n} b_j p^j (A \sin \omega t) = Ad_1 \sin \omega t + Ad_2 \cos \omega t$$

where

$$d_1 = \sum_{j=0}^{[n/2]} (-1)^j \omega^{2j} b_{2j}$$

$$d_2 = \sum_{j=0}^{n-[n/2]-1} (-1)^j \omega^{2j+1} b_{2j+1}$$

$$\sum_{j=0}^{n} b_j p^j (B \cos \omega t) = -Bd_2 \sin \omega t + Bd_1 \cos \omega t$$

Equating all cosine and sine terms,

$$Ad_1 - Bd_2 = c_1, \qquad Ad_2 + Bd_1 = c_2$$

which yield

$$A = \frac{c_1 d_1 + c_2 d_2}{d_1^2 + d_2^2}, \qquad B = \frac{-c_1 d_2 + c_2 d_1}{d_1^2 + d_2^2}$$

Now, by directly substituting $i\omega$ in the transfer function,

$$\begin{aligned} G_{m,n}(i\omega) &= \frac{\sum_{j=0}^{m} a_j (i\omega)^j}{\sum_{j=0}^{n} b_j (i\omega)^j} \\ &= \frac{\sum_{j=0}^{[m/2]} (-1)^j \omega^{2j} a_{2j} + i \sum_{j=0}^{m-[m/2]-1} (-1)^j \omega^{2j+1} a_{2j+1}}{\sum_{j=0}^{[n/2]} (-1)^j \omega^{2j} b_{2j} + i \sum_{j=0}^{n-[n/2]-1} (-1)^j \omega^{2j+1} b_{2j+1}} \\ &= \frac{c_1 + ic_2}{d_1 + id_2} \\ &= \frac{(c_1 + ic_2)(d_1 - id_2)}{d_1^2 + d_2^2} \\ &= \frac{c_1 d_1 + c_2 d_2}{d_1^2 + d_2^2} + i \frac{(-c_1 d_2 + c_2 d_1)}{d_1^2 + d_2^2} \end{aligned}$$

which is the desired result.

Corollary.

$$A \sin \omega t + B \cos \omega t = C \sin (\omega t + \phi)$$

where

$$C(\omega) = (A^2 + B^2)^{1/2}$$

$$\phi(\omega) = \arctan \left(\frac{B}{A}\right)$$

The amplitude C is called the *gain* of the transfer function, and the angle ϕ its *phase.*

Theorems 10.1 and 10.2 will be used in connection with specific techniques for mechanization on the computer in this chapter. It should be emphasized that many other important properties of transfer functions have not been described only because they are not required here.

It is clear from (10.2) that if $m = 0$, the equation is similar to (3.7). Any mechanization would otherwise seem to require differentiation of the input function $f(t)$. But early simulation equipment users and automatic control investigators recognized that this is *not* the case; it is possible to arrange circuitry so that (10.2) can be mechanized without differentiation of $f(t)$, even when the coefficients a_j and b_j are (differentiable) functions of time, or when initial conditions are not zero.

We examine here various methods to mechanize transfer functions on the computer both when the coefficients are constant (Sec. 10.2) and variable (Sec. 10.3). In addition, mechanizations of transfer functions are required in other areas of analog computer techniques. Time delays, both fixed (Sec. 10.4) and variable (Sec. 10.5), are often closely approximated by specific transfer functions. Similarly, differentiation circuits are also so approximated and are discussed in Sec. 10.6.

10.2 Transfer Functions with Constant Coefficients

If the given transfer function has constant coefficients, several ways are available to construct analog circuits that solve the equations correctly without using differentiation circuits. We describe and illustrate each separately, pointing out desirable and undesirable features.

10.2.1 Beck's method [2]–[4]

Beck's method of setting up transfer functions on the analog computer was among the earliest. It is the simplest compared with other methods, requiring no coefficient manipulation. The foundation for his method is given here as a theorem:

Theorem 10.3. If

$$\sum_{j=0}^{n} b_j p^j u = f(t) \tag{10.8}$$

and

$$x = \sum_{j=0}^{m} a_j p^j u \tag{10.9}$$

then

$$\sum_{i=0}^{n} b_i p^i x = \sum_{i=0}^{m} a_i p^i f(t) \tag{10.10}$$

Proof. Differentiate (10.8) i times, multiply by a_i and sum over i. Similarly, differentiate (10.9) i times, multiply by b_i, and sum over i. We have

$$\sum_{i=0}^{m} a_i \sum_{j=0}^{n} b_j p^{(j+i)} u = \sum_{i=0}^{m} a_i p^i f(t)$$

$$\sum_{i=0}^{n} b_i \sum_{j=0}^{m} a_j p^{(j+i)} u = \sum_{i=0}^{n} b_i p^i x$$

The left-hand sides are clearly equal since the roles of i and j can be interchanged, so that (10.10) follows.

Figure 10.2 contains a schematic for the general unscaled mechanization of a transfer function using Beck's method assuming $t = 1$. It uses the transformation defined by (3.8) for the auxiliary variable u:

$$p^j u = u_j \tag{10.11}$$

and does not include any inverters that may be required.

If initial conditions on x and its $n - 1$ derivatives are not zero, they must be taken into account. Levine [5] modified Beck's original circuit for providing initial conditions in a way requiring no additional equipment. The proof of Theorem 10.3 holds even when the n integrator variables u_j have initial conditions. We provide these to match the given $(p^i x)_0$ as follows:

Differentiate (10.8) i times for $i = 0, 1, 2, \ldots, m - 1$ and (10.9) k times for $k = 0, 1, 2, \ldots, n - 1$:

$$\begin{aligned} &\sum_{j=0}^{n} b_j p^{j+i} u = p^i f(t), \qquad i = 0, 1, \ldots, m-1 \\ &p^k x = \sum_{j=0}^{m} a_j p^{j+k} u, \qquad k = 0, 1, \ldots, n-1 \end{aligned} \tag{10.12}$$

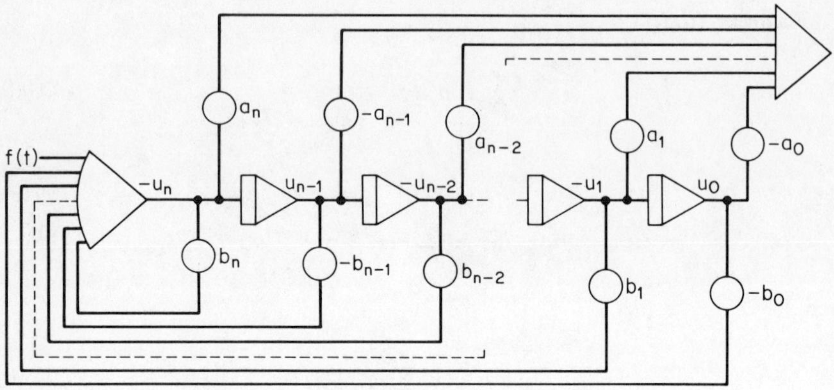

FIG. 10.2 Circuit diagram mechanizing Beck's method.

System (10.12) represents $n + m$ linear algebraic equations for $n + m$ unknowns (the functions $u_0, u_1, \ldots, u_{n+m-1}$). The values $(p^k x)_0$ and $[p^i f(t)]_0$ are presumably available so that each $(u_i)_0$ can be precalculated. Only n of them, however, are required as initial conditions.

Marzollo [6] showed that the linear system (10.12) could not be solved for the $(u_i)_0$ if the numerator and denominator of the transfer function have a common polynomial factor of degree $k > 0$. For such a case, one of the subsequent methods can be used for an analog computer mechanization, or Beck's method can be modified slightly as follows:

Suppose

$$x = G_{(m+k),(n+k)}(p)f(t) = \frac{\left(\sum_{j=0}^{m} a_j p^j\right)\left(\sum_{j=0}^{k} c_j p^j\right)}{\left(\sum_{j=0}^{n} b_j p^j\right)\left(\sum_{j=0}^{k} c_j p^j\right)} f(t) \qquad (10.13)$$

where cancellation of the common factor does not apply because the initial conditions are not zero. In (10.13), the a_j and b_j factors are relatively prime. Then we mechanize the system using two auxiliary variables, u and z, so that

$$\sum_{j=0}^{n} b_j p^j u = f(t), \qquad \sum_{j=0}^{k} c_j p^j z = 0$$
$$x = \left(\sum_{j=0}^{m} a_j p^j\right) u + z \qquad (10.14)$$

as shown in Fig. 10.3 using $i = 1$. Since we may differentiate x, and sum,

$$\begin{aligned}\sum_{i=0}^{k} c_i p^i x &= \left(\sum_{i=0}^{k} c_i p^i\right)\left(\sum_{j=0}^{m} a_j p^j\right)u + \sum_{i=0}^{k} c_i p^i z \\ &= \left(\sum_{i=0}^{k} c_i p^i\right)\left(\sum_{j=0}^{m} a_j p^j\right)u\end{aligned}$$

(10.13) can be formally shown to be satisfied.

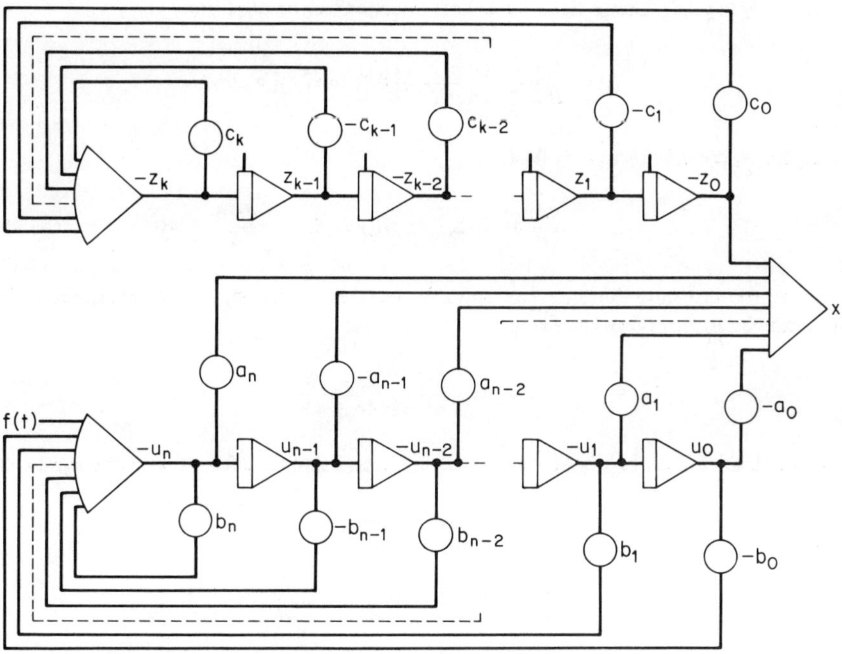

FIG. 10.3 Beck's method, modified for common factors in numerator and denominator and nonzero initial conditions.

Initial conditions are now found for the u_j variables and z_j variables ($n + k$ variables in all) by appropriately differentiating the equations in (10.14) and solving the set of $m + 2n + 2k$ linear equations

$$
\begin{aligned}
p^i x &= \sum_{j=0}^{m} a_j p^{j+i} u + p^i z, && i = 0, 1, 2, \ldots, n + k - 1 \\
p^i f(t) &= \sum_{j=0}^{n} b_j p^{j+i} u, && i = 0, 1, 2, \ldots, m + k - 1 \\
0 &= \sum_{j=0}^{k} c_j p^{j+i} z, && i = 0, 1, 2, \ldots, n - 1
\end{aligned} \tag{10.15}
$$

This partitioned system is still not adequate under one circumstance: (10.15) is not solvable when $\sum_{j=0}^{n} b_j p^j$ and $\sum_{j=0}^{k} c_j p^j$ are not relatively prime, i.e., have a common root. In that case, the root originally was a multiple root and cannot be partitioned. It is then necessary to use other methods.

Example 10.1. Using Beck's method, mechanize the equation

$$x = \frac{3(p + 1)(p + 2)}{2(p + 3)(p + 4)} f(t) = G_{2,2}(p) f(t)$$

with initial conditions at $t = 0$ of

$$x_0 = 1, \qquad \left(\frac{dx}{dt}\right)_0 = 2, \qquad f(0) = 5, \qquad \frac{df(0)}{dt} = 10 \tag{10.16}$$

Since numerator and denominator have no roots in common, we convert the transfer function by multiplying the factors:

$$G_{2,2} = \frac{3p^2 + 9p + 6}{2p^2 + 14p + 24}$$

and mechanize it as in Theorem 10.3. Initial conditions are found with the equations

$$
\begin{aligned}
x &= 3\frac{d^2u}{dt^2} + 9\frac{du}{dt} + 6u \\
\frac{dx}{dt} &= 3\frac{d^3u}{dt^3} + 9\frac{d^2u}{dt^2} + 6\frac{du}{dt} \\
f &= 2\frac{d^2u}{dt^2} + 14\frac{du}{dt} + 24u \\
\frac{df}{dt} &= 2\frac{d^3u}{dt^3} + 14\frac{d^2u}{dt^2} + 24\frac{du}{dt}
\end{aligned}
$$

By substitution of the initial conditions, it is found that

$$(u_0)_0 = -\frac{49}{36} = -1.3611, \qquad \left(\frac{du}{dt}\right)_0 = \frac{71}{18} = 3.9444$$

The final system is shown unscaled in Fig. 10.4, again using $i = 1$.

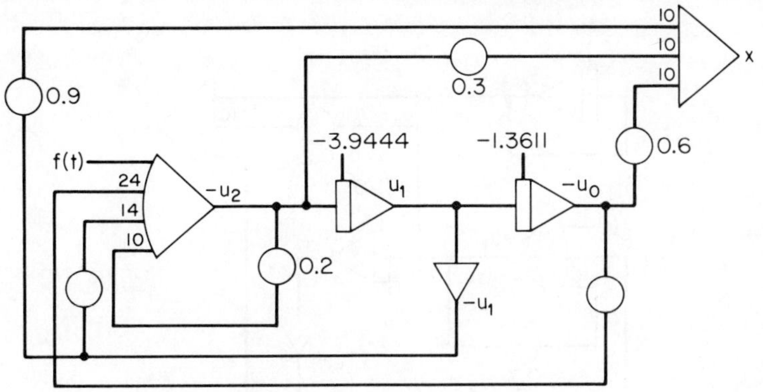

FIG. 10.4 Example 10.1 using Beck's method.

Example 10.2. Using the modification of Beck's method, mechanize the equation

$$x = \frac{2}{3}\frac{(p+1)(p+3)}{(p+1)(p+2)}f(t)$$

with the same initial conditions as in (10.16).

If initial conditions were zero, the common factor $p + 1$ could be cancelled. Otherwise we use system (10.14):

$$(3p + 6)u = f(t), \qquad (p + 1)z = 0, \qquad x = (2p + 6)u + z \tag{10.17}$$

and calculate the initial conditions for u_0 and z_0 from

$$\begin{aligned} x &= (2p + 6)u + z \\ px &= (2p^2 + 6p)u + pz \\ (p + 1)z &= 0 \\ (3p + 6)u &= f \\ (3p^2 + 6p)u &= pf \end{aligned}$$

We readily find that

$$(u_0)_0 = \frac{31}{6}, \qquad (z_0)_0 = -\frac{38}{3}$$

so that Fig. 10.5 represents an unscaled computer diagram to solve (10.17) with $\dot{\imath} = 1$.

10.2.2 Johnson's method [7], [8]

Johnson retained the simplicity of Beck's method in that the original coefficients are used in the mechanization, but he also provided a method having three other advantages: (1) Initial conditions can be easily calculated for all

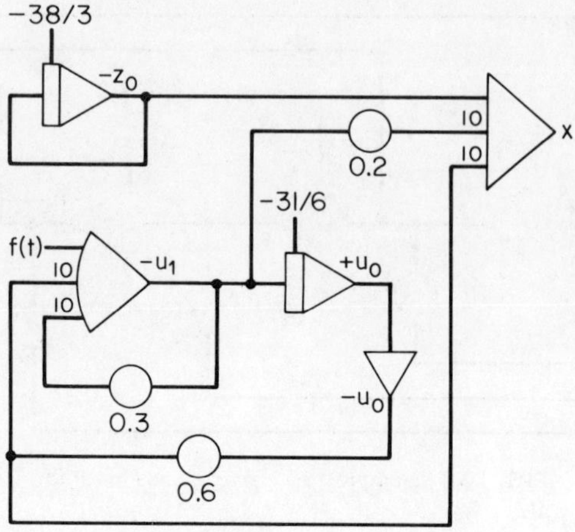

FIG. 10.5 Example 10.2: Mechanization.

cases; (2) variable coefficients are more easily mechanized; and (3) no more than $n + 3$ amplifiers are required. The basic circuit is mechanized from the following theorem:

Theorem 10.4. Let

$$y_1 = b_n x - a_n f(t), \qquad b_n \neq 0 \tag{10.18}$$

and let new functions y_j, $1 \leq j \leq n - 1$, be defined by

$$\frac{dy_j}{dt} = y_{j+1} - b_{n-j}x + a_{n-j}f(t) \tag{10.19}$$

with

$$\frac{dy_n}{dt} = -b_0 x + a_0 f(t) \tag{10.20}$$

where $a_j \neq 0$ for at least one j. Then the equation

$$x = \frac{\sum_{i=0}^{n} a_i p^i}{\sum_{i=0}^{n} b_i p^i} f(t) \tag{10.21}$$

is satisfied. Furthermore, initial conditions are found by

$$(y_j)_0 = \sum_{i=0}^{j-1} b_{n+i-j+1}(p^i x)_0 - \sum_{i=0}^{j-1} a_{n+i-j+1}[p^i f(0)] \tag{10.22}$$

Proof. Differentiate (10.19) $n - j$ times (for $j = 1, 2, \ldots, n - 1$) and sum over these values of j:

$$\sum_{j=1}^{n-1} p^{n-j+1} y_j = \sum_{j=1}^{n-1} p^{n-j} y_{j+1} - \sum_{j=1}^{n-1} p^{n-j}(b_{n-j}x) + \sum_{j=1}^{n-1} p^{n-j}[a_{n-j}f(t)]$$

Cancellation occurs for some of the derivatives of y_j, so that

$$p^n y_1 = p y_n - \sum_{j=1}^{n-1} p^{n-j}(b_{n-j}x) + \sum_{j=1}^{n-1} p^{n-j}[a_{n-j}f(t)]$$

and by substituting (10.20),

$$p^n y_1 = -\sum_{j=1}^{n} p^{n-j}(b_{n-j}x) + \sum_{j=1}^{n} p^{n-j}[a_{n-j}f(t)] \tag{10.23}$$

From (10.18),

$$p^n y_1 = p^n(b_n x) - p^n[a_n f(t)]$$

so that (10.23) becomes

$$0 = -\sum_{j=0}^{n} p^{n-j}(b_{n-j}x) + \sum_{j=0}^{n} p^{n-j}[a_{n-j}f(t)] \tag{10.24}$$

Since the a's and b's are constant, (10.21) is proved by factoring these out, letting $j = n - i$, and counting backwards. The initial conditions can be shown

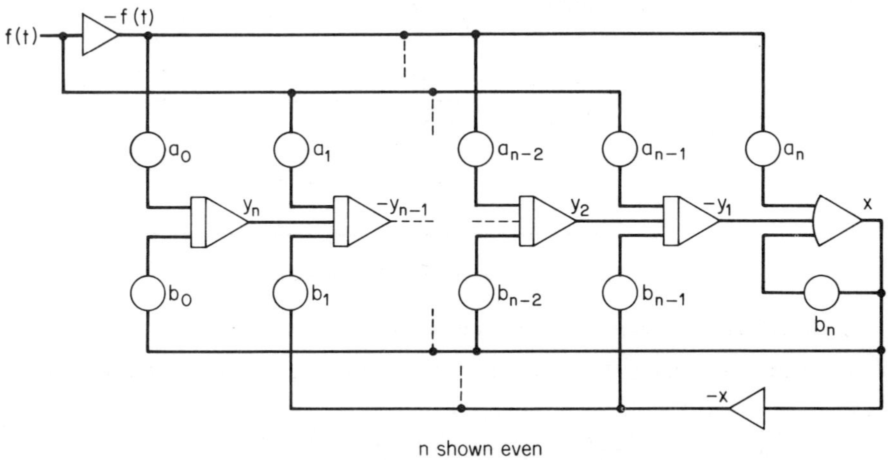

FIG. 10.6 Circuit diagram mechanizing Johnson's method.

to satisfy (10.22) by induction. A schematic for Johnson's method with $i = 1$ is shown in Fig. 10.6, with the y_i alternating in sign to avoid inverters.

Example 10.3. Using Johnson's method, mechanize the equation

$$x = \frac{2p^2 + 8p + 6}{3p^2 + 9p + 6} f(t)$$

with the same initial conditions (10.16) as previously.

This is, in fact, Example 10.2, but it is not necessary to factor the numerator or denominator to determine if there are any common ones. We proceed to set up (10.18)-(10.20):

$$y_1 = 3x - 2f(t)$$

$$\frac{dy_1}{dt} = y_2 - 9x + 8f(t)$$

$$\frac{dy_2}{dt} = -6x + 6f(t)$$

and calculate the initial conditions from

$$(y_1)_0 = b_n x_0 - a_n f(0) = 3 \times 1 - 2 \times 5 = -7$$

$$(y_2)_0 = b_{n-1} x_0 - a_{n-1} f(0) + b_n \left(\frac{dx}{dt}\right)_0 - a_n \left(\frac{df}{dt}\right)_0 = -45$$

Note the simplicity in calculating the initial conditions. Figure 10.7 shows a final unscaled circuit diagram with $i = 1$.

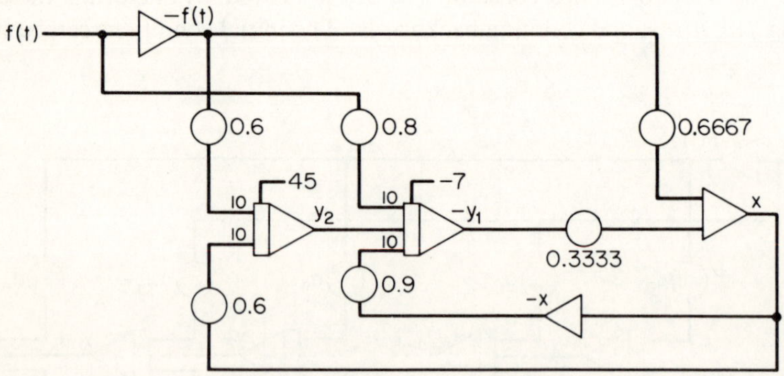

FIG. 10.7 Example 10.3: Mechanization.

10.2.3 Laning's and Battin's method [9], [10]

In control theory, it is often advantageous to express the differential system

$$\sum_{i=0}^{n} b_i p^i x = \sum_{i=0}^{n} a_i p^i f(t), \qquad b_n \neq 0$$

as a system of equations of the form

$$x = y_1 + c_n f(t) \tag{10.25}$$

$$\frac{dy_i}{dt} = y_{i+1} + c_{n-i}f(t), \qquad i = 1, 2, \ldots, n-1 \tag{10.26}$$

$$\frac{dy_n}{dt} = -\frac{1}{b_n}\left(\sum_{i=1}^{n} b_{i-1}y_i\right) + c_0 f(t) \tag{10.27}$$

To determine the proper c_i, we successively eliminate each y_i by differentiation, and we may show that

$$\sum_{i=0}^{n} b_i p^i x = \sum_{i=0}^{n}\left\{\sum_{j=i}^{n} b_j p^i[c_{n-j+i}f(t)]\right\} \tag{10.28}$$

from which

$$a_i = \sum_{j=i}^{n} b_j c_{n-j+i} \tag{10.29}$$

when the c's are constant (which occurs when the a's and b's are constant). This provides a convenient method to calculate each c_i, since (10.29) is a triangular array:

$$\begin{aligned} a_n &= b_n c_n \\ a_{n-1} &= b_n c_{n-1} + b_{n-1}c_n \\ &\vdots \\ a_0 &= b_n c_0 + b_{n-1}c_1 + \cdots + b_0 c_n \end{aligned}$$

Recursive formulas may be used:

$$c_n = \frac{a_n}{b_n}, \qquad c_i = \frac{1}{b_n}\left(a_i - \sum_{j=i}^{n-1} b_j c_{n-j+i}\right)$$

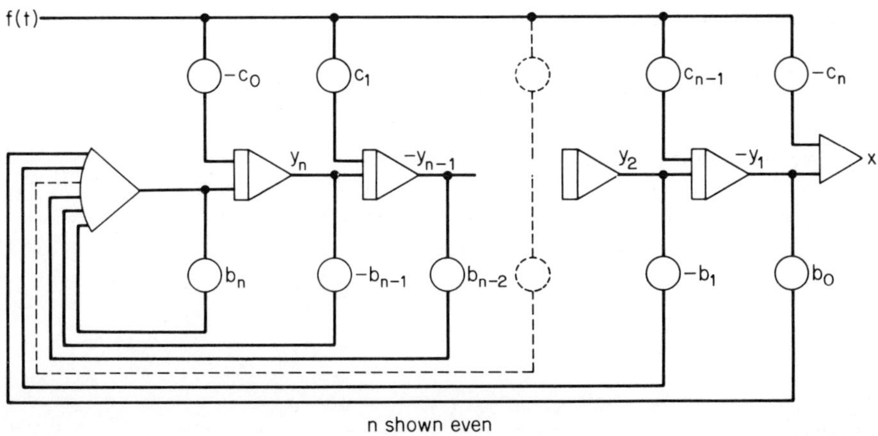

FIG. 10.8 Laning's and Battin's method.

as i decreases from $n - 1$ to 0. Figure 10.8 shows a general schematic for Laning's and Battin's method without regard for any sign inversions that may be required and using $i = 1$. Note that the y_i are again generated with alternating signs to save inverters at each integration.

Brockett [11] recognized that the c_i are the first successive $n + 1$ coefficients of a Laurent expansion of the transfer function

$$\frac{\sum_{i=0}^{n} a_i p^i}{\sum_{i=0}^{n} b_i p^i} = c_n + c_{n-1}p^{-1} + c_{n-2}p^{-2} + \cdots + c_0 p^{-n} + \cdots \tag{10.30}$$

A practical rapid way of obtaining the coefficients c_i is to perform the division as in (10.30).

Initial conditions are found by differentiation of (10.25) and substitution of (10.26). We find that recursive formulas are again convenient.

$$(y_j)_0 = (p^{j-1}x)_0 - \sum_{i=0}^{j-1} \{p^i[c_{n-j+i+1}f(t)]\}_0$$

which becomes, when the c_i's are constant

$$(y_j)_0 = (p^{j-1}x)_0 - \sum_{i=0}^{j-1} c_{n-j+i+1}[p^i f(t)]_0 \tag{10.31}$$

Example 10.4. Using Laning's and Battin's method, mechanize the transfer function

$$x = \frac{p^2 + 2p + 2}{p^3 + 3p^2 + 2p + 1} f(t)$$

with the initial conditions at $t = 0$,

$$x = \frac{dx}{dt} = \frac{d^2x}{dt^2} = 0, \qquad f = 0, \qquad \frac{df}{dt} = 1$$

Performing the division via Brockett's method,

$$\begin{array}{r|l}
 & 0 \quad + 1p^{-1} - 1p^{-2} + 3p^{-3} \ldots \\
p^3 + 3p^2 + 2p + 1 & 0p^3 + p^2 \quad + 2p \quad + 2 \\
 & \quad\; p^2 \quad + 3p \quad + 2 \quad \ldots \\
 & \quad\quad\; - p \quad\; 0 \quad\; \ldots \\
 & \quad\quad\; - p \quad - 3 \quad \ldots \\
 & \quad\quad\quad\quad + 3 \quad \ldots
\end{array}$$

so that

$$c_3 = 0, \quad c_2 = 1, \quad c_1 = -1, \quad c_0 = 3$$

The system becomes, from (10.25) to (10.27)

$$x = y_1, \qquad \frac{dy_1}{dt} = y_2 + f(t)$$

$$\frac{dy_2}{dt} = y_3 - f(t), \qquad \frac{dy_3}{dt} = -(y_1 + 2y_2 + 3y_3) + 3f(t)$$

From (10.31), the initial conditions are

$$(y_1)_0 = x_0 - c_3 f(0) = 0$$

$$(y_2)_0 = \left(\frac{dx}{dt}\right)_0 - c_2 f(0) - c_3 \frac{df(0)}{dt} = 0$$

$$(y_3)_0 = \left(\frac{d^2x}{dt^2}\right)_0 - c_1 f(0) - c_2 \frac{df(0)}{dt} - c_3 \frac{d^2 f(0)}{dt^2} = -1$$

Note that $c_3 = 0$ because $a_3 = 0$, so that only f and df/dt are required by (10.31). The final unscaled mechanization is shown in Fig. 10.9. It deviates from the general

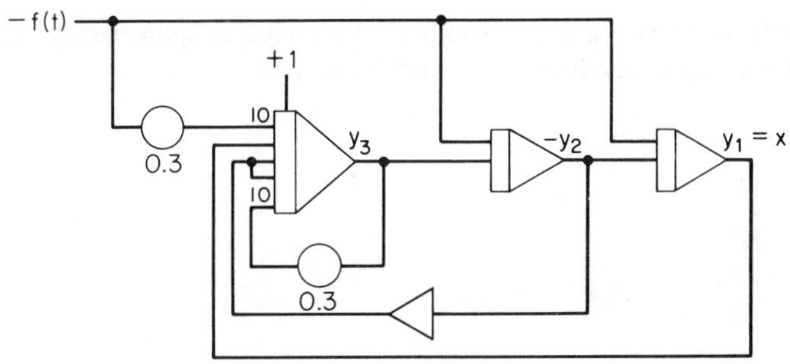

FIG. 10.9 Example 10.4: Mechanization.

case shown in Fig. 10.8 because $b_n = 1$. Most programmers divide numerator and denominator of any transfer function by b_n to normalize this leading coefficient. This leads to simpler circuits and should be done unless the effect of varying b_n is needed.

10.2.4 Other methods

Control system studies often involve low-order transfer functions in factored form, where each factor can represent one component in the system. Because of this, Johnson [7] and Heizer and Abraham [12] prepared tables of frequently used first- and second-order transfer functions showing the transfer function, circuit, and pot settings in terms of the constants involved. In general, more equipment is needed when circuits for low-order transfer functions are cascaded than if a circuit were drawn for the nonfactored high-order one. Thus, such mechanizations should be avoided when possible. There are cases, however, when the effect of varying parameters of individual blocks is to be studied, so that such tables are useful. They are easy to construct from the basic methods described and may be designed either to minimize the number of amplifiers required or to get potentiometers proportional to the system parameters. (See Problem 10-8.)

Other methods for transfer function synthesis exist [13] but are not so useful as the previous three methods described and will not be discussed here.

10.3 Variable Coefficients

There is occasionally a need to mechanize an equation of the form (10.2) when the coefficients a_j and b_j are not constants but given functions of time. Any method used to mechanize the equation when the coefficients are constant can be adapted for the nonconstant coefficient case. We emphasize that $f(t)$ may be an arbitrary signal so that the methods here involve constructing circuits that do not need any differentiating units. For the following development, we let $m = n$ but allow $a_n = 0$.

10.3.1 Matyash's method [13], [14]

Matyash adapted Johnson's method (Sec. 10.2.2) quite easily. In essence, we look for variable coefficients A_j and B_j such that

$$y_1 = B_n x - A_n f(t)$$

$$\frac{dy_j}{dt} = y_{j+1} - B_{n-j}x + A_{n-j}f(t), \qquad 1 \leq j \leq n-1$$

$$\frac{dy_n}{dt} = -B_0 x + A_0 f(t)$$

and at the same time satisfy (10.21). To determine the A_j and B_j in terms of the original a_j and b_j, we proceed as in the proof of Theorem 10.4 to obtain an equation similar to (10.24).

$$\sum_{j=0}^{n} p^{n-j}(B_{n-j}x) = \sum_{j=0}^{n} p^{n-j}[A_{n-j}f(t)]$$

Thus we must determine the A_j's and B_j's so that

$$\sum_{i=0}^{n} b_i p^i x = \sum_{i=0}^{n} p^i(B_i x) \tag{10.32}$$

$$\sum_{i=0}^{n} a_i[p^i f(t)] = \sum_{i=0}^{n} p^i[A_i f(t)]$$

are satisfied. Expanding the right-hand side of (10.32) by ordinary rules of differentiation we obtain

$$\sum_{j=0}^{n} p^j(B_j x) = \sum_{j=k}^{n} \sum_{k=0}^{j} \binom{j}{k} (p^{j-k} B_j)(p^k x)$$

where the symbols

$$\binom{j}{k} = \frac{j!}{k!\,(j-k)!}$$

are the binomial coefficients. If (10.32) is to be satisfied, we equate the components when $k = i$. Thus,

$$b_i = \sum_{j=i}^{n} \binom{j}{i} (p^{j-i} B_j)$$

This is a triangular system of n simultaneous equations

$$\begin{aligned} b_n &= B_n \\ b_{n-1} &= B_{n-1} + \binom{n}{n-1} pB_n \\ b_{n-2} &= B_{n-2} + \binom{n-1}{n-2} pB_{n-1} + \binom{n}{n-2} p^2 B_n, \\ &\text{etc.} \end{aligned}$$

which can also be shown by induction to satisfy the inverse "adjoint" relationship

$$\begin{aligned} B_n &= b_n \\ B_{n-1} &= b_{n-1} - \binom{n}{n-1} pb_n \\ B_{n-2} &= b_{n-2} - \binom{n-1}{n-2} pb_{n-1} + \binom{n}{n-2} p^2 b_n \\ &\text{etc.} \end{aligned}$$

In general,

$$B_{n-k} = \sum_{i=0}^{k} (-1)^{k-i} \binom{n-i}{n-k} \frac{d^{k-i} b_{n-i}}{dt^{k-i}} \tag{10.33}$$

In an analogous manner, we also have

$$A_{n-k} = \sum_{i=0}^{k} (-1)^{k-i} \binom{n-i}{n-k} \frac{d^{k-i} a_{n-i}}{dt^{k-i}} \tag{10.34}$$

Figure 10.6 is used with A_j and B_j replacing the a_j and b_j.

Example 10.5 [14]. Construct a circuit to solve

$$5(1 - e^{-t/5})\frac{d^2x}{dt^2} + (3 - e^{-t/5})\frac{dx}{dt} + \frac{2}{5}x = (3 - e^{-t/5})\frac{df}{dt} + \frac{2}{5}f(t)$$

We have

$$\begin{aligned} a_2 &= 0, & b_2 &= 5(1 - e^{-t/5}) \\ a_1 &= 3 - e^{-t/5}, & b_1 &= 3 - e^{-t/5} \\ a_0 &= \tfrac{2}{5}, & b_0 &= \tfrac{2}{5} \end{aligned}$$

From (10.33) we obtain (for $n = 2$)

$$B_2 = b_2 = 5(1 - e^{-t/5})$$

$$B_1 = b_1 - \binom{2}{1}\frac{db_2}{dt} = 3 - 3e^{-t/5}$$

$$B_0 = b_0 - \binom{1}{0}\frac{db_1}{dt} + \binom{2}{0}\frac{d^2b_2}{dt^2} = \frac{2}{5}(1 - e^{-t/5})$$

Similarly, from (10.34)

$$A_2 = a_2 = 0$$

$$A_1 = a_1 - \binom{2}{1}\frac{da_2}{dt} = 3 - e^{-t/5}$$

$$A_0 = a_0 - \binom{1}{0}\frac{da_1}{dt} + \binom{2}{0}\frac{d^2a_2}{dt^2} = \frac{1}{5}(2 - e^{-t/5})$$

The equations solved are then

$$y_1 = B_2x, \qquad \frac{dy_1}{dt} = y_2 - B_1x + A_1f(t)$$

$$\frac{dy_2}{dt} = -B_0x + A_0f(t)$$

Because of the initial indeterminacy ($B_2 = 0$ when $t = 0$), it is much better to substitute y_1/B_2 for x and solve

$$\frac{dy_1}{dt} = y_2 - \frac{B_1}{B_2}y_1 + A_1f(t)$$

$$\frac{dy_2}{dt} = -\frac{B_0y_1}{B_2} + A_0f(t)$$

Since the ratio of the B's are constant, much simplification results:

$$\frac{dy_1}{dt} = y_2 - \frac{3}{5}y_1 + A_1f(t), \qquad \frac{dy_2}{dt} = -\frac{2}{25}y_1 + A_0f(t) \tag{10.35}$$

An unscaled circuit diagram to solve (10.35) is shown in Fig. 10.10. Note that the division by B_2 initially is not valid. In actual practice, this circuit should be replaced by an alternate that will provide computer stability at $t = 0$. For this purpose, we

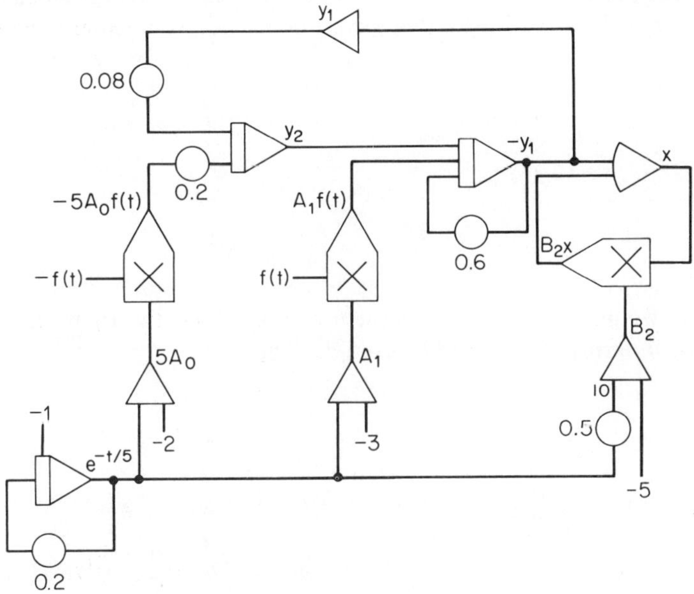

FIG. 10.10 Example 10.5: Mechanization.

may show that $x_0 = 2[f(0)]$ initially, if $(y_1)_0 = (y_2)_0 = 0$, and x remains close to this value for small t. A switching arrangement or a four-quadrant division circuit (Chapter 11) can be used instead.

10.3.2 Laning's and Battin's method [9]

Figure 10.8 applies in Laning's and Battin's method even when the coefficients are variable; the method of finding the c's is different. Equation (10.28) holds when the a's and b's are variable. We expand the derivative expression using the binomial coefficients. Hence,

$$\sum_{i=0}^{n} b_i p^i x = \sum_{k=0}^{n} \left[\sum_{j=k}^{n} \sum_{r=0}^{k} b_j \binom{k}{r} p^{k-r} c_{n-j+k} p^r f(t) \right]$$

The summation signs may now be permuted provided the variable limits are fixed to include only the terms that are meaningful. We have

$$\sum_{k=0}^{n} \sum_{j=k}^{n} \sum_{r=0}^{k} = \sum_{r=0}^{n} \sum_{k=r}^{n} \sum_{j=k}^{n}$$

so that when $r = i$, we obtain

$$a_i = \sum_{k=i}^{n} \sum_{j=k}^{n} b_j \binom{k}{i} p^{k-i} c_{n-j+k} \tag{10.36}$$

This is once again a triangular set of linear equations enabling the calculation of each c_j step by step. To see this, we introduce new summation indices s and q defined by

$$s = j - k, \qquad q = k - i$$

which transform (10.36) to

$$a_i = \sum_{q=0}^{n-i} \sum_{s=0}^{n-q-i} b_{q+s+i} \binom{q+i}{i} p^q c_{n-s}$$

With $q = 0$ and $s = n - i$, the term $b_n c_i$ appears on the right and may be solved for in terms of previously solved c_j. Thus

$$c_n = \frac{1}{b_n}(a_n)$$

$$c_{n-1} = \frac{1}{b_n}[a_{n-1} - b_{n-1}c_n - nb_n(pc_n)]$$

$$c_{n-2} = \frac{1}{b_n}\Big[a_{n-2} - b_{n-2}c_n - b_{n-1}c_{n-1} - b_{n-1}\binom{n-1}{n-2}(pc_n)$$

$$- b_n\binom{n-1}{n-2}(pc_{n-1}) - b_n\binom{n}{n-2}(p^2c_n)\Big]$$

etc.

Much calculation is required for high-order systems.

Example 10.6. Using Laning's and Battin's method, construct a circuit to solve

$$\frac{d^2x}{dt^2} + (3 - e^{-t/5})\frac{dx}{dt} + \frac{2}{5}x = (3 - e^{-t/5})\frac{df}{dt} + \frac{2}{5}f(t)$$

Since

$$a_2 = 0, \qquad b_2 = 1$$
$$a_1 = 3 - e^{-t/5}, \qquad b_1 = 3 - e^{-t/5}$$
$$a_0 = \tfrac{2}{5}, \qquad b_0 = \tfrac{2}{5}$$

we obtain

$$c_2 = a_2 = 0$$

$$c_1 = a_1 - b_1c_2 - 2b_2\frac{dc_2}{dt} = 3 - e^{-t/5}$$

$$c_0 = a_0 - b_0 c_2 - b_1 c_1 - b_1 \binom{1}{0} \frac{dc_2}{dt} - b_2 \binom{1}{0} \frac{dc_1}{dt} - b_2 \binom{2}{0} \frac{d^2 c_2}{dt^2}$$
$$= \tfrac{2}{5} - (3 - e^{-t/5})^2 - \tfrac{1}{5}e^{-t/5} = -8.6 + 5.8e^{-0.2t} - e^{-0.4t}$$

with the final system to solve being

$$x = y_1, \qquad \frac{dy_1}{dt} = y_2 + c_1 f(t), \qquad \frac{dy_2}{dt} = -(b_0 y_1 + b_1 y_2) + c_0 f(t)$$

Figure 10.11 shows an unscaled diagram for solving this system assuming $\dot{t} = 1$.

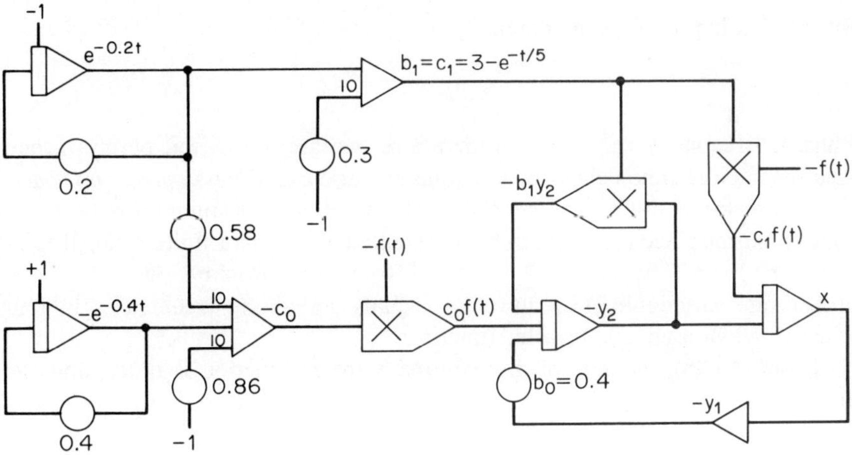

FIG. 10.11 Example 10.6: Mechanization.

10.4 Fixed Time Delays

Some problems involve transport lags or dead time, which require the generation of a fixed time delay T. Given a signal $f(t)$, we must generate

$$y = f(t - T) = G(p)f(t) \tag{10.37}$$

In control problems, for example, there is a delay between the introduction of a signal and the sensing of it by the system because of finite velocities of transmission. In some cases, the time is so small and the effect of delay is so negligible that it may be ignored. In others, however, the delay is important enough to include when the system is simulated on the computer. Although some specialized equipment to produce delays may be readily available, it is also possible to accurately delay bandwidth-limited signals with a linear system of differential equations.

The transfer function $G(p)$ associated with this delay is calculated easily. If

$$f(t) = \sin \omega t$$

then we wish

$$\begin{aligned} y &= \sin \omega(t - T) \\ &= \cos \omega T \sin \omega t - \sin \omega T \cos \omega t \end{aligned} \tag{10.38}$$

so that, from Theorem 10.2,

$$G(i\omega) = \cos \omega T - i \sin \omega T = e^{-i\omega T}$$

By substituting $i\omega = p$, we obtain

$$G(p) = e^{-pT}$$

Thus, by frequency response considerations, we are led to the purely formal notation for the time delay transfer function, also called the *Laplace shift operator*. In terms of differential equations, the notation is meaningful only for an approximation since e^{-pT} cannot be represented exactly as a finite rational fraction in p. From Theorem 10.2, a rational fraction approximation of e^{-pT} in a proper interval should properly delay signals containing frequencies in that interval, when used as a transfer function.

From (10.38), the gain of the required transfer function is unity, and the phase shift

$$\psi = -\omega T$$

is linear in ω and T. If e^{-pT} is approximated by a rational fraction

$$G_{m,n}(pT) = \frac{N(pT)}{D(pT)} = \frac{\sum_{j=0}^{m} a_j(pT)^j}{\sum_{j=0}^{n} b_j(pT)^j}$$

there are two errors to consider: a phase error ϵ_ϕ

$$\epsilon_\phi = \psi - \phi = -\omega T - \phi \tag{10.39}$$

where ϕ is the phase of $G_{m,n}(pT)$, and a gain error ϵ_g

$$\epsilon_g = 1 - C$$

where C is the gain of $G_{m,n}(pT)$. In both cases, the errors are functions of the single variable ωT. This variable is also called the *nondimensional frequency variable* or just the *frequency*.

The mechanization of $G_{m,n}(pT)$ requires n integrators. It is easy to show [15] that the maximum value of the phase shift, $\phi_{\max}$, is given by

$$\phi_{\max} = n\pi$$

Any effective approximation to the time delay should encompass a frequency range for $0 \leq \omega T \lesssim n\pi$.

10.4.1 Padé approximants

Padé first gave a rational fraction expansion of e^{-pT}:

$$e^{-pT} \simeq P_{m,n} = \frac{\sum_{j=0}^{m} (-1)^j \dfrac{\binom{m}{j}}{\binom{m+n}{j}} \dfrac{(pT)^j}{j!}}{\sum_{j=0}^{n} \dfrac{\binom{n}{j}}{\binom{m+n}{j}} \dfrac{(pT)^j}{j!}}$$

He showed that

$$\lim_{m+n\to\infty} P_{m,n} = e^{-pT}$$

The expansion has the property of agreeing with the McLaurin expansion by a maximum number of terms (at least $m + n - 1$) when a power series in pT is obtained by dividing the numerator by the denominator. The maximum number of derivatives is therefore correct at $pT = 0$. When $m = n$, the most accurate approximation can be expected with n integrators, a reason that "all-pass" approximants, as discussed below, are used so frequently for time delays.

10.4.2 All-pass approximants

Much early work on delays was done by communication engineers [16], but early attempts to simulate delays on an analog computer used Padé approximants [17], [18] with $m = n$. $P_{n,n}$ is of the form

$$G_{n,n}(pT) = \frac{\sum_{j=0}^{n} (-1)^j b_j (pT)^j}{\sum_{j=0}^{n} b_j (pT)^j} \tag{10.40}$$

which is called an "all-pass" approximant because all frequencies are transmitted with unity gain. In the case of a Padé all-pass approximation.

$$b_j = \frac{\binom{n}{j}}{\binom{2n}{j} j!} \tag{10.41}$$

but programmers by no means are limited to the use of the Padé expansion. In fact, in one of the earliest papers on time-delay simulation, Morrill [17] attempted to improve the frequency response characteristics by graphically obtaining new coefficients that successfully delayed signals containing a larger range of frequencies.

All-pass approximants have no gain error, so that only one error function ϵ_ϕ need be of concern. This fact greatly simplifies the determination of reasonable b_j. Many authors [19]–[25] have therefore obtained all-pass approximations that improve the frequency response of the Padé approximant to some degree, usually at the expense of having a greater ϵ_ϕ at the smaller frequencies.

To mechanize an all-pass approximant, Johnson's method (Sec. 10.2.2) requires the least number of amplifiers. Since $a_j = (-1)^j b_j$, (10.18)–(10.20) transform to (with $y = x$)

$$\begin{aligned} y_1 &= [y - (-1)^n f(t)]b_n T^n \\ \frac{dy_j}{dt} &= -[y - (-1)^{n-j} f(t)]b_{n-j}T^{n-j} + y_{j+1} \end{aligned} \tag{10.42}$$

for $1 \leq j \leq n - 1$, and

$$\frac{dy_n}{dt} = -[y - f(t)]b_0$$

A minimum potentiometer circuit can be constructed from (10.42) by scaling each y_j as $[y_j/(b_{n-j}T^{n-j})]$. Then

$$\begin{aligned} \frac{b_{n-1}}{b_n T}\left(\frac{y_1}{b_{n-1}T^{n-1}}\right) &= y - (-1)^n f(t) \\ \frac{d\left(\dfrac{y_j}{b_{n-j}T^{n-j}}\right)}{dt} &= -[y - (-1)^{n-j} f(t)] \\ &\quad + \left(\frac{b_{n-j-1}}{b_{n-j}T}\right)\left(\frac{y_{j+1}}{b_{n-j-1}T^{n-j-1}}\right) \\ \frac{d\left(\dfrac{y_n}{b_0}\right)}{dt} &= -y + f(t) \end{aligned} \tag{10.43}$$

With $i = 1$, this circuit is shown in Fig. 10.12. Successive y_j are generated with alternating sign to save inverters, and the shorthand notation of

$$g_j = \frac{b_{j-1}}{b_j}, \qquad j = 1, 2, \ldots, n \tag{10.44}$$

is used.

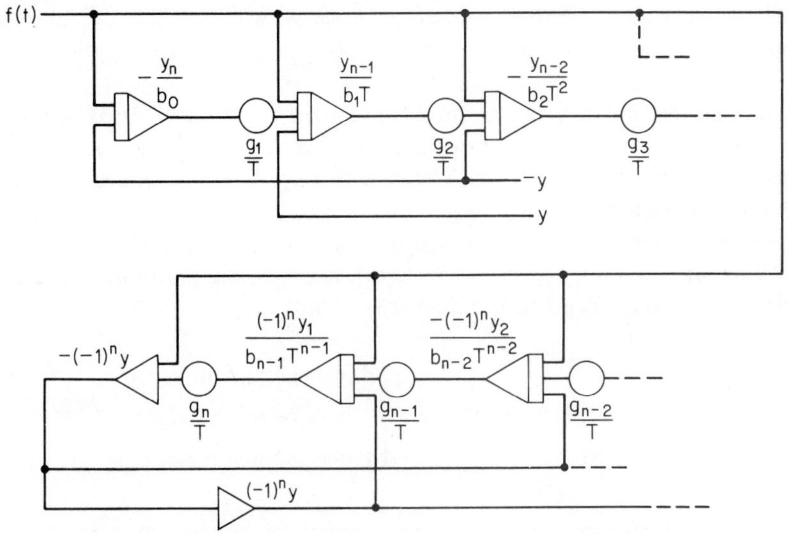

FIG. 10.12 Convenient mechanization of an all-pass approximant.

In practice, some of the amplifier outputs may have to be rescaled, or as an alternative, $f(t)$ can be reduced by a factor of 10 while y is subsequently increased by the same factor. If g_j/T is too large, the capacitors used for the integrators can be changed from 1μF to $10^{-k}\mu$F ($k > 0$) and the gains g_j/T replaced by $g_j/(10^kT)$. Excellent results can then be obtained by this circuit. A circuit using Beck's method is shown in Sec. 10.5. While an extra amplifier is then required, the scaling is considerably simplified.

10.4.3 Chebyshev all-pass approximants

All-pass approximants are often compared using the language of a Chebyshev norm (Chapter 7) such as "the maximum error is ϵ for all frequencies ω such that $0 \leq \omega T \leq B$," where B is called the *usable bandwidth* or *range* of the approximation. This practice led several authors [26]–[28] to optimize the g_j using a Chebyshev norm. The phase error of (10.39) can sometimes be made to conform to the usual error characteristics of a Chebyshev best fit, i.e., exactly n alternations. The range B of the approximation however, must first be fixed, since approximations usually have meaning only in a fixed interval. For a given B and number of integrators n, there often exists a set of n parameters g_j corresponding to exactly n alternations in the error function ϵ_ϕ with deviation $\epsilon = \max |\epsilon_\phi|$, despite the approximating form being nonlinear in each parameter. Hausner and Furlani [27] showed that this set, when it existed, provided for the best Chebyshev fit. These approximants are referred to as $C_{n,n}^{\phi,\epsilon}$.

It is also possible to define a time-delay error function

$$\epsilon_T = \frac{\epsilon_\phi}{\omega T} = -\frac{\phi}{\omega T} - 1$$

Minimizing $|\epsilon_T|$ for a given range B minimizes the relative error of the time delay ($100\epsilon_T$ is the per cent error of the time delay). Using ϵ_T as the error function permits the statement: "All frequencies x for $0 \leq x \leq B$ are delayed $T(1 \pm \epsilon)$ seconds, where $\epsilon \leq \max |\epsilon_T|$."

Approximants found by minimizing ϵ_T are referred to as $C_{n,n}^{T,\epsilon}$.

Whether $\epsilon = \max |\epsilon_T|$ or $\epsilon = \max |\epsilon_\phi|$, it is possible to optimize one of the parameters ϵ, n, or B, given the other two. Thus, if

$$\left.\begin{matrix} n, \epsilon \\ n, B \\ \epsilon, B \end{matrix}\right\} \text{ are given, then } \left\{\begin{matrix} B \text{ may be maximized,} \\ \epsilon \text{ may be minimized,} \\ n \text{ may be minimized.} \end{matrix}\right.$$

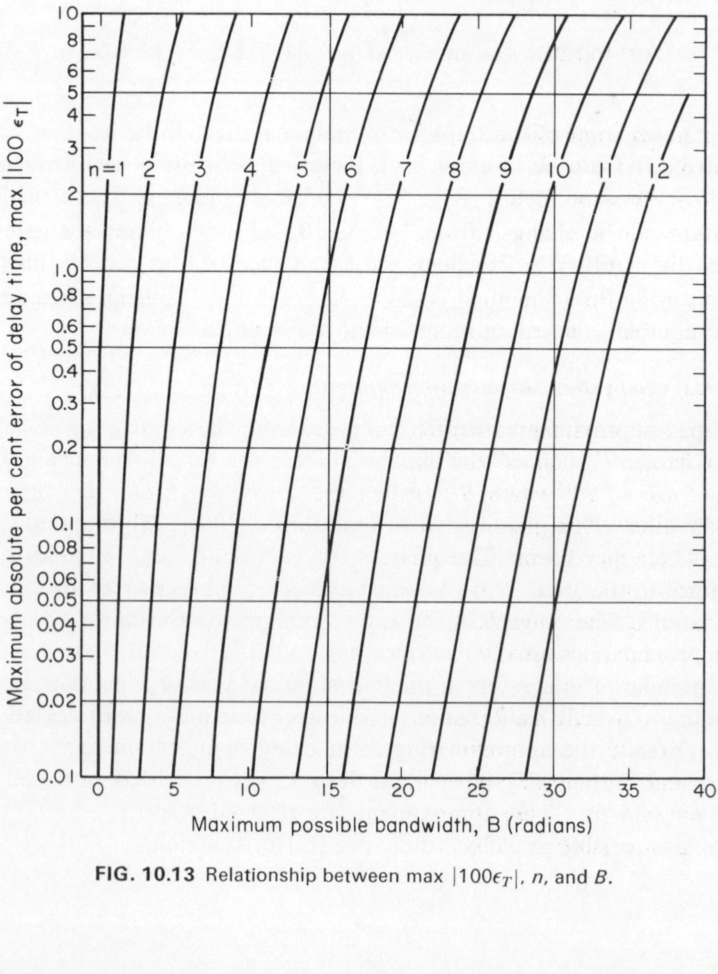

FIG. 10.13 Relationship between max $|100\epsilon_T|$, n, and B.

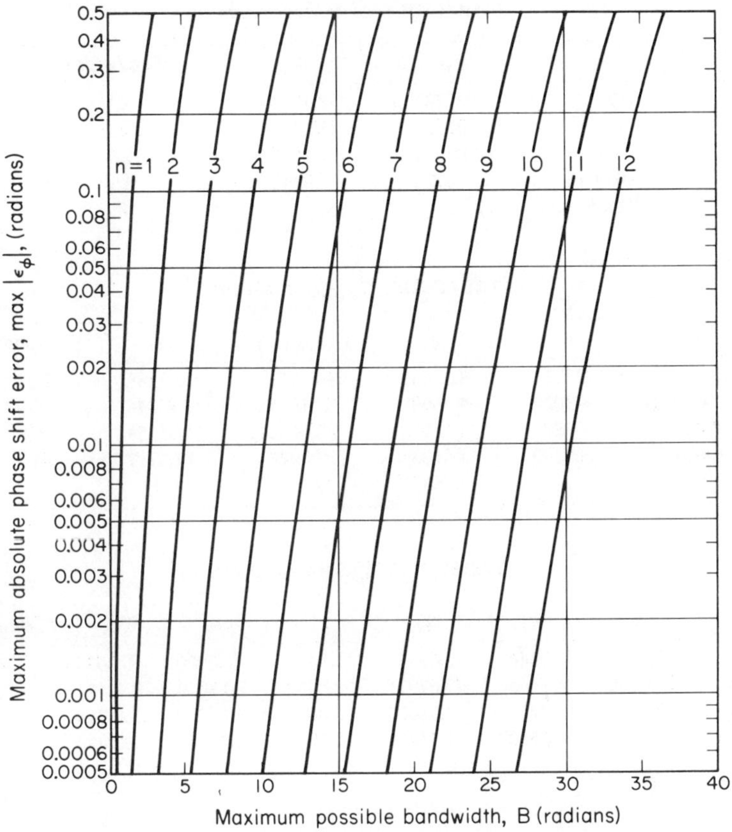

FIG. 10.14 Relationship between max $|\epsilon_\phi|$, n, and B.

Figures 10.13 and 10.14 show empirically derived relationships between max $|100\epsilon_T|$, n, and B, and max $|\epsilon_\phi|$, n, and B, respectively. They allow a programmer to determine the third parameter when any two are known or guessed. Tables 10.1–10.10 contain a partial listing of these two types of Chebyshev approximants, as well as the Padé approximants. More complete tables are given in [29]. For the Padé approximants, one may show from (10.41) that

$$g_j = j\left(\frac{2n - j + 1}{n - j + 1}\right)$$

Typical error curves $\epsilon_T(x)$ and $\epsilon_\phi(x)$ are shown in Figs. 10.15 and 10.16. Note how the oscillating error functions of the Chebyshev approximants enable the bandwidth to be extended considerably more than the Padé approximants.

Example 10.7. Determine the Chebyshev circuit for a $\frac{1}{2}$-s delay for signals containing frequencies up to $\omega = 15$ rad/s within an accuracy of 1 per cent.

TABLE 10.1 $C_{2,2}^{T,\epsilon}$ *Approximants*

ϵ	g_1	g_2	*Bandwidth*
0.1	2.22222	3.67084	5.2
0.01	2.02020	5.24359	2.8
0.001	2.00200	5.76018	1.6
0.0001	2.00020	5.92412	0.9
Padé	2	6	

TABLE 10.2 $C_{4,4}^{T,\epsilon}$ *Approximants*

ϵ	g_1	g_2	g_3	g_4	*Bandwidth*
0.1	2.22222	3.32633	14.6055	5.73881	12.7
0.01	2.02020	4.36668	9.11054	11.8144	8.9
0.001	2.00200	4.56160	8.64621	15.3553	6.7
0,0001	2.00020	4.61303	8.70427	17.3717	4.9
Padé	2	4.66667	9	20	

TABLE 10.3 $C_{6,6}^{T,\epsilon}$ *Approximants*

ϵ	g_1	g_2	g_3	g_4	g_5	g_6	*Bandwidth*
0.1	2.22222	3.22485	12.0692	5.24156	46.3020	6.60748	20.1
0.01	2.02020	4.17666	7.67971	9.56820	20.8010	17.4477	15.4
0.001	2.00200	4.32419	7.31080	10.8603	18.2222	25.1219	12.5
0.0001	2.00020	4.35503	7.31549	11.2918	18.1255	30.4078	10.2
Padé	2	4.4	7.5	12	20	42	

TABLE 10.4 $C_{8,8}^{T,\epsilon}$ *Approximants*

ϵ	g_1	g_2	g_3	g_4	g_5	g_6	g_7	g_8	*Bandwidth*
0.1	2.22222	3.17570	11.2527	5.01751	33.8387	6.15801	110.007	6.89595	27.3
0.01	2.02020	4.09416	7.21185	8.85975	15.8217	14.6883	38.0217	22.3015	21.9
0.001	2.00200	4.22495	6.87424	9.74959	14.1917	18.0771	30.6729	34.4482	18.6
0.0001	2.00020	4.24781	6.86262	9.97383	14.0939	19.3642	29.7503	43.6209	16.0
Padé	2	4.28571	7	10.4	15	22	35	72	

TABLE 10.5 $C_{10,10}^{T,\epsilon}$ *Approximants*

ϵ	g_1	g_2	g_3	g_4	g_5	g_6	g_7	g_8	g_9	g_{10}	*Ba w*
0.1	2.22222	3.14684	10.8484	4.89287	29.7180	5.92088	74.1077	6.54595	220.110	6.93212	[illegible]
0.01	2.02020	4.04803	6.98063	8.51222	14.1679	13.5604	26.8855	19.4451	61.6595	26.5647	[illegible]
0.001	2.00200	4.17079	6.65884	9.25024	12.8342	15.9156	22.6063	25.7313	46.1582	43.3118	[illegible]
0.0001	2.00020	4.18970	6.64029	9.40400	12.7343	16.5817	22.1310	28.3818	43.4655	56.6610	[illegible]
Padé	2	4.22222	6.75	9.71429	13.3333	18	24.5	34.6667	54	110	

TABLE 10.6 $C_{2,2}^{\phi,\epsilon}$ *Approximants*

ϵ	g_1	g_2	*Bandwidth*
0.5	3.70535	2.33354	5.8
0.05	2.16849	4.46434	3.6
0.005	2.02372	5.37648	2.3
0.0005	2.00361	5.74966	1.4
Padé	2	6	

TABLE 10.7 $C_{4,4}^{\phi,\epsilon}$ *Approximants*

ϵ	g_1	g_2	g_3	g_4	*Bandwidth*
0.5	3.49434	2.38575	18.3977	4.98368	11.8
0.05	2.13023	4.11098	9.74435	10.7492	8.9
0.005	2.01512	4.50022	8.75860	14.3537	6.8
0.0005	2.00185	4.59178	8.68094	16.5731	5.3
Padé	2	4.66667	9	20	

TABLE 10.8 $C_{6,6}^{\phi,\epsilon}$ *Approximants*

ϵ	g_1	g_2	g_3	g_4	g_5	g_6	*Bandwidth*
0.5	3.43000	2.39895	14.0989	4.94952	42.7032	7.69314	17.8
0.05	2.11959	4.00311	7.98190	9.31212	21.2963	17.4899	14.7
0.005	2.01301	4.29519	7.36884	10.7396	18.4593	24.5502	12.1
0.0005	2.00147	4.34875	7.31976	11.2297	18.1524	29.6094	10.1
Padé	2	4.4	7.5	12	20	42	

TABLE 10.9 $C_{8,8}^{\phi,\epsilon}$ *Approximants*

ϵ	g_1	g_2	g_3	g_4	g_5	g_6	g_7	g_8	*Bandwidth*
0.5	3.39933	2.40429	12.6875	4.92593	29.5003	7.58998	76.5590	10.4273	24.0
0.05	2.11476	3.95129	7.40893	8.77311	15.8423	15.1114	36.6499	24.4229	20.5
0.005	2.01210	4.20520	6.91334	9.72163	14.2743	18.1306	30.7750	35.2884	17.8
0.0005	2.00132	4.24533	6.87059	9.96781	14.1262	19.3565	29.8829	43.7059	15.4
Padé	2	4.28571	7	10.4	15	22	35	72	

TABLE 10.10 $C_{10,10}^{\phi,\epsilon}$ *Approximants*

	g_1	g_2	g_3	g_4	g_5	g_6	g_7	g_8	g_9	g_{10}	*Band-width*
	3.38148	2.40702	11.9891	4.90961	25.1341	7.52450	49.6771	10.2717	119.932	13.1750	30.2
	2.11203	3.92090	7.12679	8.49227	14.0424	14.0637	25.6347	21.2783	55.7283	31.4605	26.6
5	2.01161	4.15491	6.68803	9.24601	12.8714	16.0219	22.5761	26.3294	45.5523	46.3235	23.6
05	2.00124	4.18850	6.64816	9.40914	12.7616	16.6229	22.2041	28.6071	43.6192	58.3975	21.0
é	2	4.22222	6.75	9.71429	13.3333	18	24.5	34.6667	54	110	

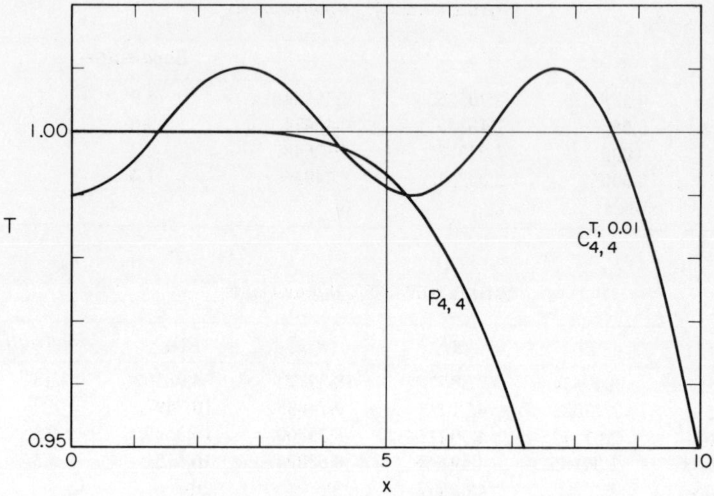

FIG. 10.15 Graph of $\epsilon_T(x)$ using $C_{4,4}^{T,0.01}$ and $P_{4,4}$.

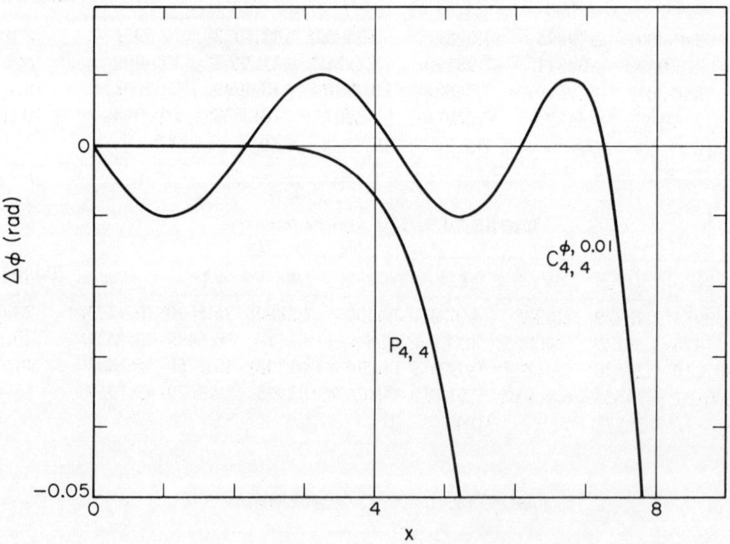

FIG. 10.16 Graph of $\epsilon_\phi(x)$ using $C_{4,4}^{\phi,0.01}$ and $P_{4,4}$.

Since $\omega_{\max} T = 7.5$ rad, it may be determined from the tables or Fig. 10.13 that a minimum of four integrators is required and that $C_{4,4}^{T,0.01}$ is the desired approximant. The g_j/T (Table 10.2) transform to (with four significant figures) 4.040, 8.733, 18.22, and 23.63. Figure 10.17 is the final circuit, where 0.1-μF capacitors are used to lower the gains by a factor of 10. Table 10.2 indicates the nondimensional frequency bandwidth is 8.9. It should be emphasized that the error can be reduced to about 0.3 per cent with four integrators and a bandwidth of 7.5 (see Fig. 10.13), but that the g_j must then be specially computed. Note that five integrators can reduce the error to 0.01 per cent. Thus, one may always trade equipment for accuracy, as well as bandwidth for accuracy or bandwidth for equipment.

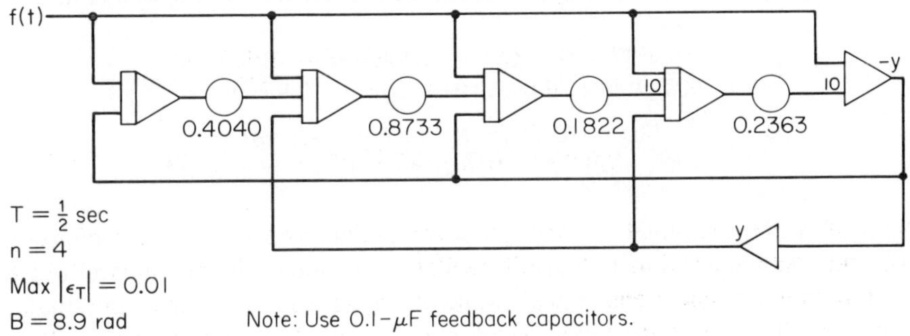

FIG. 10.17 Example 10.7: Mechanization.

To aid in determining whether or not the approximant is set up correctly, it is convenient to use a known $f(t)$ and observe the response. The unit step function $U(t)$ is a convenient known $f(t)$ because it contains all frequencies. Appendix I contains graphs of the time response of the Chebyshev and Padé approximants listed in Tables 10.1-10.10.

By comparing the actual location and magnitude of the peaks with his output, a programmer may gauge the validity of his circuit. If the outputs do not agree, he may troubleshoot the circuit by systematically checking all gains, integrator time constants, etc., until the trouble is located and corrected. It is particularly important to check the output of high-order systems since they are sensitive to small changes in the coefficients. Little difficulty was experienced in obtaining the correct output for all the approximants listed.

For very large bandwidths, circuits for $C_{n,n}^{T,\epsilon}$ can be cascaded and their listed bandwidths summed. In this way, any finite bandwidth can be obtained, although, of course, equipment limitations usually prevent too many integrators from being used. Specialized equipment should be obtained for very large delays as required in some correlation measurements.

10.4.4 Low-pass approximants

The "precursor" or "spike" common to step function responses to all-pass transfer functions (see Appendix I) is troublesome in some applications where discontinuous functions must be delayed. The bandwidth required is then infinite and cannot be simulated with differential equations. However, since some small oscillation about zero is often acceptable in a simulation, and González [30] showed that low-pass approximants are preferred in some correlation studies, some researchers have obtained low-pass approximants with good characteristics. Single [19], for example, simply cascaded a first-order low-pass filter with transfer function $1/(1 + as)$, to his all-pass approximants. Irvine [31] obtained excellent step-function responses by cascading 20 sections of second-order systems with a transfer function of the form $1/(p^2 + ap + b)$. (His passive network arrangements required only 20 amplifiers, not 60.) King and Rideout [32] modified a $P_{2,2}$ approximant and obtained two approximants

$$K_{3,4} = \frac{537 - 175.8Tp + 22.45T^2p^2 - 1.51T^3p^3}{537 + 361.2Tp + 101.9T^2p^2 + 12.93T^3p^3 + T^4p^4}$$

$$R_{3,4} = \frac{476 - 160.5Tp + 22.8T^2p^2 - 1.48T^3p^3}{476 + 309Tp + 91.8T^2p^2 + 12.1T^3p^3 + T^4p^4}$$

both of which exhibited a small precursor oscillation. Their motivation was that the Padé approximant $P_{3,4}$ still oscillated too much in the interval $0 \leq \tau \leq T$ for their applications (a peak value of 16 per cent of the final amplitude). Large amplitude oscillations can be expected for all $P_{n-1,n}$ approximants. $P_{n-2,n}$ transfer functions have not been investigated in much detail.

Budak [33], [34] provides a sophisticated method of finding a suitable approximant $G_{m,n}$ which has maximally flat phase characteristics; i.e., if ϕ is the phase of $G_{m,n}$, $d^j\phi/d\omega^j$ is exact at $\omega = 0$ for all j up to as large an integer as possible [35]. At the same time, one parameter k is available to qualitatively or quantitatively control the precursor oscillation. His method is based on Thomson's [35] and Storch's [36] observation that Bessel all-pole transfer functions

$$T_{0,n}(p) = \frac{B_{0,n}}{\sum_{i=0}^{n} B_{i,n}(pT)^i}$$

where

$$B_{i,n} = \frac{(2n - i)!}{2^{n-1}(n - i)!\, i!} \tag{10.45}$$

have maximally flat phase characteristics and represent a reasonable set of approximations to e^{-pT}. Budak forms the transfer function $B^k_{m,n}(pT)$ in terms of the parameter k:

$$e^{-pT} = \frac{e^{-kpT}}{e^{-(k-1)pT}} \simeq \frac{T_{0,n}(kpT)}{T_{0,m}[(k-1)pT]} \equiv B^k_{m,n}(pT)$$

which produces a delay of T sec. Since the phase shift of the numerator minus the phase shift of the denominator determines the final phase shift, the transfer function $B^k_{m,n}(pT)$ has a very flat phase characteristic for all k. It is interesting to note also that

$$B^{0.5}_{n,n}(pT) = P_{n,n}(pT) \tag{10.46}$$

from the coefficient definitions (10.45) and (10.41). Budak considers only cases where $m = n - 1$, but there is little reason not to consider all-pass approximants where $m = n$; in fact, such approximants are convenient because of (10.46).

To mechanize a general time delay transfer function by Johnson's method, we may again scale each y_j as in (10.43). Because the numerator coefficients a_j may not be identical with those in the denominator b_j, the final system becomes

$$\frac{b_{n-1}}{b_n T}\left(\frac{y_1}{b_{n-1}T^{n-1}}\right) = y - \frac{a_n}{b_n} f(t)$$

$$\frac{d\left(\dfrac{y_j}{b_{n-j}T^{n-j}}\right)}{dt} = -\left[y - \frac{a_{n-j}}{b_{n-j}} f(t)\right] + \left(\frac{b_{n-j-1}}{b_{n-j}T}\right)\left(\frac{y_{j+1}}{b_{n-j-1}T^{n-j-1}}\right)$$

$$\frac{d\left(\dfrac{y_n}{b_0}\right)}{dt} = -y + \frac{a_0}{b_0} f(t)$$

A more general mechanization is therefore shown in Fig. 10.18, where the notations

$$g_j = \frac{b_{j-1}}{b_j}, \qquad h_j = \frac{a_j}{b_j}$$

are used.

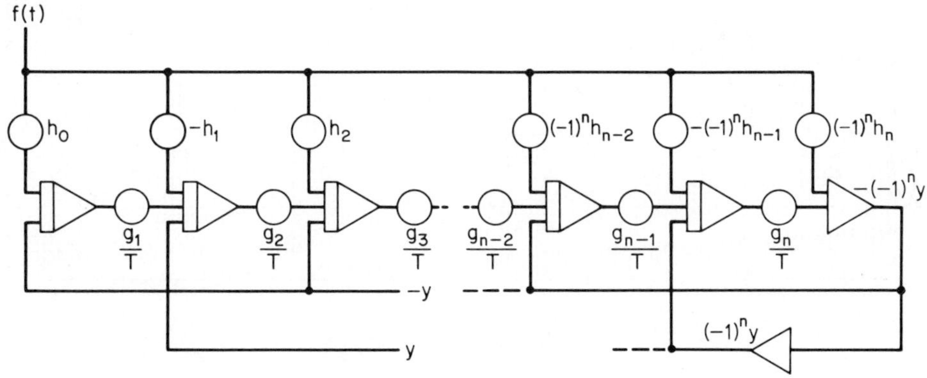

FIG. 10.18 More general time-delay mechanization.

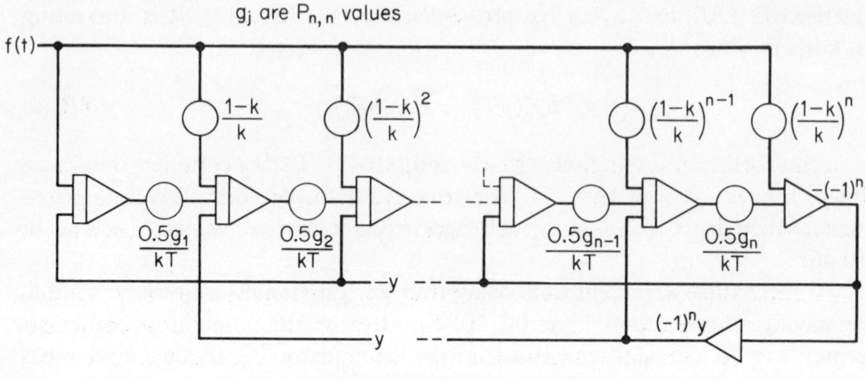

FIG. 10.19 General mechanization of $B^k_{n,n}$.

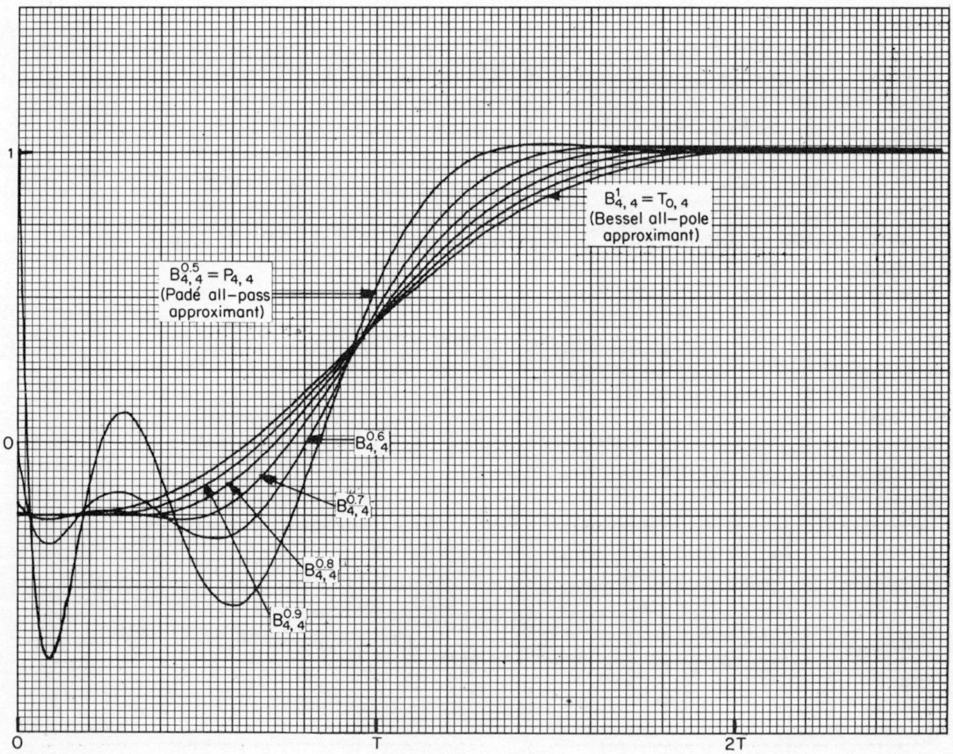

FIG. 10.20 Graph of $B^k_{4,4}$.

We now apply Fig. 10.18 to the generation of the Budak approximants $B^k_{n,n}$ by using (10.46) and (10.45). It is readily computed that

$$\frac{a_j}{b_j} = (-1)^j \left(\frac{1-k}{k}\right)^j$$

and that the g_j require an additional factor of $1/k^n$. However, we may include a factor $1/k^{n-j}$ in the scaling of y_j in order to distribute the k's in the manner that T was distributed. Thus, Fig. 10.19 evolves, where the g_j used are Padé values given in the tables. It is clear that when $k = 1$, the Bessel all-pole approximant $T_{0,n}$ results and when $k = 0.5$, the Padé all-pass approximant $P_{n,n}$ is obtained. For $0.5 < k < 1$, $B^k_{n,n}$ and its attendant flat phase characteristic is mechanized. Note that even for k close to 0.5, the initial amplitude of the precursor spike becomes $[(1-k)/k]^n$, which can be small indeed. In practice, it is necessary to rescale each y_j or to apply the tricks suggested in mechanizing all-pass approximants, to prevent overloads and large coefficients. (Or else, see a mechanization using Beck's method in Sec. 10.5.)

Figure 10.20 shows a graph for step inputs to $B^k_{4,4}$ for various k. As reported by Budak, $0.7 \leq k \leq 0.8$ represents a good compromise between the ringing produced by $P_{4,4}$ and the slow rise time of $T_{0,4}$.

It should be emphasized that low-pass transfer functions do not have a unity gain, and the phase error is, in general, greater than comparable all-pass approximants. The only reason users seek low-pass approximants is the poor qualitative step response inherent in the all-pass approximants. When low-pass approximants are used to delay bandwidth-limited signals, much noticeable error results. These should be avoided when possible.

10.5 Variable Time Delays

In the simulation of certain problems in signal transmission, chemical processes, heat transfer in nuclear reactors, etc., it is necessary to construct circuits approximating a variable time delay $T(t)$. One must not think that circuits to generate constant delays may be arbitrarily used by replacing the constant T by $T(t)$ [37]. The mechanizations use T in the scale factors, for one thing, so that the proofs of all the circuits could not be right. Another reason might be evident from the arguments in Sec. 10.3—a transfer function with time-variable coefficients requires manipulation of the coefficients before simulation. This section might suggest that the first n derivatives of $T(t)$ (for an nth-order approximation) would be required for a proper mechanization, but this is a false concept. We cannot simply expand $e^{-T(t)p}$ in a rational fraction because the simple frequency analysis of Theorem 10.2 breaks down. When a pure frequency $\sin \omega t$ is the input to the proper transfer function producing a variable delay, its output must be $\sin \omega[t - T(t)]$ within some approximation error. In (10.7), the A and B are therefore variable functions of time, which destroys the theorem's proof. A rational fraction expansion of e^{-x} does not provide a transfer function with the proper phase shifts when $pT(t)$ is substituted for x.

In this section, two methods of producing variable delays are discussed. In the first, an actual physical situation involving flow of an incompressible fluid in pipes requires a variable delay when the flow is variable but not of the simple $f[t - T(t)]$ variety. Margolis and O'Donnell [38] first used this example to justify their technique for mechanization, and Vichnevetsky [39] later rigorized it. In the second method, the formal generation of $f[t - T(t)]$ is attacked by expanding $e^{-pT(t)}$ in a power series about e^{-pT_0}, a constant delay, in the manner of Chernyshev [40]. Such an expansion gives the proper output.

10.5.1 Constant time delay using Beck's method

Beck's method is rarely used to mechanize a constant time delay circuit because an extra amplifier is required. However, the circuit lends itself nicely in the extension to the first type of variable delay, so that we now consider mechanization, by system (10.8)–(10.10), of a unity-gain all-pass time-delay approximation (10.40). Again we have $a_j = (-1)^j b_j$ with all $b_j > 0$. Reasonable scaled variables u_j [in (10.11)] are proportional to $(T^j b_j u_j)$, which transform the equations to

$$\begin{aligned}
\frac{d(b_0 u_0)}{dt} &= \frac{b_0}{Tb_1}(Tb_1 u_1) = \frac{g_1}{T}(Tb_1 u_1)\\
\frac{d(Tb_1 u_1)}{dt} &= \frac{g_2}{T}(T^2 b_2 u_2)\\
&\ \ \vdots \qquad\qquad \vdots\\
\frac{d(T^{n-2}b_{n-2}u_{n-2})}{dt} &= \frac{g_{n-1}}{T}(T^{n-1}b_{n-1}u_{n-1})\\
\frac{d(T^{n-1}b_{n-1}u_{n-1})}{dt} &= \frac{g_n}{T}(T^n b_n u_n) = \frac{g_n}{T}\left[f(t) - \sum_{j=0}^{n-1} T^j b_j u_j\right]
\end{aligned} \tag{10.47}$$

with each g_j defined as in (10.44). The required inputs to the final summer now can be taken directly from each amplifier output as derived from (10.9). The resulting circuit (Fig. 10.21) is also a minimum potentiometer circuit with only one input lead per integrator, an advantage when $\dot{t}$ is not constant. Further scaling is not required for full amplitude step inputs, a decided advantage over the mechanization of a constant delay using Johnson's method.

10.5.2 Margolis' and O'Donnell's treatment [38]

Consider the temperature $y(t)$ (or any other property) of a stream of particles in an incompressible, ideal fluid in a pipe, passing position x_2 while flowing with a uniform velocity profile $V(t)$, as shown in Fig. 10.22. Assuming the

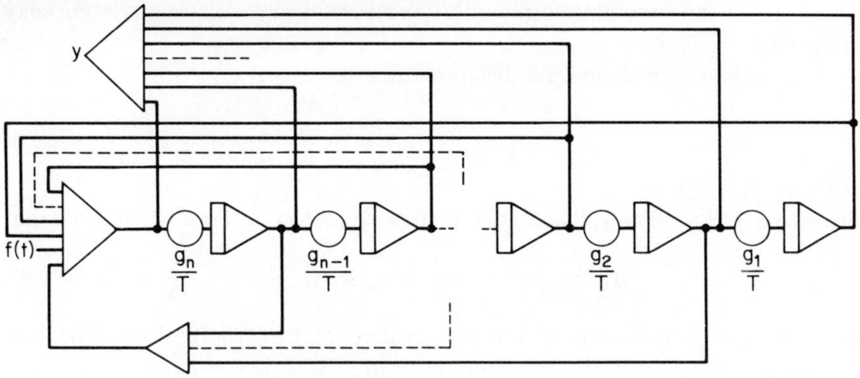

Notes: (1) $i=1$
(2) Shown for n even.
(3) Amplifier output $-(-1)^i T^i b_i u_i$ feeds pot g_i/T.

FIG. 10.21 All-pass time-delay mechanization using Beck's method.

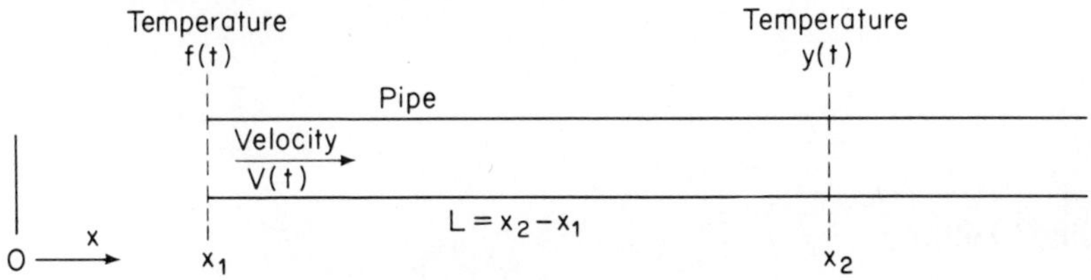

FIG. 10.22 Diagram for Margolis' and O'Donnell's argument.

temperature of individual particles does not change during the flow, we may readily conclude that

$$y(t) = f(t - T) \tag{10.48}$$

where $f(t)$ is the temperature of the stream passing position x_1 and T is the time of travel from x_1 to x_2. T changes with time because $V(t)$ is variable. In fact, an individual particle has a velocity $v(x)$ as a function of x so that

$$T = \int_{x_1}^{x_2} \frac{dx}{v(x)}$$

When stated as a function of time, this typical problem requires a variable delay in (10.48).

Let us, however, define the distance variable

$$q = \int_{t_0}^{t} V(t)\,dt \tag{10.49}$$

Let $z(q)$ correspond to $y(t)$ and $g(q)$ correspond to $f(t)$. We easily conclude that

$$y(t) = z(q) = g(q - L) = f(t - T) \tag{10.50}$$

because every particle must travel the constant length $L = x_2 - x_1$. When expressed as a function of distance, the delay in (10.50) is constant.

There is mathematically no difference in approximating, by differential equations, (10.50) with constant L or (10.37) with constant T. Accordingly, we use any suitable all-pass transfer function and write equations corresponding to (10.47):

$$\begin{aligned}
\frac{d(b_0u_0)}{dq} &= \frac{g_1}{L}(Lb_1u_1) \\
\frac{d(Lb_1u_1)}{dq} &= \frac{g_2}{L}(L^2b_2u_2) \\
\vdots \qquad & \qquad \vdots \\
\frac{d(L^{n-1}b_{n-1}u_{n-1})}{dq} &= \frac{g_n}{L}\left[g(q) - \sum_{j=0}^{n-1} L^j b_j u_j\right]
\end{aligned}$$

From (10.49),

$$\frac{dq}{dt} = V(t)$$

In terms of t,

$$\begin{aligned}
\frac{d(b_0u_0)}{dt} &= g_1\frac{V(t)}{L}(Lb_1u_1) \\
\frac{d(Lb_1u_1)}{dt} &= g_2\frac{V(t)}{L}(L^2b_2u_2) \\
\vdots \qquad & \qquad \vdots \\
\frac{d(L^{n-1}b_{n-1}u_{n-1})}{dt} &= g_n\frac{V(t)}{L}\left[f(t) - \sum_{j=0}^{n-1} L^j b_j u_j\right]
\end{aligned}$$

The function $L/V(t)$ may have units of time, but it most certainly is not equal to $T(t)$ unless V is constant. Thus, the constant time delay circuit in Fig. 10.21

can be carried over directly with each g_j/T being replaced by $g_j V(t)/L$, a process which requires n multipliers. The general schematic for each integrator is shown in Fig. 10.23. It should be pointed out that by using

$$\dot{t} = \frac{L}{V(t)}$$

the n multiplications are eliminated. This may not save equipment if too many other integrators are involved in the simulation, since inputs to all other integrators must be multiplied by $\dot{t}$, too.

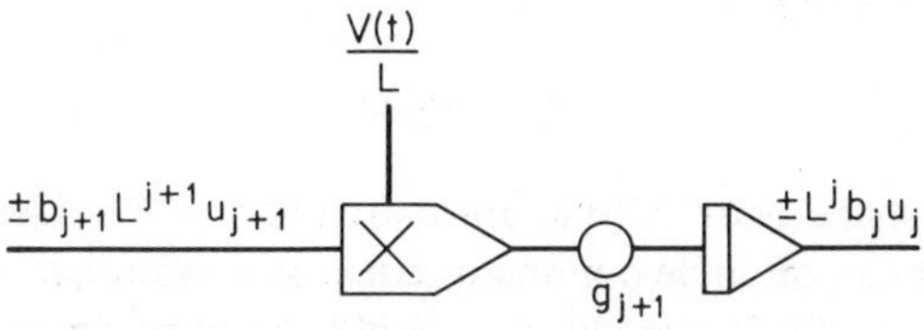

FIG. 10.23 Integrator adjustment for incompressible flows.

10.5.3 Chernyshev's expansion [40]

In cases where $T(t)$ is explicitly given, Chernyshev proposed an interesting approximation to the transfer function $e^{-pT(t)}$:

$$W(t, p) = e^{-pT(t)} = (e^{-pT_0})^{T(t)/T_0} \tag{10.51}$$

where T_0 is constant. Now e^{-pT_0} represents a constant delay and can be approximated by an all-pass transfer function

$$e^{-pT_0} = \frac{\sum_{j=0}^{n} (-1)^j b_j p^j T_0^j}{\sum_{j=0}^{n} b_j p^j T_0^j}$$

$$= 1 - \frac{2\sum_{j=1}^{n/2} b_{2j-1} p^{2j-1} T_0^{2j-1}}{\sum_{j=0}^{n} b_j p^j T_0^j}$$

for positive even n. By letting

$$a(t) = \frac{T(t)}{T_0}, \qquad G_{n-1,n}(pT_0) = \frac{2\sum_{j=1}^{n/2} b_{2j-1} p^{2j-1} T_0^{2j-1}}{\sum_{j=0}^{n} b_j p^j T_0^j}$$

(10.51) can be expanded in a binomial series,

$$W(t, p) = 1 - a(t)G_{n-1,n}(pT_0) + \frac{a(t)[a(t)-1]}{2!}[G_{n-1,n}(pT_0)]^2 + \cdots$$
$$= 1 + \sum_{j=1}^{\infty} (-1)^j \left(\prod_{k=1}^{j} [a(t) - k + 1]\right) \frac{[G_{n-1,n}(pT_0)]^j}{j!} \qquad (10.52)$$

which converges for all $|G_{n-1,n}(pT_0)| < 1$. This condition determines the choice of T_0. If we truncate (10.52) after using m terms, a modulus error M results:

$$M = \left| \prod_{k=1}^{m} [a(t) - k + 1] \frac{G_{n-1,n}(pT_0)}{m!} \right|$$

as well as a phase error

$$\phi = a(t)\psi$$

where ψ is the phase error of the approximant to e^{-pT_0}. Despite these errors, reasonable results can often be obtained with a small number of terms, especially when the maximum of ωT is small. A general schematic for mechanizing (10.52) is shown in Fig. 10.24.

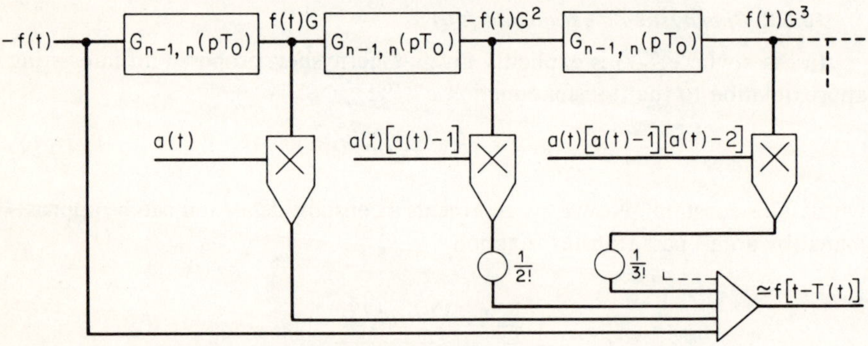

FIG. 10.24 General schematic for mechanizing Chernyshev's expansion.

Example 10.8. Construct a circuit using three terms of Chernyshev's expansion for a time delay of

$$T(t) = 0.9 + 0.1 \sin t$$

for a slowly changing signal containing frequencies no greater than 1 rad/s.

We first determine a value of T_0 that minimizes the maximum modulus error M. To do this, let us assume that the all-pass transfer function used has virtually no phase error. In that case, the modulus M of $G_{n-1,n}(pT_0)$ is seen from Fig. 10.25 to be

$$|G_{n-1,n}(pT_0)| = 2 \sin\left(\tfrac{1}{2}\,\omega T_0\right)$$

since the all-pass function has unity gain for all ω. The condition for convergence

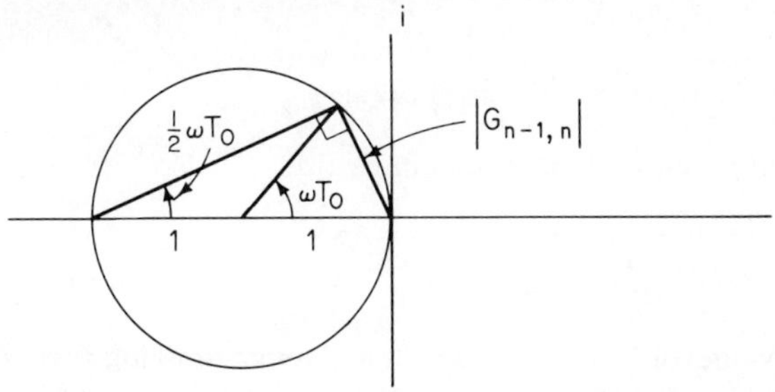

FIG. 10.25 Determining $|G_{n-1,n}(pT_0)|$ from its complex-plane graph.

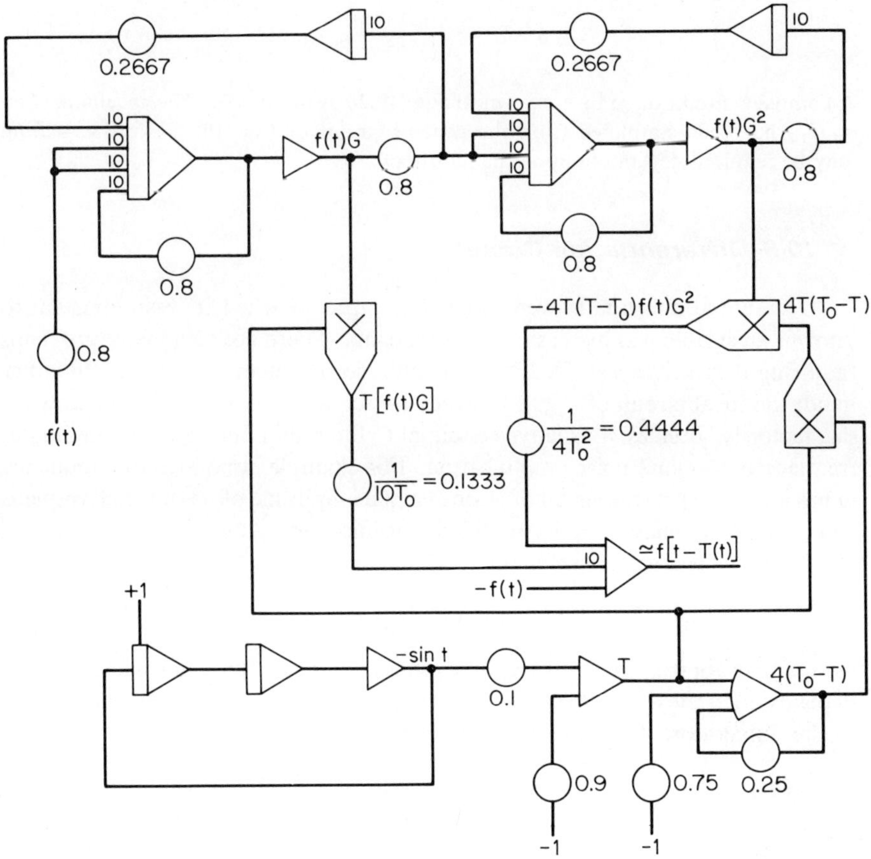

FIG. 10.26 Example 10.8.

then becomes

$$\sin\left(\tfrac{1}{2}\,\omega T_0\right) < \tfrac{1}{2}$$

For a three-term series, we thus seek to minimize

$$M = \left|\frac{4}{3}\frac{T}{T_0}\left(\frac{T}{T_0} - 1\right)\left(\frac{T}{T_0} - 2\right)\sin^3\left(\frac{\omega T_0}{2}\right)\right|$$

Using extreme values of $T = 1$ and $\omega = 1$, and approximating $\sin\left(\tfrac{1}{2}\,\omega T_0\right) = \tfrac{1}{2}\,\omega T_0$, we find, by taking $dM/dT_0 = 0$, that $T_0 = \tfrac{3}{4}$ s. Thus, M is now at worst equal to $1/48$ or about 2 per cent of the amplitude of $f(t)$ at frequencies near $\omega = 1$. Since $(\omega T_0)_{\max} = 0.75$, a second-order $P_{2,2}$ circuit has virtually no phase error and can be used, so that

$$G_{1,2}(0.75p) = \frac{9p}{\left(\frac{3}{4}\right)^2 p^2 + \frac{9}{2}p + 12}$$

A complete circuit diagram is shown in Fig. 10.26 using $i = 1$. The mechanization of $G_{1,2}$ has been simplified from the general circuit of Fig. 10.18. Further scaling may be required depending upon the nature of $f(t)$.

10.6 Differentiation Circuits

Differentiation circuits were used in Chapter 6 when it was important to know if a variable was increasing or decreasing. There are also many problems requiring derivatives as a feedback quantity (see Problem 10-18). On the other hand, the ideal circuit of Fig. 2.13 generally requires too much current to work satisfactorily. Besides, it usually is essential to filter an input signal at the higher frequencies to avoid noise amplification. For example, if noise at the frequency ω has amplitude A (the signal of $A \sin \omega t$), the amplitude of its time derivative is $A\omega$. High-frequency noise is amplified considerably.

When viewed as a transfer function,

$$G(p) = p$$

we see that approximation is required, since a transfer function mechanization depends upon the denominator being a polynomial in p of degree at least as high as the numerator. The simplest approximation is

$$G(p) = \frac{gp}{bp + 1}$$

which is shown mechanized in Fig. 10.27, based upon the standard general mechanization using Johnson's method simplified. The value of g is usually a scale factor, while b can be made as small as desired to determine the cutoff frequency. A circuit of this type was recommended as the most versatile by

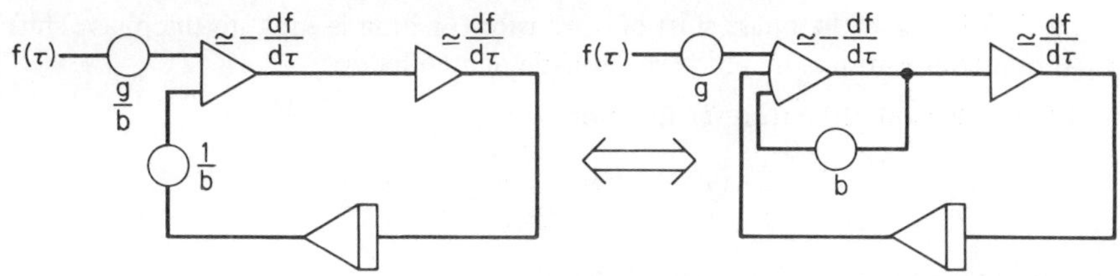

FIG. 10.27 Simple derivative circuits.

Fairchild and Krovetz [41], who examined several circuits. Besides the convenient adjustment of b and g, both $df/d\tau$ and $-df/d\tau$ are available for either an f or $-f$ input.

King [42] describes an interesting two-amplifier circuit (Fig. 10.28) applicable when external capacitors are available for patching. If R is the input and

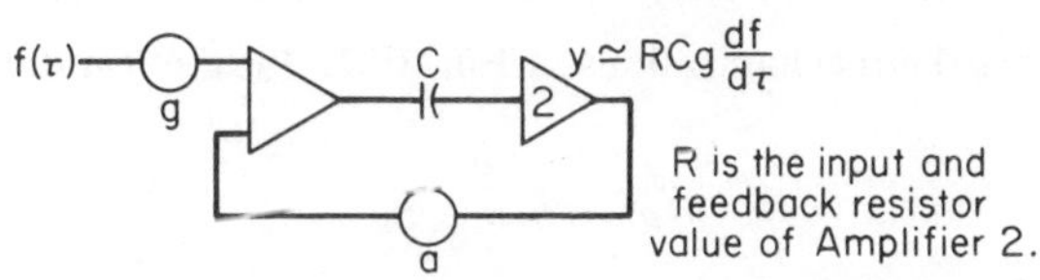

FIG. 10.28 King's circuit for differentiation.

feedback resistor values, and C is the patched capacitor value, one may show that

$$y = \frac{gp}{\frac{1}{RC} + p(1 - a)} f(\tau)$$

providing nonideal amplifier characteristics are ignored. When $a = 1$, the output equals the derivative $RCg(df/d\tau)$. King suggests that $0.995 \leq a \leq 0.999$ to reach a compromise between satisfactory transient response and noise rejection. The output can always be adjusted for an inaccurate capacitor by adjusting g.

Second-order approximations to derivative circuits requiring two integrators have also been used. In general, these are used when $f(\tau)$ is so noisy that additional filtering is desired [41]. However, care should be exercised to insure that the simulation is not sensitive to pot settings and is stable [43].

PROBLEMS

10-1. Show that the output of two cascaded transfer functions is invariant to their order.

10-2. Show that the phase shift of a transfer function is equal to the phase shift of the numerator minus the phase shift of the demoninator.

10-3. Mechanize the transfer function

$$\frac{(p+2)(p-3)(p+1)}{(p^2+p+3)(p+1)}$$

using Beck's method, assuming all required derivatives are initially zero.

10-4. Repeat Problem 10-3 with initial conditions

$$f = \frac{df}{dt} = \frac{d^2f}{dt^2} = 0, \qquad x = \frac{dx}{dt} = \frac{d_2x}{dt^2} = 1$$

10-5. Repeat Problem 10-4 using Johnson's method.

10-6. Show that the common factor in Problem 10-3 can be cancelled if all required derivatives are zero initially.

10-7. Repeat Problem 10-4 using Laning's and Battin's method.

10-8. Show that the mechanizations in Fig. 10.29 for the transfer function

$$\frac{a_1p + a_0}{b_2p^2 + b_1p + b_0}$$

are equivalent.

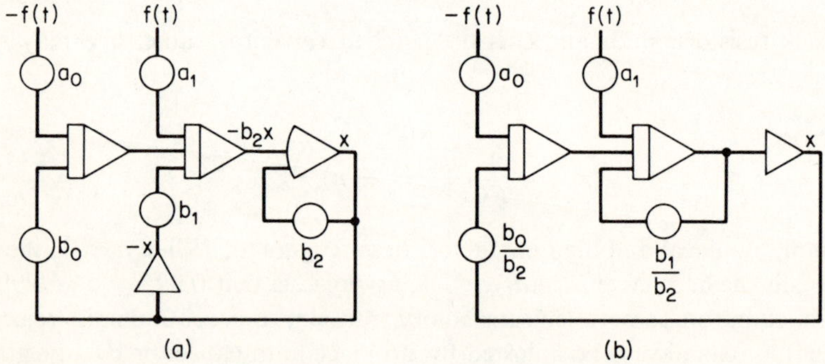

FIG. 10.29 Equivalent mechanizations.

10-9. Using Matyash's method for mechanizing a linear differential equation with variable coefficients, mechanize the equation

$$\frac{d^2x}{dt^2} + \cos t\frac{dx}{dt} + x\sin t = e^{-t}\frac{df}{dt} + e^{-2t}f$$

10-10. Repeat Problem 10-9 for Laning's and Battin's method.

10-11. Prepare a circuit for a 4-s delay when $0 \leq \omega \leq 3.75$ so that the error does not exceed 1 per cent (or 0.04 s).

10-12. Repeat Problem 10-11, but keep the phase shift error less than 0.0005 rad.

10-13. Mechanize a $B^k_{8,8}$ approximant and determine a value of k that produces a reasonably high-rise, low-oscillating step response.

10-14. Mechanize the delays in Problems 10-11 and 10-12 by Beck's method.

10-15. Construct a circuit to delay $f(t)$ for

$$T(t) = 0.5 + 0.5e^{-0.1t}$$

using as many terms in Chernyshev's expansion as required to keep the modulus error below 2 per cent. Assume $0 \leq \omega \leq 1$ rad/s. [For simplicity, take $(\omega T_0)_{\max} = 1$.]

10-16. Filter theory plays an important part in network analysis and design [44], [45]. Construct a circuit mechanizing the following fifth-order elliptic filter:

$$\frac{0.073796p^4 + 0.272353p^2 + 0.22573}{p^5 + 1.05019p^4 + 1.7549p^3 + 1.15975p^2 + 0.7129p + 0.22573}$$

10-17. A block diagram for a simplified version of a temperature controller for a single capacity process with dead time is shown in Fig. 10.30 [46]. Assuming the flow

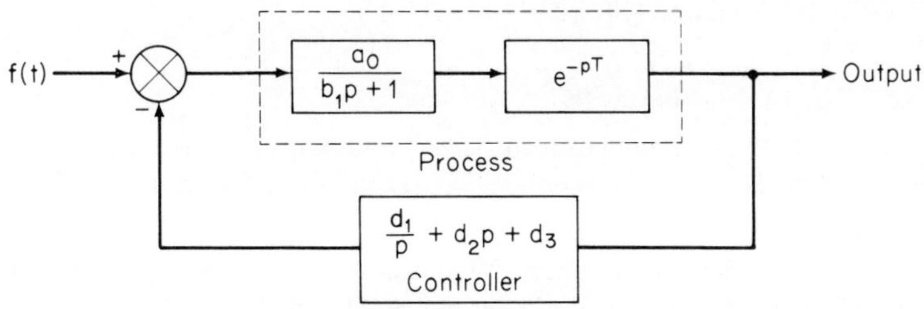

FIG. 10.30 Temperature controller block diagram.

rate is constant (so that the dead time is constant), draw an unscaled circuit diagram to simulate the process and controller on the computer, making sure that each parameter can be varied easily to ascertain its effect on the simulation. Assume a $C^{\phi,0.0005}_{4,4}$ approximant is adequate for the delay circuit.

10-18. A flood routing equation is given by the following nonlinear equation [47]:

$$\frac{dy}{dt} = C_0(y)[f(t) - y] - C_1(y)\frac{df}{dt}$$

Because C_0 and C_1 are given functions of y, not t, containing certain characteristics of the reach, a derivative circuit cannot be eliminated. Draw an unscaled computer diagram for the computer solution of the equation.

REFERENCES

[1] Coddington, E. A., *An Introduction to Ordinary Differential Equations* (Englewood Cliffs, N. J.: Prentice-Hall, 1961).

[2] Beck, C., "A Method for Solving Problems on the REAC by the Use of Transfer Functions without Passive Networks," *Symposium I on REAC Techniques*, Reeves Instr. Corp., New York, March 15-16, 1951 (ASTIA No. 108488).

[3] Beck, C., "Treating Transfer Functions on Analog Computers," *Electrical Manufacturing*, vol. 62, Oct. 1958.

[4] Langill, A. W., Jr., "An 'Idiot-Proof' Method for Analog of Transfer," *Electronic Design*, vol. 10, June 21, 1962.

[5] Levine, L., *Methods for Solving Engineering Problems Using Analog Computers* (New York: McGraw-Hill Book Co., 1964).

[6] Marzollo, A., "Nonobservable or Noncontrollable Analog Schemes for the Solution of a Special Class of Differential Equations," *IEEETEC*, vol. 15, Oct. 1966.

[7] Johnson, C. L., *Analog Computer Techniques*, 2nd ed. (New York: McGraw-Hill Book Co., 1963).

[8] Matyáš, J., "Letter to the editor," *Automatika i Telemekhanika*, vol. 21, Oct. 1960.

[9] Laning, J. H., Jr., and R. H. Battin, *Random Processes in Automatic Control* (New York: McGraw-Hill Book Co., 1956).

[10] Miller, K. S., and J. B. Walsh, "Initial Conditions in Computer Simulation," *IRETEC*, vol. 10, March 1961.

[11] Brockett, R. W., "Optimal Control Theory and Applications," *Compressed Time and Statistical Analog Computation* (course notes), GPS Instr. Co., Inc., Needham, Mass.

[12] Heizer, L. E., and S. J. Abraham, "Transfer Function Simulation by Means of Amplifiers and Potentiometers," *Journal ACM*, vol. 3, July 1956.

[13] Kogan, B. Ya., "On Methods of Simulating Rational-Fraction Functions without the Use of Differentiating Elements," *Automatika i Telemekhanika*, vol. 21, Jan. 1960.

[14] Matyash, I., "Methods of Analog Computer Solution of Linear Differential Equations with Variable Coefficients," *Automatika i Telemekhanika*, vol. 20, July 1959.

[15] Vichnevetsky, R., "An Extension of a Theorem of Linear Circuit Theory to Discontinuous Delay Simulators," *AICA*, vol. 4, Oct. 1962.

[16] Thomson, W. E., "Time-Delay Circuits (correspondence)," *IRETEC*, vol. 4, June 1955.

[17] Morrill, C. D., "A Sub-Audio Time Delay Circuit," *IRETEC*, vol. 3, June 1954.

[18] Cunningham, W. J., "Time-Delay Networks for an Analog Computer," *IRETEC*, vol. 3, Dec. 1954.

[19] Single, C. H., "An Analog for Process Lags," *Control Eng.*, vol. 3, Oct. 1956.

[20] Warfield, J. N., *Introduction to Electronic Analog Computers* (Englewood Cliffs, N. J.: Prentice-Hall, Inc., 1959).

[21] Smith, G. W., and R. C. Wood, *Principles of Analog Computation* (New York: McGraw-Hill Book Co., 1959).

[22] Wierwille, W. W., "A New Method for Obtaining Continuous Delays on the Analog Computer," *IEEETAC*, vol. 8, Jan. 1963.

[23] Hepner, C. F., "Improved Methods of Simulating Time Delays," *IEEETEC*, vol. 14, April 1965.

[24] Wierwille, W. W., "Experimental Study of a New Method of Time Delay for Analog Computers," *IEEETEC*, vol. 14, Aug. 1965.

[25] Bridgman, A., and W. W. Wierwille, "Comments on 'Experimental Study of a New Method of Time Delay for Analog Computers'," *IEEETEC*, vol. 15, Aug. 1966.

[26] Hausner, A., "Letter to the Editor (correspondence)," *IEEETEC*, vol. 14, Oct. 1965.

[27] Hausner, A., and C. M. Furlani, "Chebyshev All-Pass Approximants for Time-Delay Simulation," *IEEETEC*, vol. 15, June 1966.

[28] Herrmann, O., "About Transfer Functions with Approximately Constant Time-Delay," *IEEETEC*, vol. 15, June 1966.

[29] Hausner, A., and C. M. Furlani, "Tables of Best Tchebycheff Approximants for Time-Delay Simulation," *Harry Diamond Labs. Report TM-65-58*, Washington, D. C.

[30] González, G., "Delay Approximations for Correlation Measurements Using Analog Computers," *IEEETEC*, vol. 14, Aug. 1965.

[31] Irvine, N.L., "Simulation of Dead Time," *Proc. National Simulation Conference*, Jan. 19–21, 1956, Dallas, Texas.

[32] King, W. J., and V. C. Rideout, "Improved Transport Delay Circuits for Analog Computer Use," *Proceedings of the 3rd Inter. Analogue Comp. Meetings* (Brussels: Presses Académiques Européennes, 1962).

[33] Budak, A., "A Maximally Flat Phase and Controllable Magnitude Approximation," *IEEETCT*, vol. 12, June 1965.

[34] Budak, A., "An Improved Time Delay Simulation," *IEEETEC*, vol. 15, Feb. 1966.

[35] Thomson, W. E., "Networks with Maximally-Flat Delay," *Wireless Eng.*, vol. 29, Oct. 1952.

[36] Storch, L., "Synthesis of Constant-Time-Delay Ladder Networks Using Bessel Polynomials," *Proc. IRE*, vol. 42, Nov. 1954.

[37] Korn, G. A., "Network Approximations of Variable Time Delays Require Caution," *IRETEC*, vol. 11, Feb. 1962.

[38] Margolis, S. G., and J. J. O'Donnell, "Rigorous Treatments of Variable Time Delays," *IEEETEC*, vol. 12, June 1963.

[39] Vichnevetsky, R., "Analogue Computer Simulation of Time-Dependent Delay Using the Concept of Generalised Transfer Function," *AICA*, vol. 6, April 1964.

[40] Kogan, B. Ya., and M. K. Chernyshev, "Delay Simulation with the Aid of Operational Amplifiers," *Automatika i Telemekhanika*, vol. 27, Mar. 1966.

[41] Fairchild, B. T., and L. J. Krovetz, "10 Circuits for Differentiation on Analog Computers," *Control Eng.*, vol. 12, Feb. 1965.

[42] King, R. E., "A Dual Amplifier Differentiator," *IEEETEC*, vol. 15, Feb. 1966.

[43] Bridgeman, A., "A Note on the Simulation of the Imperfect Differentiator," *Simulation*, vol. 4, Jan. 1965.

[44] Bridgman, A., "Analog Techniques for Filter Simulation," *Simulation*, vol. 8, April 1967.

[45] Larrowe, V. L., "Band-Pass Quadrature Filters," *IEEETEC*, vol. 15, Oct. 1966.

[46] "Simulation of a Process Controller," *Goodyear Aircraft Corp. Report GER-4689*, Akron, Ohio, March 11, 1952.

[47] "GEDA Application to Flood-Control and Water-Distribution Problems," *Goodyear Aircraft Corp. Report GER-6053*, Akron, Ohio, May 4, 1954.

See also:

[48] Liu, B., "A Time Domain Approximation and its Application to Lumped Delay Lines," *IRETCT*, vol. 9, Sept. 1962.

[49] Noronha, L. G., "Analogue Simulation of Transfer Functions," *AICA*, vol. 6, Oct. 1964.

[50] Ahlbeck, D. R., "Simulating a Jet Gas Turbine with an Analog Computer," *Simulation*, vol. 7, Sept. 1966.

[51] Seddon, J. P., and R. A. Johnson, "The Simulation of a Variable Delay," *IEEETC*, vol. 17, Jan. 1968.

[52] McAvoy, T. J., "Least-Square Dead-Time Approximations," *IEEETC*, vol. 17, Feb. 1968.

[53] Thomas, C. H., "Transport Time-Delay Simulation for Transmission Line Representation," *IEEETC*, vol. 17, March 1968.

Chapter 11

ALGEBRAIC EQUATIONS AND PROGRAMMING PROBLEMS

In this chapter, we describe methods to find the roots of an explicit set of algebraic (or transcendental) equations

$$f_i \equiv f_i(\tau, x_1, x_2, \cdots, x_n) = 0, \quad i = 1, 2, \cdots, n \tag{11.1}$$

where τ implies the equations contain time-varying parameters. Two situations arise in practice:

(1) Some problems require the solution of (11.1) as an end product. There are no time-varying parameters (a *stationary* problem), but there may be run-varying parameters necessitating many solutions.

(2) Other problems require the use of the x_k as feedback in a larger problem involving other differential equations. A classic example is in the solution of Lagrange's equations (Chapter 12), where the x_k are time-varying accelerations. We must track the solutions of (11.1) as parameters change with time (a *tracking* problem).

If $n = 1$ or 2, we may find solutions of stationary problems by using elementary geometrical concepts and function plotting methods. For larger n, it is easier to use a digital computer to solve stationary problems, but we consider analog methods in case a digital computer is not available, and to build competence in solving such systems. An understanding of how solutions can be obtained will help prepare for the solutions of more general optimization problems discussed in other chapters, particularly Chapter 19.

For tracking problems, we wish to extend the implicit function generation technique of Chapter 9, where a high-gain amplifier circuit solved one algebraic equation in one variable and tracked solutions when the equation contained time-varying parameters. We require the f_i of (11.1) to be instantaneously generable from a given set of x_k by fixed function-generating equipment (no integrators). We seek to supply n inputs $\phi_i(\tau, x_1, x_2, \ldots, x_n)$ into n high-gain amplifiers so that the roots $x_1, x_2, \ldots, x_n$ appear at their outputs, as shown in Fig. 11.1.

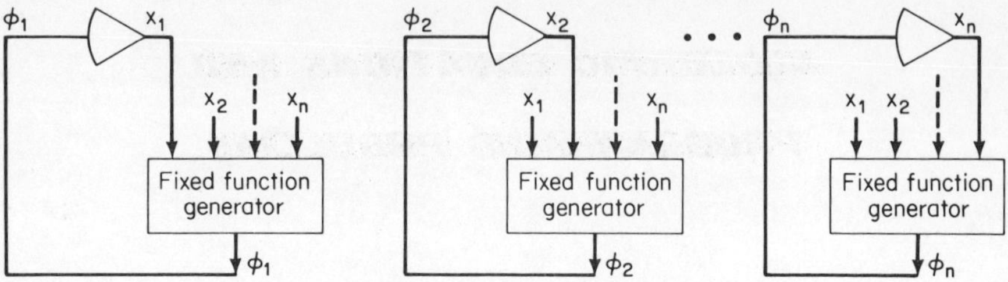

FIG. 11.1 General algebraic equation solver.

The functions ϕ_i must vanish when (11.1) are satisfied, and the arrangement must be stable or else noise will quickly upset any solution. There may be zero or more (even an infinite set of) solutions to (11.1). Because the high-gain amplifiers behave as integrators, the stability of such circuits ultimately depends upon the stability of

$$\dot{x}_i + g_i\phi_i = 0, \qquad i = 1, 2, \cdots, n \tag{11.2}$$

where $g_i \gg 0$. As in the case of one equation in one unknown, we can minimize the effects on nonideal components by literally programming (11.2), as shown in Fig. 11.2. This may be done by providing very small capacitors as feedback around each high-gain amplifier as was done in Fig. 9.3.

The problem is that of determining the $g_i\phi_i$ so that (11.2) is stable. Much attention has been devoted to this problem that may be solved by a variety of general methods.

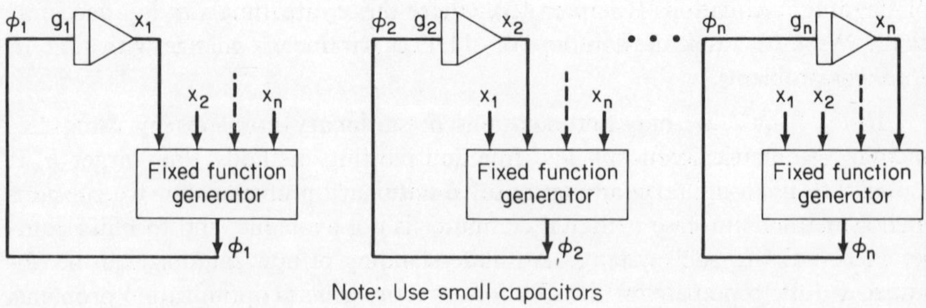

FIG.11.2 General algebraic equation solver using integrators.

In all high-gain amplifier methods, tracking errors exist when parameters change with time. It is sometimes necessary to estimate the magnitude of such errors by a technique similar to that of Theorem 9.2 for an equation in one variable.

The solution of a set of algebraic equations may be considered as a special optimization problem, called a *nondynamic* problem. This chapter also discusses solutions of programming problems. This is another nondynamic problem where there are more unknowns than equations, but additional inequality constraints are imposed on the system, and a given function of the x_i is to be maximized or minimized at the same time. The methods used to solve programming problems are similar to those for solving simple algebraic equations.

Some other nondynamic problems are discussed in other chapters. See Chapter 15, for example, for the determination of eigenvalues and eigenvectors of matrices.

Random search techniques for solving general optimization problems (including nondynamic ones) are discussed in Chapter 19.

11.1 Graphical Methods

When applicable, the simplest method to obtain solutions to $f(x) = 0$ is to draw a graph of $f(x)$ and read the roots. An advantage of this is that many complicated expressions in f are generable accurately with integrators. Of course, if the roots must be tracked for use in other circuits, graphing is inapplicable. For two equations in two variables

$$f_1(x_1, x_2) = 0 \tag{11.3}$$

$$f_2(x_1, x_2) = 0 \tag{11.4}$$

graphical methods may also be used. With x_1 considered an independent variable, (11.3) and (11.4) define two implicit functions, $x_2 = \phi_1(x_1)$ and $x_2 =$

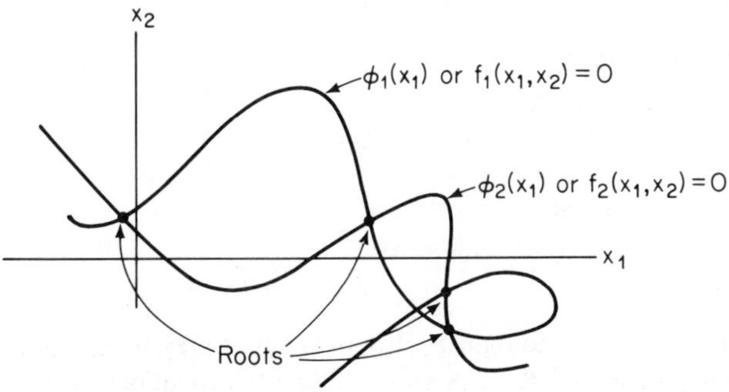

FIG. 11.3 Graphical solution of a system of two equations.

$\phi_2(x_1)$, that can be graphed on the same scale (Fig. 11.3). The coordinates of the intersection of the curves are read directly from the graph and are solutions of (11.3) and (11.4).

Another reason for obtaining graphs in such simple situations is that it is very easy to do. With the parametric technique, accurate graphs can be obtained even when subsequent methods using fixed function-generating equipment fail. The following example is a case in point, although perhaps a special case.

Example 11.1. Graph the 9th-order Chebyshev polynomial

$$T_9 = 256x^9 - 576x^7 + 432x^5 - 120x^3 + 9x \tag{11.5}$$

to determine its real roots [1]. This polynomial was chosen because of its highly oscillatory nature. We will ignore alternate schemes for plotting this function which depends upon any special characteristics of Chebyshev polynomials, in order to provide a general technique for getting accurate polynomial graphs. (All Chebyshev polynomials may actually be graphed with two oscillator circuits—see Chapter 7.)

Any method used will eventually determine that all the roots x_i are such that $-1 < x_i < 1$, which produces the difficulty. Equation (11.5) is already properly scaled, but the large gains preclude summing terms $a_i x^i$, no matter how the powers of x are generated. For example, an error in x^7 of 0.0001 results in an error of 0.0576 in T_9, which is already greater than 5 per cent. Since small errors are inevitable, meaningless results are obtained.

Using $\dot{x} = 1$, T_9 can be generated by integration with system (5.12) and (5.13). But the error polynomial due to the open-loop integration (Sec. 5.6.2) can be shown to grow so rapidly that meaningless results are also quickly obtained.

With $\dot{x} = -x$, however, linear equipment circuits with negative feedback will be developed that produce very accurate results. By using the summation Theorem 5.1, we may readily program a polynomial as the output of a set of cascaded exponentiator circuits as follows:

For any polynomial P_n,

$$P_n = \sum_{i=0}^{n} a_i x^i \tag{11.6}$$

define another polynomial

$$P_{n-1} = -\sum_{i=0}^{n} a_i(n-i)x^i \tag{11.7}$$

It follows for $\dot{x} = -x$, that

$$\dot{P}_n = -nP_n - P_{n-1} \tag{11.8}$$

Since P_{n-1} is of lower degree than P_n [from (11.7)], it may be generated similarly by defining a new polynomial P_{n-2}. A set of differential equations of the form

$$\dot{P}_i = -iP_i - P_{i-1}$$

results, where the process terminates eventually since $i \to 0$. When a polynomial linear with x is needed, it can be formed directly from the amplifier producing x. For the polynomial T_9, we have

$$\begin{aligned} P_7 &= 1152x^7 - 1728x^5 + 720x^3 - 72x \\ P_5 &= 3456x^5 - 2880x^3 + 432x \\ P_3 &= 5760x^3 - 1728x \\ P_1 &= 3456x \end{aligned} \tag{11.9}$$

with the concurrent differential equations

$$\begin{aligned} \dot{x} &= -x, & x_0 &= 1 \\ \dot{T}_9 &= -9T_9 - P_7, & (T_9)_0 &= 1 \\ \dot{P}_7 &= -7P_7 - P_5, & (P_7)_0 &= 72 \\ \dot{P}_5 &= -5P_5 - P_3, & (P_5)_0 &= 1008 \\ \dot{P}_3 &= -3P_3 - 3456x, & (P_3)_0 &= 4032 \end{aligned} \tag{11.10}$$

Note that successive polynomials were defined as P_{i-2} from P_i because we prefer the subscript of P to indicate the highest power of x. Initial conditions were calculated from (11.9) assuming $x_0 = 1$. System (11.10) obtains only positive roots. For $-1 \leq x < 0$, we use $x_0 = -1$ and recalculate the other initial conditions.

Figure 11.4 is the circuit diagram (time scaled at 0.1) that produced the graph of Fig. 11.5. Since the theoretical roots occur at sin $(20°j)$, for $j = 1 - 4$, the accuracy is remarkable, especially considering the oscillatory nature of this polynomial.

Example 11.2. By graphing the following implicity defined functions, obtain the roots to

$$f_1 \equiv y^3 - x^2y + x^3 + 1 = 0 \tag{11.11}$$

$$f_2 \equiv y^2 \cos\left(\tfrac{1}{2}x\right) + x^2e^{-y} - 1.2 = 0 \tag{11.12}$$

such that $-2 \leq x, y \leq 2$.

By differentiation with respect to x, we readily determine that the function defined by

$$\frac{dy_1}{dx} = \frac{2xy_1 - 3x^2}{3y_1^2 - x^2} \tag{11.13}$$

$$\frac{dy_2}{dx} = \frac{\frac{1}{2}y_2^2 \sin\left(\frac{1}{2}x\right) - 2xe^{-y_2}}{2y_2 \cos\left(\frac{1}{2}x\right) - x^2e^{-y_2}} \tag{11.14}$$

are the ones needed in (11.11) and (11.12). Each equation is treated as a separate system and solved on the computer. Initial conditions must satisfy the original equations so that some point solutions must initially be found separately for (11.11) and (11.12). It is easily verified that the points $(-1, -1)$, $(0, -1)$, $(-1, 0)$, and $(-1, 1)$ satisfy (11.11) and can be used as initial conditions in (11.13). Similarly, the points $(\pm\sqrt{1.2}, 0)$ and $(0, \pm\sqrt{1.2})$ are usable in (11.14).

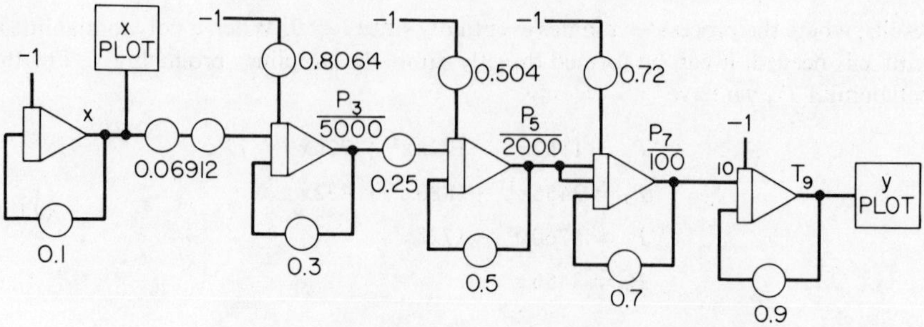

FIG. 11.4 Circuit for graphing $T_9(x)$, $0 < x \leq 1$.

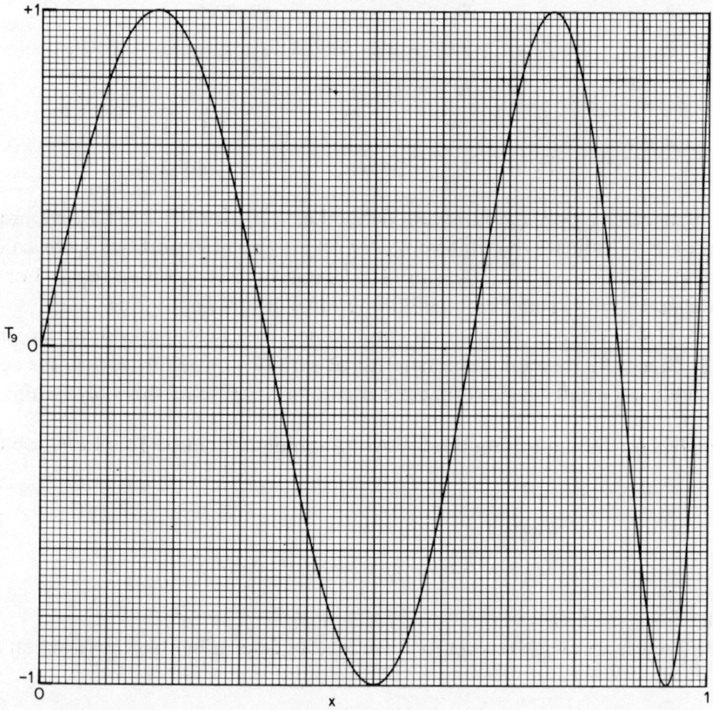

FIG. 11.5 Graph of $T_9(x)$ obtained from the circuit of Fig. 11.4.

The solution of (11.13) is straightforward using

$$\dot{y}_1 = k(2xy_1 - 3x^2), \qquad \dot{x} = k(3y_1^2 - x^2) \tag{11.15}$$

as shown in the unscaled diagram of Fig. 11.6. Notice how manual function switches are used to enable the computer operator to select both positive and negative k and control the direction of plotting.

The solution of (11.14) requires the generation of the three nonrational functions $\sin(\frac{1}{2}x)$, $\cos(\frac{1}{2}x)$, and e^{-y_2}. With

$$S = \sin(\tfrac{1}{2}x), \qquad C = \cos(\tfrac{1}{2}x), \qquad E = e^{-y_2}$$

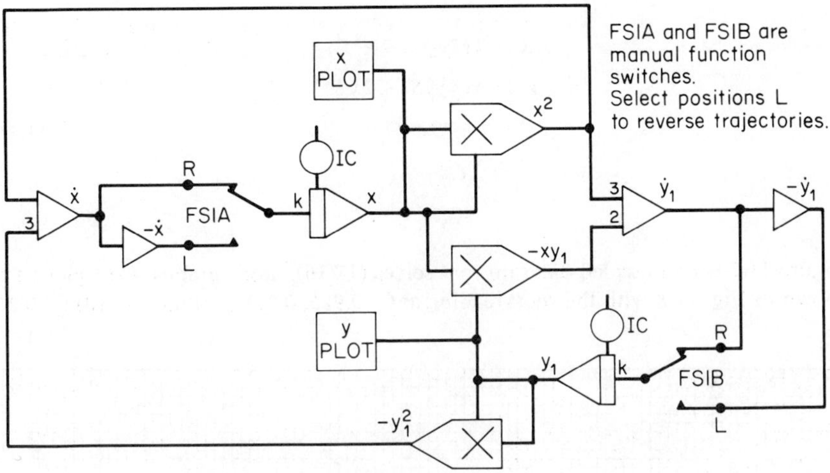

FIG. 11.6 Diagram for solution of (11.15).

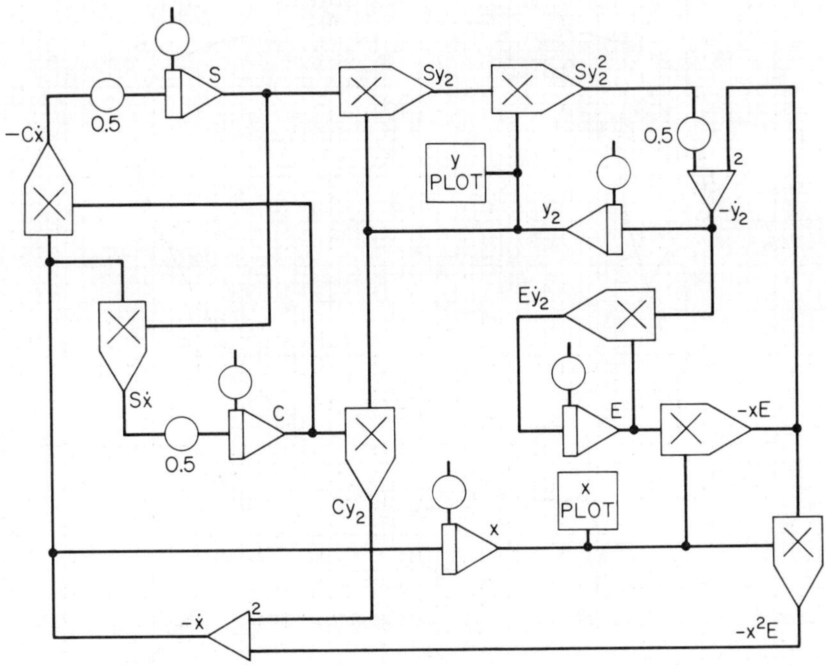

Note: Include function switches to reverse trajectories.

FIG. 11.7 Diagram for solution of (11.16).

the required differential system is found to be

$$\begin{aligned}
\dot{x} &= k(2y_2C - x^2E) \\
\dot{y}_2 &= k(\tfrac{1}{2}y_2^2S - 2xE) \\
\dot{C} &= -\tfrac{1}{2}S\dot{x} \\
\dot{S} &= \tfrac{1}{2}C\dot{x} \\
\dot{E} &= -E\dot{y}_2
\end{aligned} \tag{11.16}$$

Figure 11.7 is an unscaled diagram that solves (11.16). Both graphs were plotted as shown in Fig. 11.8 with the roots reading as (−0.985, 0.97), (−1.03, −0.10) (−0.25,

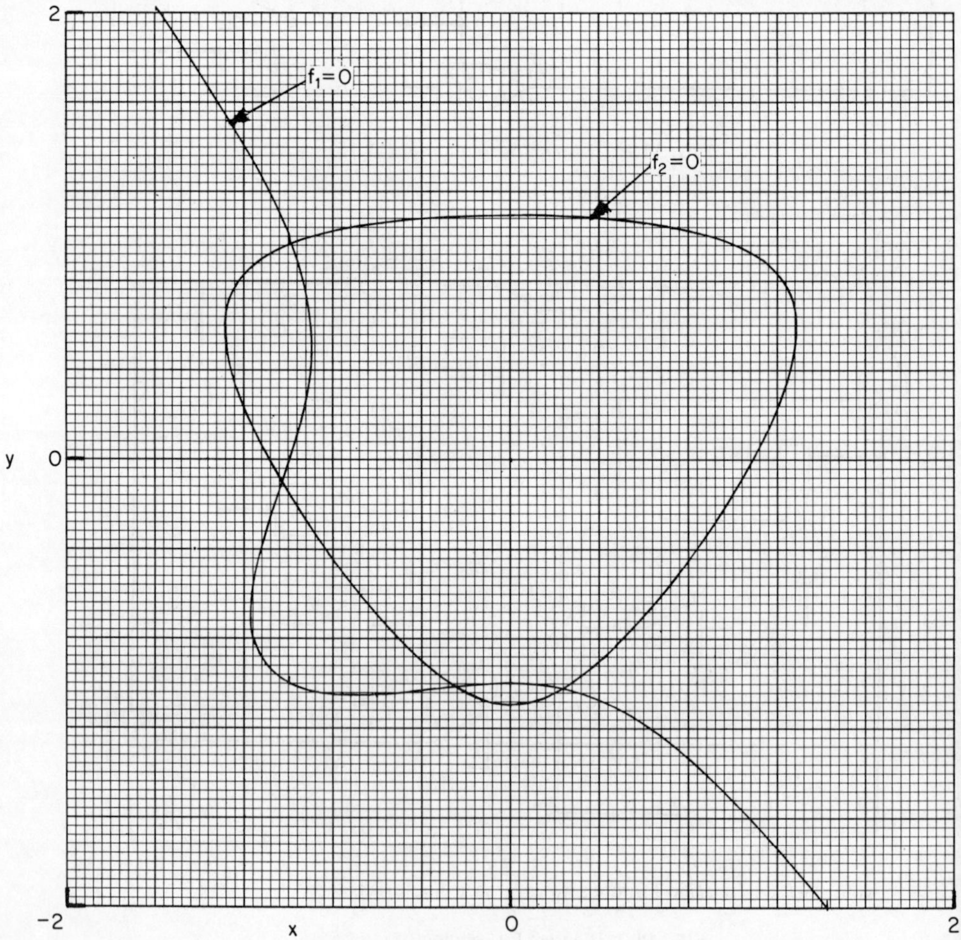

FIG. 11.8 Graphs of (11.12) and (11.11).

-1.015), and ($+0.235$, -1.02). Since the actual roots are (-0.9886, 0.9709), (-1.036, -0.1044), (-0.2525, -1.016), and (0.2396, -1.024), we see that the accuracy is good.

If better accuracy is needed, the roots can be improved by a simple numerical process called the *Newton-Raphson method*, no matter how the original estimated roots were obtained. We demonstrate the procedure with (11.11) and (11.12).

Let (x_r, y_r) be an estimated root and $(\Delta x, \Delta y)$ be the error (real root minus the estimate). Then, for (11.11),

$$(\Delta y + y_r)^3 - (\Delta x + x_r)^2(\Delta y + y_r) + (\Delta x + x_r)^3 + 1 = 0$$

However, we can assume that Δx and Δy are quite small so that all expansions can ignore powers of Δx and Δy larger than one and all cross-products of Δx and Δy. The expansion reduces to

$$3y_r^2\Delta y + y_r^3 - x_r^2\Delta y - 2x_r y_r\Delta x - x_r^2\Delta y + x_r^2 y_r + 3x_r^2\Delta x + x_r^3 + 1 = 0$$

or

$$(-2x_r y_r + 3x_r^2)\Delta x + (3y_r^2 - x_r^2)\Delta y = -f_1(x_r, y_r)$$

This has the effect of linearizing all equations and could also be obtained with a truncated Taylor series expansion of $f_1(x, y)$ in (11.11). Thus, the general case becomes

$$\frac{\partial f_1}{\partial x}\Delta x + \frac{\partial f_1}{\partial y}\Delta y = -f_1(x_r, y_r)$$

$$\frac{\partial f_2}{\partial x}\Delta x + \frac{\partial f_2}{\partial y}\Delta y = -f_2(x_r, y_r)$$

which is a set of linear equations in Δx and Δy and may be solved on the computer or with a desk calculator. This iteration process could be continued indefinitely, but convergence to the desired accuracy usually occurs quickly.

For higher-order systems, graphical methods are rarely used because searching and crossplotting techniques, which could be applied, are very inefficient. Frequently, third- or fourth-order systems can be reduced to second-order ones, where the graphical methods produce good results.

11.2 Vector-Matrix Notation and Concepts

Chapter 9 showed how to avoid setting up an unstable circuit when implicitly solving one equation in one unknown by the application of Theorem 9.1, a very simple rule. When several equations are solved simultaneously with high-gain amplifier connections, the rule for avoiding instability is not nearly so simple. Complications arise from extra feedback paths supplied from the outputs of all amplifiers to the input of any one amplifier. To find the rules for stability in a succinct manner, vector-matrix notation is practically mandatory. Thus, while such notation has been avoided in previous chapters primarily because a programmer must work with the individual scalar equations of a system, this chapter introduces vectors and matrices to enable their application to understanding the stability of differential equations, and for subsequent work.

The principal terminology and notation of vector-matrix formulations are defined in the following paragraphs. Also, the most useful theorems for our purposes will be stated. Interested readers are urged to consult more detailed texts for proofs of these and other pertinent theorems [2], [3].

An $m \times n$ matrix $\mathbf{A}$ is an array of mn complex numbers arranged rectangularly so that there are m rows and n columns:

$$\mathbf{A} \equiv \begin{pmatrix} a_{1,1} & a_{1,2} & \cdots & a_{1,n} \\ a_{2,1} & a_{2,2} & \cdots & a_{2,n} \\ \vdots & \vdots & & \vdots \\ a_{m,1} & a_{m,2} & \cdots & a_{m,n} \end{pmatrix} \equiv (a_{i,j})$$

For the most part, our interest lies here only in square matrices ($m = n$). We use bold-faced capital letters for the matrix, with the corresponding ordinary small letter to indicate the elements. By convention, the array is inclosed by brackets or parentheses.

A *vector* is a special matrix with $m = 1$ or $n = 1$. If $m = 1$, we have a *row* vector

$$\mathbf{x} \equiv (x_1, x_2, \cdots, x_n) \equiv (x_j)$$

and if $n = 1$, a *column* vector

$$\mathbf{x} \equiv \begin{pmatrix} x_1 \\ x_2 \\ \vdots \\ x_m \end{pmatrix} \equiv (x_i)$$

Vector elements are conventionally written with single subscripts, while the vectors are denoted by small bold-faced letters corresponding to the symbol used for the elements.

A *scalar* is a 1×1 matrix, or one number, conventionally written without bold-face, subscripts, or brackets.

The *magnitude* or *metric* of a vector $\mathbf{x}$ can be defined in many ways, but is usually written $||\mathbf{x}||$ in all cases. In the one instance this concept is used in this book, either the Euclidean definition

$$||\mathbf{x}|| \equiv \left(\sum_{i=1}^{n} x_i^2\right)^{1/2}$$

or the simpler definition

$$||\mathbf{x}|| \equiv \sum_{i=1}^{n} |x_i|$$

can be used, with the results stated still valid.

The derivative and integral of a matrix $\mathbf{A}$ with respect to a scalar t are defined as

$$\frac{d\mathbf{A}}{dt} \equiv \begin{pmatrix} \frac{da_{1,1}}{dt} & \frac{da_{1,2}}{dt} & \cdots & \frac{da_{1,n}}{dt} \\ \frac{da_{2,1}}{dt} & \frac{da_{2,2}}{dt} & \cdots & \frac{da_{2,n}}{dt} \\ \vdots & \vdots & & \vdots \\ \frac{da_{m,1}}{dt} & \frac{da_{m,2}}{dt} & \cdots & \frac{da_{m,n}}{dt} \end{pmatrix}$$

$$\int_{t_0}^{t_1} \mathbf{A}\,dt \equiv \begin{pmatrix} \int_{t_0}^{t_1} a_{1,1}\,dt & \int_{t_0}^{t_1} a_{1,2}\,dt & \cdots & \int_{t_0}^{t_1} a_{1,n}\,dt \\ \int_{t_0}^{t_1} a_{2,1}\,dt & \int_{t_0}^{t_1} a_{2,2}\,dt & \cdots & \int_{t_0}^{t_1} a_{2,n}\,dt \\ \vdots & \vdots & & \vdots \\ \int_{t_0}^{t_1} a_{m,1}\,dt & \int_{t_0}^{t_1} a_{m,2}\,dt & \cdots & \int_{t_0}^{t_1} a_{m,n}\,dt \end{pmatrix}$$

We write $\mathbf{A} = \mathbf{B}$ if and only if both have the same number of rows and columns and

$$a_{i,j} = b_{i,j}$$

for all i and j. Thus, for expressing the system of equations

$$\frac{dx_i}{dt} = f_i(x_1, x_2, \cdots, x_n), \qquad (x_i)_0 = c_i$$

for $i = 1, 2, \ldots, n$, we may use the vector equation

$$\frac{d\mathbf{x}}{dt} = \mathbf{f}, \qquad (\mathbf{x})_0 = \mathbf{c}$$

as a shorthand notation.

Of fundamental importance to a square matrix **A** is its *determinant*, written $|\mathbf{A}|$, which is a number found by summing the $n!$ terms of the form

$$\delta(k_1, k_2, \cdots, k_n)a_{1,k_1}a_{2,k_2}\cdots a_{n,k_n}$$

where $k_1 \neq k_2 \neq \cdots \neq k_n$ takes on all possible combinations of integers between 1 and n inclusive, and $\delta(k_1, k_2, \ldots, k_n) = 1$ (or -1) if the ordering of $k_1, k_2, \ldots, k_n$ is an even (or odd) permutation of the ordering $1, 2, \ldots, n$. An even permutation is an ordering of these integers which can be formed by an even number of interchanges between numbers. Thus, if $n = 3$, we may list all possible orderings of the integers by actually permuting the adjacent integers:

ordering	1, 2, 3	1, 3, 2	3, 1, 2	3, 2, 1	2, 3, 1	2, 1, 3
δ	$+1$	-1	$+1$	-1	$+1$	-1

We then have for the 3×3 matrix $\mathbf{A} \equiv (a_{i,j})$,

$$\begin{aligned}|\mathbf{A}| = a_{1,1}a_{2,2}a_{3,3} - a_{1,1}a_{2,3}a_{3,2} + a_{1,3}a_{2,1}a_{3,2} \\ - a_{1,3}a_{2,2}a_{3,1} + a_{1,2}a_{2,3}a_{3,1} - a_{1,2}a_{2,1}a_{3,3}\end{aligned}$$

Determinants can also be expanded in terms of *minors* $M_{i,j}$, which are determinants of the arrays formed by ignoring the ith row and jth column of **A**. It can be shown that

$$|\mathbf{A}| = \sum_{i=1}^{n} (-1)^{i+j} a_{i,j} M_{i,j}$$

for any j. The number $(-1)^{i+j}M_{i,j}$ is called the *cofactor* of $a_{i,j}$ and often written $A_{i,j}$.

A matrix is *singular* if $|\mathbf{A}| = 0$; otherwise it is *nonsingular*.

The *rank* of a matrix is the order of the largest nonzero determinant that can be formed from the rows and columns of the matrix. A nonsingular $n \times n$ matrix therefore has rank n. A singular matrix has rank less than n.

A square matrix of rank $m < n$ has m linearly independent rows; i.e., all other rows can be expressed as linear combinations of the linearly independent rows.

The *discriminants* Δ_i of a matrix **A** are the determinants of the array formed by the first i rows and i columns. Clearly, $\Delta_n = |\mathbf{A}|$ when **A** is square.

Special matrices have special names. For example:

A *null* matrix, written **0**, consists of only zero elements.

A *diagonal* matrix is square with elements $a_{i,j} = 0$ unless $i = j$. The elements $a_{i,i}$ are said to lie along the *main* diagonal and are called main diagonal elements.

A *unity* or *identity* matrix, written $\mathbf{I}$, is a diagonal matrix with all main diagonal elements $a_{i,i} = 1$.

The *transpose* of $\mathbf{A}$, written $\mathbf{A}'$, is a matrix whose elements are $(a_{j,i})$; i.e., it is obtained from $\mathbf{A}$ by interchanging rows and columns. Thus, we also have that the transpose of a column vector is a row vector and vice versa.

A square matrix $\mathbf{A}$ is *Hermitian* if the elements of $\mathbf{A}$ are such that $a_{i,j} = a^*_{j,i}$, where $a^*_{j,i}$ are complex conjugates of $a_{j,i}$.

If all $a_{i,j}$ are real, $\mathbf{A}$ is a *real* matrix.

If $\mathbf{A}' = \mathbf{A}$, they are both *symmetric* (with $a_{j,i} = a_{i,j}$). A real symmetric matrix is Hermitian, while a real Hermitian matrix must be symmetric.

If $\mathbf{A}' = -\mathbf{A}$; i.e., $(a_{j,i}) = (-a_{i,j})$, then $\mathbf{A}$ is *skew symmetric*. The diagonal elements must be 0.

The Jacobian matrix $\mathbf{J}(\mathbf{f}, \mathbf{x})$ of the set of functions $\mathbf{f}(\mathbf{x})$ is defined by

$$\mathbf{J}(\mathbf{f}, \mathbf{x}) \equiv \begin{pmatrix} \dfrac{\partial f_1}{\partial x_1} & \dfrac{\partial f_1}{\partial x_2} & \cdots & \dfrac{\partial f_1}{\partial x_n} \\ \dfrac{\partial f_2}{\partial x_1} & \dfrac{\partial f_2}{\partial x_2} & \cdots & \dfrac{\partial f_2}{\partial x_n} \\ \vdots & \vdots & & \vdots \\ \dfrac{\partial f_n}{\partial x_1} & \dfrac{\partial f_n}{\partial x_2} & \cdots & \dfrac{\partial f_n}{\partial x_n} \end{pmatrix} \tag{11.17}$$

and is important when considering system stability of nonlinear equations.

11.2.1 Matrix operations

A matrix algebra has evolved in which matrix addition, subtraction, multiplication, and division are defined in the most useful manner. Except for division, the same operation symbols are used as in ordinary algebra, but because matrices and vectors are printed in bold face, no confusion can result. We have

Addition and subtraction. $\mathbf{C} = \mathbf{A} \pm \mathbf{B}$ means that $c_{i,j} = a_{i,j} \pm b_{i,j}$ provided $\mathbf{A}$ and $\mathbf{B}$ have the same number of rows and columns. The commutative and associative laws for addition follow from this definition:

$$\mathbf{A} + \mathbf{B} = \mathbf{B} + \mathbf{A}, \qquad (\mathbf{A} + \mathbf{B}) + \mathbf{C} = \mathbf{A} + (\mathbf{B} + \mathbf{C})$$

so that parentheses can be omitted without confusion. Also, $\mathbf{A} + \mathbf{0} = \mathbf{A}$, so that $\mathbf{0}$ is a generalization of 0 in the algebra of numbers.

Multiplication. The product $\mathbf{C} = \mathbf{AB}$ is defined so that

$$c_{i,j} \equiv \sum_{k=1}^{p} a_{i,k} b_{k,j} \tag{11.18}$$

and is meaningful only if **A** is a $m \times p$ matrix and **B** is $p \times n$ matrix (they are then said to be *conformable*). This seemingly arbitrary definition is quite useful when matrices are applied in linear transformations. In general, $\mathbf{AB} \neq \mathbf{BA}$ even when both matrices are square and conformable, so that the order must be carefully preserved. However, the associative and distributive laws can be shown to apply:

$$(\mathbf{AB})\mathbf{C} = \mathbf{A}(\mathbf{BC})$$

$$\mathbf{A}(\mathbf{B} + \mathbf{C}) = \mathbf{AB} + \mathbf{AC}$$

$$(\mathbf{B} + \mathbf{C})\mathbf{A} = \mathbf{BA} + \mathbf{CA}$$

Also

$$\mathbf{AI} = \mathbf{IA} = \mathbf{A}$$

so that **I** is a generalization of 1 in the algebra of numbers. Finally

$$\mathbf{A0} = \mathbf{0A} = \mathbf{0}$$

A scalar a can multiply a matrix **A** using the definition

$$a(a_{i,j}) \equiv (aa_{i,j})$$

Note in the above definitions, that

A row vector times a matrix is a row vector.
A matrix times a column vector is a column vector.
A row vector **x** times a column vector **y** is a 1×1 matrix (a scalar), and is called the *inner product*, often written $\mathbf{x} \cdot \mathbf{y}$.

With these definitions of multiplication we may write (11.2) as

$$\dot{\mathbf{x}} + \mathbf{G}\boldsymbol{\phi} = \mathbf{0}$$

where **G** is a diagonal matrix called a *gain* matrix, with $g_{i,i} = g_i > 0$.

Division. If **A** is square, and there exists a solution **X** to the equation

$$\mathbf{AX} = \mathbf{I}$$

then **X** is called the *inverse* of **A**, written $\mathbf{A}^{-1}$. This parallels the usual algebraic definition of division, although the expression "divide by **A**" is never used (we multiply by $\mathbf{A}^{-1}$). If $\mathbf{A}^{-1}$ exists, we may show that

$$\mathbf{A}\mathbf{A}^{-1} = \mathbf{A}^{-1}\mathbf{A} = \mathbf{I}$$

and it is unique.

It can also be shown that the elements of $\mathbf{X} = \mathbf{A}^{-1}$ are given by

$$(x_{i,j}) = \frac{1}{|\mathbf{A}|}(A_{j,i}) \tag{11.19}$$

where $A_{i,j}$ is the cofactor of $a_{i,j}$. Thus $\mathbf{A}^{-1}$ exists only if $\mathbf{A}$ is nonsingular.

A linear algebraic system of equations

$$\sum_{j=1}^{n} a_{i,j}x_j - b_i = 0$$

for $i = 1, 2, \ldots, n$, can now be written simply as

$$\mathbf{A}\mathbf{x} - \mathbf{b} = \mathbf{0}$$

and its solution

$$\mathbf{x} = \mathbf{A}^{-1}\mathbf{b}$$

found by multiplying by $\mathbf{A}^{-1}$ on the left and adding $\mathbf{A}^{-1}\mathbf{b}$ to both sides. This implicitly uses these facts: If $\mathbf{A} = \mathbf{B}$, then $\mathbf{CA} = \mathbf{CB}$, $\mathbf{AC} = \mathbf{BC}$, and $\mathbf{A} + \mathbf{C} = \mathbf{B} + \mathbf{C}$.

11.2.2 Characteristic roots

The *determinantal equation*

$$|\mathbf{A} - \lambda\mathbf{I}| \equiv \begin{vmatrix} a_{1,1} - \lambda & a_{1,2} & \cdots & a_{1,n} \\ a_{2,1} & a_{2,2} - \lambda & \cdots & a_{2,n} \\ \vdots & \vdots & & \vdots \\ a_{n,1} & a_{n,2} & \cdots & a_{n,n} - \lambda \end{vmatrix} = 0$$

is called the *characteristic* equation of $\mathbf{A}$ and is a polynomial of degree n in λ having n roots λ_i, which are called the *characteristic roots* or *eigenvalues* of $\mathbf{A}$. The reason they are so important is, as we shall see, that they completely determine the stability of the set of linear differential equations

$$\dot{\mathbf{x}} + \mathbf{A}\mathbf{x} - \mathbf{b} = \mathbf{0}$$

We give names to special matrices whose characteristic roots have special qualities:

Let $\mathbf{x}$ be a row vector. A Hermitian matrix $\mathbf{A}$ is said to be *positive definite* if and only if the quadratic form

$$\mathbf{xAx}' \equiv \sum_{j=1}^{n} \sum_{i=1}^{n} a_{i,j} x_i x_j > 0$$

unless $\mathbf{x} = \mathbf{0}$. In the definition of *negative* definite, the symbol " $<$ " replaces " $>$." Clearly, if $\mathbf{A}$ is positive definite, then $-\mathbf{A}$ is negative definite and vice versa.

Theorem 11.1. The characteristic roots of a Hermitian matrix $\mathbf{A}$ are real. They are, in addition, greater (lesser) than zero if and only if $\mathbf{A}$ is positive (negative) definite.

Of the several tests to determine whether or not a matrix is positive definite, the simplest is the following:

Theorem 11.2. A Hermitian matrix $\mathbf{A}$ is positive definite if and only if all the discriminants Δ_i, $i = 1, 2, \ldots, n$ are positive.

When the characteristic roots are complex, stability considerations require knowledge of whether or not the real parts are all positive or all negative. When this is the case, we define the following:

A *positive* (*negative*) *stable* matrix $\mathbf{A}$ has characteristic roots, the real parts of which are positive (negative). Clearly, if $\mathbf{A}$ is positive stable, then $-\mathbf{A}$ is negative stable and vice versa.

Tests for determining whether or not a general matrix is positive stable, short of determining the characteristic equations, do not generally exist. Of course, a real positive definite matrix is positive stable. There are other special cases, notably

Theorem 11.3. If the diagonal elements of $\mathbf{A}$ are real and positive so that

$$a_{i,i} > \sum_{\substack{j=1 \\ j \neq i}}^{n} |a_{i,j}|$$

for all i, then $\mathbf{A}$ is positive stable.*

Another situation that may occur is expressed by the following:

Theorem 11.4. Every real matrix $\mathbf{A}$ can be expressed as the sum of a symmetric matrix $\mathbf{S}$ and a skew-symmetric matrix $\mathbf{T}$ by letting

$$s_{i,j} = \frac{a_{i,j} + a_{j,i}}{2}, \qquad t_{i,j} = \frac{a_{i,j} - a_{j,i}}{2}$$

If $\mathbf{S}$ is positive definite, $\mathbf{A}$ is positive stable.

*The symbol " $\geq$ " can replace " $>$ " in the result of this theorem under certain conditions. See Varga [4], p. 23.

11.3 Linear Algebraic Equations

In this section, we consider methods to track time-varying solutions of

$$\mathbf{f} \equiv \mathbf{Ax} - \mathbf{b} = \mathbf{0} \tag{11.20}$$

with the high-gain integrator circuits of Fig. 11.2. All elements of $\mathbf{A}$ and $\mathbf{b}$ are real. The gains g_i will be so large that solutions appear almost instantly at the outputs of each integrator. As $\mathbf{A}$ or $\mathbf{b}$ change relatively slowly, the outputs $\mathbf{x}$ track the proper solutions if the circuits are stable. We assume the g_i are not so large that nonideal components affect the stability [5], [6].

All methods of computer solution depend upon the following simple theorem:

Theorem 11.5. The equation

$$\dot{\mathbf{x}} + \mathbf{Ax} - \mathbf{b} = \mathbf{0}$$

is stable if and only if $\mathbf{A}$ is positive stable.

This is an extremely deceptive theorem, for we often find it quite difficult to determine if $\mathbf{A}$ is positive stable. The results of Theorems 11.1–11.4 provide this information only in certain cases. These, however, occur frequently enough in physical systems to warrant the following paragraph, which is the first method of solution.

11.3.1 Direct method [7]–[9]

If $\mathbf{A}$ is known to be positive stable, take $\boldsymbol{\phi} = \mathbf{f}$ and solve

$$\dot{\mathbf{x}} + g(\mathbf{Ax} - \mathbf{b}) = \mathbf{0}, \qquad g > 0$$

The rate of convergence of the solution is directly proportional to g. (The matrix $g\mathbf{A}$ is positive stable, too.)

Example 11.3. Solve on the computer

$$\begin{pmatrix} 3 & 1 & 1 & 0 \\ 5 & 10 & 2 & 2 \\ 3 & -4 & 12 & -4 \\ 1 & 1 & 2 & 5 \end{pmatrix} \begin{pmatrix} x_1 \\ x_2 \\ x_3 \\ x_4 \end{pmatrix} = \begin{pmatrix} 1 \\ 2 \\ 3 \\ 4 \end{pmatrix} \tag{11.21}$$

This matrix is positive stable by Theorem 11.3, so we may program

$$\begin{aligned}
\dot{x}_1 + g(3x_1 + x_2 + x_3 \phantom{{}+ 2x_4} - 1) &= 0 \\
\dot{x}_2 + g(5x_1 + 10x_2 + 2x_3 + 2x_4 - 2) &= 0 \\
\dot{x}_3 + g(3x_1 - 4x_2 + 12x_3 - 4x_4 - 3) &= 0 \\
\dot{x}_4 + g(x_1 + x_2 + 2x_3 + 5x_4 - 4) &= 0
\end{aligned}$$

For convenience, we scale each x_i identically and obtain a large value of g by using small feedback capacitors.

When using Theorem 11.3 in a computer solution, a gain matrix with different elements can be used to normalize the diagonal coefficients. A greater convenience is achieved by solving

$$
\begin{aligned}
\dot{x}_1 + g(\qquad x_1 + 0.3333x_2 + 0.3333x_3 \qquad\qquad\qquad - 0.3333) &= 0\\
\dot{x}_2 + g(0.5000x_1 + \qquad x_2 + 0.2000x_3 + 0.2000x_4 - 0.2000) &= 0\\
\dot{x}_3 + g(0.2500x_1 - 0.3333x_2 + \qquad x_3 - 0.3333x_4 - 0.2500) &= 0\\
\dot{x}_4 + g(0.2000x_1 + 0.2000x_2 + 0.4000x_3 + \qquad x_4 - 0.8000) &= 0
\end{aligned}
\tag{11.22}
$$

in which all coefficients are less than or equal to one, with Theorem 11.3 still applying. Figure 11.9 shows a diagram for the solution of (11.22).

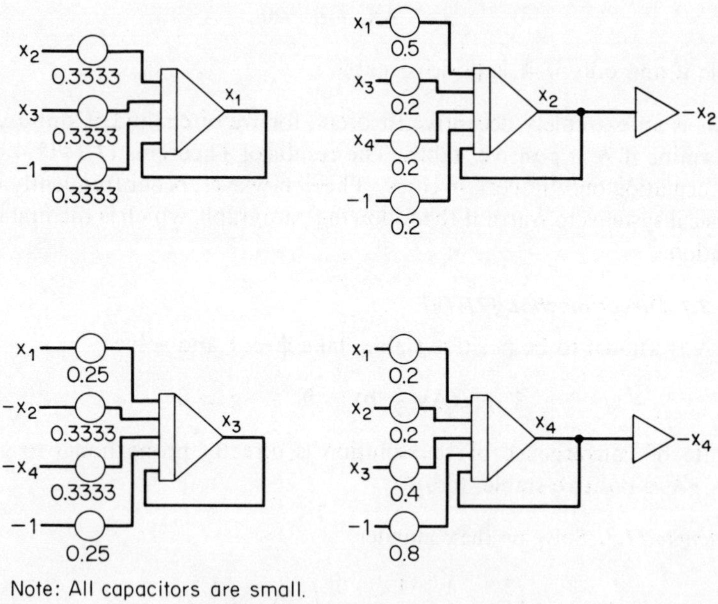

FIG. 11.9 Direct method, Example 11.3.

Example 11.4 (***The effect of gain***). Solve on the computer

$$
\mathbf{Ax} = \begin{pmatrix} 1 & -4 & -7 \\ 2 & 2 & 2 \\ 4 & 4 & 5 \end{pmatrix} \begin{pmatrix} x_1 \\ x_2 \\ x_3 \end{pmatrix} = \begin{pmatrix} 1 \\ 2 \\ 3 \end{pmatrix}
$$

The symmetric part of this matrix is

$$\mathbf{S} = \begin{pmatrix} 1 & -1 & -\frac{3}{2} \\ -1 & 2 & 3 \\ -\frac{3}{2} & 3 & 5 \end{pmatrix}$$

Is **S** positive definite? We calculate the discriminants of **S** to be

$$\Delta_1 = 1, \qquad \Delta_2 = 1, \qquad \Delta_3 = \tfrac{1}{2}$$

and find the answer is yes from Theorem 11.2. Hence **A** is positive stable from Theorem 11.4. We program

$$\begin{aligned} \dot{x}_1 + g(\ x_1 - 4x_2 - 7x_3 - 1) &= 0 \\ \dot{x}_2 + g(2x_1 + 2x_2 + 2x_3 - 2) &= 0 \\ \dot{x}_3 + g(4x_1 + 4x_2 + 5x_3 - 3) &= 0 \end{aligned}$$

for a large $g > 0$. Can we normalize the coefficients without introducing instabilities, as in Example 11.3? The answer in this case is yes, for we have the following theorem.

Theorem 11.6. If the symmetric part of **A** is positive definite, then **GA** is positive stable for any gain matrix **G**.

We may therefore program

$$\begin{aligned} \dot{x}_1 + g_1(\ x_1 - 4x_2 - 7x_3 - 1) &= 0 \\ \dot{x}_2 + g_2(2x_1 + 2x_2 + 2x_3 - 2) &= 0 \\ \dot{x}_3 + g_3(4x_1 + 4x_2 + 5x_3 - 3) &= 0 \end{aligned}$$

for any set of $g_i > 0$.

We must not think that arbitrary gains can be indiscriminantly used as in the two previous examples. The general positive stable matrix can have negative main diagonal elements. For example

$$\begin{pmatrix} 6 & -7 \\ 5 & -5 \end{pmatrix} \tag{11.23}$$

is positive stable (see Problem 11-6), but its normalized equivalent

$$\begin{pmatrix} \frac{6}{7} & -1 \\ 1 & -1 \end{pmatrix}$$

is not.

Example 11.5 (The effect of scaling). The solution of (11.21) is

$$x_1 = 0.2571, \qquad x_2 = -0.1259, \qquad x_3 = 0.3544, \qquad x_4 = 0.6320$$

Suppose we wished more optimum amplitude scaling using variables of $(2x_1)$, $(5x_2)$, $(2x_3)$, and (x_4). Equations (11.22) transform to

$$\begin{aligned}
(2\dot{x}_1) + g[\quad (2x_1) + (\tfrac{2}{15})(5x_2) + \tfrac{1}{3}(2x_3) \qquad\qquad - \tfrac{2}{3}] &= 0 \\
(5\dot{x}_2) + g[\ \tfrac{5}{4}(2x_1) + \qquad (5x_2) + \tfrac{1}{2}(2x_3) + \ (x_4) - 1] &= 0 \\
(2\dot{x}_3) + g[\ \tfrac{1}{4}(2x_1) - (\tfrac{2}{15})(5x_2) + \ (2x_3) - \tfrac{2}{3}(x_4) - \tfrac{1}{2}] &= 0 \\
(\dot{x}_4) + g[(\tfrac{1}{10}(2x_1) + \ \tfrac{1}{25})(5x_2) + \tfrac{1}{5}(2x_3) + \ (x_4) - \tfrac{4}{5}] &= 0
\end{aligned} \tag{11.24}$$

The net result must yield a stable system since the solutions to (11.22) and (11.24) are identical. The matrix

$$\mathbf{B} = \begin{pmatrix} 1 & \frac{2}{15} & \frac{1}{3} & -0 \\ \frac{5}{4} & 1 & \frac{1}{2} & 1 \\ \frac{1}{4} & -\frac{2}{15} & 1 & -\frac{2}{3} \\ \frac{1}{10} & \frac{1}{25} & \frac{1}{5} & 1 \end{pmatrix}$$

is positive stable although not immediately obvious. We therefore have another method for determining if a given matrix is positive stable.

Theorem 11.7. If a set of g_i can be found (positive or negative) so that when the ith row of $\mathbf{A}$ is multiplied by g_i and the ith column is divided by g_i and the resulting matrix is positive stable by a previous theorem, $\mathbf{A}$ is positive stable.

11.3.2 Transpose method [7]–[9]

If $\mathbf{A}$ is *not* positive stable, the equations must be manipulated before solving them on the computer. One of the earliest methods is based on the following theorem.

Theorem 11.8. The matrix $\mathbf{A'A}$ is positive definite if $\mathbf{A}$ is nonsingular. The solution of the equation

$$\mathbf{A'Ax} - \mathbf{A'b} = \mathbf{0}$$

satisfies (11.20), while the set of equations

$$\dot{\mathbf{x}} + \mathbf{G}(\mathbf{A'Ax} - \mathbf{A'b}) = \mathbf{0} \tag{11.25}$$

converges to the solution of (11.20), for any gain matrix $\mathbf{G}$.

Theorem 11.8 provides a very general method for solution of linear algebraic equations. The required multiplication can be performed with a desk calculator or by the computer. In the latter case, more equipment is needed than usual.

Example 11.6. Solve on the computer

$$\begin{pmatrix} f_1 \\ f_2 \\ f_3 \end{pmatrix} = \begin{pmatrix} 1 & 3 & 5 \\ 2 & -2 & 3 \\ 4 & 5 & -1 \end{pmatrix} \begin{pmatrix} x_1 \\ x_2 \\ x_3 \end{pmatrix} - \begin{pmatrix} 1 \\ 2 \\ 3 \end{pmatrix} = \begin{pmatrix} 0 \\ 0 \\ 0 \end{pmatrix} \tag{11.26}$$

This matrix cannot be positive stable because the *trace* is negative (see Problem 11-7).

Both $\mathbf{A'A}$ and $\mathbf{A'b}$ are found with definition (11.18). Thus we have

$$\begin{pmatrix} 1 & 2 & 4 \\ 3 & -2 & 5 \\ 5 & 3 & -1 \end{pmatrix} \begin{pmatrix} 1 & 3 & 5 \\ 2 & -2 & 3 \\ 4 & 5 & -1 \end{pmatrix} = \begin{pmatrix} 21 & 19 & 7 \\ 19 & 38 & 4 \\ 7 & 4 & 35 \end{pmatrix}$$

$$\begin{pmatrix} 1 & 2 & 4 \\ 3 & -2 & 5 \\ 5 & 3 & -1 \end{pmatrix} \begin{pmatrix} 1 \\ 2 \\ 3 \end{pmatrix} = \begin{pmatrix} 17 \\ 14 \\ 8 \end{pmatrix}$$

and we program

$$\begin{aligned} \dot{x}_1 + g_1(21x_1 + 19x_2 + 7x_3 - 17) &= 0 \\ \dot{x}_2 + g_2(19x_1 + 38x_2 + 4x_3 - 14) &= 0 \\ \dot{x}_3 + g_3(7x_1 + 4x_2 + 35x_3 - 8) &= 0 \end{aligned}$$

for any set of $g_i > 0$.

Because of the tedium in the hand calculation, many prefer to perform the matrix multiplication on the computer. This preference is virtually unanimous for problems requiring several coefficient changes, since the entire calculation would otherwise have to be performed each time. For convenience, the **f** vector is formed first, as shown in Fig. 11.10, and the product vector next, using the components of f. Thus, stability of the arrangement is assured at the expense of approximately twice the number of components as in the direct method. As we shall see, this method is identical with that using one method of steepest descent to be discussed in the next section. Many more multipliers are required, however, when the coefficients are variable functions of time.

Gephart [8] gives a variant to the above transpose method by solving

$$\dot{\mathbf{y}} = g(\mathbf{Ax} - \mathbf{b}), \qquad \mathbf{x} = -\mathbf{A'y}$$

which forms $\mathbf{x}$ at summer outputs.

Goldberg and Brown [7] also have a transpose method based on the fact that when $\mathbf{A}$ is nonsingular, the matrix of matrix arrays

$$\begin{pmatrix} \mathbf{I} & -\mathbf{A'} \\ \mathbf{A} & \mathbf{0} \end{pmatrix}$$

is positive stable. For a more general result useful in applying linear constraints, see Chapter 12.

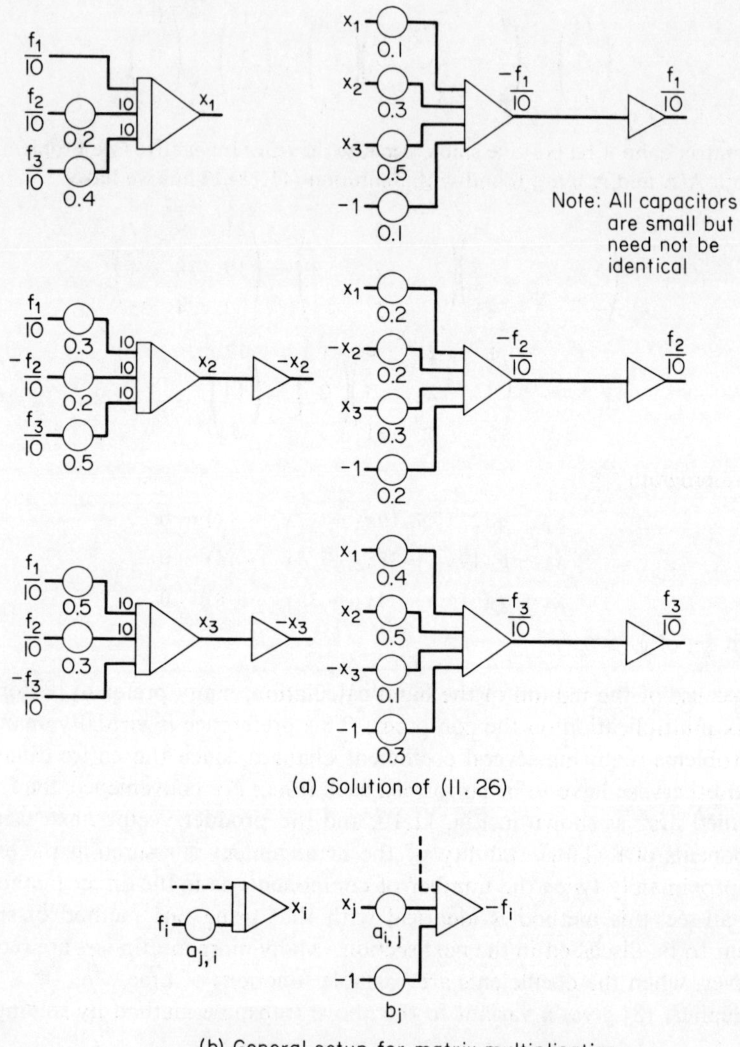

(a) Solution of (11.26)

(b) General setup for matrix multiplication

FIG. 11.10 Transpose method, Example 11.6.

11.3.3 Sign inversion method

The nuisance of matrix premultiplication or the required extra equipment for solving linear algebraic systems containing a matrix that is not positive stable have led researchers (see Rybashov, [9], for example) to examine other techniques. The sign inversion method simply attempts to transform a system to contain a positive stable matrix by multiplying each equation by $+1$ or -1.

This procedure, however, does not always work. For example, any matrix with zero main diagonal elements

$$\begin{pmatrix} 0 & a_{1,2} & \cdots & a_{1,n} \\ a_{2,1} & 0 & \cdots & a_{2,n} \\ \vdots & \vdots & \ddots & \vdots \\ a_{n,1} & a_{n,2} & \cdots & 0 \end{pmatrix}$$

cannot be so transformed, because the trace cannot be positive, a necessary condition for a matrix to be positive stable (see Problem 11-7). As another example, the matrix in (11.26) cannot be so transformed. Since the determinant of

$$\mathbf{A} = \begin{pmatrix} 1 & 3 & 5 \\ 2 & -2 & 3 \\ 4 & 5 & -1 \end{pmatrix}$$

is $|\mathbf{A}| = 119$, two sign inversions are required to maintain its positive value. The second row must be positive for the trace to be positive, hence we try

$$\mathbf{A}_1 = \begin{pmatrix} 1 & 3 & 5 \\ -2 & 2 & -3 \\ -4 & -5 & 1 \end{pmatrix} \tag{11.27}$$

which is not positive stable (see Problem 11-7). If rows 1 and 3 are sign inverted, the trace is zero. Hence no variant of this matrix is positive stable by sign inversion.

11.3.4 Reordering method

The systems of equations

$$\begin{pmatrix} 1 & 3 & 5 \\ -2 & 2 & -3 \\ 4 & 5 & -1 \end{pmatrix} \begin{pmatrix} x_1 \\ x_2 \\ x_3 \end{pmatrix} - \begin{pmatrix} 1 \\ -2 \\ 3 \end{pmatrix} = \begin{pmatrix} 0 \\ 0 \\ 0 \end{pmatrix}$$

and

$$\begin{pmatrix} 2 & -2 & 3 \\ 4 & 5 & -1 \\ 1 & 3 & 5 \end{pmatrix} \begin{pmatrix} x_1 \\ x_2 \\ x_3 \end{pmatrix} - \begin{pmatrix} 2 \\ 3 \\ 1 \end{pmatrix} = \begin{pmatrix} 0 \\ 0 \\ 0 \end{pmatrix}$$

are identical, but the latter contains a matrix easily shown to be positive stable. Similarly, the systems

$$\begin{pmatrix} 3 & 5 \\ 4 & 2 \end{pmatrix}\begin{pmatrix} x_1 \\ x_2 \end{pmatrix} - \begin{pmatrix} a \\ b \end{pmatrix} = \begin{pmatrix} 0 \\ 0 \end{pmatrix}$$

$$\begin{pmatrix} 1 & 0 \\ 0 & 2 \end{pmatrix}\begin{pmatrix} -3 & -5 \\ 4 & 2 \end{pmatrix}\begin{pmatrix} x_1 \\ x_2 \end{pmatrix} - \begin{pmatrix} 1 & 0 \\ 0 & 2 \end{pmatrix}\begin{pmatrix} -a \\ b \end{pmatrix} = \begin{pmatrix} 0 \\ 0 \end{pmatrix}$$

have identical solutions, but the matrix of the latter is positive stable. In the first situation, a simple reordering (with changing all signs in one row) accomplished stability, while in the second case, premultiplication by a special gain matrix and one row inversion produced the positive stable matrix.

The question naturally arises: Given a linear system (11.20), how can a stable system

$$\dot{\mathbf{x}} + \mathbf{G}(\mathbf{Bx} - \mathbf{g}) = \mathbf{0} \tag{11.28}$$

be mechanized by reordering (which includes changing all of the signs in some of the rows) the rows of (11.20), and choosing a **G** matrix for stability? In (11.28), the rows of $\mathbf{Bx} - \mathbf{g}$ correspond to rows of (11.20) or their negative, so that identical solutions exist. We shall speak of **B** as a *reordering* of **A**.

Gilbert and McHugh [10] provided one answer to the question. They showed that a reordering of **A** always exists when $|\mathbf{A}| \neq 0$, so that **B** contains positive discriminants and positive diagonal elements. When such an ordering is found, they provided an experimental method to determine a gain matrix **G** for stability. Unfortunately, a determination of **B** analytically itself involved partial solutions of an $n \times n$ set of linear algebraic equations, which somewhat defeats their purpose.

It is much easier to determine both **B** and **G** experimentally. The following method does not restrict the diagonal elements of **B** to be positive. It provides for a strongly stable system of equations and can sometimes be used to stabilize nonlinear systems as well (Sec. 11.5.1).

Since the **b** vector in (11.20) does not determine the final stability, we shall work with the system $\mathbf{f} \equiv \mathbf{Ax} = \mathbf{0}$, and seek a reordering of **A** so that $\dot{\mathbf{x}} + \mathbf{G}(\mathbf{Bx}) = \mathbf{0}$ is stable, with $\mathbf{x} \to \mathbf{0}$ as $\tau \to \infty$ for any set of initial conditions. Now,

(1) Normalize the matrix **A** so that the maximum absolute coefficient of any row is one.

(2) Set up the **f** vector using n integrators with no feedback, each labeled $x_1, x_2, \ldots, x_n$, using arbitrary initial conditions (say 0.2).

(3) If $a_{1,j} = +1$ (or -1), let $\dot{x}_j + 10f_1 = 0$ (or $\dot{x}_j - 10f_1 = 0$). This partial system is stable in the OPERATE mode; the other **x** components do not change, so we effectively solve

$$\dot{x}_j + 10a_{1,j}x_j = \text{other constant terms}$$

In OPERATE, f_1 is forced to 0, while the other f_j assume other values. Gains for the final system will all be relative and will be increased by a time scaling factor to produce very rapid convergence. We fix all gains initially to be arbitrarily less than or equal to 10 for recording purposes.

(4) Record the values of each f_j after $f_1 = 0$ in OPERATE. Perturb the initial conditions of the remaining components of the **x** vector one at a time by some fixed amount (say 0.01), and record the net change of all the remaining components of the **f** vector, in OPERATE, after $f_1 = 0$. Suppose a variation in $(x_i)_0$ changes f_k more (in absolute value) than a variation of any other initial condition changes any other **f** component. (Ignore ties when they occur.) "*Assign*" f_k *as a feedback to* x_i.

(5) Connect f_k through a gain pot g_2 as an input to integrator x_i. Set $g_2 \ll 1$ and place the computer in OPERATE. If $f_k \to 0$, all is well. If not, connect $-f_k$ through the pot. We shall show that the final 2×2 subsystem is stable for one case or the other if the matrix **A** has rank ≥ 2.

(6) Increase g_2 (to no greater than 10) by trial until the most rapid convergence is obtained, by recording f_1 and f_k and observing how fast each approach 0 in OPERATE. A repetitive computer, if available, is ideal for such a purpose.

(7) Repeat step 4 and successively assign one of the remaining **f** components as feedback to one of the **x** components based on the maximum sensitivity test performed in the OPERATE mode after all previously connected **f** components are nulled.

We now prove the following theorem:

Theorem 11.9. The experimental procedure described generates **B** (a reordering of **A**) and **G**, so that **GB** is positive stable when $|\mathbf{A}| \neq 0$. Further, if a partial stable system of order $m - 1$ is generated, and all changes of initial conditions of unused **x** components produce no change in any of the unused **f** components in the OPERATE mode, then **A** has rank $m - 1$.

Proof. By induction. At the $(m - 1)$st stage of construction, the system

$$\begin{pmatrix} \dot{z}_1 \\ \dot{z}_2 \\ \vdots \\ \dot{z}_{m-1} \end{pmatrix} + \begin{pmatrix} g_1 & & & \\ & g_2 & & 0 \\ & & \ddots & \\ 0 & & & g_{m-1} \end{pmatrix} \begin{pmatrix} b_{1,1} & b_{1,2} & \cdots & b_{1,m-1} \\ b_{2,1} & b_{2,2} & \cdots & b_{2,m-1} \\ \vdots & \vdots & & \vdots \\ b_{m-1,1} & b_{m-1,2} & \cdots & b_{m-1,m-1} \end{pmatrix} \begin{pmatrix} z_1 \\ z_2 \\ \vdots \\ z_{m-1} \end{pmatrix} = \begin{pmatrix} 0 \\ 0 \\ \vdots \\ 0 \end{pmatrix}$$

is stable, where each z_j is the x_j in order of selection. We call the partial gain matrix $\mathbf{G}_{m-1}$, the partial coefficient matrix $\mathbf{B}_{m-1}$, and the partial vector system

$\mathbf{z}_{m-1}$. Clearly, $|\mathbf{B}_{m-1}| \neq 0$ and therefore $|\mathbf{G}_{m-1}\mathbf{B}_{m-1}| \neq 0$. Each eigenvalue of $\mathbf{G}_{m-1}\mathbf{B}_{m-1}$ has positive real parts.

Let a new row with subscript m be added with extremely small gain g_m together with the corresponding mth column. From an expansion of the new determinantal equation, we determine that the eigenvalues $\lambda_1, \lambda_2, \ldots, \lambda_{m-1}$ are continuous with respect to g_m at $g_m = 0$; for sufficiently small g_m the previous eigenvalues will have positive real parts. Hence, we concentrate on the value of the new eigenvalue λ_m introduced by the new row. The determinantal expansion of the new matrix product $\mathbf{G}_m\mathbf{B}_m$ yields [3]

$$|\mathbf{G}_m\mathbf{B}_m - \lambda_m\mathbf{I}| = |\mathbf{G}_m\mathbf{B}_m| - a\lambda_m + \mathcal{O}(\lambda_m^2)$$

where a is the sum of the product of all combinations of $m - 1$ eigenvalues. The principal term of this sum as $g_m \to 0$ and $\lambda_m \to 0$ comes from

$$a \simeq \prod_{j=1}^{m-1} \lambda_j = |\mathbf{G}_{m-1}\mathbf{B}_{m-1}|$$

Hence, for sufficiently small g_m,

$$\lambda_m \simeq \frac{|\mathbf{G}_m\mathbf{B}_m|}{|\mathbf{G}_{m-1}\mathbf{B}_{m-1}|}$$

and is real. Now, λ_m is directly proportional to a small g_m if $|\mathbf{B}_m| \neq 0$. We control its sign by inverting the signs of the coefficients if $|\mathbf{B}_m| < 0$ as part of the construction procedure.

Suppose $|\mathbf{B}_m| = 0$. Then the last column added to $\mathbf{B}_{m-1}$ must be expressible as a linear combination of the previous columns; i.e., there exists a vector $\mathbf{z}_m$ so that

$$\mathbf{B}_m\mathbf{z}_m = \mathbf{0}$$

In the OPERATE mode, we selected f_k for row m because

$$f_k = \sum_{j=1}^{m-1} b_{k,j}z_j + b_{k,m}(z_m)_0 + \text{other constant terms}$$

changed greatly when $(z_m)_0$ was changed, after all

$$f_i = \sum_{j=1}^{m-1} b_{i,j}z_j + b_{i,m}(z_m)_0 + \text{other constant terms}$$

were forced to zero. Because the last row of $\mathbf{B}_m\mathbf{z}_m$ cannot change when the mth component $(z_m)_0$ is changed $(\mathbf{B}_m\mathbf{z}_m = \mathbf{0})$, f_k is constant for all $(z_m)_0$. Since the choice was based on maximum sensitivity, no other row would have done either.

Hence, $n - m + 1$ rows of the original **A** matrix were linearly dependent on $m - 1$ rows; the rank of **A** was $m - 1$, which proves the theorem.

By choosing the row with maximum sensitivity, we attempt to select eigenvectors which are nearly as orthogonal as possible. (Actually, any row with nonzero sensitivity could have been stabilized.) By making g_m as large as possible, we obtain relative gains which are as close as possible. The final system is strongly stable; i.e., wide variations of changes in coefficients still produce stability. It is therefore useful for some tracking systems; each g_j will be increased by a large positive factor for rapid convergence. We note that the final **B** matrix will often contain negative or zero diagonal elements, producing complex characteristic roots, but ones with large positive real parts. Because each $|\mathbf{B}_m| > 0$, the discriminants of **B** are greater than 0.

Example 11.7. Obtain a reordering of the system

$$\begin{pmatrix} 4 & 7 & 3 & 7 \\ 0 & 6 & 9 & 9 \\ 1 & 9 & 0 & 9 \\ 5 & 3 & 5 & 5 \end{pmatrix} \begin{pmatrix} x_1 \\ x_2 \\ x_3 \\ x_4 \end{pmatrix} - \begin{pmatrix} b_1 \\ b_2 \\ b_3 \\ b_4 \end{pmatrix} = \begin{pmatrix} 0 \\ 0 \\ 0 \\ 0 \end{pmatrix}$$

Ignoring the **b** vector and normalizing **A**, we set

$$\begin{pmatrix} f_1 \\ f_2 \\ f_3 \\ f_4 \end{pmatrix} = \begin{pmatrix} 0.5714 & 1 & 0.4286 & 1 \\ 0 & 0.6667 & 1 & 1 \\ 0.1111 & 1 & 0 & 1 \\ 1 & 0.6 & 1 & 1 \end{pmatrix} \begin{pmatrix} x_1 \\ x_2 \\ x_3 \\ x_4 \end{pmatrix}$$

Although the list of procedures chose f_1 as the first feedback error, any row will do. We arbitrarily set

$$\dot{x}_4 = 10 f_4$$

as the first subsystem, with $z_1 = x_4$. Using constants $(x_1)_0 = (x_2)_0 = (x_3)_0 = 0.2$, the nominal and perturbed values of all remaining f_j are measured in the OPERATE

TABLE 11.1

For $f_4 = 0$	Nominal values $x_1 = x_2 = x_3 = 0.2$	$x_1 = 0.21$ $x_2 = x_3 = 0.2$	$x_2 = 0.21$ $x_1 = x_3 = 0.2$	$x_3 = 0.21$ $x_1 = x_2 = 0.2$
$-f_1$	0.1204	0.1245	0.1163	0.1262
$\|\Delta f_1\|$		0.0041	0.0041	0.0058
$-f_2$	0.1870	0.1970	0.1861	0.1870
$\|\Delta f_2\|$		0.0100	0.0009	0
$-f_3$	0.2978	0.3069	0.2938	0.3078
$\|\Delta f_3\|$		0.0091	0.0040	0.0100

mode when $f_4 = 0$, as shown in Table 11.1. A tie exists between variations in f_2 due to perturbations in x_1 and variations in f_3 due to perturbations in x_3. We arbitrarily use $z_2 = x_3$:

$$\dot{x}_3 = g_2(-f_3)$$

and determine that $g_2 = 10$ produces the most rapid convergence of this second-order subsystem. The system $\dot{x}_3 = g_2(f_3)$ was unstable, necessitating the sign change. Proceeding, we construct Table 11.2 and find we need f_2 as a feedback to x_1. We determine by trial that we need

TABLE 11.2

For $f_3 = f_4 = 0$	Nominal values $x_1 = x_2 = 0.2$	$x_1 = 0.21$ $x_2 = 0.2$	$x_2 = 0.21$ $x_1 = 0.2$
$-f_1$	-0.0503	-0.0511	-0.0520
$\lvert\Delta f_1\rvert$		0.0008	0.0017
$-f_2$	0.1870	0.1970	0.1861
$\lvert\Delta f_2\rvert$		0.0100	0.0009

$$\dot{x}_1 = g_3(-f_2)$$

for stability with $g_3 \simeq 3$ producing the most rapid convergence. With one equation left, we determine that

$$\dot{x}_2 = 10f_1$$

is stable. In terms of the original variables, the final system (with the **B** matrix reordered in both columns and rows) is

$$\begin{pmatrix} \dot{x}_1 \\ \dot{x}_2 \\ \dot{x}_3 \\ \dot{x}_4 \end{pmatrix} + \begin{pmatrix} 3 & 0 & 0 & 0 \\ 0 & 10 & 0 & 0 \\ 0 & 0 & 10 & 0 \\ 0 & 0 & 0 & 10 \end{pmatrix}$$

$$\times \left[\begin{pmatrix} 0 & -0.6667 & -1 & -1 \\ 0.5714 & 1 & 0.4286 & 1 \\ -0.1111 & -1 & 0 & -1 \\ 1 & 0.6 & 1 & 1 \end{pmatrix} \begin{pmatrix} x_1 \\ x_2 \\ x_3 \\ x_4 \end{pmatrix} - \begin{pmatrix} -b_2/9 \\ b_1/7 \\ -b_3/9 \\ b_4/5 \end{pmatrix} \right] = \begin{pmatrix} 0 \\ 0 \\ 0 \\ 0 \end{pmatrix}$$

and remains stable for wide variations of the coefficients. The presence of 0 main diagonal terms indicates that complex characteristic roots are present. But since the real parts are large and positive, this underdamped system is strongly stable.

11.3.5 Iteration to improve accuracy

The Newton-Raphson method (Sec. 11.1) in n dimensions is best developed in terms of the Jacobian matrix $\mathbf{J}(\mathbf{f}, \mathbf{x})$. We desire the solution to

$$\mathbf{f}(\mathbf{x}) = \mathbf{0} \tag{11.29}$$

If $\mathbf{r}$ is a vector whose elements are approximate roots to (11.29), an error vector $\Delta\mathbf{x}$ must satisfy to a first approximation

$$\mathbf{J}\Delta\mathbf{x} + \mathbf{f} = \mathbf{0}$$

where $\mathbf{J}$ and $\mathbf{f}$ are evaluated at $\mathbf{r}$.

For the case of the linear equation (11.20), we determine that

$$\mathbf{J} = \mathbf{A}$$

which results in

$$\mathbf{A}\,\Delta\mathbf{x} + \mathbf{A}\mathbf{r} - \mathbf{b} = \mathbf{0}$$

This is a new linear equation with $\mathbf{Ar} - \mathbf{b}$ replacing the old $-\mathbf{b}$ vector, but with the same matrix coefficient as (11.20). Therefore, the circuit for finding $\mathbf{r}$ can be used to improve the accuracy without any change in its stability characteristics. We may accurately calculate the $\mathbf{Ar}-\mathbf{b}$ vector off the computer for insertion into the n pots. Errors in computing equipment prevent $\Delta\mathbf{x}$ from ever being zero, but this iteration method can be used to extend the accuracy indefinitely, because $\Delta\mathbf{x}$ can be scaled with as large a scale factor as needed.

11.4 The Method of Steepest Descent [11]–[13]

The method of steepest descent, or gradient method, is very popular among analog and digital programmers because of its applicability in a variety of stability problems. In this section we investigate this method for solving both linear and nonlinear algebraic equations of the form (11.1):

$$f_i \equiv f_i(\tau, x_1, x_2, \ldots, x_n) = 0, \qquad i = 1, 2, \ldots, n$$

in an implicit manner using equations of the form (11.2):

$$\dot{x}_j + g_j\phi_j = 0, \qquad j = 1, 2, \ldots, n$$

where the g_j, being much greater than zero, force each function ϕ_j to zero almost instantaneously. In turn, we require that all $f_i = 0$ when all $\phi_j = 0$. To insure the equations are independent let us also assume the Jacobian matrix $\mathbf{J}(\mathbf{f}, \mathbf{x})$ is singular at most at a finite number of points in the region of interest.

In vector-matrix notation, given the n-dimensional vector function $\mathbf{f}$ of the n variables x_j, we wish to determine values for x_j so that

$$\mathbf{f} = \mathbf{0} \tag{11.30}$$

by constructing an n-dimensional stable equation

$$\dot{\mathbf{x}} + \mathbf{G}\boldsymbol{\phi} = \mathbf{0} \tag{11.31}$$

where $\mathbf{G}$ is a gain matrix and

$$\boldsymbol{\phi}(\mathbf{x}, \tau) = \mathbf{0} \tag{11.32}$$

implies (11.30). The function $\boldsymbol{\phi}$ is an n-dimensional vector function of the x_j, and the stability of (11.31); i.e., $\boldsymbol{\phi} \to \mathbf{0}$ as $\tau \to \infty$, implies (11.32) very quickly.

The continuous gradient method attempts such a construction as follows: Let V be a positive definite scalar function of the components of $\mathbf{f}$; i.e.,

$$\begin{aligned} V > 0 \quad &\text{when} \quad \mathbf{f} \neq \mathbf{0} \\ V = 0 \quad &\text{when} \quad \mathbf{f} = \mathbf{0} \end{aligned} \tag{11.33}$$

and let us minimize V by constructing a set of differential equations such that

$$\begin{aligned} \dot{V} < 0 \quad &\text{when} \quad \mathbf{f} \neq \mathbf{0} \\ \dot{V} = 0 \quad &\text{when} \quad \mathbf{f} = \mathbf{0} \end{aligned} \tag{11.34}$$

Since V is always positive but getting smaller, we can surmise that in the neighborhood of a point where $\mathbf{f} = \mathbf{0}$, $\mathbf{f}$ will automatically be driven toward $\mathbf{0}$. Such a proposition is, in fact, supported by the well-known theorem of Lyapunov [the function V with properties (11.33) and (11.34) is called a *Lyapunov function*]:

Theorem 11.10. If a Lyapunov function V exists in a small enough region D containing the roots to (11.30), then (11.34) is stable about those x_i for which $\mathbf{f} = \mathbf{0}$.

But how can we construct the differential equations (11.34), given V? Since

$$\dot{V} = \sum_{j=1}^{n} \frac{\partial V}{\partial x_j} \dot{x}_j$$

we *choose*

$$\dot{x}_j = -g_j \frac{\partial V}{\partial x_j}, \qquad g_j > 0$$

and in this way, we guarantee at least

$$\dot{V} = -\sum_{j=1}^{n} g_j \left(\frac{\partial V}{\partial x_j}\right)^2 \leq 0$$

Equation (11.34), however, cannot be guaranteed for all points in the region of interest. It may be possible for $\mathbf{f} \neq \mathbf{0}$ and $\dot{V} = 0$. To determine conditions for such an event in terms of the f_i, let us first define the column vectors $(\partial V/\partial \mathbf{f})$ containing elements $\partial V/\partial f_k$ and $(\partial V/\partial \mathbf{x})$ containing elements $\partial V/\partial x_j$. Now we have, for $\dot{V} = 0$,

$$\frac{\partial V}{\partial x_j} = \sum_{k=1}^{n} \frac{\partial f_k}{\partial x_j} \frac{\partial V}{\partial f_k} = 0$$

which is equivalent to

$$\left(\frac{\partial V}{\partial \mathbf{x}}\right) = \mathbf{J}' \left(\frac{\partial V}{\partial \mathbf{f}}\right) = \mathbf{0}$$

A fundamental result of matrix theory requires that $|\mathbf{J}'| = 0$ or $(\partial V/\partial \mathbf{f}) = \mathbf{0}$ for this to be true. Thus, if we require V to be chosen so that $(\partial V/\partial \mathbf{f}) \neq \mathbf{0}$ anywhere in the region unless $\mathbf{f} = \mathbf{0}$, the only situations for which $\dot{V} = 0$ and $\mathbf{f} \neq \mathbf{0}$ occur are when the Jacobian $\mathbf{J}(\mathbf{f}, \mathbf{x})$ is singular. Such points are either unstable or define *false* rest points called *local minimums* of V. In any case, we have the basic theorem of the method of steepest descent:

Theorem 11.11. Let V be a positive definite function of the components of $\mathbf{f}$, and $\mathbf{G}$ an arbitrary gain matrix. Take

$$\phi_j = \frac{\partial V}{\partial x_j}, \qquad j = 1, 2, \ldots, n$$

Then the equation

$$\dot{\mathbf{x}} + \mathbf{G}\boldsymbol{\phi} = \mathbf{0}, \qquad (\mathbf{x})_0 = \mathbf{c}$$

is stable in a small enough region containing a solution of $\mathbf{f} = \mathbf{0}$. By systematically varying the initial conditions $\mathbf{c}$, solutions can be found one by one.

The two most common positive definite functions of the f_i are

(1) $V = \frac{1}{2}\sum_{i=1}^{n} f_i^2$ (sum of squares)

(2) $V = \sum_{i=1}^{n} |f_i|$ (sum of absolute values)

Both are similar to the approximation norms used in Chapter 7 because the f_i can be considered as "error functions" to be minimized. In the former case, we take

$$\phi_j = \sum_{i=1}^{n} \frac{\partial f_i}{\partial x_j} f_i$$

which results in the equations

$$\dot{x}_j + g_j \sum_{i=1}^{n} \frac{\partial f_i}{\partial x_j} f_i = 0 \tag{11.35}$$

or

$$\dot{\mathbf{x}} + \mathbf{G}\mathbf{J}'\mathbf{f} = \mathbf{0}$$

where **G** is the gain matrix and **J** the Jacobian matrix. Note the similarity to the one-dimensional case expressed by (9.10). In the latter case, we take

$$\phi_j = \sum_{i=1}^{n} \frac{\partial f_i}{\partial x_j} \operatorname{sign} f_i$$

with

$$\dot{x}_j + g_j \sum_{i=1}^{n} \frac{\partial f_i}{\partial x_j} \operatorname{sign} f_i = 0 \tag{11.36}$$

Comparison of (11.35) with (11.36) shows that multiplications are replaced by switching circuits. Both cases, however, require the generation of the f_i and all $\partial f_i/\partial x_j$, a factor that precludes the use of this method unless much equipment is available or m and n are small.

Of these two possible simple functions to use, the absolute error minimization generally converges more quickly because the input voltages remain larger. However, analog computer mechanizations of sign f_i do not include a zero value when $f_i = 0$. Thus, (11.36) produces high-frequency oscillations about the $x_j = 0$ but the results are nevertheless satisfactory in many cases.

11.4.1 Four-quadrant division circuits [31]

Usual division circuits require the denominator to be positive for stability. In solving the implicit equation

$$f \equiv a_1(\tau)x - a_2(\tau) = 0 \tag{11.37}$$

using the method of steepest descent, we obtain the equations

$$\dot{x} + ga_1 f = 0$$

or

$$\dot{x} + ga_1 \operatorname{sign} f = 0 \tag{11.38}$$

both of which are stable at the solution $x = a_2/a_1$ when $g > 0$. These circuits are shown in Fig. 11.11. It is not possible for $\dot{x} = 0$ unless $f = 0$ in these cases

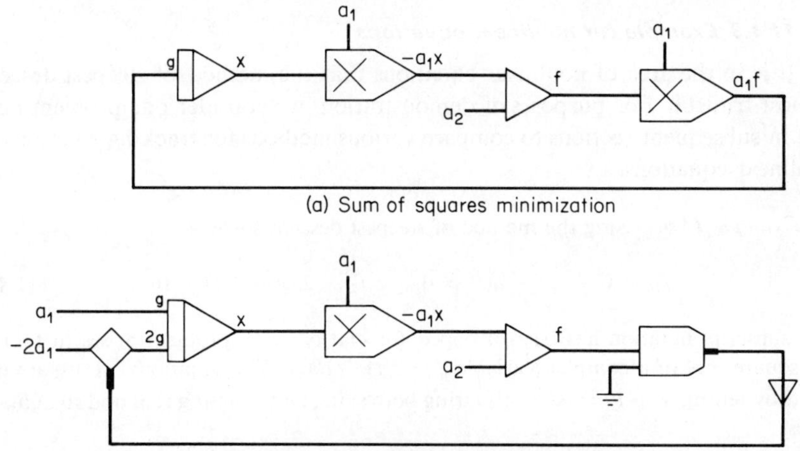

(a) Sum of squares minimization

(b) Sum of absolute error minimization

Note: High-gain amplifiers may replace integrators.

FIG. 11.11 Four-quadrant division circuits.

(allowing sign $f = 0$ when $f = 0$) so that no false stable or unstable points exist. Any high frequency oscillation in the circuit of Fig. 11.11b can be eliminated by introducing a dead zone to approximate the correct solution when $f = 0$. For we have

$$\operatorname{sign} f = \lim_{\substack{\epsilon \to 0 \\ \epsilon > 0}} [U(f - \epsilon) - U(-f - \epsilon)] \tag{11.39}$$

where the expression on the right is generable with two comparators and for small ϵ. For an interesting alternative see Problem 11-17.

11.4.2 Linear equations

Application of the method of steepest descent for a set of n linear equations using the sum squares minimization is equivalent to the transpose method (Sec. 11.3.2). To see this we need note only that for the equation

$$\mathbf{f} \equiv \mathbf{A}\mathbf{x} - \mathbf{b} = \mathbf{0}$$

the Jacobian matrix $\mathbf{J}$ is calculated to be $\mathbf{J} = \mathbf{A}$ from which (11.35) follows from (11.25). No unstable solutions are possible when $\mathbf{A}$ is nonsingular; an infinite number of unstable points exists otherwise.

There is little advantage in the use of the absolute value minimization because the multiplications are performed with potentiometers. This factor, coupled with the inherent oscillation, discourages programmers from using the absolute value minimization despite the faster convergence.

11.4.3 Example for nonlinear equations

It is in the area of nonlinear equations that the method of steepest descent is most fruitful. For purposes of demonstration, we consider one problem here and in subsequent sections to compare various methods for tracking solutions to nonlinear equations.

Example 11.8. Using the method of steepest descent solve

$$f_1 \equiv x^2 - y^2 - a(t) = 0, \qquad f_2 \equiv 2xy - b(t) = 0 \tag{11.40}$$

The subscript notation has been dropped for clarity. This problem arises in taking the square root of a complex variable $z = a(t) + ib(t)$. The equations (11.40) are derived by letting $x + iy = \sqrt{z}$, squaring both sides, and equating real and imaginary parts.

Method 1 (*Sum squares minimization*). We solve

$$\dot{x} + g_1\left(f_1\frac{\partial f_1}{\partial x} + f_2\frac{\partial f_2}{\partial x}\right) = 0, \qquad \dot{y} + g_2\left(f_1\frac{\partial f_1}{\partial y} + f_2\frac{\partial f_2}{\partial y}\right) = 0$$

which reduce to

$$\begin{aligned} \dot{x} + g_1\{\ \ 2x[x^2 - y^2 - a(t)] + 2y[2xy - b(t)]\} &= 0 \\ \dot{y} + g_2\{-2y[x^2 - y^2 - a(t)] + 2x[2xy - b(t)]\} &= 0 \end{aligned} \tag{11.41}$$

In the general case, generating f_i, $\partial f_i/\partial x_j$, multiplying, and summing is the most straightforward way of forming the $\dot{x}_j$. Here, however, further simplification is possible. We see that

$$\begin{aligned} \dot{x} + 2g_1\{x[x^2 + y^2 - a(t)] - yb(t)\} &= 0 \\ \dot{y} + 2g_2\{y[x^2 + y^2 + a(t)] - xb(t)\} &= 0 \end{aligned}$$

There is no advantage in taking $g_1 \neq g_2$. Figure 11.12 shows the mechanization using servo multipliers.

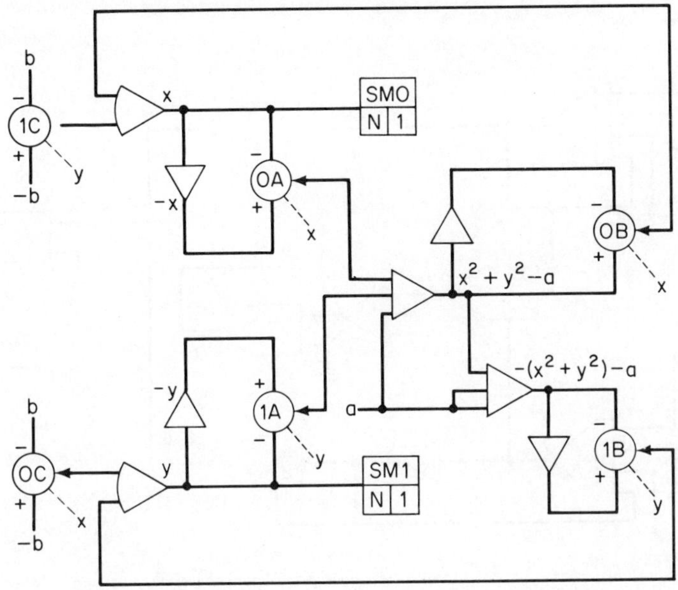

Note: Small feedback capacitors may be needed from output to grid of x and y.

FIG. 11.12 Complex square root, method 1.

Method 2 (*Absolute sum minimization*). From (11.41) we readily obtain

$$\dot{x} + 2g_1(\quad x \operatorname{sign} f_1 + y \operatorname{sign} f_2) = 0$$
$$\dot{y} + 2g_2(-y \operatorname{sign} f_1 + x \operatorname{sign} f_2) = 0$$

as stable equations. Figure 11.13 shows a circuit for these equations requiring just three multiplications. As indicated, the hold relays of any integrators used should be bypassed to obtain convergence in the RESET mode.

Since the Jacobian matrix

$$\mathbf{J} = \begin{pmatrix} \dfrac{\partial f_1}{\partial x} & \dfrac{\partial f_1}{\partial y} \\ \dfrac{\partial f_2}{\partial x} & \dfrac{\partial f_2}{\partial y} \end{pmatrix} = \begin{pmatrix} 2x & -2y \\ 2y & 2x \end{pmatrix}$$

is singular at (0, 0), $[|\mathbf{J}| = 4(x^2 + y^2)]$, an unstable or false rest point exists here when $a \neq 0$ or $b \neq 0$ for both cases. Examination of the differential equations shows it is an unstable point with noise quickly upsetting a situation where $x = y = 0$ to track the root properly. When $a = b = 0$, corresponding to the root (0, 0), convergence of the equations and accuracy are poor. This parallels the poor results of a standard square-root circuit of a real variable.

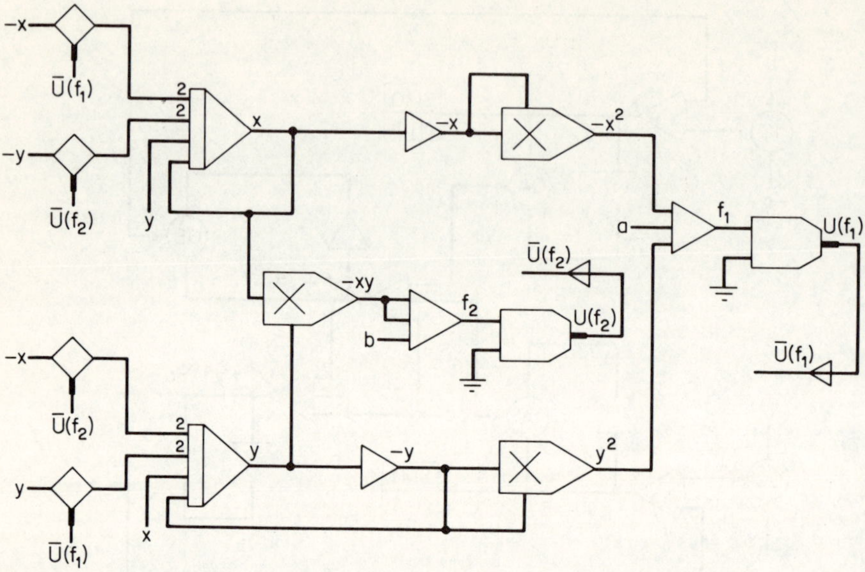

Note: Use small capacitors and bypass hold relays.

FIG. 11.13 Complex square root, method 2.

There are two solutions to (11.40) for all nonzero sets of a and b. The one found when the computer is first placed in RESET is a matter of chance, in this case, unless initial conditions can somehow be provided. Then the root which is "closest" to the initial condition is the one which is converged to. By "closest," we mean that root which can be approached by decreasing V. Once a root is singled out, it changes continuously when a and b change continuously. This is an important consideration in comparing these solutions with those constructed by a brute force method: We may solve (11.40) to find

$$x = \pm\sqrt{\frac{a + \sqrt{a^2 + b^2}}{2}}, \quad y = \pm\,(\text{sign } b)\sqrt{\frac{-a + \sqrt{a^2 + b^2}}{2}}$$

If the "+" sign is taken continuously so that x cannot be negative, y abruptly changes its sign whenever b does, provided a is negative. Thus, a brute force method requires additional switching besides much other nonlinear equipment.

11.5 Other Methods of Solutions

Just as in the transpose solution of linear equations, the equipment requirements do not make the steepest descent attractive for larger systems of nonlinear equations. In fact, the problem is much more serious since n^2 partial derivatives and n^2 nonconstant multiplications or switchings may be required. Before proceeding to a steepest descent solution, a programmer should investigate other methods of solution that do not require the generation of partial derivatives. Some of these are discussed below.

11.5.1 Reordering method

As might be expected, stability can be obtained often with a simple system

$$\dot{\mathbf{x}} + \mathbf{G}\mathbf{f} = \mathbf{0}$$

where $\mathbf{f} \equiv \mathbf{f}(\mathbf{x}, \tau)$ contains all elements f_j or $-f_j$, $\mathbf{G}$ is a diagonal gain matrix, and $\mathbf{x}$ contains the x_i. Suppose $\mathbf{x}_r$ is a root of $\mathbf{f}(\mathbf{x}, \tau) = \mathbf{0}$, and that the Jacobian $\mathbf{J}(\mathbf{f}, \mathbf{x})$ exists at $\mathbf{x} = \mathbf{x}_r$. The system can then be rewritten

$$\dot{\mathbf{x}} + \mathbf{G}[\mathbf{J}(\mathbf{x} - \mathbf{x}_r) + \mathbf{w}] = \mathbf{0} \tag{11.42}$$

where $\mathbf{w}(\mathbf{x})$ is the nonlinear part and $\mathbf{J}(\mathbf{x} - \mathbf{x}_r)$ is the linear part of $\mathbf{f}$, with $\mathbf{J}$ constant because it is evaluated at $\mathbf{x}_r$. (Time-varying parameters in $\mathbf{f}$ can be made effectively stationary by time scaling; both $\mathbf{J}$ and $\mathbf{w}$ may be considered constant in this application.) A fundamental theorem provides for the stability of (11.42) at $\mathbf{x} = \mathbf{x}_r$ when $\mathbf{J}$ is constant.

Theorem 11.12. System (11.42) is stable at $\mathbf{x} = \mathbf{x}_r$ if

(1) All solutions of $\dot{\mathbf{y}} + \mathbf{G}\mathbf{J}\mathbf{y} = \mathbf{0} \to \mathbf{0}$ as $\tau \to \infty$.
(2) $\mathbf{f}$ is continuous in a neighborhood of $\mathbf{x}_r$.
(3) The order of the nonlinear term is greater than one, i.e.,

$$\frac{||\mathbf{w}||}{||\mathbf{x} - \mathbf{x}_r||} \to \mathbf{0} \text{ as } \mathbf{x} \to \mathbf{x}_r$$

The last two conditions are satisfied if $\mathbf{f}$ is expandable in a convergent Taylor series as is usually the case in practical problems. In linear equations, conditions (2) and (3) are automatically satisfied; condition (1) is then both necessary and sufficient for stability. Autonomous nonlinear equations may possibly be stable, however, when real parts of the characteristic roots are zero at $\mathbf{x}_r$. For example, $\dot{x} + x^3 = 0$ is stable at $x = 0$. On the other hand, $\dot{x} + x^2 = 0$ is unstable at $x = 0$. If the real part of any of the characteristic roots is zero, which is called a *critical* case, stability is uncertain and an analysis is usually made by constructing Lyapunov functions.

We may also show that if at least one of the characteristic roots of $\mathbf{GJ}$ has a negative real part, system (11.42) is unstable at $\mathbf{x}_r$.

If $\mathbf{J}$ is nonsingular at $\mathbf{x}_r$, the identical reordering procedure used for linear equations can be used to insure condition (1) is satisfied. Since the elements of $\mathbf{J}$ are functions of the x_i, the trial method almost requires being at the root for sensitivity testing—a disadvantage. (If $\mathbf{J}$ is insensitive to the x_i, we need be only near the root.) But often, solutions are known for one set of parameters so that this can be done. It may also be possible to perform sensitivity tests to obtain first solutions. Let $(f_j)_0$ be the initial values of the f_j due to initial incorrect guesses $(x_i)_0$. Then

$$\bar{f}_j \equiv f_j - (f_j)_0 = 0$$

for the given $(x_i)_0$. Construct a stable system

$$\dot{\mathbf{x}} + \mathbf{G}\bar{\mathbf{f}} = \mathbf{0}$$

and consider the $(f_j)_0$ as parameters. When they are reduced to zero, we obtain a first solution to the original set of equations. Of course, the stable structure with $(f_j)_0 \neq 0$ may not be stable when they are changed. It is then necessary to restructure the system when instabilities occur—another disadvantage. But the entire process may be simpler than testing the $2^n n!$ reordering and sign-changing possibilities of the original set of equations.

If the parameters are time varying and solutions are needed as feedback to other differential equations, this reordering method is useful only if the final system is stable in all cases. It is somewhat of a nuisance to stop the solution and restructure the algebraic part of a problem because of instabilities. Section 11.5.2 discusses possible automation of such restructuring. For stationary-parameter problems, this method is applicable and desirable if many combinations of parameters need be changed and manual restructuring of the equations is tolerable.

Example 11.9. We construct a stable system for the equations

$$f_1 \equiv x^2 + y^2 + z^2 = 0$$
$$f_2 \equiv xy + yz + xz - 0.25 = 0$$
$$f_3 \equiv a_1 x + a_2 y + a_3 z - 0.25 = 0$$

for roots near the nominal values of $a_1 = 0.5$, $a_2 = -0.5$, and $a_3 = 0.5$. At the nominal values of parameters, there are four roots, of which $x \simeq 0.93$, $y \simeq 0.36$, and $z \simeq -0.07$ is the one of interest to be tracked with changing parameters.

Sensitivity testing at the root, as in Example 11.7, shows that f_1 is more sensitive to changes in x compared with other perturbations. Hence, we mechanize

$$\dot{x} + 10f_1 = 0$$

Further testing in the OPERATE mode shows that

$$\dot{z} + 10f_2 = 0, \qquad \dot{y} - 10f_3 = 0$$

are required for stability and fast convergence. Since the f_j are near zero, they must be magnified for convergence characteristics observations needed for the gain selections. These equations track the roots as the a_k are varied one at a time between the limits $0 \leq a_1 \leq 1$, $-1 \leq a_2 \leq 0$, and $0 \leq a_3 \leq 1$, while any two remain at their nominal values.

With all the general difficulties concerning stability in the range of time-varying parameters of interest, the reordering method remains powerful in many cases. We shall see one in Chapter 12, where the equations are almost linear with the nonlinear part having only a small effect. A direct or reordered system then conserves much equipment.

11.5.2 Reordered system with variable structure

It sometimes may be possible to analyze the stability of a constructed system and restructure it automatically to account for all possible cases. We illustrate this in the following example.

Example 11.10. Generate the complex square-root components, as in Example 11.8, using some direct system.

It should be pointed out that the square-root circuit for real variables (Fig. 9.8) depends on which root is desired. If both roots are needed (unusual for a given problem), some switching is required to maintain stability for both cases. For the positive root of a, we wish to solve

$$\dot{x} + g(x^2 - a) = 0$$

and for the negative root

$$\dot{x} - g(x^2 - a) = 0$$

Thus,

$$\dot{x} + g(x^2 - a)\,\text{sign}\,x = 0$$

is a direct method solution accounting for all cases.

We can expect at least a similar complication in solving (11.40). Since continuity is usually required when generating complex square roots, our circuit should account for both solutions.

Suppose we solved the equations

$$\dot{x} + x^2 - y^2 - a = 0, \qquad \dot{y} + 2xy - b = 0$$

The Jacobian of the system is

$$\begin{pmatrix} 2x & -2y \\ 2y & 2x \end{pmatrix}$$

and the equations will not be stable when $x \leq 0$. (If $a < 0$ and $b = 0$, the solution is a pure imaginary.)

On the other hand, the system

$$\dot{x} + 2xy - b = 0, \qquad \dot{y} - (x^2 - y^2 - a) = 0$$

has a Jacobian

$$\begin{pmatrix} 2y & 2x \\ -2x & 2y \end{pmatrix}$$

which must give trouble when $y \leq 0$ ($y = 0$ for $a > 0$ and $b = 0$). Thus we are led to convert from one system to another depending on the sign of a. To keep the trace positive for both possible solutions in each system, we may multiply by sign x or sign y as appropriate. The final system

$$\begin{aligned} \dot{x} + (x^2 - y^2 - a)U(a) \operatorname{sign} x + (2xy - b)U(-a) \operatorname{sign} y &= 0 \\ \dot{y} + (2xy - b)U(a) \operatorname{sign} x - (x^2 - y^2 - a)U(-a) \operatorname{sign} y &= 0 \end{aligned} \tag{11.43}$$

is stable under all possible circumstances. A diagram for solution is shown in Fig. 11.14.

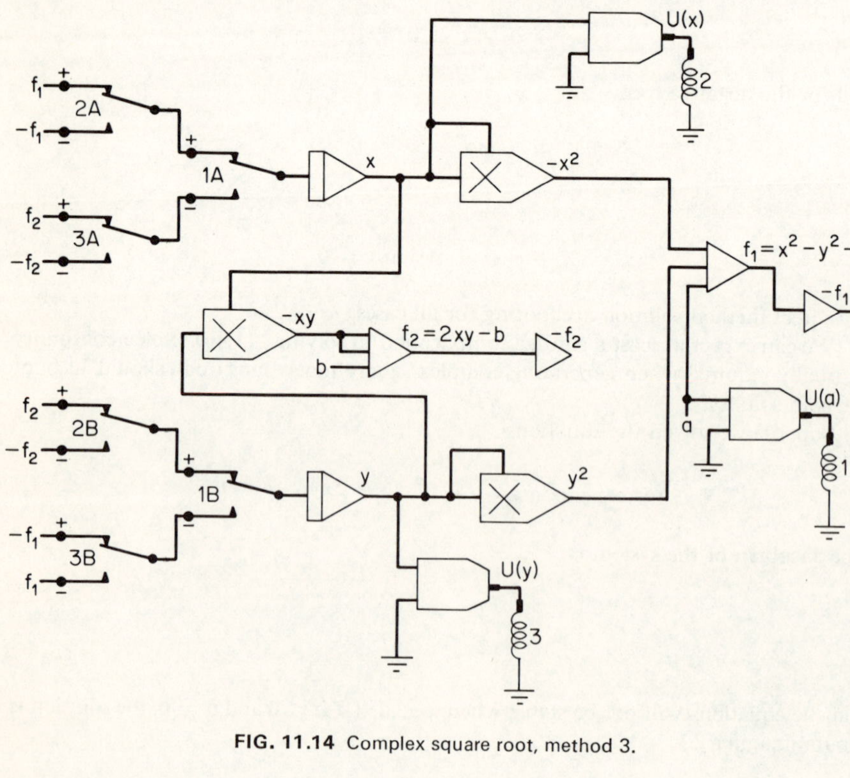

FIG. 11.14 Complex square root, method 3.

The system was constructed because we knew that x can approach zero ($y \neq 0$) only when $a < 0$. We could not hope to be so knowledgeable for more complicated systems.

11.5.3 Sign inversions with variable structure [15]

In cases where the solutions desired are stationary (no tracking because of time-varying parameters is required), Rybashov [15] proposes constructing a special type of direct system with variable structure. Let the computer solve the system

$$\dot{x}_i = \epsilon_i f_i \qquad (x_i)_0 = c_i$$

with switching provisions to let $\epsilon_i = +1$ or -1. Form a positive definite function V of the f_i on the computer and by means of a differentiator circuit, construct $\dot{V}$. If $\dot{V} > 0$, let a trigger circuit systematically search for a set of ϵ_i that makes $\dot{V} < 0$.

Two types of trajectories are now possible:

(1) After a finite number of switchings, the differential equations are stable about the points yielding $f_i = 0$ and the solutions are found.

(2) All 2^n combinations of $\epsilon_i = \pm 1$ result in unstable equations so that no solutions are found. This might be due to a condition of no solutions existing in the region of search. We must then change the initial conditions c_i and search in some other region. Or there may not be any stable solutions in the region of a root $f_i = 0$ with any set of the ϵ_i. We noted this possibility for linear equations; for example, the matrix in (11.27). A reordering of the f_i is then required. There is always a possibility that no real solutions exist and no amount of programming can discover one. Fortunately, such situations do not occur for realistic practical problems.

Example 11.11. Solve the system of algebraic equations

$$\begin{aligned} f_1 &\equiv x^2 + y^2 + z^2 - 1 = 0 \\ f_2 &\equiv x - y + z - 0.5 = 0 \\ f_3 &\equiv xy + yz + xz - 0.25 = 0 \end{aligned} \tag{11.44}$$

using a system of differential equations with a sign-changing variable structure.

A possible trigger circuit is shown in Fig. 11.15. The outputs of amplifiers 1, 2, and 3 are triangular functions of time each successively operating at full amplitude and at twice the frequency of its predecessor. When used as inputs to the comparators, all possible combinations of HI and LO levels appear as outputs during one period of the fundamental wave. These can then be used to close switches, relays, etc. Note the switch that stops the fundamental wave from oscillating when $\dot{V} < 0$.

The rest of the circuit is shown in Fig. 11.16 using the absolute function sum as V. The derivative of V is generated with noise (V should be filtered to minimize this)

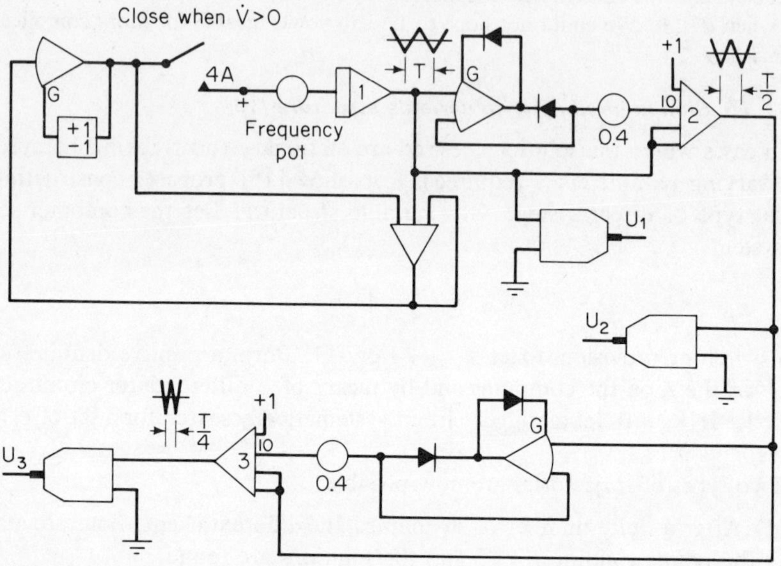

FIG. 11.15 Trigger circuit.

which must be offset by a small voltage ϵ to prevent activation of the trigger circuit when $\dot{V} = 0$. Thus it may be possible for $\dot{V}$ to be very slightly positive without activating the trigger, but this effect may be minimized by experimentally minimizing ϵ to its smallest possible value.

There is also a delay between sensing that $\dot{V} > \epsilon$ and the switching of one of the f_i. The pot controlling the frequency of the fundamental should experimentally be set to minimize any time delays while still allowing the $\dot{V}$ control to function properly ($\dot{V}$ is produced with a lag).

Figure 11.17 shows two graphs of x, y, z, and $\sum |f_i|$. In the first, two switchings produced stability for the root (0.9296, 0.3624, −0.0672) from the initial conditions shown. Eight switchings were required, however, for the root (−0.4899, −0.8624, 0.1276) from other initial conditions. The equations (11.44) are seen to be symmetric in x and z so that roots also exist at (−0.0672, 0.3624, 0.9296) and (0.1276, −0.8624, −0.4899). But they cannot be found with this circuit because the equations can never be stable at these roots. (See Problem 11-21.) Interchanging f_1 and f_3, however, can result in a stable mechanization.

The main use of a sign-changing system with variable structure occurs when some parameters are changed between runs. Then, the circuits attempt to adjust automatically to produce stable solutions. This eliminates the manual construction procedure of Sec. 11.5.2.

Rybashov [15] points out that a search for n^2 numbers $\epsilon_{i,j} = \pm 1$ can produce stable mechanizations at all the roots when solving the equation $\dot{\mathbf{x}} + \mathbf{E}\mathbf{f} = \mathbf{0}$. There are n times as many switches and comparators, but the final circuit

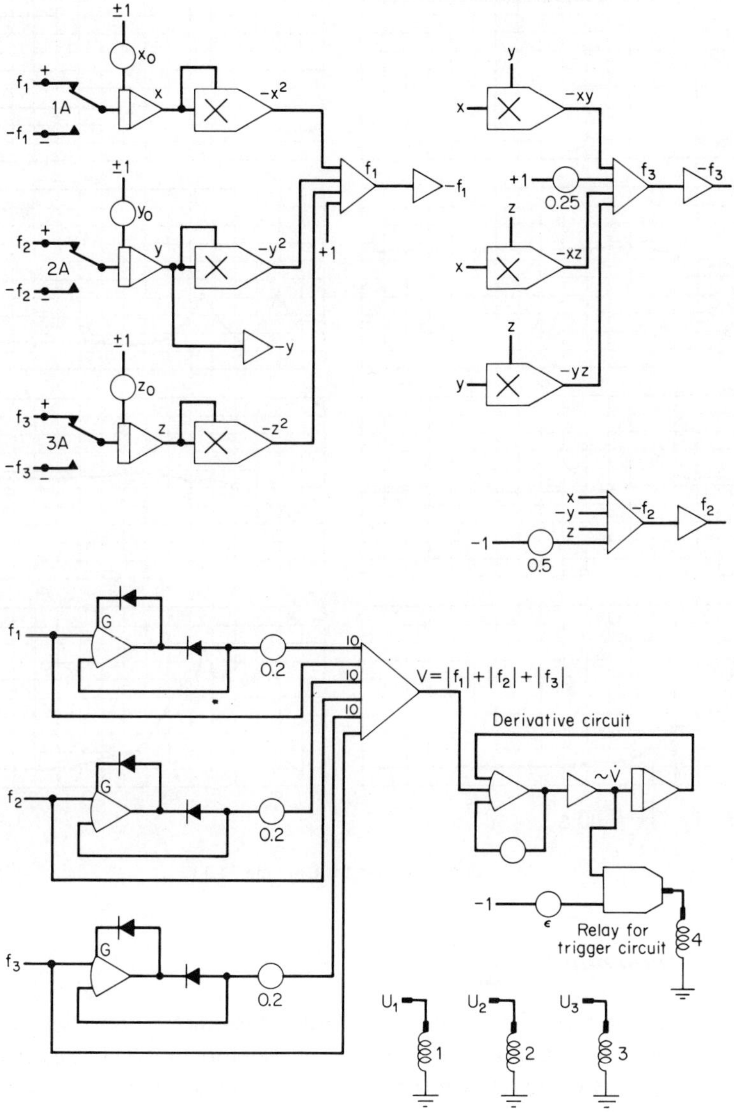

FIG. 11.16 Example 11.11.

may still be simpler than one derived by the method of steepest descent. System (11.43) is such a system where the $\epsilon_{i,j}$ were precalculated.

More sophisticated methods for driving V to zero are discussed in Chapters 19–20, using more sophisticated equipment (mode control, gated logic, and memory.)

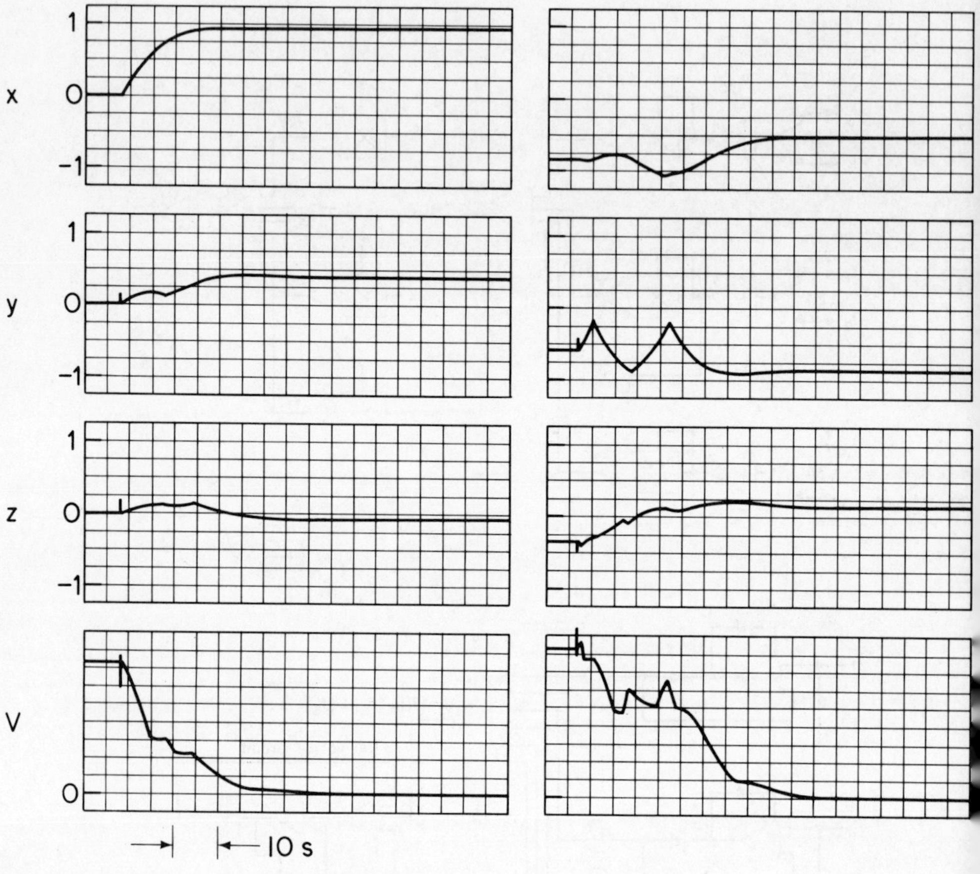

FIG. 11.17 Graphs for Example 11.11.

11.6 Dynamic Errors

Just as in the case of computing first-order dynamic errors in Sec. 9.1.2 for one function of one variable, the time-varying solution $\mathbf{x}$ of (11.32) is slightly different from the computer solution $\mathbf{x}_c$ produced by solving

$$\dot{\mathbf{x}}_c + \mathbf{G}\boldsymbol{\phi}(\mathbf{x}_c, \tau) = \mathbf{0} \tag{11.45}$$

because of the finite diagonal elements of the gain matrix $\mathbf{G}$. At any instant, a variation $\Delta\mathbf{x}$ in $\mathbf{x}$ will produce a spatial variation $\Delta\boldsymbol{\phi}$ in $\boldsymbol{\phi}$. The dynamic error vector $\Delta\mathbf{x} = \mathbf{x} - \mathbf{x}_c$ must be such that the first-order spatial relationship

$$\Delta\boldsymbol{\phi} \simeq \mathbf{J}(\boldsymbol{\phi}, \mathbf{x})\Delta\mathbf{x} \tag{11.46}$$

holds. By differentiating (11.32) with respect to τ, we see that any small variations about the nominal values of $\mathbf{x}_0$ and τ_0 must satisfy

$$\mathbf{J}\dot{\mathbf{x}} + \frac{\partial \boldsymbol{\phi}}{\partial \tau} = \mathbf{0} \quad \text{or} \quad \dot{\mathbf{x}} = -\mathbf{J}^{-1}\frac{\partial \boldsymbol{\phi}}{\partial \tau}$$

provided $\mathbf{J}$ is nonsingular and $\partial\boldsymbol{\phi}/\partial\tau$ is finite. We also expand $\boldsymbol{\phi}(\mathbf{x}_c, \tau)$ in a Taylor series, referencing it spatially about $\mathbf{x}$ as in (11.46):

$$\boldsymbol{\phi}(\mathbf{x}_c, \tau) = \boldsymbol{\phi}(\mathbf{x} - \Delta\mathbf{x}, \tau) = \boldsymbol{\phi}(\mathbf{x}) - \mathbf{J}\Delta\mathbf{x}$$

Since $\dot{\mathbf{x}}_c = \dot{\mathbf{x}} - \Delta\dot{\mathbf{x}}$, (11.45) transforms to

$$\dot{\mathbf{x}} - \Delta\dot{\mathbf{x}} + \mathbf{G}[\boldsymbol{\phi}(\mathbf{x}) - \mathbf{J}\Delta\mathbf{x}] = \mathbf{0}$$

or, by substituting the value of $\dot{\mathbf{x}}$ and noting that $\boldsymbol{\phi}(\mathbf{x}) = \mathbf{0}$,

$$\mathbf{J}^{-1}\frac{\partial \boldsymbol{\phi}}{\partial \tau} + \Delta\dot{\mathbf{x}} + \mathbf{G}\mathbf{J}\Delta\mathbf{x} \simeq \mathbf{0} \tag{11.47}$$

We now make a very important assumption: Steady-state conditions are reached almost instantly because of the large diagonal elements of the gain matrix $\mathbf{G}$. This implies that (11.31) must be stable and that there are no pure imaginary characteristic roots as in a critical case. Before $\mathbf{J}$ or $\mathbf{x}$ can significantly change, $\Delta\dot{\mathbf{x}}$ is driven approximately to $\mathbf{0}$, and $\Delta\mathbf{x}$ is constant. This is what is meant by first-order dynamic tracking errors. Thus we have the following theorem:

Theorem 11.13. Under the suitable assumption given, the instantaneous first-order dynamic tracking errors when solving a stable set of equations (11.31) are given by the error $\Delta\boldsymbol{\phi}$ in $\boldsymbol{\phi}$:

$$\Delta\boldsymbol{\phi} = -\mathbf{G}^{-1}\mathbf{J}^{-1}\frac{\partial \boldsymbol{\phi}}{\partial \tau} \tag{11.48}$$

and the error $\Delta\mathbf{x}$ in $\mathbf{x}$:

$$\Delta\mathbf{x} = -\mathbf{J}^{-1}\mathbf{G}^{-1}\mathbf{J}^{-1}\frac{\partial \boldsymbol{\phi}}{\partial \tau} \tag{11.49}$$

Equation (11.49) reduces to (9.4) in the one-dimensional case and does not include the case where $\mathbf{J}$ is singular, even if the equations are stable. Tracking is usually poor at any root where $\mathbf{J}$ is singular or $\partial\boldsymbol{\phi}/\partial\tau$ is unbounded or does not exist because $\boldsymbol{\phi}$ is discontinuous in τ.

Theorem 11.13 may be difficult to apply for an error estimation when n is large, but several important conclusions can be drawn from it. First, the vector

$\partial\boldsymbol{\phi}/\partial\tau$ is subject to time scaling on the computer. The elements of $\Delta\mathbf{x}$ and $\Delta\boldsymbol{\phi}$ are proportional to the time scale factor. Next, we see that the errors are approximately inversely proportional to the diagonal elements of $\mathbf{G}$ because $\mathbf{G}^{-1}$ has diagonal elements equal to $(1/g_{i,i})$. The $g_{i,i}$ should be as large as possible without introducing instability. Finally, the importance of the Jacobian matrix is clear. If there are several orderings of the equations that could permit stability, the one with the largest Jacobian determinant probably minimizes errors. We see this from (11.19), where the inverse elements are inversely proportional to the determinant of the matrix.

Example 11.12. Estimate the dynamic tracking errors for the equations

$$\dot{x} + g_1(x^2 - y^2 - a) = 0, \qquad \dot{y} + g_2(2xy - b) = 0 \tag{11.50}$$

in the region of stability ($x > 0$ and $a \geq 0$). With

$$\mathbf{G} = \begin{pmatrix} g_1 & 0 \\ 0 & g_2 \end{pmatrix}, \qquad \frac{\partial\boldsymbol{\phi}}{\partial\tau} = \begin{pmatrix} -\dot{a} \\ -\dot{b} \end{pmatrix}, \qquad \mathbf{J} = \begin{pmatrix} 2x & -2y \\ 2y & 2x \end{pmatrix}$$

we compute

$$\mathbf{J}^{-1} = \frac{1}{2(x^2 + y^2)} \begin{pmatrix} x & y \\ -y & x \end{pmatrix}, \qquad \mathbf{G}^{-1} = \begin{pmatrix} 1/g_1 & 0 \\ 0 & 1/g_2 \end{pmatrix}$$

and obtain

$$\Delta\boldsymbol{\phi} = \frac{1}{2(x^2 + y^2)} \begin{pmatrix} \dfrac{\dot{a}x}{g_1} + \dfrac{\dot{b}y}{g_1} \\ \dfrac{-\dot{a}y}{g_2} + \dfrac{\dot{b}x}{g_2} \end{pmatrix}$$

$$\Delta\mathbf{x} = \frac{1}{4(x^2 + y^2)^2} \begin{pmatrix} \dfrac{\dot{a}x^2}{g_1} + \dfrac{\dot{b}yx}{g_1} - \dfrac{\dot{a}y^2}{g_2} + \dfrac{\dot{b}xy}{g_2} \\ -\dfrac{\dot{a}xy}{g_1} - \dfrac{\dot{b}y^2}{g_1} - \dfrac{\dot{a}xy}{g_2} + \dfrac{\dot{b}x^2}{g_2} \end{pmatrix} \tag{11.51}$$

If we let $g_1 = g_2 = g$, and note from (11.50) that

$$x^2 - y^2 \simeq a, \qquad 2xy \simeq b, \qquad x^2 + y^2 \simeq \sqrt{a^2 + b^2}$$

(11.51) reduces to

$$\Delta\mathbf{x} = \frac{1}{4(a^2 + b^2)g} \begin{pmatrix} \dot{a}a + \dot{b}b \\ -\dot{a}b + \dot{b}a \end{pmatrix}$$

Thus, the error in the real component Δx is proportional to the rate of change of the modulus. If $a^2 + b^2$ is constant, $a\dot{a} + b\dot{b} = 0$. The error in the imaginary component Δy is proportional to the phase rate of change. If b/a is constant, $-\dot{a}b + \dot{b}a = 0$ with $M = \sqrt{a^2 + b^2}$ and $\phi = \arctan(b/a)$, so that we may calculate

$$\Delta x \simeq \frac{\dot{M}}{4Mg}, \qquad \Delta y \simeq \frac{\dot{\phi}}{4g}$$

11.7 Programming Problems [16]–[23]

Closely related to problems requiring the solution of algebraic equations are the so-called programming problems that arise in many disciplines—operations research [16], game theory and logistics [17], biological systems [19], and others [25]–[30]. The general problem is stated as follows:

Given q *inequality* relations of n variables,

$$g_k(x_1, x_2, \ldots, x_n) \leq 0, \qquad k = 1, 2, \ldots, q \tag{11.52}$$

determine the x_i so that a given function

$$\rho = \rho(x_1, x_2, \ldots, x_n)$$

is minimized. If inequalities are given in the form $h_k \geq 0$, we use $-h_k \leq 0$ as one of the g_k. Similarly, maximizing ρ is equivalent to minimizing $-\rho$. The function ρ is known by various names as a criterion, objective, utility, cost, profit function, etc.

Special cases arise in practice also. If the g_k and ρ are linear, the problem is called a *linear* (or *quadratic* if ρ is quadratic) programming problem; otherwise it is a nonlinear problem. If the g_k and ρ are *convex* functions, the problem is a convex programming problem. A convex function $\phi(x_1, x_2, \ldots, x_n)$ has the property

$$\begin{aligned}\phi[(1-\lambda)x_1 + \lambda y_1, (1-\lambda)x_2 + \lambda y_2, \ldots, (1-\lambda)x_n + \lambda y_n]\\ \leq (1-\lambda)\phi(x_1, x_2, \ldots, x_n) + \lambda\phi(y_1, y_2, \ldots, y_n)\end{aligned}$$

for all $0 \leq \lambda \leq 1$ and arbitrary x_i and y_i.

In addition to the inequalities (11.52), there may be m linear equality relations, $m < n$,

$$f_j = \sum_{i=1}^{n} a_{j,i}x_i + b_j = 0, \qquad j = 1, 2, \ldots, m \tag{11.53}$$

which may serve to reduce the number of unknowns. However, it may be convenient to keep the f_j to retain original parameters $a_{j,i}$ and b_j, so that (11.53) is included in the general problem statement.

Because of the constraints, the usual methods to minimize ρ, i.e., solving the system $\partial\rho/\partial x_i = 0$, must be modified. The customary way of dealing with this problem is to create an auxiliary *penalty* function ρ to minimize instead. This can be explained in the following intuitive way. Suppose we let

$$\rho_a = \mu\rho + a\left[\sum_{k=1}^{q} g_k^2\, U(g_k) + \sum_{j=1}^{m} f_j^2\right] \tag{11.54}$$

with $0 < \mu \leq 1$ and $a \gg 0$. Any term besides $\mu\rho$ is large and positive if the corresponding inequality or equality is not satisfied. Insertion of the step function $U(g_k)$ as a product of g_k^2 only "penalizes" ρ_a when g_k is positive, whereas f_j penalizes ρ_a when not zero. Because a is purposely made very large, we see that it "pays" to satisfy or nearly satisfy the constraints if ρ_a is to be minimized, and we intuitively feel that a minimum of ρ_a should define points x_i, which are indeed near the solution for the minimum of ρ subject to the constraints. Also, in place of g_k^2 we could have used any positive function that monotonically increases the penalty as g_k move further away from zero; $|g_k|$ for example. The same applies to f_j. In the case of linear and convex programming problems, where solutions, if they exist, are unique, we have the following theorem [23]:

Theorem 11.14. As $a \to \infty$, the absolute minimum of ρ_a defines x_i, which approach the solution of the programming problem.

The method of steepest descent may now be used to minimize ρ_a for some arbitrarily large a. By taking

$$\begin{aligned}\dot{x}_i &= -\frac{\partial \rho_a}{\partial x} \\ &= -\mu\frac{\partial \rho}{\partial x_i} - 2a\left[\sum_{k=1}^{q} g_k \frac{\partial g_k}{\partial x_i} U(g_k) + \sum_{j=1}^{m} f_j \frac{\partial f_j}{\partial x_i}\right]\end{aligned} \tag{11.55}$$

we may force $\dot{\rho}_a \leq 0$ and approach the correct solution by letting $\mu \to 0$ or $a \to \infty$ or some combination of both. For general nonlinear problems, local minimums may be reached; it is necessary to vary the initial conditions $(x_i)_0$ to search for the global minimum. For convex programming problems, however, unique rest points are always obtained [22]. These will not satisfy the constraints if no solution exists, and must be checked.

Example 11.13 (linear programming). Minimize the objective function

$$\rho = x_1 - 2x_2 + 3x_3$$

subject to the constraints

$$
\begin{aligned}
g_1 &\equiv \quad x_1 - 5x_2 - x_3 + 1 \le 0 \\
g_2 &\equiv \quad 3x_1 - x_2 + 4x_3 - \tfrac{3}{4} \le 0 \\
g_3 &\equiv -2x_1 - 2x_2 + 3x_3 - 1 \le 0 \\
g_4 &\equiv -5x_1 + 2x_2 - 4x_3 - \tfrac{1}{10} \le 0 \\
g_5 &\equiv -x_1 \le 0 \\
g_6 &\equiv -x_2 \le 0 \\
g_7 &\equiv -x_3 \le 0
\end{aligned}
$$

In linear programming problems, relationships such as $g_5 - g_7$ generally exist to force each variable to be positive. To implement these relations, the simple diode integrator limiting circuit of Fig. 6.35 can be used (without the pot, since $b = 0$), but we shall set up all the inequalities using idealized diode circuits for more precise results.

From (11.55), using the penalty function (11.54), we get

$$
\begin{aligned}
\dot{x}_1 &= -\mu - 2a[g_1 U(g_1) + 3g_2 U(g_2) - 2g_3 U(g_3) - 5g_4 U(g_4) - g_5 U(g_5)] \\
\dot{x}_2 &= 2\mu - 2a[-5g_1 U(g_1) - g_2 U(g_2) - 2g_3 U(g_3) + 2g_4 U(g_4) - g_6 U(g_6)] \qquad (11.56) \\
\dot{x}_3 &= -3\mu - 2a[-g_1 U(g_1) + 4g_2 U(g_2) + 3g_3 U(g_3) - 4g_4 U(g_4) - g_7 U(g_7)]
\end{aligned}
$$

As might be expected, the a in each equation can be different, for the matrix of coefficients inside the brackets is positive definite.

Figure 11.18 shows the kth computing amplifier for generating $g_k U(g_k)$ based on

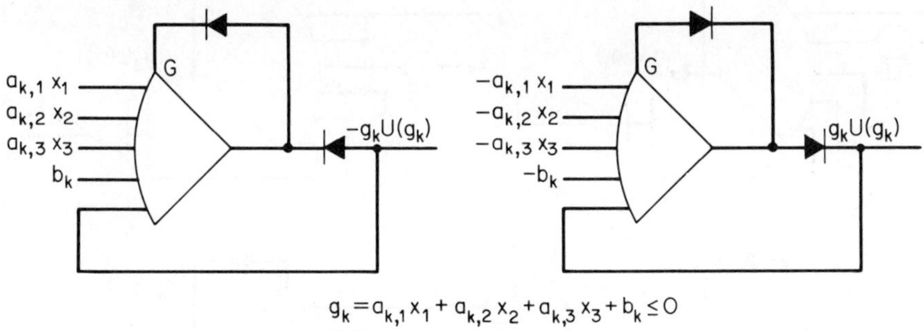

FIG. 11.18 Generation of $g_k U(g_k)$ using idealized diodes.

Fig. 6.43. Figure 11.19 is a circuit diagram that solves Example 11.13 using idealized diode circuits to implement all the U functions. Note that the individual terms in (11.56) can be summed individually at the integrator input. But to achieve very high gains with external inaccurate resistors, one input to each integrator is desirable. The low leakage of the idealized diode circuits permits gains of at least 100. With μ set to about 0.01 or lower, good values of the x_i were computed [(1.607, 4.065, -0.0001) compared with theoretical (1.6, 4.05, 0)].

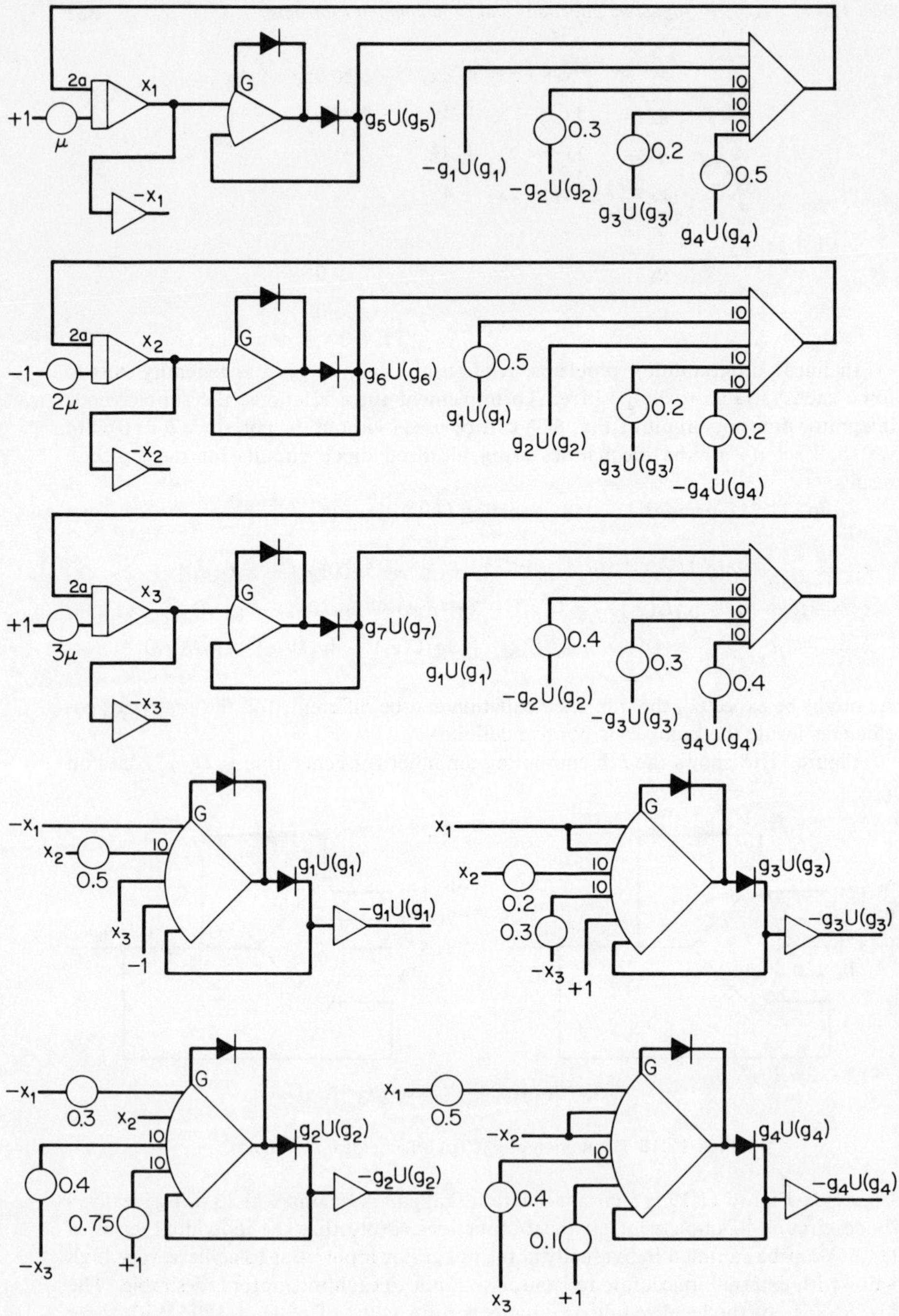

FIG. 11.19 Unscaled diagram for Example 11.13 using idealized diodes.

Theoretically, the solution of a linear programming problem must reduce n of the initial inequalities to zero. (By taking the approximate result, we see that these must be g_2, g_4, and g_7, from which the more accurate results were obtained.) Karpinskaya and Rybashov [20] use this principle directly in the solution of the problem. There are $(7!)/[(4!)(3!)]$ combinations of three equations to solve, but by systematically selecting the equations to solve, many less trials will actually be required (see [20] for details). The advantage of solving many sets of only three equations is that less equipment is required per solution.

PROBLEMS

11-1. Using $\dot{x} = -x$, determine the circuit for graphing the function

$$f(x) \equiv 4x^{10} - 6x^5 + 8x^3 - 1 = 0$$

and find the roots as in Example 11.1.

11-2. The methods of Example 11.1 are applicable to graphing sums of fractional powers of x. Using $\dot{x} = -x$, determine the (positive) roots to

$$f(x) \equiv x^{5.732} - 4x^{2.15} + 2x^{0.139} - 1 = 0$$

11-3. An extension of the system $\dot{x} = -x$ is $\dot{x} = -(x - b)$, $x_0 = a$, which permits x to sweep from a to b. For a polynomial defined by (11.6), show that the equation (11.8) holds if

$$\dot{P}_{n-1} = -nP_n + x\frac{dP_n}{dx} - b\frac{dP_n}{dx}$$

which is a polynomial of degree $n - 1$.

11-4. Determine graphically the 14 roots to the equations

$$f_1 \equiv \sin(x + y) + y = 0$$
$$f_2 \equiv x^2 + 100y^2 - 100 = 0$$

using linear equipment only.

11-5. In one dimension, the Newton-Raphson process reduces to

$$\frac{df}{dx}\Delta x = -f(x)$$

so that the value $-f(x)/(df/dx)$ must be added to the estimate of a root in order to improve it. The geometric interpretation becomes clear from Fig. 11.20 where the slope df/dx of the tangent line to $f(x)$ at x is given by $f(x)/(-\Delta x)$. Use the Newton-Raphson process to improve one of the roots in Problem 11-1.

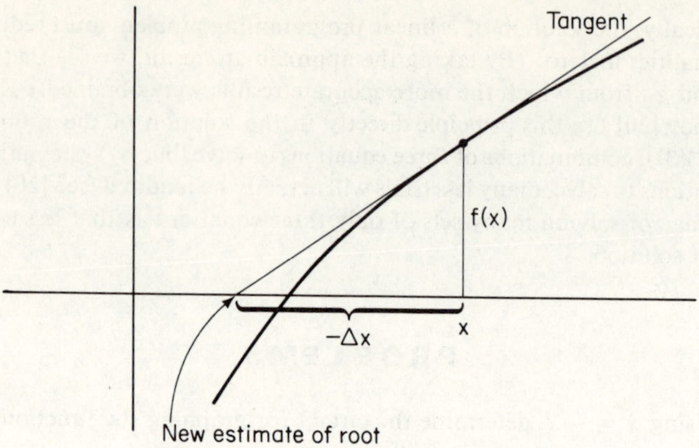

FIG. 11.20 Newton–Raphson process in one dimension.

11-6. Routh-Hurwitz stability test. Given the characteristic equation of a matrix **A**

$$f(\lambda) = \sum_{i=0}^{n} c_i \lambda^i = 0, \qquad c_n = (-1)^n$$

the Routh-Hurwitz criterion will determine if **A** is positive stable without finding the characteristic roots. Write the $n \times n$ matrix array $(r_{i,j}) = [(-1)^{n-2i+j} c_{n-2i+j}]$, i.e.,

$$\mathbf{R} = \begin{pmatrix} (-1)^{n-1}c_{n-1} & (-1)^n c_n & 0 & \cdots & 0 & 0 & 0 \\ \cdot & \cdot & & & \cdot & \cdot & \cdot \\ \cdot & \cdot & & & \cdot & \cdot & \cdot \\ \cdot & \cdot & & & \cdot & \cdot & \cdot \\ 0 & 0 & \cdots & 0 & c_0 & -c_1 & c_2 \\ 0 & 0 & \cdots & 0 & 0 & 0 & c_0 \end{pmatrix} \tag{11.57}$$

where $r_{i,j} = 0$ if $n - 2i + j < 0$ or $n - 2i + j > n$. Then **A** is positive stable if and only if the discriminants of **R** are positive. On this basis, show that the necessary and sufficient conditions that a 2×2 matrix be positive stable is that $|\mathbf{A}| > 0$ and trace of **A** [written $\text{tr}(\mathbf{A}) = \sum_{i=1}^{n} a_{i,i}] > 0$.

11-7. The chief disadvantage of the Routh-Hurwitz stability test is that it requires the characteristic equation coefficients, which entails much calculation. But for small-order matrices, equivalent tests can be developed in terms of the matrix coefficients to determine the positive stability condition more quickly. To extend the result of Problem 11-6 to a 3×3 matrix, we first note that the coefficients c_i in (11.57) can be found in terms of high-order minors. Let the ith-order minor $M_{i,k_1,k_2,\ldots,k_i}$ be the determinant of the array found by crossing out rows and columns $k_1, k_2, \ldots, k_i$, all mutually exclusive. It can then be shown that

$$c_i = (-1)^i \sum_{\substack{k_1, k_2, \ldots, k_i \\ k_1 \neq k_2 \neq \ldots \neq k_i}}^{n} M_{i,k_1,k_2,\ldots,k_i}$$

with $c_0 = |\mathbf{A}|$ and $c_{n-1} = (-1)^{n-1}\text{tr}(\mathbf{A})$. A necessary but not sufficient condition for positive stability is that the coefficients alternate in sign. Use the Routh-Hurwitz stability test to show that the necessary and sufficient conditions for positive stability of a 3 × 3 matrix is that

$$|\mathbf{A}| > 0, \qquad \text{tr}(\mathbf{A}) > 0, \qquad -c_1 > \frac{|\mathbf{A}|}{\text{tr}(\mathbf{A})}$$

11-8. Cayley-Hamilton Theorem: A matrix satisfies its own characteristic equation. Thus if

$$\sum_{i=0}^{n} c_i \lambda^i = 0$$

is the characteristic equation of **A**,

$$\sum_{i=1}^{n} c_i \mathbf{A}^i + c_0 \mathbf{I} = \mathbf{0}$$

For high-order matrices, it is easier to find the c_i by computing each $\mathbf{A}^i$ and solving a set of $n \times n$ linear equations. Verify the Cayley-Hamilton theorem for the matrix

$$\begin{pmatrix} 1 & 2 & 1 \\ 1 & 3 & 1 \\ 1 & 2 & 5 \end{pmatrix}$$

11-9. Determine if the quadratic form $2x^2 + 3y^2 + z^2 + 2xy + 2xz + 2yz$ is positive definite (Theorems 11.1 and 11.2).

11-10. Program the solution to

$$\begin{aligned} 3x + y - z &= 5 \sin t \\ -2x + 6y + 3z &= \cos t \\ x + y + 3z &= e^{-t} \end{aligned}$$

for $t_0 = 0, \dot{t} = 1$ (Theorems 11.3 and 11.5).

11-11. Program the solution to

$$\begin{aligned} 3x + 2y + 8z \sin \omega t &= 1 \\ -3xe^{-t} + 2y - 5z \cos \omega t &= 1 \\ -8x \sin \omega t + 4y \cos \omega t + 2z &= 1 \end{aligned}$$

for $\dot{t} = 1, t_0 = 0$ (Theorems 11.4 and 11.5).

11-12. Program the solution to

$$2xe^{-t} + y\cos t = f_1(t)$$
$$x\cos t + y(\sin t - 1) = f_2(t)$$

for $i = 1$, $t_0 = 0$, using the transpose method, and assuming $f_1(t)$ and $f_2(t)$ are available.

11-13. Show that the sign inversion method can always be made to work on a 2×2 nonsingular matrix containing unequal main diagonal elements.

11-14. In view of the truth of Problem 11-13, program Problem 11-12 with a switching arrangement to avoid the extra multiplications.

11-15. Indicate how to find a reordering for the matrix

$$\begin{pmatrix} -1 & 9 & 7 \\ 7 & 3 & 1 \\ 2 & 4 & -3 \end{pmatrix}$$

on the computer to produce stability.

11-16. Mechanize the dead zone expressed by (11.39) for a division circuit using (11.37) and (11.38).

11-17. Maybach [24] proposes a division circuit based on the high-gain solution of

$$\dot{x} + g(\text{sign } a_1)(\text{sign } f) = 0$$

in order to improve convergence when a_1 is near 0. Mechanize this equation ($f \equiv a_1 x - a_2 = 0$).

11-18. Mechanize the equations

$$f_1 \equiv 3x^2 + xz - z^2 - g_1(t) = 0$$
$$f_2 \equiv -x^2 + 3y^2 + xy - g_2(t) = 0$$
$$f_3 \equiv yz - y^2 + 3z^2 - g_3(t) = 0$$

by the method of steepest descent using the absolute sum minimization.

11-19. If $g_1 \simeq g_2 \simeq g_3$ in Problem 11-18, then we perceive that roots (x_1, y_1, z_1) are such that $x_1 \simeq y_1 \simeq z_1$ because of the cyclic nature of the variables. Also $(-x_1, -y_1, -z_1)$ are roots. Suppose we are interested in positive roots when $g_1, g_2, g_3 > 0$ and are approximately equal. Justify the use of the direct (reordering) method (Sec. 11.5.1).

11-20. Suppose g_1, g_2, and g_3 are run-varying constants in Problem 11-18. Set up the solution with a system of differential equations containing a variable sign-changing structure.

11-21. Prove the assertion that system (11.44) cannot be stable using the variable structure method at the root (0.1276, −0.8624, −0.4899) unless the equations are reordered.

11-22. Estimate dynamic errors inherent in solving Problem 11-18 by the direct method using

$$\dot{x} + gf_1 = 0, \qquad \dot{y} + gf_2 = 0, \qquad \dot{z} + gf_3 = 0$$

assuming $g_1 \simeq g_2 \simeq g_3$ so that $x \simeq y \simeq z > 0$ are solutions.

11-23. Do Example 11.13 with simple instead of idealized diode circuits.

11-24. Transportation problem [21]. There are m warehouses each containing an amount a_i of a given item and n stores requiring deliveries of the amount d_j of that item. In addition, there is a cost matrix, $c_{i,j}$, the cost of delivering one unit of the item from warehouse i to store j. Let $x_{i,j}$ be the amount of item delivered from warehouse i to store j. The idea is to determine the $x_{i,j}$ that minimizes the cost C of fulfilling the needs of the stores. We have to minimize

$$C = \sum_{i=1}^{m} \sum_{j=1}^{n} c_{i,j} x_{i,j}$$

under the constraints

$$-x_{i,j} \leq 0, \qquad \sum_{j=1}^{n} x_{i,j} - a_i \leq 0, \qquad \sum_{i=1}^{m} x_{i,j} - d_j = 0$$

Show that simple diode circuits are very economical of amplifiers [requiring $mn + 2(m + n)$], and propose a sample mechanization.

REFERENCES

[1] Hausner, A., "Selected Generalized Integration Techniques for Analog Computation," *Simulation*, vol. 9, July 1967.

[2] Bellman, R., *Stability Theory of Differential Equations* (New York: McGraw-Hill Book Co., 1953).

[3] Bellman, R., *Introduction to Matrix Analysis* (New York: McGraw-Hill Book Co., 1960).

[4] Varga, R. S., *Matrix Iterative Analysis* (Englewood Cliffs, N. J.: Prentice-Hall, 1962).

[5] Korn, G. A., "Stabilization of Simultaneous Equation Solvers," *Proc. IRE*, vol. 37, Sept. 1949. See also a discussion of this paper, *Proc. IRE*, vol. 38, May 1950.

[6] Cederbaum, I., and A. Fuchs, "On the Stability of Linear Algebraic Equation Solvers," *Proceedings of the 2nd Inter. Analogue Comp. Meetings* (Brussels: Presses Académiques Européennes, 1959).

[7] Goldberg, E. A., and G. W. Brown, "An Electronic Simultaneous Equation Solver," *J. Applied Physics*, vol. 19, April 1948.

[8] Gephart, L., "Linear Algebraic Systems and the REAC," *Math. Tables and Other Aids to Comp.*, vol. 6, July 1952.

[9] Rybashov, M. V., "Some Methods of Solving Systems of Linear Algebraic Equations by Electronic Simulation," *Automatika i Telemekhanika*, vol. 23, Feb. 1962.

[10] Gilbert, C. P., and B. J. J. McHugh, "Economical Circuits for the Analog Solution of Algebraic Equations," *IEEETEC*, vol. 14, April 1965.

[11] Rybashov, M. V., "Analog Solution of Algebraic and Transcendental Equations by the Gradient Method," *Automatika i Telemekhanika*, vol. 22, Jan. 1961.

[12] Brunner, W., "Steepest Ascent," *Princeton Computation Center Report No. 151*, Electronic Associates, Inc., Princeton, N. J., Dec. 29, 1959.

[13] Brunner, W., "Solution of Linear Algebraic Problems by Means of the Electronic Analog Computer," *Princeton Computation Center Report No. 176*, Electronic Associates, Inc., Princeton, N. J., Nov. 1962.

[14] Vitenberg, I. M., and E. A. Gluzberg, "Use of Electronic Simulators to Solve Boundary Problems and Algebraic, Transcendental, or Integral Equations," *Automatika i Telemekhanika*, vol. 17, July 1956.

[15] Rybashov, M. V., "Finding the Roots of Systems of Finite Equations by Means of an Electronic Simulator, Using Differential Equations With Variable Structure," *Automatika i Telemekhanika*, vol. 22, Dec. 1961.

[16] Ablow, C. M., and G. Brigham, "An Analog Solution of Programming Problems," *J. Operations Research Soc. of America*, vol. 3, Nov. 1955.

[17] Pyne, I. B., "Linear Programming on an Electronic Analogue Computer," *Trans. AIEE, Part I* (*Comm. and Elect.*), vol. 75, May 1956.

[18] DeLand, E. C., "Continuous Programming Methods on an Analog Computer," *RAND Corporation Report P-1815*, Santa Monica, Calif., Sept. 29, 1959 (ASTIA No. 616375).

[19] DeLand, E. C., "Simulation of a Biological System on an Analog Computer," *IRETEC*, vol. 11, Feb. 1962.

[20] Karpinskaya, I. N., and M. V. Rybashov, "A Method for Solving the Linear-Programming Problem with an Electronic Model," *Automatika i Telemekhanika*, vol. 24, March 1963.

[21] Longley, D., "An Analogue Computer Application in Operations Research," *Electronic Eng.*, vol. 36, June 1964.

[22] Rybashov, M. V., "The Gradient Method of Solving Convex Programming Problems on Electronic Analog Computers," *Automatika i Telemekhanika*, vol. 26, Nov. 1965.

[23] Rybashov, M. V., "Gradient Method of Solving Linear and Quadratic Programming Problems on Electronic Analog Computers," *Automatika i Telemekhanika*, vol. 26, Dec. 1965.

[24] Maybach, R. L., "Generation of Inverse Functions by the Method of Steepest Descent," *AICA*, vol. 8, Oct. 1966.

[25] Neustadt, L. W., "Applications of Linear and Non-Linear Programming Techniques," *Proceedings of the 3rd Inter. Comp. Meetings* (Brussels: Presses Académiques Européennes, 1962).

[26] Neustadt, L. W., "Simulating Money Allocation Problems on an Analog Computer," *AICA*, vol. 4, Jan. 1962.

[27] Moser, J. H., *et al.*, "Nonlinear Programming Technique for Analog Computation," *Chem. Eng. Prog.*, vol. 57, June 1961.

[28] Firth, A. W. O., "Optimization Problems, Solutions by an Analog Computer," *Computer J.*, vol. 4, April 1961.

[29] Eilon, S., and D. P. Deziel, "The Use of an Analogue Computer in Some Operational Research Problems," *Oper. Res. Quart.*, vol. 16, Sept. 1965.

[30] Rybashov, M. V., and E. E. Dudnikov, "Solution of Game Problems on Analogue Computers," *Proceedings of the 4th Inter. Analogue Comp. Meetings* (Brussels: Presses Académiques Européennes, 1966).

[31] Anderson, T. C., "Four-Quadrant Division Circuits," *Instr. and Control Systems*, vol. 35, April 1962.

See also:

[32] Chen, C. F., *et al.*, "Constructing Liapunov Functions by Means of the Analog Computer," *IEEETE*, vol. 7, June-Sept. 1964.

[33] Hay, I., "Computer Control of Cement Kiln Feed Blending," *Instr. and Control Systems*, vol. 37, July 1964.

[34] Madic, P. B., "Experiences with Analogue Computer for Solving Systems of Linear Algebraic Equations," *Proceedings of the Inter. Analogy Comp. Meeting* (Brussels: Presses Académiques Européennes, 1956).

Chapter 12

SOLUTIONS TO LAGRANGE'S EQUATIONS AND MECHANICAL PHENOMENA

Analog computers have been applied successfully to the study of particle dynamics and other mechanical phenomena, primarily because these are described by differential equations which are easily solved by the computer. In this chapter, we discuss some of the techniques used to handle such problems.

We assume some basic physical concepts are known to the reader. Ideas involving motion, such as Newton's laws, acceleration, velocity, momentum, force, work, kinetic energy, etc. can best be obtained from standard works on mechanics [1]–[4]. However, some of the subtler concepts such as the differences between holonomic and nonholonomic constraints are explained, particularly when germane to the analog technique for mechanization.

12.1 Lagrange's Equations—Development

Lagrange's formulation of the equations of motion of a dynamic system is essentially a restatement of Newton's laws of motion in terms of generalized coordinates rather than rectilinear coordinates. These are the quantities (numbers) that determine the system configuration completely at any instant of time. If known as a function of time, the generalized coordinates determine the position of all particles considered and completely determine the motion of the system. The pendulum in a plane, for example, requires the knowledge of one coordinate (an angular coordinate θ) for a complete description of its motion. There are two rectilinear coordinates x and y that can be used to describe the motion of the pendulum (see Fig. 12.1), but there is then one constraint equation

$$x^2 + y^2 = d^2 \tag{12.1}$$

so that the knowledge of one of the coordinates determines the other. Effectively there is only one coordinate needed since x and y are not independent.

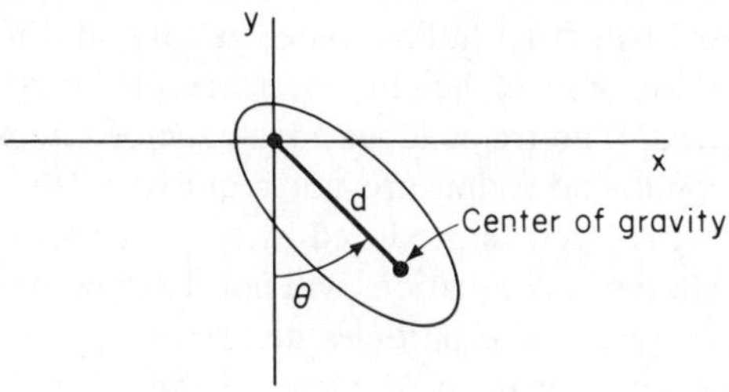

FIG. 12.1. Simple pendulum.

Suppose there are q generalized coordinates but m constraint equations used for a complete description of the motion. If they are of the type in (12.1), *i.e.*, algebraic in the coordinates and time, the system can be thought of as containing actually $n = q - m$ independent generalized coordinates and no constraints. Such a system is called *holonomic* and the constraints called *holonomic constraints.* The system is also said to have *n degrees of freedom.*

It should be emphasized that not all equality constraints are algebraic in the coordinates themselves; some depend upon the coordinate velocities and are nonintegrable. [If $x\dot{x} + y\dot{y} = 0$ were a constraint, (12.1) follows; such a constraint is integrable and holonomic.] The classic case is a billiard ball rolling on a table. Five coordinates are required to specify such a system—two distance coordinates to locate the center of gravity of the ball on the table, and three rotation coordinates to specify the orientation of the ball. If we constrain the ball to roll with perfectly rough friction (no sliding), such a constraint is nonholonomic. It is then possible to write equality constraints involving the velocities of the five coordinates. They cannot be integrable, however, since it is possible, starting from one set of five numbers, to arrive at any other set while obeying the constraint, given the suitable forces. This is different from constraint (12.1) which prohibits certain x–y pairs independent of the forces applied. In other words, positional constraints are holonomic while nonintegrable velocity constraints are not.

Another type of nonholonomic constraint involves inequalities generally used to restrict masses to given regions of space. Our billiard ball, for example, may be restricted by

$$-a \leq x \leq a, \qquad -b \leq y \leq b$$

which may represent the cushion constraint keeping it on the table. When the ball impacts a boundary, its subsequent motion depends on the nature of the impact.

Lagrange's equations of motion apply only to holonomic systems where there are n generalized coordinates for a system with n degrees of freedom. There are no explicit constraint equations that render some of the coordinates dependent upon others. This does not mean there are no constraints. The table top prevents a billiard ball from falling under gravity. But if the equations are set up without the dimension of height, the force of the table top on the ball need not be considered. The tremendous advantage of Lagrange's equations is that forces of holonomic constraints are not required in the formulation. Naturally, if applied forces (such as produced by a massé shot) cause the billiard ball to leave the table top, the equations will not describe the motion properly.

Consider, then, a system of p particles described by n generalized coordinates, q_j, all independent. Let m_k be the mass of the kth particle of the system with the transformation equations of

$$\begin{aligned} x_k &= f_k(q_1, q_2, \ldots, q_n, t) \\ y_k &= g_k(q_1, q_2, \ldots, q_n, t) \\ z_k &= h_k(q_1, q_2, \ldots, q_n, t) \end{aligned} \tag{12.2}$$

One may then show [1] from basic concepts of work, energy, forces, etc. that*

$$\frac{d}{dt}\left(\frac{\partial T}{\partial \dot{q}_j}\right) - \frac{\partial T}{\partial q_j} - Q_j = 0, \qquad j = 1, 2, \ldots, n \tag{12.3}$$

where T = kinetic energy of the system, and Q_j = external generalized forces that do work on the system. The function Q_j is positive if it tends to increase q_j and need not have units of force at all. For example, if q_j is an angular coordinate, Q_j has units of torque.

Let us evaluate (12.3) on the basis of the transformation equations (12.2) [5].† We have

*In (12.3), $\dot{q}_j \equiv dq_j/dt$. The notation violates the "dot" usage in this book to refer only to computer time derivatives, but it is customary for Lagrange's equations.

†The development here is more general than that given in [5]. It is suggested by an unpublished talk, "The Representation of Constraints in the Analysis of Physical Systems," by Dr. D. Greenwood, given at the Midwestern Simulation Council meeting on Sept. 27, 1965 [6].

$$\begin{aligned}\frac{dx_k}{dt} &= \sum_{j=1}^{n}\frac{\partial f_k}{\partial q_j}\frac{dq_j}{dt} + \frac{\partial f_k}{\partial t}\\ \frac{dy_k}{dt} &= \sum_{j=1}^{n}\frac{\partial g_k}{\partial q_j}\frac{dq_j}{dt} + \frac{\partial g_k}{\partial t}\\ \frac{dz_k}{dt} &= \sum_{j=1}^{n}\frac{\partial h_k}{\partial q_j}\frac{dq_j}{dt} + \frac{\partial h_k}{\partial t}\end{aligned} \tag{12.4}$$

with

$$T = \frac{1}{2}\sum_{k=1}^{p} m_k\left[\left(\frac{dx_k}{dt}\right)^2 + \left(\frac{dy_k}{dt}\right)^2 + \left(\frac{dz_k}{dt}\right)^2\right]$$

or

$$\begin{aligned}T = \frac{1}{2}\sum_{k=1}^{p} m_k\Bigg[&\sum_{i=1}^{n}\sum_{j=1}^{n}\frac{\partial f_k}{\partial q_j}\frac{\partial f_k}{\partial q_i}\frac{dq_j}{dt}\frac{dq_i}{dt} + 2\frac{\partial f_k}{\partial t}\sum_{j=1}^{n}\frac{\partial f_k}{\partial q_j}\frac{dq_j}{dt} + \left(\frac{\partial f_k}{\partial t}\right)^2\\ &+ \sum_{i=1}^{n}\sum_{j=1}^{n}\frac{\partial g_k}{\partial q_j}\frac{\partial g_k}{\partial q_i}\frac{dq_j}{dt}\frac{dq_i}{dt} + 2\frac{\partial g_k}{\partial t}\sum_{j=1}^{n}\frac{\partial g_k}{\partial q_j}\frac{dq_j}{dt} + \left(\frac{\partial g_k}{\partial t}\right)^2\\ &+ \sum_{i=1}^{n}\sum_{j=1}^{n}\frac{\partial h_k}{\partial q_j}\frac{\partial h_k}{\partial q_i}\frac{dq_j}{dt}\frac{dq_i}{dt} + 2\frac{\partial h_k}{\partial t}\sum_{j=1}^{n}\frac{\partial h_k}{\partial q_j}\frac{dq_j}{dt} + \left(\frac{\partial h_k}{\partial t}\right)^2\Bigg]\end{aligned} \tag{12.5}$$

This is of the form [6]

$$T = \frac{1}{2}\sum_{i=1}^{n}\sum_{j=1}^{n} a_{i,j}\frac{dq_i}{dt}\frac{dq_j}{dt} + \sum_{j=1}^{n} b_j\frac{dq_j}{dt} + c \tag{12.6}$$

where the coefficients $a_{i,j}$, b_i, and c are functions of the q_j and t. From (12.6),

$$\frac{\partial T}{\partial \dot{q}_j} = \frac{1}{2}\sum_{i=1}^{n}(a_{i,j} + a_{j,i})\frac{dq_i}{dt} + b_j$$

Since $a_{i,j} = a_{j,i}$ from comparison of (12.5) and (12.6), we write

$$\frac{\partial T}{\partial \dot{q}_j} = \sum_{i=1}^{n} a_{i,j}\frac{dq_i}{dt} + b_j \tag{12.7}$$

with (12.3) finally transforming to

$$\begin{aligned}F_j \equiv \sum_{i=1}^{n} a_{i,j}\frac{d^2q_i}{dt^2} &+ \sum_{k=1}^{n}\sum_{i=1}^{n}\frac{\partial a_{i,j}}{\partial q_k}\frac{dq_i}{dt}\frac{dq_k}{dt} + \sum_{k=1}^{n}\frac{\partial b_j}{\partial q_k}\frac{dq_k}{dt}\\ &- \frac{1}{2}\sum_{i=1}^{n}\sum_{k=1}^{n}\frac{\partial a_{i,k}}{\partial q_j}\frac{dq_i}{dt}\frac{dq_k}{dt} - \sum_{i=1}^{n}\frac{\partial b_i}{\partial q_j}\frac{dq_i}{dt} - \frac{\partial c}{\partial q_j} - Q_j = 0\end{aligned} \tag{12.8}$$

for $j = 1, 2, \ldots, n$. This system represents n equations of motion, but all the accelerations d^2q_j/dt^2 may be involved in each one. Before (12.8) can be solved on the computer, it is necessary to find expressions for these accelerations, a task which may be quite tedious if done analytically. However, the equations (12.8) are linear in the accelerations (provided Q_j contains no acceleration terms), so that we may solve for these on the computer using the techniques discussed in Chapter 11.

Before discussing the practical solution of Lagrange's equations on the computer, we note here that the coefficients $a_{i,j}$, b_i, and c, are usually found by applying the derived concept that the kinetic energy of a rigid body is equal to the sum of the translational kinetic energy of its center of mass and its rotational energy about its center of mass. Thus, the millions of individual particles and constraints making up a rigid body need not be considered directly. To write the equations of motion, we express the kinetic energy of each rigid body in the system in terms of the q_j, first by differentiating the coordinate transformation equations (12.2) of the center of mass with respect to time. If m_i is its mass, its translational kinetic energy T_i is given by

$$T_i = \frac{1}{2}m_i\left[\left(\frac{dx_i}{dt}\right)^2 + \left(\frac{dy_i}{dt}\right)^2 + \left(\frac{dz_i}{dt}\right)^2\right]$$

Next we compute the rotational energy R_i of each rigid body about its center of gravity. This is by far the most difficult task in most problems. Generalized coordinates are often chosen to simplify it. If I_i is the instantaneous moment of inertia of body i and ω is its instantaneous angular velocity, the rotational energy is given by $I_i\omega^2/2$. This is usually resolved into a sum of terms

$$\begin{aligned} R_i = {} & \tfrac{1}{2}(I_{xx}\omega_x^2 + I_{yy}\omega_y^2 + I_{zz}\omega_z^2) \\ & - (I_{yz}\omega_y\omega_z + I_{xz}\omega_x\omega_z + I_{xy}\omega_x\omega_y) \end{aligned}$$

where I_{xx}, I_{yy}, and I_{zz} are moments of inertia about the axes x, y, and z, I_{yz}, I_{xz}, and I_{xy} are products of inertia, and ω_x, ω_y, and ω_z are components of the angular velocity ω (see [1]). The net kinetic energy is then given by

$$T = \sum (T_i + R_i)$$

where the summation is over all bodies in the system

Conservative external forces (velocity-independent) are often handled by use of a potential energy function V, defined so that these components of Q_j are given by $-(\partial V/\partial q_j)$. With the Lagrangian L given by

$$L = T - V$$

Lagrange's equations can be modified to be

$$\frac{d}{dt}\left(\frac{\partial L}{\partial \dot{q}_j}\right) - \frac{\partial L}{\partial q_j} - Q_j = 0, \qquad j = 1, 2, \ldots, n \tag{12.9}$$

since $\partial V/\partial \dot{q}_j = 0$. The functions Q_j now contain only those forces not derivable from V.

12.2 Lagrange's Equations—Solutions

There are several methods that can be used to solve system (12.8) on an analog computer. If the Q_j do not contain acceleration terms, it could first be solved analytically for each individual d^2q_j/dt^2 because the equations are linear in those terms and the equations may then be mechanized in a straightforward way. Since the coefficients may be variable, however, this can be quite tedious even for a small-order system. A much better scheme is to solve the algebraic equations on the computer for d^2q_j/dt^2 using the techniques of Chapter 11, since it will then be possible to include constraints and handle acceleration-dependent external forces more easily. Both methods presented here are based upon the following theorem.

Theorem 12.1. The matrix $(a_{i,j})$ used in equations (12.6), (12.7), and (12.8) is positive definite.

Proof. From (12.5) and (12.6)

$$\frac{1}{2}\sum_{i=1}^{n}\sum_{j=1}^{n} a_{i,j}\frac{dq_i}{dt}\frac{dq_j}{dt} = \frac{1}{2}\sum_{k=1}^{p} m_k\left[\left(\frac{dx_k}{dt} - \frac{\partial f_k}{\partial t}\right)^2 + \left(\frac{dy_k}{dt} - \frac{\partial g_k}{\partial t}\right)^2 + \left(\frac{dz_k}{dt} - \frac{\partial h_k}{\partial t}\right)^2\right] \geq 0$$

with

$$a_{i,j} \equiv \sum_{k=1}^{p} m_k\left(\frac{\partial f_k}{\partial q_i}\frac{\partial f_k}{\partial q_j} + \frac{\partial g_k}{\partial q_i}\frac{\partial g_k}{\partial q_j} + \frac{\partial h_k}{\partial q_i}\frac{\partial h_k}{\partial q_j}\right)$$

Now when the generalized velocities are zero, we must have

$$\frac{dx_k}{dt} = \frac{\partial f_k}{\partial t}, \qquad \frac{dy_k}{dt} = \frac{\partial g_k}{\partial t}, \qquad \frac{dz_k}{dt} = \frac{\partial h_k}{\partial t} \tag{12.10}$$

But $(a_{i,j})$ is positive definite only if the converse is true. Suppose (12.10) holds, but not all the dq_j/dt are zero. From (12.4),

$$\sum_{j=1}^{n} \frac{\partial f_k}{\partial q_j}\frac{dq_j}{dt} = 0, \qquad \sum_{j=1}^{n} \frac{\partial g_k}{\partial q_j}\frac{dq_j}{dt} = 0, \qquad \sum_{j=1}^{n} \frac{\partial h_k}{\partial q_j}\frac{dq_j}{dt} = 0 \qquad (12.11)$$

Multiply the three equations (12.11) by $m_k(\partial f_k/\partial q_i)$, $m_k(\partial g_k/\partial q_i)$, and $m_k(\partial h_k/\partial q_i)$, and sum over k.

$$\sum_{k=1}^{p} m_k \frac{\partial f_k}{\partial q_i} \sum_{j=1}^{n} \frac{\partial f_k}{\partial q_j}\frac{dq_j}{dt} = 0$$

$$\sum_{k=1}^{p} m_k \frac{\partial g_k}{\partial q_i} \sum_{j=1}^{n} \frac{\partial g_k}{\partial q_j}\frac{dq_j}{dt} = 0$$

$$\sum_{k=1}^{p} m_k \frac{\partial h_k}{\partial q_i} \sum_{j=1}^{n} \frac{\partial h_k}{\partial q_j}\frac{dq_j}{dt} = 0$$

Interchange summation signs and sum the equations

$$\sum_{j=1}^{n} \left[\sum_{k=1}^{p} m_k \left(\frac{\partial f_k}{\partial q_i}\frac{\partial f_k}{\partial q_j} + \frac{\partial g_k}{\partial q_i}\frac{\partial g_k}{\partial q_j} + \frac{\partial h_k}{\partial q_i}\frac{\partial h_k}{\partial q_j} \right) \right] \frac{dq_j}{dt} = 0$$

or

$$\sum_{j=1}^{n} a_{i,j}\frac{dq_j}{dt} = 0$$

for each i. If not all dq_j/dt are zero, then the matrix $(a_{i,j})$ must be singular, a contradiction since this implies Lagrange's equations could not be solved at all for the d^2q_j/dt^2. Hence $(a_{i,j})$ is positive definite.

12.2.1 Solving momentum equations (Hamilton equations)

Instead of solving Lagrange's equations (12.9) directly, we may sometimes solve its first integral more efficiently:

$$\frac{\partial L}{\partial \dot{q}_j} - \int_0^t \left(\frac{\partial L}{\partial q_j} + Q_j \right) dt = (M_j)_0$$

for some constants $(M_j)_0$. From (12.7), this reduces to (since V is only a function of the q_j)

$$M_j \equiv \sum_{i=1}^{n} a_{i,j}\frac{dq_i}{dt} + b_j - \int_0^t \left(\frac{\partial L}{\partial q_j} + Q_j \right) dt - (M_j)_0 = 0 \qquad (12.12)$$

for $j = 1, 2, \ldots, n$. Since $(a_{i,j})$ is positive definite, the direct method of Chapter 11 for solving a system of linear algebraic equations can be used to solve for the dq_j/dt. Accordingly, we program to solve the algebraic equations with

$$\frac{d\left(\frac{dq_j}{dt}\right)}{d\tau} + g_j M_j = 0$$

for as large a set of g_j as possible. Note that the forces Q_j cannot involve the coordinate accelerations if this method is used, since they are not generated for feedback.

Example 12.1. Consider a variation of the simple pendulum shown in Fig. 12.2. The pendulum is constrained to move in the plane of the paper and is mounted on a projection at the center of gravity of a block, supported by a spring with spring constant k_1. The block is constrained to move in one dimension in the direction of gravity. Derive the momentum equations and solve them on the computer.

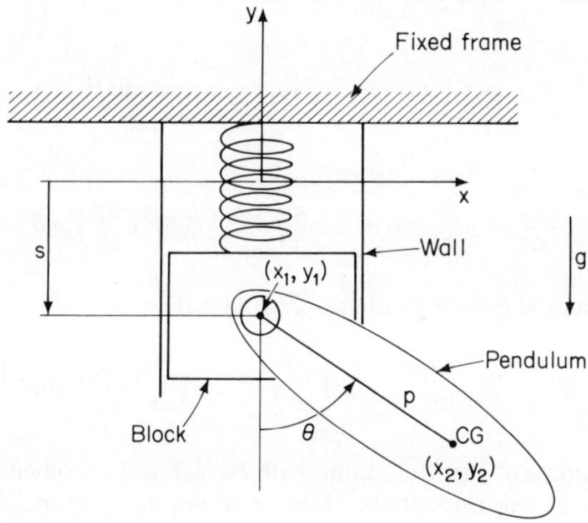

FIG. 12.2 Pendulum and block geometry.

We drop the subscript notation for the generalized coordinates in favor of simpler symbols. Two coordinates completely specify the system configuration. We choose s, a coordinate that measures the distance the center of gravity of the block is from some fixed position of the reference frame. To simplify the mathematics, we let $s = 0$ where the spring exerts no force. The other coordinate chosen is θ, a coordinate measuring the angular position of the center of gravity of the pendulum with respect to some fixed line passing through the center of rotation. To simplify the mathematics, we take the fixed line vertically in the direction of gravity.

All action takes place in the plane of the paper. The transformation equations (12.2) need not include a z coordinate. But we must fix a Cartesian reference frame from which to obtain the transformation equations for x_k and y_k. As shown in Fig. 12.2, the transformation equations for the center of gravity (x_1, y_1) of the block are given by

$$x_1 = 0, \qquad y_1 = -s$$

from which

$$\frac{dx_1}{dt} = 0, \qquad \frac{dy_1}{dt} = -\frac{ds}{dt}$$

Hence

$$T_1 = \frac{1}{2} m_b \left(\frac{ds}{dt}\right)^2$$

is the translational energy of the block, m_b being the mass of the block. The constraint prevents any rotation of the block; T_1 is the total kinetic energy of the block.

The transformation equations for the center of gravity (x_2, y_2) of the pendulum are

$$x_2 = p \sin\theta, \qquad y_2 = -s - p\cos\theta$$

from which

$$\frac{dx_2}{dt} = p\frac{d\theta}{dt}\cos\theta, \qquad \frac{dy_2}{dt} = -\frac{ds}{dt} + p\frac{d\theta}{dt}\sin\theta$$

Thus the translational energy T_2 of the pendulum is

$$T_2 = \frac{m_p}{2}\left[\left(\frac{ds}{dt}\right)^2 + p^2\left(\frac{d\theta}{dt}\right)^2 - 2p\left(\frac{d\theta}{dt}\right)\left(\frac{ds}{dt}\right)\sin\theta\right]$$

where m_p is the mass of the pendulum. With I equal to the moment of inertia of the pendulum about an axis through the center of gravity and perpendicular to the plane of the paper, the rotational energy R_2 of the pendulum is given by [1]:

$$R_2 = \frac{1}{2} I \left(\frac{d\theta}{dt}\right)^2$$

which results in a total kinetic energy of

$$T = \frac{1}{2}(m_b + m_p)\left(\frac{ds}{dt}\right)^2 + \frac{1}{2}(I + m_p p^2)\left(\frac{d\theta}{dt}\right)^2 - pm_p \frac{d\theta}{dt}\frac{ds}{dt}\sin\theta \qquad (12.13)$$

Because of gravity and the spring, two conservative forces arise from a potential function V, which may be taken as

$$V = -m_b gs - m_p g(s + p\cos\theta) + \frac{k_1}{2} s^2 \qquad (12.14)$$

The first two terms are based on the potential energy mgh of an object due to gravity, where m is its mass and h is the height as measured from any fixed reference; the gravity force is then $-(\partial V/\partial h) = -mg$. The last term is based on Hooke's law; the spring force is then $-k_1 s$.

Applying (12.12) to the Lagrangian, we obtain

$$\begin{aligned} M_s &\equiv (m_b + m_p)\left(\frac{ds}{dt}\right) - pm_p\left(\frac{d\theta}{dt}\right)\sin\theta \\ &\quad + \int_0^t [k_1 s - (m_b + m_p)g - Q_s]\,dt - (M_s)_0 = 0 \\ M_\theta &\equiv -pm_p\left(\frac{ds}{dt}\right)\sin\theta + (I + m_p p^2)\left(\frac{d\theta}{dt}\right) \\ &\quad + \int_0^t \left[pm_p\left(\frac{d\theta}{dt}\frac{ds}{dt}\cos\theta + g\sin\theta\right) - Q_\theta\right]dt - (M_\theta)_0 = 0 \end{aligned} \tag{12.15}$$

where Q_s and Q_θ are other external forces which may be applied and $(M_s)_0$ and $(M_\theta)_0$ are initial values of the momentums given by

$$\begin{aligned} (M_s)_0 &= (m_b + m_p)\left(\frac{ds}{dt}\right)_0 - pm_p\left(\frac{d\theta}{dt}\right)_0 \sin\theta_0 \\ (M_\theta)_0 &= -pm_p\left(\frac{ds}{dt}\right)_0 \sin\theta_0 + (I + m_p p^2)\left(\frac{d\theta}{dt}\right)_0 \end{aligned}$$

Assuming the Q_s and Q_θ are generable, Fig. 12.3 shows a circuit diagram solving (12.15) using electronic multipliers and generating $\sin\theta$ and $\cos\theta$ by integration. We have let

$$S = \sin\theta, \qquad C = \cos\theta$$

$$\begin{aligned} M_1 &= \int_0^t [k_1 s - (m_b + m_p)g - Q_s]\,dt - (M_s)_0 \\ M_2 &= \int_0^t \left[pm_p\left(\frac{d\theta}{dt}\frac{ds}{dt}\cos\theta + g\sin\theta\right) - Q_\theta\right]dt - (M_\theta)_0 \end{aligned} \tag{12.16}$$

$$s_1 = \frac{ds}{dt}, \qquad \theta_1 = \frac{d\theta}{dt}$$

so that

$$\begin{aligned} M_s &\equiv (m_b + m_p)s_1 - pm_p S\theta_1 + M_1 = 0 \\ M_\theta &\equiv -pm_p S s_1 + (I + m_p p^2)\theta_1 + M_2 = 0 \end{aligned}$$

Algebraically, we solve for s_1 and θ_1 by solving

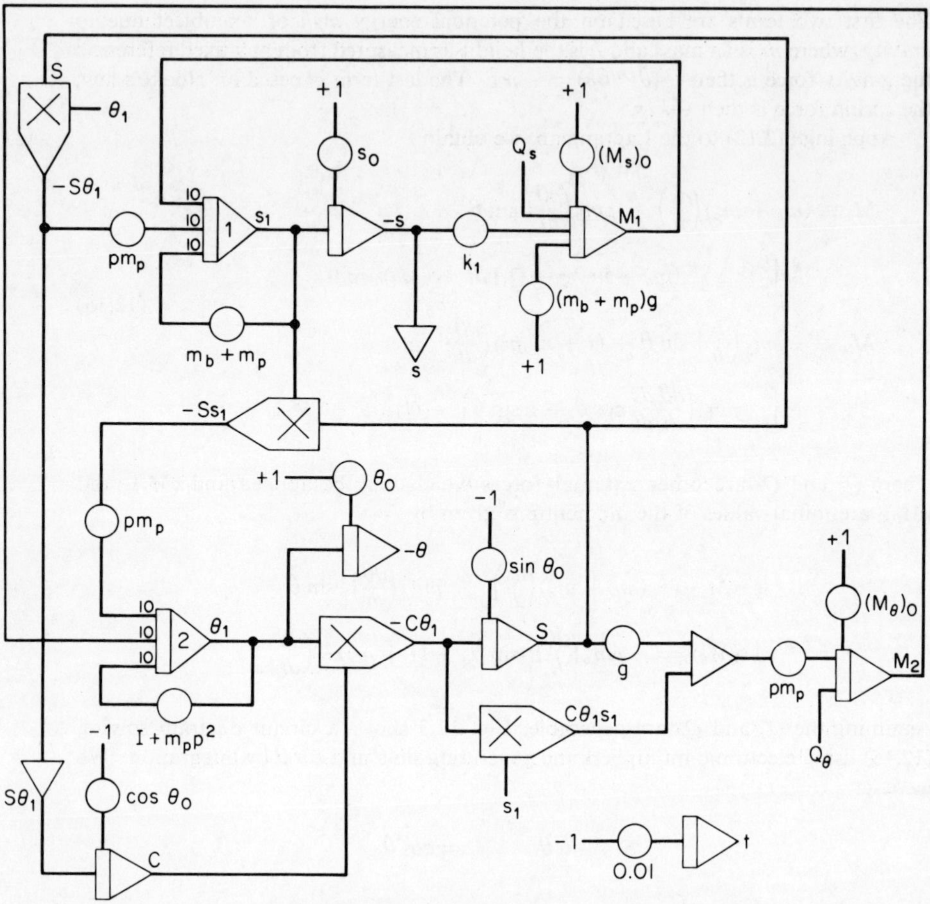

Note: Use high-gain amplifiers for 1 and 2 if possible; otherwise small capacitor feedback and bypass hold and reset relays.

FIG. 12.3 Pendulum and block, momentum equations.

$$\dot{s}_1 = -g_1 M_s, \qquad \dot{\theta}_1 = -g_2 M_\theta$$

using gains g_1 and g_2 as large as possible. With normal integrators, we solve (time scaling if necessary)

$$\begin{aligned}
\dot{t} &= 1, & t_0 &= 0\\
\dot{S} &= C\theta_1, & S_0 &= \sin\theta_0\\
\dot{C} &= -S\theta_1, & C_0 &= \cos\theta_0\\
\dot{M}_1 &= k_1 s - (m_b + m_p)g - Q_s, & (M_1)_0 &= -(M_s)_0
\end{aligned}$$

$$\dot{M}_2 = pm_p(\theta_1 s_1 C + gS) - Q_\theta, \qquad (M_2)_0 = -(M_\theta)_0$$
$$\dot{s} = s_1, \qquad (s)_0 = s_0$$
$$\dot{\theta} = \theta_1, \qquad (\theta)_0 = \theta_0$$

The first and last equations are optional since they may not be needed. (S and C determine the position of the pendulum.)

12.2.2 Solutions of the force equations [5]

Momentum equations generally involve less equipment for solving Lagrange's equations and should be used when there are no constraints or when the Q_j do not contain acceleration-dependent terms. When the latter conditions are not satisfied, the force equations (12.8) must be solved. Again, because $(a_{i,j})$ is positive definite, we use the direct method (Chapter 11) to solve the algebraic equations for the d^2q_j/dt^2. Constraints are handled by setting up an augmented system as described in Sec. 12.4. Otherwise we attempt the solution by

$$\frac{d\left(\dfrac{d^2q_j}{dt^2}\right)}{d\tau} = -g_j F_j$$

No difficulty is encountered when the external generalized forces do not contain acceleration-dependent forces.

Example 12.2. Mechanize the force equations for the pendulum-block problem as described in Example 12.1.

Using the identical development, the kinetic and potential energies of the system are still given by (12.13) and (12.14). Applying (12.9) to the Lagrangian, we obtain

$$F_s \equiv (m_b + m_p)\left(\frac{d^2s}{dt^2} - g\right) - pm_p\,(\sin\theta)\frac{d^2\theta}{dt^2} - pm_p\,(\cos\theta)\left(\frac{d\theta}{dt}\right)^2 + k_1 s - Q_s = 0 \tag{12.17}$$

$$F_\theta \equiv -pm_p\,(\sin\theta)\left(\frac{d^2s}{dt^2} - g\right) + (I + m_p p^2)\frac{d^2\theta}{dt^2} - Q_\theta = 0$$

Note that the coefficient matrix for the term d^2s/dt^2 is the same as that for $d^2s/dt^2 - g$; hence we may solve the algebraic system for the latter variable. Figure 12.4 shows a circuit diagram based on the direct method of solution using the notation

$$s_2 = \frac{d^2s}{dt^2}, \qquad \theta_2 = \frac{d^2\theta}{dt^2}$$

and the others defined by (12.16).

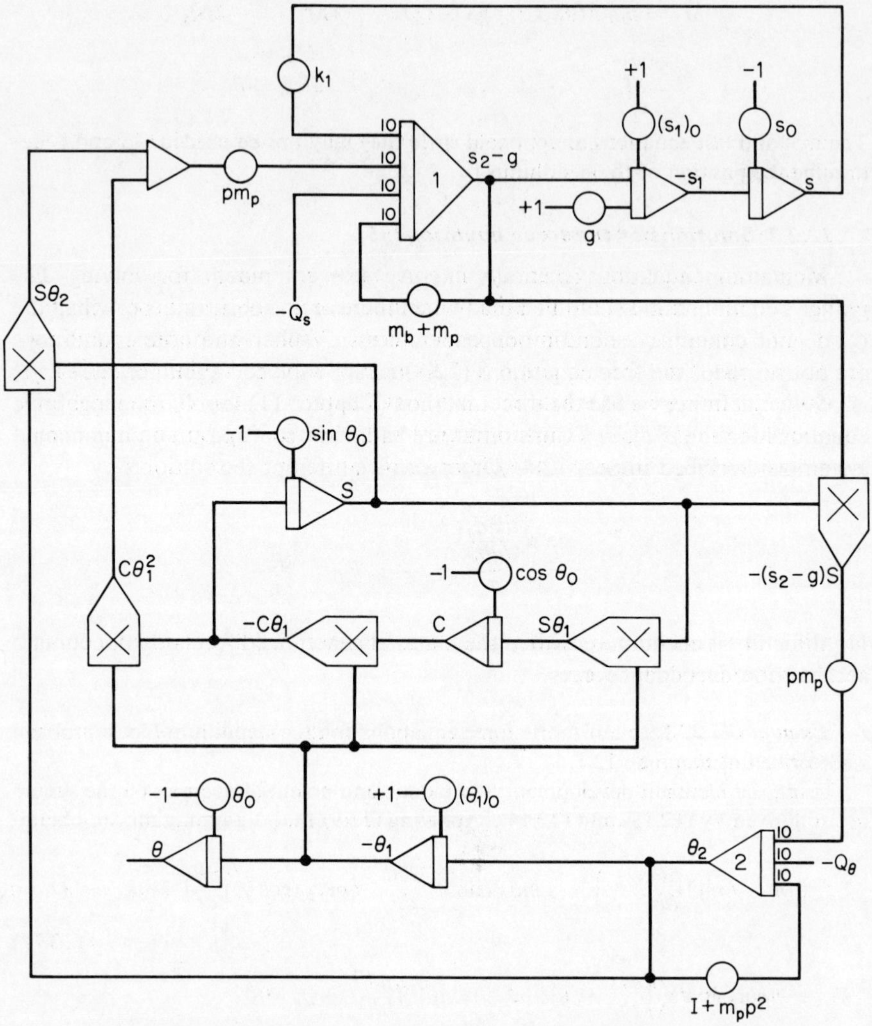

Note: Use high-gain amplifiers for 1 and 2 if possible; otherwise small capacitor feedback and bypass hold and reset relays.

FIG. 12.4 Pendulum and block, force equations.

12.3 Dissipative Forces—Friction Models

The Q_s and Q_θ in the two previous examples are external generalized forces not derivable from the potential. These may include jet reactive forces, wind forces, magnetic forces, etc. Forces that resist the motion by converting mechanical energy to some other form (e.g., heat) are called *dissipative*. Examples are

1. Motion through a medium creates drag forces that obey various laws. The most common assumes the force proportional to the square of the velocity (over a given velocity range) and can be verified experimentally.

2. A flexing spring absorbs energy with a resistive force usually assumed to be proportional to the velocity, with the constant of proportionality called the damping coefficient. Such a force is said to exhibit velocity damping and can also be verified experimentally to some degree.

3. Coulomb friction is produced when two bodies rub against one another; the resistive force is approximately proportional to the force pressing one body into the other (the normal force) with the constant of proportionality called the coefficient of friction. It is also a function of velocity and rises sharply near zero velocity. Wang and Shen [7] indicate there is also somewhat of a hysteresis effect.

When mechanical systems are simulated on a computer, dissipative forces are included if appropriate, but, unfortunately, only approximate representations are possible. Coefficients of friction, for example, vary from specimen to specimen and are functions of surface finish, impurities, aging, vibration, etc. One could not hope to simulate such friction forces to better than 10 per cent.

What is more common is a simulation that accounts for average effects. It is reasonable to hope than an average energy dissipated over a period of time in a simulation does correspond to reality. Damping coefficients of springs are found experimentally by observing the decay in the amplitude of a spring-mass system with time and using the damping coefficient corresponding to an average energy loss. It does not mean that the force resisting the motion is exactly equal to $-k_1 v$ at every instant of time. Usually, such fine differences between reality and simulation are not important to the results desired. We must always balance sophistication of a model to be simulated with its effect on the interpretation of the computer results.

12.3.1 Acceleration-dependent friction forces

The most complicated model that dissipates energy in a mechanical system is Coulomb friction. Its force is not only a function of velocity but also the normal force that may be acceleration-dependent. Thus, a toboggan sled experiences considerable variations in its normal force against the ice as dips, dives, turns, and twists. We shall examine some of the difficulties inherent in such a simulation.

Consider Fig. 12.5 showing the geometry of two sliding objects pressed together with a normal force F_n. Arrows indicate the velocity directions of each object relative to the other and the Coulomb friction force opposing the motion, the magnitude of which is given by

$$|F| = \mu|F_n|$$

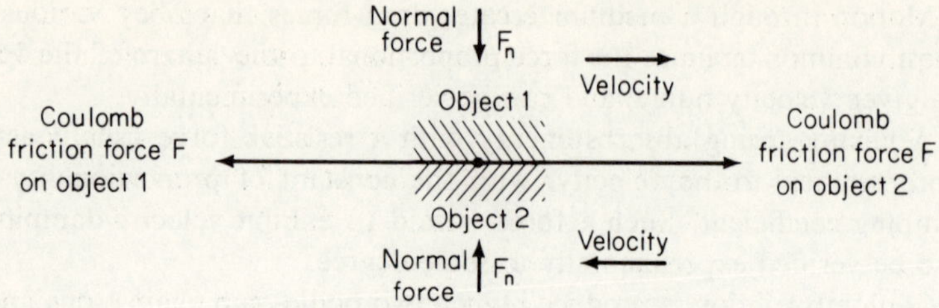

FIG. 12.5 Coulomb friction forces.

where μ is the coefficient of friction between the two surfaces. This force acts at the point of contact; for many points of contact it is distributed proportional to the pressure of contact. However, such fine details are rarely important in a simulation.

It is necessary to formulate the Q_j from the vector **F** in terms of the q_j for use in Lagrange's equations. Since the rubbing surfaces are due to constraints, the forces of the constraints are needed in the formulation. This fact diminishes the elegance of Lagrange's formulation which eliminates constraint forces in the process of getting the equations. However, the equations can help determine the forces of constraint by the concept of equivalent forces. This procedure is demonstrated in the following example.

Example 12.3. Assuming coefficients of friction of μ_s for frictional forces between the block and the wall, and μ_p for frictional forces between the block and the pendulum (the only two constraints in the plane), find the equations of motion for the system in Example 12.2.

If the pendulum were not attached to the block, the equation of motion for the block using the generalized coordinate s is given by Newton's laws

$$m_b \frac{d^2 s}{dt^2} = m_b g - k_1 s + F_\sigma \tag{12.18}$$

where F_σ is an external force applied to increase s. From (12.17), we see that

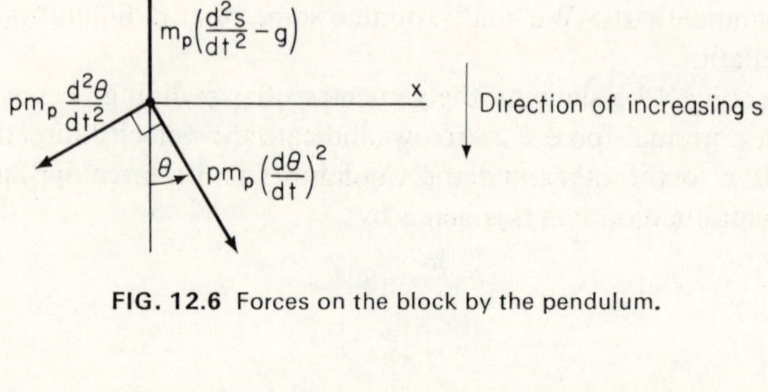

FIG. 12.6 Forces on the block by the pendulum.

$$F_\sigma = -m_p\left(\frac{d^2 s}{dt^2} - g\right) + pm_p\,(\sin\theta)\frac{d^2\theta}{dt^2} + pm_p\,(\cos\theta)\left(\frac{d\theta}{dt}\right)^2 + Q_s$$

It is clear that the first three terms are due to the inertial and centrifugal forces of the pendulum constraint. We may now construct vectorially (Fig. 12.6) these forces **F** from the pendulum that act on the block. These must be the forces of constraint. For convenience, we resolve **F** into its rectilinear components:

$$\begin{aligned} F_x &= pm_p\left(\frac{d\theta}{dt}\right)^2 \sin\theta - pm_p\left(\frac{d^2\theta}{dt^2}\right)\cos\theta \\ F_y &= m_p\left(\frac{d^2 s}{dt^2} - g\right) - pm_p\left(\frac{d\theta}{dt}\right)^2 \cos\theta - pm_p\left(\frac{d^2\theta}{dt^2}\right)\sin\theta \end{aligned} \tag{12.19}$$

and let

$$F_n = \sqrt{F_x^2 + F_y^2} \tag{12.20}$$

Let us assume the bearing is circular with radius r and no clearance (else the constraint is nonholonomic) and that the pendulum pulling on the bearing with force F_n creates a tangential friction force F_f at a point on the rim of the bearing. F_f must be included in the calculation of the normal force. Thus, a new vector diagram in Fig. 12.7 shows a normal force F_n' passing through the center of the bearing and the friction force perpendicular to it and equal in magnitude to

$$F_f = \mu_p F_n'$$

We wish to determine F_n' in terms of F_x and F_y. Resolve F_n' into its rectilinear com-

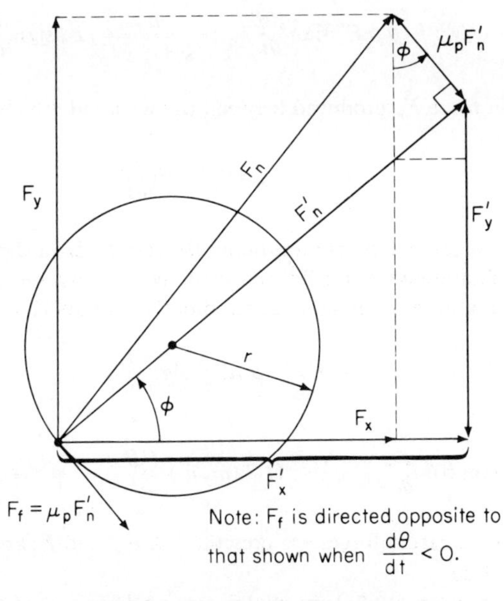

FIG. 12.7 Diagram to compute Q_θ.

ponents, defining these as F'_x and F'_y. With ϕ defined as shown we have

$$F'_x = F_x + \left(\mu_p F'_n \operatorname{sign}\frac{d\theta}{dt}\right)\sin\phi$$
$$F'_y = F_y - \left(\mu_p F'_n \operatorname{sign}\frac{d\theta}{dt}\right)\cos\phi \tag{12.21}$$

Since

$$\sin\phi = \frac{F'_y}{F'_n}, \qquad \cos\phi = \frac{F'_x}{F'_n}$$

(12.21) reduces to

$$F'_x = F_x + \mu_p F'_y \operatorname{sign}\frac{d\theta}{dt}, \qquad F'_y = F_y - \mu_p F'_x \operatorname{sign}\frac{d\theta}{dt}$$

so that we may solve for F'_x and F'_y:

$$F'_x = \frac{F_x + \mu_p F_y \operatorname{sign}\frac{d\theta}{dt}}{1+\mu_p^2}$$
$$F'_y = \frac{F_y - \mu_p F_x \operatorname{sign}\frac{d\theta}{dt}}{1+\mu_p^2} \tag{12.22}$$

The opposing torque on the pendulum is now given by

$$Q_\theta = -\mu_p r F'_n \operatorname{sign}\frac{d\theta}{dt} = -\frac{\mu_p}{\sqrt{1+\mu_p^2}} r F_n \operatorname{sign}\frac{d\theta}{dt} \tag{12.23}$$

while the friction force F_w produced between the wall and the block is taken as

$$F_w = -\mu_s |F'_x| \operatorname{sign}\frac{ds}{dt} \tag{12.24}$$

This assumes the reactive moment on the block from Q_θ is neglected. Actually, this gives rise to an additional frictional force at the wall, which prevents the rotation of the block. The final set of equations, based on (12.17) and (12.18), becomes

$$F_s \equiv m_b\left(\frac{d^2s}{dt^2} - g\right) + k_1 s + F'_y + \mu_s |F'_x| \operatorname{sign}\frac{ds}{dt} = 0$$
$$F_\theta \equiv -pm_p(\sin\theta)\left(\frac{d^2s}{dt^2} - g\right) + (I + m_p p^2)\frac{d^2\theta}{dt^2} + \frac{\mu_p}{\sqrt{1+\mu_p^2}} r F_n \operatorname{sign}\frac{d\theta}{dt} = 0 \tag{12.25}$$

assuming no other external forces are present. F'_x, F'_y, and F_n are defined in (12.22), (12.19), and (12.20).

As can be seen from (12.25), inertially dependent friction forces result in equations no longer linear in the accelerations. We could not guarantee the computer stability of

$$\dot{s}_2 = -g_1 F_s, \qquad \dot{\theta}_2 = -g_2 F_\theta \tag{12.26}$$

for all combinations of μ_p, μ_s, and r. In most physical systems, however, efforts are made to minimize the effects of inertial friction by lubrication, etc., so that the dependence of the Jacobian matrix on the friction is usually small [5]. In those cases (12.26) will be stable for large g_1 and g_2. This model also requires much equipment since it is somewhat sophisticated. The question of whether to include these friction forces or to approximate them in some other way depends upon the nature and reason of the simulation.

12.3.2 Static friction and parasitic oscillations

The friction model described in Example 12.3 suffers from one drawback. It does not account directly for the phenomenon of static friction. It is possible for friction forces, as computed in our model, to exceed the magnitude of other driving forces when the velocity is zero. In reality, relative motion between the two bodies ceases as they bind. The simulation, however, does not allow for binding. Thus, suppose the centrifugal force of the pendulum was so large when $ds/dt = 0$ that the block binds against the wall. The computer would actually compute on the basis of $ds/dt \neq 0$, since switches are either on or off. Thus

$$\frac{d^2s}{dt^2} < 0 \text{ when } \frac{ds}{dt} > 0, \qquad \frac{d^2s}{dt^2} > 0 \text{ when } \frac{ds}{dt} < 0$$

Since $ds/dt = 0$ is a switching point in the equations, *a high-frequency parasitic oscillation about $ds/dt = 0$ results* [5]. In effect, the *average* d^2s/dt^2 remains zero during the time of binding so an accurate simulation for each q_j and dq_j/dt may nevertheless be obtained. The results in [5] suggest that it is important to insure that no energy is lost during the oscillation period because static friction, which is perfectly rough friction, is not dissipative. Good results are obtained by filtering the acceleration terms d^2q_j/dt^2 before using them as inputs to the amplifier producing each d^2q_i/dt^2, with $i \neq j$. When d^2q_j/dt^2 averages zero during binding, the filtered value will remain close to zero and have little effect on the other equations. The fact that the static breakaway friction force is generally larger than the dynamic force can be accounted for by properly shaping the coefficient of friction near zero velocity.

In an effort to eliminate the parasitic oscillation inherent in using equations of the type (12.23) and (12.24), Monastyrshin [8] derives equations not solvable for second derivatives. To obtain these quantities for feedback, differentiating circuits must be used. For the cases tried, which included only ones where there were no acceleration interactions, his results were excellent. Some filtering was needed, however, to avoid the instabilities possible when using pure differentiation, so that the accuracy of the results may be similar to that obtained by simply filtering an oscillating acceleration.

For cases where static and Coulomb friction are constant in magnitude, Korn and Korn [9] program both directly with an appropriate switching circuit.

12.4 Constraints [5], [10]

We noted earlier that Lagrange's equations apply only to holonomic systems where there are the same number of generalized coordinates as degrees of freedom. If holonomic constraint equations exist, it is better (from an equipment standpoint) to eliminate as many of the coordinates as possible before solving the equations on the computer. This can be, however, quite tedious; it is also desirable at times to program the computer to allow for various constraints as variable. This section describes the modification necessary to Lagrange's equations to enable equality constraints, both holonomic and nonholonomic, to be programmed.

The method used is based on having equality constraint equations of the form

$$\sum_{j=1}^{n} A_{j,k} \frac{dq_j}{dt} + T_k = 0, \qquad k = 1, 2, \ldots, m, \qquad m \leq n \tag{12.27}$$

Such is automatically the case for nonholonomic equalities. It is necessary to differentiate holonomic constraint equations once with respect to time to arrive at the form (12.27). Thus the two types of constraints are treated essentially equal.

The additional constraints can be thought of as producing an additional external force, Q'_j, for each of the original Lagrange's equations (12.9)

$$\frac{d}{dt}\left(\frac{\partial L}{\partial \dot{q}_j}\right) - \frac{\partial L}{\partial q_j} - Q_j - Q'_j = 0 \tag{12.28}$$

for $j = 1, 2, \ldots, n$. Because these additional forces *do no work*, it can be shown [1] that

$$Q'_j = \sum_{k=1}^{m} A_{j,k} \lambda_k$$

a fact allowing us to set up a new system of $n + m$ simultaneous equations to solve on the computer. The λ_k are called Lagrange multipliers.

Only the force equations can be used. We rewrite (12.8) as

$$F_j \equiv \sum_{i=1}^{n} a_{i,j} \frac{d^2 q_i}{dt^2} - \sum_{k=1}^{m} A_{j,k} \lambda_k + F'_j = 0, \qquad j = 1, 2, \ldots, n \tag{12.29}$$

where F'_j are terms not involving d^2q_i/dt^2 or λ_k (we assume temporarily that this includes the Q_j). Now we differentiate (12.27) once with respect to time:

$$C_k \equiv \sum_{j=1}^{n} A_{j,k} \frac{d^2 q_j}{dt^2} + C'_k = 0, \qquad k = 1, 2, \ldots, m \tag{12.30}$$

where C'_k are terms not involving d^2q_j/dt^2 or λ_k.

The new system of $n + m$ linear simultaneous equations (12.29) and (12.30) contains $n + m$ unknowns, the n accelerations d^2q_j/dt^2 and the m Lagrange multipliers λ_k. Note that the matrix $\mathbf{A}_1$ of this linear system can be partitioned into four submatrices

$$\mathbf{A}_1 = \begin{pmatrix} (a_{i,j}) & (-A_{j,i}) \\ (A_{i,j}) & \mathbf{0} \end{pmatrix}$$

with $(a_{i,j})$ positive definite, and the $m \times n$ matrix $(-A_{j,i})$ being skew-symmetric to the $n \times m$ matrix $(A_{i,j})$. We may solve the system with the direct method (Chapter 11) to conserve equipment, because we prove that $\mathbf{A}_1$ is positive stable in the following theorem.

Theorem 12.2. The matrix $\mathbf{A}_1$ is positive stable at every instant of time if the constraints (12.27) are independent.

Proof. Consider the differential equation

$$\dot{\mathbf{x}} = -\mathbf{A}_1\mathbf{x} \tag{12.31}$$

where $\mathbf{x}$ is of dimension $m + n$. Multiply on the left by $\mathbf{x}'$, the transpose of $\mathbf{x}$,

$$\mathbf{x}'\dot{\mathbf{x}} = -\mathbf{x}'\mathbf{A}_1\mathbf{x}$$

The terms on the left represent half the time derivative of the nonnegative sum of squares S of the x_i:

$$\mathbf{x}'\dot{\mathbf{x}} = \frac{1}{2}\frac{d}{d\tau}\left(\sum_{i=1}^{m+n} x_i^2\right) = \sum_{i=1}^{m+n} x_i\dot{x}_i$$

The terms on the right are the quadratic form

$$-\mathbf{x}'\mathbf{A}_1\mathbf{x} = -\sum_{i=1}^{n} a_{i,j}x_ix_j$$

which is always nonpositive. Thus,

$$\frac{1}{2}\frac{d}{d\tau}\left(\sum_{i=1}^{m+n} x_i^2\right) = \frac{1}{2}\dot{S} \leq 0 \tag{12.32}$$

so the real parts of the characteristic roots of $\mathbf{A}_1$ cannot be negative. But suppose some are zero. Then these characteristic roots are either pure imaginary or zero. But they cannot be pure imaginary; if $\dot{S} = 0$, the quadratic form requires $x_i = 0$, $i = 1, 2, \ldots, n$. From (12.31) and the definition of $\mathbf{A}_1$, we must then have $\dot{x}_i = 0$, $i = n + 1, n + 2, \ldots, n + m$, so that these variables approach constants in the limit as $\tau \to 0$, not undamped sinusoids. Thus, questionable characteristic roots can only be zero. This implies that $|\mathbf{A}_1| = 0$ and that the rows of $\mathbf{A}_1$ are not linearly independent. Since $(a_{i,j})$ is positive definite, the first n rows are independent, so the last m rows cannot be independent. But these are the coefficients of the constraint equations.

A corollary of Theorem 12.2 is that (12.29) and (12.30) are solvable by

$$(\dot{q}_2)_j = -g_j F_j, \qquad \dot{\lambda}_k = -h_k C_k \tag{12.33}$$

where $(q_2)_j = d^2q_j/dt^2$ and the g_j and h_k are large gains. The direct method (12.33) will probably work also in most cases where the Q_j contain small acceleration-dependent forces.

Example 12.4. Solve the equations for a simple pendulum with point mass m in rectilinear coordinates using the constraint (12.1). (See Fig. 12.1.)

The kinetic energy is

$$T = \frac{1}{2}m\left[\left(\frac{dx}{dt}\right)^2 + \left(\frac{dy}{dt}\right)^2\right]$$

and the potential energy can be taken as

$$V = mgy$$

Application of Lagrange's equations with no constraints results in

$$m\frac{d^2x}{dt^2} = 0, \qquad m\frac{d^2y}{dt^2} + mg = 0$$

Because of the one constraint (12.1), we have one Lagrange multiplier λ. The equations are modified according to (12.27) and (12.28). Since

$$x\frac{dx}{dt} + y\frac{dy}{dt} = 0$$

we have

$$\begin{gathered} F_x \equiv m\frac{d^2x}{dt^2} - x\lambda = 0, \qquad F_y \equiv m\frac{d^2y}{dt^2} + mg - y\lambda = 0 \\ C \equiv x\frac{d^2x}{dt^2} + y\frac{d^2y}{dt^2} + \left(\frac{dx}{dt}\right)^2 + \left(\frac{dy}{dt}\right)^2 = 0 \end{gathered} \tag{12.34}$$

System (12.34) can be analytically solved for d^2x/dt^2, d^2y/dt^2, and λ, but Theorem 12.2 suggests that solutions on the computer using

$$\dot{x}_2 = -g_1F_x, \qquad \dot{y}_2 = -g_2F_y, \qquad \dot{\lambda} = -hC$$

will be stable for all sets of positive g_1, g_2, and h. To minimize dynamic errors, we make these gains as large as possible, as usual.

12.5 Impact Phenomena

Nonholonomic inequality constraints usually restrict masses to given regions of space. If violated during a computation, the inequality implies that an impact has occurred. Lagrange's equations do not apply for impacts unless the forces of the impact are specified as external forces. Although these forces are generally not known, impacts can nevertheless be simulated to some degree in terms of energy losses and velocity changes.

When two bodies impact, the basic laws governing the situation are the conservation of linear and angular momentum. In addition, a certain amount of the kinetic energy is converted into other forms (heat, vibration, etc.) depending on the nature of the materials of the bodies and the type of impact. We shall not be concerned with destructive impacts, i.e., ones in which shapes are badly deformed. Energies lost under those circumstances are highly unpredictable. For nondestructive impacts, however, a large body of experimental evidence indicates fairly regular behavior.

Consider two masses, m_1 and m_2, as shown in Fig. 12.8, traveling at velocities v_1 and v_2. Let a nondestructive impact occur (because $v_1 > v_2$) during an extremely short time compared with other phenomena of interest, resulting in two new velocities v_1' and v_2'. Then

$$k_r = -\frac{v_2' - v_1'}{v_2 - v_1} \tag{12.35}$$

is constant depending only upon the materials of each body. Coupled with the conservation of momentum, we find [1]

$$\begin{aligned} v_1' &= \frac{(m_1 - k_r m_2)v_1 + (1 + k_r)m_2 v_2}{m_1 + m_2} \\ v_2' &= \frac{(1 + k_r)m_1 v_1 + (m_2 - k_r m_1)v_2}{m_1 + m_2} \end{aligned} \tag{12.36}$$

The constant k_r, called the *coefficient of restitution*, varies between 0 and 1. When $k_r = 1$, the impact is called elastic; when $k_r = 0$, it is called inelastic.

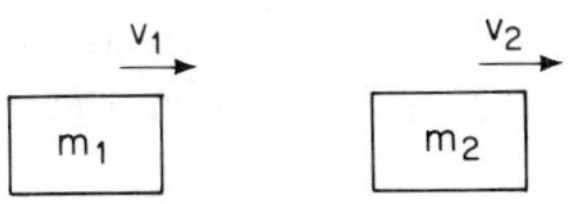

FIG. 12.8 Impacting masses.

It should be noted that in the general case only components of velocities parallel to the impacting forces are affected by (12.35). This means, since a surface reacts mostly in directions normal to it, that components of velocities normal to impacting surfaces are changed.

If one of the masses, say m_2, is attached to the inertial frame, we take $v_2 = 0$ and the limit as $m_2 \rightarrow \infty$. Then (12.36) results in

$$v_1' = -k_r v_1, \qquad v_2' = 0 \tag{12.37}$$

which are usually the equations from which experimental results are obtained.

Impact forces can also create moments that instantaneously change angular velocities [4] and should be taken into account when appropriate. Formulas similar to (12.36) result here also.

The simulation of impacts therefore requires a means to change the output of the amplifiers representing the velocities almost instantaneously. There are two methods of doing this:

1. Directly, by quickly changing the amplifier outputs as required through some switching arrangement.
2. Indirectly, by assuming the masses compress a stiff, damped spring and controlling the damping factor to absorb the proper energy [11]–[15].

12.5.1 Indirect method—stiff springs

For purposes of demonstrating circuits to simulate impact, let us assume we wish to simulate a mass M impacting a limit stop such that (12.37) holds; i.e., the stop is connected to a large inertial frame. The instant impact occurs, say at a lower limit when $x = x_L$, we assume the mass hits a very stiff spring and obeys the equation

$$M\frac{d^2x}{dt^2} + D\frac{dx}{dt} + Kx = F(t)$$

where $F(t)$ are other applied forces, D and K are the damping and force constants of the spring. The equation solved at all times is therefore

$$M\frac{d^2x}{dt^2} + \left(D\frac{dx}{dt} + Kx\right)U(x_L - x) = F(t)$$

which is easily mechanized with an idealized diode circuit. Figure 12.9 is an adaptation of Laird's circuit [12]. He also shows how the parameters D and K may be chosen to simulate any k_r by deriving the formula

$$k_r = \exp\left(\frac{-\pi D}{\sqrt{4KM - D^2}}\right)$$

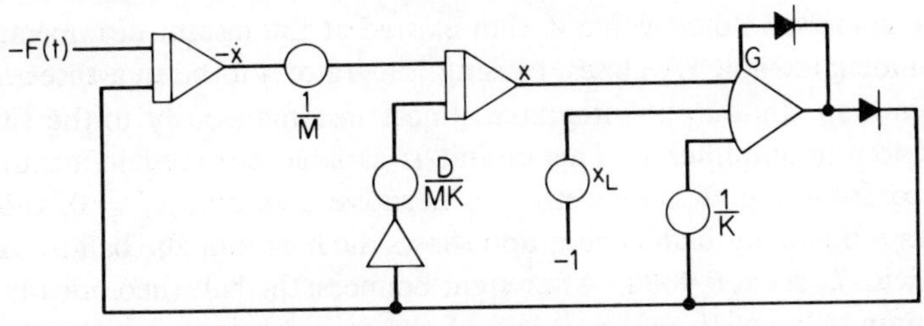

FIG. 12.9 Impact at a lower limit.

which gives an energy loss equivalent to the actual situation. We try to obtain as large a K as possible and choose D to obtain the correct k_r. Unfortunately, phase shift errors in amplifiers cause difficulty for large K. Laird recommends an experimental choice of D based on a given K, which also may be a function of the velocity. The chief advantage of such a method is the small number of amplifiers required for the simulation. On the other hand, lack of repeatability may be detrimental in some simulations. Also, a complete inelastic collision is not possible.

12.5.2 Direct method by switching

Figure 12.10 is a corresponding precision relay bounce circuit based on the idea of using a tracking circuit for memory. As long as x is in an allowable region, integrator 1 is in a three-amplifier negative feedback tracking loop. Its output is very close to v provided the external nonprecision input resistor R is sufficiently small. When $x < x_L$, the switch 1B opens so the output of inte-

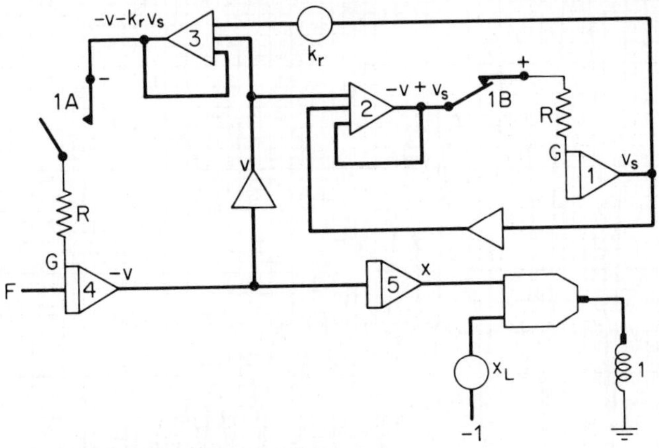

Note: Let R = 0.0001 × normal input resistor.

FIG. 12.10 Impact by switching.

grator 1 holds the stored value v_s that existed at the instant of impact. At that same moment, switch 1A closes, causing integrator 4 to be in a three-amplifier high-gain loop. Its output integrates almost instantaneously to the value $k_r v_s$ in order to null amplifier 3. This circuit is capable of excellent accuracy as is evidenced from Fig. 12.11. With F a negative constant, $x_L = 0$, the circuit simulates a bouncing ball. The graph shows the height of the ball as a function of time with k_r set at 0.9000. After eight bounces, the ball theoretically rises to a maximum value of $(0.81)^8 = 0.185$ of the initial height, which corresponds well with the graphical value.

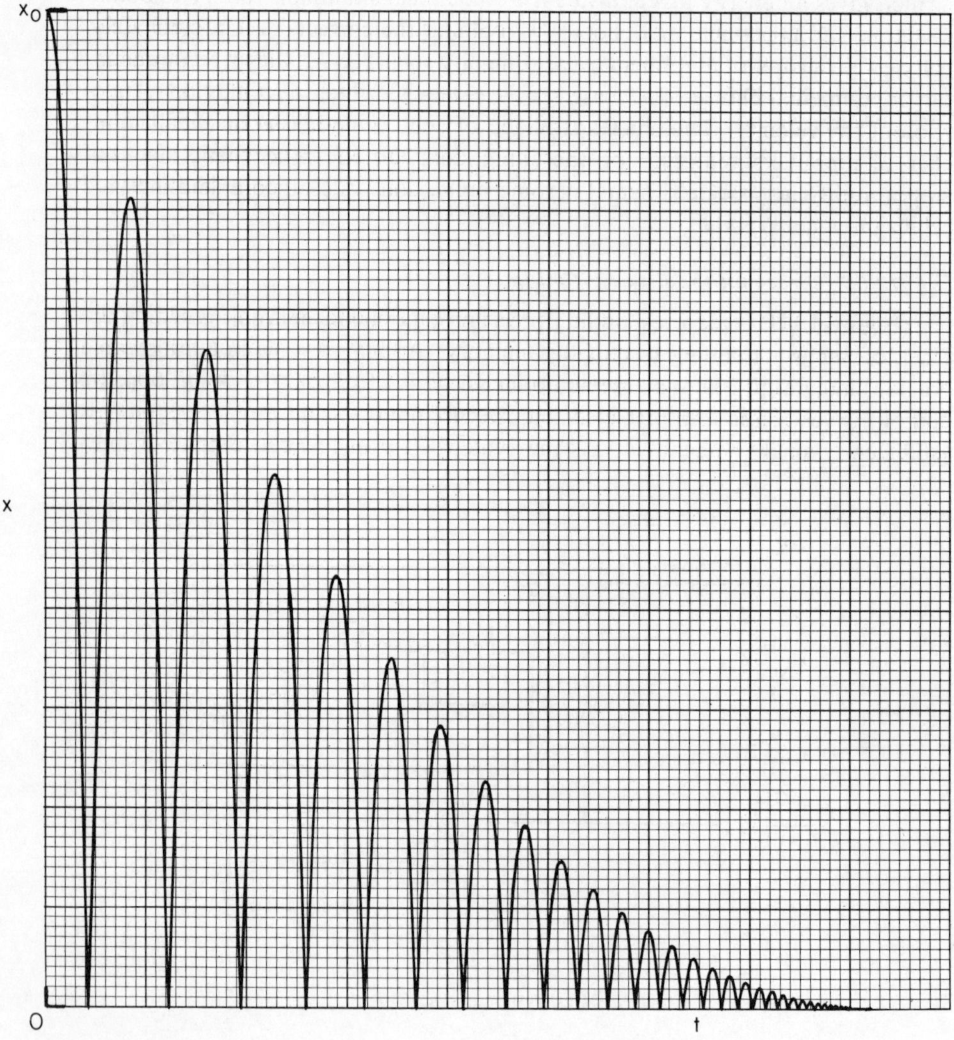

FIG. 12.11 Bouncing ball.

12.5.3 Inelastic collision with limit stop

An inelastic collision into a limit stop is a special case that can be treated more efficiently. With $k_r = 0$, the velocity must be zero after the impact. This can be achieved by shorting the capacitor of the integrator producing $-v$, as shown in Fig. 12.12. The shorting path is closed if and only if

$$[(x > x_u) \text{ AND } (F > 0)] \text{ OR } [(x < x_L) \text{ AND } (F < 0)]$$

by using three comparators and four relay switches. The lower limit x_L and the upper limit x_u as well as the driving force F can be of either polarity. Many variations to this circuit are possible.

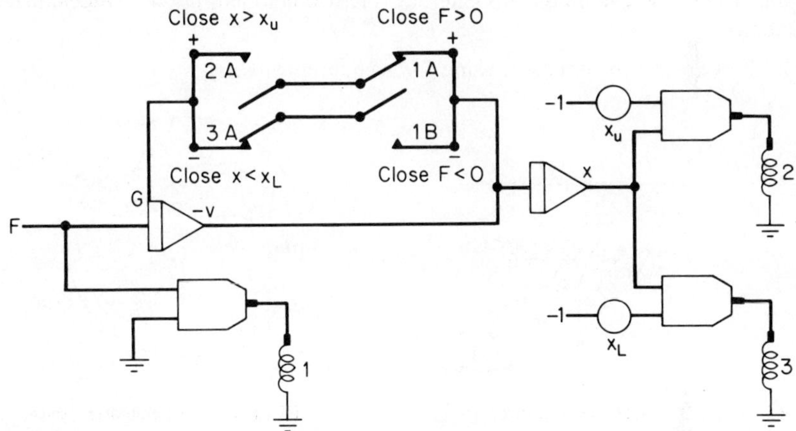

FIG. 12.12 No-drift relay circuit for inelastic limit-stop simulation.

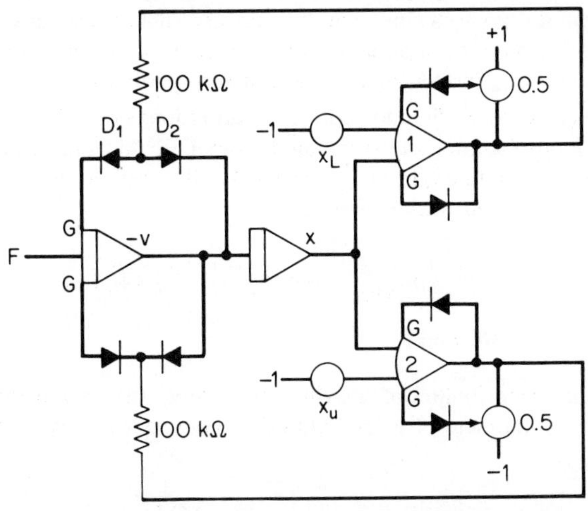

FIG. 12.13 Diode circuit for inelastic limit-stop simulation.

A diode circuit for higher-speed operation shown in Fig. 12.13 is based on one devised by Korn [16]. When x hits the lower limit stop, amplifier 1 jumps to the + reference value permitting both diodes D_1 and D_2 to conduct. This provides a shorting path across the capacitor of the amplifier producing $-v$, the output of which jumps to zero. Reversed diodes provide a similar action at the upper limit. Notwithstanding the minor disadvantages of this circuit (small drift at a limit if $F = 0$), an excellent simulation is obtained.

PROBLEMS

12-1. Set up the momentum equations to determine the motions of a double pendulum, Fig. 12.14, assuming no external forces other than gravity. Mechanize the equations.

12-2. Repeat Problem 12-1, using force equations instead.

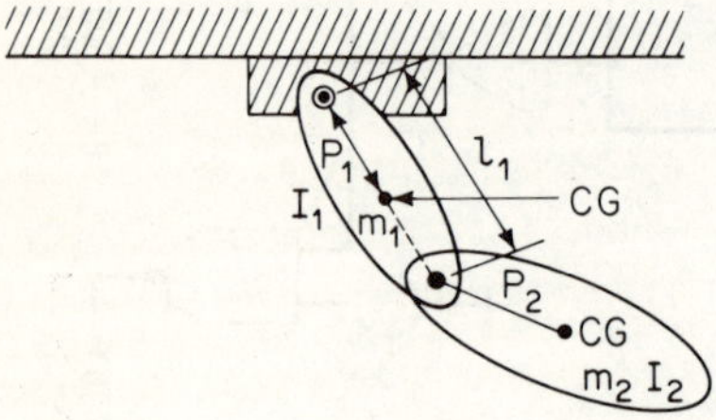

FIG. 12.14 Double pendulum.

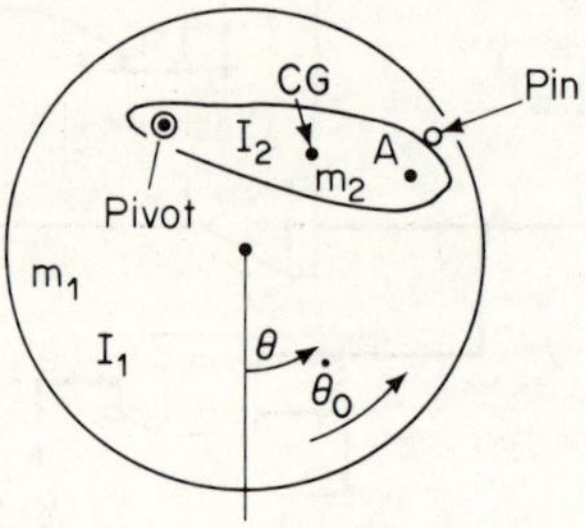

FIG. 12.15 Acceleration table.

12-3. A table is spinning at a speed $\dot{\theta}_0$, and the centrifugal force acts on a pendulum mounted horizontally (Fig. 12.15). At $t = t_0$, a pin is electrically withdrawn, allowing the pendulum to swing radially outward. Set up the equations to investigate the acceleration a mass mounted at point A receives as a function of time. Allow for varying the masses, inertias, and location of the pendulum on the table.

12-4. A good reason to include constraint equations as part of a mechanization is illustrated in the problem of a uniform ladder of length $2a$ leaning against a wall and sliding along the floor due to gravity (Fig. 12.16). If the motion is set up with the one rotation coordinate θ, we may express

$$
\begin{aligned}
T &= \frac{1}{2}m\left[\left(\frac{dx}{dt}\right)^2 + \left(\frac{dy}{dt}\right)^2\right] + \frac{1}{2}I\left(\frac{d\theta}{dt}\right)^2 \\
V &= mgy
\end{aligned}
\tag{12.38}
$$

where x and y are coordinates of the center of gravity (assumed at the center of the ladder) and I is its moment of inertia about the center of gravity. It might seem that the constraints

$$x - a\cos\theta = 0, \qquad y - a\sin\theta = 0$$

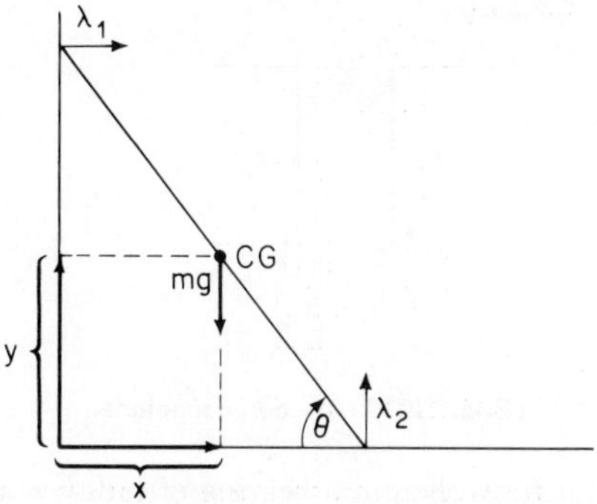

FIG. 12.16 Sliding ladder.

would allow us to obtain one equation, since T can be expressed in terms of one variable θ. For

$$\frac{dx}{dt} = -a\frac{d\theta}{dt}\sin\theta, \qquad \frac{dy}{dt} = a\frac{d\theta}{dt}\cos\theta$$

can be substituted in the expression for T. But actually the constraint is nonholonomic;

$$x - a\cos\theta \geq 0, \qquad y - a\sin\theta \geq 0$$

and the ladder leaves the wall during the motion. When this happens, two coordinates are needed to describe the motion. If the ladder leaves the floor, three coordinates are needed. By using Lagrange multipliers, we obtain from (12.38)

$$m\frac{d^2x}{dt^2} = \lambda_1, \qquad m\frac{d^2y}{dt^2} = -mg + \lambda_2$$

$$I\frac{d^2\theta}{dt^2} = a\lambda_1\sin\theta - a\lambda_2\cos\theta$$

It is clear from Fig. 12.16 that λ_1 and λ_2 must be the forces of constraint due to the wall and floor. However, these cannot be negative. Program this problem while limiting λ_1 and λ_2 to be nonnegative with ideal diode circuits.

12-5. Include rubbing friction in Problem 12-4 assuming μ to be the coefficient of friction between the ladder and both the wall and floor.

12-6. Mechanize equations (12.26) taking into account the Coulomb friction.

12-7. Set up first-order lag circuits to filter the acceleration in Problem 12-6 and thereby minimize errors due to parasitic oscillations.

12-8. Show that the gains g_j and h_k in (12.33) can be variable and do not affect the stability of the circuit.

12-9. Investigate the effects of Coulomb friction in Atwood's machine, shown in Fig. 12.17. Assume the masses m_1 and m_2 move only vertically and that the pulley,

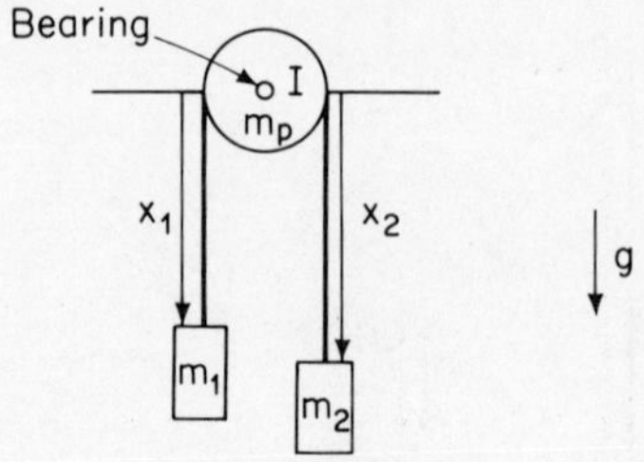

FIG. 12.17 Atwood's machine.

with mass m_p and inertia I, revolves in a bearing of radius r and coefficient of friction μ.

12-10. In the block-pendulum problem, Example 12.1, assume that a point on the pendulum is constrained to move along a curve $y = f(x)$ as shown in Fig. 12.18. Determine the equations of motion.

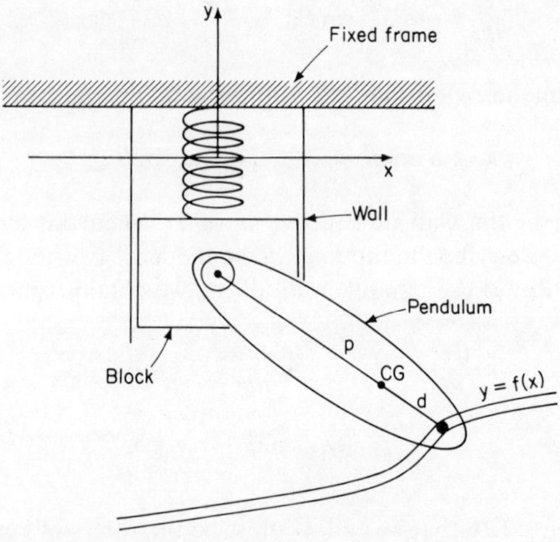

FIG. 12.18 Constraining the pendulum.

12-11. Minimize the number of amplifiers in the circuit of Fig. 12.10 by using more switches in a one-amplifier tracking loop.

12-12. Show that the theoretical error due to a nonzero F in Fig. 12.10 is of the order of RF/R_i, where R_i is the input resistor of F. (R can be made very small because the switch is not closed for long.)

12-13. Program the circuit of Fig. 12.10 to account for an upper limit also.

12-14. Mechanize Eq. (12.36) with circuits of the type in Fig. 12.10.

12-15. Mechanize the circuit of Fig. 12.12 with digital logic elements and one relay.

REFERENCES

[1] Whittaker, E. T., *A Treatise on the Analytical Dynamics of Particles and Rigid Bodies*, 4th ed. (Cambridge, England: Cambridge University Press, 1959).

[2] Goldstein, H., *Classical Mechanics* (Cambridge, Mass.: Addison-Wesley, 1951).

[3] Morse, P. M., and H. Feshback, *Methods of Theoretical Physics* (New York: McGraw-Hill Book Co., 1953).

[4] Osgood, W. F., *Mechanics* (New York: Macmillan, 1937).

[5] Hausner, A., "The Solution of Lagrange's Equations by Analog Computation," *IEEETEC*, vol. 14, Feb. 1965.

[6] Midwestern Simulation Council meeting, *Simulation*, vol. 6, March 1966.

[7] Wang, H., and C. N. Shen, "Simulating True Dry Friction on the Analog Computer," *Control Eng.*, vol. 9, Oct. 1962.

[8] Monastyrshin, G. I., "Mathematical Simulation of Dry Friction," *Automatika i Telemekhanika*, vol. 19, Dec. 1958.

[9] Korn, G. A., and T. M. Korn, *Electronic Analog and Hybrid Computers* (New York: McGraw-Hill Book Co., 1964).

[10] Greenwood, D. T., "The Representation of Constraints by Means of an Electronic Differential Analyzer," *IRETEC*, vol. 5, Sept. 1956.

[11] Ashley, J. R., "On the Analog Simulation of Mechanical Systems with Stiff Position Limit Stops," *Simulation*, vol. 2, May 1964.

[12] Laird, J. P., "Analog Simulation of Mechanical Impact," *Simulation*, vol. 4, May 1965.

[13] Ashley, J. R., "Direct Simulation of Limit Stop Damping," *Simulation*, vol. 4, May 1965.

[14] Kelly, W. A., "Mechanical Limits," *Simulation*, vol. 4, May 1965.

[15] Jenness, R. R., *Analog Computation and Simulation: Laboratory Approach* (Boston: Allyn and Bacon, 1965).

[16] Korn, G. A., "Analog-Computer Representation of Inelastic Stops," *IRETEC*, vol. 9, March 1960.

Chapter 13

CORRECTIVE SYSTEMS

A corrective system is one set up to null (or minimize) certain known errors automatically through the use of feedback circuits. We have already met such systems. The generation of implicit functions and the solution of algebraic equations by nulling inputs to high-gain amplifiers or integrators represent fundamental corrective systems. In this chapter, other useful corrective systems are described. As might be expected, the method of steepest descent plays an important role in most corrective systems because of its general applicability.

Besides the circuits themselves, the most important single consideration in any corrective system is its stability. We must insure that the errors are integrated so that they approach zero with time.

13.1 Solution Stabilization by Steepest Descent [1]–[7]

Many differential equations solved on the computer can be partially integrated analytically, and the result can be used to reduce the number of variables required to solve the problem. This is rarely done for two reasons:

1. The new variables, although fewer, are not so easily identified with physical quantities.

2. Algebraic difficulties often arise in the attempt to mechanize the fewer equations.

A classic example is the generation of $y = A \sin \omega x$ from

$$\frac{d^2 y}{dx^2} + \omega^2 y = 0 \tag{13.1}$$

using $\dot{x} = 1$, which requires two integrators for the solution. A first integral exists, however,

$$\left(\frac{dy}{dx}\right)^2 + \omega^2 y^2 = \omega^2 A^2 \tag{13.2}$$

which theoretically requires one integrator for solution, since

$$\frac{dy}{dx} = \pm\sqrt{\omega^2 A^2 - \omega^2 y^2}$$

Elaborate switching schemes must be used, which, coupled with the usually poor characteristics of a square-root circuit near zero results in an unsatisfactory generation of $A \sin \omega x$. On the other hand, if (13.1) is solved directly, we have noted that the amplitude A changes with time because of the small deficiencies of all amplifiers and associated computer equipment. The relationship (13.2) is violated eventually, preventing satisfactory results for long periods of time. Thus, with $y_1 = (dy/dx)/\omega$, the equations

$$\dot{y}_1 = -\omega y, \qquad \dot{y} = \omega y_1 \tag{13.3}$$

will not preserve the integral relationship expressed by (13.2). How then can system (13.3) be modified so that a known solution (13.2) is not violated? Or, how can the solution (13.2) be *stabilized*?

If (13.2) were part of an algebraic system

$$\epsilon \equiv y_1^2 + y^2 - A^2 = 0$$

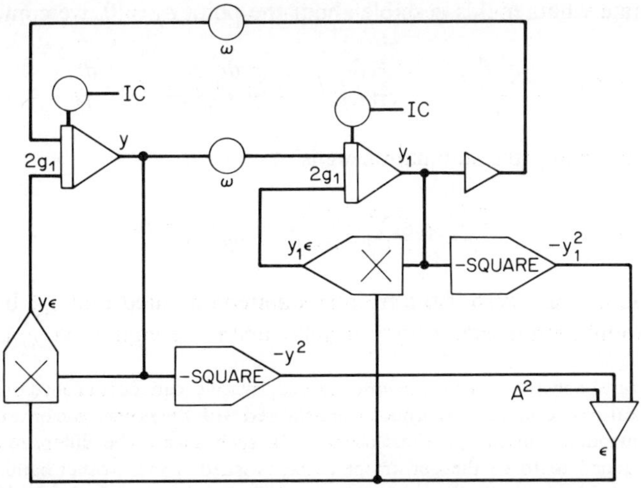

FIG. 13.1 Oscillator stabilization by steep descent by minimizing ε^2.

with y_1 and y as unknowns, and we attempted a minimization of ϵ^2, the results of Chapter 11 suggest that inputs to the high-gain amplifiers or integrators for y_1 and y contain the terms $-g_1\epsilon(\partial\epsilon/\partial y_1)$ and $-g_1\epsilon(\partial\epsilon/\partial y)$. Thus, we are led to consider the equations

$$\begin{aligned} \dot{y}_1 &= -\omega y - 2g_1 y_1(y_1^2 + y^2 - A^2) \\ \dot{y} &= \omega y_1 - 2g_1 y(y_1^2 + y^2 - A^2) \end{aligned} \tag{13.4}$$

(see Fig. 13.1) to see if this has a stabilizing influence. Indeed it does, as we readily determine generally in the next paragraph.

13.1.1 Stability of the general problem [1]

Let us consider a more general problem where there are m independent integral constraints*

$$\epsilon_j(y_1, y_2, \ldots, y_n) = 0, \qquad j = 1, 2, \ldots, m \tag{13.5}$$

and n equations ($m \leq n$)

$$\dot{y}_i = f_i(y_1, y_2, \ldots, y_n), \qquad i = 1, 2, \ldots, n \tag{13.6}$$

and where the y_i are the set of variables produced by integrators. We change system (13.6) and solve

$$\dot{y}_i = f_i - \sum_{j=1}^{m} g_j \epsilon_j \frac{\partial \epsilon_j}{\partial y_i}, \qquad i = 1, 2, \ldots, n \tag{13.7}$$

To determine whether this is stable about the point $\epsilon_j = 0$, we compute $\dot{\epsilon}_j$:

$$\dot{\epsilon}_j = \sum_{i=1}^{n} \frac{\partial \epsilon_j}{\partial y_i} \dot{y}_i = \sum_{i=1}^{n} \frac{\partial \epsilon_j}{\partial y_i} f_i - \sum_{i=1}^{n} \frac{\partial \epsilon_j}{\partial y_i} \left(\sum_{k=1}^{m} g_k \epsilon_k \frac{\partial \epsilon_k}{\partial y_i} \right)$$

or by interchanging the summation signs

$$\dot{\epsilon}_j + \sum_{k=1}^{m} g_k \epsilon_k \sum_{i=1}^{n} \frac{\partial \epsilon_k}{\partial y_i} \frac{\partial \epsilon_j}{\partial y_i} = \sum_{i=1}^{n} \frac{\partial \epsilon_j}{\partial y_i} f_i \equiv e_j \tag{13.8}$$

The functions e_j are zero if the trajectories are computed exactly, but because of small computation errors, they depend upon the computed y_j and g_k and

*An integral constraint is one whose time derivative can be deduced from the differential equations at hand. This is not to be confused with the holonomic or nonholonomic constraint introduced in the previous chapter. In such a case the differential equations must be changed to force the constraint to be satisfied. On the other hand, an integral constraint will automatically be satisfied if the differential equations are solved without error.

are not zero. Using the vectors $\boldsymbol{\epsilon} = (\epsilon_j)$, $\mathbf{y} = (y_j)$, $\mathbf{e} = (e_j)$, and the Jacobian matrix $\mathbf{J}(\boldsymbol{\epsilon}, \mathbf{y})$, we transform (13.8) to

$$\dot{\boldsymbol{\epsilon}} + \mathbf{J}(\boldsymbol{\epsilon}, \mathbf{y})\mathbf{J}'(\boldsymbol{\epsilon}, \mathbf{y})\mathbf{G}\boldsymbol{\epsilon} = \mathbf{e} \tag{13.9}$$

with $\mathbf{G}$ the diagonal gain matrix. Thus, the steep-descent procedure has a stabilizing effect, since $\mathbf{JJ}'$ is positive definite and the homogeneous equation is stable about $\boldsymbol{\epsilon} = \mathbf{0}$. By using large diagonal elements in $\mathbf{G}$, we may counter the effect of any finite $\mathbf{e}$ and force the integral relations to hold with very small error. A steep-descent stabilization procedure, however, cannot guarantee an accurate solution of (13.6). The only guarantee, for sufficiently large gains, is that (13.5) holds approximately. Because the errors expected are usually small, the gains g_k are usually made small.

Talkin [3] considers the case where the $\dot{y}_i$ themselves are found by the solution of n algebraic equations with the method of steepest descent while there are, in addition, m pure positional constraints. In that case, time scaling plays an important role, but it is always possible to time scale the problem sufficiently slow to stabilize all equations.

13.1.2 Conservative systems [5]

This stabilization procedure has direct application to mechanical systems that are conservative, i.e., the energy content $T + V$ is constant. Such is the case if all the external generalized forces $Q_j = 0$. Let

$$E = T + V = E\left(q_1, q_2, \ldots, q_n, \frac{dq_1}{dt}, \frac{dq_2}{dt}, \ldots, \frac{dq_n}{dt}\right) \tag{13.10}$$

with E_0 the constant energy defined by the initial values of position and velocity. Form

$$\epsilon_1 = E - E_0$$

and supply correction terms Δq_j and $\Delta(dq_j/dt)$ to the derivative expressions defining q_j and dq_j/dt:

$$\Delta q_j = -\epsilon_1 \frac{\partial E}{\partial q_j}, \qquad \Delta\left(\frac{dq_j}{dt}\right) = -\epsilon_1 \frac{\partial E}{\partial(dq_j/dt)}$$

In addition, any other variables formed with integrators, say because the generalized integration technique was used, could be similarly corrected.

Example 13.1. Stabilize the energy of the simple pendulum equation

$$I\frac{d^2\theta}{dt^2} = -mgd\sin\theta$$

As can be readily determined, the energy E of this system is given by

$$E = \tfrac{1}{2}I\theta_1^2 - mgd\cos\theta$$

with $\theta_1 = d\theta/dt$. Its value E_0 is constant and found from the initial values of θ_1 and θ:

$$E_0 = \tfrac{1}{2}I(\theta_1)_0^2 - mgd\cos\theta_0$$

Let us generate $S = \sin\theta$ and $C = \cos\theta$ by a generalized integration and use $\dot{t} = 1$. With $\epsilon_1 = E - E_0$, the correction terms

$$\Delta\theta_1 = -\epsilon_1\frac{\partial E}{\partial\theta_1} = -g_1(I\theta_1)\epsilon_1$$

$$\Delta C = -\epsilon_1\frac{\partial E}{\partial C} = -g_1(mgdS)\epsilon_1$$

are added to the usual expressions. Thus, the system becomes

$$\begin{aligned}\dot{\theta}_1 &= \frac{-mgdS}{I} - g_1(I\theta_1)\epsilon_1, \qquad \dot{S} = C\theta_1 \\ \dot{C} &= -S\theta_1 - g_1(mgdS)\epsilon_1\end{aligned} \tag{13.11}$$

Since small errors are expected, the gain g_1 can be taken conveniently small (say $g_1 = 1$). The oscillator amplitude should also be stabilized (as in Fig. 13.1) since the relation

$$\epsilon_2 \equiv C^2 + S^2 - 1 = 0 \tag{13.12}$$

holds, but the correction for the expression for $\dot{C}$ in (13.11) will quickly disturb it. To include the correction for (13.12), the final system solved on the computer is

$$\begin{aligned}\dot{\theta}_1 &= -\frac{mgdS}{I} - g_1(I\theta_1)\epsilon_1, \qquad \dot{S} = C\theta_1 - g_2(2S)\epsilon_2 \\ \dot{C} &= -S\theta_1 - g_1(mgdS)\epsilon_1 - g_2(2C)\epsilon_2\end{aligned}$$

13.1.3 Nonconservative systems [5]

The energy $E(t)$ of a nonconservative system can be formulated in terms of the work done on the system by the Q_j:

$$E(t) = E_0 + \int\sum_{j=1}^{n} Q_j\,dq_j = E_0 + \int_0^t\left(\sum_{j=1}^{n} Q_j\frac{dq_j}{dt}\right)dt$$

At the same time, (13.10) expresses the energy E in terms of the generalized coordinates and velocities. It is therefore possible to formulate an error function

$$\epsilon = E - E(t)$$

and stabilize the energy by considering $E(t)$ to be the instantaneous known energy. To do this, accurate multipliers for the products $Q_j(dq_j/dt)$ should be available and one accurate integrator for the formation of $E(t)$. In most instances, the errors resulting do not make the effort worthwhile. But in cases where energy is continually pumped into the system to replenish any dissipated, good results are obtained.

13.1.4 Systems with cyclic coordinates [5]

If $Q_j = 0$ and $\partial L/\partial q_j = 0$ in one of Lagrange's equations, the coordinate q_j is called *cyclic* or *ignorable*, since such information can be used to reduce the number of coordinates. From (12.9), we have

$$\frac{d}{dt}\left(\frac{\partial L}{\partial \dot{q}_j}\right) = 0$$

so that

$$\frac{\partial L}{\partial \dot{q}_j} = M_0 = \text{constant}$$

In lieu of reducing the number of coordinates, we can form

$$M(q_1, q_2, \ldots, q_n, \dot{q}_1, \dot{q}_2, \ldots, \dot{q}_n) = \frac{\partial L}{\partial \dot{q}_j}$$

and stabilize this "momentum" by applying the method of steepest descent to the constraint

$$\mu_j \equiv M - M_0$$

Again we emphasize that this procedure does not guarantee accuracy of the solution, and only guarantees the correct momentum exists throughout the solution. This is easier to see in a planetary orbit problem. We could stabilize the energy and angular momentum of the planet, which would force the solution to be an (approximate) ellipse with (approximately) fixed perigee and apogee. Other computation errors, however, will allow the ellipse to rotate slowly.

13.1.5 General integral relations

The steep-descent stabilization procedure has been used effectively in a variety of simulations where long-time errors in algebraic relationships cannot be tolerated. Most notably among these are the harmonic oscillator amplitude stabilization ([1], [6]) shown in Fig. 13.1, the stabilization of the orthonormal equations of direction cosines in coordinate transformations [1], and the enforcement of the normalization of quaternion parameters in the simulation of a spinning rigid body [7]. These examples all have one characteristic in common: The constraints force all variables to be bounded. Indeed, if this were not the case, multiplication errors in the correction terms could cause large initial errors more harmful than if the correction terms were eliminated (see Problem 13-5). The question of whether or not to include an integral constraint stabilization is often a matter of judgment, but it rarely is important in short-time problems where errors do not have time to build up to intolerable levels. Thus, in plotting continuous solutions of

$$(y')^2 + y^2 - xy - 1 = 0 \tag{13.13}$$

with initial conditions $x = 0$, $y = 1$, the poor characteristics of a square-root circuit near zero might cause us to differentiate (13.13) and solve

$$y'' = \frac{xy' + y - 2yy'}{2y'}$$

parametrically instead, with initial conditions $x = 0$, $y = 1$, and $y' = 0$. Equation (13.13) is now an integral constraint, but its solution $y \to \infty$ as $x \to \infty$ and so must terminate quickly because of scaling. It does not pay in effort and equipment to enforce this constraint.

On the other hand, suppose we wished to plot the solution of a closed curve $f(x, y) = 0$ where, because solutions are multivalued, better results are obtained by solving a differential equation. Since $\dot{x}(\partial f/\partial x) + \dot{y}(\partial f/\partial y) = 0$, the equations

$$\dot{x} = \pm \frac{\partial f}{\partial y}, \qquad \dot{y} = \mp \frac{\partial f}{\partial x}$$

are appropriate. As we have seen in the host-parasite problem (Chapter 4), errors build up sufficiently to cause the curves not to close. In that case the equations

$$\dot{x} = \pm \frac{\partial f}{\partial y} - g\frac{\partial f}{\partial x} f(x, y), \qquad \dot{y} = \mp \frac{\partial f}{\partial x} - g\frac{\partial f}{\partial y} f(x, y)$$

give much better results.

13.2 Oscillator Stabilization

The circuit of Fig. 13.1, based on the solution (13.4), is stable about the operating point

$$A^2 = y_1^2 + y^2$$

according to the theory of the previous section. But we now show that the equations are stable independent of any initial amplitude $A' \neq 0$. Let

$$B = y_1^2 + y^2$$

and form its time derivative from (13.4)

$$\dot{B} = 2y_1\dot{y}_1 + 2y\dot{y} = -4g_1B(B - A^2)$$

This is immediately integrable to

$$B = \frac{A^2}{1 - \left[\frac{(A')^2 - A^2}{(A')^2}\right]e^{-4g_1A^2\tau}}$$

where $B = (A')^2$ when $\tau = 0$. As $\tau \to \infty$, $B \to A^2$. The speed of convergence can also be increased by increasing g_1.

Although the amplitude of the oscillation can be forced to remain equal to A, the stabilization is no panacea for other types of errors. For example, suppose the frequency pots ω are set in error, perhaps because they are ganged or servo set with $\omega = \dot{x}$ in order to generate $y = A \sin x$. The equations solved are now (and we assume ω_1 and ω_2 are constant)

$$\begin{aligned} \dot{y}_1 &= -\omega_1 y - 2g_1y_1(y_1^2 + y^2 - A^2) \\ \dot{y} &= \omega_2 y_1 - 2g_2y(y_1^2 + y^2 - A^2) \end{aligned} \tag{13.14}$$

The expression $y_1\dot{y} - y\dot{y}_1$ is independent of the correction and is given by (with $g_1 = g_2$)

$$y_1\dot{y} - y\dot{y}_1 = \omega_2 y_1^2 + \omega_1 y^2$$

With $Q = y/y_1$, we deduce

$$\dot{Q} = \omega_1 Q^2 + \omega_2, \qquad Q_0 = 0$$

which results in

$$\frac{y}{y_1} = Q = \sqrt{\frac{\omega_2}{\omega_1}} \tan\left(\sqrt{\omega_1\omega_2}\,\tau\right)$$

In the limit as $g_1 \to \infty$, we also have

$$y^2 + y_1^2 = A^2$$

from which it follows that

$$y^2 = A^2 \sin^2 (\sqrt{\omega_1 \omega_2}\, \tau) \left[\frac{\omega_2}{\omega_1 + (\omega_2 - \omega_1) \sin^2 (\sqrt{\omega_1 \omega_2}\, \tau)} \right]$$

Thus, the stabilization distorts the sine wave, introducing all higher harmonics, but yields good amplitude characteristics. With $\omega_2 - \omega_1 = \Delta\omega$, and assuming $|\Delta\omega/\omega_1| \ll 1$, we further obtain

$$y \simeq A \sin (\sqrt{\omega_1 \omega_2}\, \tau) \left[1 + \frac{\Delta\omega}{2\omega_1} \cos^2 (\sqrt{\omega_1 \omega_2}\, \tau) \right]$$

$$= A \left(1 + \frac{\Delta\omega}{8\omega_1} \right) \sin (\sqrt{\omega_1 \omega_2}\, \tau) + \frac{A \Delta\omega}{8\omega_1} \sin (3\sqrt{\omega_1 \omega_2}\, \tau)$$

which is a good approximation to the generated function for large gain g_1. Without the stabilization ($g_1 = 0$), the output is a pure sine wave with amplitude approximately $A[1 + \Delta\omega/(2\omega_1)]$. The stabilization therefore decreases the error in the fundamental by a factor of four at the expense of introducing higher harmonics. Howe and Gilbert [6] discuss other types of errors, notably due to multiplier inaccuracy.

13.2.1 Another steepest-descent stabilization

Because much equipment is required for an oscillator amplitude stabilization correction based on minimizing ϵ^2, Brunner [4] suggests applying a correction based on minimizing $|\epsilon|$. The equations now become

$$\begin{aligned} \dot{y}_1 &= -\omega y - 2g_1 y_1 \operatorname{sign} (y_1^2 + y^2 - A^2) \\ \dot{y} &= \omega y_1 - 2g_1 y \operatorname{sign} (y_1^2 + y^2 - A^2) \end{aligned} \tag{13.15}$$

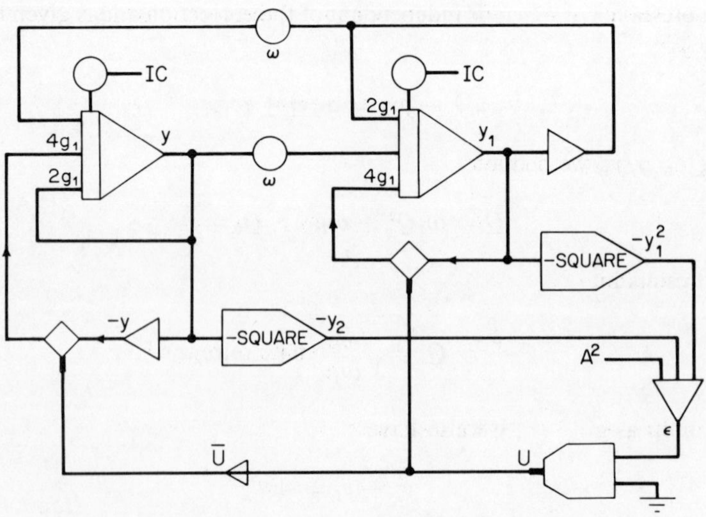

FIG. 13.2 Oscillator stabilization by steep descent by minimizing $|\varepsilon|$.

a mechanization for which is shown in Fig. 13.2 using electronic switches. In effect, two multipliers are replaced by two switches and a comparator. If the gains g_1 are small (as they should be for correction of a small error) any delay in the response of the switch does not produce any appreciable distortion; excellent results are obtained. This idea is extendable to any n-dimensional stabilization problem with m integral constraints. As many as mn multipliers are saved and should be used particularly if fast DAS's are available.

13.2.2 Stabilization with diodes

Brunner and Beadle [8] proposed the circuit in Fig. 13.3 to stabilize the amplitude of an oscillator. Only one of the outputs of each diode bridge pair transmits the voltage at one time, and this can be seen to be proportional to x sign ϵ or y sign ϵ (see Fig. 6.53). The voltage is limited to a very small value because the connection directly to the grid makes r_L effectively the forward resistance of the diode. However, this connection also integrates this voltage to zero quite rapidly. The equations solved are quite similar to (13.15) except that the gains may be dissimilar due to the mismatching of diodes.

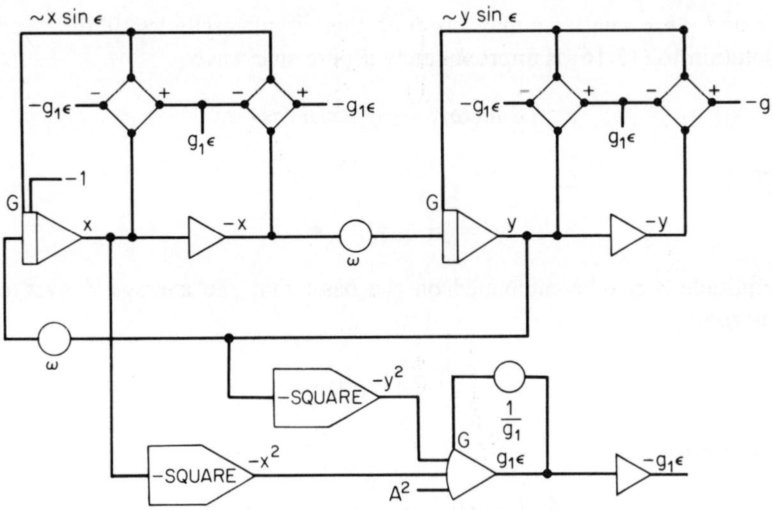

FIG. 13.3 Oscillator stabilization using diode bridges.

A simpler diode circuit was used with good success by Korte [9] and is shown in Fig. 13.4. Because one of the diodes conducts when $y_1 \neq 0$, and the voltage across the diode under these circumstances is about 0.6 or -0.6 V, for a large range of y_1, a positive feedback term of 0.6 sign y_1 is integrated by the integrator producing y_1. To keep the circuit from being unstable, a negative feedback term, $-ay_1$, is also supplied. Calling the diode conducting voltage drop v, we see that the equations solved are

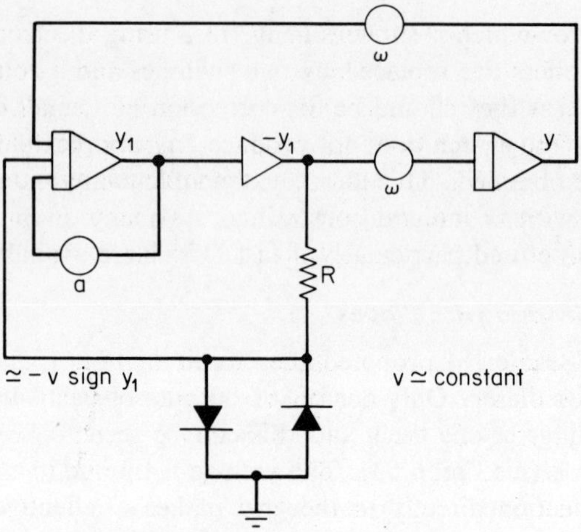

FIG. 13.4 Korte's stable, wide frequency oscillator.

$$\dot{y} = \omega y_1, \qquad \dot{y}_1 = -\omega y - a y_1 + v \operatorname{sign} y_1 \tag{13.16}$$

Since a and v are small, we may presume this circuit oscillates; if the steady-state solution to (13.16) is approximately a pure sine wave

$$y \simeq A \sin \omega\tau, \qquad y_1 \simeq A \cos \omega\tau$$

then with

$$\epsilon = y^2 + y_1^2 - A^2$$

the amplitude A can be calculated on the basis that the average $\dot{\epsilon}$ over one cycle is zero:

$$\int_0^{2\pi/\omega} \dot{\epsilon}\, d\tau = 0$$

Thus, we form

$$y\dot{y} + y_1\dot{y}_1 = -ay_1^2 + v|y_1| = -aA^2\cos^2\omega\tau + vA|\cos\omega\tau|$$

and find that

$$\int_0^{2\pi/\omega} (-aA^2\cos^2\omega\tau + vA|\cos\omega\tau|)\, d\tau = 0$$

which requires that

$$A = \frac{4v}{\pi a}$$

An energy balance exists that forces the oscillation fundamental at a specified amplitude. In practice, v is not quite constant so that A must be found by trial (a disadvantage) while a can be adjusted to vary A. This circuit is useful, despite the higher harmonics also introduced, because the waveform is independent of ω.

Another diode circuit is shown in Fig. 13.5. Pot number 1 is set at a small value, enough to make the circuit slightly unstable. Once per cycle the output

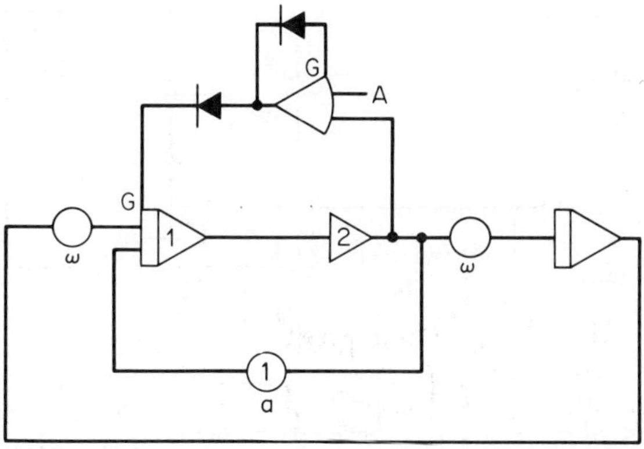

FIG. 13.5 Diode clipping circuit.

of amplifier 2 exceeds A, causing the idealized diode circuit to transmit a voltage to the grid and reduce the amplitude. This "clipping" circuit has the advantage of precision setting of A, but significant distortion can be introduced unless a is set very carefully.

13.3 Contour Tracing

An interesting contour tracing circuit was designed by Skramstad [10] and is shown in Fig. 13.6. Given

$$z = f(x, y)$$

the circuit is capable of tracing the contours of z; for any z_1, the graph $y(x)$ of

$$f(x, y) = z_1$$

is obtained. The circuit is particularly useful in on-line applications since the functional form $f(x, y)$ need not be known. It could be the output of any black box that has bounded error with x and y as inputs.

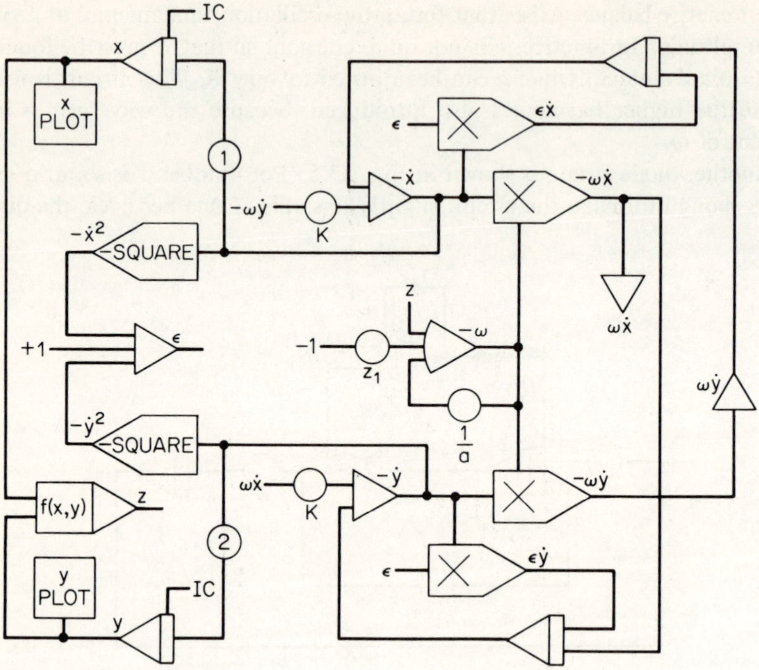

Note: pot 1 = pot 2 and both are set to control tracing speed.

FIG. 13.6 Contour tracing circuit.

The circuit is motivated by the following ideas. Let

$$\omega = -a(z - z_1)$$

be associated with a turning rate; i.e., if $z = z_1$, $\dot{x}$ and $\dot{y}$ do not change. The equations

$$\ddot{x} = -\omega\dot{y}, \qquad \ddot{y} = \omega\dot{x} \tag{13.17}$$

provide that

$$\frac{d}{dt}(\dot{x}^2 + \dot{y}^2) = 0$$

so the trajectory traces at constant velocity, constantly turning in the direction to make z closer to z_1. This produces an undamped oscillation about $z = z_1$. To introduce damping, we modify (13.17) to

$$\dot{x} = -\int \omega \dot{y}\, d\tau - K\omega\dot{y}, \qquad \dot{y} = \int \omega \dot{x}\, d\tau + K\omega\dot{x} \tag{13.18}$$

from which we may show

$$(1 + K^2\omega^2)\frac{(\dot{x}^2 + \dot{y}^2)}{2} = K_1^2 - K\int_0^\tau \omega^2(\dot{x}^2 + \dot{y}^2)\, d\tau \tag{13.19}$$

Since $\dot{x}^2 + \dot{y}^2 \to 0$ when $K > 0$, $\dot{x} \to 0$ and $\dot{y} \to 0$ and the system is damped about the curve $z = z_1$. This also means, however, that motion eventually stops, an undesirable phenomenon. To prevent this, let us introduce the constraint

$$\epsilon = \dot{x}^2 + \dot{y}^2 - 1$$

and attempt to stabilize the velocity with a small gain g (say $g = 1$) and program

$$\begin{aligned} \dot{x} &= -\int \omega\dot{y}\, d\tau - K\omega\dot{y} - g\int \epsilon\dot{x}\, d\tau \\ \dot{y} &= \int \omega\dot{x}\, d\tau + K\omega\dot{x} - g\int \epsilon\dot{y}\, d\tau \end{aligned} \tag{13.20}$$

We may solve for $\dot{x}$ and $\dot{y}$ algebraically, but since the matrix

$$\begin{pmatrix} 1 & K\omega \\ -K\omega & 1 \end{pmatrix}$$

is a stability matrix, the algebraic loop in Fig. 13.6 is stable and requires a minimum amount of equipment. Thus, programming (13.20) directly produces excellent results for a proper choice of a and K. For example, with $a \simeq 250$ and $K \simeq 5$, it is possible to track a contour of the function $|x| + |y|$ with barely noticeable error, even about the sharp right-angle corners (Fig. 13.7). Note, in Fig. 13.6, the pots 1 and 2 that control the speed of tracing.

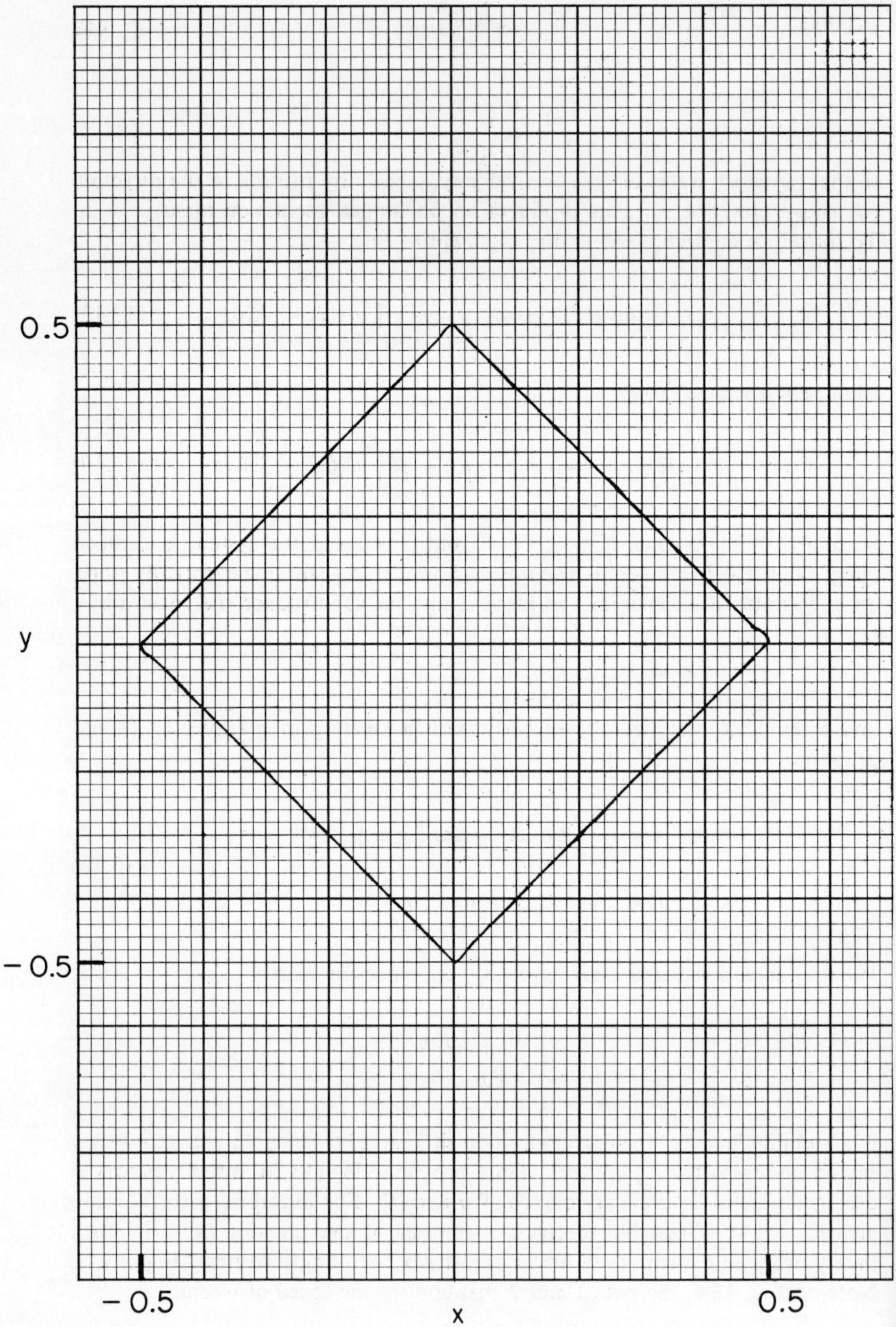

FIG. 13.7 Graph of $|x| + |y| = 0.5$ from contour tracer.

13.4 Nyquist Plotter [11]–[13], [29]

Given a linear system represented by a transfer function $G(p)$, a sinusoidal input $\sin \omega t$ results in a steady-state sinusoidal output $e_0 = A \sin \omega t + B \cos \omega t$ (Theorem 10.2). A graph of $A(\omega)$ versus $B(\omega)$ is called a Nyquist plot and is useful in a stability analysis of the system. Suppose $G(p)$ is set up on the computer and $\sin \omega t$ is used as an input. The object of a Nyquist plotter is to obtain A and B from the output waveform as a function of ω. Although several methods exist to perform this task, we show an elegant one described by Kohr [13].

Suppose A^* and B^* represent two variables that we hope approach A and B with time. Since the output is available from the system, we can consider driving the error function

$$\epsilon = (A - A^*) \sin \omega t + (B - B^*) \cos \omega t \tag{13.21}$$

to zero since ϵ can be formed on the computer once integrators representing A^* and B^* are available. If we now choose A^* and B^* to minimize

$$F = \frac{\epsilon^2}{2}$$

by a steepest-descent procedure in the usual way:

$$\begin{aligned} \dot{A}^* &= -K \frac{\partial F}{\partial A^*} = K\epsilon \sin \omega t \\ \dot{B}^* &= -K \frac{\partial F}{\partial B^*} = K\epsilon \cos \omega t \end{aligned} \tag{13.22}$$

then the resulting equations are stable, as will be shown, with $\epsilon \to 0$. It should be emphasized that this is not quite a steepest-descent procedure since the error is a function of time as well as of the space variables. For this reason, it has been called a *quasisteep-descent* procedure and its stability must be independently checked.

By differentiating (13.21) with respect to time τ, we obtain (assuming $\dot{t} = 1$)

$$\begin{aligned} \dot{\epsilon} &= \omega(A - A^*) \cos \omega t - \omega(B - B^*) \sin \omega t - \dot{A}^* \sin \omega t - \dot{B}^* \cos \omega t \\ &= \omega[(A - A^*) \cos \omega t - (B - B^*) \sin \omega t] - K\epsilon \end{aligned}$$

Differentiating with respect to time once again:

$$\begin{aligned} \ddot{\epsilon} &= -K\dot{\epsilon} + \omega^2[-(A - A^*) \sin \omega t - (B - B^*) \cos \omega t] \\ &\quad + \omega(-\dot{A}^* \cos \omega t + \dot{B}^* \sin \omega t) \\ &= -K\dot{\epsilon} - \omega^2 \epsilon \end{aligned}$$

or

$$\ddot{\epsilon} + K\dot{\epsilon} + \omega^2\epsilon = 0$$

Thus, the error ϵ is the solution of a linear differential equation. General solutions depend upon the relative values of K and 2ω. We have

$$\text{for } K < 2\omega,\ \epsilon = e^{-Kt/2}[C_1 \cos(\sqrt{\omega^2 - K^2/4}\,t) + C_2 \sin(\sqrt{\omega^2 - K^2/4}\,t)]$$
$$\text{for } K = 2\omega,\ \epsilon = e^{-Kt/2}(C_1 + C_2 t)$$
$$\text{for } K > 2\omega,\ \epsilon = C_1 e^{(-K/2+\sqrt{K^2/4-\omega^2})t} + C_2 e^{(-K/2-\sqrt{K^2/4-\omega^2})t} \tag{13.23}$$

Now it is clear that the most rapid decay of ϵ to zero occurs generally when $K = 2\omega$, since the maximum coefficient of t in the exponent is then least.

When $\epsilon = 0$, there is no direct guarantee that $A^* = A$ and $B^* = B$. From (13.21), we have only that

$$(A - A^*) \sin \omega t = -(B - B^*) \cos \omega t$$

which can hold because A^* and B^* are varying with time. But from the expression for $\dot{\epsilon} = 0$, we also have

$$(A - A^*) \cos \omega t = (B - B^*) \sin \omega t$$

Squaring each side of both equations and summing,

$$(A - A^*)^2 = (B - B^*)^2 \quad \text{or} \quad (A - A^*) = \pm(B - B^*)$$

If these were not zero, then by division we could have $\sin \omega t = \pm\cos \omega t$—an absurdity. Thus $\epsilon = \dot{\epsilon} = 0$ imply $A^* = A$ and $B^* = B$. Figure 13.8 shows the circuit diagram to mechanize (13.22) with $K = 2\omega$.

By letting ω drive the shaft of a servo multiplier, this procedure is adaptable to continuous plotting. Several difficulties, however, are encountered:

(1) The oscillator should be amplitude compensated since long times are required for the complete graph.

(2) If a pure sine wave is not generated as an input, A^* and B^* tend to oscillate. This is the case when the servo pots are misaligned and $\omega_1 \neq \omega_2$.

(3) As ω changes, it takes time for the system to reach steady-state conditions. One cannot predict generally how much time is required.

(4) The time constant of the convergence is $1/\omega$, from (13.23). Even after input steady-state conditions are reached, the rate of change of ω should be a function of ω.

Item 4 is easily arranged by taking

$$\dot{\omega} = -a\omega, \qquad \omega_0 = \omega_{\max} \tag{13.24}$$

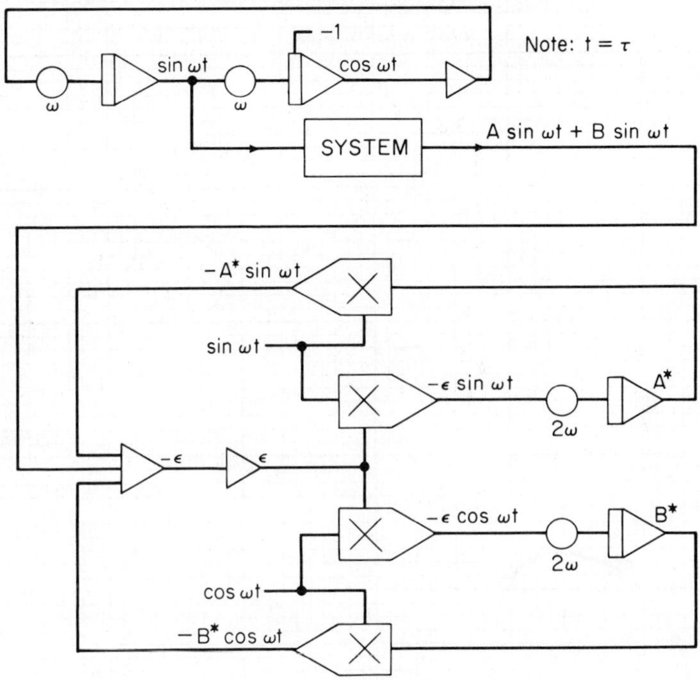

FIG. 13.8 Circuit for frequency response analyzer.

The length of time spent near a given ω is now roughly proportional to $1/\omega$ since $d\omega/\omega = -a\,dt$. If the time in item 3 is similarly inversely proportional to ω then

$$\dot{\omega} = -a\omega^2$$

could be used. One may, in general, tailor changing ω to the needs of the given system.

Figure 13.9 is a continuous plot of A^* versus B^* for the system

$$G(p) = \frac{0.8}{(p+1)^4}$$

using an amplitude stabilized oscillator as an input and (13.24) to change ω, with $\omega_0 = 10$ rad/s. Small oscillations may occur due mostly to distortions of the input sine wave.

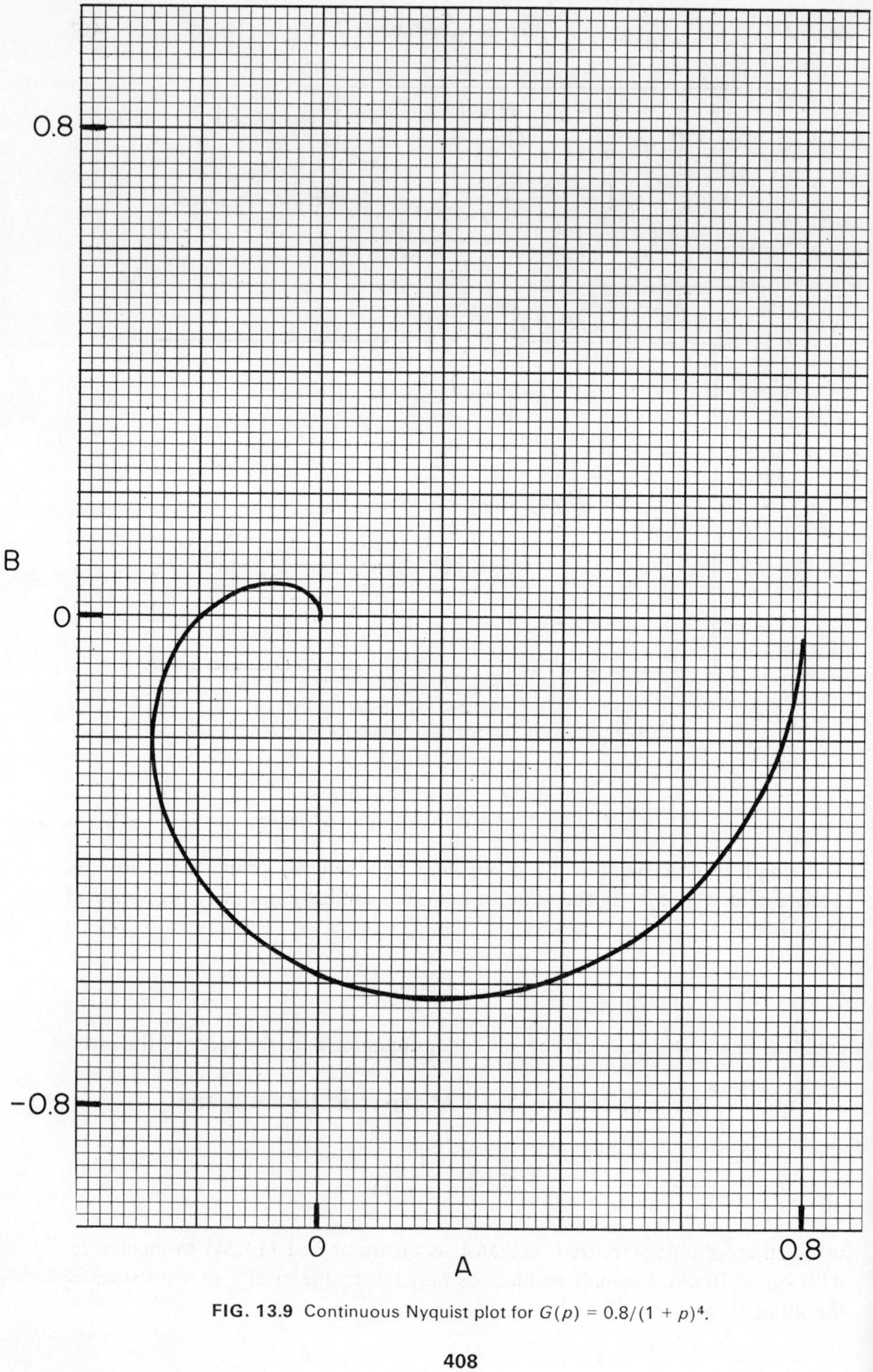

FIG. 13.9 Continuous Nyquist plot for $G(p) = 0.8/(1 + p)^4$.

13.5 Parameter Identification [14], [18]–[28]

Another quasisteep-descent application is the solution to the so-called parameter identification problem, one type of which can be described as follows:

Let a mathematical model of a physical system be described by the differential equation

$$\epsilon \equiv \sum_{i=0}^{n} \left[b_i \frac{d^i y}{dt^i} - a_i \frac{d^i f(t)}{dt^i} \right] = 0 \tag{13.25}$$

where $b_n \neq 0$, and for convenience we may take $b_n = 1$. The value n is known, but the coefficients b_i and a_i are not. Some $f(t)$ is supplied to the system and $y(t)$ is obtained experimentally. Both are available as data for $0 \leq t \leq T$. The problem is to determine the a_i and b_i.

Consider the case where a_i and b_i are constant. There are at most $2n + 1$ parameters (some of the a_i or b_i could be known). To get an insight into the nature of such a problem, let us assume that we could obtain, somehow, all the needed derivatives of $f(t)$ and $y(t)$. We need pick $2n + 1$ distinct points in time, set up $2n + 1$ equations from (13.25) in the $2n + 1$ unknowns and solve for the parameters, provided the matrix was nonsingular. The difficulty with this procedure is that the answers may easily depend upon the points chosen. In fact, because of noise and other errors, we could expect this to be so. Intuitively, we see that if many sets of $2n + 1$ points are chosen and the a_i and b_i calculated for each set are identical or close to one another, the model of the system expressed by (13.25) is very good. On the other hand, if there is wide variation between the a_i and b_i found, say for t near zero compared with the t near T, the model of the system is poor. We could not expect a constant set of parameters of force $\epsilon = 0$ for all time. This discussion applies only to the former case. We assume the model structure (13.25) is exact. Otherwise some norm must be defined to determine a best estimate of the parameters over the entire interval, and that estimate could easily change with different data.

Let a_i^* and b_i^* be variables that we hope to force dynamically to approach a_i and b_i. It would be desirable to operate on the error function

$$\epsilon^* = \sum_{i=0}^{n} \left[b_i^* \frac{d^i y}{dt^i} - a_i^* \frac{d^i f(t)}{dt^i} \right]$$

but the process of differentiation is so noise-amplifying that a direct formation of ϵ^* is precluded. Instead, we proceed by passing the ϵ in (13.25) through any linear filter $L(p)$ (a stable rational transfer function in $p \equiv d/dt$). We have

$$\sum_{i=0}^{n} \left\{ L(p) \left[b_i \frac{d^i y}{dt^i} \right] - L(p) \left[a_i \frac{d^i f(t)}{dt^i} \right] \right\} = L(p)[0]$$

The three important properties of linear filters in this situation, which are easily verified, are

1. $L(p)[ax] = aL(p)[x]$ when a is constant
2. $L(p)\left[\frac{d^i y}{dt^i}\right] = \frac{d^i[L(p)y]}{dt^i}$
3. $L(p)[0] = 0$

The definition of a linear filter requires that

$$L(p)\left[\sum_{i=1}^{n} x_i\right] = \sum_{i=1}^{n} L(p)[x_i]$$

Letting

$$x = L(p)[y], \qquad g(t) = L(p)[f(t)]$$

we obtain

$$\epsilon_T \equiv \sum_{i=0}^{n}\left[b_i \frac{d^i x}{dt^i} - a_i \frac{d^i g(t)}{dt^i}\right] = 0 \tag{13.26}$$

Now, however, the derivatives of x and g can be easily made available on the computer with a good choice of $L(p)$—namely, that *the order of $L(p)$ should be at least n and its numerator should be constant* (*an all-pole filter*):

$$L(p) = \frac{1}{\sum_{i=0}^{r} c_i p^i}, \qquad r \geq n, \qquad c_n \neq 0$$

If (13.26) is mechanized on the computer, we let a_i^* and b_i^* be variables as before,

$$\epsilon_T^* = \sum_{i=0}^{n}\left[b_i^* \frac{d^i x}{dt^i} - a_i^* \frac{d^i g(t)}{dt^i}\right]$$

and attempt to drive ϵ_T^* to zero with the descent equations

$$\dot{b}_i^* = -k_i \epsilon_T^* \frac{d^i x}{dt^i}, \qquad \dot{a}_i^* = h_i \epsilon_T^* \frac{d^i g}{dt^i}, \qquad i = 1, 2, \ldots, n \tag{13.27}$$

which minimizes $(\epsilon_T^*)^2/2$ (k_i and h_i are constant gains). Equation (13.27) requires that two filter circuits be constructed for producing the derivatives of x and g simultaneously. A final circuit is shown in Fig. 13.10 with $i = 1$. Virtually any values of c_i can be used as long as $L(p)$ is stable. We may verify that $\dot{\epsilon}_T$ has a negative feedback term, a stabilizing influence, but only a neces-

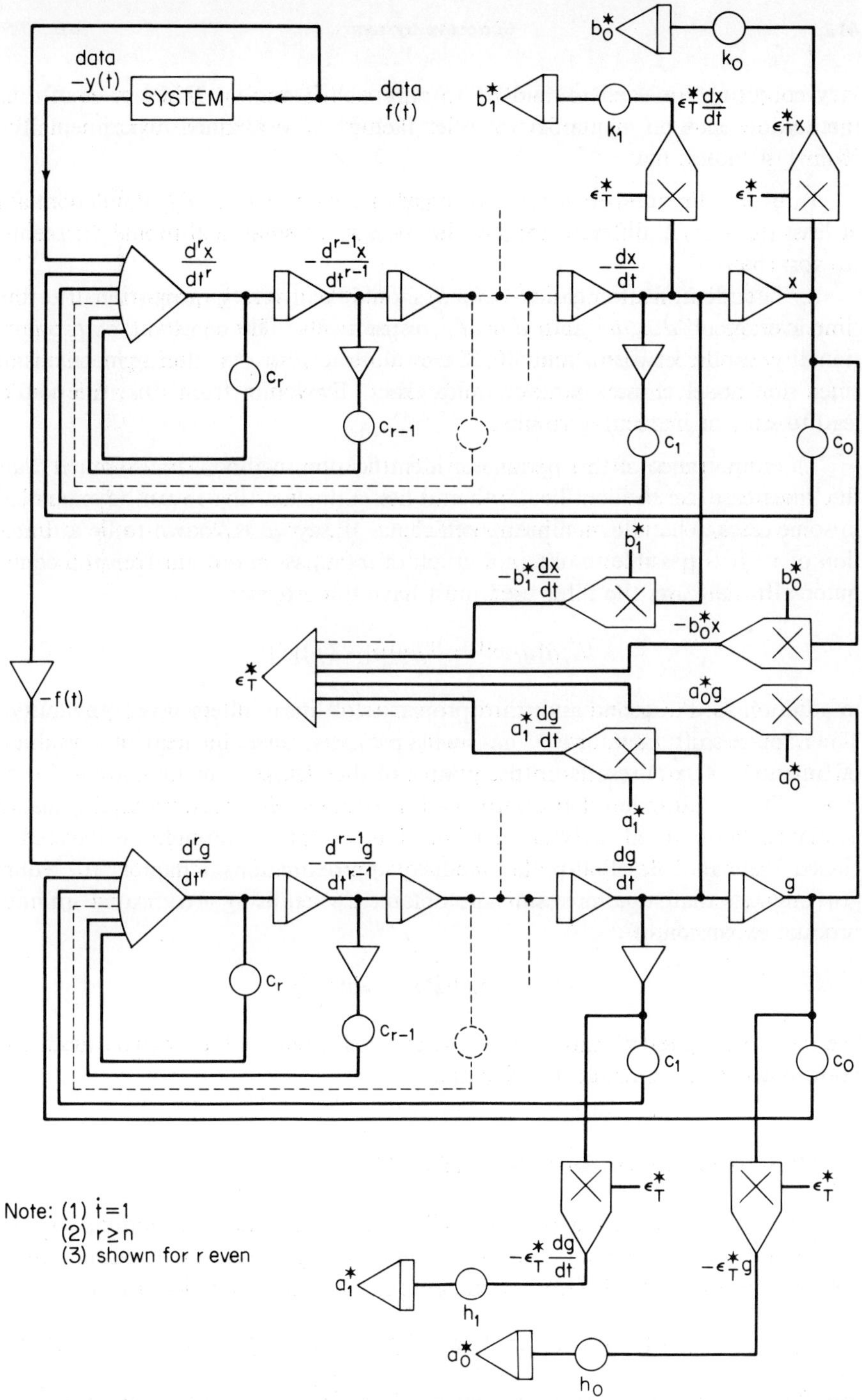

FIG. 13.10 Circuit for identification using linear filters.

sary condition for each a_i^* and b_i^* to approach a_i and b_i. The success of this method depends on a number of other factors, also obtained experimentally. Kohr [14] notes that

1. If q is the number of unknown parameters, the input $f(t)$ should contain at least $(q + 1)/2$ different frequencies. A sum of sinusoidal inputs $f(t)$ seems to work best.

2. The identification gains k_i and h_i should be inversely proportional to the time average of $|d^i x/dt^i|$ and $|d^i g/dt^i|$, respectively. The constants of proportionality can lie between 1 and 50. These values act also as a filtering mechanism since the model chosen is never quite exact. Deviation from this rule could lead to slow or inaccurate results.

The importance of this parameter identification method is not so much that the linear case can be identified; rather, it lies in the fact that it can be extended in some cases to handle nonlinear coefficients. If, say, a_0 is known to be a function of y_0, it is possible to obtain a graph of $a_0(y_0)$ as an output from the computer. In that case, the filter used must have the property

$$L(p)[a \cdot x] = L(p)[a] \cdot L(p)[x]$$

in addition to the second and third properties of linear filters used previously. The Laplace shift operator e^{-pT} has such a property, since the delay of a product of functions $a(t)x(t)$ results in the product of the delayed functions $a(t - T) \times x(t - T)$. An all-pole filter approximating a time delay must be used to make available the required derivatives of the filter output. Although we have discussed Padé and Bessel all-pole time delay approximations (Chapter 10), Kohr [14] suggests that superior results are obtained when using a truncated infinite product expansion of

$$e^{pT} = \cosh Tp + \sinh Tp$$

See [14] for suggested values of the filter coefficients, which can be used for both linear and nonlinear identification.

13.6 Regression Analysis [15]–[17]

The term regression analysis is used in statistics when determining an optimum set of parameters in data fitting while using a least squares norm. Suppose we know, from physical considerations, that some functional form

$$F[a_1, a_2, \ldots, a_n, X_1(t), X_2(t), \ldots, X_m(t)] = 0 \tag{13.28}$$

should exist between m available signals $X_j(t)$ for $0 \leq t \leq T_f$ and n constant parameters a_i. Unlike the identification problem, F does not contain derivatives of the $X_j(t)$. The problem is to determine the best set of a_i such that

$$\rho = \tfrac{1}{2}\int_0^{T_f} F^2\,dt \tag{13.29}$$

is minimized.

To attack this problem, we apply (7.4) and solve

$$\frac{\partial \rho}{\partial a_i} = 0, \qquad i = 1, 2, \ldots, n \tag{13.30}$$

for the a_i. We shall show that (13.30) sometimes can be solved by the direct method of solving algebraic equations (Chapter 11), by forming the a_i at outputs of high-gain amplifiers.

The a_i found could easily be functions of the upper limit of integration in (13.29) because of noise, possible model imperfections, etc. Thus, we consider the a_i as variables of an upper limit T, and write $a_i(T)$ in (13.28) with

$$\rho(T) = \tfrac{1}{2}\int_0^T \{F[a_1(T), a_2(T), \ldots, a_n(T), X_1(t), X_2(t), \ldots, X_m(t)]\}^2\,dt$$

By differentiating with respect to T, we see that

$$\frac{d\rho}{dT} = \int_0^T \left(\sum_{i=1}^n F\frac{\partial F}{\partial a_i}\frac{da_i}{dT}\right) dt + \tfrac{1}{2}F^2[a_1(T), a_2(T), \ldots, a_n(T), X_1(T), X_2(T), \ldots, X_m(T)] \tag{13.31}$$

Focus attention on the integral expression in (13.31). The summation and integral signs can be interchanged, and because the a_i are not functions of t, we have

$$\int_0^T \left(\sum_{i=1}^n F\frac{\partial F}{\partial a_i}\frac{da_i}{dT}\right) dt = \sum_{i=1}^n \left(\frac{da_i}{dT}\int_0^T F\frac{\partial F}{\partial a_i}\,dt\right)$$

We intend to let $\dot{T} = 1$ on the computer and would like this term to be negative as possible to minimize ρ. Thus, we program

$$\dot{a}_i = -g_i \int_0^T \left(F\frac{\partial F}{\partial a_i}\right) dt \tag{13.32}$$

and use as large a value of g_i as possible. But, from (13.29) and (13.30)

$$\frac{\partial \rho}{\partial a_i} = \int_0^{T_f} \left(F\frac{\partial F}{\partial a_i}\right) dt = 0$$

A mechanization of (13.32) produces the a_i when $T = T_f$ as well as the functions $a_i(T)$ for other upper limits.

Only one difficulty exists in programming (13.32). The required integrals

$$\int_0^T F\left(\frac{\partial F}{\partial a_i}\right) dt$$

may not be generable as continuous functions of T, since F and $\partial F/\partial a_i$ are functions of T and t. If F is linear in the a_i, we may show that these integrals are generable continuously as T increases.

Example 13.2 [15]-[17]. Suppose two signals $X(t)$ and $Y(t)$ are on tape (or are generated on-line by some process) for $0 \leq t \leq T_f$ and we wish to relate them linearly

$$Y(t) = mX(t) + b$$

With

$$\rho = \tfrac{1}{2}\int_0^T [Y(t) - mX(t) - b]^2\, dt \tag{13.33}$$

we use

$$\begin{aligned} \dot{m} &= -g\int_0^T \{-X(t)[Y(t) - mX(t) - b]\}\, dt \\ \dot{b} &= -g\int_0^T \{-[Y(t) - mX(t) - b]\}\, dt \end{aligned} \tag{13.34}$$

and $\dot{T} = 1$. Because m and b are not functions of t, (13.34) reduces to

$$\begin{aligned} \dot{m} &= g\left[-m\int_0^T X^2(t)\, dt - b\int_0^T X(t)\, dt + \int_0^T X(t)Y(t)\, dt\right] \\ \dot{b} &= g\left[-m\int_0^T X(t)\, dt - bT + \int_0^T Y(t)\, dt\right] \end{aligned} \tag{13.35}$$

A circuit to mechanize (13.35) is shown in Fig. 13.11. For very large g, we continuously force $\dot{m}$ and $\dot{b} \simeq 0$, which gives a running account of $m(T)$ and $b(T)$. If m and b are not constant for all points, the model may not be very good. If they are almost constant, a small g acts as a filter so that noise will not appreciably disturb average values. When $T = T_f$, and g is large, the solution to the original regression problem is obtained. Because $t = \tau$ (the process is in real time), the functions $X(\tau)$ and $Y(\tau)$ are assumed available in the figure.

If we had found (13.30) for arbitrary limit T:

$$\begin{aligned} m\int_0^T X^2(t)\, dt + b\int_0^T X(t)\, dt - \int_0^T X(t)Y(t)\, dt &= 0 \\ m\int_0^T X(t)\, dt + bT - \int_0^T Y(t)\, dt &= 0 \end{aligned} \tag{13.36}$$

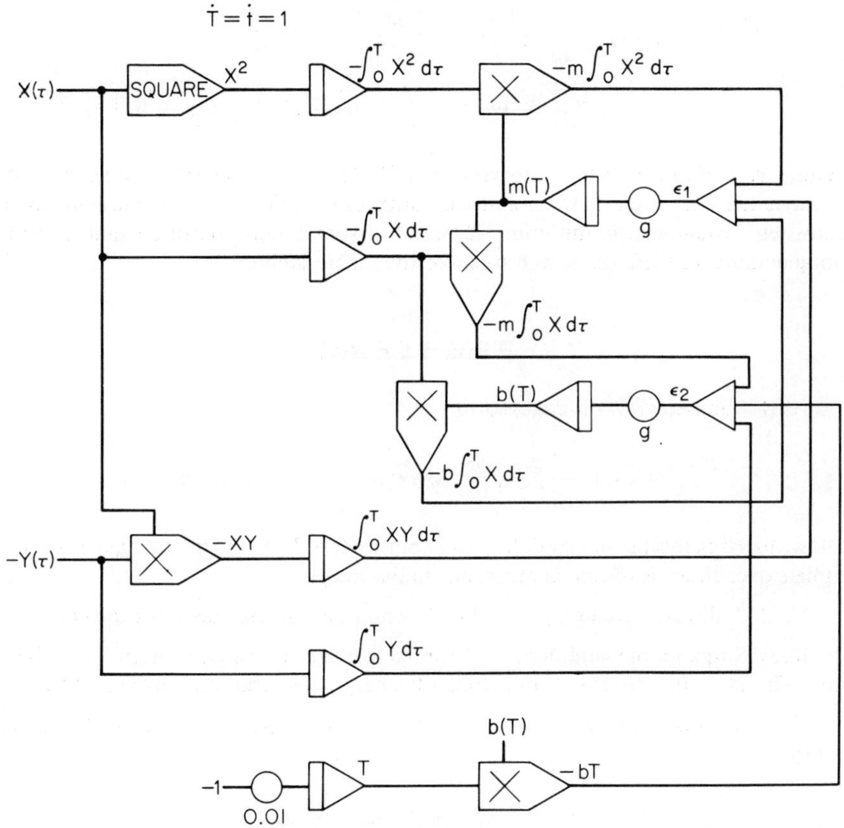

FIG. 13.11 Mechanization of the linear regression problem.

and attempted to solve these equations on the computer using the direct method, we would have had to verify that the matrix

$$\mathbf{A} = \begin{pmatrix} \int_0^T X^2(t)\, dt & \int_0^T X(t)\, dt \\ \int_0^T X(t)\, dt & T \end{pmatrix}$$

was positive stable. The method of steepest descent demonstrates that such resulting matrices will always be positive stable (actually positive definite). In this case, verification is simple as demonstrated below.

The trace of $\mathbf{A}$ is positive for $T > 0$, and

$$|\mathbf{A}| = T\int_0^T X^2(t)\, dt - \left[\int_0^T X(t)\, dt\right]^2$$

Let $\bar{X} = (1/T)\int_0^T X(t)\, dt$. Now

$$\int_0^T (X - \bar{X})^2\,dt = \int_0^T X^2\,dt - 2\bar{X}\int_0^T X\,dt + \bar{X}^2 T$$
$$= \int_0^T X^2\,dt - \bar{X}^2 T = \int_0^T X^2\,dt - \frac{1}{T}\left(\int_0^T X\,dt\right)^2 \geq 0$$

Thus, $|\mathbf{A}| \geq 0$ for $T > 0$ and nonconstant X. The direct method is stable since **A** is positive definite. Even if X remains constant for a while, (13.36) is not a set of inconsistent equations; $\dot{m}$ and $\dot{b}$ in (13.35) are driven to zero, but the m and b are not unique and can drift. (**A** is said to be positive semidefinite.)

PROBLEMS

13-1. Suppose (13.7) is changed to

$$\dot{y}_i = f_i - \sum_{j=1}^{m} g_j \frac{\partial \epsilon_j}{\partial y_i} \operatorname{sign} \epsilon_j, \qquad i = 1, 2, \ldots, n$$

Show that this results in a stabilization about each $\epsilon_j = 0$, at the savings of mn multipliers over the cost of mn comparators and switches.

13-2. Stabilize the energy of a double pendulum as described in Problem 12-1.

13-3. Suppose the pendulum in Example 13.1 has a dissipative term, $-k(d\theta/dt)$, or $-k\theta_1$ $(k > 0)$. Modify ϵ_1 to include the energy loss according to Sect. 13.1.3.

13-4. A conservative dynamical system with two degrees of freedom has kinetic energy

$$T = \frac{1}{2}\frac{\dot{q}_1^2}{a_1 + a_2 q_2^2} + \frac{1}{2}\dot{q}_2^2$$

and potential energy

$$V = a_3 + a_4 q_2^2$$

Set up the equations of motion and stabilize the energy and the cyclic coordinate q_1.

13-5. Generate the functions $\cosh x$ and $\sinh x$, $x_0 = 0$ with integrators and stabilize the relation

$$\cosh^2 x - \sinh^2 x - 1 = 0$$

If $\dot{x} = 1$, show that multiplier errors could produce inaccuracies greater than those existing without the stabilization.

13-6. Program the equations [3]

$$x^2 + y^2 + z^2 = R^2$$
$$ax + by + cz = d$$
$$\dot{x}^2 + \dot{y}^2 + \dot{z}^2 = v^2$$

which trace a circle at constant velocity in space. Assume all symbols are constant except x, y, and z.

13-7. Plot the equation

$$x^2 + \sin^2 y = a^2$$

by solving parametric differential equations and stabilize the errors by using fixed-function generators to generate $\sin y$ and $\cos y$ where needed.

13-8. Repeat the oscillator stabilization error analysis beginning with (13.14) assuming unequal stabilization gains $g_1 \neq g_2$.

13-9. Show that the solution of (13.15) approaches the proper amplitude independent of any initial amplitude.

13-10. Find the solution of (13.15) as $g_1 \to \infty$ if the ω pots are in error and are set at ω_1 and ω_2.

13-11. Show that the equations

$$\begin{aligned} \dot{y}_1 &= -\omega y - g\epsilon \operatorname{sign} y_1 \\ \dot{y} &= \omega y_1 - g\epsilon \operatorname{sign} y \\ \epsilon &= y_1^2 + y^2 - A^2, \qquad A^2 \text{ constant} \end{aligned}$$

are stable. (*Hint:* Compute $\dot{\epsilon}$.) Suggest a mechanization.

13-12. Show that the equations

$$\begin{aligned} \dot{y}_1 &= -\omega y - g(\operatorname{sign} \epsilon)(\operatorname{sign} y_1) \\ \dot{y} &= \omega y_1 - g(\operatorname{sign} \epsilon)(\operatorname{sign} y) \end{aligned}$$

are stable and suggest a mechanization.

13-13. Derive (13.19) from (13.18).

13-14. Apply the results of Problem 13-11 to Skramstad's contour-tracing circuit to prevent complete damping of the velocity and suggest equations that parallel (13.20). Mechanize the equations.

13-15. Apply the previous circuit to plot the equation in Problem 13-7. Which method would you think gives less error?

13-16. Suppose the open-loop system in Fig. 13.12 is unstable so the Nyquist plotter cannot obtain the proper results, but the closed-loop system as shown is stable.

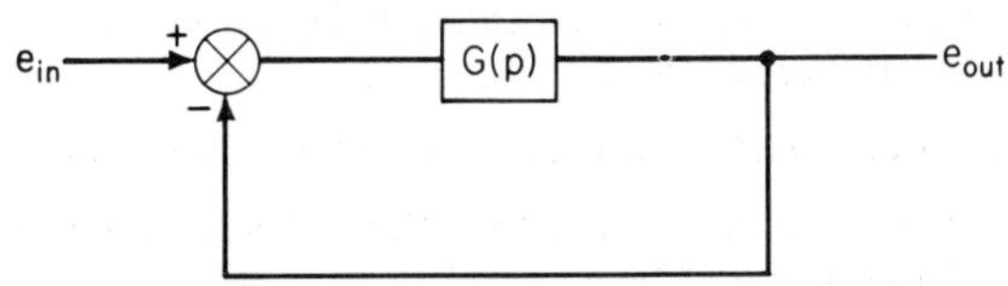

FIG. 13.12 Closed-loop system for Problem 13-16.

How may the outputs $A^*(\omega)$ and $B^*(\omega)$ of the closed-loop system be used to obtain the frequency response of $G(p)$?

13-17. Set up the circuitry to identify the parameters a_i and b_i for a model

$$b_2\ddot{y} + b_1\dot{y} + b_0 y = a_1\dot{f}(t) + a_0 f(t)$$

when $y(t)$ and $f(t)$ are available for $0 \leq t \leq T$.

13-18. Suppose a signal $r(t)$ is generated and two parameters x and y should be such that [16]

$$\epsilon = x + v^2(t - y)^2 - r(t) = 0$$

with v a known constant. Set up the equations and circuitry to obtain continuous best estimates of x and y as t increases, using the regression technique.

13-19. Repeat Problem 13-18 assuming v is unknown also. (To simplify the circuitry, let v^2 be the parameter [16].)

REFERENCES

[1] Hochfeld, E., ed., "Stabilization of Computer Circuits," *Wright Air Development Center Technical Report 57-425*, Nov. 1957 (ASTIA No. 155740).

[2] Turner, R. M., "On the Reduction of Error in Certain Analog Computer Calculations by the Use of Constraint Equations," *Proc. Western Joint Comp. Conference*, 1960.

[3] Talkin, A. I., "A Systematic Approach to Programming an Analog Computer to Generate a Large Class of Trajectories," *IEEETEC*, vol. 13, Oct. 1964.

[4] Brunner, W., "Steepest Ascent," *Princeton Computation Center Report No. 151*, Electronic Associates, Inc., Princeton, N. J., Dec. 29, 1959.

[5] Hausner, A., "The Solution of Lagrange's Equations by Analog Computation," *IEEETEC*, vol. 14, Feb. 1965.

[6] Howe, R. M., and E. G. Gilbert, "Trigonometric Resolution in Analog Computers by Means of Multiplier Elements," *IRETEC*, vol. 6, June 1957.

[7] Mitchell, E. E. L., and A. E. Rogers, "Quaternion Parameters in the Simulation of a Spinning Rigid Body," *Simulation*, vol. 4, June 1965.

[8] Brunner, W., and R. Beadle, "Amplitude Stabilization of Harmonic Oscillator with Diode Gates," *Princeton Computation Center Report No. 173*, Electronic Associates, Inc., Princeton, N. J., Sept. 1962.

[9] Korte, E. L., "Stable Wide-Frequency Oscillator," *Simulation*, vol. 2, Jan. 1964.

[10] Skramstad, H. K., "Contour Tracing with an Analog Computer," *AIEE* paper no. 59-461, Winter General Meeting, New York, N. Y., Feb. 1-6, 1959.

[11] Goldberg, E. A., "An Analog Computer Nyquist Plotter," *IRE International Convention Record, part 2,* 1960.

[12] Bekey, G. A., and L. W. Neustadt, "Analog-Computer Techniques for Plotting Bode and Nyquist Diagrams," *Proc. Western Joint Comp. Conference,* 1960.

[13] Kohr, R. H., "An Automatic Frequency Response Analyzer," *Simulation,* vol. 8, Jan. 1967.

[14] Kohr, R. H., "On the Identification of Linear and Nonlinear Systems," *Simulation,* vol. 8, March 1967.

[15] Rubin, A. I., "Continuous Data Analysis with Analog Computers Using Statistical and Regression Techniques," *Princeton Computation Center Report No. 160,* Electronic Associates, Inc., Princeton, N. J., Dec. 20, 1960.

[16] Rubin, A. I., "Continuous Regression Techniques Using Analog Computers," *IRETEC,* vol. 11, Oct. 1962.

[17] Rubin, A. I., *et al.*, "Regression Analysis and Parameter Identification," *Simulation,* vol. 9, July 1967.

[18] Murphy, C. H., "Use of REAC as a Curve Fitting Device," *Symposium II on Simulation and Computing Techniques, part 2,* Reeves Instr. Corp., New York, April 28–May 2, 1952 (ASTIA No. 170848).

[19] Brown, R. W., Sr., "A New Analog Computer Technique for Determining System Transfer Function Coefficients," *Project Typhoon Symposium III on Simulation and Computing Techniques, part 1,* Bureau of Aeronautics and U.S. Naval Air Devel. Center, Johnsville, Pa., Oct. 12–14, 1953.

[20] Clymer, A. B., "Direct System Synthesis by Means of a Computer," *Trans. AIEE, part I* (*Comm. and Elect.*), vol. 77, Jan. 1959.

[21] Clynes, M., "Respiratory Control of Heart Rate: Laws Derived from Analog Computer Simulation," *IRETME,* vol. 7, Jan. 1960.

[22] McGrath, R. J., and V. C. Rideout, "A Simulator Study of a Two-Parameter Adaptive System," *IRETAC,* vol. 6, Feb. 1961.

[23] Potts, T. F., *et al.*, "The Automatic Determination of Human and Other System Parameters," *Proc. Western Joint Computer Conference,* 1961.

[24] Kohr, R. H., "A Method for the Determination of a Differential Equation Model for Simple Nonlinear Systems," *IEEETEC,* vol. 12, Aug. 1963.

[25] Fricke, L. H., and R. A. Walsh, "The Use of a Portable Analog Computer for Process Identification, Calculation and Control," *Proc. Fall Joint Comp. Conference,* 1964.

[26] Meissinger, H. F., and G. A. Bekey, "An Analysis of Continuous Parameter Identification Methods," *Simulation,* vol. 6, Feb. 1966.

[27] Meissinger, H. F., "Parameter Optimization by an Automatic Open-Loop Computing Method," *Proceedings of the 4th Inter. Analogue Comp. Meetings* (Brussels: Presses Académiques Européennes, 1966).

[28] Jackson, G. A., "On the Stability of Continuous Parameter Tracking Systems," *Simulation*, vol. 11, July 1968.

[29] Sprague, C. H., "A Flexible Sinusoidal Response Analyzer," *Simulation*, vol. 11, Aug. 1968.

See also:

[30] Keller, R. E., "An Analog Computer Realization of the Euclidian Tools," *IRETEC*, vol. 11, Aug. 1962.

Chapter 14

SOLUTIONS TO PARTIAL DIFFERENTIAL EQUATIONS

The mathematical description of many phenomena involves the use of partial differential equations, which are differential equations characterized by the existence of more than one independent variable. Difficulties with obtaining solutions to such equations on an analog computer are instantly clear: the solutions are surfaces at the very least, whereas the analog computer produces only trajectories. Thus we can only hope that some parametrization of certain variables will generate trajectories that lie on the solution surface. Unfortunately, this cannot always be done without introducing some mathematical approximation. In turn, the answers obtained are then subject to both approximation error as well as computation error.

Much theory has been developed for analytical and numerical solutions to partial differential equations. Readers are urged to consult standard texts (e.g., Weinberger [1]) for a comprehensive view of the field. The emphasis here is how an analog computer has been used to obtain solutions to some of the problems.

14.1 First-Order Equations

First-order partial differential equations in n independent variables can always be reduced to a set of ordinary differential equations. For two independent variables x and y, for example, a general (possibly nonlinear) partial differential equation may be written

$$f(x, y, z, p, q) = 0, \qquad p \equiv \frac{\partial z}{\partial x}, \qquad q \equiv \frac{\partial z}{\partial y} \tag{14.1}$$

By Cauchy's method of characteristics [2], five ordinary differential equations are determined:

$$\begin{aligned} \frac{dx}{ds} &= \frac{\partial f}{\partial p}, \qquad \frac{dy}{ds} = \frac{\partial f}{\partial q}, \qquad \frac{dz}{ds} = p\frac{\partial f}{\partial p} + q\frac{\partial f}{\partial q} \\ \frac{dp}{ds} &= -\frac{\partial f}{\partial x} - p\frac{\partial f}{\partial z}, \qquad \frac{dq}{ds} = -\frac{\partial f}{\partial y} - q\frac{\partial f}{\partial z} \end{aligned} \tag{14.2}$$

solutions of which define the so-called *characteristic curves* or *characteristics* of the solution surface of (14.1) in terms of an arbitrary parameter s. A characteristic has the property that if one point of it is in common with the solution surface, the entire curve lies on the solution surface. Characteristics may be determined by solving (14.2) using initial conditions that satisfy (14.1) and are consistent with the space curve through which the solution must be passed. (A particular solution can be passed through a space curve just as a first-order ordinary differential equation is passed through a point.) These ideas are crystallized by the following example.

Example 14.1. Find characteristics for the equation

$$f \equiv xp + yq + pq - z = 0 \tag{14.3}$$

passing through the space curve (a parabola) parametrically given by

$$x = \lambda, \qquad y = \lambda, \qquad z = \lambda^2 \tag{14.4}$$

for the parameter λ [3].

From (14.2), (14.3) transforms to

$$\begin{aligned} &\frac{dx}{ds} = x + q, \qquad \frac{dy}{ds} = y + p \\ &\frac{dz}{ds} = px + qy + 2pq, \qquad \frac{dp}{ds} = 0, \qquad \frac{dq}{ds} = 0 \end{aligned} \tag{14.5}$$

Initial conditions must lie on the curve (14.4). With

$$x_0 = \lambda, \qquad y_0 = \lambda, \qquad z_0 = \lambda^2$$

we must determine appropriate initial conditions p_0 and q_0. The solution surface $z = z(x, y)$ must contain the initial curve. Hence

$$\frac{dz}{d\lambda} = \frac{\partial z}{\partial x}\frac{dx}{d\lambda} + \frac{\partial z}{\partial y}\frac{dy}{d\lambda} = p_0\frac{dx}{d\lambda} + q_0\frac{dy}{d\lambda}$$

must be satisfied along the initial curve. Thus,

$$2\lambda = p_0 + q_0 \tag{14.6}$$

In addition, (14.3) must be satisfied:

$$\lambda p_0 + \lambda q_0 + p_0 q_0 - \lambda^2 = 0 \tag{14.7}$$

Solving (14.6) and (14.7) for p_0 and q_0 results in

$$p_0 = \lambda(1 \pm \sqrt{2}), \qquad q_0 = \lambda(1 \mp \sqrt{2})$$

either set of which can be used. For each λ, the functions x, y, and z are generated parametrically by solving (14.5) with $\dot{s} = +1$ or -1. The totality of solutions for all λ define the solution surface, but in practice, a dense set of λ in the region of interest is obtained.

This method is extendable to a first-order partial differential equation in n independent variables by solving $2n + 1$ ordinary differential equations. When the partial differential equations are linear, however, only $n + 1$ equations are needed to generate characteristics. Little and Soudack [4] demonstrated the use of an analog computer in the case of a linear partial differential equation in two independent variables x and y. For

$$P(x, y, z)\frac{\partial z}{\partial x} + Q(x, y, z)\frac{\partial z}{\partial y} = R(x, y, z)$$

characteristic curves are generated with the system

$$\dot{x} = P(x, y, z), \qquad \dot{y} = Q(x, y, z), \qquad \dot{z} = R(x, y, z)$$

where the initial conditions x_0, y_0, and z_0 lie on the initial condition curve.

In some special instances, solution curves that are not characteristics may be generated, say for constant x; see, for example, Hemple [5].

14.2 Separation of Variables

A number of linear partial differential equations of second and higher order are separable; i.e., the solution for two independent variables is of the form

$$z = \sum_{j=1}^{\infty} a_j f_j(x) g_j(y)$$

which can be constructed from the assumption that solution components are of the form $f(x)g(y)$. This often leads to an eigenvalue system of differential equations [1] with the analog computer being capable of quickly determining the eigenvalues, corresponding eigenfunctions, and Fourier coefficients a_j. The development of such a system is best illustrated by an example.

Example 14.2 [6]. In the study of viscous compressible flow in a channel, the equation

$$\frac{\partial^2 z}{\partial y^2} = (1 - y^2)\frac{\partial z}{\partial x} \tag{14.8}$$

with the boundary conditions

$$z(x, \pm 1) = 0, \qquad z(0, y) = z_0(y), \qquad z(\infty, y) = 0 \tag{14.9}$$

must be solved for $-1 \leq y \leq 1$ and $x \geq 0$. Assuming solutions of the form $z = f(x)g(y)$ exist, we have by direct substitution into (14.8)

$$f(x)\frac{d^2 g(y)}{dy^2} = (1 - y^2)\frac{df(x)}{dx}\, g(y)$$

or

$$\frac{\left[\dfrac{d^2 g(y)}{dy^2}\right]}{g(y)(1 - y^2)} = \frac{\left[\dfrac{df(x)}{dx}\right]}{f(x)}$$

But the expression on the left is only a function of y and the expression on the right is only a function of x. Since the x and y are independent, they must equal a constant. Hence, such solutions must satisfy the equations

$$\frac{d^2 g}{dy^2} = -\lambda^2(1 - y^2)g, \qquad \frac{df}{dx} = -\lambda^2 f \tag{14.10}$$

where the constant was taken negative $(-\lambda^2)$ to force $f \to 0$ as $x \to \infty$. Another boundary condition to be satisfied [since $f(0) = 0$ implies $f(x) = 0$] is

$$g(-1) = g(1) = 0$$

It turns out that nontrivial solutions for g exist, but only for discrete values of λ. We label them λ_j, $j = 1, 2, \ldots$ in order of increasing value. These are the *eigenvalues.* To each eigenvalue there corresponds the *eigenfunction* g_j which satisfies the boundary conditions. The set of functions f_j corresponding to each λ_j is such that

$$f_j = x_0 e^{-\lambda_j^2 x}$$

where we may normalize these by requiring $x_0 = 1$. There are, in fact, an infinite number of pairs of functions f_j and g_j such that $f_j g_j$ satisfies the differential equations and some of the boundary conditions (14.9). It follows from the linearity of (14.8) that the function

$$z = \sum_{j=1}^{\infty} a_j f_j g_j$$

will be a solution if the remaining boundary condition, forcing $z = z_0(y)$ when $x = 0$, can be satisfied. We satisfy this condition by properly choosing the a_j.

The most important property of these eigenfunctions is that they are orthogonal over the interval $[-1, 1]$, which permits determining the a_j easily. To see this, let g_m and g_n be two distinct eigenfunctions corresponding to the eigenvalues λ_m and λ_n:

$$\frac{d^2 g_n}{dy^2} + \lambda_n^2 g_n(1 - y^2) = 0, \qquad \frac{d^2 g_m}{dy^2} + \lambda_m^2 g_m(1 - y^2) = 0$$

Multiply the first equation by g_m, and the second by g_n, subtract, and integrate in the interval $-1 \leq y \leq 1$:

$$\int_{-1}^{1} \left[g_m \frac{d^2 g_n}{dy^2} - g_n \frac{d^2 g_m}{dy^2} + (\lambda_n^2 - \lambda_m^2) g_n g_m (1 - y^2) \right] dy = 0$$

But

$$g_m \frac{d^2 g_n}{dy^2} - g_n \frac{d^2 g_m}{dy^2} = \frac{d}{dy} \left(g_m \frac{dg_n}{dy} - g_n \frac{dg_m}{dy} \right)$$

so the first part of the integral is zero because g_m and $g_n = 0$ at the endpoints. Thus

$$(\lambda_n^2 - \lambda_m^2) \int_{-1}^{1} g_n g_m (1 - y^2)\, dy = 0$$

which shows that the eigenfunctions g_j are orthogonal with respect to the weight function $(1 - y^2)$. We may use the theory in Chapter 7 to determine the a_j from

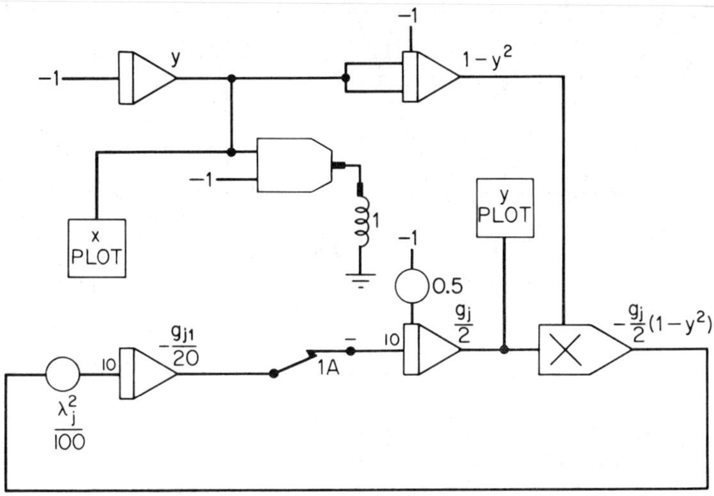

FIG. 14.1 Determination of odd eigenvalues and eigenfunctions.

$$a_j = \frac{1}{h_j}\int_{-1}^{1} (1 - y^2) z_0(y) g_j(y)\, dy$$

with

$$h_j = \int_{-1}^{1} (1 - y^2) g_j^2(y)\, dy$$

either on the computer or otherwise.

As a practical way of determining the eigenfunctions and eigenvalues for this case, we recognize from (14.10) that the eigenfunctions must be either symmetric or antisymmetric about $y = 0$. [If $g(x, y)$ is a solution, then so are $-g(x, y)$ and $g(x, -y)$.]

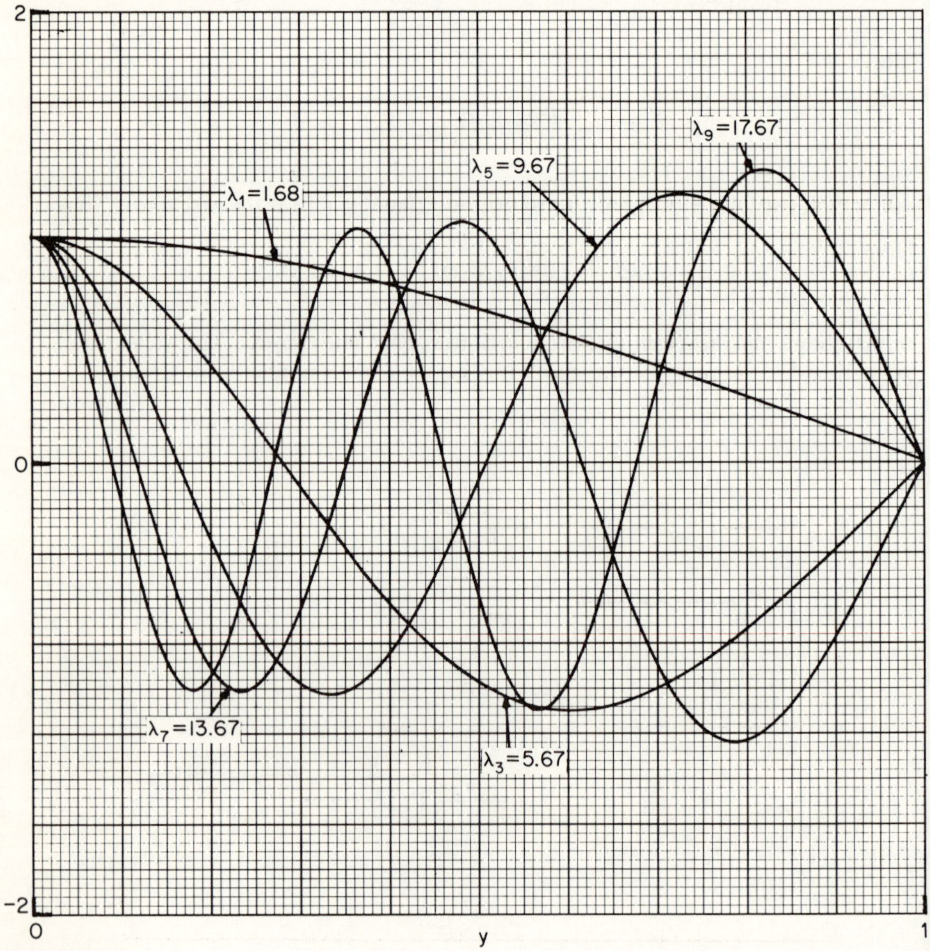

FIG. 14.2 Odd eigenfunctions $\lambda_1 - \lambda_9$.

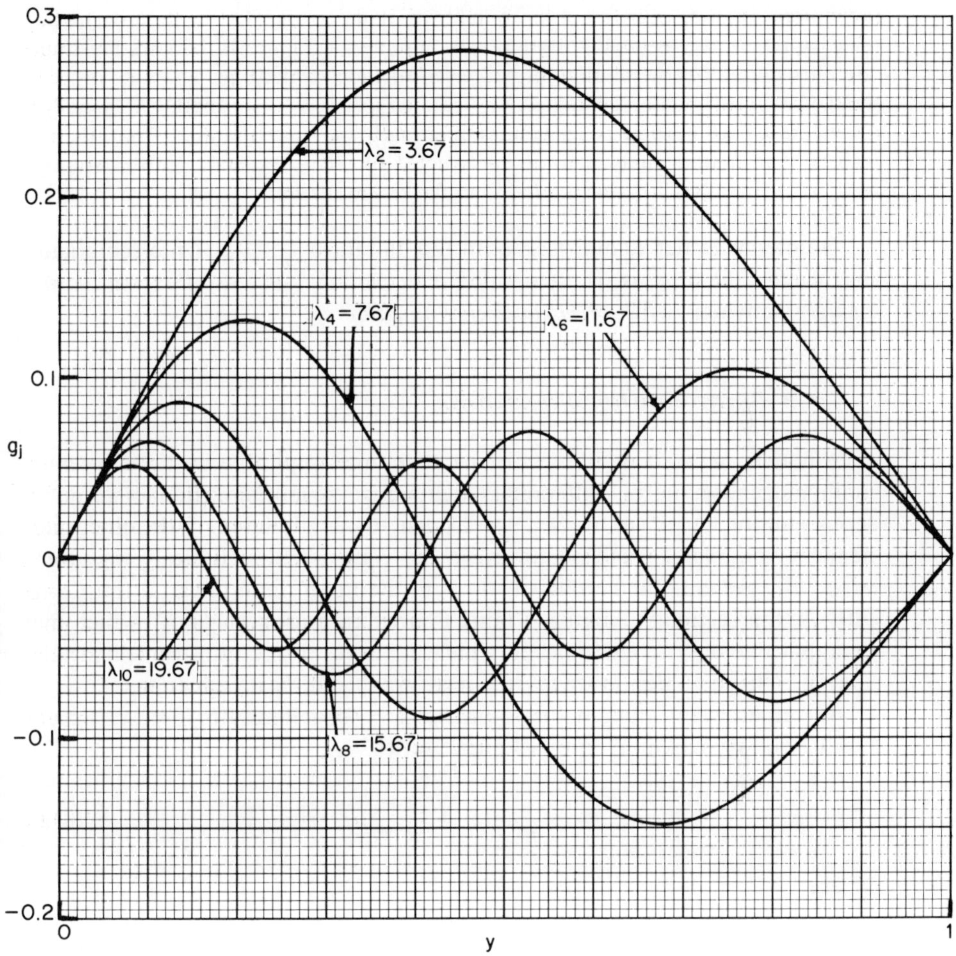

FIG. 14.3 Even eigenfunctions $\lambda_2 - \lambda_{10}$.

Thus we may solve (14.10) for g in the interval $0 \le y \le 1$, starting at $y_0 = 0$ and using $\dot{y} = 1$. For the symmetric cases, we use

$$\left(\frac{dg}{dy}\right)_0 = 0, \qquad g_0 = 1$$

to normalize the functions. For the antisymmetric cases we use

$$\left(\frac{dg}{dy}\right)_0 = 1, \qquad (g)_0 = 0$$

to simplify the mechanization. In both cases, the value of λ_j must be adjusted until the endpoint $g_j(1) = 0$—this is a boundary-value problem. If λ_j^2 is controlled by a potentiometer, these may be easily found by trial and error. (Or the iterative method of Chapter 19 may be used with iterative computers.) Figure 14.1 shows the simple circuit required for the solution of odd eigenfunctions while Figs. 14.2 and 14.3 are the actual eigenfunctions obtained. By reading g_j at $y = 1$ on the DVM, we adjust the values λ_j^2 easily to give eigenvalues differing by no more than 0.01 from the true values. The solution may then be constructed for either constant x or constant y, which is generally all that is required in a larger simulation.

Numerous other examples exist in the literature [7]–[9]. An eigenfunction and eigenvalue analysis is often called a *modal* analysis with the term *normal mode* synonymous with *orthogonal eigenfunction.*

14.3 Finite Fourier Transforms

Another analog computer technique sometimes applicable in the solution of nonhomogeneous partial differential equations is the mechanization of a Fourier series solution $z(x_0, t)$ at a given spatial point x_0 as a function of the other independent variable t. Such solutions can be constructed analytically in several ways, but perhaps the simplest is through the use of finite integral transforms [10]. In this section we discuss a finite Fourier transform [11], [12] which may be applicable only to equations in which the order of the transformed variable is even. This includes, however, the heat, wave, and beam equations.

Given an even function $F(x, t)$ (about $x = 0$) defined for $0 \leq x \leq \pi$ and $t > 0$, we may expand F into a Fourier series

$$F(x, t) = \frac{1}{\pi} a_0(t) + \frac{2}{\pi} \sum_{j=1}^{\infty} a_j(t) \cos jx \tag{14.11}$$

where

$$a_j(t) = \int_0^{\pi} F(x, t) \cos jx \, dx \tag{14.12}$$

(see Chapter 7). The set of coefficients $a_j(t)$ is called the *finite Fourier cosine transform* of $F(x, t)$. Similarly there is a sine transform, and others [11] (see Problem 14-8).

Assuming $\partial^2 F/\partial x^2$ is continuous, we integrate twice by parts and find

$$\int_0^{\pi} \frac{\partial^2 F(x, t)}{\partial x^2} \cos jx \, dx = -j^2 a_j(t) - \frac{\partial F}{\partial x}(0, t) + (-1)^j \frac{\partial F(\pi, t)}{\partial x} \tag{14.13}$$

This will help derive ordinary differential equations for determining the a_j. In turn, the a_j can be mechanized on the computer and the series constructed from (14.11).

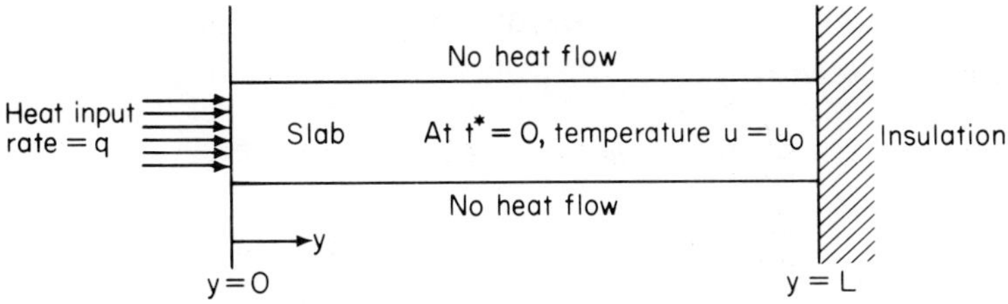

FIG. 14.4 Nose-cone heating idealization.

Example 14.3. Solution of the heat equations for a reentry vehicle [11]. We idealize a missile nose cone as a slab of length L, one section of which is shown in Fig. 14.4. Let y be the length coordinate, t^* be time, and $u(y, t^*)$ be the temperature. The differential equation of heat conduction is given by

$$\frac{\partial u}{\partial t^*} = \alpha \frac{\partial^2 u}{\partial y^2}$$

where α, the thermal diffusivity of the material, is constant. Boundary conditions are determined by what is happening to heat at the boundaries. We assume because of insulation that there is no flow at $y = L$; this requires that

$$\frac{\partial u}{\partial y} = 0 \qquad \text{at } y = L$$

Also, the heat input rate $q(t^*)$ at $y = 0$ is known experimentally:

$$k \frac{\partial u}{\partial y} = -q(t^*)$$

where k is the constant thermal conductivity of the material. Finally, we assume the temperature u_0 at $t^* = 0$ is uniform throughout the slab:

$$u(y, 0) = u_0$$

We wish to determine the temperature $u(0, t^*)$ to see if the material melts, the prime object of the study.

Because the boundary conditions involve the first derivatives $\partial u/\partial y$, the finite Fourier cosine transform is ideal for obtaining the series solution. First, we transform the variables to simplify the mathematics. Let

$$F = u - u_0, \qquad x = \frac{\pi}{L} y, \qquad t = \alpha \frac{\pi^2}{L^2} t^*$$

Then

$$\frac{\partial F}{\partial t} = \frac{\partial^2 F}{\partial x^2}, \qquad 0 \le x \le \pi, \qquad t > 0 \tag{14.14}$$

with the initial condition

$$F(x, 0) = 0$$

and boundary conditions

$$\frac{\partial F}{\partial x} = -\frac{L}{\pi k} q(t^*) = -Q(t) \qquad \text{at } x = 0$$

$$\frac{\partial F}{\partial x} = 0 \qquad \text{at } x = \pi$$

Now multiply both sides of (14.14) by $\cos jx\, dx$ and integrate from 0 to π:

$$\int_0^\pi \frac{\partial F}{\partial t} \cos jx\, dx = \int_0^\pi \frac{\partial^2 F}{\partial x^2} \cos jx\, dx \tag{14.15}$$

Using the series solution (14.11) and (14.12) it is clear that

$$\int_0^\pi \frac{\partial F}{\partial t} \cos jx\, dx = \frac{da_j}{dt}$$

so that (14.15) transforms to [from (14.13)]

$$\frac{da_j}{dt} = -j^2 a_j + Q(t), \qquad (a_j)_0 = 0$$

which are solved on the computer.

If the initial temperature $F(x, 0)$ is not zero, the initial conditions $(a_j)_0$ must be calculated from (14.12). A truncated series solution based on (14.11) can be constructed for any x in the region of interest. The coefficients behave roughly as $1/j^2$ so that usually relatively few terms are required for a given accuracy. Figure 14.5

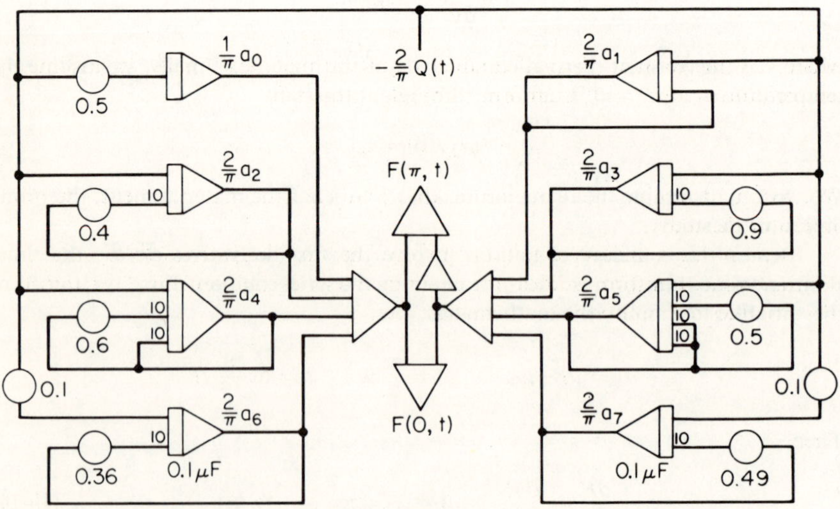

FIG. 14.5 Truncated series mechanization using the finite Fourier cosine transform.

shows a mechanization for the first eight terms of the series solution for $F(\pi, t)$ and $F(0, t)$. The temperature at other points could be mechanized by getting an appropriate weighted sum of the a_j from (14.11).

If the boundary conditions involve functional values of temperature instead of its derivative, the finite Fourier sine transform is appropriate. Similarly, when one boundary condition involves a derivative and the other, the temperature, mixed Fourier transforms can be developed. See Problems 14-8 and 14-9.

14.4 Finite Differences

An extremely powerful approximation method applicable to solving virtually all partial differential equations is the so-called difference-differential equation method first introduced by Hartree and Womersley [13] in connection with mechanical differential analyzers. A programmer must set up a grid of stations for all but one of the independent variables. He must replace derivatives in these variables by finite difference approximations, and derive ordinary differential equations for the unknown functions at these stations in the remaining continuous variable. Boundary and initial conditions must also be taken into account.

In the case of two independent variables x and y, for example, where the dependent variable $z(x, y)$ is defined by a partial differential equation, constant values of x can be distributed in the range of interest. If these are called $x_1, x_2, \ldots, x_n$, and we let

$$z_i = z(x_i, y)$$

an expression of the form $\partial^j z/\partial y^j$ at $x = x_i$ can be replaced by $d^j z_i/dy^j$, while $\partial^j z/\partial x^j$ at any x can be replaced by an appropriate finite difference approximation in terms of the ordinates $z_1, z_2, \ldots, z_n$. An extension to mixed partial derivatives can be made easily.

The primary difficulties that can arise with this scheme are

(1) Determining the amount, n, of stations and location of each x_i for a given accuracy.

(2) Choosing an appropriate finite difference scheme.

(3) Accounting for boundary conditions.

These difficulties are by no means independent. Their interrelationships will be better appreciated by studying the errors in finite difference approximations.

14.4.1 Obtaining finite difference approximations and errors for derivatives

Several methods can obtain finite difference approximations and their errors. Consider, for example, a Taylor series expansion for equally spaced data $z(x_j - h)$, $z(x_j)$, and $z(x_j + h)$, which we shall write as z_{j-1}, z_j, and z_{j+1} (Fig. 14.6),

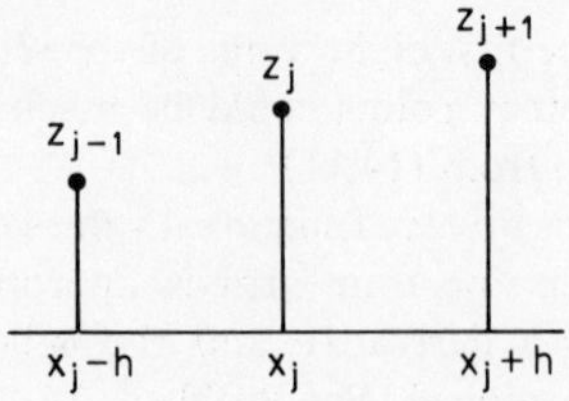

FIG. 14.6 Equally spaced data.

and let us find in terms of these ordinates a three-term approximation formula to $z' = dz/dx$ at $x = x_j$, which we write as z'_j, and its error. Expand z in a Taylor series about $x = x_j$:

$$z = z_j + z'_j(x - x_j) + \frac{z''_j(x - x_j)^2}{2!} + \frac{z'''_j(x - x_j)^3}{3!} + \cdots \tag{14.16}$$

carrying at least an extra term throughout the computation. Now substitute other known data points

$$\text{at } x = x_j + h, \qquad z = z_{j+1}$$
$$\text{at } x = x_j - h, \qquad z = z_{j-1}$$

This leads to

$$\begin{aligned} z_{j+1} &= z_j + hz'_j + \frac{h^2}{2!} z''_j + \frac{h^3}{3!} z'''_j + \cdots \\ z_{j-1} &= z_j - hz'_j + \frac{h^2}{2!} z''_j - \frac{h^3}{3!} z'''_j + \cdots \end{aligned} \tag{14.17}$$

Finally, we eliminate z''_j from the simultaneous equations and solve for z'_j. We then have

$$z'_j = \frac{-z_{j-1} + z_{j+1}}{2h} - \frac{1}{6} h^2 z'''_j + \cdots \tag{14.18}$$

The first term on the right, consisting of a weighted sum of the ordinates, is called a *Lagrangian* or *numerical differentiation* formula. The approximation in (14.18) is said to be of second order because the power of h in the leading error term is 2. It is also known as an *h^2 approximation* as well as a *three-point formula*. (The weight of the central term z_j is 0.)

An approximation based on a Taylor series expansion about the *center* of an equally spaced array is called a *central difference* approximation. If based on expansions about points to the right of center, it is called a *backward difference* approximation, while that based on points to the left of center is called a *forward difference* approximation.

Lagrangian formulas for derivatives at the known functional values for equally spaced data have been tabulated extensively. Tables 14.1–14.4 give

TABLE 14.1

Two-Point Lagrangian Formulas	Leading Error Term
$z_{j+1/2} = (\ z_j + z_{j+1})/2$	$-\frac{h^2}{8} z''_{j+1/2}$
$z'_j = (-z_j + z_{j+1})/h$	$-\frac{h}{2} z''_j$
$z'_{j+1/2} = (-z_j + z_{j+1})/h$	$-\frac{h^2}{24} z'''_{j+1/2}$
$z'_{j+1} = (-z_j + z_{j+1})/h$	$+\frac{h}{2} z''_{j+1}$

TABLE 14.2

Three-Point Lagrangian Formulas	Leading Error Term
$z_{j-1/2} = (\ 3z_{j-1} + 6z_j - z_{j+1})/8$	$+\frac{h^3}{16} z'''_{j-1/2}$
$z_{j+1/2} = (-z_{j-1} + 6z_j + 3z_{j+1})/8$	$-\frac{h^3}{16} z'''_{j+1/2}$
$z'_{j-1} = (-3z_{j-1} + 4z_j - z_{j+1})/(2h)$	$+\frac{h^2}{3} z'''_{j-1}$
$z'_{j-1/2} = (-z_{j-1} + z_j)/h$	$-\frac{h^2}{24} z'''_{j-1/2}$
$z'_j = (-z_{j-1} + z_{j+1})/(2h)$	$-\frac{h^2}{6} z'''_j$
$z'_{j+1/2} = (- z_j + z_{j+1})/h$	$-\frac{h^2}{24} z'''_{j+1/2}$
$z'_{j+1} = (\ z_{j-1} - 4z_j + 3z_{j+1})/(2h)$	$+\frac{h^2}{3} z'''_{j+1}$

TABLE 14.3

Four-Point Lagrangian Formulas	Leading Error Term
$z_{j-1/2} = (\ 5z_{j-1} + 15z_j - 5z_{j+1} + z_{j+2})/16$	$-\frac{5h^4}{128} z^{(IV)}_{j-1/2}$
$z_{j+1/2} = (-z_{j-1} + 9z_j + 9z_{j+1} - z_{j+2})/16$	$+\frac{3h^4}{128} z^{(IV)}_{j+1/2}$
$z_{j+3/2} = (\ z_{j-1} - 5z_j + 15z_{j+1} + 5z_{j+2})/16$	$-\frac{5h^4}{128} z^{(IV)}_{j+3/2}$
$z'_{j-1} = (-11z_{j-1} + 18z_j - 9z_{j+1} + 2z_{j+2})/(6h)$	$-\frac{h^3}{4} z^{(IV)}_{j-1}$
$z'_{j-1/2} = (-23z_{j-1} + 21z_j + 3z_{j+1} - z_{j+2})/(24h)$	$+\frac{h^3}{24} z^{(IV)}_{j-1/2}$
$z'_j = (-2z_{j-1} - 3z_j + 6z_{j+1} - z_{j+2})/(6h)$	$+\frac{h^3}{12} z^{(IV)}_j$
$z'_{j+1/2} = (\ z_{j-1} - 27z_j + 27z_{j+1} - z_{j+2})/(24h)$	$+\frac{3h^4}{640} z^{(V)}_{j+1/2}$
$z'_{j+1} = (\ z_{j-1} - 6z_j + 3z_{j+1} + 2z_{j+2})/(6h)$	$-\frac{h^3}{12} z^{(IV)}_{j+1}$
$z'_{j+3/2} = (\ z_{j-1} - 3z_j - 21z_{j+1} + 23z_{j+2})/(24h)$	$-\frac{h^3}{24} z^{(IV)}_{j+3/2}$
$z'_{j+2} = (-2z_{j-1} + 9z_j - 18z_{j+1} + 11z_{j+2})/(6h)$	$+\frac{h^3}{4} z^{(IV)}_{j+2}$

TABLE 14.4

Five-Point Lagrangian Formulas	Leading Error Term
$z_{j-3/2} = (35z_{j-2} + 140z_{j-1} - 70z_j + 28z_{j+1} - 5z_{j+2})/128$	$+\frac{7h^5}{256} z^{(V)}_{j-3/2}$
$z_{j-1/2} = (-5z_{j-2} + 60z_{j-1} + 90z_j - 20z_{j+1} + 3z_{j+2})/128$	$-\frac{3h^5}{256} z^{(V)}_{j-1/2}$
$z_{j+1/2} = (3z_{j-2} - 20z_{j-1} + 90z_j + 60z_{j+1} - 5z_{j+2})/128$	$+\frac{3h^5}{256} z^{(V)}_{j+1/2}$
$z_{j+3/2} = (-5z_{j-2} + 28z_{j-1} - 70z_j + 140z_{j+1} + 35z_{j+2})/128$	$-\frac{7h^5}{256} z^{(V)}_{j+3/2}$
$z'_{j-5/2} = (-93z_{j-2} + 229z_{j-1} - 225z_j + 111z_{j+1} - 22z_{j+2})/(24h)$	$+\frac{563h^4}{640} z^{(V)}_{j-5/2}$
$z'_{j-2} = (-25z_{j-2} + 48z_{j-1} - 36z_j + 16z_{j+1} - 3z_{j+2})/(12h)$	$+\frac{h^4}{5} z^{(V)}_{j-2}$
$z'_{j-3/2} = (-22z_{j-2} + 17z_{j-1} + 9z_j - 5z_{j+1} + z_{j+2})/(24h)$	$-\frac{71h^4}{1920} z^{(V)}_{j-3/2}$
$z'_{j-1} = (-3z_{j-2} - 10z_{j-1} + 18z_j - 6z_{j+1} + z_{j+2})/(12h)$	$-\frac{h^4}{12} z^{(V)}_{j-1}$
$z'_{j-1/2} = (z_{j-2} - 27z_{j-1} + 27z_j - z_{j+1})/(24h)$	$+\frac{3h^4}{640} z^{(V)}_{j-1/2}$
$z'_j = (z_{j-2} - 8z_{j-1} + 8z_{j+1} - z_{j+2})/(12h)$	$+\frac{h^4}{30} z^{(V)}_j$
$z'_{j+1/2} = (+ z_{j-1} - 27z_j + 27z_{j+1} - z_{j+2})/(24h)$	$+\frac{3h^4}{640} z^{(V)}_{j+1/2}$
$z'_{j+1} = (-z_{j-2} + 6z_{j-1} - 18z_j + 10z_{j+1} + 3z_{j+2})/(12h)$	$-\frac{h^4}{12} z^{(V)}_{j+1}$
$z'_{j+3/2} = (-z_{j-2} + 5z_{j-1} - 9z_j - 17z_{j+1} + 22z_{j+2})/(24h)$	$-\frac{71h^4}{1920} z^{(V)}_{j+3/2}$
$z'_{j+2} = (3z_{j-2} - 16z_{j-1} + 36z_j - 48z_{j+1} + 25z_{j+2})/(12h)$	$+\frac{h^4}{5} z^{(V)}_{j+2}$
$z'_{j+5/2} = (22z_{j-2} - 111z_{j-1} + 225z_j - 229z_{j+1} + 93z_{j+2})/(24h)$	$+\frac{563h^4}{640} z^{(V)}_{j+5/2}$

first derivative and ordinate formulas and errors for 2-, 3-, 4-, and 5-point data. Higher derivative expressions could also be obtained but will not be needed in this chapter. Note that the order of the error need not be the same for all formulas within one group—central difference formulas are, in general, more accurate than either forward or backward formulas. Note also that approximations can be obtained at noninteger stations by expanding the Taylor series about these stations.

Approximations will also be useful for functional values and their derivatives for data not equally spaced. We obtain a 3-point formula for $z'_{1/2}$ for a configuration shown in Fig. 14.7. Such a configuration is useful to approximate left-hand boundary conditions in some analog problems. Using (14.16) with $j = \frac{1}{2}$, we obtain

$$z_1 = z_{1/2} + \frac{h}{2} z'_{1/2} + \left(\frac{h}{2}\right)^2 \frac{z''_{1/2}}{2!} + \left(\frac{h}{2}\right)^3 \frac{z'''_{1/2}}{3!} + \cdots$$

$$z_2 = z_{1/2} + \frac{3h}{2} z'_{1/2} + \left(\frac{3h}{2}\right)^2 \frac{z''_{1/2}}{2!} + \left(\frac{3h}{2}\right)^3 \frac{z'''_{1/2}}{3!} + \cdots$$

and calculate, by multiplying the first row by 9 and subtracting, that

$$z'_{1/2} = \frac{-8z_{1/2} + 9z_1 - z_2}{3h} + \frac{h^2}{8} z'''_{1/2} + \cdots \tag{14.19}$$

The error is of order two.

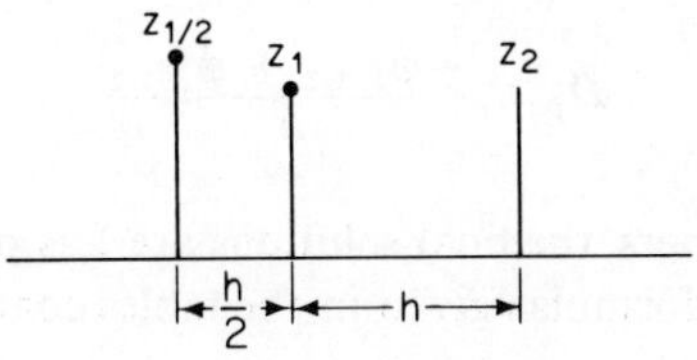

FIG. 14.7 Arrangement typical for some boundary conditions.

14.4.2 Rules for mechanization

Before considering examples solved on the computer by finite difference schemes, we list three rules to aid the mechanization. These evolved through analysis and experience and will be clarified in the examples presented.

Rule 14.1. To obtain a kth-order solution, all approximations must be of kth order including those accounting for boundary conditions.

This is actually a definition of a kth-order solution. It stems from considering how an error introduced is propagated when replacing a derivative (say in an ordinary differential equation) by a difference approximation of order k. From a Taylor expansion about the initial conditions, we find the error term is also of order k.

Rule 14.2. If only even derivatives of a dependent variable (such as u, u'', etc.) are specified at a boundary, arrange the grid stations so that an integer station (say x_0 or x_1) appears at the boundary. If at least one odd derivative (u', u''', etc.) is specified at a boundary, a half-integer station (say $x_{1/2}$) should be placed at the boundary.

The reasons for Rule 14.2 are convenience and better accuracy as will become apparent in subsequent examples.

The next rule listed may seem strange to readers oriented in digital computer applications of finite difference schemes. In approximating say a second derivative $D = u''$ by finite differences, we may show that a second-order approximation is

$$D_j = u''_j \simeq \frac{u_{j-1} - 2u_j + u_{j+1}}{h^2}$$

which could be implemented directly on an analog computer also. However, Clymer [14] and Fisher [15] found that computer solutions are less sensitive to error when high-order derivatives are not formed directly as summer outputs. Better results are obtained by first obtaining all lower-order derivatives as summer outputs using first-difference formulas. Thus, if we define $\phi = u'$ and mechanize

$$\phi_{j-1/2} \simeq \frac{-u_{j-1} + u_j}{h}, \qquad \phi_{j+1/2} \simeq \frac{-u_j + u_{j+1}}{h}$$

$$D_j \simeq \frac{-\phi_{j-1/2} + \phi_{j+1/2}}{h}$$

as outputs of three summers, the final solutions are less prone to error. For this reason, the Lagrangian formulas given in the tables contain only first derivative approximations.

Rule 14.3. Generate high-order derivatives with first-order derivative approximations, mechanizing all lower-order derivatives as summer outputs.

Rule 14.3 supersedes the general equipment rule (Chapter 4) since more amplifiers will be required. The rule is broken when not enough amplifiers are available for a given problem or if it is known that solutions are insensitive to error.

We shall now examine second-order solutions to three classical problems and obtain more insight into the insertion of boundary conditions.

14.4.3 Heat equation [16]–[20]

A general form of the heat equation in one dimension is given by

$$C(x)\frac{\partial u}{\partial t} = \frac{\partial}{\partial x}\left[K(x)\frac{\partial u}{\partial x}\right] + q(x, t) \tag{14.20}$$

where

$u(x, t)$ = temperature

$C(x)$ = heat capacity

$K(x)$ = thermal conductivity

$q(x, t)$ = rate of heat supplied by sources in the medium

$K(x)(\partial u/\partial x)$ = heat flux

To solve this or any equation by a finite difference approach, we must consider first the boundary conditions of the problem. Figure 14.4 shows a case where du/dt is known at each boundary. But to demonstrate the use of Rule 14.2, we choose an example that contains an even-order boundary condition at one end and an odd-order boundary condition at the other end.

Example 14.4. Let a slab of nonuniform material of length L, initially at uniform temperature T_0, be insulated at the right end and contact a constant temperature heat sink (say ice) at 0° at the left end. Determine the temperature distribution by a finite difference approach using 12 stations and approximation formulas with error of order 2. Assume $C(x)$ and $K(x)$ vary continuously, and that $q(x, t)$ is a known heat rate function.

The boundary conditions reduce to

$$u = 0 \text{ at the left end}$$

$$u' = 0 \text{ at the right end}$$

According to Rule 14.2, an integer station is placed at the left end, and a half-integer one at the right end.

Figure 14.8 shows the grid placement of the 12 x_j's. We first compute $h = L/12.5$

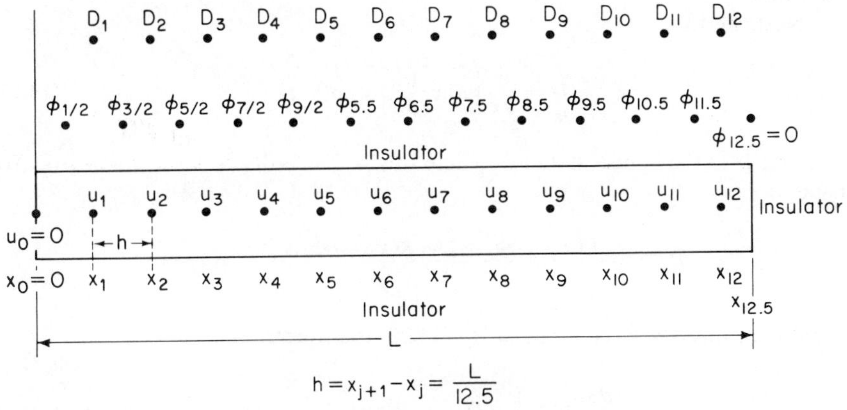

FIG. 14.8 Example 14.4.

to locate u_{12} at a distance $h/2$ from the right-hand boundary and u_1 at a distance h from the left-hand boundary. Now, we let

$$\phi = \frac{\partial u}{\partial x} \tag{14.21}$$

and determine the ϕ stations at half-integer values to take advantage of slightly greater accuracy than for other methods. Figure 14.9 now shows how $\phi_{j+1/2}$ can be computed with second-order error from Table 14.1 or 14.2. We have

$$u'_{j+1/2} = \phi_{j+1/2} \simeq \frac{-u_j + u_{j+1}}{h}$$

Since $u_0 = 0$ is the given left-end boundary condition, $\phi_{1/2}$ can be computed in terms of u_1. Also $\phi_{12.5}$ is known to be zero. Thus, our placing of the right-hand boundary

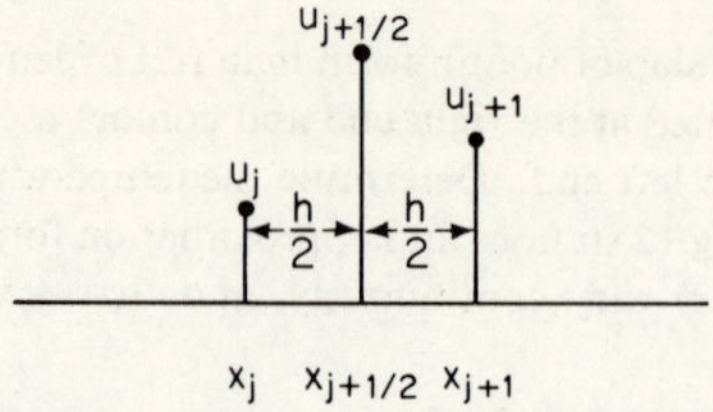

FIG. 14.9 Computing $\phi = \partial u/\partial x$ at half-integer stations.

at a half-integer station simplifies computing all required ϕ variables with second-order error. Without this arrangement, we would need either a fictitious point beyond the boundary or change to a less accurate backward difference formula at the boundary.

Next, we let

$$D = \frac{\partial}{\partial x}\left[K(x)\frac{\partial u}{\partial x}\right] + q(x, t) \tag{14.22}$$

and find D at the integer stations because this is the feedback term in (14.20). We have

$$D_j \simeq \frac{-K(x_{j-1/2})\phi_{j-1/2} + K(x_{j+1/2})\phi_{j+1/2}}{h} + q(x_j, t)$$

Finally, because $\partial u/\partial t$ becomes du/dt, (14.20) transforms to

$$\frac{du_j}{dt} \simeq \frac{1}{C(x_j)} D_j, \qquad j = 1, 2, \ldots, 12$$

a set which is mechanized easily. Figure 14.10 is one variation of such a mechanization. The initial temperature need not have been constant but could have just as easily been taken into account. This method produces more accurate results than do alternative methods (see Problem 14-10).

Howe and Haneman [20] analyzed the accuracy of the above method when $C(x)$ and $K(x)$ where constant, and $q(x, t) = 0$. By separation of variables, we may determine the exact eigenvalues of the resulting sinusoidal eigenfunctions. On the other hand, the solution of the difference equations contains only the same number of eigenvalues as there are stations, and these are slightly larger than the theoretical ones. The per cent error increases with the order of the eigenvalue. Since high-order terms are rarely important in practical problems, Howe and Haneman conclude that 8 to 16 stations should be adequate for most applications. Actually, the accuracy could be improved or the number of stations decreased by obtaining fourth-order solutions as discussed subsequently.

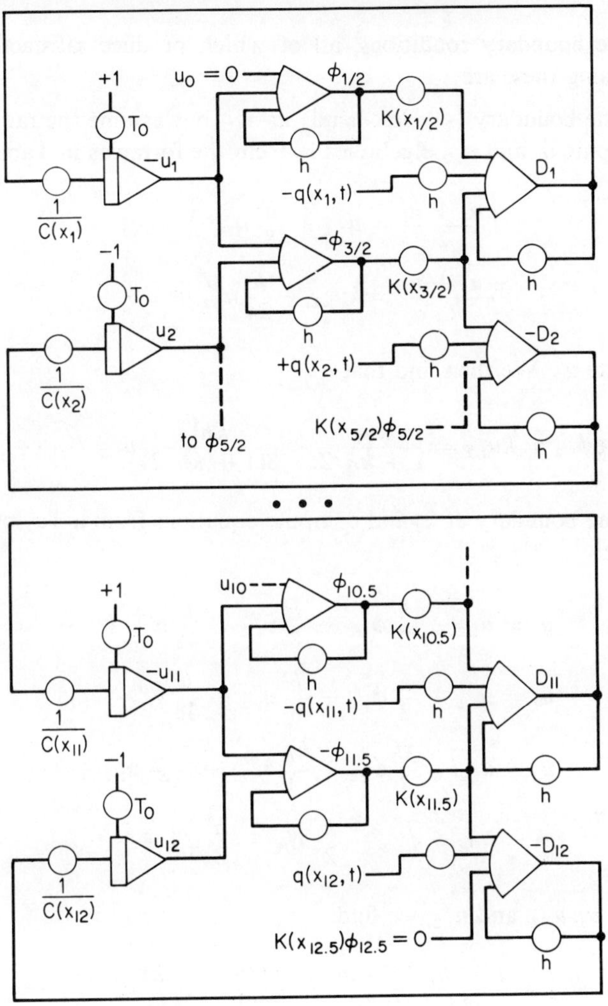

FIG. 14.10 A 12-station mechanization of the one-dimensional heat equation.

14.4.4 Boundary conditions

Example 14.4 illustrated the station placement if either u or $\partial u/\partial x$ is given at a boundary. There are cases, however, both in heat transfer problems and in connection with other partial differential equations, in which both functions are needed at a boundary. Or, in the general case, an even- and odd-order derivative must be computed for the feedback term. In a heat transfer problem, for example, the heat flux may be proportional to the temperature (convective transfer) or proportional to u^4 (radiation transfer) at the boundary. Let us assume the former: $\partial u/\partial x = ku$ at the boundary. There are then several ways to insert the boundary conditions, all of which produce satisfactory results. Notable among these are:

1. Set the boundary (say left-hand) at $x_{1/2}$ but assume the range extends* to x_0. Compute u_0 and $u_{1/2}$ algebraically from the formulas in Table 14.1

$$\frac{u_0 + u_1}{2} = u_{1/2} + \frac{h^2}{8} u''_{1/2} + \cdots$$

$$\frac{-u_0 + u_1}{h} = k u_{1/2} + \frac{h^2}{24} u'''_{1/2} + \cdots$$

and eliminate u_0. We then find that

$$\phi_{1/2} = k u_{1/2} = \frac{k u_1}{1 + kh/2} - \frac{kh^2}{8(1 + kh/2)} u''_{1/2} + \cdots \qquad (14.23)$$

2. Set the boundary at x_0 and compute equations from a Taylor expansion at station $\frac{1}{2}$:

$$u_0 = u_{1/2} - \frac{h}{2}\phi_{1/2} + \frac{h^2}{8} u''_{1/2} - \frac{h^3}{48} u'''_{1/2} + \cdots$$

$$u_1 = u_{1/2} + \frac{h}{2}\phi_{1/2} + \frac{h^2}{8} u''_{1/2} + \frac{h^3}{48} u'''_{1/2} + \cdots$$

$$u_2 = u_{1/2} + \frac{3h}{2}\phi_{1/2} + \frac{9h^2}{8} u''_{1/2} + \frac{27h^3}{48} u'''_{1/2} + \cdots$$

$$\phi = k u_0 = \phi_{1/2} - \frac{h}{2} u''_{1/2} + \frac{h^2}{8} u'''_{1/2} + \cdots$$

Eliminating u_0, $u_{1/2}$, and $u''_{1/2}$, we find

$$\phi_{1/2} = \frac{u_2 - u_1 + 2khu_1}{h(3 + 2kh)} - \frac{h^2(19 + 2kh)}{24(3 + 2kh)} u'''_{1/2} + \cdots \qquad (14.24)$$

3. Set the boundary at $x_{1/2}$ and compute $\phi_{1/2}$ from (14.19)

$$\phi_{1/2} = k u_{1/2} = \frac{-8u_{1/2} + 9u_1 - u_2}{3h} + \frac{h^2}{8} u'''_{1/2} + \cdots$$

or

$$\phi_{1/2} = \frac{k(9u_1 - u_2)}{3kh + 8} + \frac{3kh^3}{8(3kh + 8)} u'''_{1/2} + \cdots \qquad (14.25)$$

* The use of a fictitious point x_0 is somewhat artificial, but identical results can be obtained more simply with a Taylor expansion about $x = x_{1/2}$.

In all cases, the solution of the equations approaches the actual solution as $h \to 0$. But (14.23) and (14.24) are not quite second-order errors. Accordingly, we use method 3 consistently whenever mixed boundary conditions are given. We set the boundary at a half-integer station according to Rule 14.2 and compute the required feedback variables algebraically at that station using (14.19) or its equivalent right-hand boundary expression

$$u'_{j+1/2} = \frac{8u_{j+1/2} - 9u_j + u_{j-1}}{3h} - \frac{h^2}{8} u'''_{j+1/2} + \cdots$$

This procedure is applicable for a general partial differential equation and for maintaining continuity equations for discontinuous materials.

Example 14.5. The slab consists of material with discontinuous $K(x)$. Set up continuity condition at the boundary of discontinuity.

We prearrange the boundary at a half-integer station, say $n+\frac{1}{2}$. The continuity equation takes the form that the heat flux is constant over the boundary:

$$K_L\left(\frac{\partial u}{\partial x}\right)_L = K_R\left(\frac{\partial u}{\partial x}\right)_R \tag{14.26}$$

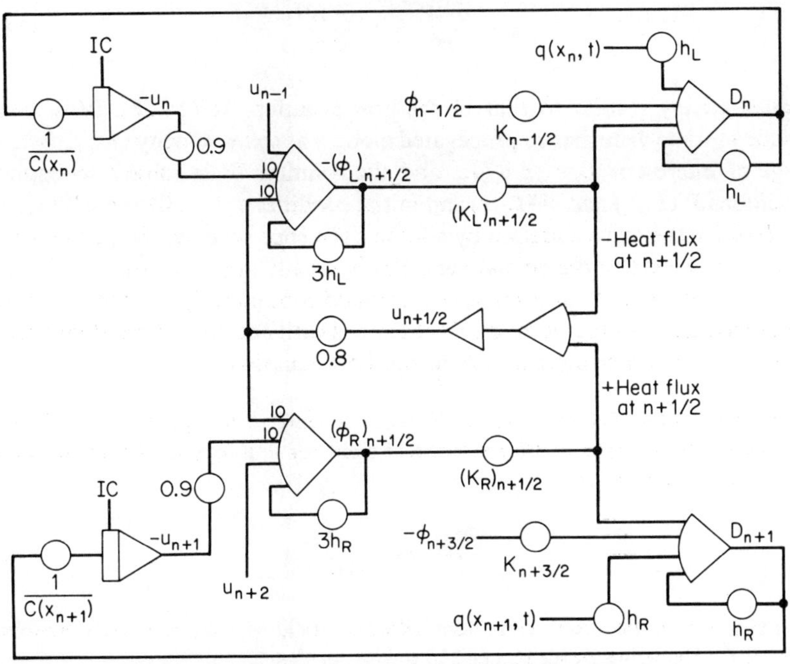

FIG. 14.11 Heat flux diagram at a discontinuous boundary.

where the subscripts L and R stand for values just to the left- and right-hand sides of the boundary. Hence

$$(K_L)_{n+1/2}\left(\frac{8u_{n+1/2}-9u_n+u_{n-1}}{3h_L}\right)\simeq(K_R)_{n+1/2}\left(\frac{-8u_{n+1/2}+9u_{n+1}-u_{n+2}}{3h_R}\right) \tag{14.27}$$

is algebraically used to solve for $u_{n+1/2}$ while either side can be taken as the heat flux at station $n+\frac{1}{2}$. Different station spacings can be used on either side of the boundary. Figure 14.11 shows an implicit circuit for mechanizing $u_{n+1/2}$ in (14.27) together with the variables at adjacent stations.

The application of finite difference techniques to solving heat transfer problems is easily extended to symmetrical situations involving polar and spherical coordinates as well as more than one spatial coordinate [17]. In the latter case, a grid of stations in two or more dimensions must be used, which can tax the amplifier supply of even large installations. For this reason, two-dimensional heat transfer problems, requiring $m \times n$ stations, are done with relatively small m and n, giving just minimal engineering information.

14.4.5 Wave equation [20]

Treatment of the wave equation in one dimension

$$\frac{\partial^2 W}{\partial t^2}=v^2\frac{\partial^2 W}{\partial x^2} \tag{14.28}$$

is qualitatively similar to that of the heat equation. In (14.28), W is the amplitude of some disturbance propagated along x at some velocity $v(x, t, W)$. The range of interest is $L_1 \leq x \leq L_2$, and the solution W is subject to boundary conditions $W(L_1, t)$ and $W(L_2, t)$ and initial conditions $W(x, 0)$ and $\partial W(x, 0)/\partial t$.

To solve the wave equation by a finite difference scheme, the partial derivatives with respect to the spatial variables is usually approximated by finite differences while the W_i at each station is found continuously in time t. This leads to coupled second-order differential equations with t as the independent variable, instead of first-order equations as in the heat equation.

Example 14.6. Set up a finite difference scheme to solve for the displacement $D(x, t)$ of a stretched nonuniform string of mass per unit length $m(x)$ under constant tension T. The equation is

$$\frac{\partial^2 D}{\partial t^2}=\frac{T}{m(x)}\frac{\partial^2 D}{\partial x^2}$$

with $0 \leq x \leq L$, boundary conditions $D(0, t) = D(L, t) = 0$ and initial conditions to provide for the string being plucked in the center

$$D(x, 0) = \frac{D_0 x}{L} \qquad \text{for } 0 \le x \le L/2$$

$$= D_0\left(1 - \frac{x}{L}\right) \qquad \text{for } L/2 \le x \le L$$

$$\frac{\partial D(x, 0)}{\partial t} = 0, \qquad D_0 \ll 1$$

We proceed as in the heat equation by placing stations at $x = 0$ and $x = L$ making it easy to find all variables needed (Rule 14.2). Let

$$V = \frac{\partial D}{\partial x}, \qquad S = \frac{\partial V}{\partial x}$$

We now have

$$V_{j+1/2} \simeq \frac{-D_j + D_{j+1}}{h}, \qquad S_j \simeq \frac{-V_{j-1/2} + V_{j+1/2}}{h}, \qquad \frac{d^2 D_j}{dt^2} \simeq \frac{T}{m(x_j)} S_j$$

for all stations except at the boundaries. These equations require about five amplifiers per station (Fig. 14.12). Only three per station are required, however, if we choose to violate Rule 14.3 and sum at the integrator input instead of forming the S_j. More computation error can then be expected.

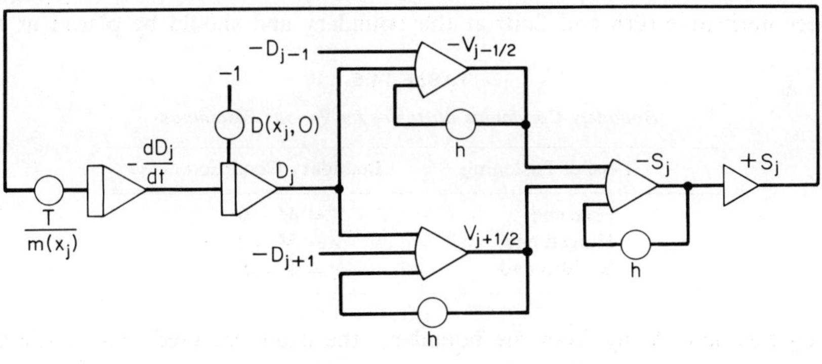

FIG. 14.12 Wave equation loop.

The errors of the normal mode frequencies are qualitatively similar to decay constant errors obtained for the heat equation [20].

14.4.6 Vibrating beam equation [14], [15], [20]–[22]

The equation describing the dynamic transverse displacement y of a beam of length L is given by

$$\rho(x)\frac{\partial^2 y}{\partial t^2} + \frac{\partial^2}{\partial x^2}\left[EI(x)\frac{\partial^2 y}{\partial x^2}\right] = f(x, t) \tag{14.29}$$

where

$$\begin{aligned} x &= \text{distance along the beam} \\ t &= \text{time} \\ EI(x) &= \text{flexural rigidity at } x \\ \rho(x) &= \text{mass per unit length at } x \\ f(x, t) &= \text{external load (force per unit length along the beam)} \\ M(x, t) &= EI(x)(\partial^2 y/\partial x^2) = \text{bending moment} \\ V(x, t) &= \partial M/\partial x = \text{shear force} \end{aligned}$$

Initial conditions $y(x, 0)$ and $\partial y(x, 0)/\partial t$ are given, and the boundary conditions depend upon how the beam is fastened. Table 14.5 shows these conditions for various fastenings with the definition

$$\phi(x, t) = \frac{\partial y(x, t)}{\partial x}$$

A free end or a built-in end fastening requires mixed boundary conditions so the end should be set up at a half-integer station. A hinged end specifies even-order derivative (0th and 2nd) at the boundary and should be placed at an integer station. Away from the boundary, the equations used to approximate a solution to (14.29) with second-order error are

TABLE 14.5

Boundary Conditions of Beams for Various Fastenings

Type of Fastening	Boundary Condition
Free end	$V = M = 0$
Hinged end	$y = M = 0$
Built-in end	$y = \phi = 0$

$$\phi_{j+1/2} \simeq \frac{-y_j + y_{j+1}}{h}, \qquad M_j \simeq EI(x_j)\frac{-\phi_{j-1/2} + \phi_{j+1/2}}{h}$$

$$V_{j+1/2} \simeq \frac{-M_j + M_{j+1}}{h}, \qquad D_j \simeq \frac{-V_{j-1/2} + V_{j+1/2}}{h}$$

$$\frac{d^2 y_j}{dt^2} \simeq \frac{1}{\rho(x_j)}[f(x_j, t) - D_j]$$

Near a boundary, however, some of the required terms are not available. For

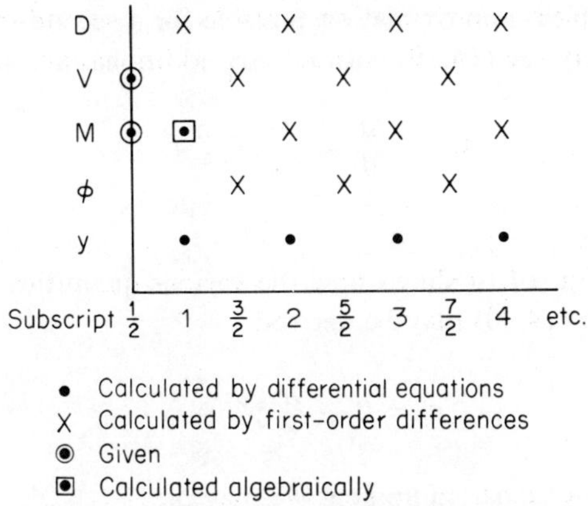

FIG. 14.13 Calculations for a free end boundary.

example, if the left-hand end is free, we place it at station $x_{1/2}$ because $V_{1/2} = M_{1/2} = 0$ are mixed boundary conditions. Now, if D_1 were calculated with the above formulas, the fictitious point y_0 (unavailable) would be needed. To see better which variables can and cannot be generated by the above formulas, it is a good idea to draw a map of calculations. Such a map is shown in Fig. 14.13. Subscripts (or stations) are used for the abscissa, and the hierarchy of successive derivative symbols for the ordinate. The value represented by each cross in the diagram is obtained by central difference formulas using two adjacent symbols below it. The values of the other symbols are obtained as indicated in the diagram.

As suggested by the map, if M_1 is calculated algebraically in some special way, y_0 is not needed. This is the approach taken in this case. It is emphasized that any given map is not unique. We could generate $\phi_{1/2}$ with second-order error using a one-sided formula involving y_1, y_2, and y_3. The following method, however, uses the derivative information and is slightly more accurate.

Let $x_{1/2}$ be the station of the end for either a free or built-in end. We then have

Free end. $V_{1/2} = M_{1/2} = 0$

Make a Taylor series expansion about $x = x_{1/2}$.

$$M_1 = M_{1/2} + \frac{h}{2}M'_{1/2} + \frac{h^2}{8}M''_{1/2} + \cdots$$

Since $M'_{1/2} = V_{1/2} = 0$, we have

$$M_1 = 0 + \frac{h^2}{8}M''_{1/2} + \cdots \tag{14.30}$$

which is the simplest approximation possible for a second-order computation. We may alternately use (14.19) without any additional amplifiers and find

$$M_1 = \frac{M_2}{9} - \frac{h^3}{24} M'''_{1/2} + \cdots \tag{14.31}$$

Built-in end. $\phi_{1/2} = y_{1/2} = 0$

The map in Fig. 14.14 shows how the various quantities are calculated. A formula similar to (14.30) may be derived

$$y_1 = 0 + \frac{h^2}{8} y''_1 + \cdots \tag{14.32}$$

or, without any additional equipment we may use

$$y_1 = \frac{y_2}{9} - \frac{h^3}{24} y'''_{1/2} + \cdots \tag{14.33}$$

The variable D_1 is not required because y_1 is calculated algebraically, not by a differential equation. Mirror images of each map are required for right-hand boundaries.

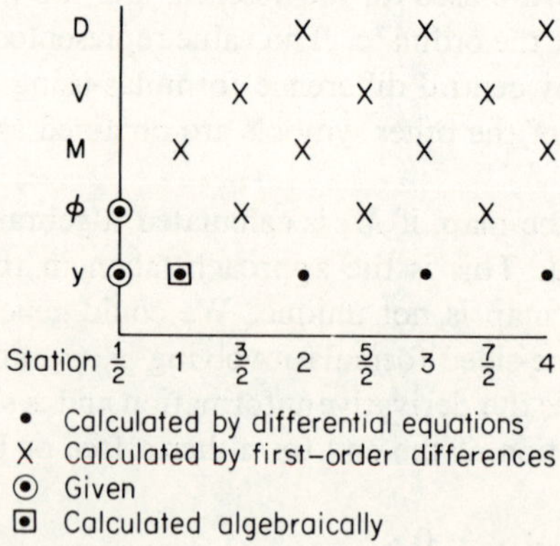

FIG. 14.14 Calculations for a built-in boundary.

The amplitude for each central station is generated according to Fig. 14.15 by using five amplifiers per station. These may be reduced to three amplifiers per station [21], but the solution is then somewhat more sensitive to small errors (Rule 14.3 is violated).

The accuracy of the finite difference solution can be compared with the true solution for simple situations that are easy to calculate. Theoretical normal mode frequencies λ_j for a continuous uniform cantilever beam [one end built-in and

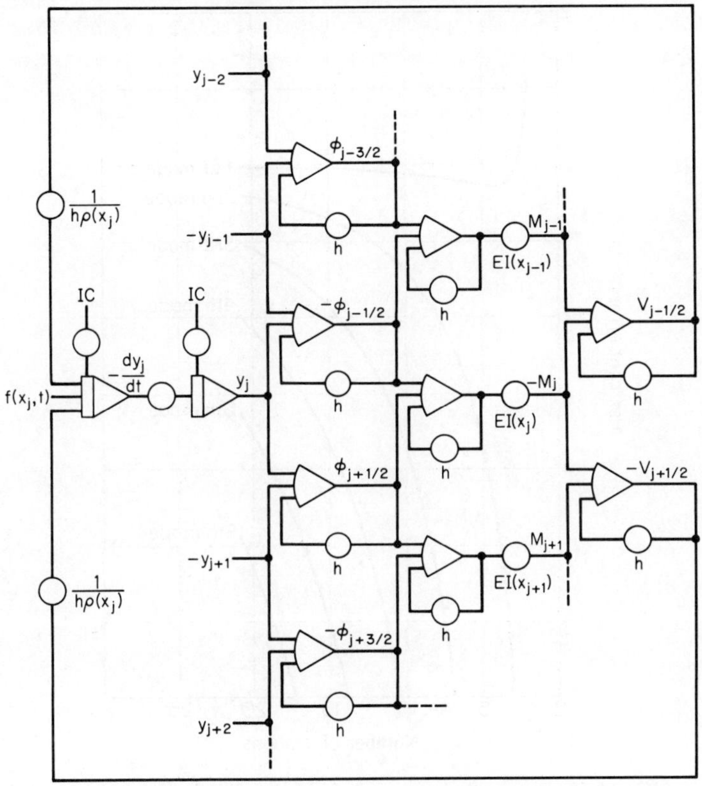

FIG. 14.15 Vibrating beam circuit for one station.

the other free with $EI(x) = 1$ and $\rho(x) = 1$] are given by the solution to the transcendental equation [21]

$$\cos \sqrt{\lambda_j} \cosh \sqrt{\lambda_j} + 1 = 0$$

The first six modes are found to be 3.51602, 22.0345, 61.6972, 120.902, 199.860, and 298.556. On the other hand, we may determine the $n - 1$ eigenvalues of the matrix $\mathbf{E}$ for the linear system

$$\frac{d^2\mathbf{y}}{dt^2} + \mathbf{E}\mathbf{y} = \mathbf{0} \tag{14.34}$$

where

$$\mathbf{y} = \begin{pmatrix} y_2 \\ y_3 \\ \cdot \\ \cdot \\ \cdot \\ y_n \end{pmatrix}$$

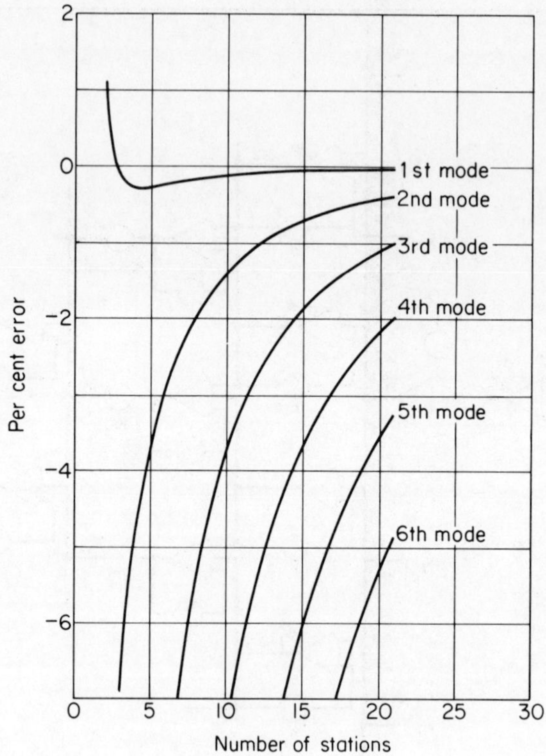

FIG. 14.16 Per cent error in mode frequency of a cantilever beam, second-order approximation.

which results from the finite difference mechanization with n stations. These eigenvalues are proportional to the square of the natural frequencies of the system and are compared readily with those of the continuous case. Figure 14.16 shows how the error of the finite difference mechanization changes with n for the first six normal modes using the third-order boundary equations (14.31) and (14.33). The mechanization of the boundary conditions differs in this graph from those of Howe and Howe [21] who used second-order boundary equations (14.30) and (14.32), so the error graph is different. The mechanization described here produces much better results for the fundamental frequency—an error of no worse than 0.3 per cent for $n \geq 3$,—but slightly poorer results for the higher modes.

14.5 Fourth-Order Error Mechanizations

Tables 14.3 and 14.4 show approximations for ordinates and first derivatives using 4- and 5-point equally spaced data. These are rarely required for computations on a digital computer, since it is too easy to reduce h and compute with more stations to improve accuracy. On an analog computer, however, such a procedure leads to excessive equipment requirements. Fisher [23], and then Greenwood [24], investigated the use of higher-order approximations. Three items are of concern when using higher-order approximations:

1. Spurious solutions are introduced into the computer solution [23]. These can be made negligible by a suitable choice of boundary equations.

2. The first-derivative approximation for points away from a boundary must be determined. Fourth-order first-derivative approximations, in general, require five-point formulas. But the central difference four-point formula at the half-integer station (Table 14.3) is equivalent to one-sided five-point formulas near the center. As Greenwood [24] noted, the terms of the four-point formula successively alternate in sign, which makes their mechanization very efficient. Their accuracy is also almost ten times as great as the central five-point formula. For this reason the third-order solutions are rarely computed—fourth-order solutions are just as easy to compute.* We therefore take as the basic approximation

$$u'_{j+1/2} \simeq \frac{u_{j-1} - 27u_j + 27u_{j+1} - u_{j+2}}{24h} + \frac{3h^4}{640} u^{(V)}_{j+1/2} \tag{14.35}$$

3. Required ordinates in (14.35) are not available near boundaries so that one-sided differences must be used. Also, special boundary conditions can be taken into account by specially computing the required approximation formula. Care must be taken at all times to use approximations no worse than fourth order (Rule 14.1). This is discussed in greater detail later.

Because (14.35) computes derivatives at half-integer stations, it is still convenient to compute derivatives successively by staggering: even-order finite differences are computed at integer stations, and odd-order derivatives at half-integer stations. Also, boundaries having these derivatives specified are placed at integer or half-integer stations with a half-integer station being slightly more accurate in the case of mixed boundary conditions.

Example 14.7. Repeat the heat transfer problem (Example 14.4) obtaining fourth-order solutions.

With the same definitions of ϕ as in (14.21) and D as in (14.22), we again let $h = L/12.5$ to enable the right-hand boundary to be placed at a half-integer station with $\phi_{12.5} = 0$, and an integer station $x = 0$ at the left-hand boundary with $u_0 = 0$. Figure 14.17 shows a map of how each point is calculated. Four symmetrical points below a cross are needed now for central difference formulas, and five points for one-sided formulas. For example, since $u_0 = 0$, the five-point formula (Table 14.4) reduces to

$$\phi_{1/2} \simeq \frac{17u_1 + 9u_2 - 5u_3 + u_4}{24h}$$

* Provided an even-order derivative is approximated by finite differences.

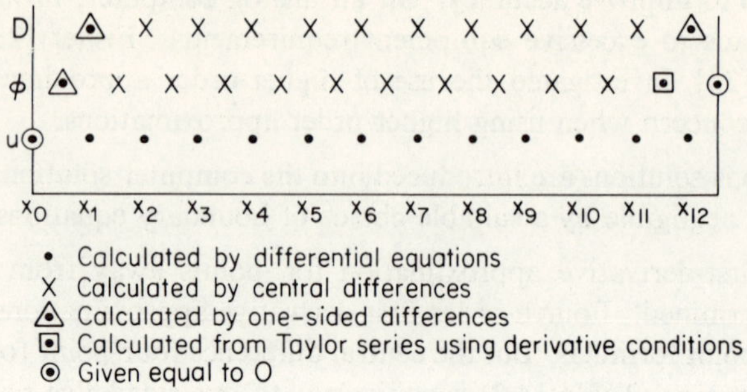

FIG. 14.17 Map of fourth-order heat transfer computation.

which is needed for a fourth-order error. Because the terms do not alternate in sign, extra inverters are required near the boundary. Away from the boundary, we have

$$\phi_{j+1/2} \simeq \frac{u_{j-1} - 27u_j + 27u_{j+1} - u_{j+2}}{24h}$$

$$D_j \simeq \frac{K_{j-3/2}\phi_{j-3/2} - 27K_{j-1/2}\phi_{j-1/2} + 27K_{j+1/2}\phi_{j+1/2} - K_{j+3/2}\phi_{j+3/2}}{24h}$$

The variable $\phi_{11.5}$ can be calculated in a special nonstandard way. We could use a one-sided formula involving u_8 through u_{12}. But slightly more accuracy is obtained if the formula used involves u_9 through u_{12} *and* with the condition that $u'_{12.5} = 0$. Accordingly, we set up a truncated Taylor series about $x = x_{11.5}$ using primes to denote derivatives:

$$u \simeq u_{11.5} + u'_{11.5}(x - x_{11.5}) + \frac{u''_{11.5}(x - x_{11.5})^2}{2}$$
$$+ \frac{u'''_{11.5}(x - x_{11.5})^3}{6} + \frac{u^{(\text{IV})}_{11.5}(x - x_{11.5})^4}{24} + \frac{u^{(\text{V})}_{11.5}(x - x_{11.5})^5}{120}$$

and set up five equations from the assumed five known conditions

$$u'_{11.5} = 0 \simeq u'_{11.5} + u''_{11.5}h + u'''_{11.5}\frac{h^2}{2} + u^{(\text{IV})}_{11.5}\frac{h^3}{6} + u^{(\text{V})}_{11.5}\frac{h^4}{24}$$

$$u_{12} \simeq u_{11.5} + u'_{11.5}\frac{h}{2} + u''_{11.5}\frac{h^2}{8} + u'''_{11.5}\frac{h^3}{48} + u^{(\text{IV})}_{11.5}\frac{h^4}{384} + u^{(\text{V})}_{11.5}\frac{h^5}{3840}$$

$$u_{11} \simeq u_{11.5} - u'_{11.5}\frac{h}{2} + u''_{11.5}\frac{h^2}{8} - u'''_{11.5}\frac{h^3}{48} + u^{(\text{IV})}_{11.5}\frac{h^4}{384} - u^{(\text{V})}_{11.5}\frac{h^5}{3840}$$

$$u_{10} \simeq u_{11.5} - 3u'_{11.5}\frac{h}{2} + 9u''_{11.5}\frac{h^2}{8} - 27u'''_{11.5}\frac{h^3}{48} + 81u^{(\text{IV})}_{11.5}\frac{h^4}{384} - 243u^{(\text{V})}_{11.5}\frac{h^5}{3840}$$

$$u_9 \simeq u_{11.5} - 5u'_{11.5}\frac{h}{2} + 25u''_{11.5}\frac{h^2}{8} - 125u'''_{11.5}\frac{h^3}{48} + 625u^{(\text{IV})}_{11.5}\frac{h^4}{384}$$
$$- 3125u^{(\text{V})}_{11.5}\frac{h^5}{3840} \tag{14.36}$$

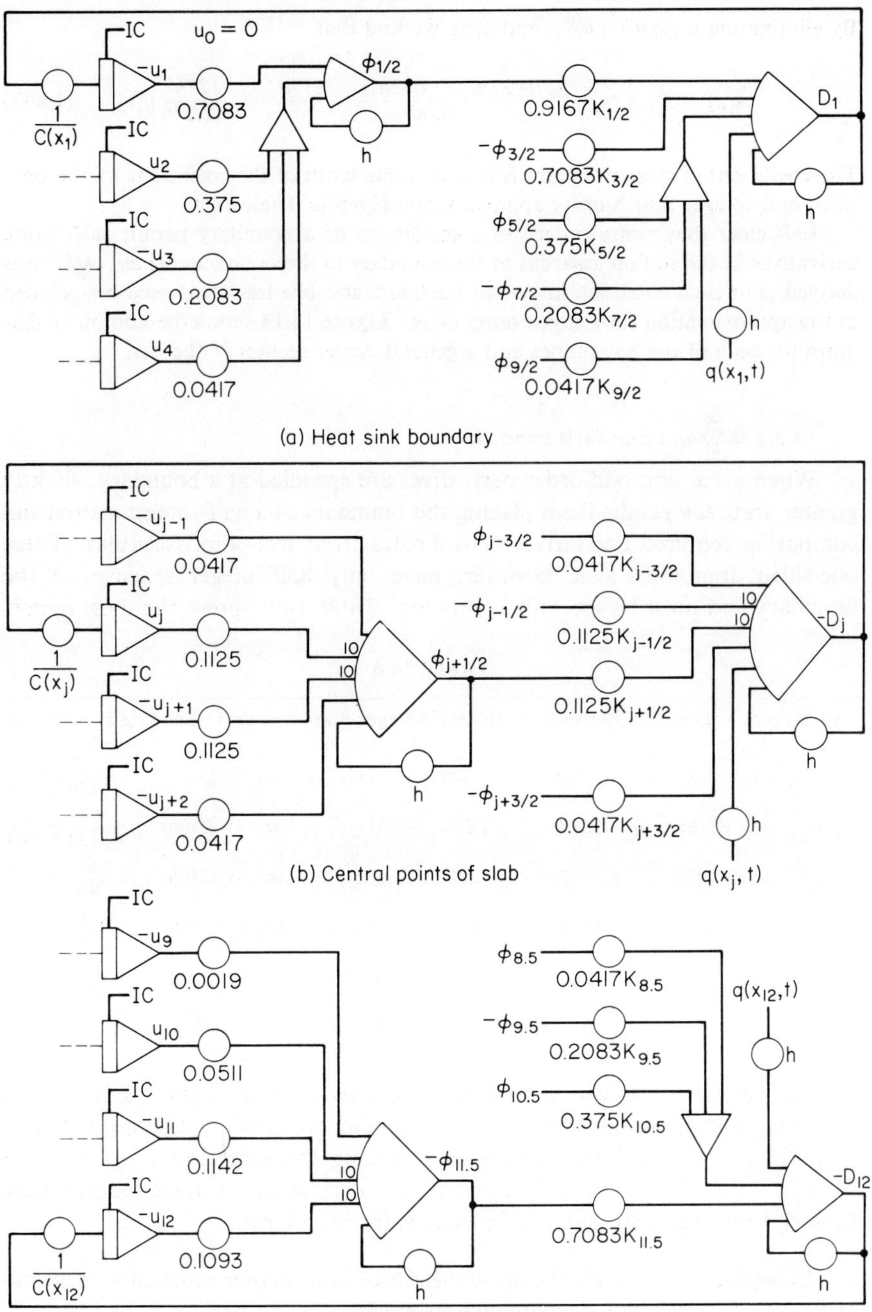

FIG. 14.18 Diagram for a fourth-order heat transfer computation.

By eliminating $u_{11.5}$, $u''_{11.5}$, $u'''_{11.5}$, and $u^{(IV)}_{11.5}$, we find that

$$u'_{11.5} = \phi_{11.5} \simeq \frac{-u_9 + 27u_{10} - 603u_{11} + 577u_{12}}{528h} + \frac{127h^4}{42{,}240} u^{(V)}_{11.5} \quad (14.37)$$

The coefficient of this error term is less than one-tenth of the coefficient of the one-sided half-integer fourth-order approximation given in Table 14.4.

It is clear that nonzero derivative conditions at a boundary permit calculating derivatives at the station adjacent to the boundary in the same manner as (14.37) was derived. For each condition known at the boundary, one less point need be included in the approximation for a given order error. Figure 14.18 shows the computer diagram for each of the boundaries and a general center section of the bar.

14.5.1 Mixed boundary conditions

When even- and odd-order derivatives are specified at a boundary, slightly greater accuracy results from placing the boundary at a half-integer station and computing required derivatives or ordinates from five-point formulas. These one-sided approximations, however, have only half-integer spacings at the boundary and must be specially computed. Table 14.6 shows the most important of these formulas with a half-integer ordinate endpoint at the left. For the half-integer endpoint at the right, the formulas are similar. The ordinate formula for $j + \frac{1}{2}$ is derivable simply by replacing the subscripts $j + i$ by $j - i$ for $i = -\frac{3}{2}, -1, 0, +1, +2$ while the first derivative formulas require those subscript replacements and sign inversions for each term.

TABLE 14.6

Five-Point Lagrangian Formulas with Half-Integer Station at Left	Leading Error Term
$z_{j-1/2} = (\ \ -8z_{j-3/2} + 35z_{j-1} + 35z_j - 7z_{j+1} + z_{j+2})/56$	$+\frac{h^5}{128} z^{(V)}_{j-1/2}$
$z'_{j-3/2} = (-2816z_{j-3/2} + 3675z_{j-1} - 1225z_j + 441z_{j+1} - 75z_{j+2})/(840h)$	$+\frac{7h^4}{128} z^{(V)}_{j-3/2}$
$z'_{j-1} = (\ -192z_{j-3/2} + 35z_{j-1} + 210z_j - 63z_{j+1} + 10z_{j+2})/(210h)$	$-\frac{h^4}{40} z^{(V)}_{j-1}$
$z'_{j-1/2} = (\ 128z_{j-3/2} - 1085z_{j-1} + 1015z_j - 63z_{j+1} + 5z_{j+2})/(840h)$	$+\frac{h^4}{1920} z^{(V)}_{j-1/2}$
$z'_j = (\ 64z_{j-3/2} - 210z_{j-1} + 35z_j + 126z_{j+1} - 15z_{j+2})/(210h)$	$+\frac{h^4}{40} z^{(V)}_j$
$h = x_{j+1} - x_j$	

Example 14.8. Program the one-dimensional heat transfer equation with fourth-order error in a slab with discontinuous materials.

As in Example 14.5, we place the discontinuous boundary at a half-integer station, say $n + \frac{1}{2}$, and program to find $u_{n+1/2}$ from (14.26) and Table 14.6

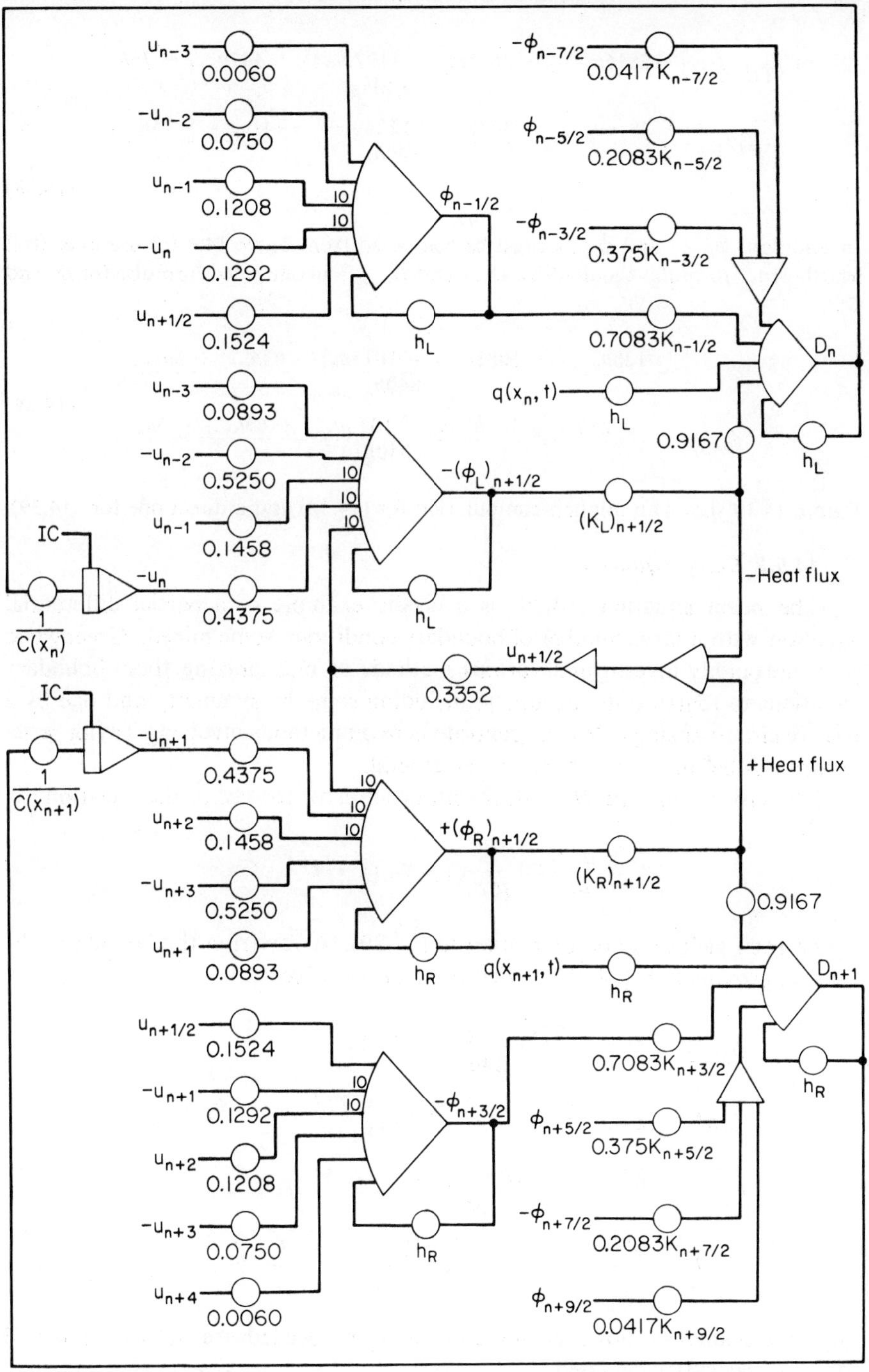

FIG. 14.19 Fourth-order computation at a discontinuous boundary.

$$(K_R)_{n+1/2}\,\frac{-2816u_{n+1/2}+3675u_{n+1}-1225u_{n+2}+441u_{n+3}-75u_{n+4}}{840h_R}$$
$$\simeq (K_L)_{n+1/2}\,\frac{+2816u_{n+1/2}-3675u_{n}+1225u_{n-1}-441u_{n-2}+75u_{n-3}}{840h_L} \tag{14.38}$$

In addition, $\phi_{n+3/2}$ and $\phi_{n-1/2}$ must be computed from Table 14.6 for use in central fourth-order formulas required by D_{n-1} and D_{n+2}, and one-sided formulas for D_n and D_{n+1}

$$\begin{aligned}\phi_{n+3/2} &\simeq \frac{128u_{n+1/2}-1085u_{n+1}+1015u_{n+2}-63u_{n+3}+5u_{n+4}}{840h_R}\\ \phi_{n-1/2} &\simeq \frac{-128u_{n+1/2}+1085u_{n}-1015u_{n-1}+63u_{n-2}-5u_{n-3}}{840h_L}\end{aligned} \tag{14.39}$$

Figure 14.19 shows an implicit computation for (14.38) and a direct one for (14.39).

14.5.2 Beam equation

The beam equation (14.29) is a classic example of a partial differential equation with a large number of boundary conditions, some mixed. Greenwood [24] thoroughly investigated various methods of mechanizing these boundary equations to fourth-order accuracy, including some by symmetry and one by a passive circuit analogy. We concentrate here on methods involving Taylor series expansions because they are the most general.

We wish to find the D_j with fourth-order error for use in the equation

$$\frac{d^2y_j}{dt^2} \simeq \frac{1}{\rho(x_j)}[f(x_j, t) - D_j]$$

which then yields an approximation to (14.29). Away from the boundary, the basic approximation (14.35) is used for each variable:

$$\begin{aligned}\phi_{j+1/2} &\simeq \frac{y_{j-1}-27y_j+27y_{j+1}-y_{j+2}}{24h}\\ M_j &\simeq EI(x_j)\,\frac{\phi_{j-3/2}-27\phi_{j-1/2}+27\phi_{j+1/2}-\phi_{j+3/2}}{24h}\\ V_{j+1/2} &\simeq \frac{M_{j-1}-27M_j+27M_{j+1}-M_{j+2}}{24h}\\ D_j &\simeq \frac{V_{j-3/2}-27V_{j-1/2}+27V_{j+1/2}-V_{j+3/2}}{24h}\end{aligned}$$

Near the boundary, however, some of the terms needed are not available and must be specially calculated.

Figure 14.20 is a possible map for including the various boundary condi-

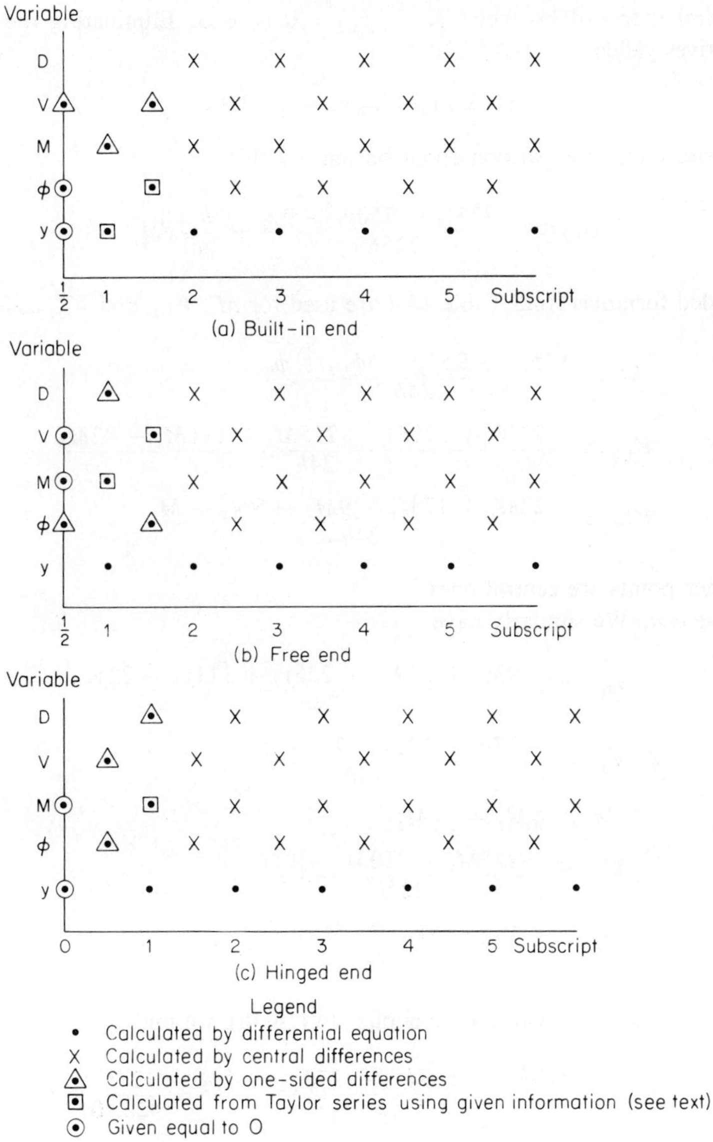

FIG. 14.20 Map of fourth-order beam end conditions.

tions to fourth order. It is by no means a unique map for each type of fastening. For the built-in and free ends, the maps were made similar to accommodate similar formulas. The roles of M and V are interchanged with y and ϕ.

Built-in end. A Taylor series expansion of y about station $\frac{1}{2}$ results in equations similar to (14.36) when $\phi_{1/2} = y_{1/2} = 0$ is used. Eliminating the lowest derivatives yields

$$y_1 \simeq \tfrac{2}{9}y_2 - \tfrac{1}{25}y_3 + \tfrac{1}{48}h^4 y_{1/2}^{(\mathrm{IV})}$$

Similarly, such an expansion about station $\frac{3}{2}$ yields

$$\phi_{3/2} \simeq \frac{-225y_1 + 250y_2 - 9y_3}{225h} + \frac{h^4}{240}y_{3/2}^{(\mathrm{V})}$$

One-sided formulas from Table 14.4 are used for M_1, $V_{1/2}$, and $V_{3/2}$ as follows:

$$M_1 \simeq \frac{17\phi_{3/2} + 9\phi_{5/2} - 5\phi_{7/2} + \phi_{9/2}}{24h}$$

$$V_{1/2} \simeq \frac{-93M_1 + 229M_2 - 225M_3 + 111M_4 - 22M_5}{24h}$$

$$V_{3/2} \simeq \frac{-22M_1 + 17M_2 + 9M_3 - 5M_4 + M_5}{24h}$$

All other points are central ones.

Free end. We similarly have

$$\phi_{1/2} \simeq \frac{-93y_1 + 229y_2 - 225y_3 + 111y_4 - 22y_5}{24h}$$

$$\phi_{3/2} \simeq \frac{-22y_1 + 17y_2 + 9y_3 - 5y_4 + y_5}{24h}$$

$$M_1 \simeq \tfrac{2}{9}M_2 - \tfrac{1}{25}M_3$$

$$V_{3/2} \simeq \frac{-225M_1 + 250M_2 - 9M_3}{225h}$$

$$D_1 \simeq \frac{17V_{3/2} + 9V_{5/2} - 5V_{7/2} + V_{9/2}}{24h}$$

Hinged end. By a calculation similar to (14.36), we find

$$M_1 \simeq \frac{-577\phi_{1/2} + 603\phi_{3/2} - 27\phi_{5/2} + \phi_{9/2}}{528h} + \frac{127h^4}{42{,}240}\phi_1^{(\mathrm{V})}$$

The one-sided formulas are standard:

$$\phi_{1/2} \simeq \frac{17y_1 + 9y_2 - 5y_3 + y_4}{24h}$$

$$V_{1/2} \simeq \frac{17M_1 + 9M_2 - 5M_3 + M_4}{24h}$$

$$D_1 \simeq \frac{-22V_{1/2} + 17V_{3/2} + 9V_{5/2} - 5V_{7/2} + V_{9/2}}{24h}$$

Solutions with the above approximations result in errors proportional to h^4. In the case of a uniform cantilever beam, Greenwood reports that the error in the first four modes is reduced almost to zero for $n = 12$.

Boundary conditions can be included in a variety of ways, as long as the final matrix **E** in (14.34) contains characteristic roots that are positive real.

14.6 Serial Method [25]

We have set up difference differential equations in what could be termed the "natural" way, in discussing previous examples. Finite difference approximations were used for derivatives with respect to a spatial coordinate, while "computer time" derivatives were used for derivatives with respect to a time coordinate. Theoretically, this procedure could be reversed; the choice of which direction to integrate continuously is not predetermined. If an equation is written, for example, as

$$\frac{\partial^2 u}{\partial x^2} = \frac{\partial u}{\partial y}$$

there is no initial bias in which direction, x or y, to integrate continuously. Many partial differential equations, in fact, do not involve time, but only spatial coordinates, although the above equation is just a form of the diffusion equation where y replaces t.

The serial method refers to a finite difference method where $u(x, t_{i+1})$ depends only upon $u(x, t_i)$ and possibly previously found curves $u(x, t_{i-1})$, etc. Continuous integration is done in the x direction. The major advantage is that the same computer program can be used to find each curve successively (or serially). (The previous method discussed is often referred to as the parallel method.) As we shall see, the disadvantages are that the equations are usually unstable, that a boundary-value problem results, and that curve storage is needed.

Let us solve the actual diffusion equation

$$\frac{\partial^2 u}{\partial x^2} = \frac{\partial u}{\partial t} \tag{14.40}$$

by the serial method and see what is involved.

Example 14.9. Assuming an initial temperature profile $u(x, 0) = u_0(x)$ and boundary conditions

$$u(0, t) = 0, \qquad u(L, t) = 0 \tag{14.41}$$

(ice at both ends of the bar), solve (14.40) by the serial method.

If we let, for fixed t_1,

$$u_1(x) = u(x, t_1)$$

with $\Delta t = t_1 - t_0$, then a backward difference approximation results:

$$\frac{d^2u_1}{dx^2} \simeq \frac{u_1 - u_0}{\Delta t} \tag{14.42}$$

which has errors proportional to Δt. High-order difference equations are prone to instabilities when using the serial method [23]. Thus, one source of difficulty is error control.

A circuit to solve (14.42) is shown in Fig. 14.21 with $\dot{x} = 1$. We see that the circuit contains positive feedback because (14.42) is unstable. Also, the initial conditions from (14.41) require that $u_1(0) = 0$, but du_1/dt must be chosen so that $u_1(L) = 0$. If Δt is made small, this can become a very sensitive value—another source of difficulty.

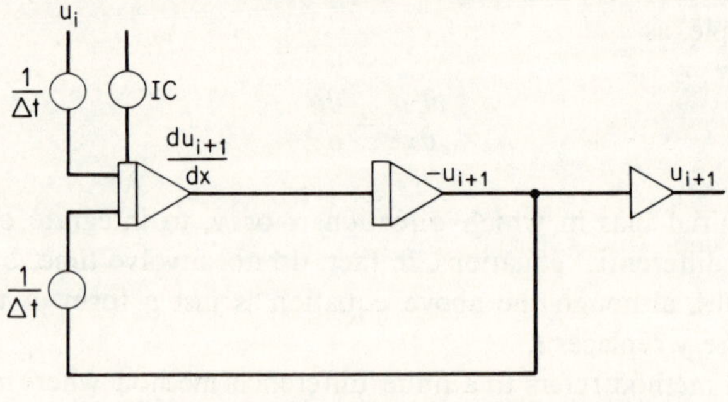

FIG. 14.21 Heat equation by the serial method.

The major difficulty is also apparent from Fig. 14.21. Assuming the boundary conditions are satisfied so that $u_1(x)$ is generated, it must be stored so that it could be used as an input to find $u_2(x)$ [or $u(x, t_2)$]. One method is to record the solution and approximate it on a DFG. But this is somewhat of a nuisance if many curves are needed.

The serial method can be useful when a convenient function storage device is available. The digital part of a hybrid computer is such a device—hybrid computation offers an excellent way to make use of the low analog equipment requirements of the serial method. See Chapter 18 for a simple storage circuit based on point memory.

Other methods of solving partial differential equations have been attempted on an analog computer. These are the Monte Carlo methods [26], [27], and a method that converts the equation into an equivalent integral equation [28]. Because much associated equipment is required, these will not be discussed here.

PROBLEMS

14-1. Mechanize the final equations in Example 14.1.

14-2. Given the equation [4]

$$y\frac{\partial z}{\partial x} + x\frac{\partial z}{\partial y} = z - 1$$

with an initial condition curve

$$x = \lambda, \qquad y = 2\lambda, \qquad z = (\lambda + 1)^2$$

derive the differential equations of the characteristic curves and mechanize them.

14-3. Modify Fig. 14.1 to include the determination of a_j assuming $z_0(y)$ is available.

14-4. Set up Example 14.4 to find eigenvalues and eigenfunctions from equations found by separating variables. Show that by generating $K(x)$ and $K(x)C(x)$ no divisions are necessary.

14-5. Set up the wave equation in Example 14.6 as an eigenvalue problem.

14-6. If the homogeneous vibrating beam equation [(14.29) with $f(x, t) = 0$] is set up as an eigenvalue problem, the spatial equation is of fourth order, and two boundary conditions are needed at each end. Instead of a one-parameter boundary-value problem, two initial conditions must be searched. Although automatic procedures are developed in Chapter 19, mechanize the equations to determine the eigenvalues and eigenfunctions of a cantilever beam by a simple searching procedure.

14-7. Instead of an insulated end, assume convective heating at $y = L$ in Example 14.3. This means that

$$k\frac{\partial u}{\partial y} = h_1[u(L, t^*) - u_i]$$

where h_1 is a convective heat transfer coefficient, and u_i, the constant temperature inside the vehicle [11].

14-8. The finite μ-transform is the set of coefficients d_μ defined from

$$d_\mu(t) = \int_0^\pi F(x, t)\cos\left[(\mu - \tfrac{1}{2})x\right] dx$$

for integral μ. Show that

$$\int_0^\pi \frac{\partial^2 F}{\partial x^2}\cos\left[(\mu - \tfrac{1}{2})x\right] dx = -(\mu - \tfrac{1}{2})^2 d_\mu(t) - \frac{\partial F(0, t)}{\partial x} + (-1)^{\mu+1}(\mu - \tfrac{1}{2})F(\pi, t)$$

and is therefore useful when boundary conditions involve a derivative at one end and an ordinate at the other.

14-9. Use the finite μ-transform in setting up the analog solution of the following problem [12]: The pressure $p(x, t)$ and flow rate $f(x, t)$ of a fluid in a pipe of length S with a first-order dissipation term are given by

$$L\frac{\partial f}{\partial t} + Rf = -\frac{\partial p}{\partial x}, \qquad C\frac{\partial p}{\partial t} = -\frac{\partial f}{\partial x}$$

for constants L, R, and C. These are equivalent to

$$LC\frac{\partial^2 p}{\partial t^2} + RC\frac{\partial p}{\partial t} = \frac{\partial^2 p}{\partial x^2}, \qquad LC\frac{\partial^2 f}{\partial t^2} + RC\frac{\partial p}{\partial t} = \frac{\partial^2 f}{\partial x^2}$$

Given $p(0, t)$ and a terminal relationship

$$p(S, t) = g[f(S, t)]$$

where $g(x)$ is a known function, determine $p(S, t)$, $f(S, t)$, and $f(0, t)$.

14-10. If successive second-order first differences are taken at integer stations instead of staggering them alternately at half-integer and integer stations, the results are less accurate (from the central difference formula of Table 14.2). Nevertheless, a scheme of this sort is needed when an odd derivative is to be approximated. How can the boundary condition of, say, Example 14.4, be included in such a mechanization?

14-11. Set up an implicit computer circuit to determine $u_{1/2}$ and $\phi_{1/2}$ in the case of radiation transfer at a boundary. Assume u is expressed in °R and that

$$\phi_{1/2} = ku_{1/2}^4$$

14-12. A disc of homogeneous material and radius R is at initial temperature u_0 and has its outer circumference in contact with a heat sink at 0°. Because of symmetry, the temperature u is a function only of the radius r from the center, and the heat flow is radial. The equation governing such a case is given by

$$C\frac{\partial u}{\partial t} = K\left[\frac{1}{r}\frac{\partial}{\partial r}\left(r\frac{\partial u}{\partial r}\right)\right]$$

Set up a finite difference approximation using five stations. *Hint:* $(\partial u/\partial r)_0 = 0$.

14-13. If rapid fluctuations in temperature are expected in one area of the bar, better accuracy is obtained by using closely spaced stations near that interval. How may the equations to the left and right of a point be coupled when that point defines an interval spacing change?

14-14. In two dimensions, the heat transfer equation can be written [17]

$$C(x, y)\frac{\partial u}{\partial t} = \frac{\partial}{\partial x}\left[K(x, y)\frac{\partial u}{\partial x}\right] + \frac{\partial}{\partial y}\left[K(x, y)\frac{\partial u}{\partial y}\right] + q(x, y, t)$$

for Cartesian coordinates x and y. Discuss implementing this equation on the computer for various boundary conditions on the rim of a rectangular plate.

14-15. Modify Fig. 14.12 so that only three amplifiers per station are required for the solution of the stretched string equation.

14-16. Show that

$$y_j \simeq \frac{141y_{j-1} - 93y_{j-2} + 23y_{j-3}}{71} + \frac{22}{71}h^4 y_{j+1/2}^{(IV)}$$

provided $y'_{j+1/2} = 0$. Use this formula in an alternate method of setting up the boundary condition in Example 14.7.

14-17. Show that

$$\phi_{1/2} \simeq \frac{-34y_1 + 39y_2 - 5y_3}{29h} + \frac{3091}{55,680}h^4 y_{1/2}^{(V)}$$

$$\phi_{3/2} \simeq \frac{-30y_1 + 31y_2 - y_3}{29h} + \frac{2983}{1,336,320}h^4 y_{1/2}^{(V)}$$

provided $M_{1/2} = V_{1/2} = 0$ and suggest another method for providing fourth-order boundary equations for a free end of a beam.

14-18. Show that computing D_2 from a one-sided formula in the fourth-order mechanization of the built-in end of a beam (excluding the $V_{1/2}$ term) is equivalent to the central formula.

14-19. The approximate equations governing the heat exchange of a thin-walled tube of length L through which flows a fluid at velocity $v(t)$ and steam surrounding the tube are given by

$$\frac{\partial u_T}{\partial t} = \alpha_1(u_s - u_T) - \alpha_2(u_T - u_w)$$

$$\frac{\partial u_w}{\partial t} + v\frac{\partial u_w}{\partial x} = \beta(u_T - u_w)$$

where

$u_s(t) =$ temperature of the steam (known)

$u_T(x, t) =$ temperature of the tube wall

$u_w(x, t) =$ temperature of the fluid

$x =$ distance along the tube

$t =$ time

$\alpha_1, \alpha_2, \beta =$ constants

The initial temperature of the wall $u_T(x, 0)$ and fluid $u_w(x, 0)$ are known, as is the entrance temperature $u_T(0, t)$ of the wall and fluid $u_w(0, t)$. Set up a second-order finite difference mechanization.

REFERENCES

[1] Weinberger, H. F., *Partial Differential Equations* (New York: Blaisdell, 1965).

[2] El'sgol'ts, L. E., *Differential Equations* (Delhi, India: Hindustan Publishing Corp., 1961).

[3] Evans, G. W., II, "Some Partial Differential Equations Solvable by the REAC," *Symposium I on REAC Techniques*, Reeves Instr. Corp., New York, March 15-16, 1951 (ASTIA No. 108488).

[4] Little, W. D., and A. C. Soudack, "On the Analog Computer Solution of First-Order Partial Differential Equations," *AICA*, vol. 7, Oct. 1965.

[5] Hemple, A. B., "Exact Simulation of a Special First Order Partial Differential Equation," *AICA*, vol. 5, Oct. 1963.

[6] Corcos, G. M., *et al.*, "Application of the Electronic Differential Analyzer to Eigenvalue Problems," *Symposium II on Simulation and Computing Techniques*, Reeves Instr. Corp., New York, April 28-May 2, 1952 (ASTIA No. 170488).

[7] Howe, C. E., *et al.*, "Solution of Linear Differential Equations with Time-Varying Coefficients by the Electronic Differential Analyzer," *Symposium II on Simulation and Computing Techniques*, Reeves Instr. Corp., New York, April 28-May 2, 1952 (ASTIA No. 170488).

[8] Howe, C. E., and R. M. Howe, "Application of the Electronic Differential Analyzer to the Oscillation of Beams Including Shear and Rotary Inertia," *J. Applied Mechanics*, vol. 22, March 1955.

[9] Powell, R. V., "Distributed Parameter Vibration with Structural Damping and Noise Excitation," *IRETEC*, vol. 8, June 1959.

[10] Karplus, W. J., and P. A. Stephens, Jr., "Application of Finite Integral Transforms to Analog Simulations," *Trans. AIEE, Part 1* (*Comm. and Elect.*), vol. 78, May 1959.

[11] Liban, E., "The Application of Finite Fourier Transforms to Analog Computer Simulations," *Proc. Spring Joint Comp. Conference*, 1962.

[12] Schiesser, W. E., "An Alternate Derivation of Fourier Series Solutions in the Modal Simulation of Unidirectional Fluid Dynamics," *Simulation*, vol. 5, Dec. 1965.

[13] Hartree, D. R. and J. R. Womersley, "A Method for the Numerical or Mechanical Solution of Certain Types of Partial Differential Equations," *Proc. Royal Soc. of London*, vol. 161, Aug. 1937.

[14] Clymer, A. B., "Operational Analog Simulation of the Vibration of a Beam and a Rectangular Multicellular Structure," *IRETEC*, vol. 8, Sept. 1959.

[15] Fisher, M. E., "Limitations Due to Noise, Stability, and Component Tolerance on the Solution of Partial Differential Equations by Differential Analysers," *J. Electronics and Control*, vol. 8, Feb. 1960.

[16] Presson, A. G., "Solution of a Heat Transfer Problem with the Electronic Differential Analyzer," *Project Typhoon Symposium III on Simulation and Comput-*

ing Techniques, Part 1, Bureau of Aeronautics and U.S. Air Devel. Center, Johnsville, Pa., Oct. 12–14, 1953.

[17] Howe, R. M., "Application of Difference Techniques to Heat Flow Problems Using the Electronic Differential Analyzer," *University of Michigan Report 2115-3-T*, Ann Arbor, Michigan, May 1954 (ASTIA No. 35220).

[18] "Analysis of Transient Aerodynamic Heating Problems with GEDA Equipment," *Goodyear Aircraft Corp., Report GER-5434*, Akron, Ohio, Jan. 7, 1955.

[19] Hermann, P. J., *et al.*, "Application of GEDA Analog Computers to Study Temperature Transients in a Plastic Windshield," *Proc. National Simulation Conference*, January 19–21, 1956, Dallas, Texas.

[20] Howe, R. M., and V. S. Haneman, Jr., "The Solution of Partial Differential Equations by Difference Methods Using the Electronic Differential Analyzer," *Proc. IRE*, vol. 41, Oct. 1953.

[21] Howe, C. E., and R. M. Howe, "Application of Difference Techniques to the Lateral Vibration of Beams Using the Electronic Differential Analyzer," *University of Michigan Report 2115-1-T*, Ann Arbor, Michigan, February 1954 (ASTIA No. 35219).

[22] Clymer, A. B., "Methods of Simulating Structural Dynamics," *Simulation*, vol. 7, Oct. 1966.

[23] Fisher, M. E., "Higher Order Differences in the Analogue Solution of Partial Differential Equations," *Journal ACM*, vol. 3, 1956. Also in *Proceedings of the Inter. Analogy Comp. Meeting* (Brussels: Presses Académiques Européennes, 1956).

[24] Greenwood, D. T., "A Comparison of Higher-Order Difference Methods in the Solution of Beam-Vibration Problems," *IRETEC*, vol. 11, Feb. 1962.

[25] Brunner, W., "Serial Solution of Partial Differential Equations on an Analog Computer," *Princeton Computation Center Report No. 180*, Electronic Associates, Inc., Princeton, N. J., June 21, 1961.

[26] Chuang, K., *et al.*, "A Stochastic Method of Solving Partial Differential Equations Using an Electronic Analog Computer," *University of Michigan Report 2900-91-T*, Willow Run Labs., Ann Arbor, Michigan, June 1960 (ASTIA No. 238424).

[27] Little, W. D., "Hybrid Computer Solutions of Partial Differential Equations by Monte Carlo Methods," *Proc. Fall Joint Comp. Conference*, 1966.

[28] Witsenhausen, H., "Survey of Non-Standard Techniques for Analog Solution of Partial Differential Equations," *Princeton Computation Center Report*, Electronic Associates, Inc., Princeton, N. J., June 20, 1961.

For other interesting applications and information, see:

[29] Yanowitch, M., "Solution of a Flutter Problem," *Reeves Instr. Corp. Report* (reprinted from project Cyclone report), New York, June 1955.

[30] Henning, Tore, "Comparison of Two Ways of Simulating a Simple Tubular Heat Exchanger," *AICA*, vol. 3, April 1961.

[31] Jones, L. G., and W. C. Skinner, "Analog Solutions to Unsteady State Radial Gas Flow Problems," *Proceedings of the 3rd Inter. Analogue Comp. Meetings* (Brussels: Presses Académiques Européennes, 1962).

[32] Miura, T., and J. Iwata, "Effect of the Initial Voltage Error on the Solution of the Vibrations of Beams by Analog Computers," *Proceedings of the 3rd Inter. Analogue Comp. Meetings* (Brussels: Presses Académiques Européennes, 1962).

[33] Butler, R., *et al.*, "Differential Analyser Treatment of the Problem of Solidifying Steel," *AICA*, vol. 4, Jan. 1962.

[34] Gilliland, M. C., "A Mutually Coupled Finite Difference Operator for Diffusion Processes," *AICA*, vol. 4, April 1962.

[35] Prawitz, J., "Simulation of an Atmospheric Transport Problem," *AICA*, vol. 4, April 1962.

[36] Wiig, K. M., "An Evaluation of Static and Dynamic Errors in the Solution of a First-Order Partial Differential Equation on an Analog Computer," *AICA*, vol. 6, Oct. 1964.

[37] Otoba, K., *et al.*, "Flood Simulator for the River Kitakami," *Simulation*, vol. 4, Feb. 1965.

[38] Miura, T., and J. Iwata, "A Consideration on the Solution of Partial Differential Equations by Analog Computer," *Simulation*, vol. 6, Feb. 1966.

[39] Vichnevetsky, R., "A New Method of Analysing the Truncation Error in the Finite Difference Representation of Partial Differential Operators," *Proceedings of the 3rd Inter. Analogue Comp. Meetings* (Brussels: Presses Académiques Européennes, 1962).

[40] Childs, D., "Modal Simulation of Unidirectional Fluid Dynamics (Water Hammer)," *Simulation*, vol. 5, Sept. 1965.

Chapter 15

PROBLEMS IN COMPLEX VARIABLES

The theory of functions of a complex variable is useful in a wide variety of practical engineering problems. In investigations of linear systems, for example, we have seen (Chapter 11) that the real parts of the complex eigenvalues of the matrix of the system completely determine its stability. Finding these eigenvalues is equivalent to finding the roots (zeros) of the characteristic polynomial equation, when the latter is explicitly stated. Naturally, engineers having access to analog computers have devised means of determining both eigenvalues of matrices and complex roots of polynomials, as well as solutions to other complex variable problems.

15.1 Basic Concepts

We consistently use the notation

$$z = x + iy$$

where x and y are real and $i^2 = -1$ as an independent variable in the problems considered. Any expression of the type

$$w = u + iv$$

automatically implies u and v are real functions of x and y. We also write

$$\text{Re}(w) = u, \qquad \text{Im}(w) = v$$

as well as

$$w^* = u - iv$$

where w^* is called the *conjugate* of w. Geometrically, any complex variable such as w can be graphed in the u-v plane (or w plane) by the point (u, v) taking u as the abscissa and v as the ordinate. This instantly suggests a polar coordinate form of w:

$$w = r(\cos\theta + i\sin\theta) \tag{15.1}$$

where

$$r = \sqrt{u^2 + v^2}, \qquad \theta = \arctan\left(\frac{v}{u}\right)$$

and are called the *modulus* and *argument* of w, respectively. A function

$$w = f(z) = u + iv$$

in a region R is defined as in real numbers; for all x, y in R, there is defined a u, v in the w plane. This can be interpreted as a mapping, further discussed in Sec. 15.4.

Derivatives of a complex function w

$$\frac{dw}{dz} = r + is$$

are defined in the same way as for real functions:

$$\frac{df}{dz} = \lim_{\Delta z \to 0} \frac{f(z + \Delta z) - f(z)}{\Delta z}$$

provided the limit exists independent of how Δz approaches zero. If the derivative exists in an entire region R, the function is said to be *analytic* in R. In this chapter, any function $f(z)$ specified [except its conjugate $f^*(z)$] is taken as analytic. We now have the following theorem.

Theorem 15.1. A necessary and sufficient condition that $w = f(z) = u + iv$ is analytic in R is that the Cauchy–Riemann equations

$$\frac{\partial u}{\partial x} = \frac{\partial v}{\partial y}, \qquad \frac{\partial u}{\partial y} = -\frac{\partial v}{\partial x} \tag{15.2}$$

hold in R. Furthermore,

$$\operatorname{Re}\left(\frac{dw}{dz}\right) = r = \frac{\partial u}{\partial x}, \qquad \operatorname{Im}\left(\frac{dw}{dz}\right) = s = \frac{\partial v}{\partial x} \tag{15.3}$$

We frequently like to make use of the power of an analog computer for solving differential equations rapidly. Many techniques can be extended to the case of complex variables by using the important Cauchy integral theorem:

Theorem 15.2. If $f(z)$ is analytic in an open simply connected* region R, then

$$\int_{z_0}^{z_1} f(z)\,dz$$

is independent of the path of integration connecting z_0 and z_1.

As implied by Theorem 15.2, definite integrals are always taken along some path; the indefinite integral

$$g(z) = \int_{z_0}^{z} f(z)\,dz$$

being independent of the path, is an unambiguous function of z under the terms of the theorem. This is a remarkable and useful property enabling us to generate functions of complex variables by integration.

15.1.1 Complex amplifiers [1]

The properties of a complex function $f(z) = u + iv$ of a complex variable $x + iy$ expressed in Theorems 15.1, 15.2, and others, make for more efficient circuits for many problems than if we just thought of u and v as two functions of two independent variables x and y. To help take advantage of these theorems, it is convenient to pair off the two amplifiers representing a complex variable in a manner that parallels the basic idea that a complex number is an ordered pair of real numbers. We shall inclose the two amplifiers in a rectangular box, with the upper amplifier always representing the real part of the variable. The rectangle itself is labeled with the letter or the expression that the variable represents. This procedure helps keep track of the variables and also helps to adjust one's thinking in terms of complex rather than real variables. We call such a configuration a *complex amplifier* and the amplifiers therein real and imaginary components of the complex amplifier.

15.1.2 Complex operations

Complex amplifiers are patched according to the rules of complex numbers:

*A region R is simply connected when every simple closed curve in R has its interior in R.

$$(a + bi) \pm (c + di) = (a \pm c) + (b \pm d)i$$
$$(a + bi) \times (c + di) = (ac - bd) + (bc + ad)i \tag{15.4}$$
$$\frac{a + bi}{c + di} = \frac{ac + bd}{c^2 + d^2} + \frac{bc - ad}{c^2 + d^2}i$$

These may always be performed in terms of the real operations that the computer can handle. Functions of a complex variable can similarly be generated when the real and imaginary parts are explicitly or implicitly known by using implicit function techniques and function generation techniques. As we shall see, generalized integration by the Cauchy integral theorem is also possible. Thus, virtually all techniques of real variable function generation covered in Chapters 5–9 have their complex counterpart.

From (15.4), we see that complex operations can involve much nonlinear equipment. On the other hand, it is often possible to reduce the required equipment by noting special relationships. For example, the product relationship in (15.4) is generable with three real multiplications:

$$ac - bd = \frac{(a + b)(c + d) + (a - b)(c - d) - 4bd}{2}$$

$$ad + bc = \frac{(a + b)(c + d) - (a - b)(c - d)}{2}$$

Similarly, squaring a variable requires only two multiplications

$$(a + bi)^2 = a^2 - b^2 + (2ab)i = (a + b)(a - b) + (2ab)i$$

These procedures require more summers and are not recommended unless multipliers are at a premium. We see, however, that it may be possible to balance the need of equipment to fit the available supply.

15.2 Generalized Integration [1]

Because of the Cauchy integral theorem 15.2, many functions of z are generable by an integration process in virtually the exact manner as real functions were in Chapter 5. We must first obtain a complex differential equation (or a set of equations) to be solved on the computer. What does it mean to solve a complex differential equation on an analog computer? Suppose

$$\frac{df}{dz} = g(z) = p(x, y) + iq(x, y) \tag{15.5}$$

is the differential equation from which we wish to generate f as z changes. As in the case of real variables, we use the dot notation

$$\dot{z} = \dot{x} + i\dot{y} = u + iv, \qquad (z_0) = z_0 = (x)_0 + i(y)_0$$

whenever we wish

$$\dot{x} \equiv u, \qquad (x)_0 = x_0; \qquad \dot{y} \equiv v, \qquad (y)_0 = y_0$$

This indicates the change in z parametrically in terms of τ. Because the chain rule applies to complex variables also, (15.5) is transformed to

$$\dot{f} = g(z)\dot{z} = (p + iq)(u + iv) = (up - qv) + i(qu + pv)$$

Thus, f is generated by the equations

$$\dot{\text{Re}}(f) = up - qv, \qquad \dot{\text{Im}}(f) = qu + pv$$

with the proper initial conditions.

In many problems, we may specify $\dot{z}$ to simplify resulting expressions. The specification of any $\dot{z}$, say

$$\dot{z} = f(z), \qquad (z)_0 = z_0$$

determines a path* in the z plane, since it represents two real differential equations for the generation of x and y.

Example 15.1. Find the value of $F = e^z$ at $z = 1 + i$ by solving the proper differential equation using a linear path for z starting from the origin ($z_0 = 0$).

To pass through the point $1 + i$, we must have

$$\dot{z} = k(1 + i), \qquad z_0 = 0$$

In general, the power series for functions of a complex variable is identical with that function of the same name in real variables. This not only gives the same functional values along the real axis but results in the identical differential equation. From the power series definition of e^z, we may easily show that

$$\frac{dF}{dz} = e^z = F$$

Thus

$$\dot{F} = F\dot{z} = kF(1 + i)$$

or expanding

*The solution curve of a differential equation is often called a trajectory. However, we usually speak of varying z along a path. For this reason, the term path is used instead of trajectory even though x and y are varied by solving differential equations.

$$\dot{F} = k\{\text{Re}(F) - \text{Im}(F) + i[\text{Re}(F) + \text{Im}(F)]\}, \qquad F_0 = 1$$
$$\dot{z} = k(1 + i), \qquad z_0 = 0$$

The circuit could look like Fig. 15.1, shown unscaled in amplitude but with a time scale factor of $1/k$. To obtain e^{1+i}, we need not have started from $(0, 0)$. If we

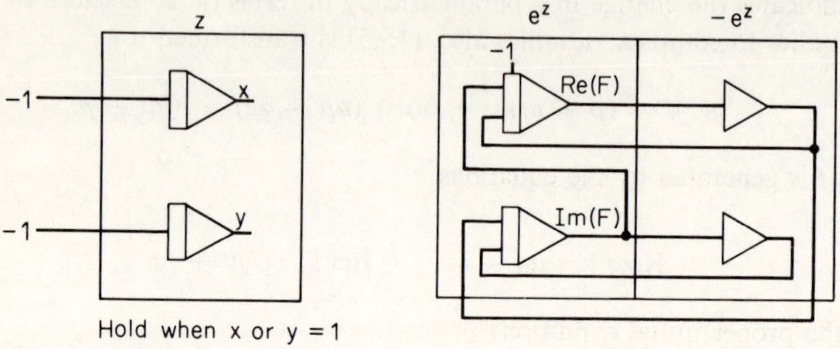

FIG. 15.1 Determining e^{1+i} from the differential equations.

started from $(1, 0)$, we would have required

$$\dot{z} = i, \qquad z_0 = 1; \qquad \dot{F} = iF, \qquad F_0 = e$$

The equation for F is then mechanized by an oscillator circuit, for

$$\dot{\text{Re}}(F) + i\dot{\text{Im}}(F) = \dot{F} = i\{\text{Re}(F) + i\text{Im}(F)\}$$

or

$$\dot{\text{Re}}(F) = -\text{Im}(F), \qquad \dot{\text{Im}}(F) = \text{Re}(F)$$

Similarly, if $\dot{z} = 1$, then $\dot{F} = F$ leads to the fact that, for constant y, the real and imaginary parts of F increase exponentially. These considerations lead to the explicit formula

$$e^z = e^x (\cos y + i \sin y)$$

which need not, however, be known or calculated to generate the function by integration.

From the above discussion, we see that much of real variable function generation with integrators carries over to complex functions.

Theorem 15.3. The linear equipment and summation theorems (5.1 and 5.2) apply to complex variables: Group I functions ($\sin cz$, $\cos cz$, and e^{cz}) are generable with linear equipment when $\dot{z} = (a + bi)$; Group II functions [$\ln z^c$, z^c, $\sin (\ln z^c)$, $\cos (\ln z^c)$] when $\dot{z} = (a + bi)z$; and similar formulas apply in all other cases.

The proof of Theorem 15.3 is as straightforward as those for Theorems 5.1 and 5.2.

The path defined by

$$\dot{z} = a + bi \tag{15.6}$$

is a linear path in the z plane. If $a = 0$, the path is parallel to the y axis, while $b = 0$ means the path is parallel to the x axis. The path

$$\dot{z} = (a + bi)z \tag{15.7}$$

is a spiral in the z plane. If $b = 0$, the path is radial ($y = cx$) while $a = 0$ means the path is circular.

It will not always be possible to use only linear equipment to generate functions with integrators. Even the case where nonlinear equipment is required, generating functions by integration can reduce the equipment requirements considerably.

Example 15.2. Generate the functions $1/z$, $1/z^2$, and $1/z^3$, which are required in the Joukowsky airfoil transformation [2]. In this case, we must generate the functions while the z-plane path is an off-center circle:

$$x = x_0 + R\cos t, \qquad y = y_0 + R\sin t \tag{15.8}$$

because of the properties of a conformal mapping in this problem. However, we easily verify that (15.8) is the solution of

$$\dot{z} = i(z - z_0), \qquad (z)_0 = x_0 + R + iy_0 = z_0 + R$$

if we define z_0 to be the center of the circle; $z_0 = x_0 + iy_0$. Now we let

$$F_1 = \frac{1}{z}, \qquad F_2 = \frac{1}{z^2}, \qquad F_3 = \frac{1}{z^3}$$

and we have

$$\dot{F}_1 = -\frac{\dot{z}}{z^2} = -\frac{i(z - z_0)}{z^2} = -iF_1 + iz_0F_2$$

$$\dot{F}_2 = -\frac{2\dot{z}}{z^3} = -\frac{2i(z - z_0)}{z^3} = -2iF_2 + 2iz_0F_3$$

$$\dot{F}_3 = -\frac{3\dot{z}}{z^4} = -\frac{3i(z - z_0)}{z^4} = -3iF_3 + 3iz_0F_2^2$$

which can be solved on the computer. Since z_0 is constant for any run, these functions are generable with one complex squaring circuit and linear equipment. Figure 15.2 shows an unscaled diagram requiring much less equipment than required by a

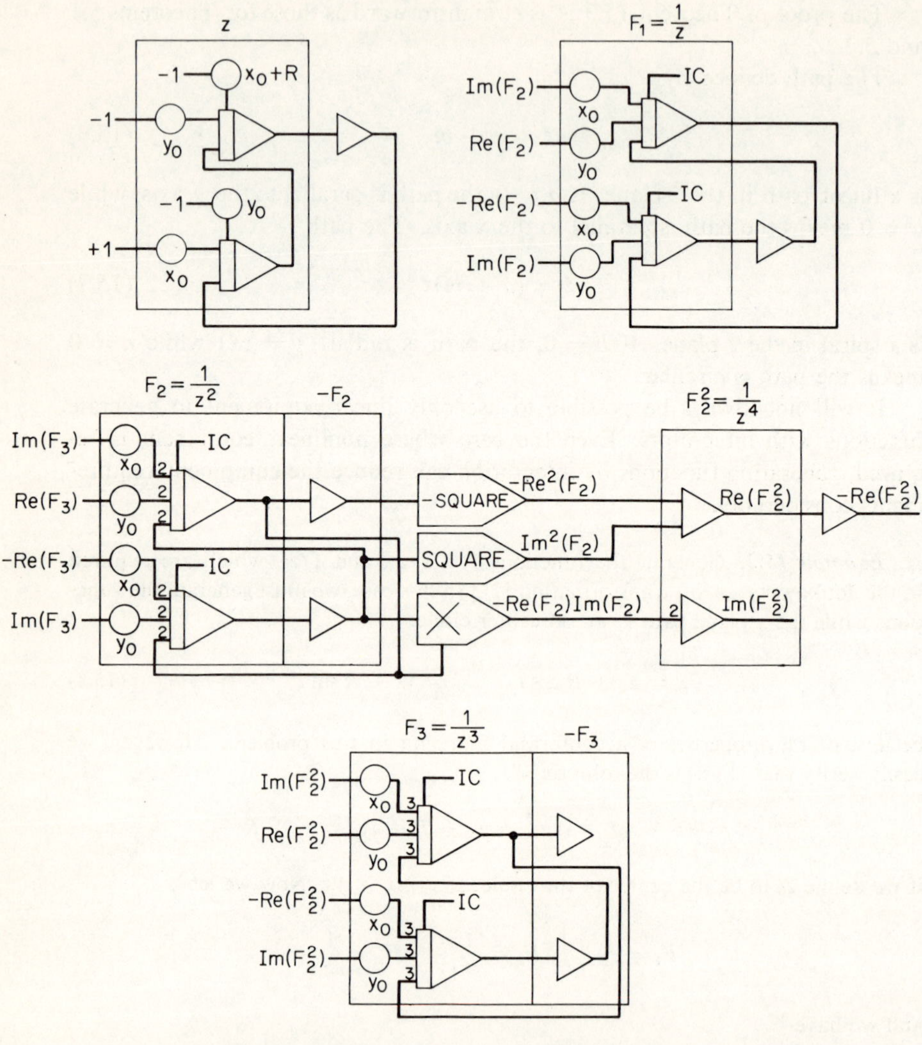

FIG. 15.2 Generation of $1/z$, $1/z^2$, and $1/z^3$, using $\dot{z} = z - z_0$.

direct method [2]. The initial conditions of each integrator must be calculated from the initial condition of z. Many pots control the circle center x_0, y_0, an inconvenience unless accurate ganged pots or rapid automatic pot-setting equipment is available.

Since complex division requires much equipment [Eq. (15.4)], it often pays to eliminate any division needed in right-hand expressions for differential equations. The method depends on the purpose for generating the function. Suppose, for example, we were solving the equations

$$\dot{z} = i, \qquad \dot{F} = \frac{f(F, z)}{g(F, z)}$$

in a *searching* problem; i.e., we were looking for certain properties of F to be satisfied. The most common example of this is finding the roots of F. Then we can use the system

$$\dot{z} = ig(F, z), \qquad \dot{F} = f(F, z)$$

which generates $F(z)$ without divisions. The z-plane path, however, is radically changed if g is not real. In searching problems this is not important since the search can be accomplished by any number of paths covering the region of interest.

On the other hand, we may be involved in a *mapping* problem where the z-plane path must be traced along lines parallel to the y axis. To eliminate division in such a case, we must use

$$\dot{z} = igg^*, \qquad \dot{F} = fg^*$$

Since gg^* is positive real, the resulting *real* differential equations are multiplied by values that effectively only time scale the solutions. Thus, for any denominator D, we multiply all right-hand sides by DD^* to eliminate division in a mapping problem, but by D in a searching problem. The resulting equations are usually simpler in the latter case.

15.3 Polynomials

The most frequent search problem in complex variables consists of finding the roots of polynomials. Such information is required, for example, in determining the analytic solution of a linear system.

Several methods have been devised to determine these roots by a searching procedure. When searching for roots of $w = f(z)$ by scanning the z plane, the following two important theorems are useful:

Theorem 15.4. Let C_z be a simple closed curve in the z plane such that $|f(z)| > 0$ for points on C_z. As z traces curve C_z, another closed curve C_w is traced in the w plane. The number of zeros in the interior of C_z equals the number of times that C_w encircles $w = 0$. (This theorem is applicable to all analytic functions containing no poles.) In practice, Theorem 15.4 is usually used by taking a circle about the z-plane origin.

Theorem 15.5 (DeMoivre's theorem). In the polar coordinate form of z (15.1):

$$z = r(\cos \theta + i \sin \theta)$$

then

$$z^j = r^j(\cos j\theta + i \sin j\theta)$$

for all real j.

In most cases, the polynomial is formed by generating powers of z and finding

$$w(z) = \sum_{j=0}^{n} a_j z^j \tag{15.9}$$

at the output of a complex summer.

15.3.1 Linear scan ($\dot{z} = i$) [1]

Simpler circuits result if paths parallel to either the x or y axis are used for a linear scan. Since initial conditions are easier to calculate for real z, the most convenient set of paths should originate from the real axis. Hence, we use (15.6) with $a = 0$. The value of b is absorbed in time scaling so we refer to these paths as

$$\dot{z} = i, \qquad z_0 = x_0$$

This scans only the upper half plane of z; the equation $\dot{z} = -i$, $z_0 = x_0$, must be used separately to investigate the lower half plane. It should be remembered that when the polynomial (15.9) has *real* coefficients, complex zeros must come in conjugate pairs. Examination of the half plane is sufficient to determine all zeros in that case.

The integral powers of z,

$$F_j = z^j, \qquad j = 1, 2, \ldots \tag{15.10}$$

are now generated by

$$\dot{F}_j = jz^{j-1}\dot{z} = ijF_{j-1}, \qquad (F_j)_0 = x_0^j$$

Thus, all powers of z through z^j must be generated. To save on inverter requirements, the four successive functions generated instead consist of

$$z^k, \qquad (z^{k+1})^*, \qquad -z^{k+2}, \qquad -(z^{k+3})^*$$

with initial conditions of x_0^k, x_0^{k+1}, $-x_0^{k+2}$, and $-x_0^{k+3}$, as shown mechanized in Fig. 15.3. Generating the polynomial requires two additional summers plus

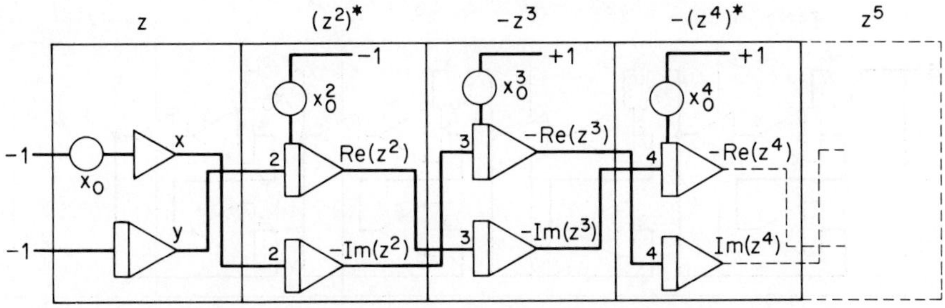

FIG. 15.3 Powers of z with $\dot{z} = i$, $z_0 = x_0$.

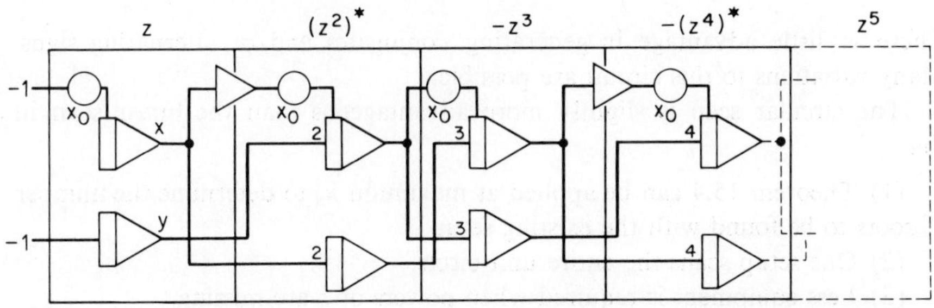

FIG. 15.4 Powers of z with $\dot{z} = i$, $z_0 = x_0$, using ganged pots.

any inverters needed because of the signs of the coefficients. The initial conditions are supplied by pots after being calculated for a given constant x_0.

For a repetitive computer or to automate a sequence of runs without calculating initial conditions each time x_0 is changed, ganged or servo pots can be used. Here, inverters are required for every other initial condition as shown in Fig. 15.4. It is convenient to normalize $|z| \leq 1$ to avoid any loading problems. Simply scale each amplifier as z^k after transforming the original equation (see Problem 15-6).

15.3.2 Circular scan ($\dot{z} = iz$) [1], [3], [4]

Instead of a spiral scan (15.7), we let $a = 0$ to keep integration times short. One revolution of the independent variable terminates the run with any set of initial conditions. Again, we let $b = 1$ to simplify notation and $z_0 = x_0$ to ease the calculation of initial conditions. Now, for F_j defined in (15.10), we have

$$\dot{F}_j = jz^{j-1}\dot{z} = ijF_j, \qquad (F_j)_0 = x_0^j \tag{15.11}$$

Thus, powers of z are generable independent of other powers; (15.11) represents an oscillator circuit that could also be derived from Theorem 15.5. Figure 15.5 shows an unscaled diagram for the generation of the first four powers of z.

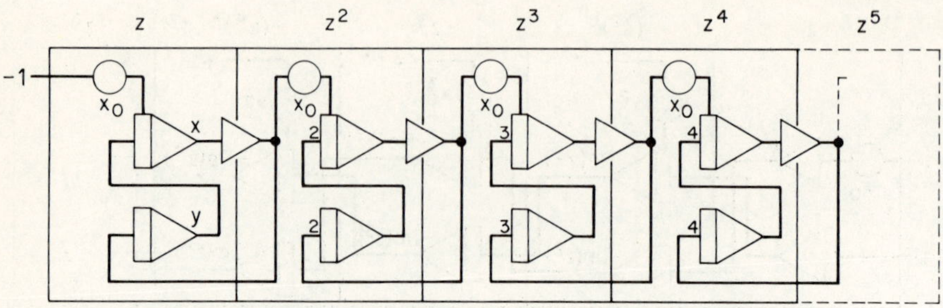

FIG. 15.5 Powers of z with $\dot{z} = iz$, $z_0 = x_0$, using ganged pots.

There is little advantage in generating conjugates and/or alternating signs. Many variations to this circuit are possible.

The circular scan is slightly more advantageous than the linear scan in that

(1) Theorem 15.4 can be applied at maximum x_0 to determine the number of roots to be found with the existing setup.

(2) One setup scans the entire unit circle.

(3) Less equipment is required when powers of z are missing.

(4) Linear scans involve open-loop integrations which are more prone to error buildup.

15.3.3 Radial scan ($\dot{z} = -z$) [3]

If many resolvers are available, it may be advantageous to use a radial scan. With

$$\dot{z} = -z, \qquad z_0 = r_0 \cos\theta_0 + ir_0 \sin\theta_0$$

we get

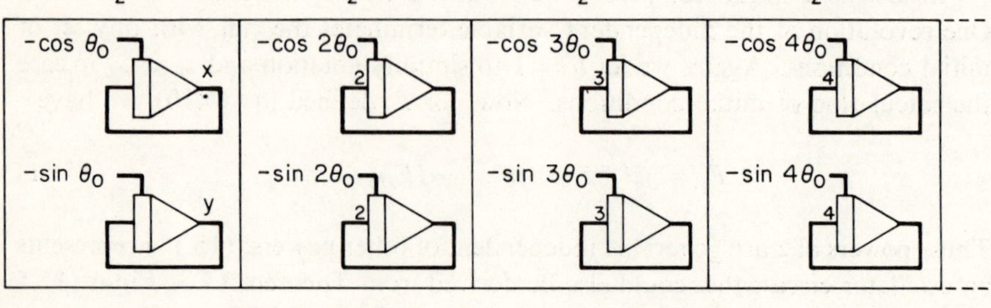

FIG. 15.6 Powers of z with $\dot{z} = -z$, using resolvers.

$$\dot{F}_j = -jF_j, \qquad (F_j)_0 = r_0^j \cos j\theta_0 + ir_0^j \sin j\theta_0$$

If we take $r_0 = 1$, corresponding to the maximum value of the modulus of z during any run, the resolvers are needed only to supply initial conditions for constant θ_0 as shown in Fig. 15.6. Alternately, the values can be computed and supplied by potentiometers. The main advantage of this mechanization is the high accuracy of the negative feedback circuits.

Example 15.3. Find the roots to the polynomial

$$w = f(z) = 5z^4 + 4z^3 + 3z^2 + 2z + 1 \tag{15.12}$$

We use the circular scan to generate powers of z as in Fig. 15.5. The functions $u = \text{Re}(w)$ and $-v = -\text{Im}(w)$ are then constructed with two summers as shown in Fig.

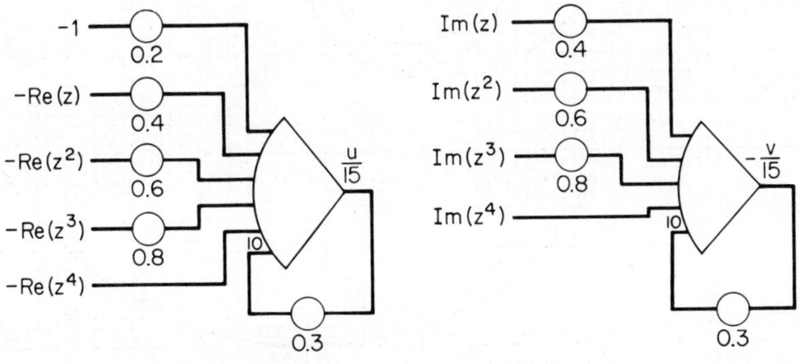

FIG. 15.7 Generating $f(z)$ at summer outputs.

15.7. With $x_0 = 1$, Fig. 15.8 shows the w-plane closed curve C_w produced as z traces out the unit circle. All radial lines from the origin intersect C_w four times indicating all roots lie within the unit circle $|z| < 1$. By varying x, the w plane is observed easily. A succession of runs shows pairs of roots occur with $x_0 \simeq 0.65$ and 0.69. For these values, we plot u, v, and x as a function of y and determine graphically that the roots are

$$z_{1,2} \simeq -0.54 \pm 0.36i, \qquad z_{3,4} \simeq +0.15 \pm 0.67i$$

It is possible to iterate upon these values (see Problem 15-9) to obtain

$$z_{1,2} \simeq 0.537832 \pm 0.358283i, \qquad z_{3,4} \simeq 0.137832 \pm 0.678155i$$

or any desired accuracy. The errors are due mostly to inaccurate ganged pots.

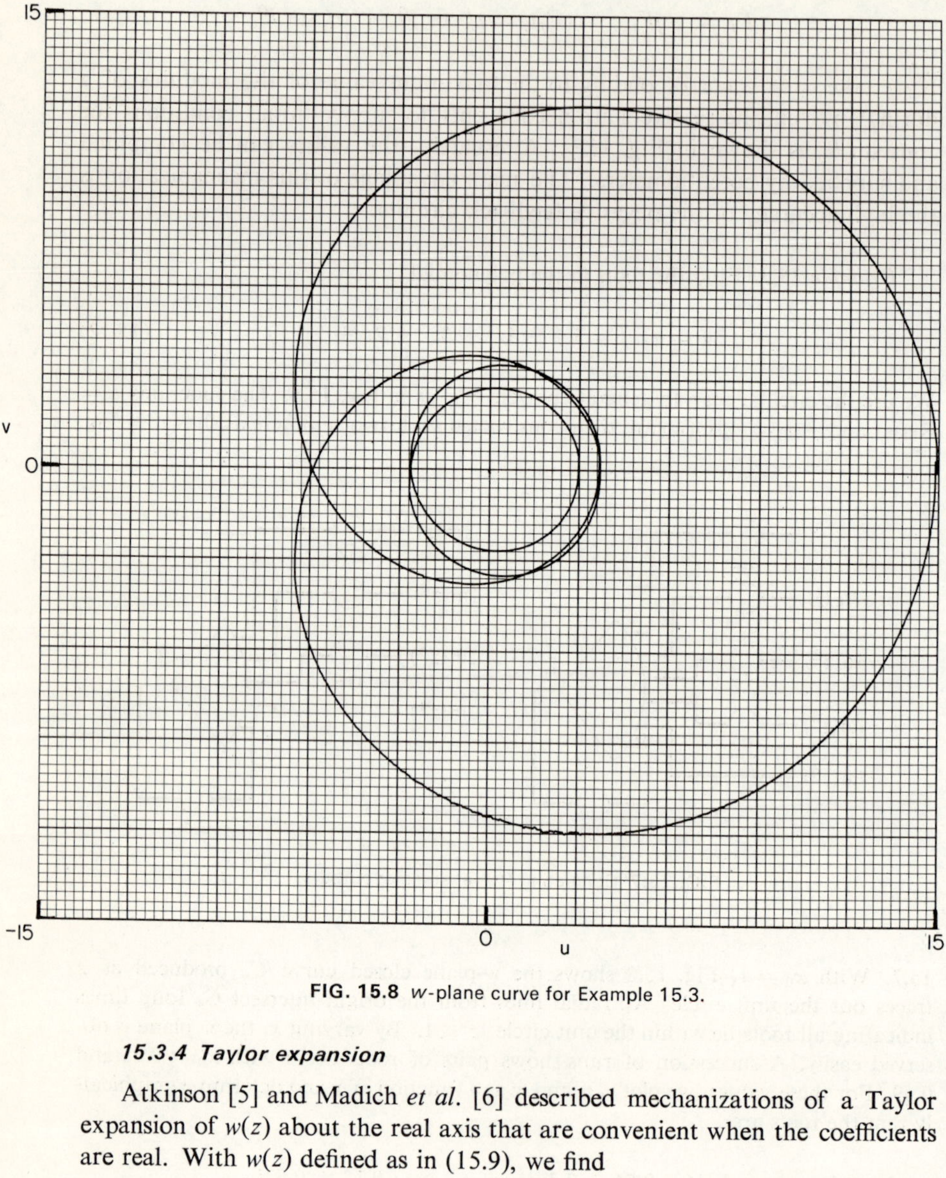

FIG. 15.8 w-plane curve for Example 15.3.

15.3.4 Taylor expansion

Atkinson [5] and Madich *et al.* [6] described mechanizations of a Taylor expansion of $w(z)$ about the real axis that are convenient when the coefficients are real. With $w(z)$ defined as in (15.9), we find

$$w(z) = w(x) + iyw'(x) - \frac{y^2}{2!}w''(x) + \cdots + \frac{i^n y^n w^{(n)}(x)}{n!}$$

where the primes are derivatives with respect to z evaluated at $z = x$. A more convenient form is

$$\operatorname{Re}(w) = w(x) - y^2\left\{\frac{w''(x)}{2!} - y^2\left[\frac{w^{(IV)}(x)}{4!} - y^2\left(\frac{w^{(VI)}(x)}{6!} - \cdots\right)\right]\right\}$$

$$\frac{\operatorname{Im}(w)}{y} = w'(x) - y^2\left\{\frac{w'''(x)}{3!} - y^2\left[\frac{w^{(V)}(x)}{5!} - y^2\left(\frac{w^{(VII)}(x)}{7!} - \cdots\right)\right]\right\} \tag{15.13}$$

If $w(x)$ is generated with $\dot{x} = 1$ by (5.12) and (5.13), all derivatives needed in (15.13) are made available. With y^2 set on a ganged pot, Fig. 15.9 shows a general circuit diagram for its mechanization. The open-loop integration for determining $w(x)$ and its derivatives is prone to error as usual.

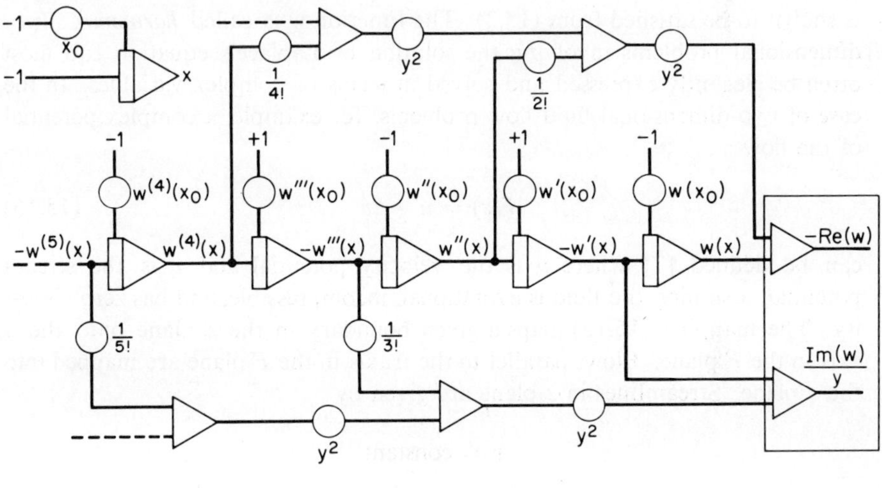

FIG. 15.9 Taylor series mechanization of a polynomial.

Of all the methods discussed, the circular scan combines simplicity, convenience, and accuracy to make it preferable over other methods. For finding real roots, the method in Sec. 11.1 is recommended.

15.4 Conformal Mapping [8]–[10]

In a strict sense, a conformal mapping for x, y in region R by

$$u = f(x, y), \qquad v = g(x, y) \tag{15.14}$$

onto a region R' is one that preserves angular measure; i.e., if two curves in the x-y plane intersect with an angle α between them, the corresponding curves in the u-v plane also intersect, forming the angle α in the same sense. We may show, however, that if the first partial derivatives are continuous in R, the necessary and sufficient conditions for the mapping (15.14) to be conformal is that the Cauchy–Riemann equations (15.2) hold. In turn, this means that an analytic complex function $w(z)$ is defined in R. Thus, the theory of conformal mapping is intimately associated with the theory of complex variables.

Practical applications of conformal mapping depend upon the fact that Laplace's equation

$$\frac{\partial^2 w}{\partial x^2} + \frac{\partial^2 w}{\partial y^2} = 0$$

is shown to be satisfied from (15.2). The function w is called *harmonic*. Two-dimensional problems involving the solution of Laplace's equation can most often be elegantly expressed and solved in terms of complex variables. In the case of two-dimensional fluid flow problems, for example, a complex potential of the flow

$$F(z) = u + iv \tag{15.15}$$

can be defined [7] where u is the velocity potential and v is the stream potential, assuming the fluid is irrotational, incompressible, and has zero viscosity. The mapping (15.15) maps a given boundary in the z plane onto the u axis in the F plane. Flows parallel to the u axis in the F plane are mapped into the z plane. Streamlines in z plane are given by

$$v = \text{constant}$$

but $(dF/dz)^*$ can be shown [7] to be a velocity vector, so that streamlines can also be plotted with the equation

$$\dot{z} = \left(\frac{dF}{dz}\right)^* \tag{15.16}$$

Similarly,

$$u = \text{constant}$$

defines equipotential lines orthogonal to the streamlines. Thus, these can be plotted with the equation

$$\dot{z} = i\left(\frac{dF}{dz}\right)^*$$

If the fluid has uniform density, pressure distributions needed in a study can be calculated from Bernoulli's equation. Here, we consider just obtaining the graphs of streamlines.

Example 15.4. The two-dimensional complex potential of flow about a circle of radius R is given by [7]

$$F(z) = A\left(z + \frac{R^2}{z}\right) \tag{15.17}$$

where A is the free stream velocity. Plot a set of streamlines.

From (15.16), the streamlines can be graphed by using

$$\dot{z} = A\left(1 - \frac{R^2}{z^2}\right)^* \tag{15.18}$$

Conjugate properties useful in separating (15.18) into real and imaginary parts are

$$(a \pm b)^* = a^* \pm b^*, \qquad (ab)^* = a^* b^*, \qquad \left(\frac{a}{b}\right)^* = \frac{a^*}{b^*}$$

which leads to

$$\dot{z} = A\left[1 - \frac{R^2}{(z^*)^2}\right]$$

To eliminate division we multiply the right-hand side by $z^2 z^{*2}$:

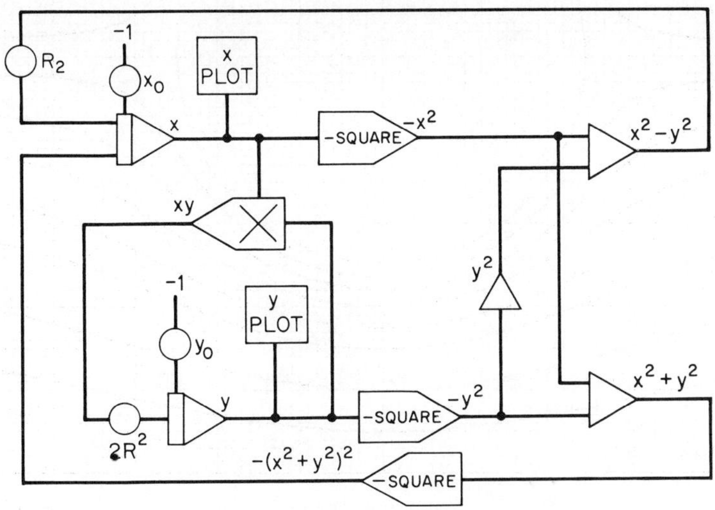

FIG. 15.10 Circuit for streamlines about a circle.

$$\dot{z} = A[(zz^*)^2 - R^2 z^2] \tag{15.19}$$

It is simple to separate (15.19) because $zz^* = x^2 + y^2$. Since A is absorbed in time scaling we use

$$\dot{x} = (x^2 + y^2)^2 - R^2(x^2 - y^2), \qquad \dot{y} = -2R^2 xy$$

Figure 15. 10 shows the circuit that plots these streamlines for $R = \frac{1}{2}$, and Fig. 15.11 shows the streamlines. To minimize errors, x_0 was made zero, but provision to change the signs of the inputs to both integrators was made through the use of function switches.

The equations derived are not always so easily separated into real and imaginary parts in terms of x and y. The following example illustrates how generalized integration is often useful in these problems.

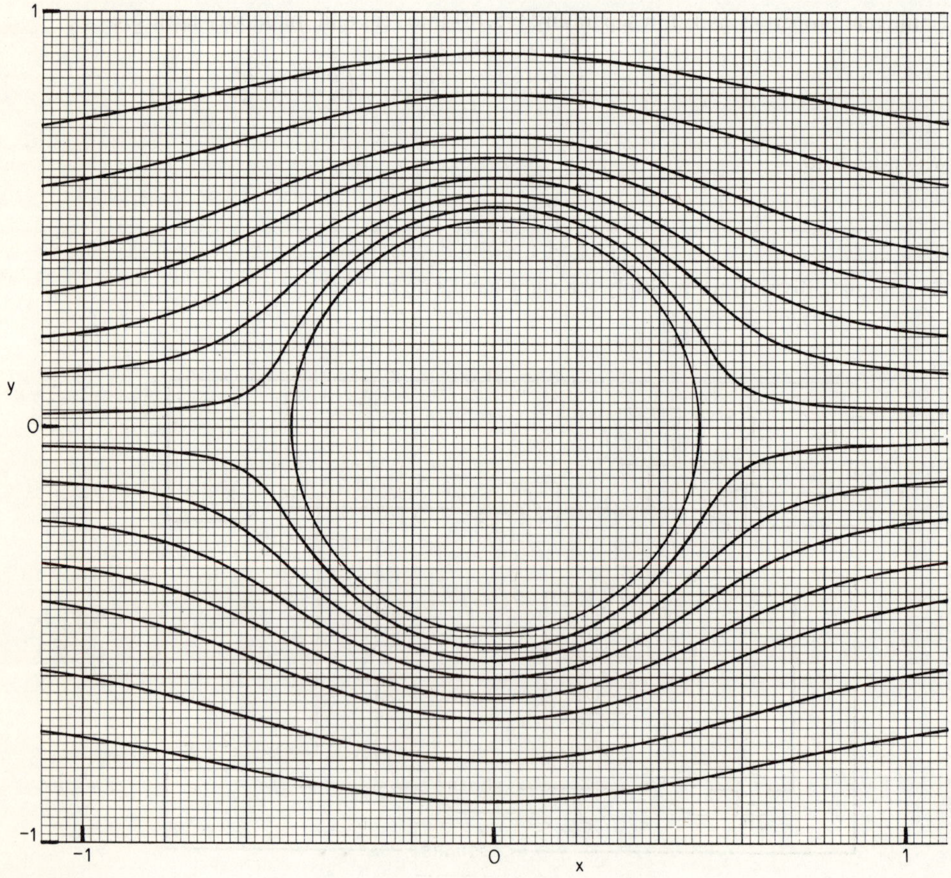

FIG. 15.11 Streamlines about a circle.

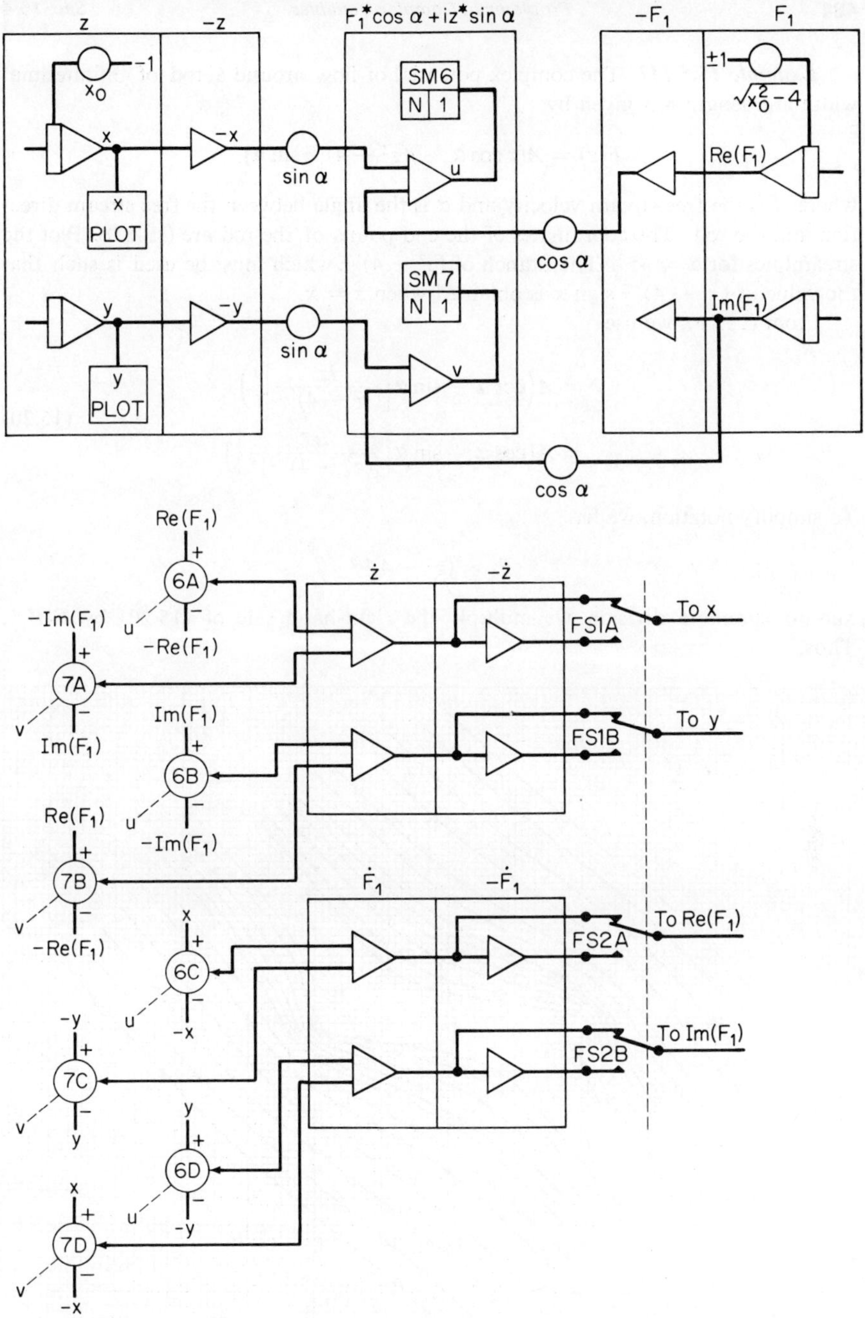

Note: Select all function switches either up or down to plot trajectories in both directions.

FIG. 15.12 Circuit for streamlines about a rod.

Example 15.5 [1]. The complex potential of flow around a rod of infinitesimal width and length 4 is given by

$$F(z) = A[z \cos \alpha - i(z^2 - 4)^{1/2} \sin \alpha]$$

where A is the free stream velocity and α is the angle between the free stream direction and the rod. The coordinates of the end points of the rod are $(\pm 2, 0)$. Plot the streamlines for $\alpha = 45°$. The branch of $(z^2 - 4)^{1/2}$ which must be used is such that the value of $(x^2 - 4)^{1/2}$ sign x is obtained when $z = x$.

From (15.16), we use

$$\begin{aligned} \dot{z} &= A\left(\cos \alpha - \sin \alpha \left[\frac{iz}{(z^2 - 4)^{1/2}}\right]^*\right) \\ &= A\left(\cos \alpha + \sin \alpha \left[\frac{iz^*}{\{(z^2 - 4)^{1/2}\}^*}\right]\right) \end{aligned} \tag{15.20}$$

To simplify notation, we let

$$F_1 = (z^2 - 4)^{1/2}$$

and to eliminate division we multiply the right-hand side of (15.20) by $F_1 F_1^*$. Thus,

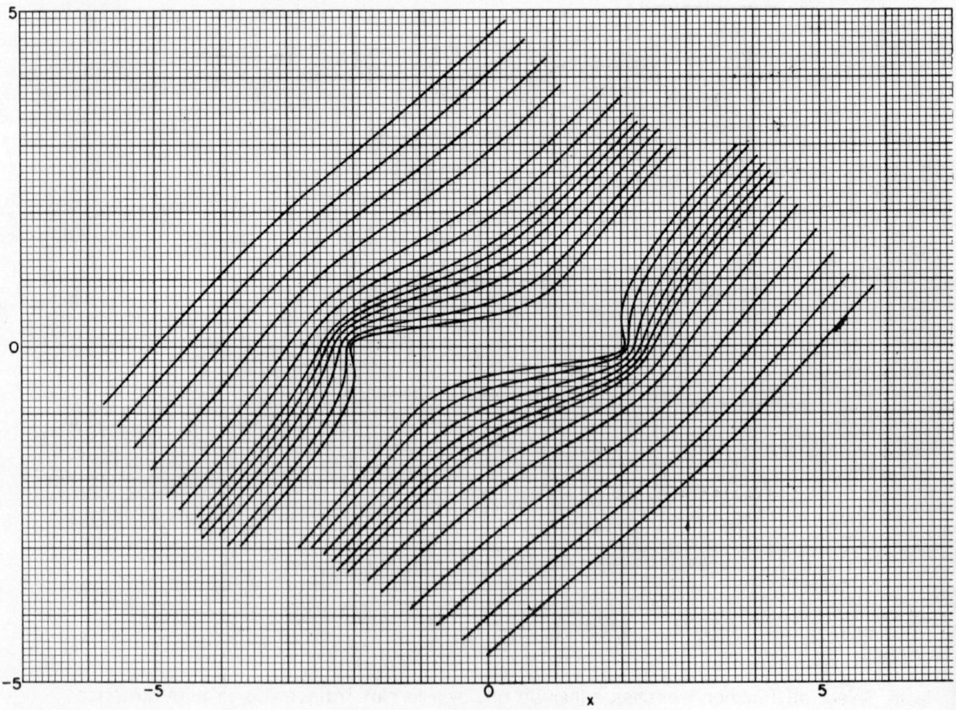

FIG. 15.13 Streamlines about a rod.

$$\dot{z} = F_1(F_1^* \cos \alpha + iz^* \sin \alpha) \tag{15.21}$$

But now, F_1 is easily generated by differential equations, since

$$\dot{F}_1 = \frac{z\dot{z}}{F_1} = z(F_1^* \cos \alpha + iz^* \sin \alpha) \tag{15.22}$$

Thus, the entire process of obtaining the streamlines requires just two complex multiplications, since the term $F_1^* \cos \alpha + iz^* \sin \alpha$ is generable with potentiometers. Figure 15.12 shows a circuit diagram that mechanizes (15.21) and (15.22) using servo multipliers while Fig. 15.13 contains the streamlines obtained. Runs were started on the real axis to make it simpler to calculate the initial conditions for F_1.

15.5 The Method of Steepest Descent

The method of steepest descent can be applied to track complex roots of n complex equations in n complex unknowns. Although such problems may be considered as consisting of $2n$ real equations in terms of $2n$ real unknowns, with real variable steepest-descent methods (Sec. 11.4) applicable, the Cauchy–Riemann relations simplify the circuitry in most cases. In this section, we examine the general method of steepest descent as applied to complex variables and look at specific applications and examples.

Suppose we have one function f of one variable z and wish to track the root for

$$f(z, \tau) \equiv u(x, y, \tau) + iv(x, y, \tau) = 0 \tag{15.23}$$

where f contains parameters changing with τ. This is equivalent to tracking the roots to

$$u(x, y, \tau) = 0, \qquad v(x, y, \tau) = 0$$

and using real variable methods described in Sec. 11.4. Whether we take the positive definite function as the sum of squares or the sum of absolute values, the partial derivatives of u and v with respect to x and y must be obtained. In the former case, we mechanize according to (11.35)

$$\dot{x} = -g_1\left(\frac{\partial u}{\partial x}u + \frac{\partial v}{\partial x}v\right), \qquad \dot{y} = -g_2\left(\frac{\partial u}{\partial y}u + \frac{\partial v}{\partial y}v\right)$$

Now we take $g_1 = g_2 = g$ and let

$$\frac{df}{dz} = r + is$$

From the Cauchy–Riemann conditions (15.2) and (15.3) we have

$$\dot{x} = -g(ru + vs), \qquad \dot{y} = -g(su - vr) \tag{15.24}$$

which can be concisely stated by

$$\dot{z} = -gf\left(\frac{df}{dz}\right)^* \tag{15.25}$$

Rybashov [11] has shown there are no false rest points; (15.25) will yield roots of (15.23) if they exist.

If we attempt to minimize the sum of absolute values:

$$V = |u| + |v|$$

then from (11.36), we must solve (taking $g_1 = g_2 = g$)

$$\dot{x} = -g(r \operatorname{sign} u + s \operatorname{sign} v), \qquad \dot{y} = -g(s \operatorname{sign} u - r \operatorname{sign} v) \tag{15.26}$$

Since sign u and sign v replace u and v in (15.26) we are motivated to define the complex sign function

$$\operatorname{sign} f = \operatorname{sign} u + i \operatorname{sign} v \tag{15.27}$$

to express (15.26) concisely as

$$\dot{z} = -g\left(\frac{df}{dz}\right)^* \operatorname{sign} f$$

Thus, the formulas derived are very similar to those for real variables.

Each situation is easily extendable to n complex functions of n complex variables:

Theorem 15.6. Given n analytic functions $f_j = u_j + iv_j$ of n complex variables $z_j = x_j + iy_j$, $j = 1, 2, \ldots, n$, the steep-descent equations for minimizing the function

$$V = \tfrac{1}{2}\sum_{j=1}^{n} f_j f_j^* = \tfrac{1}{2}\sum_{j=1}^{n} (u_j^2 + v_j^2)$$

are given by

$$\dot{z}_j = -g_j \sum_{k=1}^{n} \left(\frac{\partial f_k}{\partial z_j}\right)^* f_k, \qquad j = 1, 2, \ldots, n \tag{15.28}$$

where the g_j are positive real.

In matrix notation we have

$$\dot{\mathbf{z}} = -\mathbf{G}(\mathbf{J}')^*\mathbf{f}$$

where $\mathbf{G}$ is the gain matrix, $(\mathbf{J}')^*$ is the conjugate transpose of the Jacobian matrix, and $\mathbf{f}$ is the vector with components f_k. Similarly, to minimize (for faster convergence)

$$V = \sum_{j=1}^{n} (|u_j| + |v_j|)$$

we program

$$\dot{z}_j = -g_j \sum_{k=1}^{n} \left(\frac{\partial f_k}{\partial z_j}\right)^* \text{sign } f_k, \qquad j = 1, 2, \ldots, n$$

where the complex sign function is defined as in (15.27). Both methods reduce to the corresponding real formulas if the z_j are taken to be real. It must be emphasized, as in the case of real functions, the right-hand expressions must be generated by fixed function-generating equipment and not with integrators.

15.5.1 Polynomial with complex coefficients

Theorem 15.6 has direct application to polynomials with complex coefficients

$$f(z) = \sum_{j=0}^{n} a_j(\tau)z^j = u + iv \tag{15.29}$$

With a_j complex functions of τ, we also have

$$\frac{df}{dz} = \sum_{j=0}^{n} ja_j z^{j-1} = r + is \tag{15.30}$$

Thus, the components of f and df/dz, both needed for $\dot{z}$, can be formed by summing individual terms, once complex amplifiers representing the z^j are available. The z^j can be formed from the relationship

$$z^j = z^{j-1}z \tag{15.31}$$

when j is even or odd, or

$$z^j = (z^{j/2})^2 \tag{15.32}$$

when j is even. The latter method requires less multipliers and is generally preferable. The mechanization of $a_j z^j$ requires four pots for constant a_j or four multipliers for variable a_j (only two if a_j is real or pure imaginary). Finally, $f(df/dz)^*$ is formed from the basic rule for complex multiplication

$$f\left(\frac{df}{dz}\right)^* = \mathrm{Re}(f)\,\mathrm{Re}\left(\frac{df}{dz}\right) + \mathrm{Im}(f)\,\mathrm{Im}\left(\frac{df}{dz}\right) + i\left[\mathrm{Im}(f)\,\mathrm{Re}\left(\frac{df}{dz}\right) - \mathrm{Re}(f)\,\mathrm{Im}\left(\frac{df}{dz}\right)\right]$$

This procedure is direct and has been successfully used even for polynomials with real coefficients [1]. The following method, however, is more efficient in terms of equipment when the coefficients are real.

15.5.2 Polynomials with real coefficients

Polynomials with real coefficients arise frequently in practice; tracking their complex roots is a simpler matter. Although first attacked in polar coordinates, requiring the use of resolvers [12], an efficient and accurate circuit was subsequently developed by Kokotović and Šiljak [13] based on a transformation of Mitrović. The method is simple to mechanize. With definitions (15.29) and (15.30), we also define the variable

$$R = x^2 + y^2$$

and a set of functions $\psi_k(x, R)$ defined recursively

$$\psi_0 = 0, \qquad \psi_1 = -1, \qquad \psi_k = +2x\psi_{k-1} - R\psi_{k-2}$$

We may now show that

$$u = a_0 + R\sum_{k=1}^{n} a_k\psi_{k-1} - x\sum_{k=1}^{n} a_k\psi_k$$

$$v = -y\sum_{k=1}^{n} a_k\psi_k$$

$$r = \frac{\partial u}{\partial x} = x\sum_{k=1}^{n} ka_k\psi_{k-1} - \sum_{k=1}^{n} ka_k\psi_k$$

$$s = \frac{\partial v}{\partial x} = -y\sum_{k=1}^{n} ka_k\psi_{k-1}$$

from which either system (15.24) or (15.26) may be mechanized. A general diagram for the latter case is shown in Fig. 15.14 using servo multipliers to perform the needed multiplications for R and x, when generating the ψ_k.

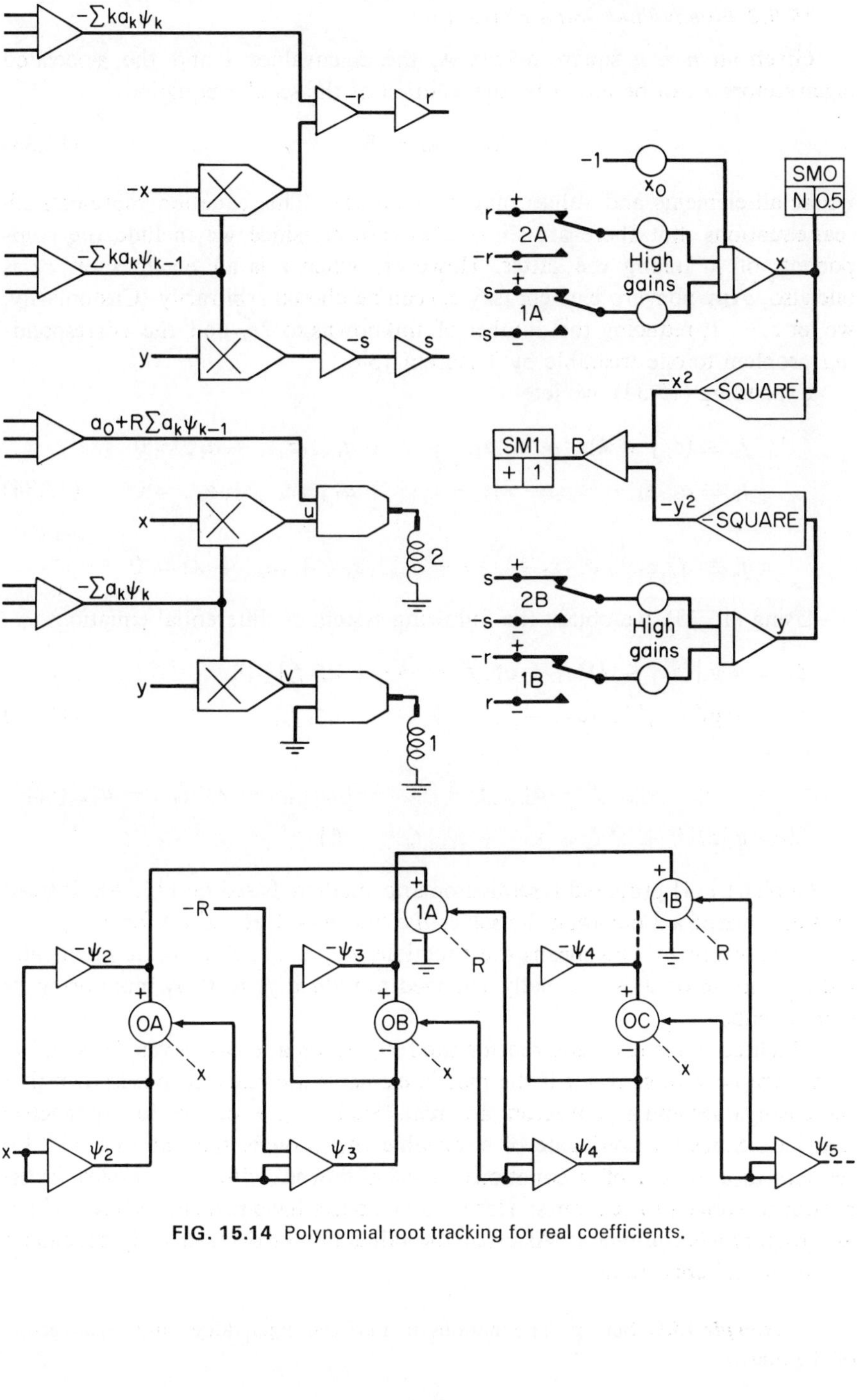

FIG. 15.14 Polynomial root tracking for real coefficients.

15.5.3 Eigenvalues and eigenvectors

Given an $n \times n$ square matrix $\mathbf{A}$, the eigenvalues λ and the associated eigenvectors $\mathbf{z}$ can be found by the solution of the secular equation

$$\mathbf{Az} - \lambda\mathbf{z} = \mathbf{0} \tag{15.33}$$

where all elements and values may be complex. This equation represents $2n$ real equations, but there are $2n + 2$ unknowns, since we include the components of λ among the latter. However, when $\mathbf{z}$ is an eigenvector, $c\mathbf{z}$ is one also. Any nonzero element (say z_n) can be chosen arbitrarily. Customarily, we let $z_n = 1$, reducing the number of unknowns to $2n$, and the corresponding problem to one treatable by Theorem 15.6.

Expanding (15.33), we let

$$\begin{aligned} f_1 &\equiv (a_{1,1} - \lambda)z_1 + a_{1,2}z_2 + \cdots + a_{1,n-1}z_{n-1} + a_{1,n} = 0 \\ f_2 &\equiv a_{2,1}z_1 + (a_{2,2} - \lambda)z_2 + \cdots + a_{2,n-1}z_{n-1} + a_{2,n} = 0 \\ &\vdots \\ f_n &\equiv a_{n,1}z_1 + a_{n,2}z_2 + \cdots + a_{n,n-1}z_{n-1} + (a_{n,n} - \lambda) = 0 \end{aligned} \tag{15.34}$$

Using (15.28), we obtain the following system of differential equations

$$\begin{aligned} \dot{z}_1 &= -g_1[(a_{1,1} - \lambda)^* f_1 + a_{2,1}^* f_2 + \cdots + a_{n,1}^* f_n] \\ \dot{z}_2 &= -g_2[a_{1,2}^* f_1 + (a_{2,2} - \lambda)^* f_2 + \cdots + a_{n,2}^* f_n] \\ &\vdots \\ \dot{z}_{n-1} &= -g_{n-1}[a_{1,n-1}^* f_1 + a_{2,n-1}^* f_2 + \cdots + (a_{n-1,n-1} - \lambda)^* f_{n-1} + a_{n,n-1}^* f_n] \\ \dot{\lambda} &= g_n[z_1^* f_1 + z_2^* f_2 + \cdots + z_{n-1}^* f_{n-1} + f_n] \end{aligned} \tag{15.35}$$

Gephart [14] proposed a semiautomatic method based on (15.28). Instead of mechanizing a $\dot{\lambda}$ equation, he solves the first $n - 1$ equations for $z_1, z_2, \ldots, z_{n-1}$ for a given λ. This means the functions $f_1, f_2, \ldots, f_{n-1}$ are automatically reduced to zero. λ is manually changed to reduce f_n to 0 by monitoring its components.

Much equipment is required for the fully automatic case, even for small n. Considerable savings result if the matrix elements are real and it is known that the eigenvalues and eigenvectors are real. Such a case occurs for symmetric matrices. Since the conjugate of a variable then equals the variable, (15.35) reduces to a system of n equations in n unknowns, with all complex cross-products reduced to real ones. Hermitian matrices have real eigenvalues, which also require less equipment. As usual, the values obtained are strongly dependent on the initial conditions.

Example 15.6. Set up the equations to find the eigenvalues and eigenvectors of the matrix

$$\begin{pmatrix} 0.1+i & 0.5 & 0.2 \\ -0.3 & 0.2-i & 0.8 \\ 0.3 & -0.6 & 0.3+0.2i \end{pmatrix}$$

Letting $z_3 = 1$, we have from (15.34)

$$f_1 \equiv (0.1+i)z_1 + 0.5z_2 + 0.2 - \lambda z_1 = 0$$
$$f_2 \equiv -0.3z_1 + (0.2-i)z_2 + 0.8 - \lambda z_2 = 0$$
$$f_3 \equiv 0.3z_1 - 0.6z_2 + 0.3 + 0.2i - \lambda = 0$$

and taking all $g_j = 1$, from (15.35),

$$\dot{z}_1 = -[(0.1-i)f_1 - 0.3f_2 + 0.3f_3] + \lambda^* f_1$$
$$\dot{z}_2 = -[0.5f_1 + (0.2+i)f_2 - 0.6f_3] + \lambda^* f_2$$
$$\dot{\lambda} = z_1^* f_1 + z_2^* f_2 + f_3$$

Thus, six complex products are required, or a minimum of 18 multipliers.

PROBLEMS

15-1. Given variables a, b, c, d, mechanize the circuit for complex division as given in (15.4) by a direct method.

15-2. Letting

$$x + iy = \frac{a+bi}{c+di}$$

discuss an implicit mechanization of x and y from

$$(x+iy)(c+di) - (a+bi) = 0$$

using some direct method.

15-3. Demonstrate that the function ln z must be multivalued by generating it from Theorem 15.3 using $\dot{z} = iz$.

15-4. Using a linear scan, set up the circuit to find the roots to

$$f(z) \equiv z^4 - 3z^3 + 3z^2 + 2i = 0$$

15-5. Repeat Problem 15-4 for a circular scan.

15-6. When roots to a polynomial lie outside the unit circle, two courses of action are possible. We may scale the roots with the transformation

$$z_1 = az, \qquad 0 < a < 1$$

compute $f(z_1/a)$ and find the roots z_1, or we may simply use

$$z_2 = \frac{1}{z}$$

and find the roots to $f(1/z_2)$. In the latter case, a new polynomial $w(z_2) = (z_2)^n f(1/z_2)$ can be investigated, since it has the same roots as $f(1/z_2)$. Apply both methods to find the roots of

$$f(z) \equiv z^4 + 0.1z^3 - 0.01z^2 + 0.003z + 0.0005 = 0$$

using the circular scan. Which method seems more reasonable for this problem?

15-7. Generate the functions $a_j z^j$ for use in the radial scan (Sec. 15.3.3).

15-8. Show that only half the integrators shown in Fig. 15.6 are really needed for the radial scan, since the real parts of z^j are always linearly related to the imaginary parts for any θ_0. Suggest an alternate mechanization using deMoivre's theorem.

15-9. The Newton-Raphson method for improving estimates of real roots of real functions is directly applicable to complex roots of analytic functions, because the Taylor series is also applicable. Using the value

$$z_0 = 0.15 + 0.67i$$

as an estimate of the root to (15.12), find the actual root to four significant places by iteration with

$$z_{j+1} = z_j - \frac{f(z_j)}{\left[\dfrac{df(z_j)}{dz}\right]}$$

15-10. Mechanize Example 15.3 with the Taylor series method of Sec. 15.3.4.

15-11. Petrich [15] uses the Taylor expansion in a special way to test for the stability of a polynomial (i.e., to see if the roots have negative real parts). Given

$$w(z) = \sum_{j=0}^{n} a_j z^j, \qquad a_j \text{ real}$$

the roots all have negative real parts if and only if the change in the argument of w

$$\Delta \arg w(z) = -\tfrac{1}{2}n\pi$$

as z varies from ∞ to 0 along the y axis. To mechanize for this condition, he noted that, since $x = 0$,

$$\frac{w^{(k)}(0)}{k!} = a_k$$

so that (15.13) becomes

$$\mathrm{Re}(iy) = a_0 - y^2\{a_2 - y^2[a_4 - y^2(a_6 - \cdots)]\}$$

$$\frac{\mathrm{Im}(iy)}{y} = a_1 - y^2\{a_3 - y^2[a_5 - y^2(a_7 - \cdots)]\}$$

Set up a circuit mechanizing these equations with ganged potentiometers.

15-12. Repeat Problem 15-11 using integrators to obtain the even powers of y, forming the equation at the output of a summer.

15-13. In another paper, Petrich and Madich [16] find analytical solutions of a linear differential equation of the form

$$\sum_{j=0}^{n} a_j \frac{d^j y}{dt^j} = 0$$

by finding the n roots of the corresponding algebraic equation

$$\sum_{j=0}^{n} a_j z^j = 0$$

on the analog computer. The exact solution, if all roots are distinct, is then

$$y = \sum_{j=1}^{n} c_j e^{z_j t}$$

where z_j are the roots, and c_j are constants determined by the initial conditions. Apply this method to the equation

$$\frac{d^4 y}{dt^4} + 0.1\frac{d^3 y}{dt^3} - 0.01\frac{d^2 y}{dt^2} + 0.003\frac{dy}{dt} + 0.0005y = 0$$

using the results of Problem 15-6. Initial conditions at $t = 0$ are $y = dy/dt = d^2y/dt^2 = d^3y/dt^3 = 1$.

15-14. For frequency analyses of transfer functions

$$f(z) = \frac{p(z)}{q(z)} = \frac{\sum_{j=0}^{n} a_j z^j}{\sum_{j=0}^{n} b_j z^j}$$

where a_j, b_j are real, Ogar [17] generates the real and imaginary parts of each polynomial with $\dot{z} = i$, $z_0 = 0$, in order to obtain $p(iy)$ and $q(iy)$. Nyquist and Bode plots can then be obtained by use of inverse resolvers and multipliers using the relationships

$$|f| = \frac{|p|}{|q|}, \qquad \arg f = \arg p - \arg q$$

Assuming $p(iy)$ and $q(iy)$ are available, draw a circuit diagram for $|f|$ and $\arg f$.

15-15. Very efficient circuits for $p(iy)$ and $q(iy)$ can be obtained using the radial scan in Problem 15-14. Since $x = 0$, this amounts to $\dot{y} = -y$, $y_0 = y_{\max}$. Generate all four polynomials with n integrators and show that, if two multipliers are available, only $n/2$ integrators need be used.

15-16. Since p and q become unbounded when $n \neq 0$, it might be advantageous to generate $f(z)$ directly with integrators, using $\dot{z} = i$, in almost the same way that rational functions were generated in Chapter 5. Let

$$z_j = \frac{z^j}{q}$$

We now have

$$f(z) = \sum_{j=0}^{n} a_j z_j$$

and the system of differential equations to generate z:

$$\dot{z}_j = \frac{jz^{j-1} - i\dfrac{dq}{dz} z_j}{q}$$

But dq/dz is another polynomial

$$\frac{dq}{dz} = \sum_{j=0}^{n} jb_j z^{j-1}$$

so that

$$\frac{\left(\dfrac{dq}{dz}\right)}{q} = \sum_{j=0}^{n} jb_j z_{j-1}$$

Hence

$$\dot{z}_j = jz_{j-1} - iz_j\left(\sum_{j=0}^{n} jb_j z_{j-1}\right)$$

The advantage of such a system is that y can be allowed to go to infinity while the functions remain bounded. Generate the real and imaginary parts of the transfer function

$$f(z) = \frac{3z - 4}{8z^2 + 5z + 2}$$

in this way.

15-17. Obtain the equations to plot the streamlines of flow around a corner from the complex potential

$$F(z) = Az^2$$

and mechanize the equations.

15-18. Repeat Problem 15-17 for flow in an angular wedge $\theta = \pi/n$ from the complex potential

$$F(z) = Az^n$$

15-19. As a method of testing the stability of a polynomial, Petrić [18] notes that the mapping

$$w = \frac{z+1}{z-1}$$

maps the left-hand plane onto the unit circle. The stability of

$$f(z) = \sum_{j=0}^{n} a_j z^j$$

can be tested by finding n roots of

$$f_1(z) = (z-1)^n f\left(\frac{z+1}{z-1}\right)$$

by using a circular scan at unit radius. Check the polynomial in Problem 15-4 for stability in this way.

15-20. Circulation can be added to the flow by adding to the potential of (15.17) the term of $i\Gamma \ln (z/R)$, Γ constant. How will this affect Fig. 15.10 for obtaining streamlines?

15-21. Using the method of steepest descent, set up a circuit to track the roots of $F(z) = e^{-z} + az$ as a changes with time.

15-22. Mechanize Eqs. (15.31) and (15.32).

15-23. Set up the circuit to track the roots of

$$f(z) = z^3 + 8z^2 + 3z + a(t)$$

using the theory of Sec. 15.5.2.

REFERENCES

[1] Hausner, A., "Analog Computer Techniques for Problems in Complex Variables," *IEEETEC*, vol. 14, Dec. 1965.

[2] "GEDA Construction of Conformal Maps in Fluid Dynamics," *Goodyear Aircraft Corp. Report GER-5672*, Akron, Ohio, Nov. 12, 1953.

[3] Bauer, L., and S. Fifer, "The Solution of Polynomial Equations on the REAC," *Symposium I on REAC Techniques*, Reeves Instr. Corp., New York, March 15-16, 1951 (ASTIA No. 108488).

[4] Petrich, J., "Trigonometric Method for Conformal Mapping with Algebraic Polynomials by the Use of a Repetitive Differential Analyser," *AICA*, vol. 3, Jan. 1961.

[5] Atkinson, C., "Polynomial Root Solving on the Electronic Differential Analyser," *Math. Tables and Other Aids to Comp.*, vol. 9, Oct. 1955.

[6] Madich, P., *et al.*, "The Use of a Repetitive Differential Analyzer for Finding Roots of Polynomial Equations," *IRETEC*, vol. 8, June 1959.

[7] Churchill, R. V., *Complex Variables and Applications*, 2nd ed. (New York: McGraw-Hill Book Co., 1960).

[8] Tomlinson, N. P., *et al.*, "Analog Computer Construction of Conformal Maps in Fluid Dynamics," *J. Applied Physics*, vol. 26, Feb. 1955.

[9] Heinhold, J., and T. H. München, "Konforme Abbildung Mittels Elektronischer Analogrechner," *Mathematik, Technik, Wirtschaft*, vol. 6, No. 2, 1959.

[10] Kulisch, U., "Behandlung von Differentialgleichungen im Komplexen auf dem Elektronischen Analogrechner," *AICA*, vol. 4, Jan. 1962.

[11] Rybashov, M. V., "Analog Solution of Algebraic and Transcendental Equations by the Gradient Method," *Automatika i Telemekhanika*, vol. 22, Jan. 1961.

[12] Levine, L., and H. F. Meissinger, "An Automatic Analog Computer Method for Solving Polynomials and Finding Root Loci," *IRE National Convention Record, Part 4*, 1957.

[13] Kokotović, P., and D. D. Šiljak, "Automatic Analog Solution of Algebraic Equations and Plotting of Root Loci by Generalized Mitrović's Method," *IEEE International Convention Record, Part 1*, 1964.

[14] Gephart, L., "Linear Algebraic Systems and the REAC," *Math. Tables and Other Aids to Comp.*, vol. 6, July 1952.

[15] Petrich, I. I., "Application of the Mikhailov Criterion in Investigating the Stability of Linear Systems by Means of Simulating Devices," *Automatika i Telemekhanika*, vol. 24, Nov. 1963.

[16] Petrich, J., and P. Madich, "Analytical Solution of Differential Equations with the Use of Differential Analyser," *AICA*, vol. 2, July 1960.

[17] Ogar, G. W., "Obtaining the Frequency Response of Physical Systems by Analog Computer Techniques," *IRE National Convention Record, Part 2*, 1961.

[18] Petrić, J., "A New Method of Examining the Stability of Linear Systems Using a Repetitive Differential Analyzer," *IRETEC*, vol. 10, Dec. 1961.

Other methods for obtaining eigenvalues and eigenvectors have been discussed by the following:

[19] Dekker, Ir. L., and W. Heins, "Resonance Method for the Determination of the Eigenvalues of Matrices on an Analog Computer," *AICA*, vol. 5, Oct. 1963.

[20] Neustadt, L., "A Method of Computing Eigenvectors and Eigenvalues on an Analog Computer," *Math. Tables and Other Aids to Comp.*, vol. 13, July 1959.

Chapter 16

PASSIVE NETWORK ANALYSIS

16.1 Basic Concepts

Almost since their inception, analog computers have been applied successfully to analyze passive electrical networks containing resistances, capacitances, and inductances. The equations involved require no multipliers when these are constant. Three procedures to mechanize an analog circuit have evolved: the nodal method (Sec. 16.2), the loop method (Sec. 16.3), and the branch method (Sec. 16.4). When ideal diodes are also included in the circuit, the branch mechanization can be extended easily to include them. Section 16.5 contains a method of dealing with ideal diodes.

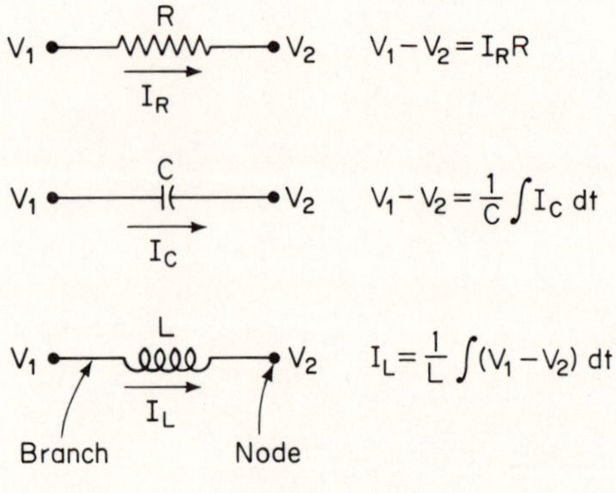

FIG. 16.1 Fundamental relations and symbols.

It is assumed the reader is familiar with the basic physics of electrical networks. Any of the following concepts can be reinforced by a standard text on circuit theory (e.g. [1]).

We take voltage drops in the direction of current flow. The basic relations, independent of whether or not L, R, or C are constant, are shown in Fig. 16.1. Independent nodes will be labeled V, while loop or branch currents will be labeled I, both with appropriate subscripts. Using operator notation, $p \equiv d/dt$, these relationships can be solved for currents

$$I_R = \frac{V_1 - V_2}{R}, \qquad I_C = p[C(V_1 - V_2)],$$
$$I_L = \frac{1}{L} p^{-1}(V_1 - V_2) \tag{16.1}$$

or voltage drops,

$$V_1 - V_2 = I_R R, \qquad V_1 - V_2 = \frac{1}{C} p^{-1}(I_C),$$
$$V_1 - V_2 = p(LI_L) \tag{16.2}$$

By convention, one of the nodes is taken as a reference node and is labeled 0 or with the symbol for ground.

16.1.1 Kirchhoff's laws

All equations of a circuit are derived from a fundamental set of rules, physically verified:

Kirchhoff's voltage law. The algebraic sum of voltage drops around any closed loop of branches is zero. In traversing the path around the loop, $-I$ is used for an expression when the direction of the path is not the same direction as the current. In other words, the sense of a voltage drop must be preserved.

Kirchhoff's current law. The algebraic sum of currents emanating from any node is zero. The sense of the current must also be preserved here.

To those familiar with analytical circuit analysis, it might seem, at this point, that the matter is ended. Simply write the equations and solve them on the computer or otherwise. That this is not so is evidenced by the early papers of Walters [2] and Scott [3]. It is possible to unknowingly mechanize a set of equations and get poor results because the mechanization is unstable. Besides describing the various mechanizations, the purpose of this chapter is to show how to avoid such unstable mechanizations.

16.1.2 The order of the characteristic equation

If computer instabilities exist in any of the three mechanizations to be described, it can usually be traced to one situation: The number of integrators used exceeds that minimum actually required, resulting in a matrix coefficient that has a zero determinant. The question of the order of the circuit's characteristic equation (which is the minimum number of integrators needed) is therefore quite important and should be predetermined by the programmer. Otterman [4] provides a systematic method for doing this, although the number is readily available by inspection for most networks.

To understand more fully how this number can be obtained, consider the circuit in Fig. 16.2, where each capacitor, resistor, and inductor, as well as each

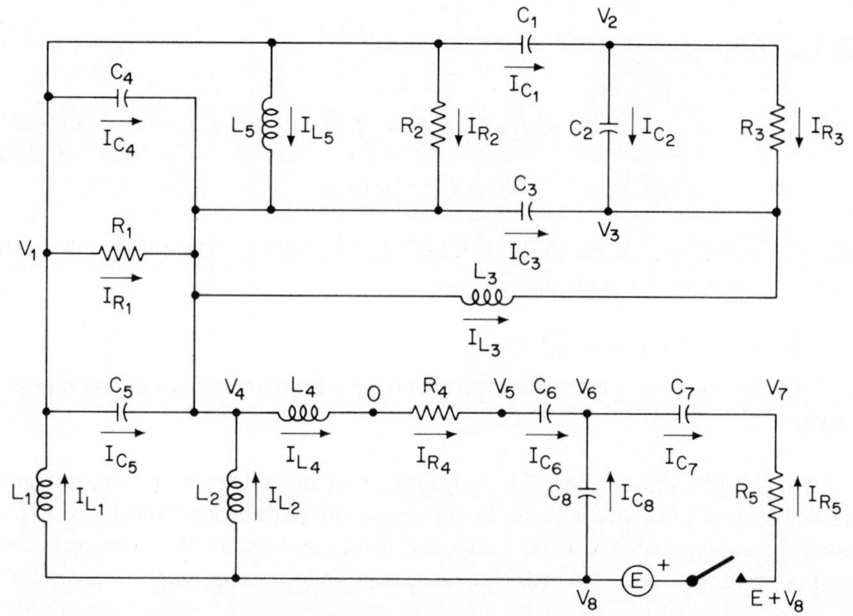

FIG. 16.2 Sample circuit.

branch current and each independent node has been properly labeled. The reference node marked with 0 was taken arbitrarily. There are five inductors and eight capacitors. These are the *energy storage* elements. From the relationships in (16.1) and (16.2), at most 13 integrators should be required in any mechanization. But we know from elementary considerations that capacitors in parallel could be replaced by a single capacitor without changing voltages or other currents in the circuit. The capacitors C_4 and C_5 could be replaced by one labeled $C_4 + C_5$, reducing the number of energy storage elements to 12 quickly. Thus, the order of the characteristic equation of the circuit can be less than the number of storage elements. But how much less?

Focus attention now on the loop of capacitors C_1, C_2, C_3, and C_4 in Fig. 16.2. From Kirchhoff's voltage law, we see that

$$V_1 - V_2 = \frac{1}{C_1}\int I_{C_1}\,dt$$

$$V_2 - V_3 = \frac{1}{C_2}\int I_{C_2}\,dt$$

$$V_4 - V_3 = \frac{1}{C_3}\int I_{C_3}\,dt$$

$$V_1 - V_4 = \frac{1}{C_4}\int I_{C_4}\,dt$$

Since these voltage drops form a closed loop,

$$\frac{1}{C_1}\int I_{C_1}\,dt + \frac{1}{C_2}\int I_{C_2}\,dt - \frac{1}{C_3}\int I_{C_3}\,dt - \frac{1}{C_4}\int I_{C_4}\,dt = 0 \qquad (16.3)$$

These integrals are therefore not independent. Two capacitors in parallel, in fact, is a special case of a loop of capacitors. It is clear that any loop of capacitors can reduce the required number of integrals by one. If the capacitors are constant, we also have

$$\frac{1}{C_1}I_{C_1} + \frac{1}{C_2}I_{C_2} - \frac{1}{C_3}I_{C_3} - \frac{1}{C_4}I_{C_4} = 0$$

Thus,

Theorem 16.1. Each independent loop of capacitors results in one algebraic equation relating the integrals of the currents through those capacitors. Any one of the integrals may be found in terms of the others. If the capacitors are constant in a loop, there exists a time-independent algebraic equation relating those currents.

The emphasis on independent loops is necessary to insure that the equations are independent (see Problem 16-1).

Similarly, there is a loop of inductors L_1, L_5, and L_2. We have

$$I_{L_1} = \frac{1}{L_1}\int (V_8 - V_1)\,dt$$

$$I_{L_5} = \frac{1}{L_5}\int (V_1 - V_4)\,dt \qquad (16.4)$$

$$I_{L_2} = \frac{1}{L_2}\int (V_8 - V_4)\,dt$$

so that

$$L_1 I_{L_1} + L_5 I_{L_5} - L_2 I_{L_2} = K \tag{16.5}$$

where $K = 0$ since there are no voltage or current sources in the loop (Problem 16-3). Thus, one of the currents may be calculated from the other two algebraically, so that one of the integrals in (16.4) need not be implemented. Again, we have the following theorem.

Theorem 16.2. Each independent loop of inductors results in one independent algebraic equation relating the currents through those inductors. Any one of the currents may be found in terms of the others thereby reducing the required integrals by one.

Loops of capacitors or inductors are not the only way algebraic equations may be found to reduce the number of integrators in a mechanization. Other ways involve the concept of a *cut set* defined in a simple, intuitive way, from the topology of the circuit.

Definition. Remove any n elements of the circuit (or imagine them cut in two). These elements form a cut set if the resulting circuit is divided into two unconnected parts, and each element contributes to this division. For purposes of this definition, a node is considered a part of a circuit, while current and voltage sources are shorted out.

The significance of a cut set lies in the following theorem, provable directly from Kirchhoff's law of currents.

Theorem 16.3. The algebraic sum of currents of elements forming a cut set, with sense preserved in the direction of one part, is zero.

In Fig. 16.2, the inductors L_1, L_2, and L_4 form a cut set. We therefore have

$$I_{L_1} + I_{L_2} - I_{L_4} = 0$$

from which

$$\frac{1}{L_1}\int (V_8 - V_1)\, dt + \frac{1}{L_2}\int (V_8 - V_4)\, dt - \frac{1}{L_4}\int V_4\, dt = 0$$

If the inductors are constant, we also have

$$\frac{1}{L_1}(V_8 - V_1) + \frac{1}{L_2}(V_8 - V_4) - \frac{V_4}{L_4} = 0$$

Thus,

Theorem 16.4. Each independent cut set of inductors results in one algebraic equation relating the integrals of the voltages across those inductors. Any one of the integrals may be found in terms of the others. If the inductors are constant, there exists a time-independent algebraic equation relating those voltages.

Note that L_1, L_2, L_3, and L_4 do not form a cut set because L_3 does not contribute to the division when each are "cut."

Similarly, capacitors C_6, C_7, and C_8 form a cut set since V_6 is isolated when they are cut. Hence,

$$I_{C_6} - I_{C_7} + I_{C_8} = 0$$

(known directly from Kirchhoff's law of currents), so that

$$\int I_{C_6}\,dt - \int I_{C_7}\,dt + \int I_{C_8}\,dt = K \tag{16.6}$$

where $K = 0$ if no initial charge on the capacitors is present. From (16.6) and (16.2) we have

$$C_6(V_5 - V_6) - C_7(V_6 - V_7) + C_8(V_8 - V_6) = K$$

Theorem 16.5. Each independent cut set of capacitors results in one independent algebraic equation relating the voltages around those capacitors. Any one of the voltages may be found in terms of the others thereby reducing the required integrals by one.

This section concludes with the final answer concerning the order of the characteristic equation of a network, as first described by Otterman [4].

Theorem 16.6. The order of the differential equation describing a network is equal to the sum of capacitors and inductors in the circuit less the sum of the number of independent loops of capacitors, loops of inductors, cut sets of capacitors, and cut sets of inductors.

An analog circuit can always, and indeed should, be arranged using this number of integrators.

We turn our attention in the following sections to specific examples for mechanizing the analog circuits using the three major methods. However, a word first about simplifying the diagram of electrical circuit on hand. Whenever possible, combine parallel or series resistors, capacitors, or inductors. In Fig. 16.2, for example, R_2 and R_1 can be replaced by a single resistor as can the parallel capacitors C_4 and C_5. Similarly, if branch currents are used, the law of currents requires that elements in series have the same current, if no other element is connected to their common node. In Fig. 16.2, we may write

$$I_{L_4} = I_{R_4} = I_{C_6}$$

It is simplest to label each current with the same symbol. In this regard, simple numerical subscripts to I are preferred over all others.

16.2 Nodal Mechanizations

The nodal method of circuit analysis is sometimes useful when the voltage at the nodes is of interest. Equations are obtained by writing the voltages at each nonreference node in terms of the currents and using Kirchhoff's current law to sum these to zero. For n nonreference nodes, this yields n equations in terms of the n unknown voltages. Care must be exercised here, however, because direct mechanization of these equations can lead to instabilities. To see this, we consider an *LRC* circuit with inductive coupling [2], as shown in Fig. 16.3.

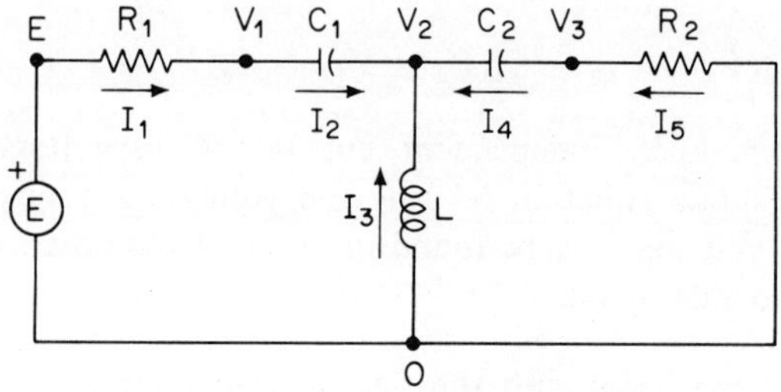

FIG. 16.3 LRC circuit with inductive coupling.

Example 16.1. Write the nodal equations of the *LRC* circuit in Fig. 16.3, assuming constant components. There are three nonreference nodes, V_1, V_2, V_3. We have at $V_1, I_1 = I_2$ or

$$\frac{E - V_1}{R_1} = C_1 \frac{d}{dt}(V_1 - V_2) \tag{16.7}$$

At $V_2, I_2 + I_3 + I_4 = 0$ or

$$C_1 \frac{d}{dt}(V_1 - V_2) - \frac{1}{L}\int V_2\,dt + C_2 \frac{d}{dt}(V_3 - V_2) = 0 \tag{16.8}$$

and at V_3, $I_4 = I_5$ or

$$C_2 \frac{d}{dt}(V_3 - V_2) = -\frac{V_3}{R_2} \tag{16.9}$$

In matrix form, this yields

$$\begin{pmatrix} C_1 & -C_1 & 0 \\ -C_1 & C_1 + C_2 & -C_2 \\ 0 & -C_2 & C_2 \end{pmatrix} \begin{pmatrix} \dfrac{dV_1}{dt} \\ \dfrac{dV_2}{dt} \\ \dfrac{dV_3}{dt} \end{pmatrix} = - \begin{pmatrix} \dfrac{V_1 - E}{R_1} \\ \dfrac{1}{L}\int V_2\, dt \\ \dfrac{V_3}{R_2} \end{pmatrix}$$

Any attempt to mechanize these equations directly by solving the algebraic equations on the computer is doomed to failure because the determinant of the coefficient matrix is 0. Note also that if dV_1/dt, dV_2/dt, and dV_3/dt are found by any method, four integrators are needed for the mechanization because V_1, V_3, and $\int V_2\, dt$ are needed as feedback. This number exceeds the order of the characteristic equation, which is three. Thus, the equations (16.7)-(16.9) could not be solved without further manipulation.

As an example of a circuit that can easily be solved by a nodal mechanization, consider the resistances and capacitors of the circuit of Fig. 16.3 interchanged [2], as shown in Fig. 16.4.

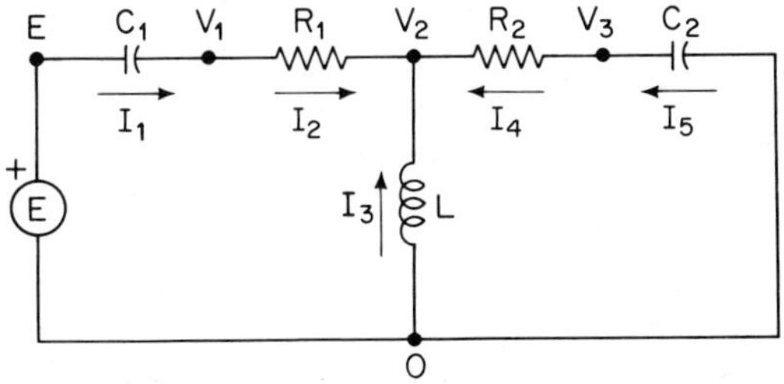

FIG. 16.4 Interchanging the capacitors and inductors.

Example 16.2. Write the nodal equations for the circuit in Fig. 16.4 and mechanize the solution.

We have

$$\frac{V_1 - V_2}{R_1} = C_1 \frac{d}{dt}(E - V_1)$$

$$\frac{V_3 - V_2}{R_2} = -C_2 \frac{dV_3}{dt}$$

and

$$\frac{V_1 - V_2}{R_1} + \frac{V_3 - V_2}{R_2} - \frac{1}{L}\int V_2\, dt = 0 \tag{16.10}$$

Now, only dV_1/dt and dV_3/dt appear in the equations with V_2 found by solving (16.10). The equations mechanized are

$$\frac{d}{dt}(E - V_1) = \frac{V_1 - V_2}{R_1 C_1}$$

$$\frac{dV_3}{dt} = -\frac{V_3 - V_2}{R_2 C_2}$$

$$V_2 = -\frac{R_1 R_2}{(R_1 + R_2)L}\int V_2\,dt + \frac{R_2}{R_1 + R_2}V_1 + \frac{R_1}{R_1 + R_2}V_3$$

which is straightforward as shown in Fig. 16.5.

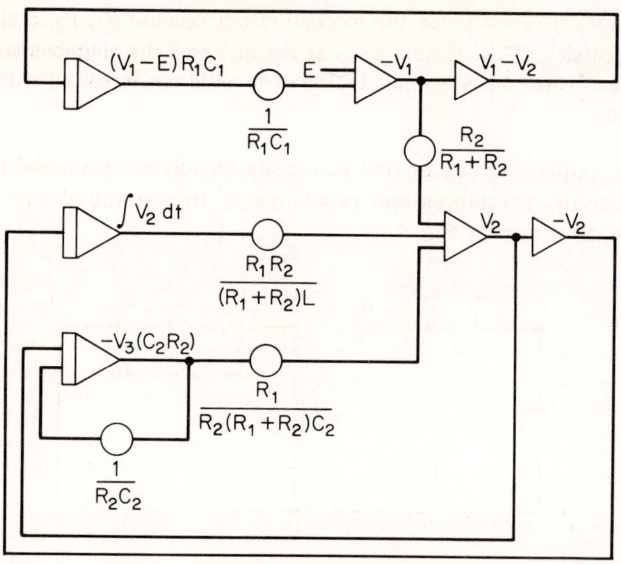

FIG. 16.5 Nodal mechanization.

This question naturally arises: When will a nodal mechanization be straightforward and when will difficulties arise in the derivative matrix? The answer was provided by Otterman [5], [6].

Theorem 16.7. Excessive integrators can be avoided in a nodal analysis if the network is *L-C continuous*, i.e., if lines can be traced from a node to every capacitor in the circuit by traversing only capacitors and lines can be traced from this same node to every inductor by traversing only inductors (short circuit all current and voltage sources in the trace). Use this node as the reference node when writing the equations.

The nodal analysis is extendible to circuits that are not *L-C* continuous [5], but such a technique involves linear transformations into new voltage variables. Most networks are more easily mechanized using the loop or branch analysis methods. Unless there are relatively few nodes compared with loops, or the circuit is L-C continuous (this includes resistance networks), the nodal analysis mechanization is not recommended.

16.3 Loop Mechanizations

Equations may also be found in terms of loop currents by applying Kirchhoff's voltage law and using the principle of superposition in linear systems. If there are n loops formed by the circuits, n equations can be written in terms of n loop current variables, and solved on the computer. Care must be exercised, however, that the loop current variables are chosen to minimize the number of integrators required and that this number is the order of the characteristic equation of the circuit. To see that difficulties can arise, we look at the following:

***Example 16.3** [6].* Mechanize the analog circuit of the network in Fig. 16.6 using the loop currents as indicated.

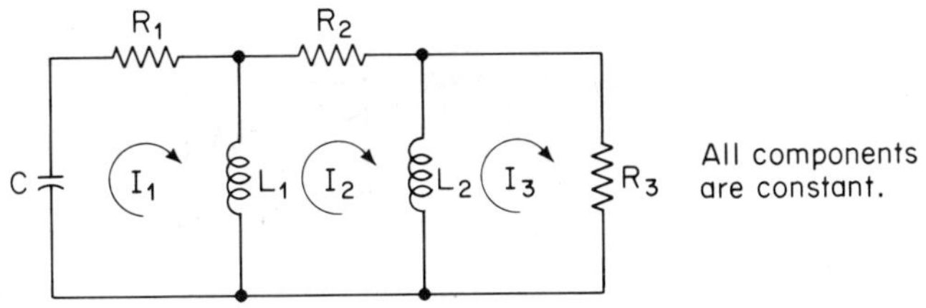

FIG. 16.6 Wrong choice of loop currents.

Applying Kirchhoff's law of voltage about each loop, we get

$$I_1 R_1 + L_1 \frac{dI_1}{dt} + \frac{1}{C}\int I_1\,dt - L_1 \frac{dI_2}{dt} = 0$$

$$I_2 R_2 + (L_1 + L_2)\frac{dI_2}{dt} - L_1 \frac{dI_1}{dt} - L_2 \frac{dI_3}{dt} = 0$$

$$I_3 R_3 + L_2 \frac{dI_3}{dt} - L_2 \frac{dI_2}{dt} = 0$$

or, in matrix notation

$$\begin{pmatrix} L_1 & -L_1 & 0 \\ -L_1 & L_1 + L_2 & -L_2 \\ 0 & -L_2 & L_2 \end{pmatrix} \begin{pmatrix} \dfrac{dI_1}{dt} \\ \dfrac{dI_2}{dt} \\ \dfrac{dI_3}{dt} \end{pmatrix} = \begin{pmatrix} -I_1 R_1 - \dfrac{1}{C}\int I_1\,dt \\ -I_2 R_2 \\ -I_3 R_3 \end{pmatrix}$$

Since the determinant of the coefficient matrix is 0, any direct mechanization to solve for the derivatives of the I_j on computer is doomed to failure. Note also that four integrators would be required to find I_1, I_2, I_3, and $\int I_1\,dt$ if the derivatives were available, while the order of the characteristic equation is only three.

On the other hand, if the loop currents are specified as shown in Fig. 16.7, the resulting equations are

$$L_2 \frac{dI_1}{dt} + R_3(I_1 + I_2 + I_3) = 0$$
$$L_1 \frac{dI_2}{dt} + (R_2 + R_3)I_2 + (R_2 + R_3)I_3 + R_3 I_1 = 0 \qquad (16.11)$$
$$\frac{1}{C}\int I_3\,dt + (R_1 + R_2 + R_3)I_3 + (R_2 + R_3)I_2 + R_3 I_1 = 0$$

One typical analog circuit for the solution of (16.11) with $\dot{\imath} = 1$ is shown in Fig. 16.8.

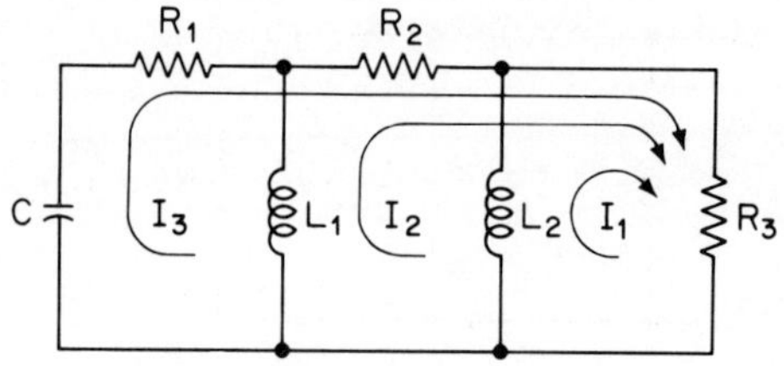

FIG. 16.7 Right choice of loop currents.

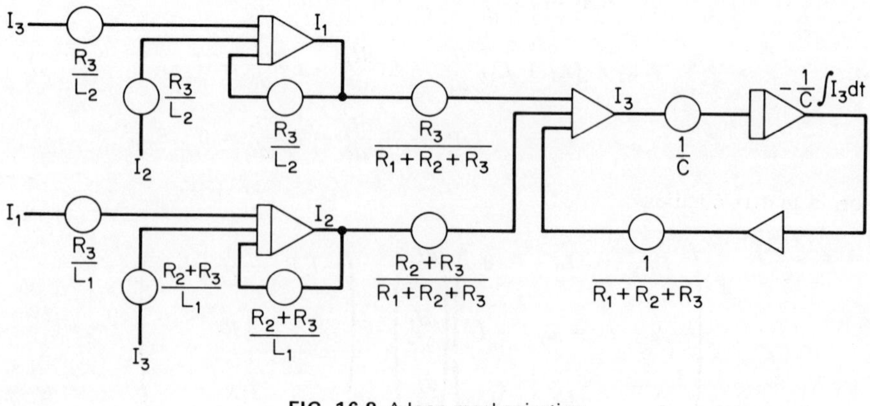

FIG. 16.8 A loop mechanization.

Note that

(1) Relatively few amplifiers are needed for the mechanization. In general, *loop mechanizations are amenable to minimum amplifier circuits.*

(2) All loop currents are made available. *Loop mechanizations are convenient if currents, their derivatives, or integrals are of interest.*

(3) No voltages are indicated. Additional equations must be mechanized to obtain voltage-time graphs.

(4) The mechanization is wrong if L_1, L_2, or C is variable. The voltage drop across a variable inductor L with current I is $d(LI)/dt$. The circuit, however, can be modified to include this change (see Problem 16-7).

(5) If component variables change from run to run, say R_3, it is quite inconvenient to make the changes. *Loop mechanizations are not easily amenable to minimum potentiometer circuits.*

The italicized statements above characterize the advantages and disadvantages of a loop mechanization.

16.3.1. Choosing loop current variables

Example 16.3 showed how important it was to choose loop current variables properly. A general method for choosing these variables has been determined by Otterman [6] and is summarized below.

(1) Start to redraw the circuit. Convert current sources to voltage sources and short out all voltage sources. Draw all resistors in the circuit. If any closed loops have been formed, trace loops through these elements.

(2) Add inductors or capacitors, one at a time. The term "new" elements now refers to those elements in any new loop formed through which no previous loop has been traced.

(3) If the new elements are inductors only, trace the new loop current through inductors only, if possible.

(4) If the new elements are inductors (where step 3 does not apply) or inductors and resistors, trace the new loop through inductors and resistors only, if possible. Otherwise, a loop with the smallest number of elements will suffice.

(5) If the new elements are capacitors only, trace the new loop through capacitors only, if possible.

(6) If the new elements are capacitors (where step 5 does not apply), or capacitors and resistors, trace the new loop through capacitors and resistors only, if possible. Otherwise, a loop with the smallest number of elements will suffice.

(7) If the new elements contain both capacitors and inductors, any loop will suffice.

The entire procedure is summarized in one sentence: *Maximize the independent number of loops containing like elements and minimize the independent number of loops containing both capacitors and inductors.*

In applying the above procedure, complete all loops first that do not contain both capacitors and inductors, to avoid "false starts" [5]. In addition, such a procedure allows using the same loops for writing Kirchhoff's law of voltages.

16.3.2 The equations for solution

The n equations obtained from Kirchhoff's voltage law now consist of the following (assuming each L is constant):

(1) If m distinct loops contain inductors, label the loop currents with subscripts 1 to m. Then there are m equations containing derivative terms dI_j/dt, $j = 1, 2, \ldots, m$. These are called *inductance equations*. With

$$\frac{d\mathbf{I}_L}{dt} = \begin{pmatrix} \dfrac{dI_1}{dt} \\ \dfrac{dI_2}{dt} \\ \vdots \\ \dfrac{dI_m}{dt} \end{pmatrix}$$

the equations have the form

$$\mathbf{L}\frac{d\mathbf{I}_L}{dt} - \mathbf{b} = \mathbf{0}$$

where $\mathbf{L}$ is an inductance matrix and $\mathbf{b}$ contains no derivative terms. The matrix $\mathbf{L}$ can be shown to be symmetric and positive definite and from Theorem 11.4, the direct method of solving algebraic equations can be used for its solution on the computer.

(2) If q distinct loops contain resistors and no inductors, label the loop currents with subscripts $m + 1$ to $m + q$. These are called *resistance equations*. Then there are q equations containing current terms I_k, $k = m + 1, m + 2, \ldots, m + q$. With

$$\mathbf{I}_R = \begin{pmatrix} I_{m+1} \\ I_{m+2} \\ \vdots \\ I_{m+q} \end{pmatrix}$$

these equations have the form

$$\mathbf{R}\mathbf{I}_R - \mathbf{d} = \mathbf{0}$$

where $\mathbf{R}$ is a resistance matrix and $\mathbf{d}$ contains no derivative terms dI_j/dt, $j = 1, 2, \ldots, m$ and no current terms I_k, $k = m + 1, \ldots, m + q$. The matrix $\mathbf{R}$ can also be shown to be symmetric and positive definite, so that the direct method for solving these equations can also be used.

(3) There remain $n - m - q$ equations from loops of capacitors. These are called *capacitance equations*. Label these loop currents I_i, $i = m + q + 1, \ldots, n$. With

$$\int \mathbf{I}_C \, dt = \begin{pmatrix} \int I_{m+q+1} \, dt \\ \int I_{m+q+2} \, dt \\ \vdots \\ \int I_n \, dt \end{pmatrix}$$

the equations have the form

$$\mathbf{C} \int \mathbf{I}_C \, dt - \mathbf{f} = \mathbf{0}$$

where $\mathbf{C}$ is a positive definite capacitance matrix and $\mathbf{f}$ contains no terms $\int I_i \, dt$, $i = m + q + 1, \ldots, n$ or I_k, $k = m + 1, \ldots, m + q$, or dI_j/dt, $j = 1, 2, \ldots, m$. The direct method for solving this system of algebraic equations is likewise used.

Thus, the equations contain three loosely coupled sets of algebraic equations for finding the highest-order derivative of the currents needed. These are solved by the direct method, integrated to produce needed feedback terms, in much the same way as Lagrange's equations were solved. Stability of the mechanization is assured because the coefficient matrix of each set of equations is positive definite.

Example 16.4. Mechanize the analog circuit for the network in Fig. 16.9.

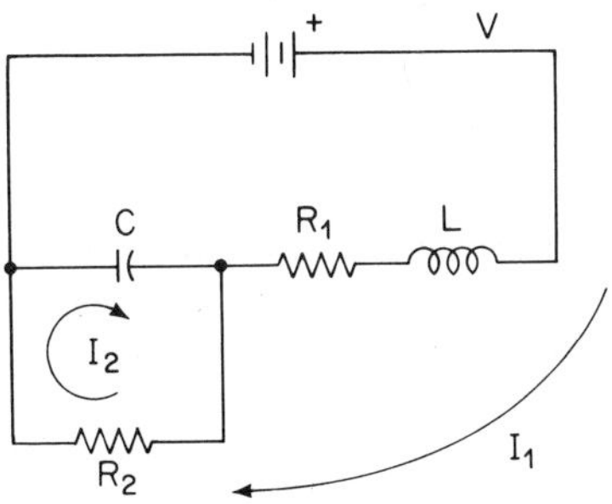

FIG. 16.9 Example 16.4.

If the inner loop containing C and L were completed first, a false start would have been made, since a loop can be traced through L that does not pass through C. Thus, I_1 is traced through L and R_2 (step 4). Similarly, I_2 is traced through C and R_2 (step 6). Now, we have

Inductance equation.

$$(1)\ L\frac{dI_1}{dt} + (R_1 + R_2)I_1 + R_2I_2 = V$$

Resistance equation.

$$(2)\ R_2I_1 + R_2I_2 + \frac{1}{C}\int I_2\,dt = 0$$

both of which are easily mechanized as shown in Fig. 16.10. A high-gain amplifier for the resistance equation need not have been used.

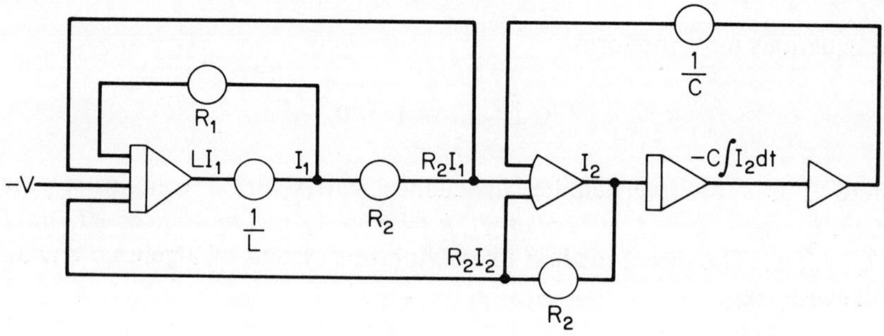

FIG. 16.10 Mechanization of Example 16.4.

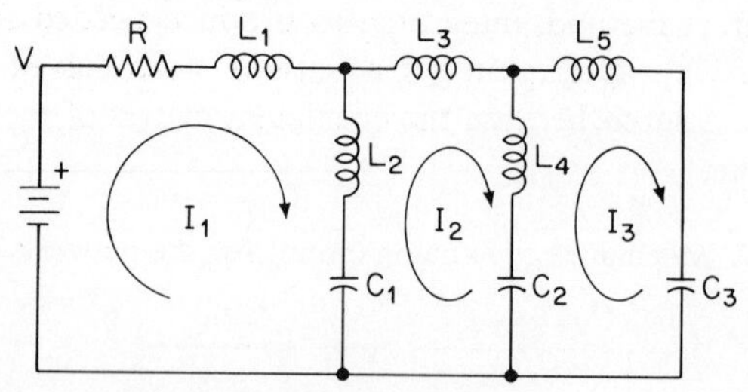

FIG. 16.11 Example 16.5.

Example 16.5 *[7].* Mechanize the analog circuit for the network in Fig. 16.11.

No loops can be completed without tracing at least one capacitor and one inductor. Hence, the simplest possible loops are chosen (step 7). We now have

Inductance equations.

$$(1)\ RI_1 + (L_1 + L_2)\frac{dI_1}{dt} - L_2\frac{dI_2}{dt} + \frac{1}{C_1}\int (I_1 - I_2)\,dt - V = 0$$

$$(2)\ (L_2 + L_3 + L_4)\frac{dI_2}{dt} - L_2\frac{dI_1}{dt} - L_4\frac{dI_3}{dt} + \frac{1}{C_1}\int (I_2 - I_1)\,dt + \frac{1}{C_2}\int (I_2 - I_3)\,dt = 0$$

$$(3)\ (L_4 + L_5)\frac{dI_3}{dt} - L_4\frac{dI_2}{dt} + \frac{1}{C_2}\int (I_3 - I_2)\,dt + \frac{1}{C_2}\int I_3\,dt = 0$$

There are no resistance or capacitance equations. These equations are solved algebraically for the dI_j/dt as shown in Fig. 16.12. Six integrators are needed, corresponding to the order of the characteristic equation for this circuit. The two inductance cut sets are automatically taken care of in the mechanization. In general, all loops and cut sets of both capacitors and inductors are automatically accounted for in loop mechanizations.

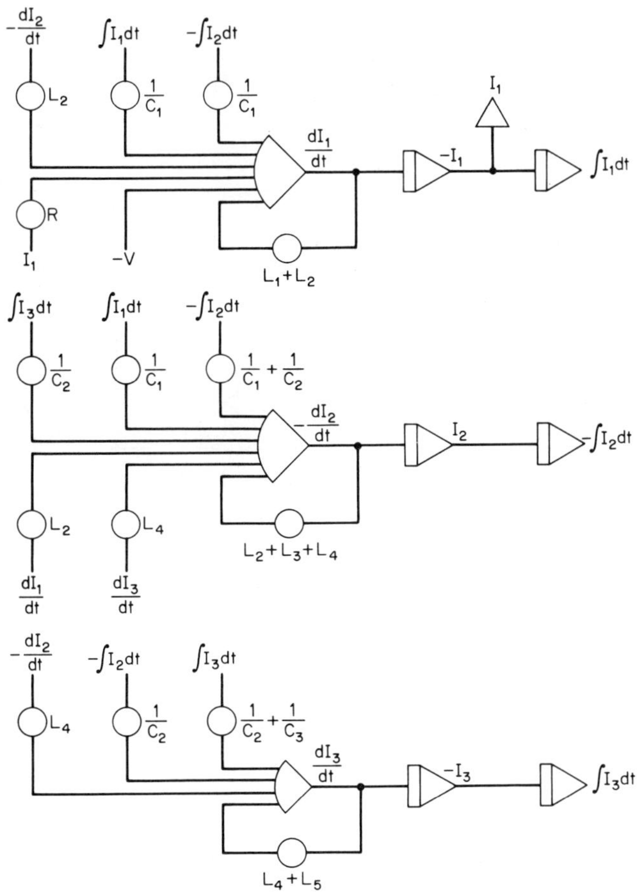

FIG. 16.12 A solution to Example 16.5.

Example 16.6. Mechanize the analog circuit of the network in Fig. 16.13.

There are two loops of capacitors, one of a resistor and capacitor, and a fourth with a resistor, capacitor, and inductor. The choice of loops is straightforward. We have

Inductance equation.

$$(1)\ R_1I_1 + L\frac{dI_1}{dt} + \frac{1}{C_1}\int (I_1 - I_3)\,dt = 0$$

Resistance equation.

$$(2)\ R_2I_2 + \frac{1}{C_7}\int (I_2 - I_4)\,dt - E = 0$$

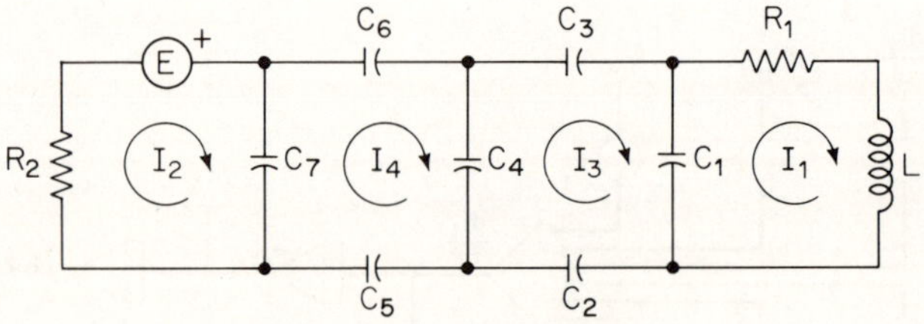

FIG. 16.13 Example 16.6.

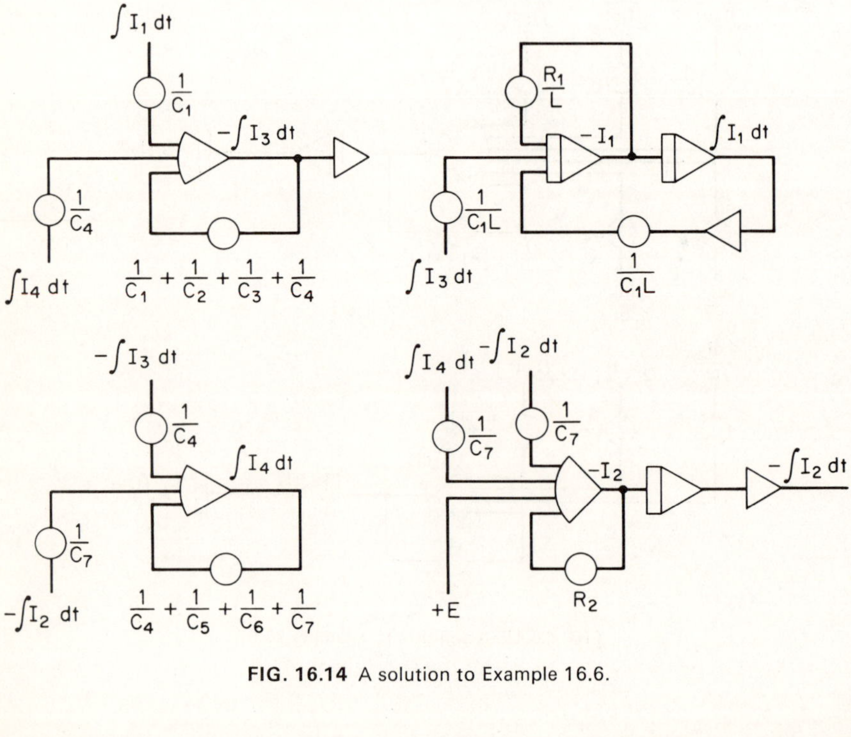

FIG. 16.14 A solution to Example 16.6.

Capacitance equations.

$$(3)\quad \left(\frac{1}{C_1}+\frac{1}{C_2}+\frac{1}{C_3}+\frac{1}{C_4}\right)\int I_3\,dt-\frac{1}{C_4}\int I_4\,dt-\frac{1}{C_1}\int I_1\,dt=0$$

$$(4)\quad \left(\frac{1}{C_4}+\frac{1}{C_5}+\frac{1}{C_6}+\frac{1}{C_7}\right)\int I_4\,dt-\frac{1}{C_4}\int I_3\,dt-\frac{1}{C_7}\int I_2\,dt=0$$

A mechanization with three integrators is shown in Fig. 16.14, where $\int I_3\,dt$ and $\int I_4\,dt$ are found algebraically from the capacitance equations. This network is quite difficult to analyze by other methods, and it tends to show up the strong advantage of loop mechanizations; namely, that there need be no concern about cut sets. Although there are eight energy storage components, there are three independent cut sets of capacitors (C_2 and C_3, C_5 and C_6, and C_3, C_4, and C_5) as well as the two loops. Thus, the characteristic equation of the circuit is only three.

Example 16.7. To show the application of the loop method to more complicated networks, we shall write a set of loop equations for the solution of the network in Fig. 16.2. It is more difficult to show loop currents in more complicated networks, since the procedure for assigning them would clutter up the diagram. Accordingly, we prepare Table 16.1 as elements are added to the partially drawn network consisting of resistors only. The number of integrators needed in the mechanization of any loop is found easily:

TABLE 16.1

Assignment of Loops

"New" elements added	*Loops chosen*	*By step*	*Integrators needed*
R_1, R_2 (Initial loop)	R_1, R_2	1	0
R_3, C_2	R_3, C_2	6	1
C_5	R_1, C_5	6	1
C_4	C_4, C_5	5	0
L_5	R_2, L_5	4	1
C_1, C_3	$C_1, C_2, C_3\, C_4$	5	0
L_1, L_2	$L_1, L_5\, L_2$	3	0
R_5, C_7, C_8	C_7, R_5, C_8	6	1
L_3	C_3, L_3	4	2
L_4, R_4, C_6	L_2, L_4, R_4, C_6, C_8	7	2
			8 Total

(1) Loops with like elements require no integrators since they are involved only in algebraic equations.

(2) Loops with resistors and inductors require only one integrator. Both dI/dt and I appear in the equations.

(3) Loops with resistors and capacitors require only one integrator. Both I and $\int I\,dt$ appear in the equations.

(4) Loops with inductors and capacitors require two integrators because dI/dt, I, and $\int I\,dt$ appear in the equations.

Now, rearrange the list of loops in the order for labeling of the currents. All loops with an inductor are listed first, those remaining with a resistor listed next, and the remaining capacitor loops last (Table 16.2).

TABLE 16.2

Labeling of Loops

Loop	Elements	
I_1	R_2, L_5	For inductance equations
I_2	C_3, L_3	
I_3	L_1, L_5, L_2	
I_4	L_2, L_4, R_4, C_6, C_8	
I_5	R_1, R_2	For resistance equations
I_6	R_3, C_2	
I_7	R_1, C_5	
I_8	R_5, C_7, C_8	
I_9	C_4, C_5	For capacitance equations
I_{10}	C_1, C_2, C_3, C_4	

Using this list, write the loop equations, assuming all currents are traced clockwise. If an element is involved with more than one loop current, refer to the diagram to determine the proper sign. We have

Inductance equations.

$$
\begin{aligned}
I_1:&\quad R_2I_1 + L_5\frac{dI_1}{dt} + R_2I_5 - L_5\frac{dI_3}{dt} = 0\\
I_2:&\quad L_3\frac{dI_2}{dt} + \frac{1}{C_3}\int I_2\,dt - \frac{1}{C_3}\int I_{10}\,dt = 0\\
I_3:&\quad (L_1 + L_2 + L_5)\frac{dI_3}{dt} - L_5\frac{dI_1}{dt} - L_2\frac{dI_4}{dt} = 0\\
I_4:&\quad (L_2 + L_4)\frac{dI_4}{dt} + R_4I_4 + \left(\frac{1}{C_6} + \frac{1}{C_8}\right)\int I_4\,dt - L_2\frac{dI_3}{dt} - \frac{1}{C_8}\int I_8\,dt = 0
\end{aligned}
\tag{16.12}
$$

Resistance equations.

$$
\begin{aligned}
I_5:&\quad (R_1 + R_2)I_5 - R_1I_7 + R_2I_1 = 0\\
I_6:&\quad R_3I_6 + \frac{1}{C_2}\int I_6\,dt - \frac{1}{C_2}\int I_{10}\,dt = 0\\
I_7:&\quad R_1I_7 + \frac{1}{C_5}\int I_7\,dt - R_1I_5 + \frac{1}{C_5}\int I_9\,dt = 0\\
I_8:&\quad R_5I_8 + \left(\frac{1}{C_7} + \frac{1}{C_8}\right)\int I_8\,dt - \frac{1}{C_8}\int I_4\,dt - E = 0
\end{aligned}
$$

Capacitance equations.

$$I_9: \quad \left(\frac{1}{C_4} + \frac{1}{C_5}\right)\int I_9\,dt - \frac{1}{C_4}\int I_{10}\,dt + \frac{1}{C_5}\int I_7\,dt = 0$$

$$I_{10}: \quad \left(\frac{1}{C_1} + \frac{1}{C_2} + \frac{1}{C_3} + \frac{1}{C_4}\right)\int I_{10}\,dt - \frac{1}{C_2}\int I_6\,dt - \frac{1}{C_4}\int I_9\,dt - \frac{1}{C_3}\int I_2\,dt = 0$$

As a check on the correctness of the equations, particularly with regard to signs, we observe the rule: *Any term in the equation for I_i involving element E_j and $f(I_k)$ (with $i \neq k$) must appear in the equation for I_k as a term involving E_j and $f(I_i)$ with like sign. If $i = k$, the term is positive.*

Mechanization of the equations is straightforward. Using high-gain amplifiers, solve for dI_1/dt through dI_4/dt, integrating these for any required integrals. Similarly, solve for I_5 through I_8, integrating once for any required integrals. Finally, solve for $\int I_9\,dt$ and $\int I_{10}\,dt$ algebraically. The outputs of all amplifiers including the ones producing the derivative terms must be scaled.

For cases where the inductances are variable, see Problem 16-8.

If an initial voltage is placed across any capacitor, the right-hand expression of any equation containing that capacitor is set equal to that voltage with sense preserved.

16.4 Branch Mechanizations

A serious disadvantage of the loop mechanization can be the lack of simple control of the parameters of the network. If the inductor L_5 in (16.12), for example, had to be changed from run to run, three potentiometers must be changed each time. When solutions are observed on an oscilliscope to ascertain the effect of any parameter, it is much more convenient to have one potentiometer control the value of one parameter. Such an analog circuit is referred to as a *minimum potentiometer* circuit. Of course, loop mechanizations can be arranged in that way but would require many multipliers.

A branch mechanization can produce minimum potentiometer circuits at the expense of adding more amplifiers to the circuit as well as somewhat more effort to set up the circuit. There are other advantages to branch mechanizations. All voltages and branch currents are produced as outputs on the computer without additional mechanizations. It is much easier to mechanize a circuit with ideal diodes as will be seen in the next section. Finally, it is often possible (for simple networks) to draw the analog circuit without equations in a procedure called "direct simulation [8]". This is not recommended until a programmer has sufficient experience setting up analog circuits for networks.

To set up a branch mechanization for a network in a minimum potentiometer circuit, each node is labeled with a voltage variable with respect to a reference node, and each branch is labeled with a branch current which passes through it, in the same way as was done in Fig. 16.2. We note here that equations are simpler to write if the reference node is chosen to be the one that has the most elements connected to it. Suppose there are m elements and n nonreference nodes. Then there are $m + n$ variable voltages and currents. At each nonreference node, write the node current equation from Kirchhoff's current law. For each element, write the branch voltage equation [from (16.1) or (16.2)]. There are $m + n$ equations and $m + n$ unknowns, which can be mechanized on the computer. The matter is not ended here, however. If the $m + n$ equations are mechanized directly, there will be as many integrators used as there are storage elements. But we have seen the difficulties possible with such a setup. Accordingly, if there are q algebraic equations possible from cut sets of capacitors and cut sets of inductors (from Theorems 16.4 and 16.5), these equations are also mechanized. This means that q equations of the original $m + n$ are not mechanized. As will be seen, a selective process determines which they are. If there are loops of capacitors or loops of inductors, a special technique is required that will be discussed separately.

16.4.1 Procedures when there are no loops of capacitors or of inductors

Each branch equation of the original m contains two voltage variables and a branch current variable. Because we do not wish to use derivative circuits, two rules are followed in the process of mechanizing the equations (for minimum potentiometer circuits with no loops of like energy storage elements).

(1) Use a branch equation for an inductor to generate the branch current. If the inductor is part of a cut set of inductors, all but one equation is so used.

(2) Use a branch equation for a capacitor to generate one of the voltages on either side of the capacitor. A selection process determines which one to generate. If the capacitor is part of a cut set of capacitors, all but one equation is so used.

In addition, we have

(3) Use equations for a cut set to generate a voltage variable. If the cut set is formed by the junction of the elements, the voltage at the junction must be generated. Otherwise, any voltage at either end of one of the elements is satisfactory.

(4) Use some of the branch equations for resistors to generate the remaining voltage variables. A selective process determines which they are.

(5) The remaining equations generate the rest of the current variables. Be sure that a cut-set current equation replaces one of the nodal current equations.

When followed in the order listed, the above process uses each equation to generate one of the variables in the network, and the mechanization is straightforward.

In the following examples, the equations are first segregated by type and then rearranged to simplify the selection by which variables are generated. Although not strictly necessary, this procedure provides a systematic means for drawing the analog circuit.

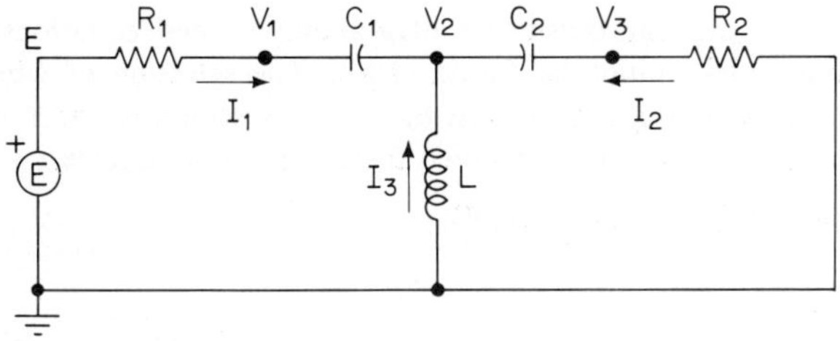

FIG. 16.15 Example 16.8.

Example 16.8. We begin by mechanizing the circuit in Fig. 16.3 by the branch method, first by simplifying the diagram to show only three branch currents as in Fig. 16.15. We need show only nodal equations at nodes where there are junctions of three or more elements. We have

Branch equations.

(1) $E - V_1 = I_1 R_1$

(2) $V_1 - V_2 = \frac{1}{C_1}\int I_1\,dt$

(3) $V_3 - V_2 = \frac{1}{C_2}\int I_2\,dt$

(4) $V_3 = -I_2 R_2$

(5) $I_3 = -\frac{1}{L}\int V_2\,dt$

Nodal equation.

(6) $I_1 + I_2 + I_3 = 0$

These are rearranged and the variable generated by each equation is specified.

Inductor equation (step 1).

(1) I_3: $\quad I_3 = -\frac{1}{L}\int V_2\,dt$

Capacitor equations (step 2).

(2) V_2: $$V_2 = V_1 - \frac{1}{C_1}\int I_1\,dt$$

(3) V_3: $$V_3 = V_2 + \frac{1}{C_2}\int I_2\,dt$$

Note here that V_2 appears outside an integral only in these equations. If not found from one of them, V_2 cannot be found at all. The selection of which voltage to generate from a capacitor equation must be made by scanning the remaining equations and choosing first the ones for which there is no other choice.

Resistor equation for voltage (step 4).

(4) V_1: $$V_1 = E - I_1 R_1$$

There was no other choice for generating V_1.

Remaining equations for current (step 5).

(5) I_2: $$I_2 = -\frac{V_3}{R_2}$$

(9) I_1: $$I_1 = -I_2 - I_3$$

The mechanization for these six equations is now straightforward, and the final circuit is shown in Fig. 16.16. Unlike the loop mechanization, *all amplifier loops contain an odd number of amplifiers*. Also, *all variables are available on the computer* and *each parameter is controlled by one and only one pot*. These italicized state-

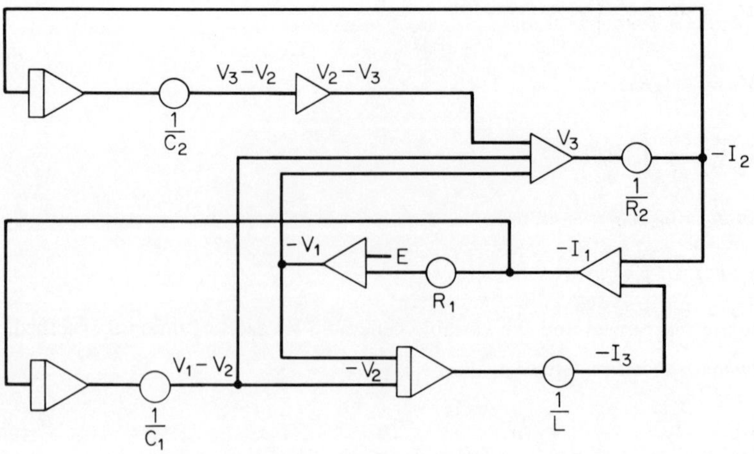

FIG. 16.16 A mechanization of Example 16.8.

ments are characteristic of branch mechanizations when no loops of like storage elements are present.

Example 16.9. Mechanize the network shown in Fig. 16.11 by the branch method.

This network is redrawn in Fig. 16.17 to show branch currents and node voltages. We have

Branch equations.

$$(1)\ V_5 = -\frac{1}{C_1}\int I_2\, dt$$

$$(2)\ V_6 = -\frac{1}{C_2}\int I_4\, dt$$

$$(3)\ V_4 = \frac{1}{C_3}\int I_5\, dt$$

$$(4)\ V - V_1 = RI_1$$

$$(5)\ I_1 = \frac{1}{L_1}\int (V_1 - V_2)\, dt \qquad (16.13)$$

$$(6)\ I_2 = \frac{1}{L_2}\int (V_5 - V_2)\, dt$$

$$(7)\ I_3 = \frac{1}{L_3}\int (V_3 - V_2)\, dt$$

$$(8)\ I_4 = \frac{1}{L_4}\int (V_6 - V_3)\, dt$$

$$(9)\ I_5 = \frac{1}{L_5}\int (V_3 - V_4)\, dt$$

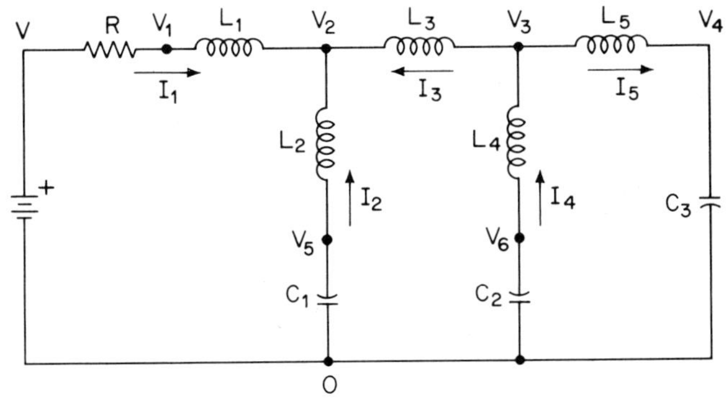

FIG. 16.17 Example 16.9

Nodal equations.

(10) $I_1 + I_2 + I_3 = 0$

(11) $I_4 - I_3 - I_5 = 0$

Note that V_2 and V_3 do not appear outside of integrals but are needed for feedback. This occurs whenever cut sets are formed from the junction of inductors. They are generated, however, from the equations derived by differentiating the nodal equations at the junctions.

Cut-set equations.

(12) $\frac{1}{L_1}(V_2 - V_1) + \frac{1}{L_2}(V_2 - V_5) + \frac{1}{L_3}(V_2 - V_3) = 0$

(13) $\frac{1}{L_4}(V_3 - V_6) + \frac{1}{L_3}(V_3 - V_2) + \frac{1}{L_5}(V_3 - V_4) = 0$

These two extra equations replace two of the integrals in (16.13). Arbitrarily, we choose to eliminate the equations for I_1 and I_5. Rearranging, we have

Inductor equations (step 1).

(1) I_2: $\quad I_2 = \frac{1}{L_2}\int (V_5 - V_2)\,dt$

(2) I_3: $\quad I_3 = \frac{1}{L_3}\int (V_3 - V_2)\,dt$

(3) I_4: $\quad I_4 = \frac{1}{L_4}\int (V_6 - V_3)\,dt$

Capacitor equations (step 2).

(4) V_5: $\quad V_5 = -\frac{1}{C_1}\int I_2\,dt$

(5) V_6: $\quad V_6 = -\frac{1}{C_2}\int I_4\,dt$

(6) V_4: $\quad V_4 = \frac{1}{C_3}\int I_5\,dt$

Cut-set equations (step 3).

(7) V_2: $\quad \frac{1}{L_1}(V_2 - V_1) + \frac{1}{L_2}(V_2 - V_5) + \frac{1}{L_3}(V_2 - V_3) = 0$

(8) V_3: $\quad \frac{1}{L_4}(V_3 - V_6) + \frac{1}{L_3}(V_3 - V_2) + \frac{1}{L_5}(V_3 - V_4) = 0$

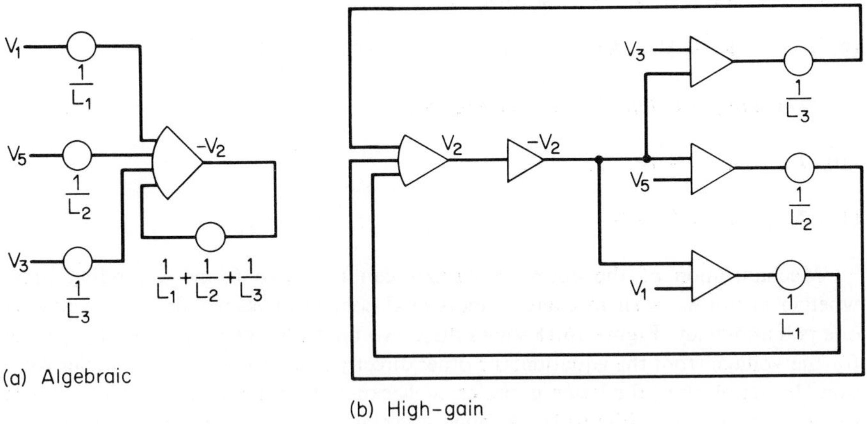

FIG. 16.18 Two methods of mechanizing a cut-set relation.

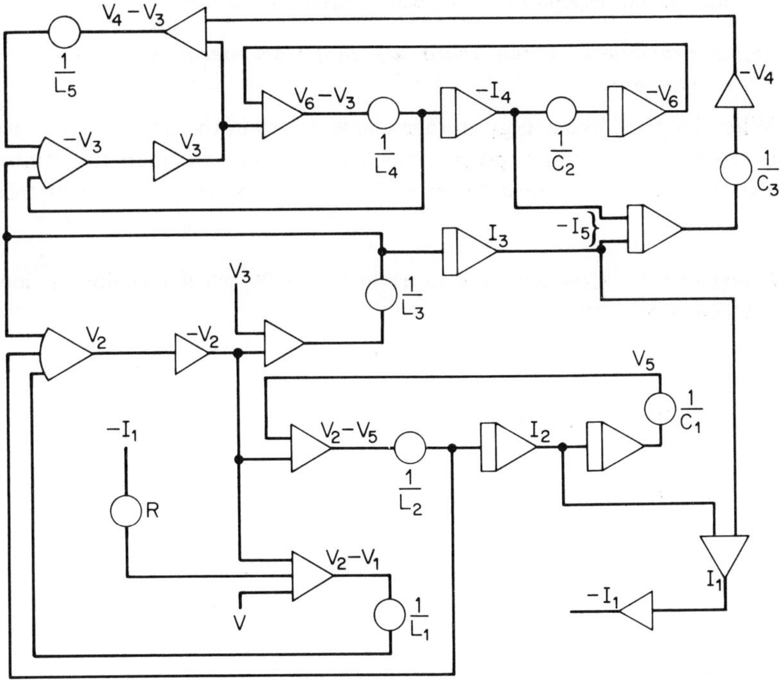

FIG. 16.19 A mechanization of Example 16.9.

Resistor equation for voltage (step 4).

(9) V_1: $V_1 = V - RI_1$

Remaining equations for current (step 5).

(10) I_1: $I_1 = -I_2 - I_3$

(11) I_5: $I_5 = I_4 - I_3$

Mechanization of the cut-set equations can take two forms depending upon whether or not we wish to exercise individual control of each inductance value by one potentiometer. Figure 16.18 shows these two methods—one algebraic, by solving for the voltage from the equation; the other directly from a high-gain amplifier solution. In many cases, the latter is preferred because the generated variables are also needed for feedback. Figure 16.19 shows a sample mechanization. Note that V_3 and V_2 are each involved in one algebraic loop, but an inspection of the matrix shows it to be positive definite as long as the generation of each nodal voltage is solved from the current equation from that node. Note also that many more amplifiers are needed than the corresponding loop mechanization in Fig. 16.12.

16.4.2 Procedures when there are loops of energy storage elements

When loops of like energy storage elements are in the network, the procedures to obtain a minimum potentiometer circuit and maintain the minimum number of integrators must be modified. That this is so is evident from a simple example.

Example 16.10. Consider the network in Fig. 16.20, which contains one loop of four inductors. We have

Branch equations.

(1) $I_1 = \frac{1}{L_1}\int (V_1 - E)\,dt$

(2) $I_2 = \frac{1}{L_2}\int (E - V_3)\,dt$

(3) $I_3 = \frac{1}{L_3}\int (V_3 - V_2)\,dt$

(4) $I_4 = \frac{1}{L_4}\int (V_2 - V_1)\,dt$

(5) $V_3 - V_1 = \frac{1}{C}\int I_5\,dt$

(6) $V_2 = -I_6 R$

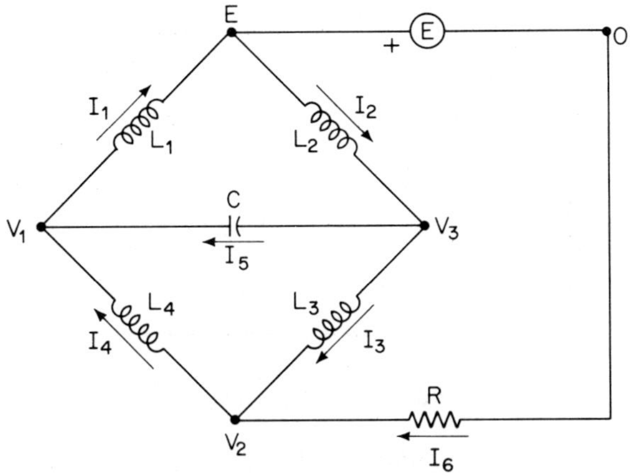

FIG. 16.20 Example 16.10.

Nodal equations.

(7) $I_4 + I_5 - I_1 = 0$

(8) $I_3 + I_6 - I_4 = 0$ (16.14)

(9) $I_3 + I_5 - I_2 = 0$

We note first that only two of the equations contain voltage variables outside an integral. The third comes from the cut set formed by L_1, L_2, L_3, and L_4:

Cut-set equation.

(10) $I_1 - I_2 + I_3 - I_4 = 0 \Rightarrow$ (by differentiation)

$$\frac{1}{L_1}(V_1 - E) - \frac{1}{L_2}(E - V_3) + \frac{1}{L_3}(V_3 - V_2) - \frac{1}{L_4}(V_2 - V_1) = 0$$

This equation must be used to generate either V_1 or V_3. If one pot per component is to be mechanized, a partial circuit can be drawn as in Fig. 16.21, using the cut-set equation to generate V_3. Now we also have a

Loop equation.

$$L_1 I_1 + L_2 I_2 + L_3 I_3 + L_4 I_4 = 0 \tag{16.15}$$

The best procedure to properly use the inductor equations is to first rearrange nodal equations to include any cut-set relation as required in step 5. Thus, (16.14) is changed to

$$\begin{aligned} I_1 - I_2 + I_3 - I_4 &= 0 \\ I_3 + I_6 - I_4 &= 0 \\ I_3 + I_5 - I_2 &= 0 \end{aligned}$$

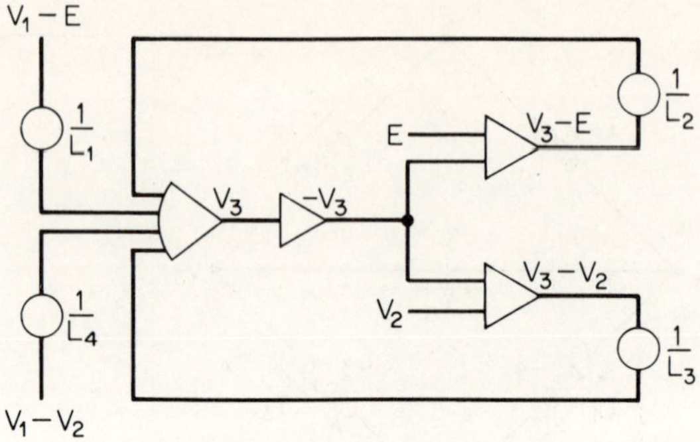

FIG. 16.21 A mechanization for V_3.

Next differentiate those equations involving the loop inductor currents (in this case, all of them):

$$\frac{dI_1}{dt} - \frac{dI_2}{dt} + \frac{dI_3}{dt} - \frac{dI_4}{dt} = 0$$

$$\frac{dI_3}{dt} + \frac{dI_6}{dt} - \frac{dI_4}{dt} = 0 \qquad (16.16)$$

$$\frac{dI_3}{dt} + \frac{dI_5}{dt} - \frac{dI_2}{dt} = 0$$

Substitute the derivative expressions for each inductor equation in (16.16):

$$\frac{1}{L_1}(V_1 - E) - \frac{1}{L_2}(E - V_3) + \frac{1}{L_3}(V_3 - V_2) - \frac{1}{L_4}(V_2 - V_1) = 0 \qquad (16.17)$$

and

$$\frac{1}{L_3}(V_3 - V_2) + \frac{dI_6}{dt} - \frac{1}{L_4}(V_2 - V_1) = 0$$

$$\frac{1}{L_3}(V_3 - V_2) + \frac{dI_5}{dt} - \frac{1}{L_2}(E - V_3) = 0 \qquad (16.18)$$

Equation (16.17), of course, is a restatement of the cut-set equation, but the two in (16.18) are used to generate I_5 and I_6. Now we may classify the equations and indicate how each variable should be generated.

Inductor equations.

I_6: $\quad I_6 = \frac{1}{L_4}\int (V_2 - V_1)\,dt - \frac{1}{L_3}\int (V_3 - V_2)\,dt$

I_5: $\quad I_5 = \frac{1}{L_2}\int (E - V_3)\,dt - \frac{1}{L_3}\int (V_3 - V_2)\,dt$

Capacitor equation (step 2).

V_1: $\quad V_1 = V_3 - \frac{1}{C}\int I_5\,dt$

Cut-set equation (step 3).

$$V_3: \quad \frac{1}{L_1}(V_1 - E) - \frac{1}{L_2}(E - V_3) + \frac{1}{L_3}(V_3 - V_2) - \frac{1}{L_4}(V_2 - V_1) = 0$$

Resistor equation (step 4).

$$V_2: \quad V_2 = -I_6 R$$

Remaining current equations (step 5).

$$I_1 - I_2 + I_3 + I_4 = 0$$
$$I_3 + I_6 - I_4 = 0$$
$$I_3 + I_5 - I_2 = 0$$
$$L_1 I_1 + L_2 I_2 + L_3 I_3 + L_4 I_4 = 0$$

The remaining current equations are not mechanized unless these currents are needed. They require the use of additional pots set to the inductance values. A final mechanization is shown in Fig. 16.22. All loops contain an odd number of amplifiers.

If we had generated two of the inductor currents with integrators, inductance values would have to be set on more than one pot, since we would have been forced to use (16.15) in the mechanization. The trick proposed here avoids this by generating only $I_3 - I_4$ and $I_3 - I_2$, which are needed for feedback.

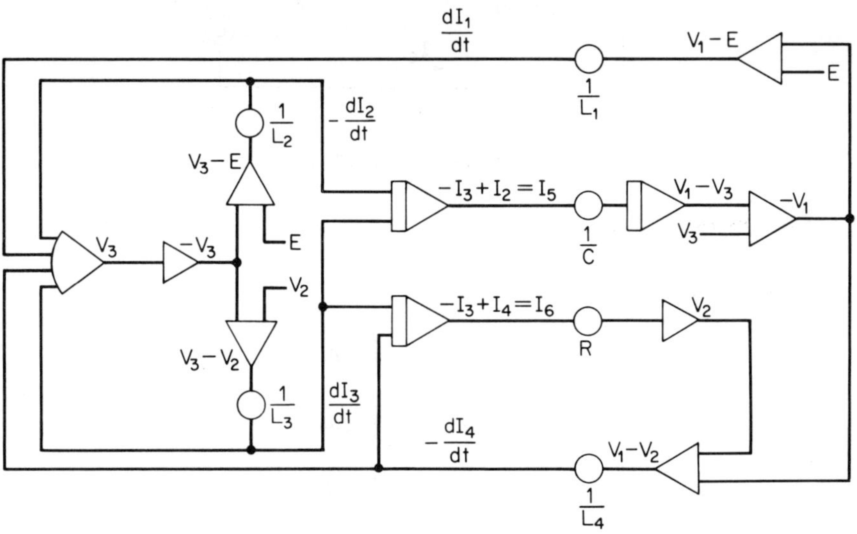

FIG. 16.22 A final circuit for Example 16.10.

We may formulate a general rule:

Step 6. When a loop of inductors is in a network, minimize the number of nodal equations containing the current through those elements. Then differentiate those nodal equations involving the inductor currents and substitute the voltage expressions for those current derivatives in the loop. Use these equations to find currents other than the inductor currents. In this way, the currents through other elements connected to the loop are made available for feedback while preserving the one-pot-per-inductor concept. If the currents in the inductors are needed, use the loop current equations (Theorem 16.1) and solve a matrix equation.

This rule is particularly important when some of the inductors in the loop form part of a cut set of inductors, but it need not be followed if the inductors just form a loop. (See Problem 16-14.)

When loops of capacitors are in a network, a similar rule exists. To help formulate this, we consider a relatively simple network as shown in Fig. 16.23.

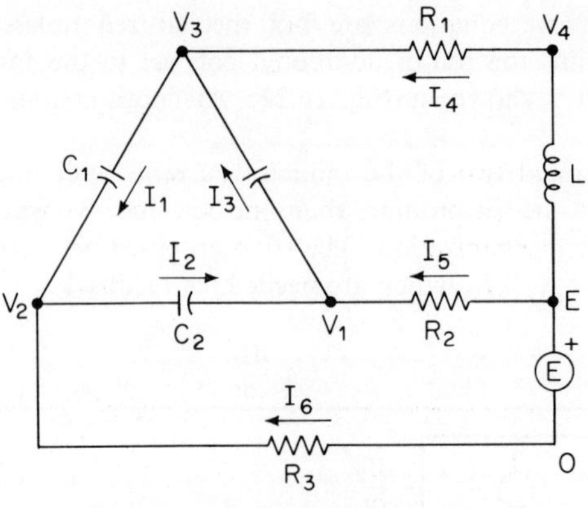

FIG. 16.23 Example 16.11.

Example 16.11. We have

Branch equations.

$$V_3 - V_2 = \frac{1}{C_1}\int I_1\, dt$$

$$V_2 - V_1 = \frac{1}{C_2}\int I_2\, dt$$

$$V_1 - V_3 = \frac{1}{C_3}\int I_3\, dt$$

V_4: $\quad V_4 - V_3 = I_4 R_1$

I_4: $$I_4 = \frac{1}{L}\int (E - V_4)\,dt$$

$$E - V_1 = R_2 I_5$$
$$V_2 = -R_3 I_6$$

Nodal equations.

$$I_3 - I_1 + I_4 = 0$$
$$-I_3 + I_2 + I_5 = 0$$
$$I_1 - I_2 + I_6 = 0$$

First, I_1, I_2, or I_3 cannot be obtained from the nodal equations since its matrix has zero determinant. Therefore, the loop equation

$$\frac{1}{C_1}\int I_1\,dt + \frac{1}{C_2}\int I_2\,dt + \frac{1}{C_3}\int I_3\,dt = 0$$

or

$$\frac{1}{C_1}I_1 + \frac{1}{C_2}I_2 + \frac{1}{C_3}I_3 = 0$$

would have to be used to obtain those currents. The same problem exists as in the case of loops of inductors; namely, performing a matrix inversion after integration leads to using more than one pot for some of the capacitance values. Our experience in Example 16.10 leads us to consider integrating values other than the basic currents in the capacitor branches. This is most easily done if the capacitor equations are changed to

$$C_1(V_3 - V_2) = \int I_1\,dt$$
$$C_2(V_2 - V_1) = \int I_2\,dt$$
$$C_3(V_1 - V_3) = \int I_3\,dt$$

and the nodal equations are rearranged so that the currents in the loop appear in one less equation

$$I_3 - I_1 + I_4 = 0$$
$$-I_3 + I_2 + I_5 = 0$$
$$I_4 + I_5 + I_6 = 0$$

Now we may integrate the nodal equations involving those currents in the loop of capacitors, substituting the proper voltage expressions.

Capacitor equations (step 2 modified).

$$V_3: \qquad C_3(V_1 - V_3) - C_1(V_3 - V_2) + \int I_4\,dt = 0$$

$$V_1: \qquad -C_3(V_1 - V_3) + C_2(V_2 - V_1) + \int I_5\,dt = 0$$

We also have

Inductor equation (step 1).

$$I_4: \qquad I_4 = \frac{1}{L}\int (E - V_4)\,dt$$

Resistor equation for voltage (step 4).

$$V_2: \qquad V_2 = -R_3 I_6$$

$$V_4: \qquad V_4 = V_3 + I_4 R_1$$

Remaining equations for current (step 5).

$$I_5: \qquad I_5 = \frac{E - V_1}{R_2}$$

$$I_6: \qquad I_4 + I_5 + I_6 = 0$$

$$I_3 - I_1 + I_4 = 0$$

$$-I_3 + I_2 + I_5 = 0$$

$$\frac{1}{C_1}I_1 + \frac{1}{C_2}I_2 + \frac{1}{C_3}I_3 = 0$$

A final circuit is shown in Fig. 16.24 without showing the mechanization for the currents in the loop of capacitors. Unlike the case for a loop of inductors this procedure leads to a circuit applicable even when the capacitors are not constant.

Step 7. When a loop of capacitors is in a network, minimize the number of nodal equations containing the current through those elements. Then integrate those nodal equations involving the capacitor currents and substitute the voltage expressions for those current integrals in the loop. Use these equations to find voltages at some of the nodes of the loop. In this way, the voltages at all but one of the nodes of the capacitors involved are made available for feedback to other elements of the circuit, while preserving the one-pot-per-capacitor mechanization.

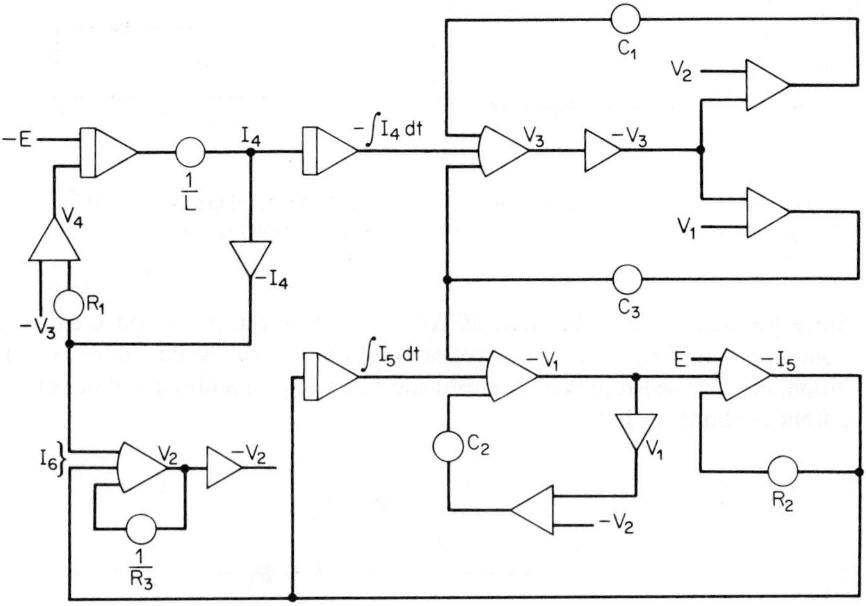

FIG. 16.24 A mechanization for Example 16.11.

16.5 Inclusion of Ideal Diodes in Network Analysis [9]

When networks containing diodes are intended for low frequency use, a good simulation can often be obtained by considering the diode to be ideal—the current through the diode is zero when it does not conduct and the voltage across the diode is zero when it does conduct. Branch mechanizations are used solely in such a case, since it is fairly easy on the computer to generate the current through the diode with feedback from the voltage across it. To demonstrate this, we analyze the behavior of the elements in Fig. 16.25, where a small inductance ΔL has been placed in series with a diode. This is not done to simulate any inductance in the diode; it is rather an artificial trick to help set up the analog circuit to simulate ideal diodes in a network.

As long as current is flowing through ΔL, $V = V_j$ regardless of the relative magnitude of V_i and V_j. That is, the voltage drop across ΔL can be of either sign as long as $I \geq 0$. To set up such a case on the computer, we may program

$$I = \frac{1}{\Delta L}\int (V_i - V_j)\,dt, \qquad I \geq 0 \tag{16.19}$$

Its corresponding mechanization is shown in nonideal fashion in Fig. 16.26.

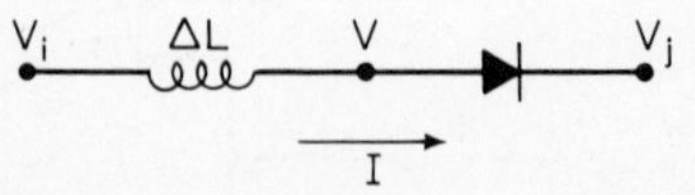

FIG. 16.25 Inductance in series with a diode.

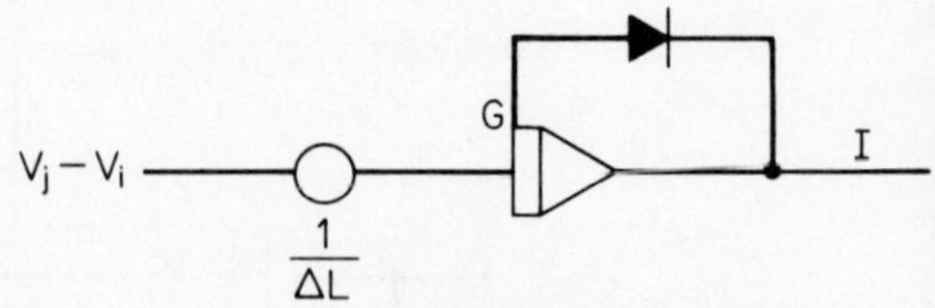

FIG. 16.26 Nonideal mechanization of the elements in Fig. 16.25.

Since we wish to take the limit as $\Delta L \to 0$, the integrator is converted to a high-gain amplifier, which we have learned behaves as a fast integrator. In addition, an idealized representation is made by adding an additional diode at the output as shown in Fig. 16.27.

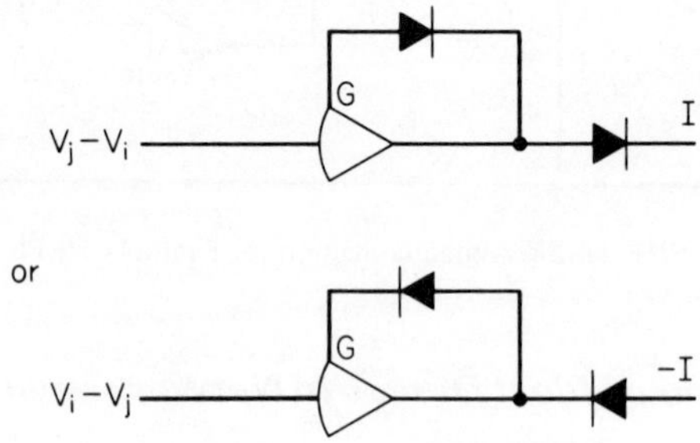

FIG. 16.27 Limit as $\Delta L \to 0$ (idealized).

The basic method can now be stated:

To include a diode between any two junctions V_i and V_j in a network, include the diode current I in all pertinent equations. Program (16.19) and let $\Delta L \to 0$. The amplifier representing the current I is then involved in a feedback loop that preserves all voltage and current equations in the network. Only one exception exists to this rule, as described later in Example 16.13.

As might be expected, the amplifier for diode current can be involved only in a feedback loop which, effectively, is in the solution of algebraic equations. It may be necessary to place a small capacitor from output to grid to prevent instabilities, as discussed in Chapter 11. The method of arriving at the circuit shows this equivalent to having a small inductance in series with the diode in the network. In addition, if the diode forms a loop with capacitors in the network, the current through it could be infinite theoretically if the capacitors discharged instantly. In such a case, the analog circuit operates satisfactorily, but because of delays in amplifier response, it may be necessary to limit the positive current through the diode in the circuit.

The basic rule is effective in all cases. Clearly, the order of the characteristic equation may change during the computation, say, for example, if a cut set of energy storage elements is destroyed when the diode conducts. We must always program for the maximum number of integrators possible, but as we shall see, it is always possible to program a constraint equation when less integrators are needed than used. Furthermore, constraint equations do not involve the use of additional pots set to any parameter values, so the one-pot-per-parameter programming may be preserved.

Example 16.12. We look first at an elementary network, shown in Fig. 16.28,

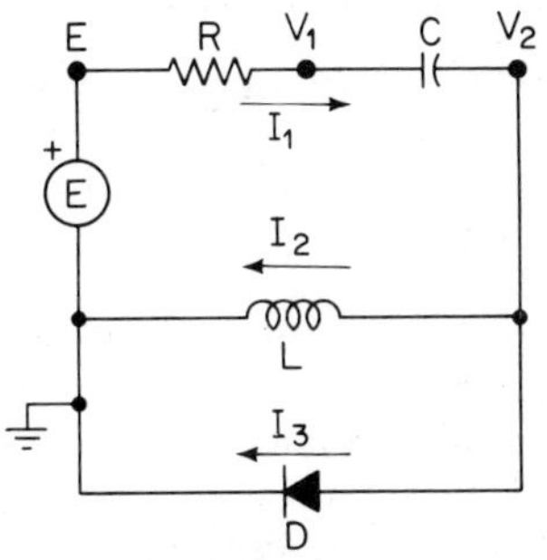

FIG. 16.28 Example 16.12.

where a single diode D is placed across L in a series LRC network. We proceed as before, writing branch and nodal equations but including the diode current and pretending a small inductance ΔL is in series with the diode:

Branch equations.

V_1: $$E - V_1 = I_1 R$$

V_2: $$V_1 - V_2 = \frac{1}{C}\int I_1\,dt$$

I_2: $$I_2 = \frac{1}{L}\int V_2\,dt$$

I_3: $$I_3 = \frac{1}{\Delta L}\int V_2\,dt, \qquad I_3 \geq 0$$

Nodal equation.

I_1: $$I_3 + I_2 = I_1$$

Assignment of variables generated from each equation is straightforward and, in this case, unique as indicated above. A simple analog circuit follows in Fig. 16.29. Note that I_3 is in a three-amplifier summing loop and automatically keeps $V_2 \leq 0$. If the

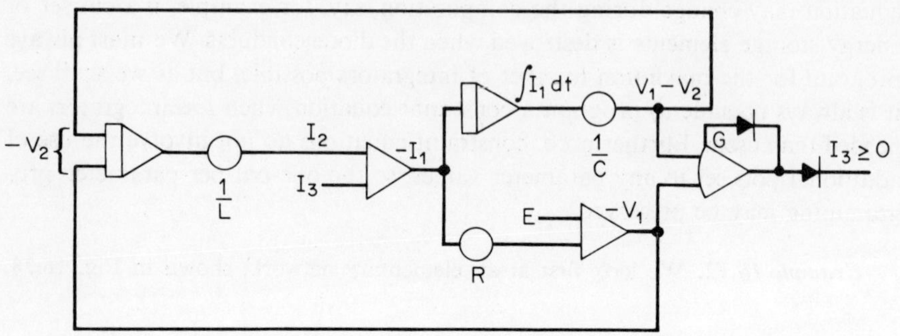

FIG. 16.29 Mechanization of Example 16.12.

diode does not conduct, ($V_2 < 0$), then I_3 is 0 and the amplifier has no effect on the circuit. All equations are satisfied.

Example 16.13. Figure 16.30 is part of a network that covers the exception noted earlier in setting up a circuit for diode current. The voltage V_0 cannot be obtained as an input to a high-gain amplifier because the node is a junction of inductors only. Furthermore, the inductors form a cut set when the diode does not conduct. Presumably the voltages V_1, V_2, and V_3 are generated from other equations, but V_0 is needed to generate the currents.

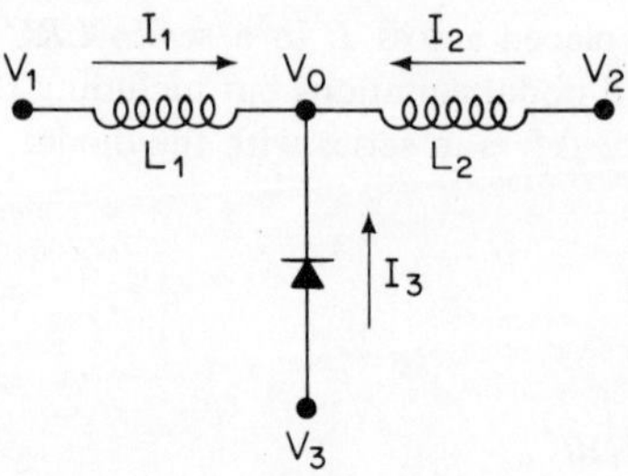

FIG. 16.30 Example 16.13—destruction of an inductor cut set.

To generate V_0, we proceed according to step 3, assuming a small inductor ΔL is in series with the diode. Since

$$I_1 + I_2 + I_3 = 0, \qquad I_3 \geq 0 \tag{16.20}$$

we have

$$\frac{1}{L_1}(V_1 - V_0) + \frac{1}{L_2}(V_2 - V_0) + \frac{1}{\Delta L}(V_3 - V_0) = 0 \tag{16.21}$$

When $(V_3 - V_0)/\Delta L < 0$, we would like a diode circuit to make it zero so that it does not affect the mechanization of (16.21). A mechanization in Fig. 16.31 meets these requirements. When the diode is conducting, V_0 is forced to equal V_3 by the action of the idealized diode amplifier. When it is not conducting, it has no effect on

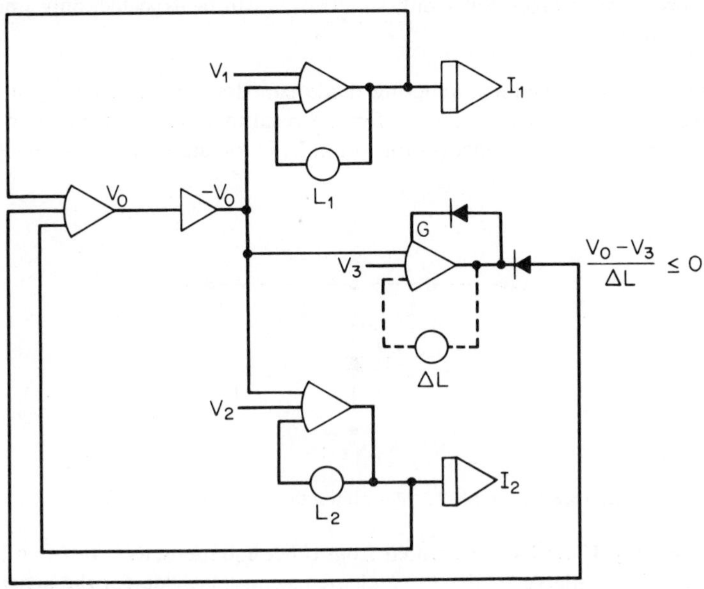

FIG. 16.31 Mechanization of Example 16.13.

the circuit. Then $dI_1/dt = -dI_2/dt$ and $I_1 = -I_2$. In this case, two integrators are not needed, but we are forced to use them. However there is an inequality constraint equation derivable from (16.20)

$$-I_3 = I_1 + I_2 \leq 0$$

which can be enforced in long term runs, if necessary, to insure that small computing errors do not build up (Fig. 16.32). If I_3 is ever negative, the feedback path

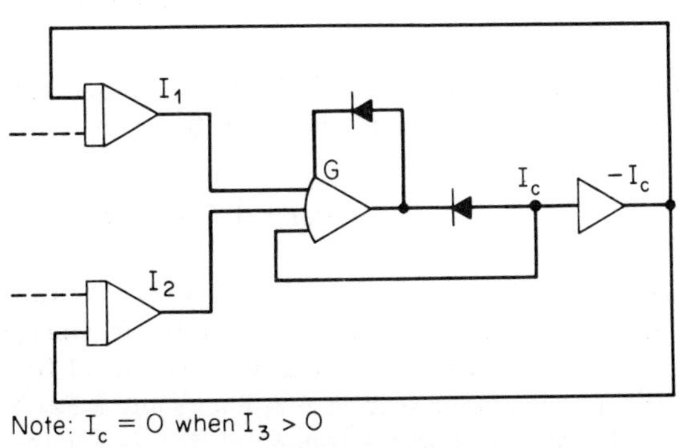

FIG. 16.32 Enforcing the constraint.

supplies corrective voltages. Since only small errors can be expected, only small gains need be patched.

Example 16.14. When a conducting diode can destroy a capacitor cut set, programming can automatically account for the resulting constraint equation. Figure 16.33 shows that part of a network, for example, involving such a cut set of capac-

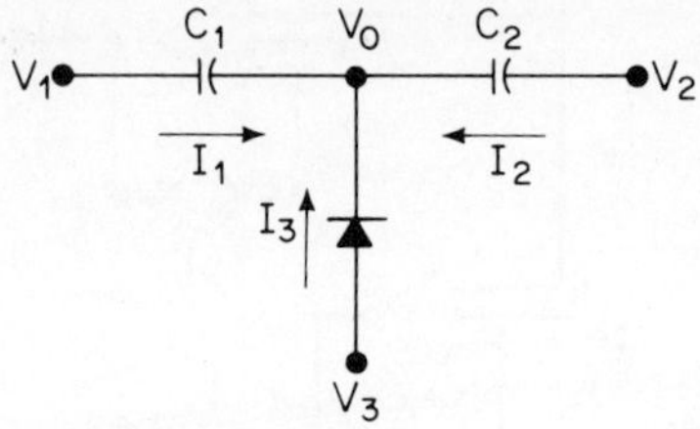

FIG. 16.33 Example 16.14—destruction of a capacitor cut set.

itors. Presumably V_1 or V_2 is generated from other equations, as is I_1, I_2, and V_3. Let us suppose V_1 is available so the three elements must provide for the generation of I_3, V_0, and V_2 from the equations

$$V_1 - V_0 = \frac{1}{C_1}\int I_1\,dt, \qquad V_2 - V_0 = \frac{1}{C_2}\int I_2\,dt$$

$$I_3 = \frac{1}{\Delta L}\int (V_3 - V_0)\,dt, \qquad I_3 \geq 0$$

The nodal equation

$$I_1 + I_2 + I_3 = 0$$

ordinarily provides for the generation of V_0 as well as one of the currents, say I_2. Now we must have

$$\int I_1\,dt + \int I_2\,dt + \int I_3\,dt = K = \text{constant} \tag{16.22}$$

where K is determined from the initial conditions. Equation (16.22) can be programmed for the algebraic solution of V_2, however, by noting that

$$\int I_1\,dt + C_2(V_2 - V_0) + \int I_3\,dt = K$$

holds, while using previous principles. Figure 16.34 shows the resulting mechanization. It is clear that all equations are satisfied when the *IC* voltages make $V_2 - V_0$ and $V_1 - V_0$ correspond to any initial conditions (determined from K). When the diode conducts, the amplifier representing $-I_3$ enters into a feedback loop through the inputs V_3 and $-V_1$. When it does not conduct, $I_1 = -I_2$, and the integrator $\int I_3\,dt$

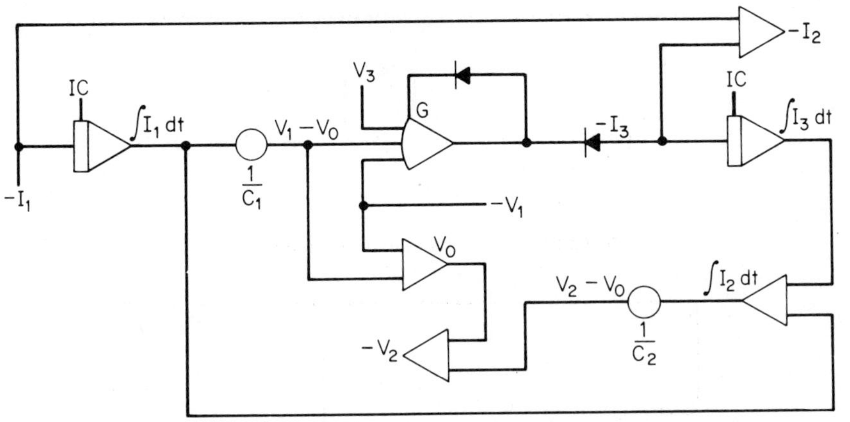

FIG. 16.34 Mechanization of Example 16.14.

remains constant, effectively removing it from the circuit. Thus, no constraint equations need be mechanized. Two integrators are used when they are needed, but one is automatically constant when the cut set is in force.

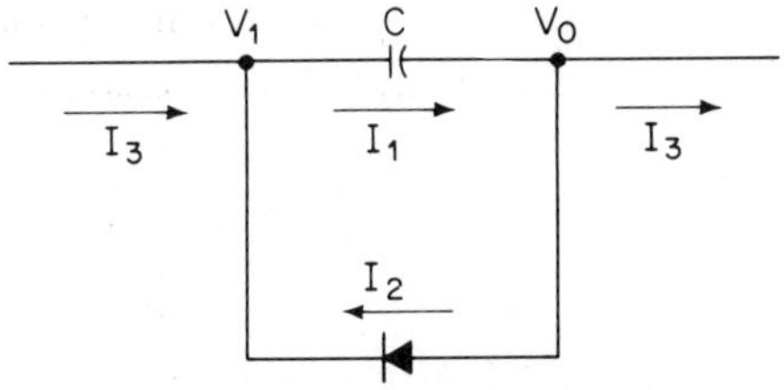

FIG. 16.35 Example 16.15—a diode across a capacitor.

Example 16.15. A diode across a capacitor is the simplest example of a case where the diode current may have to be limited. In Fig. 16.35, we note that

$$I_1 = I_2 + I_3, \qquad V_1 - V_0 = \frac{1}{C}\int I_1\,dt$$

$$I_2 = \frac{1}{\Delta L}\int (V_0 - V_1)\,dt, \qquad I_2 \geq 0$$

must be used to generate I_1, I_2, and one of the voltages V_1 or V_0. Presumably, the other voltage and I_3 are generated from other equations in the network. The analog circuit must be drawn as shown in Fig. 16.36 when previous principles are used. Since I_1 and I_2 are not needed elsewhere, the only feedback path for I_2 is through the integrator, which is then limited essentially to negative values only. There can be a small lag in response in the high-gain amplifier allowing the integrator to be slightly positive causing the amplifiers I_2 and $-I_2$ to saturate. Alternate circuits are shown in Fig. 16.37. In (a) and (b), the limiting is still ideal but scaling and response is

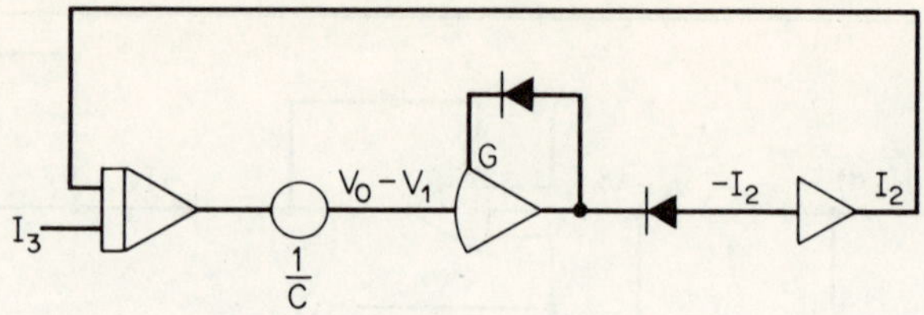

FIG. 16.36 The mechanization of Example 16.15.

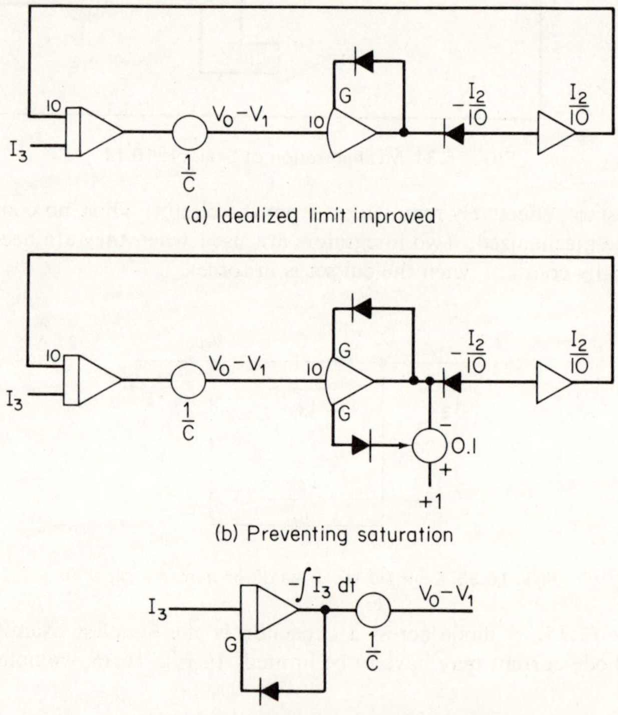

FIG. 16.37 Alternate methods of limiting the integrator.

somewhat improved. In both cases, a series of positive pulses is emitted to buck out the input I_3, in the same way as in a derivative circuit. I_2 can be expected to be noisy, but $V_0 - V_1$ is limited to zero within a very small tolerance (usually less than 0.01 per cent of full scale). In (c) the limit is not ideal and $-\int I_3\, dt$ can be as great as 0.5V, but the circuit is simpler and may be satisfactory.

Active circuits have also been simulated on the analog computer by simulating equivalent networks ([10]–[12]), sometimes with nonlinear characteristics.

PROBLEMS

16-1. A part of a network consisting of five capacitors is shown in Fig. 16.38. There are three loops (C_1, C_2, C_3), (C_3, C_4, C_5), and (C_1, C_2, C_4, C_5). Write the equation for the integrals of the current for each loop and show they are not independent (Theorem 16.1).

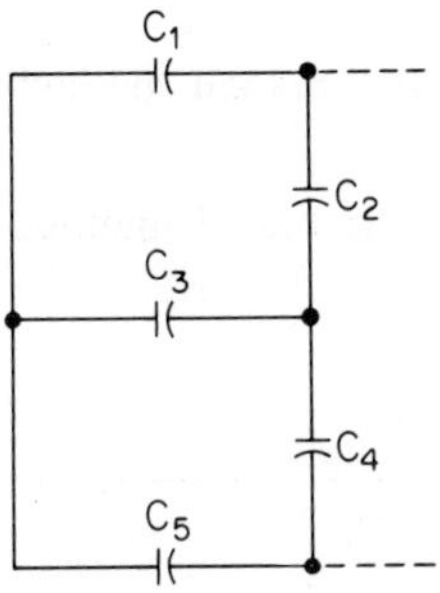

FIG. 16.38 Loops of capacitors.

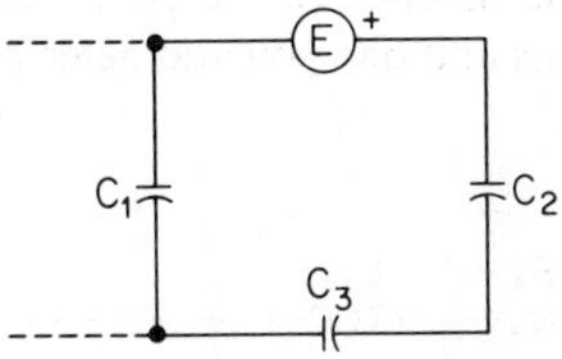

FIG. 16.39 Loop of capacitors with voltage source.

16-2. When a loop contains capacitors and voltage sources only, Theorem 16.1 also applies in modified form. Write the equation for the integrals of the currents for the loop in Fig. 16.39 and discuss how Theorem 16.1 is affected.

16-3. Discuss any modifications in Theorem 16.2 when there are voltage sources in a loop of inductors.

16-4. Discuss any modifications necessary in Theorem 16.4 when a cut set of inductors includes a known current source, such as shown in Fig. 16.40.

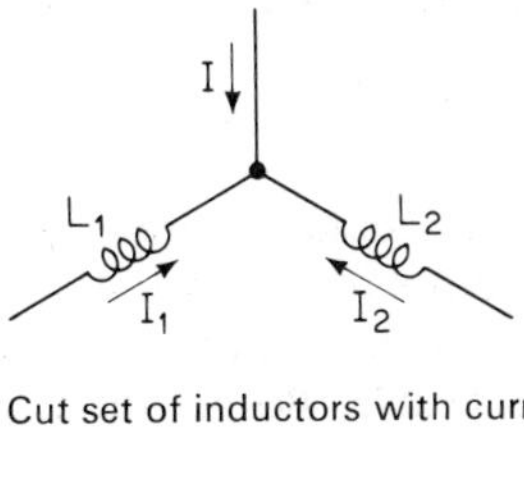

FIG. 16.40 Cut set of inductors with current source.

16-5. Discuss any modifications necessary in Theorem 16.5 when a cut set of capacitors includes a known current source.

16-6. Write down all cut sets of capacitors in Fig. 16.13 and show only three are independent.

16-7. Modify the analog circuit of Fig. 16.8 to include the possibility that the components are variable.

16-8. Rewrite the inductance equations (16.12) to include the possibility the inductances L are variable and set up a circuit to solve for the currents. (*Hint*: solve a matrix equation after integration.)

16-9. Solve the network shown in Fig. 16.3 with a loop mechanization.

16-10. Solve the network shown in Fig. 16.15 with a loop mechanization.

16-11. Loop and branch mechanizations are not unique. Describe why, using the networks of Figs. 16.6 and 16.9 as examples.

16-12. Mechanize the circuit in Fig. 16.9 with the branch method using one potentiometer per component value.

16-13. Mechanize the circuit in Fig. 16.41 with the branch method using two integrators and one potentiometer per component.

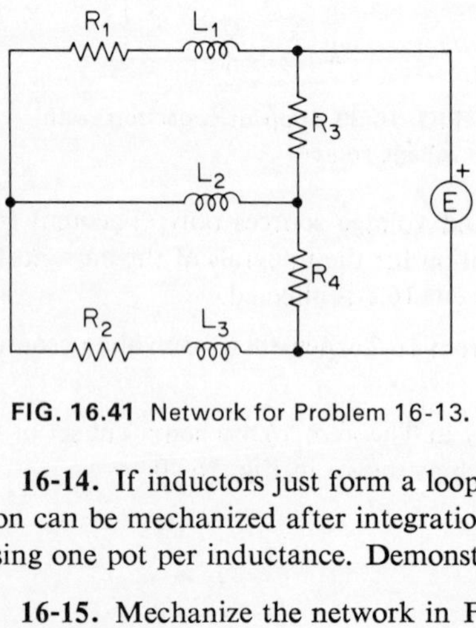

FIG. 16.41 Network for Problem 16-13.

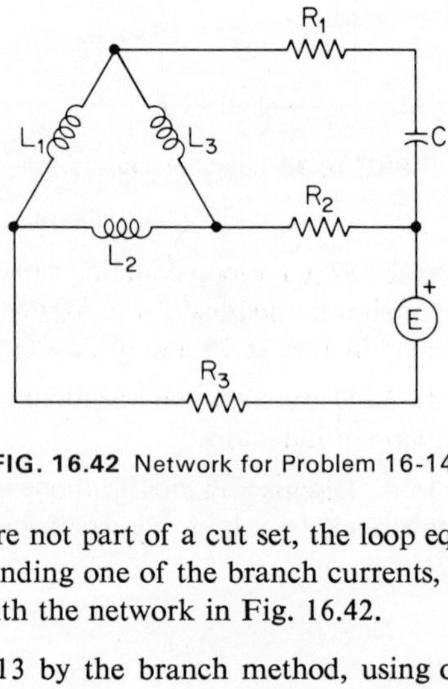

FIG. 16.42 Network for Problem 16-14.

16-14. If inductors just form a loop and are not part of a cut set, the loop equation can be mechanized after integration for finding one of the branch currents, still using one pot per inductance. Demonstrate with the network in Fig. 16.42.

16-15. Mechanize the network in Fig. 16.13 by the branch method, using only three integrators.

16-16. An equivalent mechanization of ideal diodes in a network can be obtained by including a small resistance ΔR in series with the diode and by letting $\Delta R \to 0$ in the limit while insuring the diode current or voltage drop is positive only. In many respects, this concept is more flexible. Besides the circuitry shown in Fig. 16.27, we

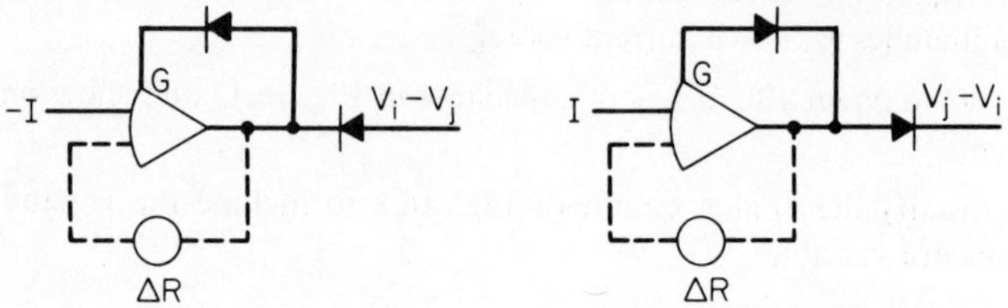

FIG. 16.43 Alternate method of including ideal diodes in a network.

may use the circuits in Fig. 16.43. If I is negative, the amplifier is in a feedback loop to force I to 0; if I is positive, $V_i - V_j$ is forced to zero. Use one of these circuits to mechanize the network in Fig. 16.44 and program a constraint equation also. Note that a simpler analog circuit is obtained than the one in Fig. 16.31, but comment on its preference if inductors are also part of a loop.

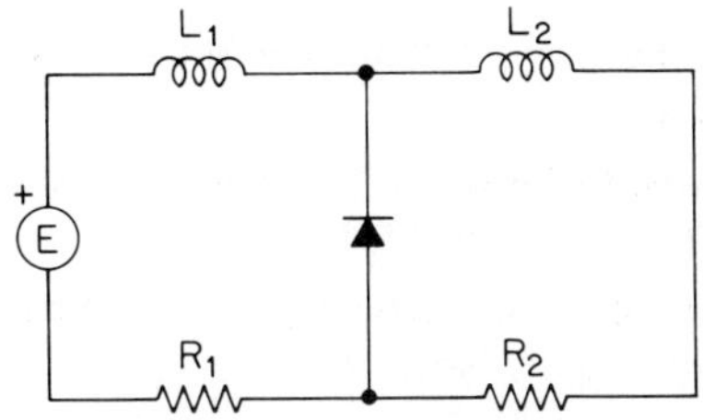

FIG. 16.44 Problem 16-16.

16-17. Mechanize the circuit shown in Fig. 16.45 by the branch method [9].

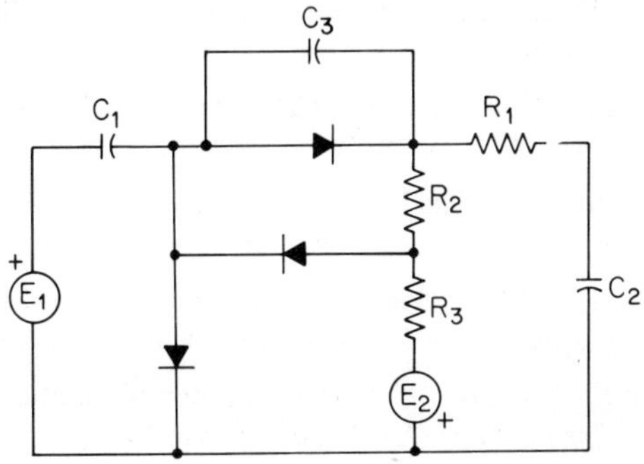

FIG. 16.45 Problem 16-17.

16-18. Filters, or other common circuits, can be mechanized directly from the network rather than the transfer function. Mechanize the circuit of Fig. 16.46, which can be a standard Butterworth, Tchebycheff, or Bessel filter, depending on the choice of element values [13]. (Note the loop of capacitors.)

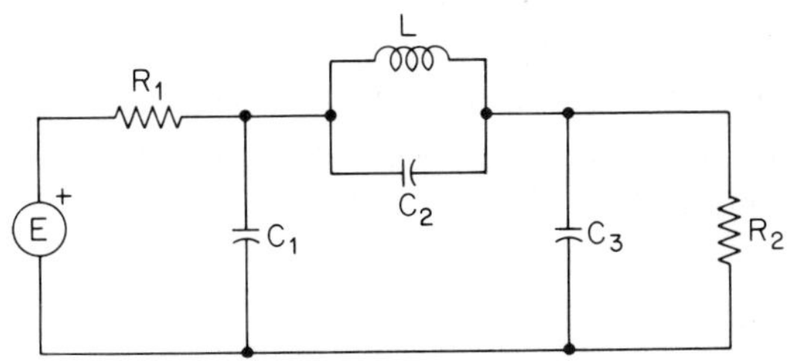

FIG. 16.46 Third-order filter.

REFERENCES

[1] Guillemin, E. A., *Introductory Circuit Theory* (New York: John Wiley & Sons, Inc. 1953).

[2] Walters, L. G., "Hidden Regenerative Loops in Electronic Analog Computers," *IRETEC*, vol. 2, June 1953.

[3] Scott, N. R., "On the Use of Redundant Integrators in Analog Computers," *IRETEC*, vol. 6, Dec. 1957.

[4] Otterman, J., "On the Order of the Differential Equation Describing an Electrical Network," *Proc. IRE*, vol. 45, July 1957.

[5] Otterman, J., "How to Avoid Extra Integrators When Simulating *RLC* Networks," *Control Eng.*, vol. 4, Nov. 1957.

[6] Otterman, J., "On the Loop- and Node-Analysis Approaches to the Simulation of Electrical Networks," *IRETEC*, vol. 7, Sept. 1958.

[7] "GEDA Analysis of an Electrical Circuit," *Goodyear Aircraft Corp. Report GER-5455*, Akron, Ohio, March 18, 1953.

[8] Larrowe, V. L., "Direct Simulation Bypasses Mathematics, Simplifies Analysis," *Control Eng.*, vol. 1, Nov. 1954.

[9] Hausner, A., "Analyzing Networks Containing Diodes by Analog Computation," *Simulation*, vol. 6, June 1966.

[10] Lambert, J. M., and A. J. Heidrich, "Radar Systems Simulation Techniques," *IRE National Convention Record, part 4*, 1959.

[11] Berger, E. L., and R. M. Taylor, "Optimization of a Radar in its Environment by GEESE Techniques," *Proc. Western Joint Comp. Conference*, 1961.

[12] Balaban, P., and J. Logan, "Analog Computer Simulation of Semiconductor Circuits," *Proc. Spring Joint Comp. Conference*, 1968.

[13] Bridgman, A. D., "Analog Techniques for Filter Simulation," *Simulation*, vol. 8, April 1967.

For an application of signal flow graph theory to obtain an analog circuit for a network, see

[14] Yates, B. H., "Analog Simulation of Passive Networks," *Instr. and Control Systems*, vol. 38, Feb. 1965.

For other papers on network simulation, see

[15] Norris, A., "The Analog Method in Electronics," *Proceedings of the 3rd Inter. Analogue Comp. Meetings* (Brussels: Presses Académiques Européennes, 1962).

[16] Steber, G. R., and G. L. Elmergreen, "The Simulation of Null Networks," *Simulation*, vol. 11, Aug. 1968.

Chapter 17

ERROR ANALYSIS AND ERROR REDUCTION

Of fundamental importance in all computations, whether done by an analog or digital computer, a desk calculator, a slide rule, or simply by pencil and paper, is whether or not the answer is in error, by how much, and its effect on the conclusions drawn. If errors are not known, there may be little confidence in the solution, and the information provided will be used cautiously, if at all; indeed, it should be so used. On the other hand, if confidence is generated either by experience, experiment, or analysis, the answer and conclusions will be trusted more and the computation will have been worthwhile.

In analog computations, any distrust of a computer solution is aggravated by the inherent error of all components. This is not to say that the solutions cannot be trusted or that conclusions cannot be drawn from them. The amount of error in each solution will simply affect the conclusions drawn. As an analog computer programmer, it is important to understand how the errors arise, how they affect the solution, and, if it is important, how to reduce them when possible.

17.1 Maintenance and Reliability

As a start in reducing errors generally, it is best to adhere to a strict maintenance program where defective equipment is repaired or replaced periodically and operating equipment is adjusted to meet manufacturer's specifications. Such a program must be emphasized, for it instills in a programmer confidence in his equipment and will generally tend to minimize errors.

Computer manuals generally provide periodic schedules for preventive maintenance, where adjustable elements of a component are carefully adjusted to within the proper limits. These include capacitor and resistor values, amplifier offset and drift, servo pot alignment, electronic multiplier breakpoint and slope adjustment of the squaring cards, etc. Such schedules must be strictly adhered to and even modified as experience dictates. Frequent balancing of amplifiers by the operator is also necessary, particularly for a sensitive problem.

The problem of reliability of a computer is intimately associated with the maintenance programs but can be considered separately in terms of component failures with use. Loveman [1] tabulated component failures for an early analog computer installation. Notwithstanding a 75-per cent down time (there was no patch panel) for repairs and maintenance, his conclusion was that the simulator had a high reliability. Sixty-three per cent of all failures in a year were vacuum tubes, with an average catastrophic failure requiring less than 20 minutes to repair. Such a down-time rate in terms of more modern equipment is extremely poor. The addition of a patch panel allowed special prewired test boards [2] to be developed; plug-in units allowed instant replacement of amplifiers or other components that failed. These could be repaired at leisure off the machine. Most important of all, reliable transistorized circuitry was developed and incorporated in computers manufactured in the 1960s thus reducing the amount of failures and increasing reliability considerably. Modern analog computers are highly reliable in the sense that catastrophic component failure rates are low.

17.2 Checks on Solutions

The first formal symposium on analog computer techniques produced no papers to determine if the computer solutions were correct [5]. There was, however, a discussion of various quick methods used for checks on the solution. When a program is run initially, we must perform such checks to promote self-trust in the error limits of a solution. There are three items of immediate concern:

(1) Are there patching errors and are the potentiometers set correctly?
(2) Are there any equipment malfunctions?
(3) Is the frequency range of the computer components exceeded?

The first is the fault of the operator and might be labeled "mistakes" rather than unavoidable errors. The operator has little control over the second item, but both generally produce profound effects that are easy to detect. The last item is inherent in the computer, but the operator can control its error by not running the problem too fast.

Many qualitative checks can and should be made to insure that any initial difficulties are discovered and corrected. A programmer should assume that he has made mistakes and that there exists faulty equipment unless he can be convinced otherwise. Checks frequently performed consist of the following [6]–[9]:

1. *Patching check.* Inputs and outputs of all components in a diagram are literally checked with a mark (✓) on the diagram as they are patched. When patching is completed, it is cross checked (✗) by comparing the input and output leads on the patchboard with the diagram. A similar scheme is used to check the diagram against the equations mechanized.

2. *Static check.* Supply nonzero initial conditions to all components, even if they are false. Compute the outputs of all components and the inputs to all integrators based on these values and compare them with the measured values. If they agree, there is reasonable assurance that there are no component malfunctions and no mistakes. If not, locate the source of the difficulty and correct it. Computers often have a special mode, the ST (STATIC TEST) mode, to do this. In some cases, there are static test reference voltages on the computer that are live in the ST mode, but zero in IC, HOLD, or OPERATE. These can be used to patch nonzero initial conditions into integrators that are zero at the start of a problem, without having to remove the leads after the check. This check should never be omitted, except, perhaps, in the simplest problem.

3. *Dynamic check.* Run the problem at two speeds, separated by a factor of at least two. If no apparent change in the solution occurs, there is reasonable assurance that the dynamic limitations are not exceeded at either speed. This check is particularly important if servos or other electromechanical components are used, or if there are high gains in some integrators.

Even if the computer seems to operate well and results are repeatable, there is still no assurance the solution is obtained with acceptable error. Further checks—some rapid, others more costly—can be made. These include

4. *Engineering insight.* The results should seem reasonable to the investigator.

5. *Special solution.* Special cases sometimes can be calculated quickly. For example, in a trajectory problem, eliminate drag and run a vacuum trajectory that is easily calculated.

6. *Function generation check.* Be sure that all functions generated, particularly those by integration (Chapter 5), are within acceptable error bounds.

7. *Experimental check.* If the equations are based on a mathematical model of a device, and data are available on the device, the computer solution can be checked against the data. A discrepancy here, however, can just mean a poor model.

8. *Linear equation check.* Solutions run at half the initial conditions should be half scale. Solutions run with the sign of initial conditions changed

will change the sign of all amplifier outputs. This is useful for determining the effect of offsets and unequal negative and positive reference voltages. A selected root transient check [9] can be made if the characteristic roots are obtained. This consists of setting proper initial conditions to null all terms of the solution corresponding to each root except one and comparing the computer damping and period with those predicted by the characteristic root. A frequency response check [9] can also be made if the frequency response of a system is available.

The above checks can often be made with a desk calculator and without any special knowledge of numerical analysis. They do not answer the question "How much error exists in the solution?" Their primary function is to determine quickly that the circuits are operating properly. As much information as is available should be applied to determine this. Any further checking usually requires computer aid as discussed in the next four subsections.

17.2.1 Dynamic substitution check

Let us assume the computer is solving the equations in canonical form

$$\dot{x}_i = f_i(x_1, x_2, \ldots, x_n), \qquad i = 1, 2, \ldots, n \tag{17.1}$$

with appropriate initial conditions a_i. As described by Richmond and Loveman [9]–[10], this check requires that the program make all $\dot{x}_i$ available throughout the computation. All variables, including outputs of summers, are recorded in ink by recorders. At periodic intervals of τ, these recordings are sampled with the data being transcribed to punched cards for statistical analysis on digital computer equipment. In particular, the output errors are formed by differencing the input and output traces. For the case of integrators, the errors e_i have the form

$$e_i = \dot{x}_i - f_i(x_1, x_2, \ldots, x_n)$$

where the f_i have accumulated error. Integrations are checked by differentiating the output with simple two-point formulas.

Statistical analysis is necessary only because there are reading errors superimposed upon the computer errors. If the reading errors are normally distributed, their mean should be near zero and standard deviation, near constant for all traces. By systematic inspection of all statistical variables of the traces, biases and gross errors can be detected.

The dynamic substitution check has not found widespread use because it suffers on several accounts. First, the readings are generally not normally distributed due to nonlinearities in recording equipment. Second, small error cannot be detected. These errors could propagate and result in a rather large final error. The check is effectively a series of static checks throughout the solution. Finally, if a digital computer is available, accurate check solutions are more easily obtained by a Picard iteration (below) or by solving the equations from scratch (Sec. 17.2.4).

17.2.2 Picard iteration

An excellent method of checking analog computer solutions for errors, while simultaneously reducing the solution error, was described by Meissinger [11]. It is based on the Picard iteration process often used in theoretical proofs of existence and uniqueness of solutions. Given equations (17.1), suppose analog solutions $(x_i)_0$ are obtained. Then, the relations

$$(x_i)_0 = a_i + \int_0^\tau f_i[(x_1)_0, (x_2)_0, \ldots, (x_n)_0]\, dt \tag{17.2}$$

should hold, but because of errors, an evaluation of the right-hand side of (17.2) produces new functions $(x_i)_1$. Sucessive iterations using

$$(x_i)_{k+1} = a_i + \int_0^\tau f_i[(x_1)_k, (x_2)_k, \ldots, (x_n)_k]\, dt$$

must converge to the exact solution as $k \to \infty$ if it exists and is unique. This procedure is almost never used to solve differential equations even with a digital computer, because its convergence rate can be very slow. But if the analog computer solution has small error, then few iterations, sometimes only one, can provide convergence. Not only are the original errors obtained for all τ but also more accurate solutions. The inherent advantage of using a Picard iteration scheme over numerically solving the differential equations is that the former is easier to do. It only requires substitution and quadrature and can therefore be done with a desk calculator with simple integration schemes.

17.2.3 Variational method [10]

Again, let us use the subscript 0 for the analog solution $(x_i)_0$, and introduce the error function δx_i:

$$x_i = (x_i)_0 + \delta x_i$$

Thus

$$(\dot{x}_i)_0 + \delta \dot{x}_i = f_i[(x_1)_0 + \delta x_1, (x_2)_0 + \delta x_2, \ldots, (x_n)_0 + \delta x_n]$$

But we also have

$$(\dot{x}_i)_0 = f_i[(x_1)_0, (x_2)_0, \ldots, (x_n)_0] + e_i$$

where the e_i are measured from the analog solution by substituting the $(x_i)_0$ in each f_i just as in the dynamic substitution check. By subtracting, a set of variational equations is obtained.

$$\delta\dot{x}_i - f_i[(x_1)_0 + \delta x_1, (x_2)_0 + \delta x_2, \ldots, (x_n)_0 + \delta x_n] + f_i[(x_1)_0, (x_2)_0, \ldots, (x_n)_0] = -e_i \tag{17.3}$$

These are new differential equations containing the analog solutions as time-varying functions and using the e_i as forcing functions.

If the equations are linear, (17.3) reduces to

$$\delta\dot{x}_i - f_i(\delta x_1, \delta x_2, \ldots, \delta x_n) = -e_i \tag{17.4}$$

which is of the same form as the original set. Thus, the same circuit used to investigate (17.1) can be used for (17.4) provided the e_i are conveniently generated. If the δx_i are small, the linearized version of (17.3) could also be mechanized; a new circuit is needed if the equations are nonlinear.

The major disadvantage of mechanizing variational equations lies in determining the $e_i(\tau)$ from the analog solution. Only gross errors (from 1 to 10 per cent) could be detected and these are still subject to reading errors. But a Picard iteration can quickly detect such errors as well as smaller errors. Vichnevetsky later (see Sec. 17.3) used the variational approach to qualitatively examine error propagation, a more fruitful purpose for the variational equations in error analysis.

17.2.4 Digital simulation

An independent check solution is easily the best method to check for errors. The double entry system in accounting, for example, checks two sums. If they agree, it is highly unlikely that either is in error.

In the early 1950s, an independent digital computer check solution was a costly affair. Programming and debugging, error control, and running times made much simulation easier on an analog computer. Nevertheless, check solutions were obtained for one or several cases to insure that the analog program gave acceptable accuracy. Faster machines, simpler programming in languages such as FORTRAN, debugging routines, and the introduction of subroutines, all made digital computer programming simple and efficient. It is possible today, with fixed standard subroutines, to quickly program and solve a set of differential equations much more rapidly than can be done on an analog computer, with little of the scaling problems and inaccuracies. To be sure, the analog computer still has its place—in real time simulation, optimization, and exploratory studies—because of its inherent faster speed of computation.

In 1955, Selfridge [12] proposed a special computer language to simulate an analog computer with a digital computer. Its main feature was that the coding was very simple to learn and its format followed closely analog computer programming. Thus, it was geared to analog computer programmers. In 1959, Stein *et al.*, [13]–[14], followed with another simulation language with which a digital program could be obtained easily from an analog computer diagram. A more sophisticated integration routine was also used, producing more accurate results. A host of other simulation languages followed [15]–[37] (making this a subject in itself [39]), each attempting to make up for certain deficiencies inherent in previous ones or written for different digital computers. Significant among these are:

APACHE (*A*nalog *P*rogramming *A*nd *CHE*cking) [16]–[17]. Equations are fed into the program in virtually the same notation as they are formulated. The program returns scaled equations, static check voltages, *an analog program* (patching required and pot settings), and results for test cases. It is limited, however, in the digital and analog computer required for its use.

MIDAS (*M*odified *I*ntegration *D*igital *A*nalog *S*imulator) [22], [25], [26]. This is a block oriented language used solely for checking analog results. From an analog diagram, a block diagram can be drawn, and a program written easily from this. Blocks include INTEGRATE, SUM, NEGATIVE (inverter), MULTIPLY, DIVIDE, RESOLVER (sin and cos generation) as well as functions not so easily generable on an analog computer—EXPONENTIAL, NATURAL LOGARITHM, ARC TANGENT. Switching elements such as OUTPUT RELAY [two outputs $C = AU(B)$ and $D = AU(\bar{B})$], INPUT RELAY [SPDT—output is $AU(C) + BU(\bar{C})$], and LIMITER were included to appeal to the analog programmer with little or no digital programming experience.

MIMIC [29]. A continuation of MIDAS, this language is also algebraically oriented. Subroutine blocks can be included and provision to handle logic variables is also provided.

DSL/90 [30]. Besides standard simulation blocks, the user can include FORTRAN statements or use the language as a subroutine. It is the most flexible language of the four listed, but a knowledge of FORTRAN is required to make use of this flexibility. If FORTRAN is not used, a program written in DSL/90 resembles one written in MIMIC.

The choice of which language to use in a given situation depends on several factors [38]. For large-scale systems, APACHE provides useful output for scaling and patching. If an analog diagram is available, a MIDAS program is easily written. MIMIC or DSL/90 could be used when the analog program contains circuits not so easily represented by MIDAS (for example, solving a set of algebraic equations with high-gain amplifiers). DSL/90 is preferable if the programmer has experience with FORTRAN.

Simulation languages have been written in large numbers because of the wide variety of both digital and analog machines. New ideas will continue to find expression in even more languages, although some effort has been made in obtaining one standard language for continuous systems simulation [37].

17.3 Classical Error Theories

In this section, classical theories of errors for continuous system computers are discussed and consolidated. The basis for all theories is the first one, evolved by Miller and Murray [40]–[42]. They introduced the concept of *sensitivity equations*, the solutions of which determine the sensitivities of the original solution to small errors. Nelson [43], [44], however, found that the "true error" is conceptually more important—indeed, we have used *true error equations* in Chapter 5 to discuss certain static errors in generating functions with integrators, in Chapter 11 to discuss dynamic errors in generating implicit functions with high-gain amplifiers, and in numerous other places. They differ from the sensitivity equations in that variations to all errors are found simultaneously and without linearization. On the other hand, Vichnevetsky [45] was concerned with error propagation in a qualitative way and used approximate linearized true error equations (called *variational equations*) from which the stability often can be determined.

The three error theories are related. In a sense, we introduce error terms in the equations corresponding to those we can expect and evaluate their effect mathematically. The approach taken depends upon what one hopes to gain from the analysis. True error equations can be converted to variational equations by dropping small effect terms. Variational equations can be converted to sensitivity equations by looking at the effects of one parameter at a time and using a simple scale factor.

17.3.1 Miller–Murray theory—sensitivity equations

In the Miller–Murray theory, there are three classifications of errors:

α errors. These refer to zero memory errors (errors in pot settings, amplifier offsets, function generators, etc.) introduced into a computer solution because of the inability of computing elements to produce required input functions exactly. Instead of solving

$$F_i(\dot{x}_1, \dot{x}_2, \ldots, \dot{x}_n, x_1, x_2, \ldots, x_n) = 0 \tag{17.5}$$

the machine solves

$$G_i(\dot{x}_1, \dot{x}_2, \ldots, \dot{x}_n, x_1, x_2, \ldots, x_n, \alpha_1, \alpha_2, \ldots, \alpha_m) = 0 \tag{17.6}$$

for $i = 1, 2, \ldots, n$, where each $F_i = G_i$ when all $\alpha_j = 0$. This is a procedure we have used often. α errors are often called *static* errors.

β errors. These result from instantaneous disturbances of the integrator outputs (the memory). These errors jar the machine from one solution to another. Examples of β errors are errors in initial conditions, and integrator noise. β errors are often called *memory* errors.

Both α and β errors do not increase the order of the equations.

λ errors. These result in the machine solving equations of higher order than the F_i. All summers, potentiometers, and integrators are not ideal, with the transfer function of each tending to increase the order of the equations solved. In addition, high-gain solutions of algebraic equations, which are used for feedback, sometimes require a small capacitance from grid to output. This also increases the order of the equations, and its effect is usually greater than that produced by imperfections of amplifiers. λ errors are often called *dynamic* errors.

We can usually write out the equations with λ errors by including terms such as $\lambda_i \ddot{x}_i$ so that these do not enter when each λ_i is zero.

A comprehensive list of the major error sources in analog computers is

TABLE 17.1

α and β Errors

Component	*Source of error*	*Ideal performance*	*Approximate actual value*
Potentiometer	Cannot be set exactly	Set to value a $V_0 = aV_{in}$	$V_0 = a(1 + \alpha_0)V_{in}$
Summer	Resistor mismatches, noise, offset, finite gain, nonzero grid current	$V_0 = -\sum_{i=1}^{n} V_i g_i$	$V_0 = -\sum_{i=1}^{n} V_i g_i(1 + \alpha_i) + \alpha_0$
Integrator	Time constant mismatches, noise, drift, finite gain, nonzero grid current, switching errors at start	$V_0 = V_{IC} - \int_0^\tau \left(\sum_{i=1}^{n} V_i g_i\right) d\tau$	$V_0 = V_{IC} + \beta_1 - \int_{\beta_2}^{\tau} \left[\sum_{i=1}^{n} V_i g_i(1 + \alpha_i) + \alpha_0\right] d\tau$
High-gain amplifier	Same as summer	$\sum_{i=1}^{n} V_i g_i = 0$	$\sum_{i=1}^{n} V_i g_i(1 + \alpha_i) = \alpha_0$
Servo multiplier	Nonlinear pots, misalignment of pots, backlash, dead zone	$V_0 = V_1 V_2$	$V_0 = V_1(V_2 + \alpha_0)$
Quarter-square multipler	Approximation, noise	$V_0 = V_1 V_2$	$V_0 = V_1 V_2 + \alpha_0$
DFG and fixed function generator	Approximation, noise	$V_0 = f(V_1)$	$V_0 = f(V_1) + \alpha_0$
Relay	Nonzero contact resistance (almost negligible), comparator errors (hysteresis)	$V_0 = V_1 U(t_2 - t_1) + V_2 U(t_1 - t_2)$	$V_0 = V_1(1 + \alpha_1)U(t_2 - t_1 \pm \alpha_2) + V_2(1 + \alpha_1) \times U(t_1 - t_2 \mp \alpha_2)$
Electronic switch	Offset, comparator errors (hysteresis), mismatched components	$V_0 = V_1 U(t_2 - t_1)$	$V_0 = (V_1 + \alpha_1) \times U(t_2 - t_1 \pm \alpha_2) + \alpha_3$

NOTE: Each α_i may not be constant and varies from component to component.

TABLE 17.2

λ (Dynamic) Errors

Component	*Source of error*	*Ideal performance*	*First-order error*
Potentiometer	Parasitic capacitance, resistance, and inductance	$V_0 = aV_{\text{in}}$	Transfer function changed to $\frac{V_0}{V_{\text{in}}} = a\left(\frac{1}{\lambda_1 s + 1}\right)$
Summer	Parasitic capacitance, resistance, and inductance	$V_0 = -\sum_{i=1}^{n} V_i g_i$	Transfer function changed to $\frac{V_0}{\sum_{i=1}^{n} V_i g_i} = -\frac{1}{1+\lambda_2 s}$
Integrator	Parasitic capacitance, inductance, and resistance*	Transfer function $\frac{V_0}{\sum_{i=1}^{n} V_i g_i} = -\frac{1}{s}$	Transfer function changed to $\frac{V_0}{\sum_{i=1}^{n} V_i g_i} = -\frac{1}{\alpha_0 + (1+\alpha_1)s + \lambda_1 s_2}$
High-gain amplifier	Parasitic or actual capacitance	$\sum_{i=1}^{n} V_i g_i = 0$	$\sum_{i=1}^{n} V_i g_i = \frac{\dot{V}_0}{\lambda_1}$ (Chapter 11)
Servo multiplier	Mechanical lag	$\frac{V_0}{V_1 V_2} = 1$	Transfer function changed to $\frac{V_0}{V_1 V_2} = \frac{1}{1+\lambda_1 s}$ Velocity and acceleration of motor is limited. Friction can also be important.
Quarter-square multiplier	Parasitic capacitance, inductance, and resistance	$\frac{V_0}{V_1 V_2} = 1$	Transfer function changed to $\frac{V_0}{V_1 V_2} = \frac{1}{1+\lambda_1 s}$
DFG and fixed function generator	Parasitic capacitance, inductance, and resistance	$\frac{V_0}{f(V_1)} = 1$	Transfer function changed to $\frac{V_0}{f(V_1)} = \frac{1}{1+\lambda_1 s}$
Relay	Mechanical lag, dead-space, bouncing	$V_0 = V_1\, U(t_2 - t_1) + V_2\, U(t_1 - t_2)$	$V_0 = V_1\, U(x+\alpha_0) + V_2\, U(-x-\alpha_0)$ where $x/(t_2 - t_1) = 1/(1+\lambda_1 s)$ Bouncing not included.
Electronic switch	Parasitic capacitance, inductance, and resistance, and hysteresis	$V_0 = V_1\, U(t_2 - t_1)$	Transfer function changed to $\frac{V_0}{V_1} = \frac{1+\lambda_1 s}{1+\lambda_2 s}\, U(x \pm \alpha_0)$ where $x/(t_2 - t_1) = 1/(1+\lambda_3 s)$

*Does not include the effects of capacitor dielectric absorption.

NOTE: Each λ_i and α_i are approximately constant, but vary from component to component.

given in Tables 17.1 and 17.2. More thorough discussions of minimization of component errors through design are given in various texts (e.g. [3] and also occasionally in papers [4]).

Table 17.1 lists the major α and β errors that arise during the normal course of computation. Their values can be functions of problem variables, computer design, age, and operator skill, as well as other factors. The errors are not necessarily constant during any run, although they are usually quite small. They could be measured for any given setup, but, fortunately this is necessary only if a specific quantitative error analysis must take place.

Table 17.2 shows how an ideal performance is changed because of λ errors. The errors in both tables must be combined in any given case to determine an actual equation solved by the computer. This is really only an approximate equation, for only first-order errors are included. To estimate an error in the computer solution, we must compare it with the actual solution to the problem. Here we reach an impasse, since only the computer solution is available. There are ways around this impasse, as discussed, usually time- and cost-consuming.

Miller and Murray essentially generalized usual existence theorems for ordinary differential equations when no λ errors are present and showed the computer solution to depend analytically on the α_j and β_k. The computer solution

$$x_i \equiv x_i(\alpha_1, \alpha_2, \ldots, \alpha_m, \beta_1, \beta_2, \ldots, \beta_p)$$

can therefore be written as a power series with a Taylor expansion:

$$x_i = x_i(0, 0, \ldots, 0, 0, 0, \ldots, 0) + \sum_{j=1}^{m} \frac{\partial x_i}{\partial \alpha_j} \alpha_j + \sum_{k=1}^{p} \frac{\partial x_i}{\partial \beta_k} \beta_k + \cdots \qquad (17.7)$$

where each partial derivative of the x_i with each parameter exists.

Let γ be either an α or β parameter. Differentiate (17.6) with respect to γ:

$$\sum_{j=1}^{n} \frac{\partial G_i}{\partial \dot{x}_j} \frac{\partial \dot{x}_j}{\partial \gamma} + \sum_{j=1}^{n} \frac{\partial G_i}{\partial x_j} \frac{\partial x_j}{\partial \gamma} + \frac{\partial G_i}{\partial \gamma} = 0 \qquad (17.8)$$

(If γ is a β parameter, $\partial G_i/\partial \gamma = 0$). Now let

$$y_j = \frac{\partial x_j}{\partial \gamma}$$

Then

$$\frac{\partial \dot{x}_j}{\partial \gamma} = \frac{\partial^2 x_j}{\partial \gamma \partial \tau} = \frac{\partial y_j}{\partial \tau} = \dot{y}_j$$

so that (17.8) reduces to

$$\sum_{j=1}^{n} \frac{\partial G_i}{\partial \dot{x}_j} \dot{y}_j + \sum_{j=1}^{n} \frac{\partial G_i}{\partial x_j} y_j + \frac{\partial G_i}{\partial \gamma} = 0 \tag{17.9}$$

These are called the *first-order sensitivity equations*. The y_j are called *sensitivity functions* or *parameter influence coefficients*. If the solutions $x_j(\tau)$ are known, the coefficients $\partial G_i/\partial \dot{x}_j$, $\partial G_i/\partial x_j$, and $\partial G_i/\partial \gamma$ are all functions of time τ. Thus the equations are linear. Initial conditions depend upon whether γ is an α or β parameter. If γ is an α, $(y_j)_0 = 0$ since it has no effect on the initial value of the integrator producing x_j. If γ is a β, $y_j = 1$ at $\tau = \tau_1$ if the perturbation β occurs when $\tau = \tau_1$.

In a similar manner, higher-order sensitivity equations can be derived, although they are rarely used. It is difficult enough to obtain solutions to (17.9); higher-order sensitivities simply compound those difficulties. We assume that the α_j and β_k are small enough so that (17.7) provides the essential information needed. Thus, when there are no λ errors, an error investigation can be referred to a system of linear differential equations, the sensitivity equations. Solutions of the sensitivity equations, however, require the solution variables $x_i(t)$ and $\dot{x}_i(t)$, so that difficulties still exist.

Double subscripts are normally used for specific computations; *i.e.*, we let $y_{i,j} = \partial x_i/\partial \alpha_j$ and $z_{i,k} = \partial x_i/\partial \beta_k$.

A similar theory can be derived for λ errors, although care must be taken because solutions can be nonanalytic in λ at $\lambda = 0$. But we shall see that a λ error can be converted often to an approximate equivalent α error in stable systems.

Example 17.1. Find the first-order sensitivity equations for the system

$$\dot{x} + (\alpha_1 + 1)(x^2 + \alpha_2) = 0, \qquad x_0 = 1 + \beta_1 \tag{17.10}$$

where α_1, α_2, and β_1 are nominally zero. These static errors exist when the function $x = 1/(1 + \tau)$ is generated with an integrator using a quarter-square multiplier (Sec. 5.6.4). α_1 and β_1 are constants and we also consider α_2 constant to simplify the analysis.

For this simple one-dimensional example, we use single subscript notation and let

$$y_1 = \frac{\partial x}{\partial \alpha_1}, \qquad y_2 = \frac{\partial x}{\partial \alpha_2}, \qquad z_1 = \frac{\partial x}{\partial \beta_1}$$

Differentiating, we obtain from (17.9)

$$\begin{aligned} \dot{y}_1 + (x^2 + \alpha_2) + (\alpha_1 + 1)(2xy_1) &= 0, \qquad (y_1)_0 = 0 \\ \dot{y}_2 + (\alpha_1 + 1)(2xy_2 + 1) &= 0, \qquad (y_2)_0 = 0 \\ \dot{z}_1 + (\alpha_1 + 1)(2xz_1) &= 0, \qquad (z_1)_0 = 1 \end{aligned}$$

which reduce to

$$\begin{aligned} \dot{y}_1 + x^2 + 2xy_1 &= 0, \qquad (y_1)_0 = 0 \\ \dot{y}_2 + 2xy_2 + 1 &= 0, \qquad (y_2)_0 = 0 \\ \dot{z}_1 + 2xz_1 &= 0, \qquad (z_1)_0 = 1 \end{aligned} \tag{17.11}$$

about the nominal values of the errors. These are uncoupled equations; the effect of each error can be considered separately, as is generally true. Since x is known here to be $1/(1 + \tau)$, it can be substituted in (17.11) and solutions found analytically. With these solutions, small value errors are found from (17.7).

Solutions of the first-order sensitivity equations (17.9) require the solutions $x_i(\tau)$ for feedback in the equations. If the computer solution is stable, Meissinger [64]-[65] suggests using *it* as feedback in the final sensitivity equations. For example, to solve (17.11) on the computer, x is supplied by solving (17.10) about the nominal values of α_1, α_2, and $\beta_1 = 0$:

$$\dot{x} + x^2 = 0, \qquad x_0 = 1$$

Although reasonable answers can be expected, the price paid for a simultaneous solution is about four times the equipment needed for the solution of (17.10). If there are m parameters for n first-order differential equations, $n(m + 1)$ equations must be solved, a gigantic undertaking when n and m are not small. Even if sensitivity functions for each parameter are obtained separately, about twice the amount of equipment is required. (A possible exception to this occurs with linear systems—see Problem 17-4.)

Another method to obtain sensitivity functions for stable systems is to solve the original equations on the computer with and without a finite error γ, as suggested by Murray and Miller [42]. From (17.7)

$$y_i = \frac{\partial x_i}{\partial \gamma} \simeq \frac{x_i - x_i(0, 0, \ldots, 0)}{\gamma}$$

The disadvantage of this procedure for analog computers is clear; since the errors are small, $x_i - x_i(0, 0, \ldots, 0)$ appear as small voltage traces. Magnifying these voltages also magnifies other small errors including noise. Nevertheless, the procedure is useful because essentially the same amount of equipment is required to get x_i and $x_i(0, 0, \ldots, 0)$, since they are not found simultaneously. Korn [62] and O'Grady [63] use this method to provide first-order sensitivity coefficients automatically in connection with certain parameter optimization studies on an iterative analog computer (Chapter 19).

For finding sensitivity functions for a λ error, the best procedure, when applicable, is to convert the error to an approximate α error and use the procedures of Sec. 17.3.4. This can be shown to be equivalent to the method discussed by Tomovic *et al.* [66], for stable systems.

Besides error analysis and parameter optimization already mentioned, Meissinger [64] broadened the applications of sensitivity functions to include statistical analyses of system outputs due to randomly distributed, known input disturbances. This can be more accurately mechanized also from using exact variational equations with perturbation techniques (Sec. 17.5).

17.3.2 Nelson's theory—true error

The solutions to sensitivity equations can be misleading unless the entire infinite series (17.7) is constructed, an impossibility. To see this, we analytically solve y_2 from (17.11) and find

$$y_2 = -\frac{1}{3}\left[1 + \tau - \frac{1}{(1+\tau)^2}\right]$$

From (17.7), the small error produced by this term is approximately $-\alpha_2[1 + \tau - 1/(1+\tau)^2]/3$, which implies its amplitude grows without bounds. But this is not always true. For we may also analytically solve the equation

$$\dot{x} + x^2 + \alpha_2 = 0, \qquad x_0 = 1$$

to determine the effects of α_2 on the solution. If α_2 is negative, we find

$$x = \sqrt{-\alpha_2}\tanh\left[C + \sqrt{-\alpha_2}(1+\tau)\right]$$

where C is found from the initial condition $x_0 = 1$. But the hyperbolic tangent function approaches 1 as its argument approaches infinity, so that $x \to \sqrt{-\alpha_2}$ as $\tau \to \infty$.

The discrepancy in the results comes from neglect of the higher-order terms in the Taylor expansion (17.7). As Nelson [43]–[44] points out, a "true" error analysis, when possible, leads to a more accurate error analysis and a more accurate concept of the error.

To perform a true error analysis, it is most convenient to introduce the errors u_i to the true solutions x_i, where the computer solution x_i^* is given by

$$x_i^* = x_i + u_i$$

Since

$$\dot{x}_i^* = \dot{x}_i + \dot{u}_i$$

differential equations for the $\dot{u}_i$ can be found by differencing the expressions for $\dot{x}_i^*$ and $\dot{x}_i$:

$$\dot{u}_i = \dot{x}_i^* - \dot{x}_i$$

assuming the original equations [(17.5) and (17.6)] can be solved for the $\dot{x}_i$ and the $\dot{x}_i^*$. The final equations take the form

$$\dot{u}_i = g_i(u_1, u_2, \ldots, u_n, x_1, x_2, \ldots, x_n, \alpha_1, \alpha_2, \ldots, \alpha_m) \tag{17.12}$$

from which the errors can be investigated directly, when the x_i are known functions of τ. If preferred, the x_i can be replaced by $x_i^* - u_i$ in order to use computer outputs for a computer analysis. The error model must be quite accurate.

Example 17.2. Derive the true error equation for the differential equation in Example 17.1. We have

$$\begin{aligned} \dot{x} + x^2 &= 0, \qquad x_0 = 1 \\ \dot{x}^* + (1 + \alpha_1)(x^{*2} + \alpha_2) &= 0, \qquad x_0^* = 1 + \beta_1 \end{aligned} \tag{17.13}$$

This leads to

$$\dot{u} + 2ux^{*2} - u^2 + \alpha_2 + \alpha_1(x^{*2} + \alpha_2) = 0, \qquad u_0 = \beta_1 \tag{17.14}$$

or

$$\dot{u} + 2ux + u^2 + \alpha_2 + \alpha_1[(x + u)^2 + \alpha_2] = 0, \qquad u_0 = \beta_1 \tag{17.15}$$

with $u = x^* - x$. Quantitative results can be found from (17.14) or (17.15) for any magnitude of errors.

In stable systems, where errors tend to be small, the computer solution can be taken to be the actual solution just as in sensitivity analysis. Computer errors may

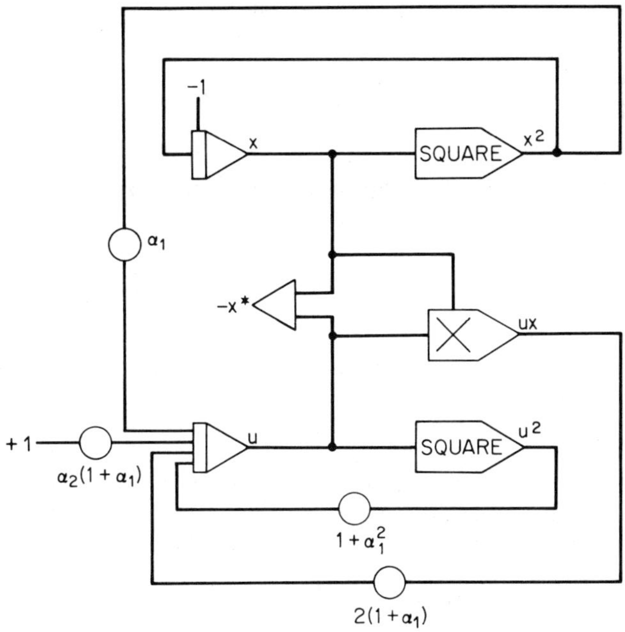

FIG. 17.1 True error determination for Example 17.2.

then be predicted by using the computer solution as feedback in the error equation. Figure 17.1 shows a mechanization of (17.15) to find u. It assumes there is little error in x (which may not be true here). The solution x^* can then be constructed by summing x and u in a technique known as *extrapolation.* Note that extrapolation with a first-order sensitivity function is linear, while true error extrapolation is exact. Care must be exercised when extrapolating effects of error to insure the actual solution has little error.

Example 17.3 (***Linear Systems***). Most fruitful results are obtained from a true error analysis of α and β errors in linear systems [44]. Given a system

$$\dot{\mathbf{x}} + \mathbf{A}(\tau)\mathbf{x} + \mathbf{f}(\tau) = \mathbf{0}, \qquad \mathbf{x}_0 = \mathbf{b}$$

determine the errors of the solution due to errors $\alpha_{i,j}$ in pot settings for the components in $\mathbf{A}$ and the errors β_i in the initial conditions $\mathbf{b}$. Let

$$\mathbf{E} = (\alpha_{i,j}), \qquad \boldsymbol{\beta} = (\beta_i)$$

be the matrix and vector for these errors. The computer solves

$$\dot{\mathbf{x}}^* + (\mathbf{A} + \mathbf{E})\mathbf{x}^* + \mathbf{f}(\tau) = \mathbf{0}, \qquad \mathbf{x}_0 = \mathbf{b} + \boldsymbol{\beta}$$

With $\mathbf{u}$ the error vector such that

$$\mathbf{x}^* = \mathbf{x} + \mathbf{u}$$

we obtain

$$\dot{\mathbf{x}} + \dot{\mathbf{u}} + (\mathbf{A} + \mathbf{E})(\mathbf{x} + \mathbf{u}) + \mathbf{f}(\tau) = \mathbf{0}$$

and

$$\dot{\mathbf{u}} + \mathbf{A}\mathbf{u} + \mathbf{E}(\mathbf{x} + \mathbf{u}) = \mathbf{0}, \qquad \mathbf{u}_0 = \boldsymbol{\beta}$$

or

$$\dot{\mathbf{u}} + (\mathbf{A} + \mathbf{E})\mathbf{u} + \mathbf{E}\mathbf{x} = \mathbf{0}, \qquad \mathbf{u}_0 = \boldsymbol{\beta} \tag{17.16}$$

We instantly perceive

(a) If $\mathbf{A}$ is constant, the characteristic roots are slightly perturbed from those of the actual solution.

(b) The driving vector $\mathbf{f}(\tau)$ in the actual equation is replaced by $\mathbf{Ex}$ in the error equation.

(c) If $\mathbf{E} \equiv \mathbf{0}$, errors in initial conditions are determined without driving functions and with the identical coefficient matrix. In addition, if $\mathbf{A}$ is constant, errors in initial conditions cannot affect the stability of the solution.

17.3.3 Variational approach [45]

Qualitative information concerning the error propagation can be more often obtained quickly from an analysis of first-order variational equations than from other methods. Assume the equations to be solved on the computer can be solved for the first derivatives

$$\dot{x}_i + f_i(\tau, x_1, x_2, \ldots, x_n) = 0, \qquad i = 1, 2, \ldots, n$$

or in vector notation

$$\dot{\mathbf{x}} + \mathbf{f}(\mathbf{x}, \tau) = \mathbf{0}$$

with appropriate initial conditions. Now, if α and β errors exist in the mechanization, the machine solves

$$\dot{\mathbf{x}}^* + \mathbf{f}^*(\mathbf{x}^*, \boldsymbol{\alpha}, \tau) = \mathbf{0}$$

where $\boldsymbol{\alpha}$ is an m-dimensional α vector such that

$$\mathbf{f}^*(\mathbf{x}^*, \mathbf{0}, \tau) = \mathbf{f}(\mathbf{x}^*, \tau)$$

From a Taylor series expansion about $\boldsymbol{\alpha} = \mathbf{0}$ and $\mathbf{x}^* = \mathbf{x}$, we have

$$\mathbf{f}^*(\mathbf{x}^*, \boldsymbol{\alpha}, \tau) \simeq \mathbf{f}(\mathbf{x}, \tau) + \mathbf{J}^*(\mathbf{f}^*, \mathbf{x}^*)(\mathbf{x}^* - \mathbf{x}) + \mathbf{J}^*(\mathbf{f}^*, \boldsymbol{\alpha})\boldsymbol{\alpha}$$

where $\mathbf{J}^*(\mathbf{f}^*, \mathbf{x}^*)$ is the $n \times n$ Jacobian of $\mathbf{f}^*$ with respect to $\mathbf{x}^*$ and $\mathbf{J}^*(\mathbf{f}^*, \boldsymbol{\alpha})$ is the $n \times m$ Jacobian of $\mathbf{f}^*$ with respect to $\boldsymbol{\alpha}$, both evaluated at $\mathbf{x}$ and $\mathbf{0}$. Thus, $\mathbf{J}^*(\mathbf{f}^*, \mathbf{x}^*) = \mathbf{J}(\mathbf{f}, \mathbf{x})$. With

$$\mathbf{u} = \mathbf{x}^* - \mathbf{x} \tag{17.17}$$

(so that $\dot{\mathbf{u}} = \dot{\mathbf{x}}^* - \dot{\mathbf{x}}$) we obtain a linearized variational equation called the *error-propagation equation*:

$$\dot{\mathbf{u}} + \mathbf{J}(\mathbf{f}, \mathbf{x})\mathbf{u} + \boldsymbol{\epsilon}(\mathbf{x}, \boldsymbol{\alpha}) \simeq \mathbf{0} \tag{17.18}$$

where $\boldsymbol{\epsilon}(\mathbf{x}, \boldsymbol{\alpha}) = \mathbf{J}^*(\mathbf{f}^*, \boldsymbol{\alpha})\boldsymbol{\alpha}$. The $\boldsymbol{\beta}$ vector jars $\mathbf{u}$ throughout the solution. Equation (17.18) can be used to determine quantitative effects of α and β errors. But since $\boldsymbol{\epsilon}$ depends on the actual solution $\mathbf{x}(\tau)$, it is strictly a function of τ for any fixed $\boldsymbol{\alpha}$. The stability of (17.18) thus depends upon the stability of the homogeneous equation

$$\dot{\mathbf{u}} + \mathbf{J}(\mathbf{f}, \mathbf{x})\mathbf{u} = \mathbf{0} \tag{17.19}$$

The Jacobian $\mathbf{J}(\mathbf{f}, \mathbf{x})$ is a function of τ as well as $\mathbf{x}$ and $\boldsymbol{\alpha}$, so that Theorem 11.12 may not apply. However, sufficient conditions for the stability of (17.19) can sometimes be derived when a Lyapunov function (Sec. 11.4) can be constructed from it. Multiply (17.19) by $\mathbf{u}'$, the transpose of $\mathbf{u}$:

$$\mathbf{u}'\dot{\mathbf{u}} + \mathbf{u}'\mathbf{J}\mathbf{u} = \mathbf{0}$$

It is easily shown that

$$\mathbf{u}'\dot{\mathbf{u}} \equiv \frac{1}{2}\frac{d}{d\tau}\left(\sum_{i=1}^{n} u_i^2\right)$$

where u_i are the components of $\mathbf{u}$. Thus, $\mathbf{u}'\mathbf{u}$ is a Lyapunov function if $\mathbf{u}'\mathbf{J}\mathbf{u}$ is a positive definite quadratic form. This requires that the symmetric part of $\mathbf{J}$ be positive definite. We therefore have the main result:

Theorem 17.1. A sufficient condition for the stability of (17.19) is that $\mathbf{J}' + \mathbf{J}$ be positive definite.

Therefore, we often can get information concerning the stability of the error propagation without specifically knowing the α or β errors. In fact, because λ errors are usually small also, the stability of (17.19) usually means that an equation will be stable for all three kinds of errors.

Example 17.4. Investigate the stability of

$$\dot{x} + x + y \sin \tau + f_1(\tau) = 0$$

$$\dot{y} + x \cos \tau + (\tfrac{1}{2} + e^{-\tau})y + f_2(\tau) = 0$$

Since

$$\mathbf{J} = \begin{pmatrix} 1 & \sin \tau \\ \cos \tau & \frac{1}{2} + e^{-\tau} \end{pmatrix}$$

$$\mathbf{J} + \mathbf{J}' = \begin{pmatrix} 2 & \sin \tau + \cos \tau \\ \sin \tau + \cos \tau & 1 + 2e^{-\tau} \end{pmatrix}$$

the equations are stable. $\mathbf{J} + \mathbf{J}'$ is positive definite for all τ.

17.3.4 Reduction of λ errors to approximate α errors for stable systems

An approximate evaluation of λ errors in stable systems can be obtained from the dynamic error ideas given in Sec. 11.6. Let us assume that instead of solving

$$\dot{\mathbf{x}} + \mathbf{f}(\mathbf{x}, \tau) = \mathbf{0}$$

the machine solves the stable equation

$$\mathbf{\Lambda}\ddot{\mathbf{x}}^* + \dot{\mathbf{x}}^* + \mathbf{f}(\mathbf{x}^*, \tau) = \mathbf{0} \tag{17.20}$$

for some constant gain matrix $\mathbf{\Lambda} = (\lambda_{i,j})$, which corresponds to a first-order λ-error analysis. The first-order effect of these λ errors can be found as an error $\boldsymbol{\epsilon}(\mathbf{x}^*, \tau)$ in $\dot{\mathbf{x}}^*$. By first-order effect, we mean essentially that $\mathbf{\Lambda}\ddot{\mathbf{x}}^*$ is negligible in estimating the value of the vector $\ddot{\mathbf{x}}^*$. Differentiating (17.20) with τ, we have

$$\ddot{\mathbf{x}}^* + \dot{\mathbf{f}}(\mathbf{x}^*, \tau) \simeq \mathbf{0}$$

Since

$$\dot{\mathbf{f}} \simeq \mathbf{J}\dot{\mathbf{x}}^* + \frac{\partial \mathbf{f}}{\partial \tau}$$

(17.20) transforms to

$$\dot{\mathbf{x}}^* + \mathbf{f}(\mathbf{x}^*, \tau) - \mathbf{\Lambda}\left(\mathbf{J}\dot{\mathbf{x}}^* + \frac{\partial \mathbf{f}}{\partial \tau}\right) \simeq \mathbf{0}$$

Using the same definition of $\mathbf{u}$ as in (17.17), we finally obtain

$$\dot{\mathbf{u}} + \mathbf{J}\mathbf{u} - \mathbf{\Lambda}\left(-\mathbf{J}\mathbf{f} + \frac{\partial \mathbf{f}}{\partial \tau}\right) \simeq \mathbf{0} \tag{17.21}$$

for stable systems, where $\dot{\mathbf{x}}^* \simeq -\mathbf{f}(\mathbf{x}^*, \tau)$ has also been substituted because this introduces only a second-order error for stable systems. Essentially, the $\boldsymbol{\epsilon}$ vector in (17.18) has been evaluated (in terms of α errors) making it possible to obtain quantitative estimates of the effects of a $\mathbf{\Lambda}$ matrix directly on the computer. Static errors must also be superimposed on these λ errors for a more accurate picture of the total error effect. This procedure gives an approximate error but is simpler than a direct evaluation of the true error involving $\ddot{\mathbf{u}}$. Note that scale factors could be taken according to $\mathbf{\Lambda}^{-1}\mathbf{u}$ as a first order of magnitude when $\mathbf{\Lambda}$ is diagonal, thus eliminating the $\mathbf{\Lambda}$ matrix elements entirely.

Example 17.5. Investigate the λ error of the one-dimensional case

$$\dot{x} + \left(\frac{1}{1+\tau}\right)x = 0, \qquad x_0 = 1 \tag{17.22}$$

where the computer solves

$$\lambda\ddot{x}^* + \dot{x}^* + \left(\frac{1}{1+\tau}\right)x^* = 0, \qquad x_0^* = 1 \tag{17.23}$$

Note that only one initial condition is specified since the computer solves (17.22), but the λ term is present because of parasitic capacitance. Thus $(\dot{x})_0^* = -1$ is computed exactly. We have

$$\mathbf{f} = \left(\frac{x}{1+\tau}\right), \qquad \mathbf{J} = \left(\frac{1}{1+\tau}\right), \qquad \frac{\partial \mathbf{f}}{\partial \tau} = -\left[\frac{x}{(1+\tau)^2}\right]$$

The error equation (17.21) reduces to one scalar equation

$$\dot{u} + \left(\frac{1}{1+\tau}\right)u + \lambda\left[\frac{2x}{(1+\tau)^2}\right] = 0, \qquad u_0 = 0 \tag{17.24}$$

Since an actual solution is found to be $x = 1/(1+\tau)$ we may investigate the error using

$$\dot{u} + \left(\frac{1}{1+\tau}\right)u + \lambda\left[\frac{2}{(1+\tau)^3}\right] = 0, \qquad u_0 = 0$$

Its solution can be shown to be

$$u = -\frac{2\lambda\tau}{(1+\tau)^2}$$

so that a computer solution is approximately

$$x^* = \frac{1}{1+\tau} - \frac{2\lambda\tau}{(1+\tau)^2} \tag{17.25}$$

The function $-2\tau/(1+\tau)^2$ may not be the first-order sensitivity function as defined by $\partial x^*/\partial\lambda$ at $\lambda = 0$. In this case [41], there are no analytic solutions at $\lambda = 0$; i.e., the power series in λ diverges for all λ except $\lambda = 0$.

To compare (17.25) with an exact solution of (17.23), we note first that $(1+\tau) \times \exp(-\tau/\lambda)$ is a solution to (17.23). By the method of variation of parameters, the exact solution is found to be

$$x^* = (1+\tau)e^{-\tau/\lambda} + \left(\frac{1}{\lambda} - 2\right)(1+\tau)e^{-\tau/\lambda}\int_0^\tau \frac{e^{u/\lambda}}{(1+u)^2}\,du$$

or

$$x^* = (1+\tau)y - 2(1+\tau)\lambda(y - e^{-\tau/\lambda})$$

where y is the solution of

$$\lambda\dot{y} + y = \frac{1}{(1+\tau)^2}, \qquad y_0 = 1 \tag{17.26}$$

Since the solution of (17.26) is approximately

$$y \simeq \frac{1}{(1+\tau)^2}$$

for small λ, we let

$$y = \frac{1}{(1+\tau)^2} + \lambda[a_1(\tau) + a_2(\tau)]$$

and by using

$$a_1 = \frac{2}{(1+\tau)^3}$$

we find that a_2 satisfies the equation

$$\dot{a}_2 + \frac{a_2}{\lambda} = \frac{6}{(1+\tau)^4}, \qquad (a_2)_0 = -2$$

This ultimately reduces to the exact equation

$$x^* = \frac{1}{1+\tau} - \frac{2\lambda\tau}{(1+\tau)^2} - \frac{4\lambda^2}{(1+\tau)^2} + 4\lambda^2(1+\tau)e^{-\tau/\lambda}$$
$$-(12\lambda^2 - 6\lambda)(1+\tau)e^{-\tau/\lambda}\int_0^\tau \frac{e^{u/\lambda}}{(1+u)^4}\,du \tag{17.27}$$

(For $\lambda = \frac{1}{2}$, this gives the exact solution without quadratures.) By noting that

$$\int_0^\tau \frac{e^{u/\lambda}}{(1+u)^3}\,du \geq \int_0^\tau \frac{e^{u/\lambda}}{(1+u)^4}\,du$$

and integrating the left-hand expression by parts, we may show that

$$(1+\tau)e^{-\tau/\lambda}\int_0^\tau \frac{e^{u/\lambda}}{(1+u)^4} < \left(\frac{\lambda}{1-3\lambda}\right)\left[\frac{1}{(1+\tau)^2}\right]$$

for $\lambda < \frac{1}{3}$, so the last term in (17.27) is of order λ^2 for small λ. The maximum error term in using (17.25) is easily calculated. Thus, although no power series in λ exists about $\lambda = 0$, we may evaluate its effect on the solution by the method suggested and obtain a "quasi-sensitivity function."

The method of obtaining the quasi-sensitivity function by converting the λ error into an approximate α error is extremely useful. Note that if (17.21) could not be solved analytically, it could be mechanized on the computer using u/λ as the computer variable. In this way, the outputs are of normal size, since u/λ is the first-order quasi-sensitivity function.

17.4 Analysis of Linear Systems with Constant Coefficients

Linear systems with constant coefficients are most frequently analyzed by determining shifts in the characteristic roots caused by small error [46]–[61]. There are two ways to do this. First, by means of Table 17.2 or by some other models, determine the equation solved by the computer and find its characteristic roots. For each summer, potentiometer, and integrator in the circuit, the order of the equation is increased by one. If the mechanization is stable, all but the original n roots damp out of the solution quickly. The original n roots are shifted slightly and the shift may be computed. The second method, adopted here, is to instantly eliminate any increases in order of the equation by converting λ errors to approximate equivalent α errors. Consider, for example, a summer whose first-order transfer function is given by

$$S(s) = \frac{-1}{1 + \lambda_2 s}$$

If the input is V_i and the output is V_0, we have

$$\lambda_2 \dot{V}_0 + V_0 + V_i = 0 \tag{17.28}$$

This represents an instantaneous small lag obtainable by the methods of Secs. 11.6 or 17.3. Assume $\lambda_2 \ddot{V}_0$ is negligible (produces second-order errors), differentiate (17.28) and find that

$$V_0 + V_i - \lambda_2 \dot{V}_i \simeq 0 \tag{17.29}$$

Now, substitute the problem expression for $\dot{V}_i$. It need not contain any perturbation terms because products of the form $\lambda_i \lambda_j$ simply produce second-order effects.

For an integrator, the transfer function is taken as

$$I(s) = \frac{-1}{\alpha_0 + (1 + \alpha_1)s + \lambda_1 s^2}$$

and its equivalent α error representation is found to be

$$(1 + 2\alpha_1)\dot{V}_0 + \alpha_0 V_0 + (1 + \alpha_1)V_i - \lambda_1 \dot{V}_i \simeq 0$$

which reduces to

$$\dot{V}_0 + \alpha_0 V_0 + (1 - \alpha_1)V_i - \lambda_1 \dot{V}_i \simeq 0 \tag{17.30}$$

for $0 \leq |\alpha_1| \ll \frac{1}{2}$.

The advantage of this method is obvious; a final equation of the same order as the original is obtained. The final matrix of coefficients for a many-integrator problem then contains small perturbation terms. It is even possible to attempt to compensate for these errors by adjusting the gains to null these errors [70].

Example 17.6 (***Harmonic Oscillator***). Determine the effect of summer and integrator delays on the harmonic oscillator circuit shown in Fig. 17.2 (ignore potentiometer errors). The ideal equations are

$$\dot{x}_2 + \omega x_1 = 0, \qquad \dot{x}_1 - \omega x_2 = 0 \tag{17.31}$$

but the computer solves

$$\lambda_1 \ddot{x}_1 + (1 + \alpha_1)\dot{x}_1 + \alpha_0 x_1 + \omega x_3 = 0$$
$$\lambda_1 \ddot{x}_2 + (1 + \alpha_1)\dot{x}_2 + \alpha_0 x_2 + \omega x_1 = 0$$

with

$$\lambda_2 \dot{x}_3 + x_3 + x_2 = 0 \tag{17.32}$$

By (17.29), (17.32) is transformed to

$$x_3 + x_2 - \lambda_2 \dot{x}_2 = x_3 + x_2 + \lambda_2 \omega x_1 \simeq 0$$

Similarly, by using (17.30) we finally obtain

$$\dot{x}_1 + \alpha_0 x_1 - \omega(1 - \alpha_1)x_2 - \lambda_2 \omega^2 x_1 - \lambda_1 \omega^2 x_1 \simeq 0$$
$$\dot{x}_2 + \alpha_0 x_2 + \omega(1 - \alpha_1)x_1 - \lambda_1 \omega^2 x_2 \simeq 0$$

The perturbed matrix is

$$\mathbf{A} = \begin{pmatrix} \alpha_0 - \omega^2(\lambda_1 + \lambda_2) & \alpha_1 \omega - \omega \\ \omega - \alpha_1 \omega & \alpha_0 - \omega^2 \lambda_1 \end{pmatrix}$$

Ignoring cross products and powers of the error terms, the characteristic roots μ can be shown to be

$$\mu \simeq \alpha_0 - \omega^2 \left(\lambda_1 + \frac{\lambda_2}{2}\right) \pm i\omega(1 - \alpha_1)$$

Since the roots from (17.31) are $\pm i\omega$, this represents a shift

$$\Delta\mu \simeq \alpha_0 - \omega^2 \left(\lambda_1 + \frac{\lambda_2}{2}\right) \mp i\omega\alpha_1$$

Thus, the amplitude of the oscillator decays for sufficiently small ω where α_0 predominates, but expands for sufficiently large ω. This result essentially agrees with

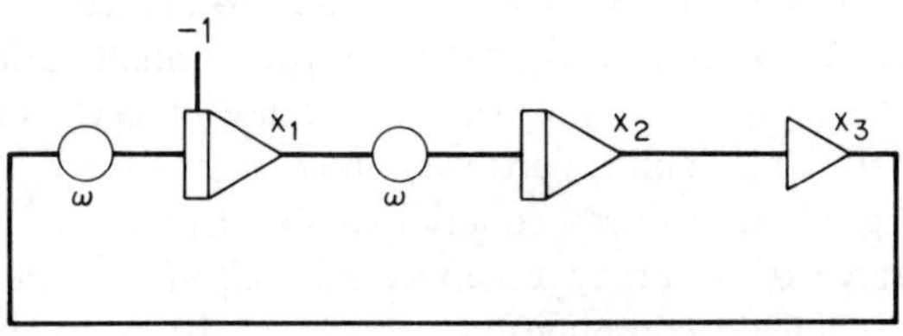

FIG. 17.2 Harmonic oscillator circuit.

Macnee's [46]: α_1 is usually much smaller than the other errors. However, this term could include the effects of incorrect integrator gains or pot settings.

The above procedure is useful for obtaining first-order effects only. Much more tedious calculations are involved if more sophisticated models are assumed or if second- and higher-order effects are to be determined. The latter is rarely necessary, since a dynamic check can determine if the solution is completely out of the small error range.

For higher-order systems, the original characteristic roots must be obtained for comparison with the perturbed roots. It is then convenient to examine the effect of each error element of the perturbing matrix separately, in a manner described by Nelson [44]. This involves solving the polynomial obtained for the shift in roots in terms of the error and original matrix elements. The process can be tedious—we see why a dynamic error check is so much more convenient.

Another source of error not yet discussed is the capacitor dielectric absorption phenomenon [51]–[54]. This is fundamental to the nature of materials from which capacitors are made; it results in requiring an accurate model of an integrator to increase the order of the equation by much more than one. Errors produced depend upon its previous operation—how long the computer is in the RESET mode, for example. The major effect is to instantaneously change the initial conditions when the computer is placed in OPERATE. This can be crucial in high-speed repetitive computers or in sensitive or unstable problems, often resulting in nonrepeatable results. In such cases, be sure to allow sufficient reset time. During operation, the major effect is to introduce other terms that can cause lags; but these can be included approximately in the small errors of the integrator as given in Table 17.2.

17.5 Perturbation Techniques for Improving Accuracy [67]–[69]

The results of the previous sections concerning error can be applied in some cases to improve solution accuracy. In obtaining linearized variational differential equations (17.18) or true error differential equations (17.12), a mechanization producing the elements u_i is quite accurate because these elements can be scaled properly. If a problem involves small variations from run to run, about a nominal solution $x(t)$ (e.g., because of perturbations in initial conditions), it pays to mechanize perturbation equations for an accurate picture of the changes produced. These equations can be linearized or not, depending upon the degree of accuracy involved. Similarly, if a variable $x(t)$ does not change much during computation, say 0.1 to 10 per cent of its initial value x_0, an improved mechanization results from using

$$\delta x = x - x_0$$

as a fundamental variable. We give two examples to illustrate these principles.

Example 17.7 (Trajectory Perturbations). The trajectory of an artillery shell in two dimensions is given approximately by

$$\begin{aligned}
\dot{V} &= -g\sin\theta - kV^2, && (V)_0 = V_0 \\
\dot{\theta} &= -\frac{g\cos\theta}{V}, && (\theta)_0 = \theta_0 > 0 \\
\dot{x} &= V\cos\theta, && (x)_0 = 0 \\
\dot{y} &= V\sin\theta, && (y)_0 = 0
\end{aligned} \tag{17.33}$$

where x and y are instantaneous horizontal and vertical positions of the shell, g is value for gravity, and k is a drag coefficient. The trajectory suffers from small perturbations in the initial firing velocity V_0 and initial firing angle θ_0. How do these affect the range R, which is the value of x when y returns to 0 after its initial rise?

Many runs could be made varying V_0 and θ_0 to determine this effect, but the resolution is poor for small changes in these parameters so that accuracy will certainly be lacking. The process of covering a grid of expected variations from statistical data is also time consuming.

True error equations can be found by the appropriate substitutions as given in Sec. 17.3.2. In this case, the functions to the right are not polynomials. Taylor expansions are needed for higher-order effects to eliminate differencing two expressions, each large in magnitude (see Problem 17-14). The simplest set of perturbation equations obtainable is the linearized set, i.e., the variational equations. This is found by taking simple partial derivatives (the Jacobian elements). Using δ to denote the variation, we have

$$\begin{aligned}
\delta\dot{V} &= -g\cos\theta\,\delta\theta - 2kV\,\delta V, && (\delta V)_0 = \text{initial perturbation in } V_0 \\
\delta\dot{\theta} &= \frac{g\sin\theta}{V}\,\delta\theta + \frac{g\cos\theta}{V^2}\,\delta V, && (\delta\theta)_0 = \text{initial perturbation in } \theta_0 \\
\delta\dot{x} &= -V\sin\theta\,\delta\theta + \cos\theta\,\delta V, && (\delta x)_0 = 0 \\
\delta\dot{y} &= V\cos\theta\,\delta\theta + \sin\theta\,\delta V, && (\delta y)_0 = 0
\end{aligned} \tag{17.34}$$

Systems (17.33) and (17.34) are solved simultaneously with an automatic hold circuit measuring δx when $y + \delta y = 0$. Statistical information on the initial perturbations of θ_0 and V_0 can be converted easily to range error statistics.

Example 17.8. A simple passive network (Fig. 17.3) contains a variable resistor $R_1(t)$ that changes very little with time and is known. Such is the case with strain gauge and other measurements. At $t = 0$, no variables are changing: $\dot{V}_1 = \dot{I}_1 = \dot{I}_2 = \dot{I}_3 = 0$. Variations in all variables are small, but these variations are needed for the study. The equations to be solved are

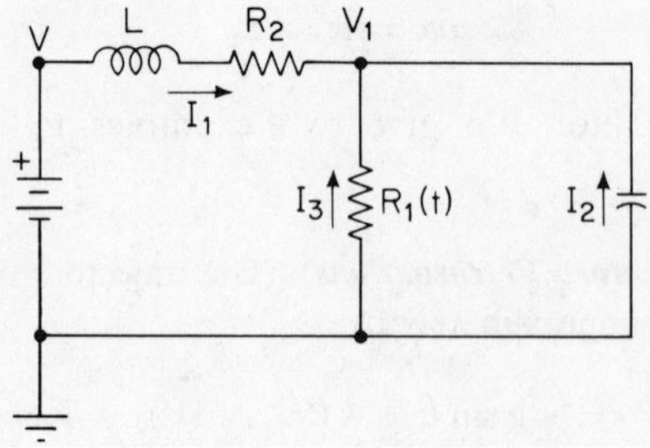

FIG. 17.3 Network with variable resistance.

$$C\frac{dV_1}{dt} = I_1 - \frac{V_1}{R_1(t)}, \qquad L\frac{dI_1}{dt} = V - V_1 - R_2 I_1 \tag{17.35}$$

with the initial conditions $(I_1)_0$ and $(V_1)_0$ determined by the algebraic equations

$$(I_1)_0 - \frac{(V_1)_0}{R_1(0)} = 0, \qquad V - (V_1)_0 - R_2(I_1)_0 = 0$$

We find that

$$(I_1)_0 = \frac{V}{R_1(0) + R_2}, \qquad (V_1)_0 = \frac{VR_1(0)}{R_1(0) + R_2}$$

These initial conditions can be calculated exactly, off the computer. Let

$$\Delta I_1 = I_1 - (I_1)_0$$

$$\Delta V_1 = V_1 - (V_1)_0$$

$$\Delta R_1(t) = R_1(t) - R_1(0)$$

By substitution in (17.35), the final exact equations are

$$L\frac{d(\Delta I_1)}{dt} = -\Delta V_1 - R_2 \Delta I_1, \qquad (\Delta I_1)_0 = 0$$

$$C\frac{d(\Delta V_1)}{dt} = \Delta I_1 - \frac{\Delta V_1}{R_1(t)} + \frac{(I_1)_0 \Delta R(t)}{R_1(t)}, \qquad (\Delta V_1)_0 = 0$$

The equations are similar to the original; the forcing function V is replaced by a new forcing function $(I_1)_0 \Delta R(t)/R_1(t)$. This is generally true for linear systems (see Problem 17-16).

17.6 Simple Static Error Reducing Schemes

When components of an analog computer were discussed in previous chapters, and in the areas of scaling, function generation, program planning, etc., some care was taken to make the programmer aware that some schemes have better error characteristics than others. Indeed, the entire art of analog computing may well depend upon recognizing potential error sources and knowing how to eliminate them. The basics of computing are quite simple; only the ability to anticipate, recognize, interpret, and troubleshoot errors can distinguish an effective programmer from others. A new programmer frequently runs into trouble and unfortuately this tends to discourage further efforts. But this ability must be developed with experience, although it may be supplemented by analysis.

In this section experiences and ideas for reducing static errors, not covered in earlier chapters, are discussed. Its primary aim is not only to provide information but to show how the knowledge of the error itself can suggest ways to reduce it.

Generating explicit functions with integrators is one example. The error itself can easily be determined, but the cause might be elusive. Often, a simple remedy is to make small adjustments in coefficient pots to attempt to compensate for mismatched resistors or capacitors, in a technique called "pot tweaking." Although perhaps an inelegant procedure, a function sometimes can be changed from one which has unacceptable error characteristics to one which is quite acceptable just by tweaking a few pots. The choice of which pots to adjust can be influenced by measuring the value of the time constant for each integrator input or simply by arbitrary trial-and-error tests.

Another example is to compensate for backlash in servo multiplication. Most of the error (Table 17.1) comes from backlash and pot misalignment (and not from nonlinearities in the pot). These can be measured easily and compensated by driving the servo by voltages $V_2 - \alpha_0$ instead of V_2. Because of backlash, α_0 has different values for $\dot{V}_2 > 0$ than for $\dot{V}_2 < 0$, but these can easily be supplied by relays.

There are often various choices for mechanizing a given algebraic equation. A simple analysis of static errors may sometimes determine which produces the least error. Kindle [71], for example, demonstrated how improved error characteristics resulted from generating the function

$$f(x) = \frac{1}{1 + x^2} \tag{17.36}$$

by an implicit technique with servo multipliers rather than by a conventional division circuit. In (17.36), the variable x had large negative and positive ranges, and the accuracy near zero was important in the application for which it was required.

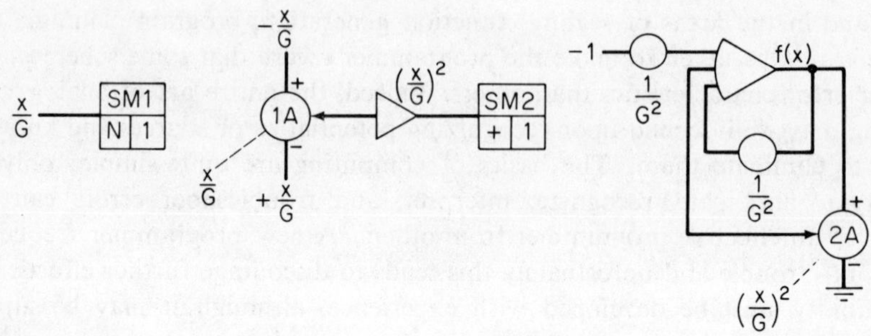

FIG. 17.4 Conventional generation of $1/(1 + x^2)$.

Suppose x is scaled as x/G, $G > 1$. The conventional circuit (Fig. 17.4) is based on the implicit equation

$$\frac{1}{G^2} f(x) + f(x)\left(\frac{x}{G}\right)^2 - \frac{1}{G^2} = 0$$

Let α_1 be the error of pot 1A and α_2 of pot 2A. The squaring circuit actually produces

$$\frac{x}{G}\left(\frac{x}{G} + \alpha_1\right)$$

and the multiplication $f(x)\, x^2/G^2$ becomes

$$f(x)\left[\frac{x}{G}\left(\frac{x}{G} + \alpha_1\right) + \alpha_2\right]$$

Even if the pots $1/G^2$ were exact, the function produced is

$$f(x) = \frac{1}{1 + G^2\left(\frac{x}{G}\right)^2 + \alpha_1 G^2\left(\frac{x}{G}\right) + \alpha_2 G^2} \tag{17.37}$$

On the other hand, an improved circuit based on the implicit equation is shown in Fig. 17.5. Not only is one servo required (using two cups), but the error characteristics are much improved. The output of the first product is

$$-f(x)\left|\frac{x}{G} + \alpha_1\right|$$

and that of the second

$$f(x)\left|\frac{x}{G}+\alpha_1\right|\left|\frac{x}{G}+\alpha_2\right|$$

Thus, the function generated is

$$f(x)=\frac{1}{1+G^2\left|\dfrac{x}{G}+\alpha_1\right|\left|\dfrac{x}{G}+\alpha_2\right|} \tag{17.38}$$

A comparison of (17.37) with (17.38) shows when $x = 0$, the denominator of f is in error by $\alpha_2 G^2$ in the conventional circuit, but by $|\alpha_1||\alpha_2|G^2$ in the improved circuit. Since each α_i is about 0.001, the improved accuracy is clear.

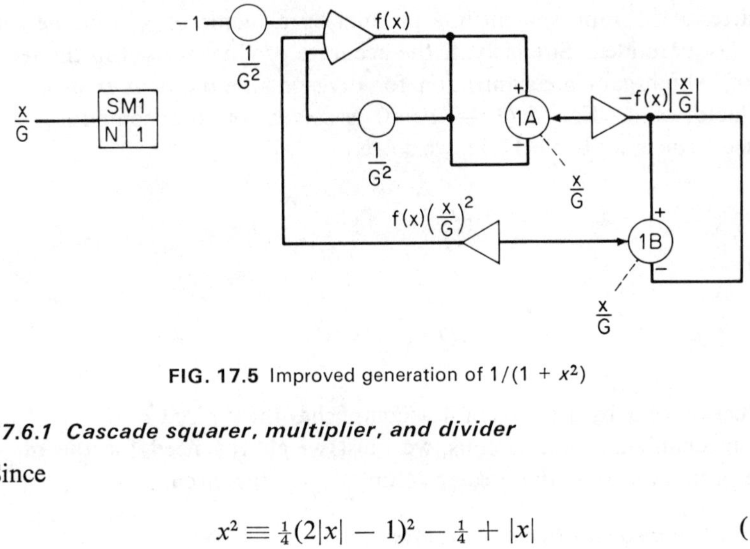

FIG. 17.5 Improved generation of $1/(1 + x^2)$

17.6.1 Cascade squarer, multiplier, and divider

Since

$$x^2 \equiv \tfrac{1}{4}(2|x| - 1)^2 - \tfrac{1}{4} + |x| \tag{17.39}$$

some effort can be made to reduce the static error of a squarer [72]. In (17.39), the function $|x|$ can be made quite precise; only the square function is subject to considerable error from the piecewise linear approximation of a diode function generator. Since summers and pot settings are subject only to small linear component errors, a mechanization of (17.39) reduces the static error of a squarer by a factor of about 4 (Fig. 17.6). Absolute-value circuits can be cascaded more than once (Problem 17-19), each time reducing the nonlinear error by a factor of 4. Of course, successive cascading eventually yields only linear component errors. Howe [72] reports achieving a square function in error by only 10 parts per million by carefully controlling the linear component errors and cascading absolute-value circuits twice with 37-segment squarers.

The same principle can be applied to multipliers by noting

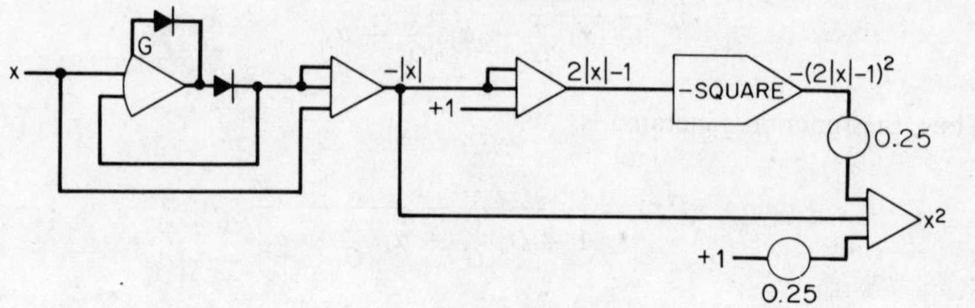

FIG. 17.6 Cascade squarer.

$$xy \equiv \tfrac{1}{4}(1 - |x + y|)^2 - \tfrac{1}{4}(1 - |x - y|)^2 + \frac{|x + y|}{2} - \frac{|x - y|}{2} \tag{17.40}$$

with a resultant improvement in a product by a factor of 4 over the conventional $\frac{1}{4}$-square idea. Similarly, if the accurate product is used in the feedback path of a high-gain mechanization for division, the quotient error is reduced by a factor of 4. For if $Qy - x = 0$ is solved on the computer, and the product error is α (Table 17.1), we solve

$$Qy + \alpha - x = 0 \tag{17.41}$$

so that

$$Q = \frac{x}{y} - \frac{\alpha}{y}$$

A reduction of α by a factor of 4 accomplishes the desired result.

In mechanizing these circuits, we must weigh the need for the more accurate product against the greater complexity of the circuit.

17.6.2 Improved division circuits

Howe [72] suggests a simple scheme for reducing quotient errors by a factor of 2 while requiring only three additional amplifiers at most. It is based on a high-gain solution of the equation

$$Q + Q(2y - 1) - 2x = 0$$

With the same error α in the product as in (17.41) we see that

$$Q = \frac{x}{y} - \frac{\alpha}{2y}$$

This circuit is shown in Fig. 17.7, but it is applicable only when $y > 0$.

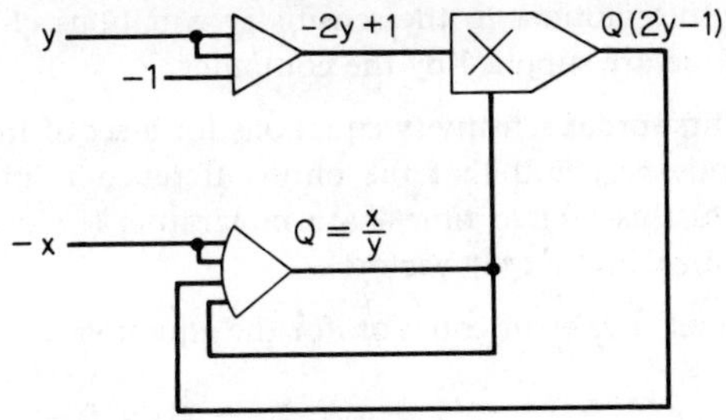

FIG. 17.7 Improved division circuit.

If y does not change much during computation

$$y = y_0 + \Delta y(t)$$

then an improved division circuit can be mechanized because the variable $\Delta y(t)$ can be scaled up. We implicitly solve the equation

$$Q[y_0 + \Delta y(t)] - x = 0 \quad \text{or} \quad y_0 Q + \Delta y Q - x = 0$$

If the scale factor of Δy is G, $G > 1$, we mechanize

$$y_0 Q + \frac{(G\Delta y)Q}{G} - x = 0$$

which reduces the multiplication error by the factor of $1/G$ (Fig. 17.8).

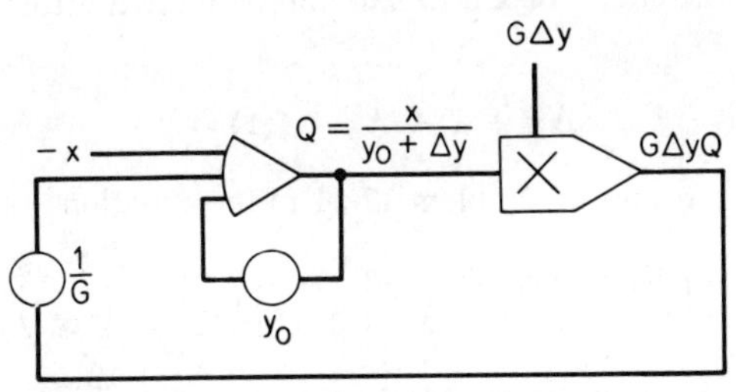

FIG. 17.8 Division circuit for an almost constant denominator.

PROBLEMS

17-1. Obtain the expressions for the second-order sensitivity functions $\partial y_j/\partial \gamma$ from (17.9).

17-2. Find the solutions analytically to the sensitivity equations (17.11).

17-3. Mechanize the solutions to the sensitivity equations (17.11) when the required functions of time are supplied by the computer.

17-4. Show that first-order sensitivity equations for a set of linear equations have the same homogeneous part and that the only difference is in the forcing functions. Kahne [73] uses this fact to time-share integrators (i.e., uses the same integrators to produce both $\mathbf{x}$ and $\partial\mathbf{x}/\partial\gamma$ vectors).

17-5. Find the exact true error equation for the equation

$$\lambda\ddot{x}^* + \dot{x}^* + (1 + \alpha_1)(x^{*2} + \alpha_2) = 0$$

$$x_0^* = 1 + \beta_1, \qquad \dot{x}_0^* = -(1 + \alpha_1)(x_0^{*2} + \alpha_2)$$

17-6. Find the true error equations for the nonzero α in the system [44]

$$\dot{x}_1 = ax_1x_2 + bx_2, \qquad \dot{x}_2\,(c + \alpha)x_1x_2 + ex_1$$

17-7. Show that extrapolating errors in linear systems with a first-order sensitivity function is equivalent to ignoring the term $\mathbf{Eu}$ in (17.16).

17-8. Given a matrix $\mathbf{A}$ with characteristic roots λ_i, derive the determinantal equation for the true change u_i in λ_i due to a small error matrix $\mathbf{E}$.

17-9. Show that if $\mathbf{B}$ is positive definite and the symmetric part of $\mathbf{BJ}$ is positive definite, the error-propagation equation

$$\dot{\mathbf{u}} + \mathbf{Ju} = \mathbf{0}$$

is stable [45].

17-10. Show that (11.43) is stable under all possible circumstances.

17-11. Convert the effects of λ errors to approximate α errors in the stable linear system

$$\mathbf{\Lambda}\ddot{\mathbf{x}} + \dot{\mathbf{x}} + \mathbf{Ax} + \mathbf{f}(\tau) = \mathbf{0}$$

17-12. Apply the results of Problem 17-11 to the equation

$$\begin{pmatrix} \lambda_1 & 0 \\ 0 & \lambda_2 \end{pmatrix}\begin{pmatrix} \ddot{x}_1 \\ \ddot{x}_2 \end{pmatrix} + \begin{pmatrix} \dot{x}_1 \\ \dot{x}_2 \end{pmatrix} + \begin{pmatrix} -3 & -4 \\ 4 & 5 \end{pmatrix}\begin{pmatrix} x_1 \\ x_2 \end{pmatrix} + \begin{pmatrix} e^{-\tau} \\ \tau e^{-\tau} \end{pmatrix} = \begin{pmatrix} 0 \\ 0 \end{pmatrix}$$

to determine the approximate error. Mechanize the final equations and indicate how to find the four quasi-sensitivity functions u_1/λ_1, u_2/λ_2, u_1/λ_2, and u_2/λ_1.

17-13. Repeat Example 17.6 taking into account lags produced by the two potentiometers, using the transfer function from Table 17.2.

17-14. Compute the exact perturbation equations of (17.33) and show they reduce to (17.34) for small variations. Then approximate the perturbation equations to the second order of the perturbation variables $\delta\theta$, δV, δx, and δy.

17-15. Convert system (17.34) to sensitivity equations for finding the change in each variable with respect to a small perturbation in the initial velocity V_0.

17-16. Determine the perturbation equation for the linear system

$$\dot{\mathbf{x}} + \mathbf{A}(t)\mathbf{x} + \mathbf{f}(t) = \mathbf{0}$$

when the initial conditions are such that $\dot{\mathbf{x}}_0 = \mathbf{0}$, i.e.,

$$\mathbf{A}(0)\mathbf{x}_0 + \mathbf{f}(0) = \mathbf{0}$$

and both **A** and **f** change very little from these nominal values.

17-17. Show that the static error in generating the function

$$f = x_1^2 - x_2^2$$

can be much improved by mechanizing the product

$$f = (x_1 + x_2)(x_1 - x_2)$$

when x_1 remains close to x_2 [71].

17-18. Because of backlash and misalignment in servo multipliers, there exists a position error α_1 when the driving variable x is increasing, and α_2 when it is decreasing. Mechanize a circuit with relays to compensate for these errors assuming $\dot{x}$ is available.

17-19. Show that

$$x^2 \equiv \frac{1}{16}\left(2\Big|2|x| - 1\Big| - 1\right)^2 + |x| + \frac{1}{4}\Big|2|x| - 1\Big| - \frac{5}{16}$$

and mechanize the equation assuming x is available. This is a double cascade squarer reducing errors in a DFG square by a factor of 16.

17-20. Mechanize (17.40) with ideal diode circuits for the absolute values.

REFERENCES

[1] Loveman, B., "Reliability of a Large REAC Installation," *Proc. Eastern Joint Comp. Conference*, 1953.

[2] Braun, W. G., and L. M. Warshawsky, "Verifier Prints Out Patchboard Locations," *Control Eng.*, vol. 2, Dec. 1955.

[3] Huskey, H.D., and G.A. Korn, eds., *Computer Handbook* (New York: McGraw-Hill Book Co., 1962).

[4] Single, C. H., and E. M. Billinghurst, "Optimization of Analog Computer Linear System Dynamic Characteristics," *Proc. Western Joint Comp. Conference*, 1961.

[5] Brock, P., "Introduction" in the session on numerical checks, *Symposium I on REAC Techniques*, Reeves Instr. Corp., New York, March 15-16, 1951 (ASTIA No. 108488).

[6] Lakatos, E., "Checking Analogue Computer Solutions," *Proc. IRE* (correspondence), vol. 39, Dec. 1951.

[7] Harris, H. E., "New Techniques for Analog Computation," *Instr. and Automation*, vol. 30, May 1957.

[8] Grosswald E., and M. Plotkin, "Systematic Tracing of Discrepancies in Analog Computers," *IRE National Convention Record, Part 4*, 1957.

[9] Richmond, W. F., Jr., and B. D. Loveman, "Checking Analogue Computer Solutions," *Symposium II on Simulation and Computing Techniques, part 2*, Reeves Instr. Corp., New York, Apr. 28–May 2, 1952 (ASTIA No. 170848).

[10] Richmond, W. F., Jr., "Checking Analogue Solutions of Unstable and Complex Systems," *Project Typhoon Symposium III on Simulation and Computing Techniques, Part 1*, Bureau of Aeronautics and U.S. Naval Air Devel. Center, Johnsville, Pa., Oct. 12–14, 1953.

[11] Meissinger, H. F., "Substitution Methods for the Verification of Analog Solutions," *Proc. National Simulation Conference*, Jan. 19–21, 1956, Dallas, Texas.

[12] Selfridge, R. G., "Coding a General-Purpose Digital Computer to Operate as a Differential Analyzer," *Proc. Western Joint Comp. Conference*, 1955.

[13] Stein, M. L., *et al.*, "A Compiler with an Analog-Oriented Input Language," *Proc. Western Joint Comp. Conference*, 1959.

[14] Stein, M. L., and J. Rose, "Changing from Analog to Digital Programming by Digital Techniques," *Journal ACM*, vol. 7, Jan. 1960.

[15] Ohlinger, L. A., "'Anatran'—First Step in Breeding the 'Diganalog'," *Proc. Western Joint Comp. Conference*, 1960.

[16] Green, C., *et al.*, "APACHE—A Breakthrough in Analog Computing," *IRETEC*, vol. 11, Oct. 1962.

[17] Green, C., *et al.*, "The Code APACHE—Intended for the Programming of an Analogue Problem by Means of a Digital Computer," *AICA*, vol. 5, April 1963.

[18] Stein, M. L., "Automatic Digital Programming of Analog Computers," *IEEETEC*, vol. 12, April 1963.

[19] Hurley, J. R., and J. J. Skiles, "DYSAC: a Digitally Simulated Analog Computer," *Proc. Spring Joint Comp. Conference*, 1963.

[20] Gaskill, R. A., *et al.*, "DAS—A Digital Analog Simulator," *Proc. Spring Joint Comp. Conference*, 1963.

[21] Gaskill, R. A., "A Versatile Problem-Oriented Language for Engineers," *IEEETEC*, vol. 13, Aug. 1964.

[22] Peterson, H, *et al.*, "MIDAS—How It Works and How It's Worked," *Proc. Fall Joint Comp. Conference*, 1964.

[23] Brennan, R. D., and H. Sano, "PACTOLUS—A Digital Analog Simulator Program for the IBM 1620," *Proc. Fall Joint Comp. Conference*, 1964.

[24] Ocker, W., and S. Teger, "HYTRAN—A Software System to Aid the Analog Programmer," *Proc. Fall Joint Comp. Conference*, 1964.

[25] Hartnett, R. T., *et al.*, "MIDAS—an Analog Approach to Digital Computation," *Simulation*, vol. 3, Sept. 1964.

[26] Blechman, G. E., "An Enlarged Version of MIDAS," *Simulation*, vol. 3, Oct. 1964.

[27] Brennan, R. D., and R. N. Linebarger, "A Survey of Digital Simulation: Digital Analog Simulator Programs," *Simulation*, vol. 3, Dec. 1964.

[28] Rideout, V. C., and L. Tavernini, "MADBLOC, A Program for Digital Simulation of a Hybrid Computer," *Simulation*, vol. 4, Jan. 1965.

[29] Peterson, H. E., *et al.*, "MIMIC—A Digital Simulator Program," *SESCA Internal Memo 65-12*, Systems Engineering Group, Wright-Patterson Air Force Base, Ohio, May 1965.

[30] Syn, W. M., and D. G. Wyman, "7090/94 DSL/90 Digital Simulation Language User's Guide," *San Jose TR 02.355*, IBM Systems Devel. Div., Devel. Lab., San Jose, Calif., July 1, 1965.

[31] Janoski, R. M., *et al.*, "COBLOC—A Program for All-Digital Simulation of a Hybrid Computer," *IEEETEC*, vol. 15, Feb. 1966.

[32] Schlesinger, S. I., and L. Sashkin, "EASL. . . A Digital Computer Language for 'Hands-on' Simulation," *Simulation*, vol. 6, Feb. 1966.

[33] Burgin, G. H., "MIDAS III. . . A Compiler Version of MIDAS," *Simulation*, vol. 6, Mar. 1966.

[34] Farris, G. J., and L. E. Burkhart, "The DIAN Digital Simulation Program," *Simulation*, vol. 6, May 1966.

[35] Tavernini, L., "On UNIP and the Construction of Digital Simulation Programs," *Simulation*, vol. 7, Nov. 1966.

[36] d'Hoop, H., and R. Monterosso, "SAHYB—A Program for Simulation of Analog and Hybrid Computers," *IEEETEC*, vol. 15, June 1966. See also "SAHYB-2: A Continuous System Simulation Language Compatible with FORTRAN-IV," *IEEETC*, vol. 17, Feb. 1968.

[37] Strauss, J.C., *et al.*, "The SCi Continuous System Simulation Language (CSSL)," *Simulation*, vol. 9, Dec. 1967.

[38] Bloom, H. M., "Problem Solving by Digital-Analog Simulation," *Harry Diamond Labs. Report TR-1357*, Wash., D. C., Oct. 1967.

[39] Chu, Y., *Digital Simulation of Continuous Systems* (New York: McGraw-Hill Book Co., 1969).

[40] Murray, F. J., "Mathematical Error Analysis for Continuous Computers," *Symposium II on Simulation and Computing Techniques, Part 2*, Reeves Instr., Corp., New York, Apr. 28–May 2, 1952 (ASTIA No. 170848).

[41] Wright, F. B., Jr., "Dynamic System Studies: Error Studies," *Wright Air Development Center Report 54-250, Part 13*, Sept. 1956 (ASTIA No. 97266).

[42] Miller, K. S., and F. J. Murray, "Dynamic System Studies: Error Analysis for Differential Analyzers," *Wright Air Development Center Report 54-250, Part 14*, Sept. 1956 (ASTIA No. 97315).

[43] Nelson, D. J., "A Foundation for the Analysis of Analog-Oriented Combined Computer Systems," *Stanford U. Tech. Report No. 1002-1*, April 1962 (ASTIA No. 278507).

[44] Nelson, D. J., "A Fundamental Error Theory for Analog Computers," *IEEETEC*, vol. 12, Oct. 1963.

[45] Vichnevetsky, R., "Error Analysis in the Computer Simulation of Dynamic Systems: Variational Aspects of the Problem," *IEEETEC*, vol. 16, Aug. 1967.

[46] Macnee, A. B., "Some Limitations on the Accuracy of Electronic Differential Analyzers," *Proc. IRE*, vol. 40, March 1952.

[47] Paul, R. J. A., "Some Factors Affecting the Accuracy of Electronic Analogue Computers," *Proceedings of the Inter. Analogy Comp. Meeting* (Brussels: Presses Académiques Européennes, 1956).

[48] Fuchs, H., "Some Considerations of the Accuracy of Linear Analog Computers," *Proceedings of the Inter. Analogy Comp. Meeting* (Brussels: Presses Académiques Européennes, 1956).

[49] Marsocci, V.A., "An Error Analysis of Electronic Analog Computers," *IRETEC*, vol. 5, Dec. 1956.

[50] Dow, P. C., Jr., "An Analysis of Certain Errors in Electronic Differential Analyzers I—Bandwidth Limitations," *IRETEC*, vol. 6, Dec. 1957.

[51] Dow, P. C., Jr. "An Analysis of Certain Errors in Electronic Differential Analyzers II—Capacitor Dielectric Absorption," *IRETEC*, vol. 7, March 1958.

[52] Miura, T., and M. Nagata, "Theoretical Consideration on Computing Errors of Slow Type Electronic Analogue Computers," *Proceedings of the 2nd Inter. Analogue Comp. Meetings* (Brussels: Presses Académiques Européennes, 1959).

[53] Miura, T., *et al.*, "Some Considerations of the Error Due to Absorption Phenomena of Integrating Capacitor for the Slow Type of Electronic Analogue Computers," *Proceedings of the 2nd Inter. Analogue Comp. Meetings* (Brussels: Presses Académiques Européennes, 1959).

[54] Miura, T., and M. Nagata, "Theoretical Consideration of Computing Errors of a Slow Type Electronic Analog Computer," *IRETEC*, vol. 7, Dec. 1958.

[55] de Backer, Ir. W., "Error Analysis of a D-C Integrator," *AICA*, vol. 2, Jan. 1960.

[56] Popovic, D. B., "On the Errors in Solving Differential Equations Caused by the Inaccurate Initial Values," *Proceedings of the 3rd Inter. Analogue Comp. Meetings* (Brussels: Presses Académiques Européennes, 1962).

[57] Korn, G. A., "Exact Design Equations for Operational Amplifiers with Fours Terminal Computing Networks," *IRETEC*, vol. 11, Feb. 1962.

[58] Kogan, B. Ja., "On the Steady-State Operational Amplifier Errors Due to the Passband Limitations," *AICA*, vol. 4, April 1962.

[59] Wadhwa, L. K., "Phase Distortion in Operational Amplifiers," *AICA*, vol. 4, April 1962.

[60] Gal'perin, M. V., "Analysis of Analog-Computer Error Due to Finite Passbands of Operational Amplifiers," *Automatika i Telemekhanika*, vol. 25, May 1964.

[61] Marsocci, V. A., "Dynamic Errors in Electronic Analog Computers," *IEEETEC*, vol. 16, April 1967.

[62] Korn, G. A., "Parameter-Perturbation Generator for Analog-Computer Error and Optimization Studies," *AICA*, vol. 5, April 1963.

[63] O'Grady, E. P., "Correlation Method for Computing Sensitivity Functions on a High-Speed Iterative Analog Computer," *IEEETEC*, vol. 16, April 1967.

[64] Meissinger, H. F., "The Use of Parameter Influence Coefficients in Computer Analysis of Dynamic Systems," *Proc. Western Joint Comp. Conference*, 1960.

[65] Meissinger, H. F., "Parameter Influence Coefficients and Weighting Functions Applied to Perturbation Analysis of Dynamic Systems," *Proceedings of the 3rd Inter. Analogue Comp. Meetings* (Brussels: Presses Académiques Européennes, 1962).

[66] Tomovic, R., *et al.*, "Sensitivity of Dynamic Systems to Parameters Which Increase the Order of Mathematical Models," *IEEETEC*, vol. 14, Dec. 1965.

[67] Bush, L., and P. Orlando, "A Perturbation Technique for Analog Computers," *IRETEC*, vol. 7, June 1958.

[68] Rubin, A. I., and G. F. Graber, "Acoustic Ray Tracing on the General-Purpose Electronic-Analog Computer," *IEEETEC*, vol. 14, June 1965.

[69] Fogarty, L. E., and R. M. Howe, "Flight Simulation of Orbital and Re-Entry Vehicles," *IRETEC*, vol. 11, Aug. 1962.

[70] Driban, S., "Programming, Setting Up, and Checking Out of One-Dimensional Multispring and Damper Systems," *Simulation*, vol. 5, Sept. 1965.

[71] Kindle, B., "Accuracy Considerations in Using Servos," *Instr. and Control Systems*, vol. 33, Jan. 1960.

[72] Howe, R. M., "Some Techniques for Accuracy Improvement in Analog Computation," *Simulation*, vol. 9, Oct. 1967.

[73] Kahne, S. J., "Sensitivity-Function Calculation in Linear Systems Using Time-Shared Analog Integration," *IEEETC*, vol. 17, April 1968.

For other information, see

[74] Cahn, L., "Accuracy of an Analog Computer," *IRETEC*, vol. 2, Dec. 1953.

[75] Hamer, H., and A. Lupinski, "A Method for Evaluating Amplifier Phase Shift at Low Frequencies," *IRETEC*, vol. 6, Sept. 1957.

[76] Meilander, W. C., "A New Method of Verifying Analog Computer Problems and Performances," *Proc. Western Joint Comp. Conference*, 1957.

[77] Holst, P. A., "Dynamic Accuracy and Error in Analog Computations," *IEEETEC*, vol. 12, June 1963.

[78] Bell, H., Jr., and V. C. Rideout, "Precision in High-Speed Electronic Differential Analyzers," *Symposium II on Simulation and Computing Techniques*, Reeves Instr. Corp., New York, April 28–May 2, 1952 (ASTIA No. 170848).

[79] Allen, W. R., "Dynamic System Studies—Operation and Maintenance Procedures for Analog Computers," *Wright Air Development Center Report 54-250, Part 6*, 1956 (ASTIA No. 110625).

[80] Murphy, B., "A Recommended Preparation, Programming and Verification Procedure of Universal Applicability to General Purpose Electronic Analog Computers," *Proceedings of the 2nd Inter. Analogue Comp. Meetings* (Brussels: Presses Académiques Européennes, 1959).

[81] Sosenskii, N. L., "On the Determination of Certain Errors of Analog Computers," *Automatika i Telemekhanika*, vol. 20, Oct. 1959.

[82] Starr, D. A., and J. J. Jonsson, "The Design of an Automatic Patching System," *Simulation*, vol. 10, June 1968.

Chapter 18

ANALOG/HYBRID COMPUTERS I: MODE CONTROL AND MEMORY

Previous chapters have emphasized "slow" analog computers that produce graphs of voltage traces directly during computation. In such computers, the mode of all integrators at any instant is uniform: the OPERATE, RESET, or HOLD mode selected by the operator automatically places all integrators into the same mode. Chronologically, such computers came first.

Repetitive computers appeared in the 1950s. They contain the means to place the computer in OPERATE and RESET automatically so that, by using a scope, rapid manual searches for parameters in a given problem can be made. But without any further equipment, a repetitive computer behaves precisely as a slow computer. It just computes much more rapidly without requiring an operator to depress the RESET or OPERATE buttons.

As early as 1954, Wadel [1]–[3] noted that simple iterations could be programmed if every integrator was not in the same mode simultaneously, and that other nonconventional applications of analog computers involving sampled data could be analyzed as well. These papers helped redirect the design of analog computers to provide such flexibility. In the late 1950s and early 1960s, convenient integrator mode control was incorporated in analog computer designs. Relay mode control was also replaced, by and large, by electronic mode control (EMC), allowing for iterative computations at high speed. With EMC, digital signals control electronic switches that control the RESET, HOLD, and OPERATE modes of individual integrators. This allows point memory to be programmed for use in iterative computations and for other applications.

18.1 Repetitive Computers

We have indicated how a repetitive computer with its virtually instantaneous feedback information can help immeasurably in searching for parameters to satisfy requirements of a problem. Algebraic root finding, simple two-point boundary-value problems, approximate curve fitting, etc., are amenable to rapid manual solutions when the operator uses this feedback information, even without memory units, automatic iterative capability, or optimization concepts.

The means for automatically placing the computer in the RESET and OPERATE modes vary with design. In any case, if the times the computer is in these modes are constants t_r and t_o, the sum

$$t_p = t_r + t_o$$

is called the *period of the rep-op cycle*, while its reciprocal $1/t_p$ gives the *frequency of solutions*. (The term "rep-op" is short for "repetitive operation.")

Small periods (say ≤ 0.1 s) are invariably obtained by using smaller feedback capacitors for the integrators than usual, so that the scaling procedures for a rep-op solution are straightforward. Amplitude scaling does not change. To time scale, pretend the problem will be solved in normal time, but decrease each capacitor by the orders of magnitude necessary to fit within the period of the rep-op cycle desired.

Scopes are used to display rep-op solutions. When frequencies of the solution are 10 or more, changing a pot or parameter slowly changes each solution slightly, but its display appears to change continuously for the same reasons a rapid sequence of still photos appears as a motion picture. Despite this tremendous advantage, there are some disadvantages. Higher solution frequencies must also increase dynamic errors. Servo multipliers could not be driven by a rapidly changing variable; multiplications must be done electrically. Even amplifiers driven at the higher frequencies are affected. One may notice that the decay of an oscillator amplitude per revolution changes radically with frequency (Chapter 17) and for higher frequencies, the amplitude expands. These effects are becoming less noticeable in computers as equipment design improves, but we must still be guided by the rule: *Never compute faster than consistent with one's purpose.*

To illustrate the versatility of a repetitive computer for obtaining a desired response without any sophisticated analytical methods, we consider a simple curve-fitting problem, one early application of such computers [4].

Example 18.1 An acceleration-time curve needed in a mechanism analysis is available in discrete data shown normalized below, and containing experimental error.

A	0	0.2	0.3	0.475	0.67	0.75	0.775	0.742
t	0	0.015	0.045	0.06	0.1	0.15	0.2	0.26
A	0.72	0.575	0.462	0.34	0.235	0.17	0.12	0.083
t	0.3	0.4	0.5	0.6	0.7	0.8	0.9	1.0

Although this could be set up on a DFG, the shape of the curve suggests it may be possible to get an acceptable approximation using the form

$$A = a(e^{-b_1 t} - e^{-b_2 t})$$

which would require two integrators (Chapter 5). Fitting with a least squares or some other norm requires the solution of three nonlinear algebraic equations. Much equipment is required to do this exactly on an analog computer by automatic means, and it should be avoided if an acceptable approximation can be obtained by adjusting the three parameters manually and comparing the solution with the data. To do this, the easiest method is to mark the data points on the screen face of a scope with washable ink and superimpose the generated function on these points. With a repetitive computer, we quickly gain insight on how the function changes with each parameter change and can adjust the pots to get a good fit. If

$$F = -e^{-b_1 t}$$

the differential system solved becomes (Chapter 5)

$$\begin{aligned} \dot{F} &= -kb_1 F, & F_0 &= -1 \\ \dot{A} &= -k(b_2 - b_1)aF - kb_2 A, & A_0 &= 0 \\ \dot{t} &= k, & t_0 &= 0 \end{aligned}$$

In normal time, the computation lasts 8 s if k is chosen to be 0.125. For a rep-op

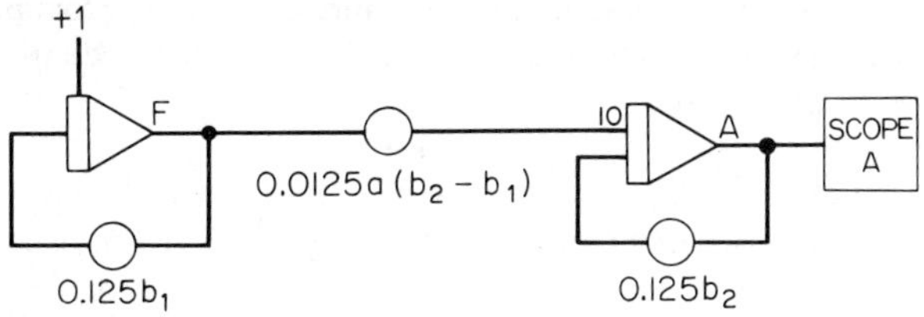

Note: Repetitive operation with $t_0 = 0.08$ s, feedback capacitors of 0.01 μF

FIG. 18.1 Diagram for Example 18.1.

solution, we use the same value, but also feedback capacitors of 0.01 μF to reduce the OPERATE time t_o to 80 ms. Figure 18.1 shows how a diagram may look for such a case. With a RESET time of 20 ms, ten solutions are obtained per second.

The fit obtained in this case was suprisingly good. Final pot settings were

$$0.125b_1 = 0.6537, \qquad 0.125b_2 = 0.5700$$
$$0.0125(b_1 - b_2)a = 0.1398$$

from which we calculate that the approximation is

$$A = 15.514(e^{-4.56t} - e^{-5.2288t})$$

However, Fig. 18.1 could be used directly in the generation of A without the need to find a, b_1, or b_2. Figure 18.2 shows how the fitted curve compares with the data points.

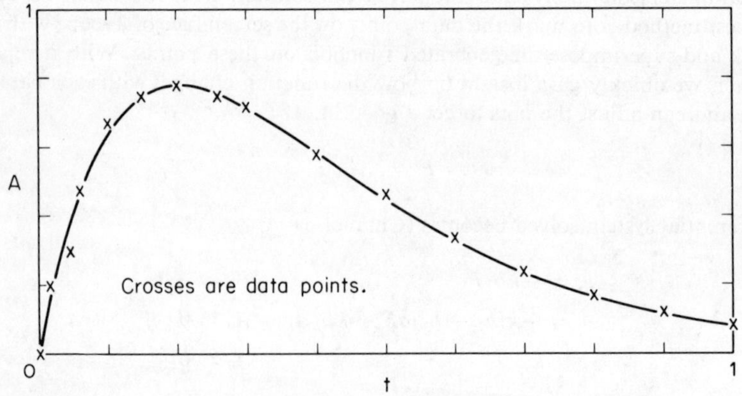

FIG. 18.2 The fitted curve.

Because there were only three parameters and the final approximation was not critical, this procedure is quite satisfactory. It can become quite tedious if many more parameters have to be adjusted.

18.2 Programmed Mode Control

Common to rep-op systems is a *master periodic signal* (Sec. 18.2.3), which alternately flips the mode relays shown in Fig. 2.11 or activates electronic switches to provide high speed rep-op. For computers with relay controls, we assume an integrator is placed in the correct computer mode if*

*This system is based on the EAI 231R-V computer, which is also the basis for all discussions of mode control.

(a) Both relay coils are not energized in the OPERATE mode.

(b) The hold relay coil is energized and the IC relay coil is not energized in the HOLD mode.

(c) Both relay coils are energized in the RESET mode.

Such computers are designed with relay coil terminations on the analog patchboard, making it possible to change the mode response of any given integrator. The advantage of such a design, as we shall see, is that point memory can be created, allowing for slow speed iterative computations in the manner suggested by Wadel [1].

18.2.1 Programmed relay mode control (RMC)

By depressing the computer mode buttons, the mode relays are energized with a high surge voltage, which levels off at some lower value. The initial high surge provides a rapid switching action, while the lower value generates enough of a magnetic field to retain the relay arm in the proper position. These connections are bottle-plugged on the analog board. If the coils are to be energized in some other way, a similar voltage-time input must be generated. In the EAI 231R-V computer, this can be done by mode relay drivers, which are designed as part of the computer system to respond to digital signals on the MLG board. A high signal (1) energizes the relay, while a low one (0) allows the arm to return to its normal state. The patching, operation, and symbolism of such a unit are shown in Fig. 18.3. The lower digital signal shown on the

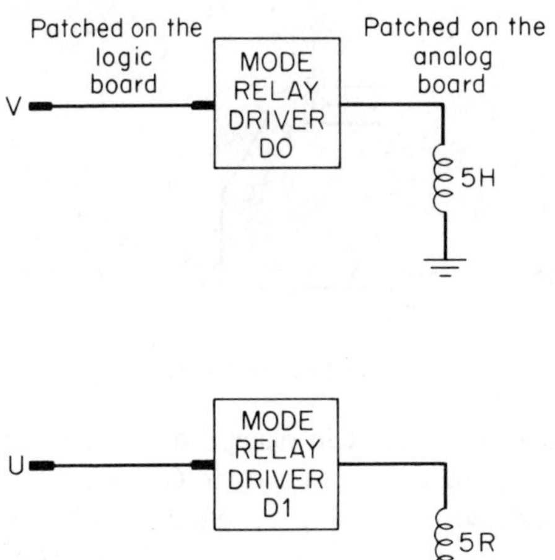

(a) Patching

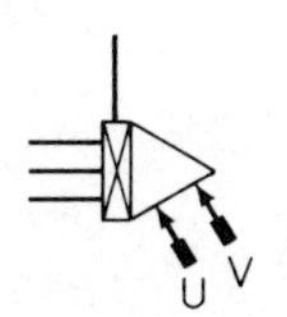

(b) Symbolism

MODE	V	U
OPERATE	0	0
HOLD	1	0
RESET	1	1

(c) Operation

FIG. 18.3 Mode relay driver patching, operation, and symbolism.

diagram for the integrator corresponds to the reset relay, and the upper, to the hold relay. A cross is placed within the rectangular portion of the integrator to remind the programmer that its mode is controlled by patched logic. Whenever two digital signals are shown controlling an integrator mode, it will be assumed that the mode relay drivers are patched as shown, with the exceptions described below.

On the logic board, digital signals IC, H, and OP corresponding to the computer modes are available and are digital 1 or 0 according to the table below.

Digital Signals

Computer mode	IC	H	OP
RESET	1	1	0
HOLD	0	1	0
OPERATE	0	0	1

These symbols and their complements $\overline{IC}$, $\bar{H}$, and $\overline{OP}$ are used whenever needed on a diagram.

As a matter of convenience, we occasionally indicate in a diagram that the IC logic signal controls the reset relay or that H controls the hold relay. Either case need not be patched through mode relay drivers, since the normal voltage energizing these relays is patchable on the analog board. In Fig. 18.4, for

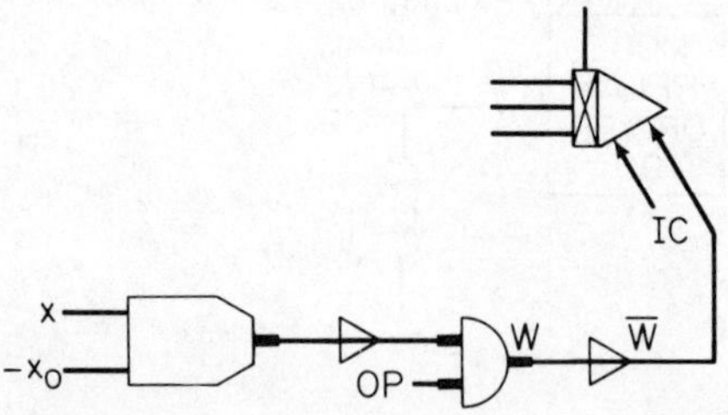

FIG. 18.4 Hold circuit when the computer is in OPERATE AND $x > x_0$.

example, a simple hold circuit is shown, which can place the integrator in the hold mode when the computer is not in OPERATE or in OPERATE AND $x > x_0$. The reset relay control can be bottle-plugged on the analog board, but $\overline{W}$ must be the input to a mode relay driver patched to hold relay. Incidentally, many of the relay circuits involving hold action discussed in Chapter 6 can be accomplished by driving the hold relay with similar circuits (see Problems 18-1 and 18-2) resulting in a saving of amplifiers. The advantage of using the hold relay instead of an externally patched one is that the integrator can sum the voltages at its inputs.

An entry of 0 shown driving a relay coil means no connection; the plug is removed from the patchboard. An entry of 1 driving a relay coil means the coil is always energized. This must be patched through a mode relay driver.

Although all iterative computations can be done at slow speed, we assume that electronic mode control (below) is available. All examples given are done at high speed.

18.2.2 Programmed electronic mode control (EMC)

Although some repetitive computers are alternately placed in OPERATE and RESET by automatically switching appropriate relays, faster computation almost demands electronic switching. Many design arrangements are possible. We assume the integrator modes can be individually programmed to either integrate or reset depending upon whether a digital signal is either 0 or 1. The signal controlling the electronic switches is programmed on the digital board and is indicated on a diagram as shown in Fig. 18.5. The computer must be in the OPERATE mode for these switches to be operable.*

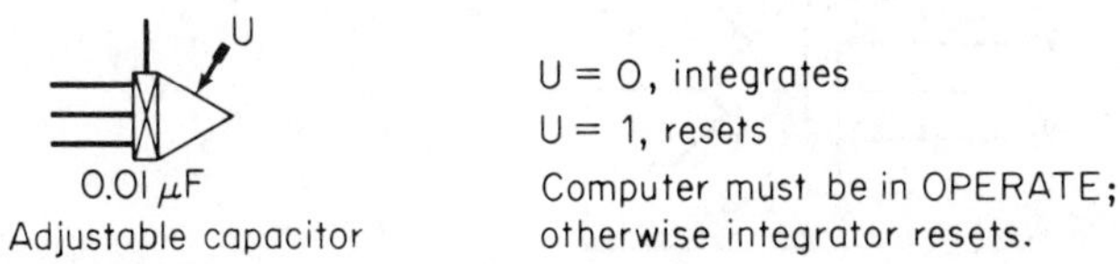

FIG. 18.5 Integrator symbol when mode is controlled by electronic switching.

Some computers provide for a HOLD mode for electronically switched integrators, as well as other conveniences, such as summing more than one *IC* voltage. However, we assume these integrators have no electronic hold mode. If holding is needed, an electronic switch is placed before any input. Some of the circuits to be shown could be simplified with a hold mode, but we leave this as an exercise for readers who have access to such equipment.

There are occasions when we wish these integrators to respond to the digital signals when the computer is placed in HOLD and RESET as well as OPERATE. (For example, when EMC integrators and RMC integrators are used simultaneously.) The EAI 231R-V computer contains integrators that are convertible from RMC to EMC by patching, but the reset and hold relays are not bypassed in the process. To enable an EMC integrator to perform in all three active computer modes, the hold and reset relays must be placed in the operate

*In subsequent discussions, we may speak of a reset or operate mode of mode-controlled integrators. These terms are independent of the RESET or OPERATE modes of the entire computer, which refer to control of the reset and hold relays. No confusion should result when we say, for example, that a rep-op integrator resets. The computer is still in the OPERATE mode.

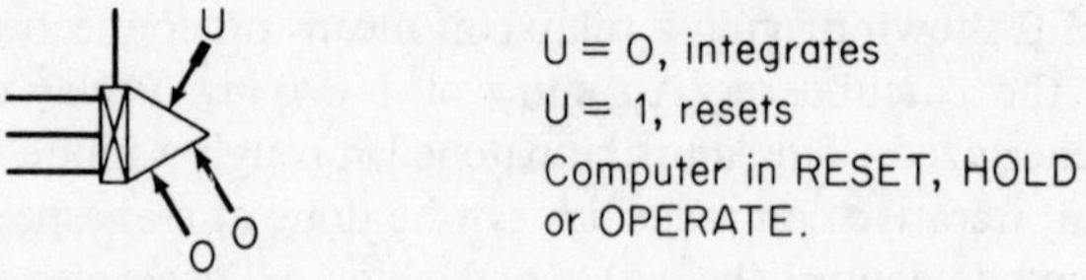

FIG. 18.6 Operation for all three active states.

position by removing the controlling voltages from the coils. Digital 0's are indicated on the diagram as shown in Fig. 18.6. For other arrangements, any convenient symbolism will do.

One use of an EMC integrator is as a high-speed electronic switch [5] as shown in Fig. 18.7. This arrangement can be used if electronic switches are in short supply.

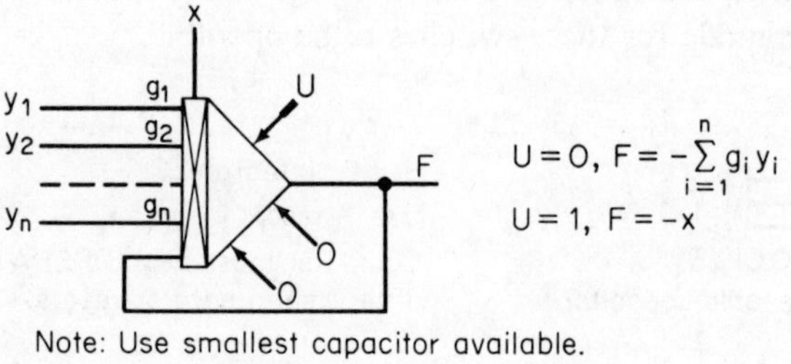

FIG. 18.7 Electronic mode controlled integrator as a switch.

18.2.3 Oscillators and counters

As stated earlier, it is necessary to obtain a periodic signal to place integrators alternately in RESET and OPERATE for repetitive operation. In the EAI 231R-V computer, there is a high-frequency (10-kHz) master square-wave oscillator used as the basic periodic source. This can be fed as an input into specially designed presettable counters that produce periodic digital drive signals used to drive the electronic switches or mode relays. Both master oscillator output and master drive input and output (there could be more than one of both) are available on the digital board. Figure 18.8 shows the schematic of

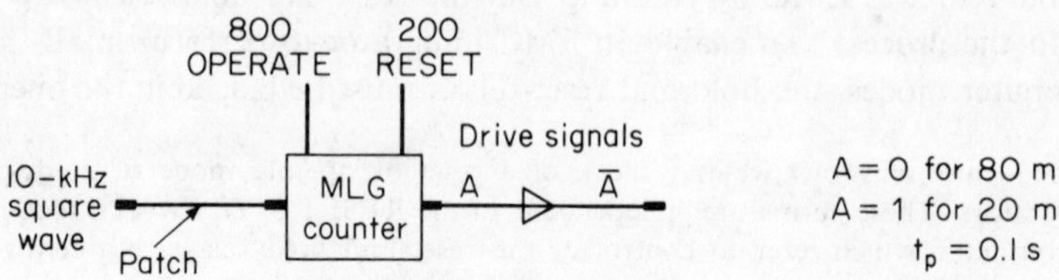

FIG. 18.8 Obtaining a drive signal for EMC integrators.

the arrangement to produce a periodic digital signal A that will be 0 for 80 ms and 1 for 20 ms. The counter values are set to 800 and 200 counts for the state of A to change as required. It counts the number of cycles of the oscillator, each one elapsing in 0.0001 s. Each counter value is presettable from 1 to 999 in steps of 1.

The symbol A is used exclusively to indicate the periodic counter output drive signal used to control the OPERATE mode of integrators in a rep-op cycle. Since the means to obtain A varies from computer to computer, a notation on the diagram showing the values of t_o and t_r is sufficient when these values are constant.

18.2.4 Variable operate periods

For computing with the parametric technique, a variable operate time circuit is shown in Fig. 18.9, where the 10-kHz oscillator is gated before it is

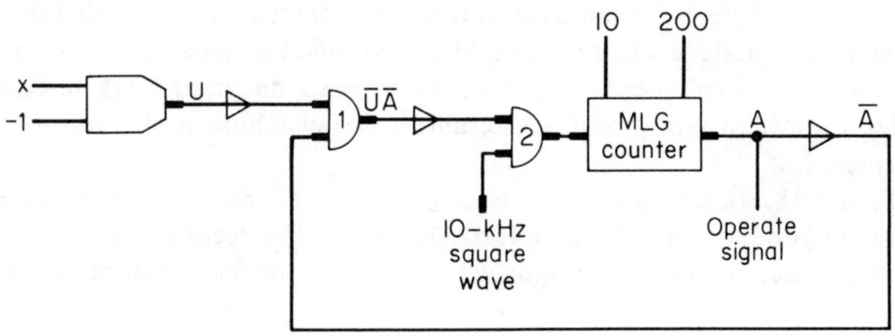

FIG. 18.9 Variable operate time for use with the parametric technique.

fed to the counter. Assume x is initially 0, and we wish to compute until $x = 1$. When $A = 1$, all integrators reset and the counter counts (for 20 ms as shown) until $A = 0$. Since $x_0 = 0$, AND gate 1 becomes high, so the counter is inhibited by AND gate 2. All integrators operate until $x = 1$. The counter resumes counting allowing the integrators to integrate for an additional 1 ms (10 counts) as shown, and then resetting the integrators for 20 ms. The action repeats itself. Thus, we allow for a variable time of operation while obtaining signals usable in track-and-store circuits described in the next section.

18.2.5 Point memory and track-store amplifiers

The principal reason for having individual mode control for each integrator is that a point memory unit can be patched easily. Such a unit stores the value of some $f(x)$ at $x = x_0$ during one run for use in the succeeding run. It is the basis of all iterative computations on an analog computer.

Consider Fig. 18.10 where $f(x)$ is a function varying with time. When $U = 1$, the output of integrator 1 essentially tracks $f(x)$ (it also inverts the

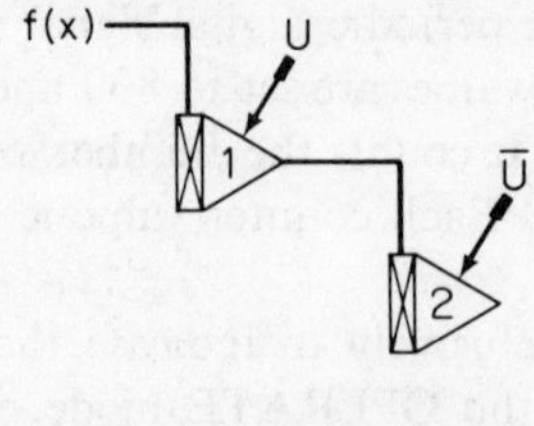

Note: Use small feedback capacitors
for accurate tracking.

FIG. 18.10 Point memory pair (zero-order hold).

sign) while integrator 2 integrates. Since integrator 2 has zero input it "holds" any voltage existing on its capacitor. Suppose when $x = x_0$, U suddenly becomes 0 and $\bar{U} = 1$. Integrator 1 holds at the value $-f(x_0)$, and integrator 2 resets to become $f(x_0)$. Integrator 2 stores the value of $f(x)$ when $\bar{U}$ changes its state from 0 to 1. When used in tandem as shown and controlled by complementary digital signals, the integrators are called a *memory pair*, or a *zero-order hold*. A small dead time zone when both integrators are resetting is needed to prevent false transference, unless the switching is simultaneous and extremely fast.

Figure 18.11a shows a circuit equivalent to Fig. 18.10 except that the pair are both operable in the IC and HOLD modes. The feedback resistor causes the integrators to track the input when the switch is on and store the voltage

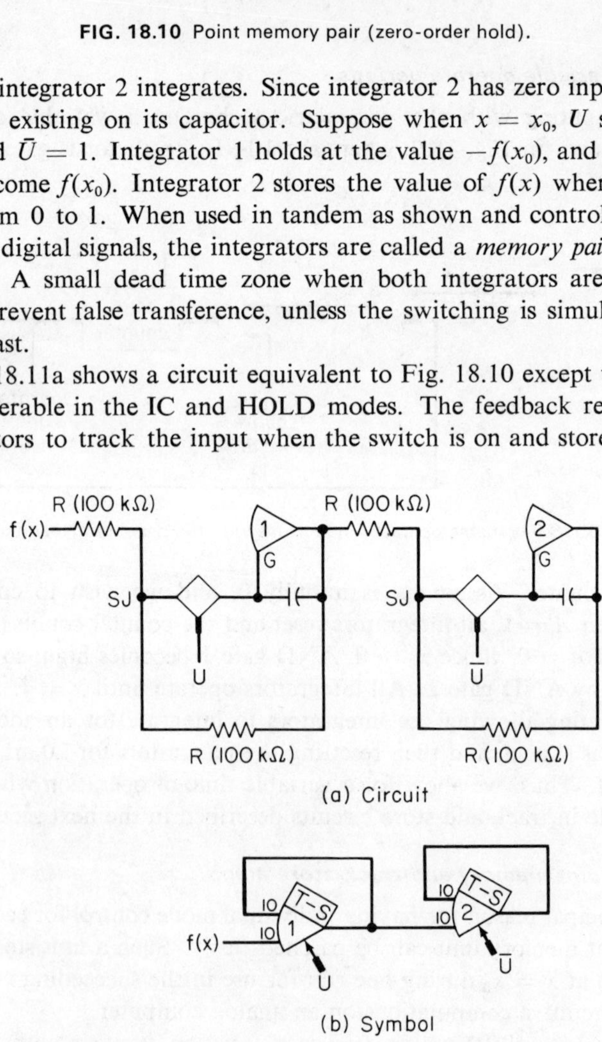

FIG. 18.11 Equivalent memory pair using summing amplifiers.

when it is off. Some computers provide a simple means to convert a standard summer to a track-and-store amplifier. Figure 18.11b shows the symbol used when summers are used as a track-and-store unit. No SJ-G connection is made externally; the electronic switch controls this connection. Track-and-store amplifiers are used hereafter for point memory units.

With zero-order hold circuitry, we may simulate the solution of difference equations [2] and sampled data systems, either by themselves [6] or combined with continuous variables [3]. The digital signal U must then be made to correspond in frequency to the time-scaled sampling frequency.

18.3 Point Memory Techniques with Rep-Op Solutions

Although rep-op solutions are observed on scopes, it is possible to obtain variplotter graphs of high-speed solutions using point memory. The following paragraphs illustrate several situations that arise.

18.3.1 Plotting high-speed rep-op variables

Suppose we are solving a set of equations with rep-op integrators and wish to obtain a graph of one rep-op variable y_r versus another x_r, where x_r is monotonically increasing with τ. (We consistently use the subscript r for variables generated in the rep-op mode when slow-speed variables are also used.) Figure 18.12 shows a circuit to accomplish this. The purpose of the normal slow-time

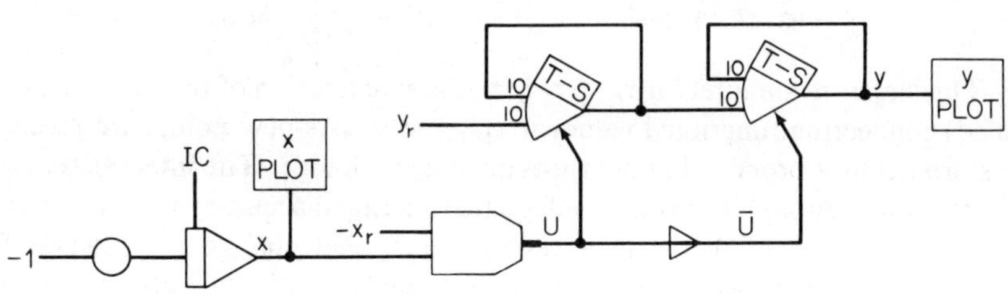

FIG. 18.12 Plotting repetitive high-speed variables.

integrator is to change x slowly. During one rep-op run, x_r increases until it is equal to x. At that instant, the value of the y integrator is updated, i.e., $y = y_r$. This integrator holds this value until it is updated again in the next rep-op cycle. By this time x has increased slightly; the new value of y corresponds to the new value $y_r(x_r)$ when $x_r = x$. *Thus the output y is a piecewise constant (staircase) function of x.* To make the abrupt changes in y unnoticeable in the graph, either a high rep-op frequency must be used in generating x_r and y_r, or x must be made to change very slowly. This is characteristic of all zero-order hold circuits.

18.3.2 First-order hold

High rep-op frequencies introduce more error, while a slowly changing x requires a long time for plotting the graph. For eliminating the staircase nature of zero-order holds in graphing and other applications, predictor circuits are used. These reconstruct the function by extrapolation, using previous values of the stored variable to predict subsequent values. The number of previous points used in the extrapolating scheme is then defined as the *order* of the hold. In Fig. 18.13, a first-order hold is shown as a linear approximation in each interval. The slope in the interval $\tau_i \leq \tau \leq \tau_{i+1}$ is an extension of the secant (shown dashed) connecting functional values at τ_{i-1} and τ_i. As new points are received and stored, a first-order hold undergoes an abrupt change. The interval $\Delta\tau = t_p$ is usually small enough to produce adequate functional accuracy with first-order holds—the figure greatly exaggerates the functional change per interval for clarity. Note that no secant information is available until the second interval. A zero-order hold (constant) must be used for the first interval approximation.

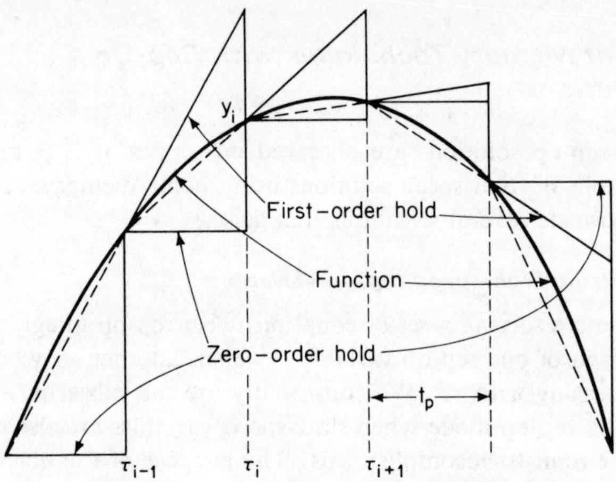

FIG. 18.13 Comparison of zero- and first-order holds.

To mechanize a first-order hold, we need first, two memory pairs to store $y(\tau_i)$ and $y(\tau_{i-1})$. Since the slope s_{i+1} of the secant is given by

$$s_{i+1} = \frac{y(\tau_i) - y(\tau_{i-1})}{t_p}$$

we must next use an integrator to mechanize

$$\dot{y} = s_{i+1}, \qquad y_0 = y_i$$

in each interval. A circuit to perform this task is shown in Fig. 18.14 assuming

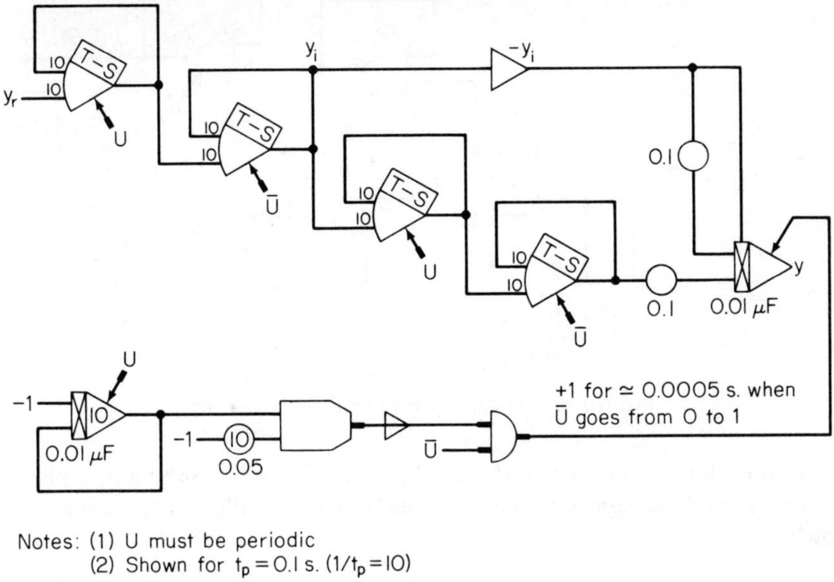

FIG. 18.14 First-order hold.

$t_p = 0.1$ s. The only difficulty is the requirement that the integrator producing the y approximation must reset in an instant when $\bar{U}$ goes from 0 to 1. In practice, a finite reset time must be used, but good results are obtained if it is a small fraction of the sampling period t_p. When $\bar{U}$ goes from 0 to 1, the AND gate output becomes 1 to reset y. But integrator 10 starts integrating at the rate of 10,000 V/s forcing the AND gate back to 0 when its output reaches 5 V (in about 0.0005 s). As we shall see in Chapter 20, *differentiating* units simplify this auxiliary circuit. Note that the time the AND gate remains 1 is controllable by the setting of pot 10 and the capacitor choice for integrator 10.

18.3.3 Plotting effects of parameter variations

These plotting circuits are extremely useful for graphing the effects of parameter variations. Just solve the equations in rep-op while varying the parameter slowly; store and plot any variable of interest. For example, suppose the solution of a set of equations depends upon one parameter β. It is varied in slow time, say between 0 and 1, while some other variable, say $y_r(x_r, \beta)$, is generated in rep-op. Suppose we wish to plot the value of $y_r(1, \beta)$. Figure 18.15 shows a circuit to do this, assuming x_r is monotonic, increasing from 0 to 1. A variable operate time circuit is used as shown, if necessary, with the signal $\bar{U}\bar{A}$ controlling the tracking unit during the compute period, and A

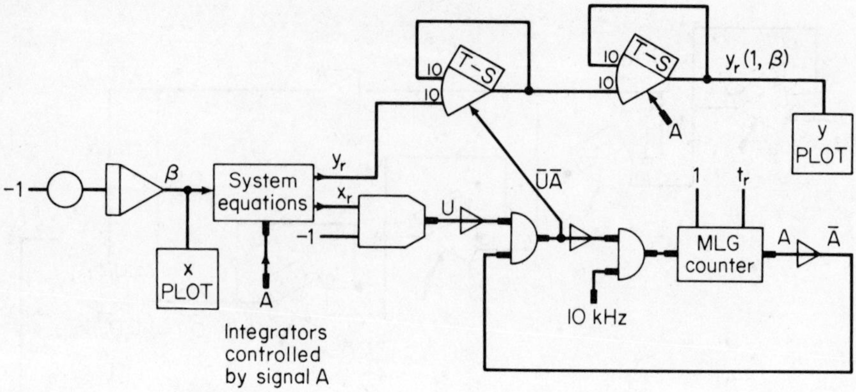

FIG. 18.15 Plotting effects of parameter variations.

controlling storing unit during the reset period. The staircase function plotted is usually good enough for such purposes when β is allowed to change very slowly.

18.4 Multispeed Operation with Memory

We consider here the uses of point memory as an aid to slow-speed computation. Units at all speeds do a substantial amount of computation, but the primary computation is done at slow speed.

18.4.1 Multiplexing and time sharing [7]–[9]

These terms refer to the use of memory circuits to reduce the amount of other equipment required in the solution of a problem. Consider a case, for example, where $f(x_1)$, $f(x_2)$, $f(x_3)$, and $f(x_4)$ are required in a problem simultaneously, and f must be set up on diode function generators. Instead of using four DFG's, a multiplexing circuit such as shown in Fig. 18.16 could be used. The amplifier x_i successively takes on values x_1, x_2, x_3, and x_4 each for one-quarter of the rep-op period, provided t_o is chosen so that t sweeps from 0 to 1 during the rep-op compute time ($A = 0$). Enough settling time for $f(x_i)$ should be permitted by this choice. At the same time each of the first three DAS closes, the first of the memory integrators track, and the second then stores the value of $f(x_i)$ when U_i becomes 0. The reset time t_r should be made large enough to allow t to return to zero with $t_o = 3t_r$. At the end of the reset time, $f(x_4)$ is stored. A rep-op period of 0.002 s is not too small for such an application. With 500 updatings per second, zero-order holds are usually sufficient for good accuracy, provided the x_i do not change radically in one rep-op period. The technique is extendible to any number of similar functions required.

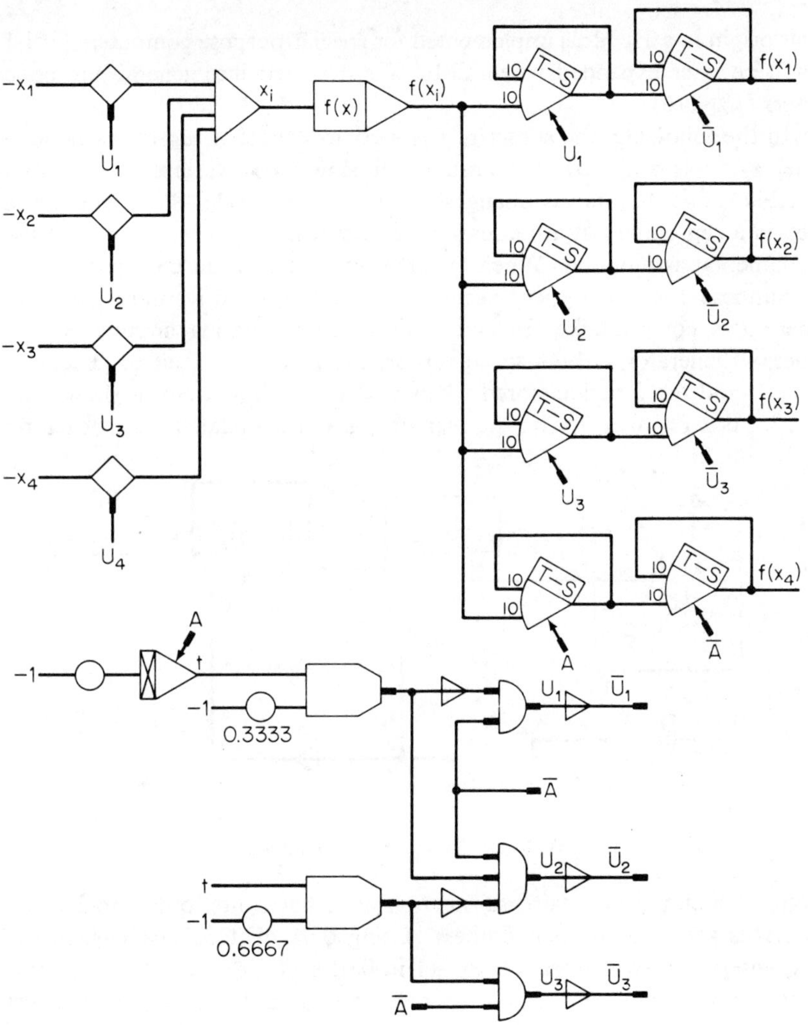

Notes: (1) Make $t_0 = 3t_r$

(2) Select capacitor and input pot to insure t ranges from 0 to 1 in time t_0

FIG. 18.16 Multiplexing circuit.

18.4.2 Function generation [10]–[16]

When solving problems in slow speed, as done in the first 16 chapters, one big difficulty is the accurate generation of analytic functions of dependent variables. Use of integrators for such a purpose (Chapter 5) may involve multipliers, and, in any case, produces functions that may have large errors when stability characteristics are poor. High-speed rep-op with memory may be used as a general-purpose analytic function generator. The following techniques had their origin in early ideas implemented for special-purpose computers [10]–[12], but were later expanded when EMC was designed into general-purpose computers [13]–[16].

In the following, the subscript r is used to denote a rep-op variable while other symbols are used for constants or slow-speed variables. A slow-speed variable is one that hardly changes during a rep-op cycle. Slow-speed variables used in a rep-op circuit are essentially constant during a rep-op run. This is a fundamental assumption. When violated, poor results can be expected.

Suppose x is a slow-speed variable ($0 \leq x \leq 1$) and we must generate e^{-x}, where $\dot{x}$ is not available. In lieu of using differentiating circuits or a diode function generator, a high-speed rep-op circuit can be used to generate e^{-x_r}, $\dot{x}_r = 1$, and the function stored when $x = x_r$. Such a circuit is shown in Fig. 18.17. As x changes slowly, the memory output is updated during each rep-op

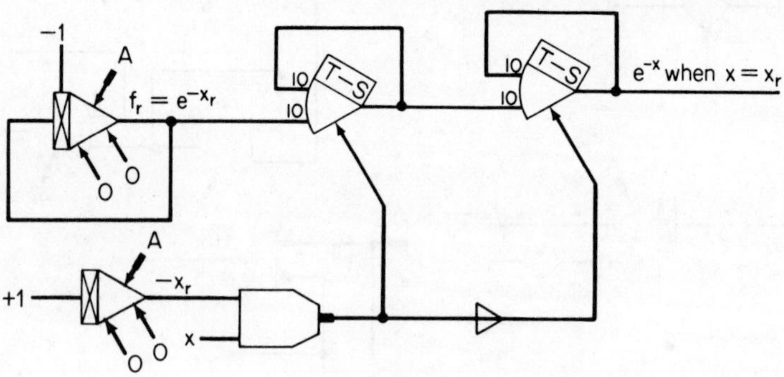

FIG. 18.17 Generating e^{-x} with rep-op.

cycle. The output is a staircase function of x, but a first-order hold could increase its accuracy and smoothness. Using 0.01-μF feedback capacitors, the $-x_r$ integrator sweeps from 0 to -1 in 0.01 s. There are then about 100 updatings per second. A variable operate time (Fig. 18.9) can increase this to the point where no first-order hold may be necessary (with a variable operate time, first-order holds are more difficult to mechanize—see Problem 18-8).

We also see that generating the function ye^{-x} is just as easy—the f_r integrator must just have an initial condition y. Remember that y is essentially constant during the rep-op cycle so that $\dot{f}_r = -f_r$ still holds. Also we shall show that no increase in equipment is needed to generate $ye^{-(x/z)}$.

The static accuracy of functions generated in this way depends upon whether or not the rep-op equations are stable. *If the rep-op equations can be solved accurately, the static accuracy of the functions generated usually exceeds that of commercial fixed or arbitrary function generators.* The ramifications of such a fact are clear. When computing in slow speed, it often pays to generate critical functions in this way, particularly if the problem terminates statically (steep descent for roots of algebraic equations, for example). Dynamic accuracy, on the other hand, can suffer severely with zero-order holds. The error is the product of the rate of change of the function generated and the rep-op period t_p. In using this technique, we must estimate if the dynamic error is tolerable, using first-order holds when necessary.

In general terms, the track-and-store method for generating analytic functions is this:

(1) Generate repetitively and track a function f_r

$$f_r(x_r, V_1, V_2, \ldots, V_n) \tag{18.1}$$

at high speed, where V_i are slow-speed variables.

(2) Store f_r when another generated function g_r is such that

$$g_r(x_r, V_1, V_2, \ldots, V_n) = 0 \tag{18.2}$$

The function f stored is obtainable by solving (18.2) for x_r

$$x_r = h(V_1, V_2, \ldots, V_n)$$

and substituting into the expression (18.1) for f_r

$$f = f_r(h, V_1, V_2, \ldots, V_n) \tag{18.3}$$

Thus f is a function of slow-speed variables and is updated each time (18.2) is satisfied during the rep-op cycle. Note that x_r is not involved in (18.3) and is, in essence, a parameter.

In the example given previously,

$$f_r = e^{-x_r}, \qquad g_r = x_r - x \tag{18.4}$$

It does not matter how the f_r or g_r are generated. But the results of Chapter 5 indicate why integration methods play such a large role in generating functions in this way. Any function generable with linear equipment can be generated with this track-and-store method using linear equipment, regardless of how the argument is obtained.

In setting up the rep-op circuit, there are three problems to be solved:

(1) Finding the f_r and g_r so that the required f is generated. The choice is usually obvious. Differential equations for f_r and x_r are written to simplify their generation.

(2) Selecting initial conditions for the integrators used in rep-op. Simplicity again makes the choice obvious.

(3) Insuring that g_r passes through 0 during the rep-op computation, and initially is always the same sign for every run.

From (18.4), for example, we have

$$\dot{f}_r = -f_r \dot{x}_r$$

so the clear choice is $\dot{x}_r =$ constant. Since $0 \leq x \leq 1$, we could choose

$$\dot{x}_r = 1, \qquad (x_r)_0 = 0 \tag{18.5}$$

or

$$\dot{x}_r = -1, \qquad (x_r)_0 = 1$$

Both insure that g_r passes through zero. But (18.5) results in

$$\dot{f}_r = -f_r, \qquad (f_r)_0 = 1$$

a stable equation which is far preferable.

Group I functions (Chapter 5) have special properties allowing certain functions of three variables to be generated with linear equipment and memory, as covered by the following theorem [16].

Theorem 18.1. Let $g_i(x)$ be Group I functions generable with $\dot{x} = 1$ as part of the system

$$\dot{g}_i(x) = \sum_{j=1}^{n} a_{i,j} g_j + b_i, \qquad i = 1, 2, \ldots, n$$

Then the functions

$$f_i = V_1 g_i\left(\frac{V_3}{V_2}\right)$$

are generable with linear equipment and memory units.

We illustrate the use of this theorem for two cases.

Example 18.2. Since x is generable in any linear system, Theorem 18.1 shows that a combination multiplication and division circuit is possible:

$$f = \frac{V_1 V_3}{V_2}$$

Assume $V_2 > 0$ and $V_2 > V_1$. Take

$$f_r = \frac{x_r V_3}{V_2}, \qquad g_r = x_r - V_1 \tag{18.6}$$

Now,

$$\dot{f}_r = \frac{\dot{x}_r V_3}{V_2}$$

so that a convenient system for rep-op is

$$\begin{aligned} \dot{x}_r &= -V_2, \qquad (x_r)_0 = V_2 \\ \dot{f}_r &= -V_3, \qquad (f_r)_0 = V_3 \end{aligned} \tag{18.7}$$

where the initial conditions are found from (18.6). We see that

$$(g_r)_0 = V_2 - V_1$$

is always positive. Further, x_r decreases, insuring that g_r goes negative. The final circuit is shown in Fig. 18.18. We could have just as easily taken the system to be

$$\begin{aligned} \dot{x}_r &= V_2, \qquad (x_r)_0 = -V_2 \\ \dot{f}_r &= V_3, \qquad (f_r)_0 = -V_3 \end{aligned}$$

to obtain the identical properties, except that $(g_r)_0$ is now initially negative but goes to 0. System (18.7) produces good results even when V_1 is negative but larger in magnitude than V_2.

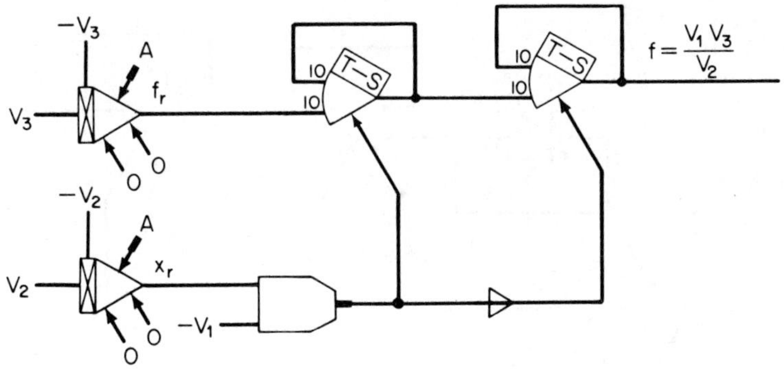

FIG. 18.18 Combination multiplication and division circuit.

Example 18.3. From Theorem 18.1, we could expect to generate with linear equipment, the function

$$f = V_1 e^{V_3/V_2}$$

where the maximum value of the exponent is, say, 2, and $V_2 > 0$. Let

$$f_r = V_1 e^{x_r/V_2}, \qquad g_r = x_r - V_3$$

Then

$$\dot{f}_r = \frac{f_r \dot{x}_r}{V_2}$$

Take

$$\dot{x}_r = -V_2, \qquad (x_r)_0 = 2V_2$$
$$\dot{f}_r = -f_r, \qquad (f_r)_0 = V_1 e^2$$

Now, $(g_r)_0$ must be positive initially (by the assumption that $V_3/V_2 < 2$) but must go through 0 eventually, even for V_3 negative. These equations must be amplitude scaled:

$$\left(\frac{\dot{x}_r}{2}\right) = -0.5V_2, \qquad \left(\frac{x_r}{2}\right)_0 = V_2$$
$$\left(\frac{\dot{f}_r}{e^2}\right) = -\left(\frac{f_r}{e^2}\right), \qquad \left(\frac{f_r}{e^2}\right)_0 = V_1$$
$$\left(\frac{g_r}{2}\right) = \left(\frac{x_r}{2}\right) - 0.5V_3$$

A scaled circuit diagram is shown in Fig. 18.19. A time scale factor of 10 was included

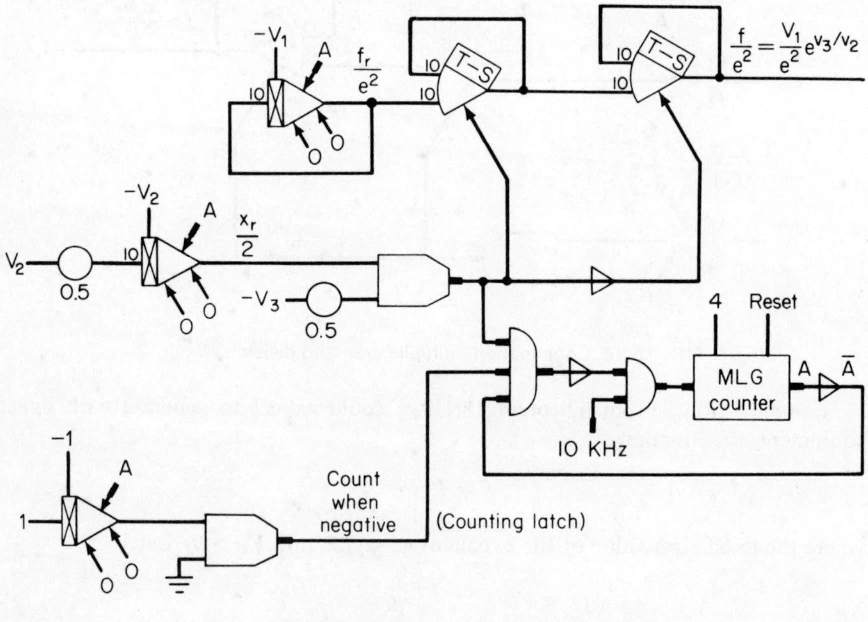

FIG. 18.19 Generation of $V_1 e^{V_3/V_2}$.

to obtain small operate times. Also, an additional counting latch provided an end of the variable operate time when the exponential became $|V_1|e^{-10} < 0.0001$. There are approximately 100 updatings per second at least, using 0.01-μF feedback capacitors.

Example 18.4. An interesting application of the track-and-store method of generating functions is an inverse resolver circuit [15]. Generate the functions

$$\begin{aligned} R &= (x^2 + y^2)^{1/2} = x \cos \theta + y \sin \theta \\ \theta &= \arctan\left(\frac{y}{x}\right) \end{aligned} \tag{18.8}$$

From (18.8) we must have

$$y \cos \theta - x \sin \theta = 0$$

so we take

$$\begin{aligned} g_r &= y \cos \theta_r - x \sin \theta_r \\ R_r &= x \cos \theta_r + y \sin \theta_r \end{aligned}$$

with no parameter x_r needed. If $y > 0$, these functions satisfy the equations

$$\begin{aligned} \dot{R}_r &= g_r, & (R_r)_0 &= x \\ \dot{g}_r &= -R_r, & (g_r)_0 &= y \\ \dot{\theta}_r &= 1, & (\theta_r)_0 &= 0 \end{aligned} \tag{18.9}$$

with $(g_r)_0$ positive. If $y < 0$, we may take

$$\begin{aligned} \dot{R}_r &= -g_r, & (R_r)_0 &= x \\ \dot{g}_r &= R_r, & (g_r)_0 &= y \\ \dot{\theta}_r &= -1, & (\theta_r)_0 &= 0 \end{aligned}$$

But now, the structure of the equations has changed and $(g_r)_0$ is negative. We are thus motivated to define

$$g'_r = -g_r$$

and solve

$$\begin{aligned} \dot{R}_r &= g'_r, & (R_r)_0 &= x \\ \dot{g}'_r &= -R_r, & (g'_r)_0 &= -y \\ \dot{\theta}_r &= -1, & (\theta_r)_0 &= 0 \end{aligned} \tag{18.10}$$

Systems (18.9) and (18.10) can be solved to generate the functions R and θ using relays (since y is a slowly changing variable) to provide the structure changes when

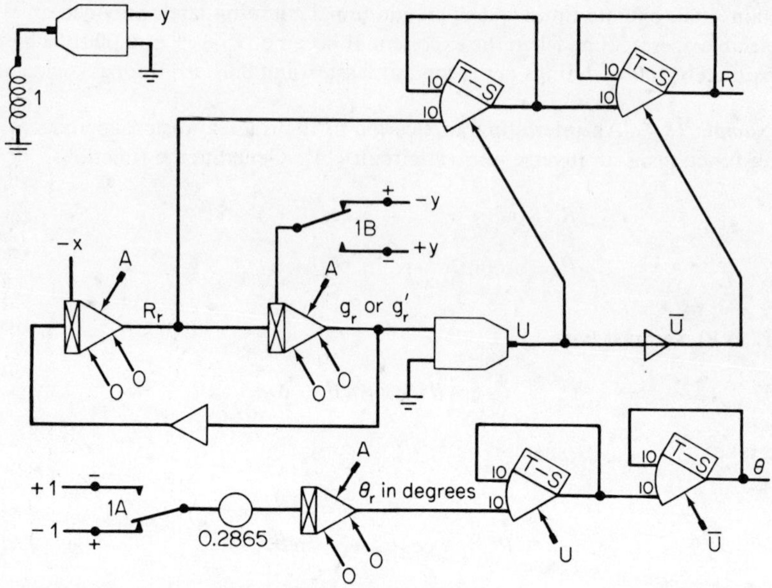

FIG. 18.20 Inverse resolver circuit.

y changes sign. The circuit shown in Fig. 18.20 provides excellent static accuracy even for small x and y. The θ_r integrator is scaled to produce θ in degrees. When $\theta = 0.9$, this corresponds to 180°. Thus, the output is equivalent to a $\pm 180°$ resolver with discontinuities at 180° and $-180°$.

18.4.3 Multiple integrals [13]

Multispeed operation with memory considerably simplifies the computation of multiple integrals. Consider, for example, the double integral

$$I = \int_a^b \int_{y_1(x)}^{y_2(x)} f(x, y)\, dy\, dx \qquad (18.11)$$

Let

$$h(x) = \int_{y_1(x)}^{y_2(x)} f(x, y)\, dy$$

If $h(x)$ is found by high-speed computation in a short time, x does not change much and can be considered constant. We solve

$$\dot{h}_r = f(x, y)\dot{y}_r, \qquad (h_r)_0 = y_1(x)$$

by selecting the most convenient $\dot{y}_r$, and store the value of h_r when $y_r = y_2(x)$. The stored value $h(x)$ is then used in (18.11)

$$I = \int_a^b h(x)\, dx$$

to find I at the slower speed. Extension to higher-order integrals requires more computing speeds.

Example 18.5. We illustrate this technique by computing the integral

$$I = \int_0^1 \int_x^1 \sqrt{y^2 - x^2 + 0.01}\, dy\, dx$$

Let

$$h(x) = \int_x^1 \sqrt{y^2 - x^2 + 0.01}\, dy$$

and generate in rep-op

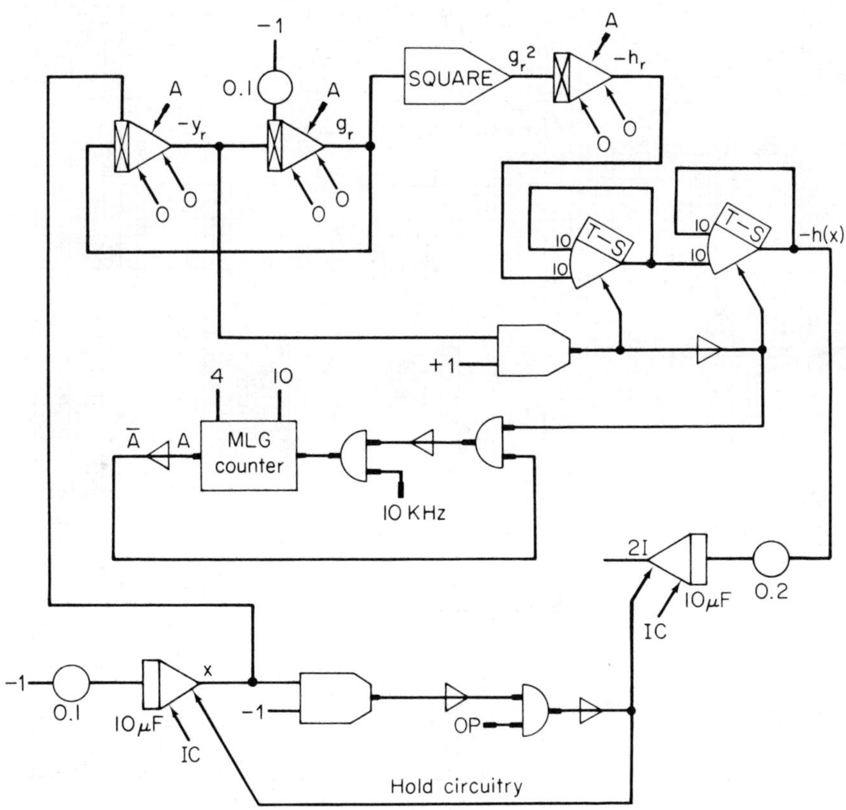

FIG. 18.21 Double integral example.

$$h_r = \int_x^1 \sqrt{y_r^2 - x^2 + 0.01}\, dy_r$$

where x is considered constant. The parametric technique can be used to generate the integrand. Let

$$g_r = \sqrt{y_r^2 - x^2 + 0.01} \qquad \text{so that} \qquad \dot{g}_r = y_r \frac{\dot{y}_r}{g_r}$$

A final convenient system for rep-op is

$$\begin{aligned} \dot{y}_r &= g_r, \qquad (y_r)_0 = x \\ \dot{g}_r &= y_r, \qquad (g_r)_0 = 0.1 \\ \dot{h}_r &= g_r^2, \qquad (h_r)_0 = 0 \end{aligned} \tag{18.12}$$

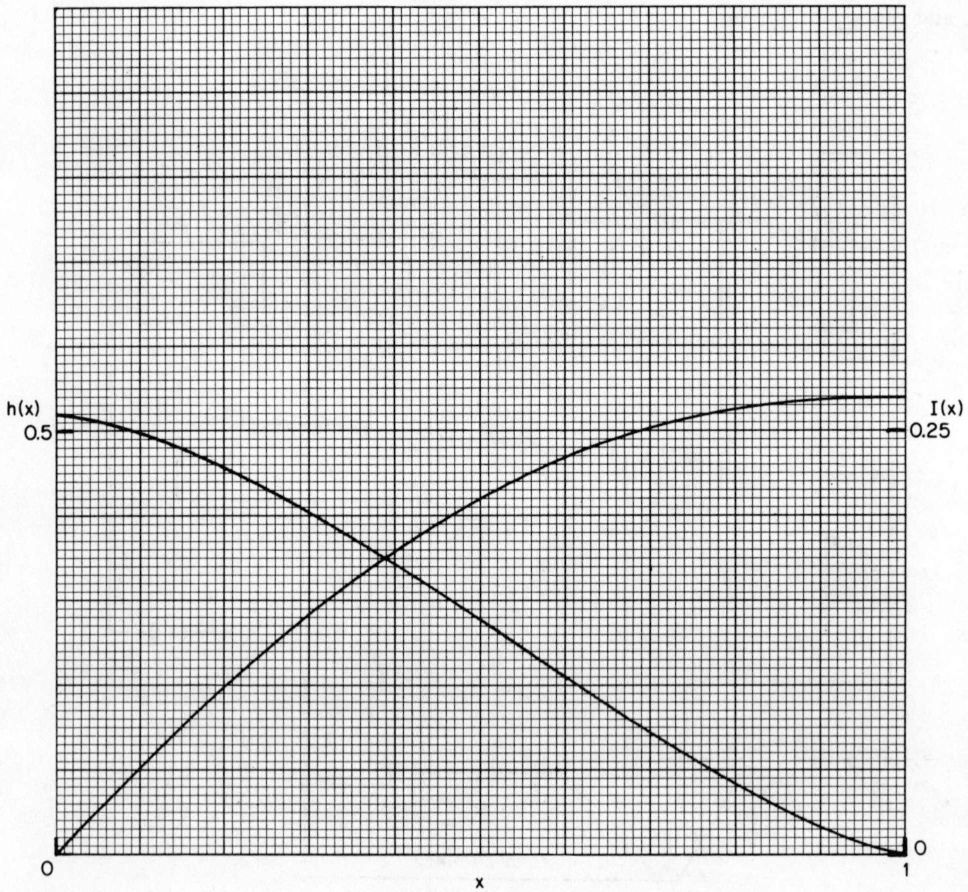

FIG. 18.22 Graphs of $h(x)$ and $I(x)$.

where h_r is stored and equals $h(x)$ when $y_r = 1$. In slow time we solve

$$\begin{aligned} \dot{x} &= 1, \qquad x_0 = 0 \\ \dot{I} &= h(x), \qquad I_0 = 0 \end{aligned} \tag{18.13}$$

and stop when $x = 1$. Figure 18.21 shows the simple circuitry required for the solution of (18.12) and (18.13), while Fig. 18.22 shows the graph of $h(x)$ and the output of the I integrator. Note how easy it was to generate $h(x)$ in rep-op! When the integrand is generable by differential equations, it is a function of one variable, not two.

18.5 Iteration and Continuous Memory [13], [15], [17], [18]

In this section, we begin a discussion on iterative techniques which will continue through the next two chapters. An iteration can be defined as a process by which a sequence of runs is made, each one being dictated by some recursive formula. Given n initial parameters x_j, a recursive formula after run $k - 1$ can have the form

$$(x_j)_k = f_j[(x_1)_{k-1}, (x_2)_{k-1}, \ldots, (x_n)_{k-1}]$$

The process *converges* if each $(x_j)_k \to (x_j)_{k-1}$ as $k \to \infty$, and they represent a solution (or approximation) to a given problem.

Continuous memory refers to the storage of an entire function produced in one rep-op run and needed for the succeeding run. Although special equipment has been designed for this purpose (see [17] for a good summary), point memory information can be used to approximate the function also. Most frequently a polynomial approximation is used because of its simplicity.

Assume $n + 1$ equally spaced values y_i of the generated function are stored during one rep-op run. We can generate the polynomial approximation $f(x)$ by integration in the succeeding run by using the system

$$\dot{f}_i = f_{i+1}, \qquad \dot{x} = 1, \qquad i = 0, 1, \ldots, n \tag{18.14}$$

where $f_i = df_i/dx$, $f_0 = f$, and $f_{n+1} = 0$. Initial conditions for (18.14) are found from the $n + 1$ values of the function from an appropriate Lagrangian formula for derivatives (Chapter 14).

Before proceeding with mechanization of this procedure, it is well to discuss the limitations of it. First, system (18.14) is unstable with errors in f_i propagating as a polynomial in τ of degree i. We could not expect good accuracy for any but small values of n. Second, system (18.14) is useful only for a continuous function with a modification necessary when the function is discontinuous. Finally, the frequencies of the function must be small enough so that an nth degree polynomial approximation is justified. Despite these limitations, the technique is quite adequate for some problems. Also, the limitations can be somewhat offset by splitting the basic interval (in which the approximation should hold) into two or more smaller subintervals, approximating segments of the function by polynomials in each subinterval, and combining the approximations by switching.

18.5.1 Mechanizations of continuous memory

We consider here only polynomials of degree three or less, although extensions to higher degree follow along the same lines.

As in Chapter 14, the symbol

$$y_j^{(i)} \equiv \frac{d^i y}{dx^i} \quad \text{evaluated at } x_j$$

is used. The independent variable x is taken to range from 0 to 1 for simplicity, and the functional values $y(x_j)$ are written y_j.

First-degree approximation. If the generated function is sufficiently linear to warrant a first-degree approximation, we take

$$f_0 = y_0, \qquad y_0' = (f_1)_0 = \frac{y_1 - y_0}{\Delta x}$$

from the two-point formulas given in Table 14.1. Δx here is 1. A sample mechanization is shown in Fig. 18.23. The y_j alternate in sign to conform with an economical method used for higher-order mechanizations. At the end of the operate period, new values of the y_j must have been generated as outputs. During reset, the integrators reset to the proper initial conditions, but the polynomial is also generated by these integrators during the operate period. Presumably, y_0 and y_1 will change because of some iteration process. If they do

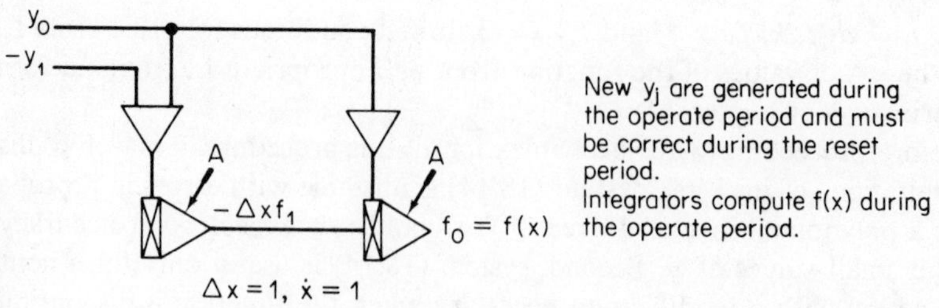

FIG. 18.23 Continuous memory—first degree.

not change, the same function will be generated every cycle. Note that the scaling permits maximum slopes of 1 and could be changed, if necessary.

Second-degree approximation. The principle of operation remains the same. However, we may calculate that the Lagrangian formulas for derivatives of f at $x = x_0$ are

$$(f_0)_0 = y_0$$

$$y_0' = (f_1)_0 = \frac{-\frac{3}{2}y_0 + 2y_1 - \frac{1}{2}y_2}{\Delta x}$$

$$y_0'' = (f_2)_0 = \frac{y_0 - 2y_1 + y_2}{(\Delta x)^2}$$

where Δx is now $\frac{1}{2}$. Figure 18.24 shows a convenient mechanization.

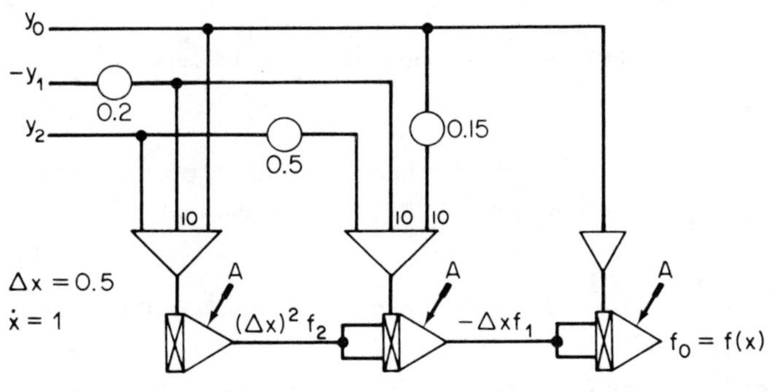

FIG. 18.24 Continuous memory—second degree.

Third-degree approximation. The derivative expressions now are

$$(f_0)_0 = y_0$$

$$y_0' = (f_1)_0 = \frac{-\frac{11}{6}y_0 + 3y_1 - \frac{3}{2}y_2 + \frac{1}{3}y_3}{\Delta x}$$

$$y_0'' = (f_2)_0 = \frac{2y_0 - 5y_1 + 4y_2 - y_3}{(\Delta x)^2}$$

$$y_0''' = (f_3)_0 = \frac{-y_0 + 3y_1 - 3y_2 + y_3}{(\Delta x)^3}$$

where Δx is now equal to $\frac{1}{3}$. The mechanization now appears as in Fig. 18.25.

In the above mechanizations, any storage device that contains the y_j during

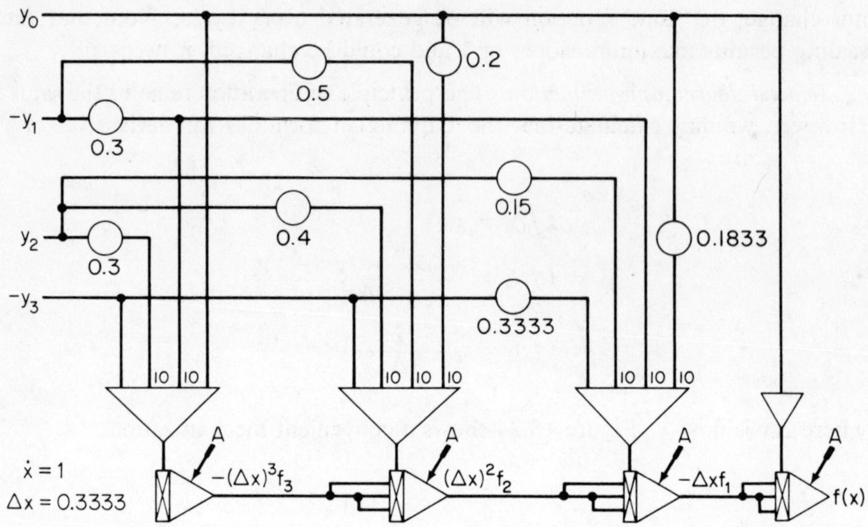

FIG. 18.25 Continuous memory—third degree.

the reset period may be allowed to change during the operate period. This fact permits an economical use of amplifiers, as demonstrated below.

18.5.2 Application to the solution of integral equations [19]–[22]

Integral equations were solved quite early in the analog computer art, but the solution involved special-purpose function generators and were manual in operation [19]–[21]. With point memory programmed for a continuous memory, however, an automatic procedure for their solution is straightforward [22]. Two types of integral equations are common.

(a) Fredholm's equation of the second kind:

$$y(x) = g(x) + \lambda \int_a^b K(x, t)y(t)\,dt \tag{18.15}$$

and

(b) Volterra's equation of the second kind:

$$y(x) = g(x) + \lambda \int_a^x K(x, t)y(t)\,dt \tag{18.16}$$

The functions $g(x)$ and the kernel $K(x, t)$ are given. It is required to find $y(x)$ satisfying the relationships (18.15) or (18.16). If $y(x)$ is not outside the integral, equations are said to be of the first kind.

Several iteration procedures are possible. If an arbitrary function $h_0(x)$ is assumed as a solution to, say, (18.15), new functions may be calculated by

$$h_{i+1}(x) = g(x) + \lambda \int_a^b K(x, t)h_i(t)\, dt$$

where $h_i(x)$ can be shown to converge to $y(x)$ as $i \to \infty$ for sufficiently small λ. It can also be shown that the procedure always converges to the $y(x)$ in Volterra's equation of the second kind (18.16). For purposes of computation, the variable x is quantized into $n + 1$ values x_j, $j = 0, 1, \ldots, n$, and the $n + 1$ relationships

$$h_{i+1}(x_j) = g(x_j) + \lambda \int_a^b K(x_j, t)h_i(t)\, dt \tag{18.17}$$

are found simultaneously. It is necessary to construct a smooth function $h_i(t)$ from the previously stored $n + 1$ values of $h_i(x_j)$ to perform the integration. This is where the continuous memory will play its part. Note that (18.17) reduces the kernel to $n + 1$ functions of one variable, simplifying its generation. We now use the notation y_j for $h_i(x_j)$ understanding that the variables change in discrete steps at the end of each iteration cycle.

Example 18.6. We compute the solution to

$$y(x) = x - \lambda \int_0^1 e^{-xt} y(t)\, dt$$

by approximating y as a cubic polynomial in x, leaving λ as variable. As λ is changed, new solutions are tracked and appear on the scope. Quantization of the range of x into three equally spaced intervals requires that

$$x_j = \frac{j}{3}, \qquad j = 0, 1, 2, 3$$

We generate the y_j values simultaneously from

$$y_j = x_j - \lambda \int_0^1 e^{-x_j t} h_i(t)\, dt$$

and use the circuit of Fig. 18.25 to generate $h_i(t)$. A final mechanization is shown in Fig. 18.26 using $\dot{t} = 1$ in rep-op, with 0.01-μF feedback capacitors and $t_o = 10$ ms. (Amplitude scaling could be much improved.) When the computer is in RESET, all integrators reset, forcing the initial conditions of the continuous memory integrators to produce $h_0(t) = t$ for the first run. Convergence to the solution then takes place in about ten runs or $\simeq 0.1$ s. Note the track-store amplifiers track the proper integrals during the entire operate period and are the correct values of y_j during the reset period. This is the only requirement that the continuous memory function generator

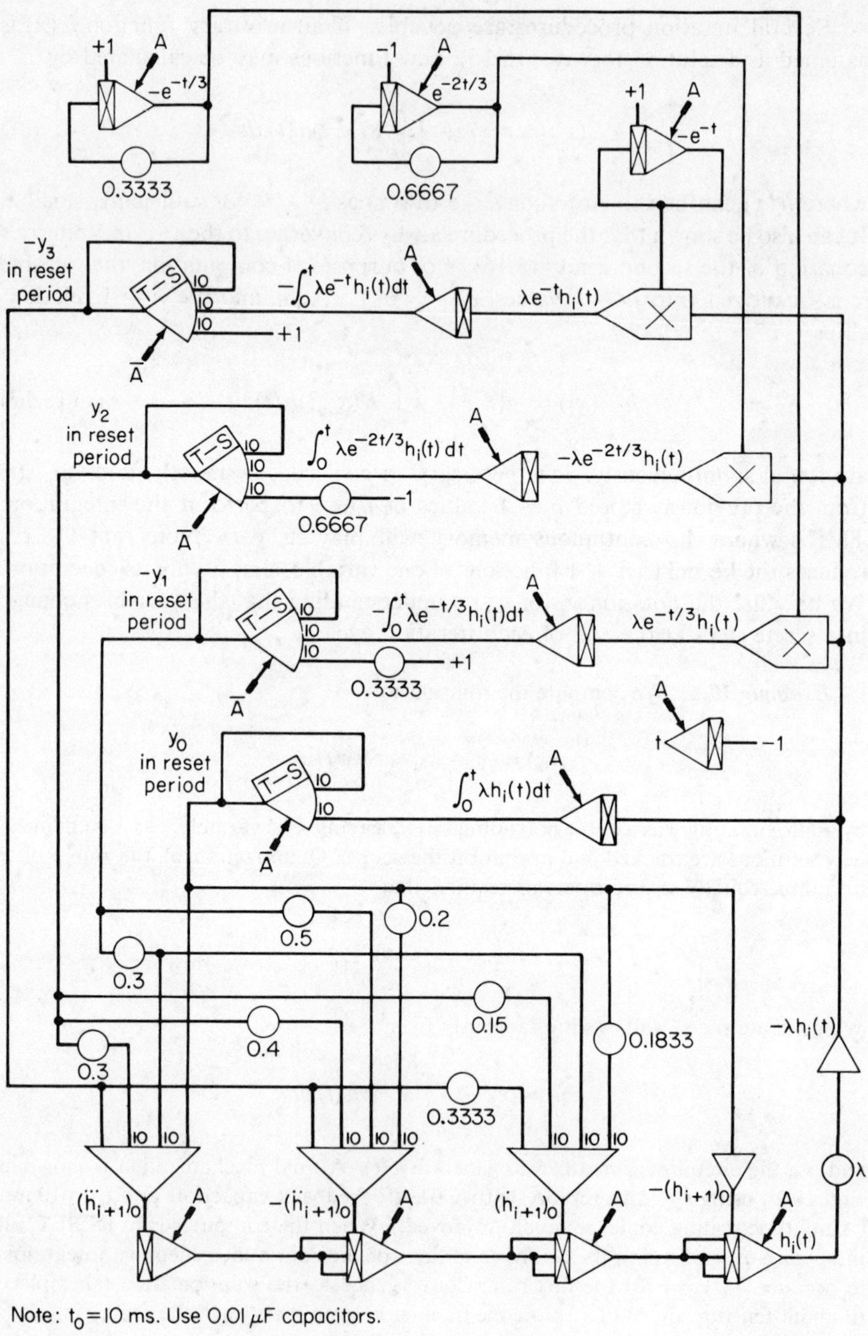

Note: $t_0 = 10$ ms. Use 0.01 μF capacitors.

FIG. 18.26 Solution for Example 18.6.

will work properly. Thus, memory pairs need not be used. For Volterra's equation of the second kind, the $\tilde{A}$ output must be AND gated with a comparator output so that only the integrals from a to x_j are stored in the track-store amplifiers during the reset period.

Figure 18.27 shows the graphs of $y(x)$ obtained for $\lambda = 0, \frac{1}{2}$, and 1 after convergence had taken place, using the plotting circuit of Fig. 18.12. During the run, d^3y_i/dx^3 remained near zero, indicating that a quadratic approximation would have been sufficient.

This was the first example of an all rep-op iteration. A continuous memory circuit will also be used to mechanize a serial solution to a partial differential

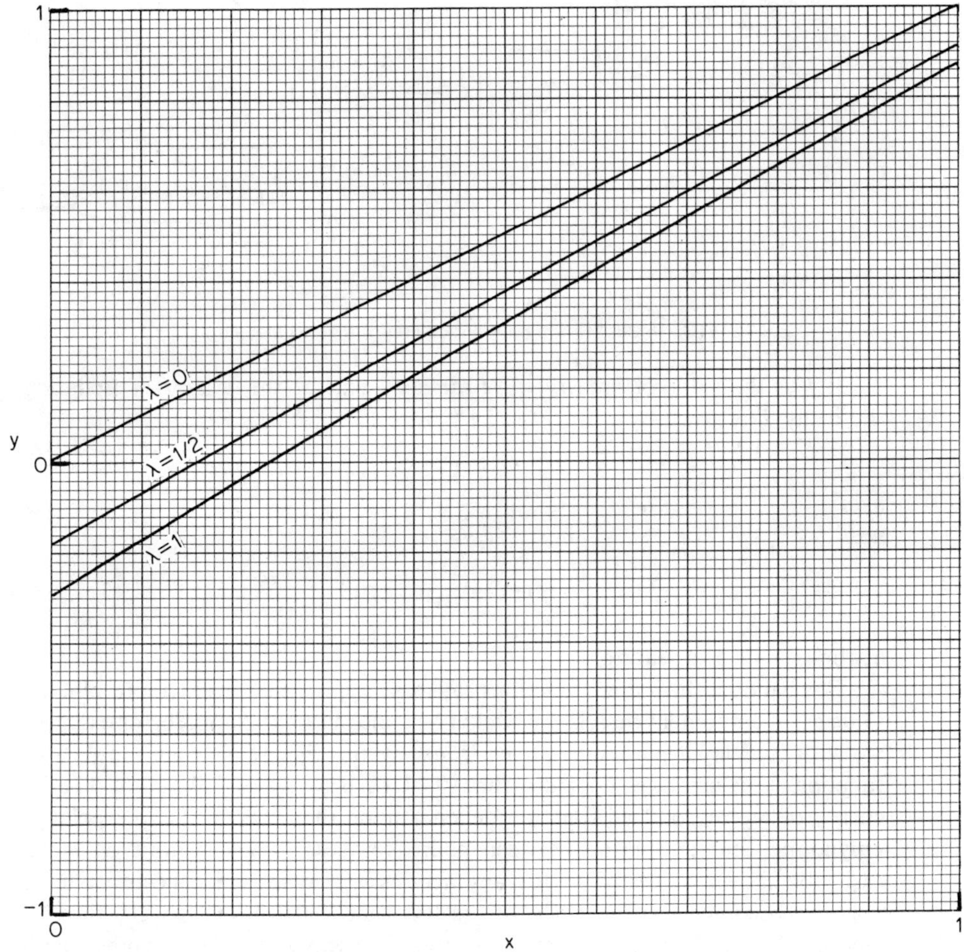

FIG. 18.27 Graphs of solution for various λ.

equation, after a discussion of boundary-value solutions in Chapter 19, and the use of gated logic for counting in Chapter 20.

18.5.3 Time delay [15]

By stacking $n + 1$ memory pairs in tandem (Fig. 18.28) a sequence of points f_i can be stored and updated every rep-op period. These in turn can be used to reproduce $f(\tau)$ at some future time by some appropriate polynomial interpolation scheme. This solves, once again, the problem (Chapter 10) of obtaining a time delay in what is called a "bucket brigade" manner.

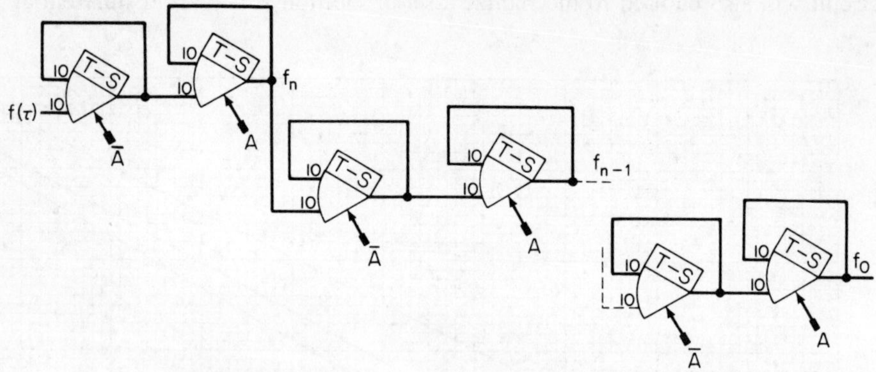

FIG. 18.28 Stacking memory pairs for a time delay circuit.

If f_0 is the value of the $(n + 1)$st memory pair at the instant of updating, the delay T is given by

$$T = nt_p$$

i.e.,

$$f_0(\tau) = f_n(\tau - T)$$

at the instant of updating.

For bandwidth-limited functions, all-pass approximants are more efficient in terms of equipment than is the bucket brigade method. Suppose, for example, we wish to delay a pure sine wave sin ωt by the time T so that the delayed function is in error by no more than 0.01 (1 per cent of full scale). The accuracy of the reproduced function depends upon the order of the interpolation polynomial and the period between point storages. A lower limit on n can easily be shown to be $\omega T/\pi$ for any accuracy, and for a second-order interpolation becomes about ωT for errors less than 0.01. If ωT is 10, then 22 amplifiers $[2(n + 1)]$ are needed for storage elements alone. On the other hand, Table 10.3 shows that a sixth-order Chebyshev all-pass approximant (requiring eight amplifiers) yields a max $|\epsilon_T| = 0.0001$ for $\omega T = 10.2$. Since the total error E of the delayed function is approximately

$$E \simeq \omega T \max |\epsilon_T|$$

its use produces errors of no more than 0.00102 for all frequencies less than 10.

A good use for the bucket brigade delay, however, is for an expected step function. First-order interpolation results in a ripple-free delayed function with the delay in error no more than t_p. The circuit for such a delay is shown in Fig. 18.29. As in first-order holds, resetting the integrator for a short time after each updating prevents drift in the produced function.

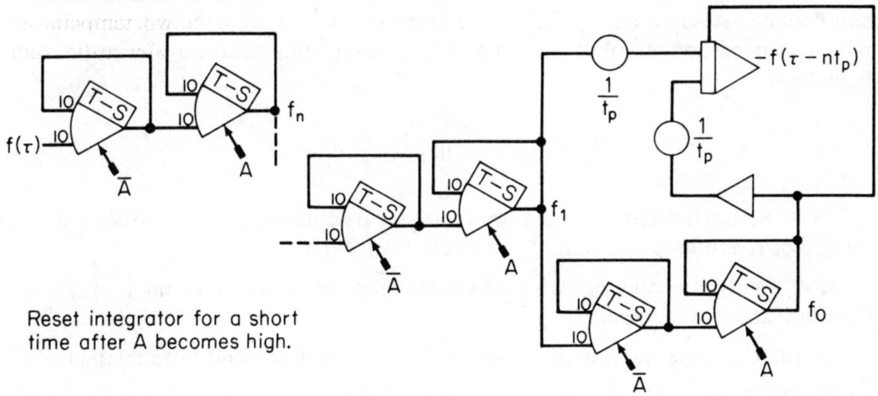

FIG. 18.29 First-order interpolation for a bucket brigade delay.

PROBLEMS

18-1. Program the min-max circuit of Fig. 6.20 with mode relay drivers using two integrators, three comparators, and AND-gate logic.

18-2. Program the counting circuit of Fig. 6.62 using mode relay drivers. (*Hint:* Generate $-y$ at an integrator output for an economical circuit.)

18-3. Difference equations are easily programmed with memory [2]. Using the notation $\Delta y_i \equiv y_i - y_{i-1}$, the outputs of memory units can form functions from which the units themselves can be updated. Program the equation

$$\Delta y_i = -0.1 y_i, \qquad y_0 = 1$$

in this way, using integrators for the memory.

18-4. In Fig. 18.12, it was assumed that x_r was monotonic in τ, but suppose it is not. How may $y_r(x_r)$ then be plotted?

18-5. When multiplexing two functions, the circuit of Fig. 18.16 can be considerably simplified. Program such a case with no comparators or AND gates.

18-6. If a slave counter is available, the integrator and comparators in Fig. 18.16 can be eliminated with four control signals and their complements generated by four AND gates. Demonstrate this procedure.

18-7. A fruitful application of multiplexing is time-sharing nonlinear equipment when solving partial differential equations by the parallel method. Then, the equations at each station are identical except for subscripts. Indicate how to solve a heat conduction problem

$$\frac{\partial u}{\partial t} = \alpha \frac{\partial^2 u}{\partial x^2} - k_1(u + a)^4 + k_2$$

by this method, using 10 stations across a bar. The equation includes radiation heating and cooling. Assume any initial temperature profile $u(0, x)$, a known temperature boundary condition $u(t, 0)$ at the left and just the radiation heat transfer at the right boundary:

$$\frac{\partial u(t, L)}{\partial t} = -k_1 [u(t, L) + a]^4 + k_2$$

18-8. Set up a first-order hold for a variable operate time. *Hint:* Using a timing integrator resetting to the value t_r, compute $1/(t_r + t_o)$.

18-9. With x, y, and z slow-speed variables, generate the function $y \ln x$ using the track-and-store method.

18-10. Generate the function $z \sin (x/y)$ using the track-and-store method.

18-11. Generate the function $y \arcsin x$ using the track-and-store method.

18-12. In Fig. 18.18, we require $V_2 > V_1$ to keep $(g_r)_0$ the same sign. Suppose all variables are positive. Arrange the circuit to produce $V_1 V_3/V_2$ independent of the relative magnitudes. (This can be convenient if V_1 is small when V_3 is large and vice versa.)

18-13. In generating functions with track-and-store, dynamic errors play a large part. In addition, there are comparator errors and capacitor mismatches. Suggest methods for compensating for these when generating the function $y = \sin x$ for $0 \leq x \leq 10$.

18-14. Generate the function $\int_0^x e^{-xu^2}\,du$ for use with a slow-speed variable x.

18-15. Mechanize a circuit to find the volume of revolution of a function $y(x)$ defined for $0 \leq x \leq 1$.

18-16. Mechanize a fourth-order continuous memory circuit based on the five points, y_0-y_4, using the fact that

$$(y')_0 = \frac{-\frac{25}{12}y_0 + 4y_1 - 3y_2 + \frac{4}{3}y_3 - \frac{1}{4}y_4}{\Delta x}$$

$$(y'')_0 = \frac{\frac{35}{12}y_0 - \frac{28}{3}y_1 + \frac{19}{2}y_2 - \frac{16}{3}y_3 + \frac{11}{12}y_4}{(\Delta x)^2}$$

$$(y''')_0 = \frac{-\frac{5}{2}y_0 + 9y_1 - 12y_2 + 7y_3 - \frac{3}{2}y_4}{(\Delta x)^3}$$

$$(y^{\text{IV}})_0 = \frac{y_0 - 4y_1 + 6y_2 - 4y_3 + y_4}{(\Delta x)^4}$$

18-17. Compute the solution $y(x)$ of

$$y(x) = \sin x - \lambda \int_0^x te^{-0.2xt} y(t)\, dt$$

for $0 \leq x \leq \pi/2$ using a cubic approximation in a continuous memory.

18-18. When there are not too many data points, curve fitting can be done by relaxation by literally mechanizing the required norm and adjusting the parameters to minimize the value observed on a DVM. Fit the data [23]

t	0	0.25	0.50	0.75	1.0
y	0	49.23	28.28	19.96	12.85

to the functional form $y = A(e^{-at} - e^{-bt})$ in this way, using a least squares norm.

18-19. Program a maximum-value circuit with an EMC integrator using a comparator output to track the input when it is larger than the integrator output [24].

18-20. Use an EMC circuit to generate a general hysteresis curve by allowing an integrator to track $f_1(x)$ or $f_2(x)$ if its output is outside these limits, where $f_2 > f_1$. (This would be an extension of an idea by Schulte [25].)

18-21. Hedges [26] uses a steep-descent circuit to compensate an oscillator and constructs an automatic Nyquist plotter with EMC integrators and memory by repetitively evaluating the integrals

$$\int_0^{\pi/\omega} \omega K \sin(\omega t - \theta)\, dt = 2 \cos \theta$$

$$\int_{-\pi/2\omega}^{\pi/2\omega} \omega K \sin(\omega t - \theta)\, dt = 2 \sin \theta$$

where $\omega \sin \omega t$ is the input to a test transfer function and $\omega K \sin(\omega t - \theta)$ is its output. Construct a circuit to perform this task.

REFERENCES

[1] Wadel, L. B., "Automatic Iteration on an Electronic Analog Computer," *IRE Wescon Convention Record*, 1954.

[2] Wadel, L. B., "An Electronic Differential Analyzer as a Difference Analyzer," *Journal ACM*, vol. 1, July 1954.

[3] Wadel, L. B., "Analysis of Combined Sampled and Continuous-Data Systems on an Electronic Analog Computer," *IRE National Convention Record, Part 4*, 1955.

[4] Teague, D. S., Jr., and R. D. Gilpin, Jr., "Preliminary Investigation of Suitability of REAC for Experimental Curve Fitting," *Symposium I on REAC Techniques*, Reeves Instr. Corp., New York, March 15-16, 1951 (ASTIA No. 108488).

[5] Stricker, J. L., "The Application of Integrators to Switching Functions," *Simulation*, vol. 6, April 1966.

[6] Saucedo, R., and T. W. Sze, "Analog Simulation of Digital Computer Programs," *Trans. AIEE, part I* (*Comm. and Elect.*), vol. 80, Jan. 1962, pp. 703-709.

[7] Rawden, E., "Time Multiplexing as Applied to Analog Computation," *IRETEC*, vol. 8, March 1959.

[8] Miura, T., and J. Iwata, "Time-Sharing Computation Utilizing Analog Memory," *AICA*, vol. 5, July 1963.

[9] Miura, T., and J. Iwata, "Study on Time-Shared Analog Computation Technique," *Simulation*, vol. 5, Nov. 1965.

[10] Broomal, J., and L. Riebman, "A Sampling Analogue Computer," *Proc. IRE*, vol. 40, May 1952.

[11] Freeman, H., and E. Parsons, "A Time-Sharing Analog Multiplier," *IRETEC*, vol. 3, March 1954.

[12] Hirsch, C. J., and F. C. Hallden, "Computing Techniques for the Sampling Parametric Computer," *IRETEC*, vol. 6, June 1957.

[13] Andrews, J. M., "Mathematical Applications of the Dynamic Storage Analog Computer," *Proc. Western Joint Comp. Conference*, 1960.

[14] Miura, T., and J. Iwata, "Combined Computation with Slow-Type and Repetitive Analog Computers," *AICA*, vol. 6, July 1964.

[15] "5800 DYSTAC® Iterative Analog Computer, Applications and Specifications," manual published by Computer Systems, Inc., Monmouth Junction, New Jersey.

[16] Hausner, A., "Generating Functions with Integrators," *Harry Diamond Labs. Report TR-1341*, Wash., D. C., March 1967.

[17] Jury, S. H., "High Speed Memory Analog Computer," *Ind. and Eng. Chem.*, vol. 53, March 1961.

[18] Jury, S. H., "Memory and Function Generation," *Military Systems Design*, vol. 6, Jan.-Feb. 1962.

[19] Fisher, M. E., "On the Continuous Solution of Integral Equations by an Electronic Analogue. I," *Proc. Cambridge Phil. Soc.*, vol. 53, part 1, Jan. 1957.

[20] Fisher, M. E., "Proposed Methods for the Analog Solution of Fredholm's Integral Equation," *Journal ACM*, vol. 5, Oct. 1958.

[21] Tomović, R., and N. Parezanović, "Solving Integral Equations on a Repetitive Differential Analyzer," *IRETEC*, vol. 9, Dec. 1960.

[22] Herschel, R., and A. Kley, "The Solution of Integral Equations of the Second Kind by Iterative Analog Computation," *Proceedings of the 4th Inter. Analogue Comp. Meetings* (Brussels: Presses Académiques Européennes, 1964).

[23] Berthiaume, P., "Analog Computer Storage Techniques and Applications," *Princeton Computation Center Report*, Electronics Associates, Inc., Princeton, N. J., 1961.

[24] Holst, P. A., "Iterative Analog Computations," *AICA*, vol. 6, April 1964.

[25] Schulte, C. W., "Use of Logical Elements to Simulate Hysteresis," *Simulation*, vol. 4, June 1965; corrected vol. 5, Nov. 1965.

[26] Hedges, F., "An Automatic Nyquist Plotter Using Track and Store Techniques," *Simulation*, vol. 6, April 1966.

For other information and applications, see:

[27] Elgerd, O. I., "An Analog Computer Study of the Transient Behavior and Stability Characteristics of Serial-Type Digital Data Systems," *Trans. AIEE, Part I* (*Comm. and Elect.*), vol. 77, July 1958.

[28] de Winne, A. L., "On the Use of Memory Circuits in Analogue Computers," *AICA*, vol. 4, April 1962.

[29] Anderson, T. C., "A Method for Evaluating Stieltjes Integrals on the Analog Computer," *IRETEC*, vol. 11, Aug. 1962.

[30] Godfrey, W. L., and R. D. Benham, "Analog Simulation of a Counter-Current Crystallization Process," *Simulation*, vol. 4, Jan. 1965.

[31] Brock, P., and S. Sherman, "Application of REAC Equipment to the Solution of Problems Involving Integral Operators, I," *Symposium I on REAC Techniques*, Reeves Instr. Corp., New York, March 15–16, 1951 (ASTIA No. 108488).

[32] Nelson, J. M., and S. Fried, "Solution of Bounded Fourier Integrals on the DYSTAC® Computer," *Eastern Simulation Council* talk, Dayton, N. J., Dec. 12, 1960.

[33] Hausner, A., "Simple Analog Computer Oscilloscope Displays," *Simulation*, vol. 3, Dec. 1964.

[34] Gilly, W. A., "A Reliable, Straightforward Technique for Generating Bode and Nyquist Diagrams," *Simulation*, vol. 8, May 1967.

[35] Gilliland, M. C., and H. H. Hara, "Use of Analog Memory for Simulation of a Melting Slab," *Instr. and Control Systems*, vol. 33, Sept. 1960.

Chapter 19

ANALOG/HYBRID COMPUTERS II: OPTIMIZATION TECHNIQUES

This chapter is a continuation of the previous one in that rep-op with mode control and memory will be used but with the primary aim of parameter optimization. It is the most fruitful use of iterative techniques. We shall see that a wide variety of iterative computations can be done with such computers, even without the gated digital elements to be discussed in Chapter 20.

19.1 Ordinary OptimizationTechniques [1]–[3]

The concept of seeking "best" values of parameters $a_1, a_2, \ldots, a_n$ in a given system has been used frequently in this book. In Chapters 7–10, approximation theory was applied to generating functions by selecting parameters to minimize some distance function. Chapter 11 was concerned with the solution of algebraic equations; the variables themselves were considered parameters, and an artificial norm was created from each equation—one that could be expected to go to zero—and the norm was minimized continuously to find the solution. Programming problems were similar except that constraints were included and the norm instead was some positive function of the variables. The method of steepest descent was used frequently there and in Chapter 13 where continuous systems were constantly corrected to minimize some norm. Included was the general problem of identification of physical systems. Eigenfunction determination (Chapter 14) and complex variable roots (Chapter 15) are other examples of forcing an artificial norm to zero to get "best" answers.

It is clear that such problems permeate all physical systems. It is rare to use a computer for the sole purpose of solving equations; optimizing the system is the more fundamental task. Further, the examples given previously form a subset of the types of optimization problems existing as well as the techniques to solve them. This section is included before we attempt to use memory or gated logic to optimize systems, to better understand the scope of the general field known as optimization. It is by no means a comprehensive discussion. Many texts are available that present a broader view in depth (e.g., [3]).

19.1.1 Problem statement

In broad terms, there are two classes of optimization problems. In the first, the parameters sought must be all constant (*ordinary* optimization) while in the second, at least one parameter is functionally related to other variables in the system (*functional* optimization). Both classes of problems require minimizing a distance function $\rho(a_1, a_2, \ldots, a_n)$ of n parameters a_i. If ρ_1 and ρ_2 are found to be functional values for distinct sets A_1 and A_2 of parameters and $\rho_1 < \rho_2$, we say that the set of parameters A_1 produces a better system than the set A_2. A best system is one in which the set of parameters A^* produces ρ^* such that $\rho^*(A^*) \leq \rho(A)$ for all other $A \neq A^*$. Existence or uniqueness of ρ^* and A^* is not generally guaranteed. If ρ is to be maximized, we shall equivalently minimize $-\rho$. The function ρ is known by the terms criterion, objective, error, utility, cost, profit, payoff, system, etc.

In addition to ρ, there may be $m(<n)$ equality relations

$$f_j(a_1, a_2, \ldots, a_n) = 0, \qquad j = 1, 2, \ldots, m \tag{19.1}$$

and q inequality constraints (region limitations)

$$g_k(a_1, a_2, \ldots, a_n) \leq 0, \qquad k = 1, 2, \ldots, q \tag{19.2}$$

as in linear and nonlinear programming problems (Chapter 11).

Optimization problems are further delineated into types. If ρ, f_j, and g_k can all be evaluated instantaneously for a given set of a_i, it is a *nondynamic* problem—otherwise a *dynamic* problem.

The problem is a *stationary* one if uncontrollable parameters do not change with time. Otherwise it is a *tracking* problem.

We consider ordinary optimization here; functional optimization is treated separately in Sec. 19.6.

19.1.2 Eliminating constraints

Constraint equations (19.1) and (19.2) pose the most difficulty in optimization solutions. Before an optimization problem is solved, careful consideration should be given to methods of insuring that the constraint equations, if any, are satisfied. Two major methods of doing this are widely used:

1. Reduction of parameters (equality constraints only). It may be possible to solve for some of the n parameters in terms of the others from the m constraint equations, particularly in nondynamic problems. By substitution into ρ and the g_k, the dimension of the problem is reduced, possibly as small as $n - m$. This is the most desirable method of treating equality constraints but often the least tractable one. If any of the parameters can be eliminated easily, it is usually advisable to do so. Sometimes it is possible to reduce ρ sequentially by making changes in $n - m$ parameters while solving for the remaining m parameters from (19.1) simultaneously.

2. Penalty function method. Change the payoff function to

$$\rho' = \mu\rho + a\{p(f_1, f_2, \ldots, f_m) + r[g_1 U(g_1), g_2 U(g_2), \ldots, g_q U(g_q)]\}$$

where p and r are positive definite functions of their arguments and $a > 0$ with $a/\mu \gg 0$. In many cases, we may show

$$\min\,(\lim_{a/\mu \to \infty} \rho') = \min \rho$$

such as in linear programming problems where $p + r = \sum_{j=1}^{m} f_j^2 + \sum_{k=1}^{q} g_k^2 U(g_k)$, or always if $q = 0$ and $p = \sum_{j=1}^{m} f_j^2$. This has the tremendous advantage of reducing a constrained problem to an unconstrained one, but requires a sequence of minimization problem solutions allowing $a/\mu \to \infty$. This is too time consuming for tracking problems.

A third method involving Lagrange multipliers is also useful when inequality constraints have been eliminated. We have used such a method when solving Lagrange's equations with constraints in Chapter 12. It was tractable there because the $n + m$ resulting simultaneous algebraic equations were linear and could be solved by the direct method. That was also a continuous tracking problem in which method 1 was assumed not applicable. We shall discuss this method more generally when looking at the gradient technique for solution.

19.1.3 Techniques of solution for nondynamic problems

There are three basic techniques for solution, each modifiable into very many subtechniques.

1. Random search [4]–[5]. Parameters are randomly varied within a given region in a sequence of trials, starting with initial values. If $\rho_k < \rho_{k-1}$ after trial k, the new $(a_i)_k$ are retained; otherwise the old $(a_i)_{k-1}$ are stored. Variations to this basic method include using different types of distributions, randomly varying a small change $(\Delta a_i)_k$ in the $(a_i)_k$, randomly selecting the sign for a constant small change $(\Delta)_k$ in the $(a_i)_k$ (random walk), reducing ρ by varying one parameter at a time (relaxation), biasing the randomness in directions of past successes (quasirandom), etc., together with combinations of these.

All equality constraints must be accounted for prior to a random search. The method is not efficient in that many trials are wasted. It cannot be used for tracking problems since convergence to solutions is too slow. But among the advantages of this method, which should be considered before its rejection for a particular problem, are its complete unconcern for bumps, ridges, discontinuities, etc., that may be present in the hypersurface of ρ in the parameter space and its ease of mechanization. Both cannot be overlooked. The former difficulty defeats many other schemes, including the powerful gradient method, while the latter is an important time saver in problem preparation. Many schemes start off with a random search to insure an initial closeness of the solution when other techniques are used.

Techniques that compare the payoff function produced by two sets of parameters and select the set producing the better one are called *direct* techniques. Random search is a pure direct technique.

Note also that changes in the parameters are made in *discrete* steps, as opposed to *continuous* changes possible with the gradient method (method of steepest descent), discussed below.

Finally, we note that testing ρ for many combinations of parameter sets in a systematic way (covering the n-dimensional space with a grid of points) is in essence a random search. This technique has been used when n is small. It is extremely inefficient but does not require random number generators and is even simpler to mechanize than a genuine random search.

2. *The continuous gradient method* (*the method of steepest descent*). This method owes its existence to the following important but simple theorem [3].

Theorem 19.1. A minimum of a continuous function $\rho(a_1, a_2, \ldots, a_n)$ occurs only at an interior point of a region where

$$\frac{\partial \rho}{\partial a_i} = 0, \qquad i = 1, 2, \ldots, n \tag{19.3}$$

(called a *stationary* point), or where at least one of the partial derivatives is not continuous or does not exist, or at a boundary of the region.

Note that Theorem 19.1 is only a necessary condition for a minimum. Stationary points could define maximums, relative or local minimums or maximums (extremums in a small region about the point), or saddle points (similar to points of inflection). It is usually obvious whether or not a stationary point is a maximum, minimum, or saddle point, but there are analytical tests to confirm intuition [3].

If we find ourselves at the "bottom of a hill," we can only be sure that we are at a local minimum. In an optimization problem, interest is usually in obtaining a global minimum (the minimum of all local minimums and all points of nonexistent derivatives). This usually complicates procedures.

Note that the Newton–Raphson procedure (Sec. 11.3.5) for solving for the set of stationary interior points with (19.3) requires the evaluation of the Jacobian of the partial derivative vector. This contains n^2 elements $\partial^2 \rho/\partial a_i \partial a_j$, which we very much wish to avoid calculating.

Inequality constraints must be accounted for. This can be done by the penalty function method as part of the problem mechanization or can be dealt with specially when violated during the problem solution. The penalty function is probably the best way to handle these constraints. After the solution is obtained, we may observe which constraints are slightly violated and which are not. Since the solutions lie on those edges where constraints are slightly violated, we may reprogram the problem to treat violated inequality constraints as equalities (equal to 0), ignoring the others.

Equality constraints, on the other hand, must be accounted for before the gradient method is mechanized. Theorem 19.1 assumes all a_i independent—not true when equality constraints exist. The method of Lagrange multipliers can produce an augmented set of necessary conditions [3].

Theorem 19.2. A minimum of a continuous function $\rho(a_1, a_2, \ldots, a_n)$ subject to the continuous function constraints

$$f_j(a_1, a_2, \ldots, a_n) = 0, \qquad j = 1, 2, \ldots, m \tag{19.4}$$

occurs only at points of the interior of the region where there exists a set of m constants λ_j (called Lagrange multipliers) such that

$$\frac{\partial \rho}{\partial a_i} + \sum_{j=1}^{m} \lambda_j \frac{\partial f_j}{\partial a_i} = 0, \qquad i = 1, 2, \ldots, n \tag{19.5}$$

or where at least one of the partial derivatives $\partial \rho/\partial a_i$ or $\partial f_j/\partial a_i$ does not exist or is not continuous, or at a boundary of the region where (19.4) is satisfied.

Equations (19.4) and (19.5) represent $n + m$ equations (possibly nonlinear), an added complexity. But (19.5) implies we may attempt to minimize an augmented payoff function

$$\rho' = \rho + \sum_{j=1}^{m} \lambda_j f_j$$

simply by considering the λ_j as additional parameters. We have attacked many nondynamic minimization problems (both stationary and tracking) by a continuous steep-descent path, using the equations

$$\dot{a}_i = -g_i \frac{\partial \rho}{\partial a_i}, \qquad i = 1, 2, \ldots, n \tag{19.6}$$

$g_i > 0$, when no equality constraints exist. When equality constraints are present, we can use the modified form

$$\begin{aligned} \dot{a}_i &= -g_i\left(\frac{\partial \rho}{\partial a_i} + \sum_{j=1}^{m} \lambda_j \frac{\partial f_j}{\partial a_i}\right), \qquad & i = 1, 2, \ldots, n \\ \dot{\lambda}_j &= -h_j f_j, & j = 1, 2, \ldots, m \end{aligned} \tag{19.7}$$

$g_i, h_j > 0$. Both systems (19.6) and (19.7) insure "rolling down the hill" along the continuous gradient vector of ρ and ρ'.

The process of minimizing ρ by solving for the interior stationary points is called an *indirect* method. The continuous gradient method is a pure indirect method, except for comparing the final ρ's obtained by varying initial conditions to ascertain the global minimum.

3. *Reordering method.* When ρ was a positive function of a more fundamental set of "errors," $f_j(a_1, a_2, \ldots, a_n)$, $j = 1, 2, \ldots, n$, as in the solution of algebraic equations (Chapter 11), we found it more economical of equipment to program some direct or reordered method

$$\dot{a}_i = g_i f_j \qquad \text{or} \qquad \dot{a}_i = g_i(-f_j), \qquad g_i \gg 0$$

for n pairs of i and j, each selected once from the integers between 1 and n. This procedure is applicable because we knew each f_j had to go to zero, but unlike the gradient or random search methods, stability is a problem. The method is attractive because it requires the least amount of equipment.

19.1.4 Types of and solutions for dynamic problems

By definition, a dynamic optimization problem necessarily involves delay in evaluating the payoff function. But the same three methods, sometimes modified, can be used for such problems.

1. *Random search [6]–[7].* No change is required for random search mechanizations essentially sequential and discrete anyway. We shall mechanize a random search in this chapter, using point memories.

2. *Discrete gradient method [8]–[10].* Any delay in evaluating ρ usually means a delay in evaluating its derivatives. For example, if

$$\rho = \int_a^b f(x, a_1, a_2, \ldots, a_n)\, dx$$

then

$$\frac{\partial \rho}{\partial a_i} = \int_a^b \frac{\partial f(x, a_1, a_2, \ldots, a_n)}{\partial a_i} \, dx \tag{19.8}$$

which is another integral. Another situation frequently arising is a payoff function of the form

$$\rho = \int_a^b f(x_1, x_2, \ldots, x_s) \, dt$$

where $x_1, x_2, \ldots, x_s$ are part of the system of s differential equations

$$\begin{aligned} \frac{dx_1}{dt} &= f_1(x_1, x_2, \ldots, x_s, a_1, a_2, \ldots, a_n) \\ \frac{dx_2}{dt} &= f_2(x_1, x_2, \ldots, x_s, a_1, a_2, \ldots, a_n) \\ &\vdots \\ \frac{dx_s}{dt} &= f_s(x_1, x_2, \ldots, x_s, a_1, a_2, \ldots, a_n) \end{aligned} \tag{19.9}$$

The gradients may still be evaluated exactly, since

$$\frac{\partial \rho}{\partial a_i} = \int_a^b \left(\sum_{j=1}^{s} \frac{\partial f}{\partial x_j} \frac{\partial x_j}{\partial a_i} \right) dt, \qquad i = 1, 2, \ldots, n \tag{19.10}$$

where $\partial x_j / \partial a_i$ are recognized as the parameter influence coefficients discussed in Chapter 17, and could be found simultaneously, using much equipment. In such cases, system (19.6) can still be used, provided memory was used to store each $\partial \rho / \partial a_i$ after each run and g_i was sufficiently small. Difficulties arise, however, because the a_i change during the solution of the differential equations. It is more common to keep each a_i constant while the differential equations are being solved and change a_i in the direction of the gradient by discrete amounts after the solution is obtained. The idea is this:

(a) Hold all a_i constant, determine and store the required $\partial \rho / \partial a_i$ one or all at a time.

(b) Change the a_i at step k so that ρ decreases in the direction of its gradient, i.e., let

$$(a_i)_k = (a_i)_{k-1} - g \left(\frac{\partial \rho}{\partial a_i} \right)_{k-1}, \qquad \begin{aligned} i &= 1, 2, \ldots, n \\ k &= 1, 2, \ldots \end{aligned} \tag{19.11}$$

where $g > 0$ is called the *step size*. Instead of "rolling down a hill," we "jump" in the direction we know is locally the steepest.

Step-size selection is the most crucial aspect of using this technique. If we take small jumps, a large number of iterations may be required for convergence.

If we take large jumps, we may overshoot the bottom of the hill, finding ourselves at a higher point than where we started. Unfortunately, some trial procedure is generally necessary except if it is known that ρ will go to 0 (see Sec. 19.5.3). There are, however, limits to the rate of convergence [33].

A major variation of this technique consists of relaxation, finding one gradient component $\partial\rho/\partial a_i$ at a time and minimizing ρ by adjusting a_i [11]. When this is accomplished, the process is repeated for a new parameter and cycled over all parameters. For such a case, the gradient really is not needed, since only one of two directions decreases ρ when $\partial\rho/\partial a_i \neq 0$. This process is therefore called a *discrete quasigradient method by relaxation*, but might equally be called a *quasirandom search by relaxation.* We shall illustrate manual methods in this chapter; automatic methods are deferred until Chapter 20 because much logic is needed besides memory.

Other variations depend strongly on how the gradient vector is obtained. If all components are obtained in one run through the use of sensitivity equations, no gated logic is needed and rapid convergence results. The price paid is the requirement of an exorbitant amount of equipment, except in the case of linear systems.

If gradient components are obtained one at a time and stored until all are available, less equipment but more logic is required and convergence is slower. Because it takes more time to obtain the gradients, more efficient use of the information is made with the "optimum gradient method" [12]. This selects that $g = g_k$ which minimizes ρ during the kth step using the partial derivatives stored. It requires an iteration within an iteration but guarantees convergence to a local minimum in the overall problem while reducing the problem to a sequence of one-dimensional searches. Each $\partial\rho/\partial a_i$ is kept constant during the search for g_k. The optimum gradient method is illustrated in Chapter 20, where the gradients are obtained by parameter perturbations using the difference approximation

$$\frac{\partial\rho}{\partial a_i} \simeq \frac{\rho(a_i + \Delta a) - \rho(a_i - \Delta a)}{2\Delta a} \tag{19.12}$$

which has errors of order $(\Delta a)^2$. Alternately, a first-order error approximation

$$\frac{\partial\rho}{\partial a_i} \simeq \frac{\rho(a_i + \Delta a) - \rho(a_i)}{\Delta a}$$

can be used.

Other methods, notably a correlation method [13]–[14] for computing the gradients have been devised but are not discussed in this book.

3. Discrete reordering method [15]–[20]. In boundary-value problems, there are n functions f_j to be driven to 0, but each can be determined only by solving the set of differential equations. But there is a direct analogy between this and the continuous case. In the discrete reordering method we attempt a scheme

$$(a_i)_k = (a_i)_{k-1} + g_i(\pm f_j)_{k-1} \tag{19.13}$$

to avoid generation of partial derivatives. Stability remains a problem, but the (i, j) pairs and the g_i can be determined by trial just as in the continuous case discussed in Chapter 11. The advantage is that no logic is required and minimal equipment is used. Once a stable arrangement is obtained, it remains stable for wide variations of parameters involved in the f_j. This method is discussed thoroughly in this chapter.

19.2 Manual Optimization by Relaxation with Memory

For large numbers of parameters, automatic optimization of parameters requires much equipment. If the problem is stationary, the process can be carried out manually. Suppose an approximation $f_a(t, a_1, \ldots, a_n)$ to a function $f(t)$ is required for $0 \leq t \leq t_f$, and both f_a and f are generable with good accuracy on the computer. An absolute error distance function is best suited for this purpose, since the voltages remain large. If we let

$$\rho = \int_0^{t_f} |f - f_a| \, dt$$

each parameter a_i is successively adjusted to reduce ρ. The process is repeated until little or no change results, i.e., $\partial\rho/\partial a_i = 0$ for all i. Ridges and valleys in the hypersurface $\rho(a_1, a_2, \ldots, a_n)$ can produce local minimums as usual. Also, the convergence rate is often slow. Nevertheless, it remains a powerful technique because of its simplicity.

By doing the problem in rep-op, ρ may be stored after each rep-op cycle and read directly from a DVM. The operator adjusts each parameter cyclically from this observed value. Ten solutions per second are sufficient for this purpose.

Example 19.1. To illustrate the manual relaxation procedure, we attempt to fit a fourth-order transfer function to yield an optimum time delay for a step function. Using Beck's method and variable scaling, the circuit is programmed as shown in Fig. 19.1. There are nine parameters which may be adjusted initially to yield a Padé all-pass approximant:

$$a_1 = a_2 = a_3 = a_4 = a_5 = 1$$

$$a_6 = \frac{20}{T}, \qquad a_7 = \frac{9}{T}, \qquad a_8 = \frac{4.6667}{T}, \qquad a_9 = \frac{2}{T}$$

For convenience, we choose $T = 2$ and we let $t_f = 8$ s. If done at slow speed, 8 s of computation time is required. With 0.01-μF feedback capacitors, 80 ms are needed for t_o—just about right.

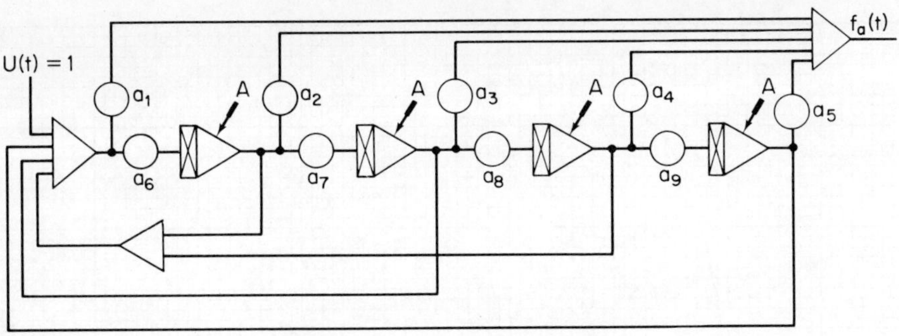

FIG. 19.1 Time delay model.

The rest of the associated circuitry is shown in Fig. 19.2. Since the input is $U(t)$, $f(t) = U(t - 2)$ is generated by switching from a t ramp after 20 ms. Generating the absolute value of $U(t - 2) - f_a$, integrating, and storing are then straightforward. Because of noise and other errors, a simple first-order, real-time filter may be used as shown.

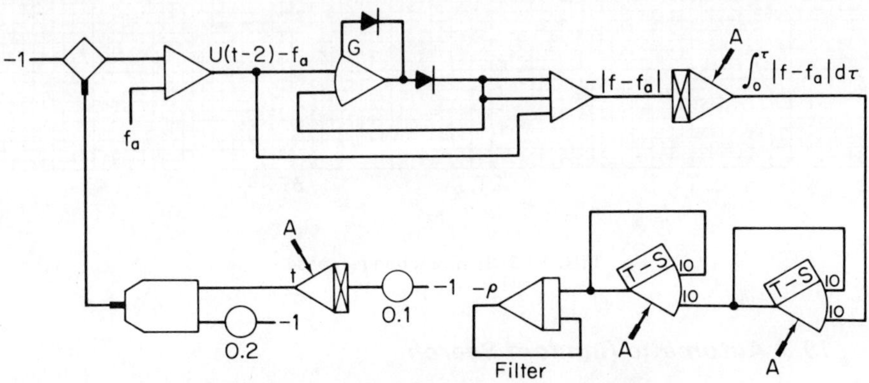

FIG. 19.2 Associated circuitry for manual relaxation.

The manual relaxation procedure took about 1 hr in this case, requiring about 25 adjustments for each parameter. Final pot values became

$$a_1 = 0.1752, \quad a_2 = 0.3142, \quad a_3 = 0.4566$$
$$a_4 = 0.6979, \quad a_5 = 1.0000, \quad a_6 = 0.4945$$
$$a_7 = 0.4028, \quad a_8 = 0.1880, \quad a_9 = 0.8795$$

with the output of the delayed step function shown in Fig. 19.3.

Manual relaxation as an optimization procedure, although tedious and time-consuming, requires the least amount of equipment of all methods. We shall return to this method and program it for an automatic variations in Chapter 20.

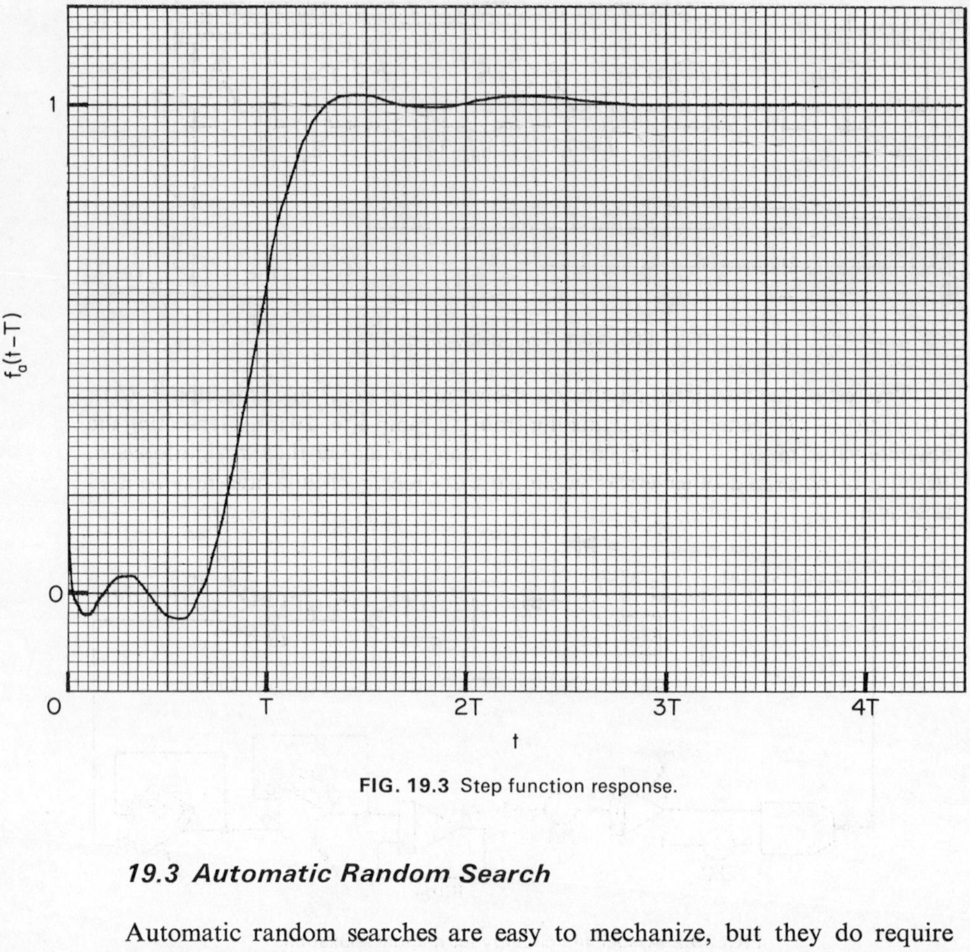

FIG. 19.3 Step function response.

19.3 Automatic Random Search

Automatic random searches are easy to mechanize, but they do require noise sources for generating random numbers, and special circuits that update the values of each parameter when successful runs are detected. A relaxation process using a random search requires more logic and will not be discussed. Besides, it is just as easy to mechanize an n-parameter random search by varying all parameters at a time, provided n random noise sources are available.

Noise generators, with selectable normally distributed outputs are usually available as part of the computer system. Lowe [21] suggests, however, that a reasonably good, uniformly distributed, pseudo-random set of numbers can be generated by sampling a high-frequency sweep generator. This procedure will be used for obtaining random excursions from a nominally existing set of $(a_i)_{k-1}$. In Fig. 19.4, a sampled noise and parameter perturbation scheme is shown.

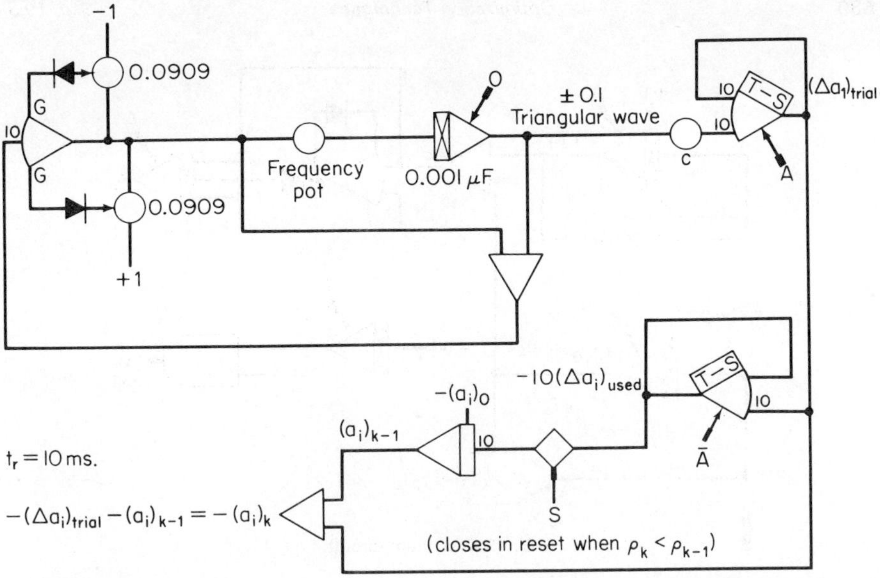

FIG. 19.4 Noise generator and parameter variation circuit.

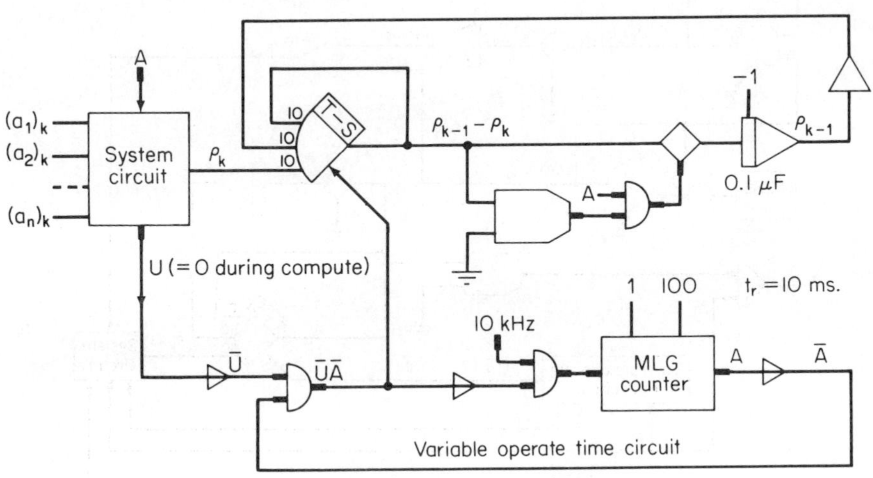

FIG. 19.5 Timing and ρ-updating circuit.

The ratio of the period of the triangular wave and the rep-op period should be irrational for a truly nonrepeating set of stored values of trial Δa_i. In practice, good results are obtained as long as the set of stored values are reasonably "jumpy" and this can be trial-set with the frequency pot. As shown, Δa_i can be as large as 0.1 and made less by setting $c < 1$. Its value is held constant during the compute period ($A = 0$) and 10 times its value is stored in the reset

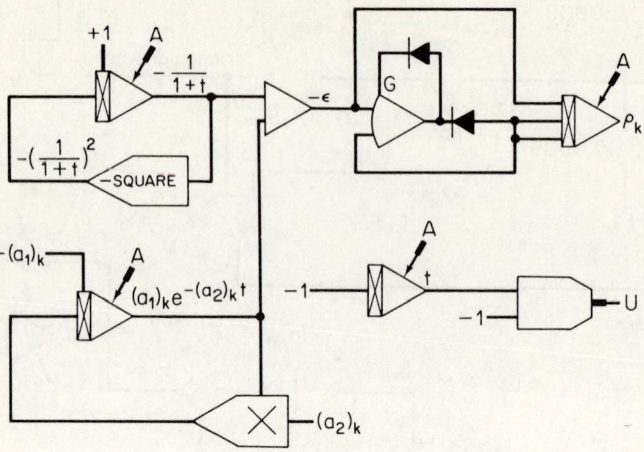

FIG. 19.6 System circuit.

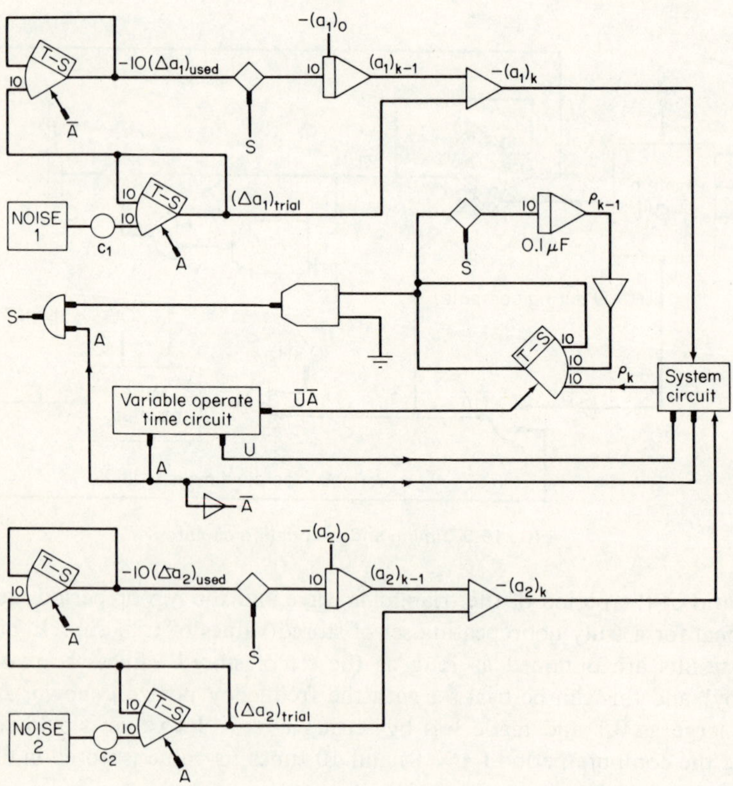

FIG. 19.7 Random search circuit for Example 19.2.

period for updating. We shall create a signal S, high in reset whenever $\rho_k < \rho_{k-1}$, and shall choose the reset time to be 10 ms. This procedure precisely updates the $(a_i)_k$ to the value $(a_i)_{k-1} + (\Delta a_i)_{\text{trial}}$ in Fig. 19.4. The constant $(\Delta a_i)_{\text{trial}}$ is integrated for 10 ms at a gain of 100, and effectively adds the constant to $(a_i)_{k-1}$. Such a circuit is called an *accumulator* circuit, and is useful for other optimization schemes.

Accumulator circuits use integrators with large capacitors (1 or 0.1 μF) when their outputs are not updated frequently. This prevents serious drift from capacitor leakage.

Figure 19.5 shows the rest of the general schematic for a random search mechanization. A variable operate time circuit (Fig. 18.9) produces the drive signal A, which controls all integrators in the system circuit, and the signal $\bar{U}\bar{A}$ which allows the track-store unit to track $\rho_{k-1} - \rho_k$ during t_o until $U = 1$, and forces it to store the value during reset. If $\rho_{k-1} - \rho_k$ is positive during reset, a successful run has been made that forces S high and updates each $(a_i)_{k-1}$ to the used value of $(a_i)_k$. The distance function ρ is also updated with an accumulator circuit; its initial condition of 1 guarantees an updating after the first run.

Example 19.2. Random Search. Let us determine the parameters a_1 and a_2 so that

$$\rho = \int_0^1 \left| a_1 e^{-a_2 t} - \frac{1}{1+t} \right| dt$$

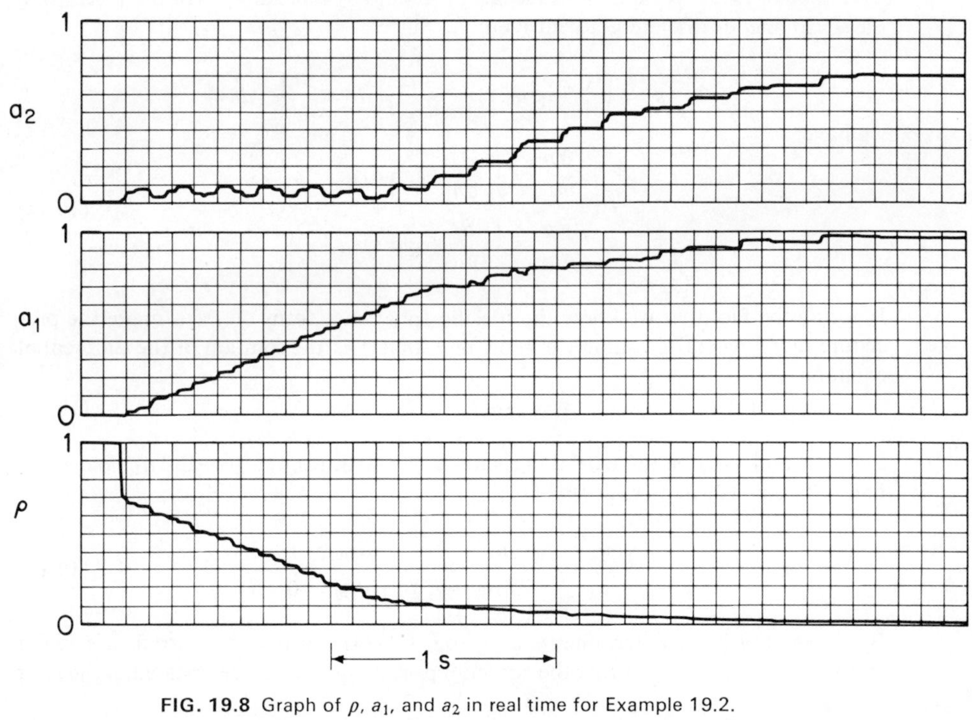

FIG. 19.8 Graph of ρ, a_1, and a_2 in real time for Example 19.2.

is minimized. The system circuit is trivial enough, as shown in Fig. 19.6. It must be supplied $(a_1)_k$ and $(a_2)_k$ to produce ρ_k.

Figure 19.7 shows the circuit for the automatic random search, while Fig. 19.8 shows the graph of ρ, a_1, and a_2 in real time when $|\Delta a_i|$ was kept less than 0.05. The method is extremely powerful for analog computers because the convergence is fairly rapid with very little equipment.

19.4 Simultaneous Discrete Gradient Optimization

Very fast convergence and high accuracy are obtained by finding the partial derivative vector simultaneously with the system. For the general case (Sec. 19.4.1), this requires about $n + 1$ times the analog computer components for an n-parameter system. For linear feedback control systems (Sec. 19.4.2), exactly twice the number of analog computer components are often required, independent of the number of parameters.

19.4.1 The general case

In the general case, we find $(\partial\rho/\partial a_i)_{k-1}$ simultaneously for all i during the operate period and mechanize (19.11) during the reset period with accumulator circuits, finding g by trial.

Example 19.3. We demonstrate this procedure by repeating Example 19.2, using the simultaneous discrete gradient method. Since

$$\rho = \int_0^1 \left| f(a_1, a_2, t) - \frac{1}{1+t} \right| dt = \int_0^1 |\epsilon|\, dt$$

we have

$$\begin{aligned} \frac{\partial\rho}{\partial a_1} &= \int_0^1 \frac{\partial f}{\partial a_1}(\text{sign } \epsilon)\, dt \\ \frac{\partial\rho}{\partial a_2} &= \int_0^1 \frac{\partial f}{\partial a_2}(\text{sign } \epsilon)\, dt \end{aligned} \tag{19.14}$$

Ignoring the fact that we know the specific functional form of f (to make the procedure more general), assume we know only that f is the solution of the differential equation

$$\dot{f} = -a_2 f, \qquad f_0 = a_1 \tag{19.15}$$

Let $u_1(t)$ and $u_2(t)$ be the partial derivatives of f with respect to a_1 and a_2. We easily find (Chapter 17) that (with $i = 1$)

$$\begin{aligned} \dot{u}_1 &= -a_2 u_1, \qquad & (u_1)_0 &= 1 \\ \dot{u}_2 &= -a_2 u_2, - f, \qquad & (u_2)_0 &= 0 \end{aligned} \tag{19.16}$$

By solving (19.15), f determines ϵ, and from (19.14)-(19.16), the partials are found and stored when $t = 1$ during the compute period, and the a_i are updated in the reset period. New partials are found after every cycle, making it unnecessary to worry about optimumly choosing g. If the a_i oscillate, g is too large. If convergence takes a long time, g is too small. We may easily control g by varying the reset period and choose one by trial—even varying it manually during the computation. Note that ρ need not be calculated; this is a pure indirect method that drives the partials to zero.

Figure 19.9 is the circuit diagram to solve the problem by generating the parameter influence coefficients. Figure 19.10 shows three runs with $g = 0.05$, 0.06, and 0.07 and includes the partials $\partial \rho/\partial a_i$ which were stored (not shown in Fig. 19.9). We see why the choice of g is so important in this problem. Initially large g (about 0.25)

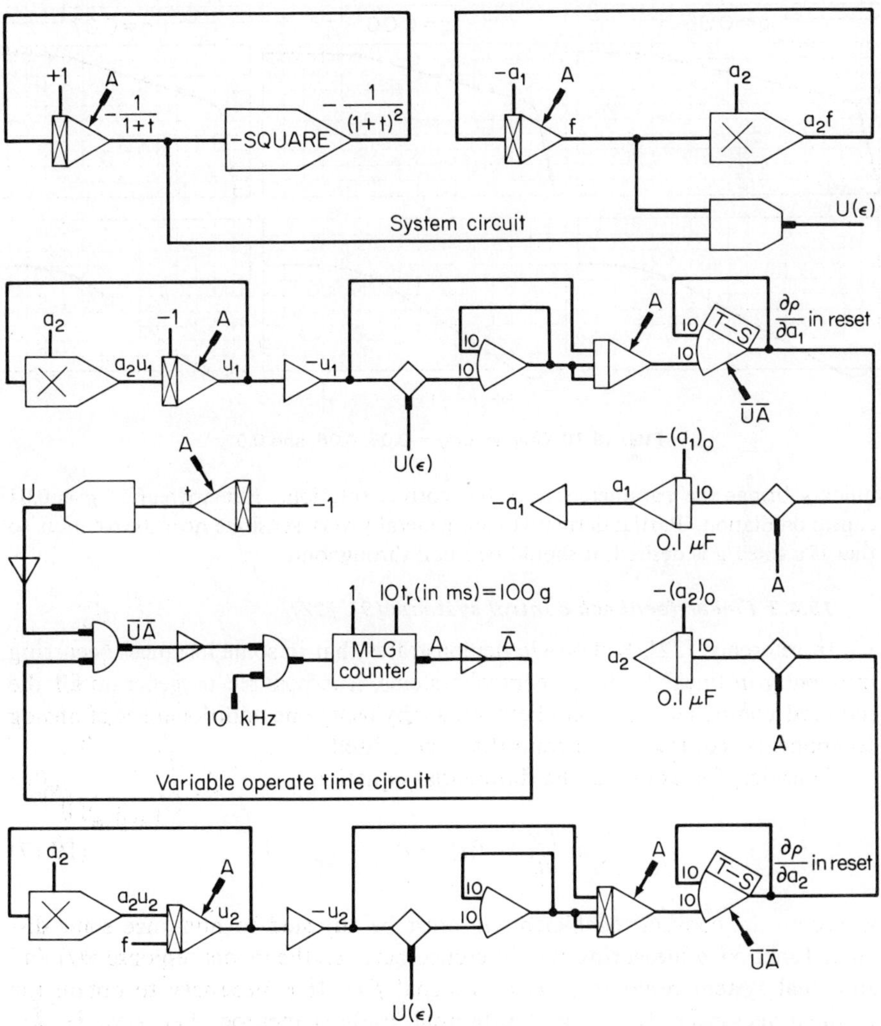

FIG. 19.9 Diagram for using parameter influence coefficients.

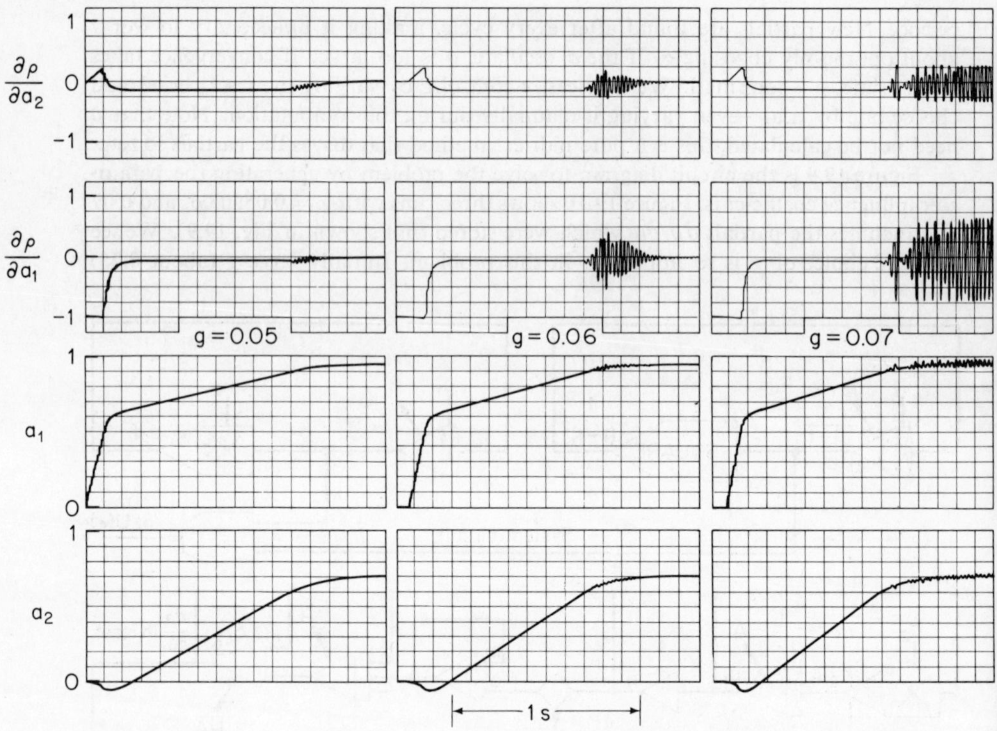

FIG. 19.10 Graphs for $g = 0.05$, 0.06, and 0.07.

quickly brings the parameters near the correct solution. But thereafter $g = 0.07$ causes oscillation. Partial derivatives are generally very sensitive near the solution, so that if a fixed g is desired, it should be small throughout.

19.4.2 Linear feedback control systems [9], [22]

In reference [22], Kokotović demonstrates that in some instances occurring frequently in linear feedback control systems, it is possible to generate all the required components of a gradient vector by using one additional set of analog components required to generate the system itself.

Consider, for example, the differential equation

$$\sum_{i=0}^{n} a_i \frac{d^i x}{dt^i} + f(t) = 0, \qquad a_n = 1 \tag{19.17}$$

where all the constant coefficients a_i must be adjusted to minimize some distance function ρ measuring the difference between the model response $x(t)$ and an actual system response $x_a(t)$ to a signal $f(t)$. It is necessary to obtain the gradient vector $\partial x/\partial a_i$ to use the discrete gradient method. Let $u_j = \partial x/\partial a_j$. Then, by differentiating (19.17) with respect to a given a_j, we have

$$\sum_{i=0}^{n} a_i \frac{d^i u_j}{dt^i} + \frac{d^j x}{dt^j} = 0 \tag{19.18}$$

for each j. Now, consider the solution of

$$\sum_{i=0}^{n} a_i \frac{d^i z}{dt^i} + x(t) = 0 \tag{19.19}$$

Differentiate (19.19) j times with respect to t:

$$\sum_{i=0}^{n} a_i \frac{d^{(i+j)} z}{dt^{(i+j)}} + \frac{d^j x}{dt^j} = 0 \tag{19.20}$$

and compare (19.18) and (19.20). We see that

$$\frac{d^j z}{dt^j} = u_j$$

But, a standard mechanization of (19.19) makes all derivatives of z available as integrator outputs, as shown in Fig. 19.11. Of course, the pots in this figure must be replaced by multipliers in any optimization scheme.

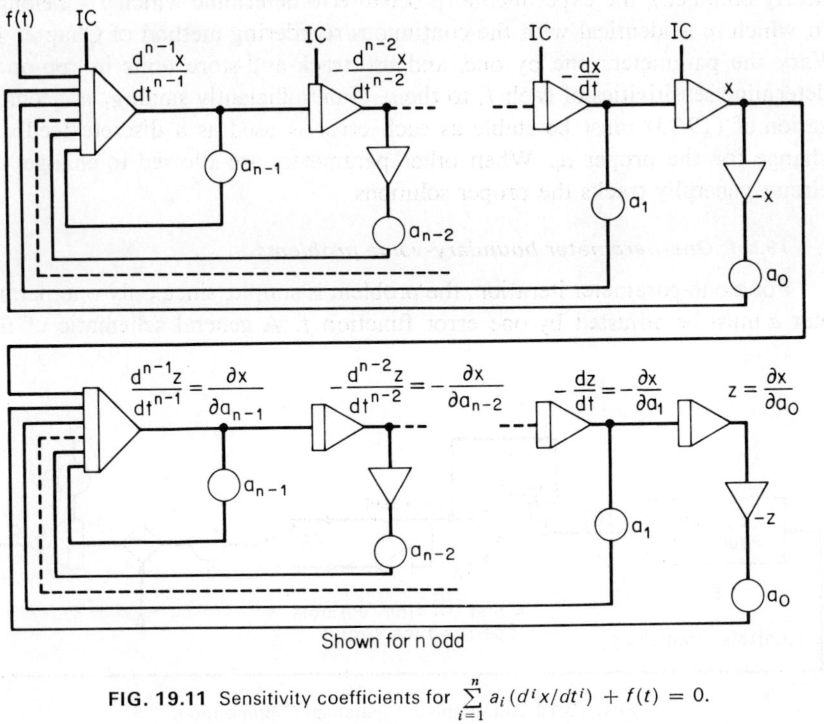

FIG. 19.11 Sensitivity coefficients for $\sum_{i=1}^{n} a_i (d^i x/dt^i) + f(t) = 0$.

The derivation holds for any initial conditions on x and its derivatives, but each $(u_j)_0$ must equal zero. Note that this procedure is not applicable for initial condition parameters of (19.17).

The general multiloop linear feedback system may require additional equipment, depending on how the parameters enter in each transfer function block. Interested readers are referred to Kokotović's development in reference [22].

19.5 Discrete Reordering Iteration for Boundary-Value Problems

A special case in ordinary optimization problems arises in the solution of algebraic equations, boundary-value problems, etc.; we artificially create a distance function that goes to 0 when the solution is obtained. Remember that ρ then consists of a sum of positive terms each involving one of a more fundamental set of errors f_j, all of which must go to zero. Previous methods described in Secs. 19.3 and 19.4 have been used for such problems, but it is reasonable to expect a simpler system by not introducing a gradient vector. Instead, use the f_j directly to optimize the a_i, according to (19.13). The major difficulty with this method is in obtaining a first solution. But if one is somehow obtained (or nearly obtained), the experimental procedure to determine which f_j "belongs" to which a_i is identical with the continuous reordering method of Chapter 11. Vary the parameters one by one, and use track-and-store units in rep-op to determine sensitivities of each f_j to the a_i. For sufficiently small g_i, a mechanization of (19.13) must be stable as each error is used as a discrete feedback change for the proper a_i. When other parameters are allowed to change, the circuit generally tracks the proper solutions.

19.5.1 One-parameter boundary-value problems

For a one-parameter iteration, the problem is simple, since only one parameter a must be adjusted by one error function f. A general schematic of this

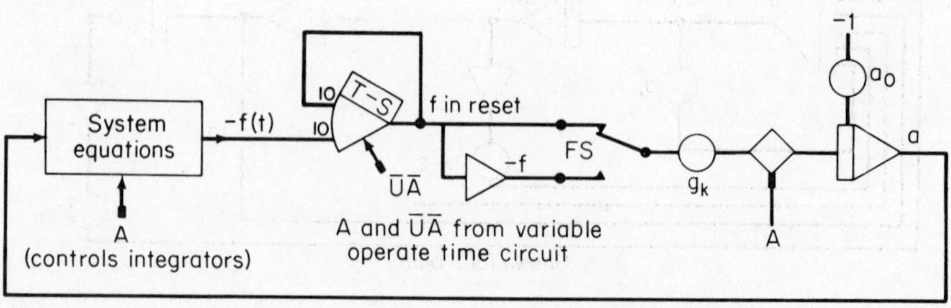

FIG. 19.12 Automatic one-parameter optimization.

is shown in Fig. 19.12. Either $+f$ or $-f$ is connected through the function switch to the DAS, depending upon which produces stability. This can usually be determined beforehand and need not be programmed.

Example 19.4. One-parameter boundary-value problem. To illustrate the method described we set up and solve the following nonlinear boundary-value problem. We are given

$$\frac{d^2V}{dx^2} = \frac{a}{(1 + K\cos\pi x)V^{1/2}}$$

In this case the initial conditions are known

$$V_0 = \left(\frac{dV}{dx}\right)_0 = 0$$

but we are required to determine a so that

$$V = 1 \quad \text{at} \quad x = 1$$

Further, a is a function of K with $-1 < K < 1$. The problem is to find this functional relationship and plot a graph of $a(K)$.

There is an initial singularity at the start of the problem that can be eliminated by the parametric technique while generating $V^{1/2}$ by integration. Let

$$V_1 = \frac{dV}{dx}, \qquad F = V^{1/2}$$

so that the equations become

$$\dot{V} = V_1\dot{x}$$

$$\dot{V}_1 = \frac{a\dot{x}}{(1 + K\cos\pi x)F}$$

$$\dot{F} = \frac{1}{2}\frac{V_1\dot{x}}{F}$$

Now we choose $\dot{x}$ to eliminate division:

$$\begin{aligned}\dot{x} &= F(1 + K\cos\pi x)\\ \dot{V} &= V_1F(1 + K\cos\pi x)\\ \dot{V}_1 &= a\\ \dot{F} &= \tfrac{1}{2}V_1(1 + K\cos\pi x)\end{aligned}$$

Further, we choose to generate the function $K\cos\pi x$ by integration. Let

$$C = K\cos\pi x, \qquad S = K\sin\pi x$$

The final set of equations is now

$$\begin{aligned}
\dot{x} &= kF(1 + C), & x_0 &= 0\\
\dot{V} &= kV_1F(1 + C), & V_0 &= 0\\
\dot{V}_1 &= ka, & (V_1)_0 &= 0\\
\dot{F} &= \tfrac{1}{2}kV_1(1 + C), & F_0 &= 0\\
\dot{C} &= -\pi kSF(1 + C), & C_0 &= K\\
\dot{S} &= \pi kCF(1 + C), & S_0 &= 0
\end{aligned} \tag{19.21}$$

with k the time scale factor. Since V is not needed for feedback, it need not be generated. The error function f' at $x = 1$:

$$f' = V(1) - 1$$

is equivalent to

$$f = F(1) - 1$$

since F and V are equal when either is one.

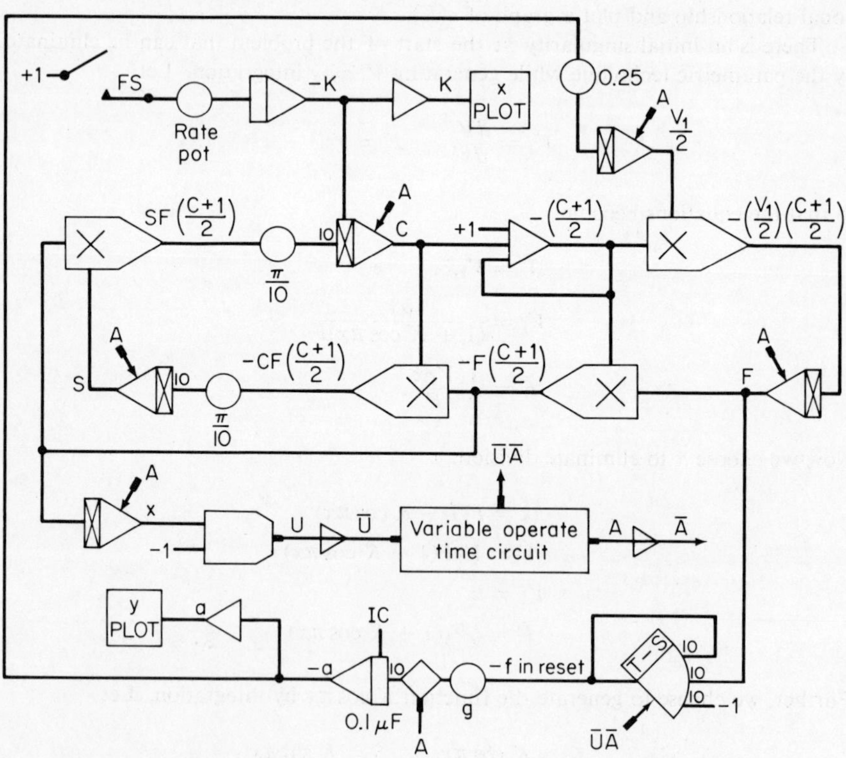

FIG. 19.13 Scaled circuit diagram to obtain $a(K)$.

From (19.21), we see that a must be positive for V_1 and F to increase. If $F(1) > 1$, a is too large; $\partial f/\partial a > 0$. This fact simplifies the circuit in Fig. 19.12 since $-f$ must always be connected to the g pot.

Figure 19.13 is a scaled circuit diagram applicable for $-1 < K \lesssim 0.85$, and using $k = \frac{1}{2}$. Since the operate time is variable, a variable operate time circuit is used. For larger values of K, the slope V_1 increases its range and should be rescaled. Figure 19.14 shows the graph of $a(K)$ obtained by sweeping K in the needed range very slowly. This can be done with a slow-time integrator ramp. Since the stability of the convergence depends upon g, and is critical for small a, we also took g proportional to a by using another multiplier.

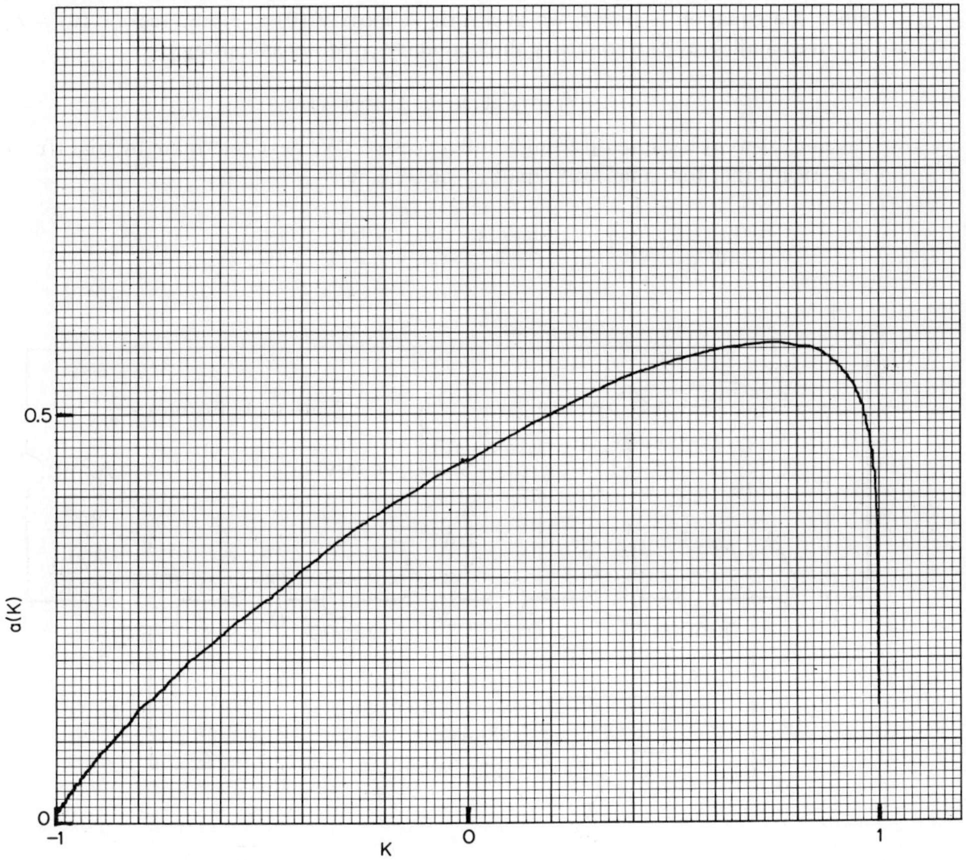

FIG. 19.14 Graph of $a(K)$.

19.5.2 Two-parameter boundary-value problems

The scheme is essentially similar to one-parameter problems, except that the influence of each parameter on the error functions may have to be determined by trial. Sometimes this is unnecessary, since the influences may be intuitively guessed.

Example 19.5. Two-parameter boundary-value problem. We solve the normalized uniform cantilever beam equation for the fundamental eigenvalue and eigenfunction by this direct method:

$$\frac{d^4y}{dx^4} - a_1 y = 0, \qquad 0 \leq x \leq 1 \tag{19.22}$$

with

$$y(0) = y'(0) = y''(1) = y'''(1) = 0$$

Since the equation is linear, we arbitrarily choose $y''(0) = 1$, and seek $y'''(0) = a_2$ and the smallest $a_1 > 0$ to satisfy the boundary conditions. Letting

$$f_1 = y'''(1), \qquad f_2 = y''(1)$$

"assign" f_2 to a_2 because a_2 strongly influences y''. We next determine by trial (from Fig. 19.15) that for sufficiently small g, the scheme

$$\left(\frac{a_1}{20}\right)_k = \left(\frac{a_1}{20}\right)_{k-1} - g\left(\frac{f_1}{2}\right)_{k-1}, \qquad \left(\frac{a_2}{2}\right)_k = \left(\frac{a_2}{2}\right)_{k-1} - g(f_2)_{k-1}$$

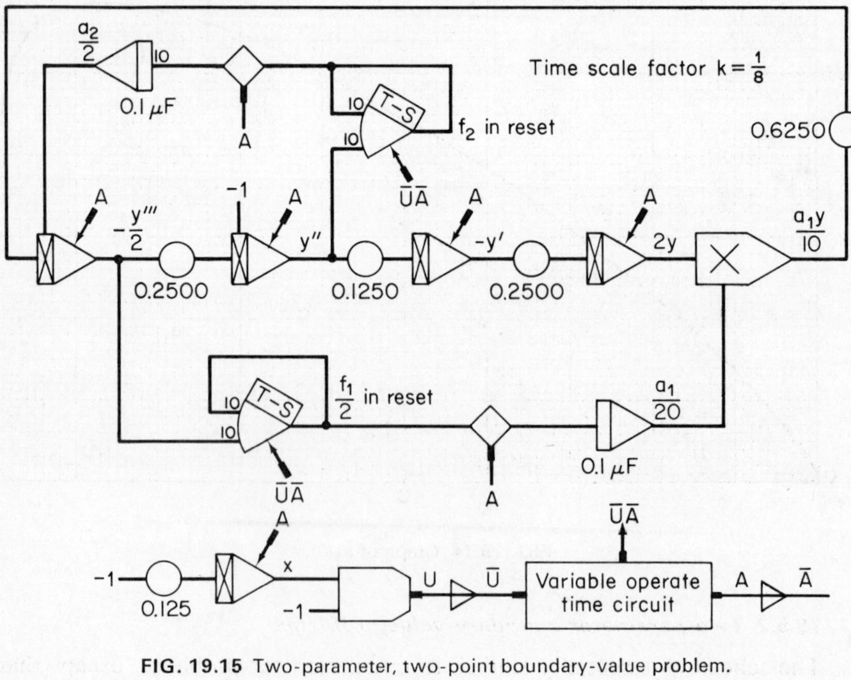

FIG. 19.15 Two-parameter, two-point boundary-value problem.

converges to each $f_j = 0$. Figure 19.16 shows three runs using $(a_1)_0 = (a_2)_0 = 0$ with $g = 0.5$, 0.6, and 0.7 (reset times set to 5, 6, and 7 ms) to show the effects of varying g. (For $g = 0.72$, the solutions did not converge.) The value of the first eigenvalue $(a_1)^{1/2}$ agreed within 0.1 per cent with the theoretical value as found in Chapter 14.

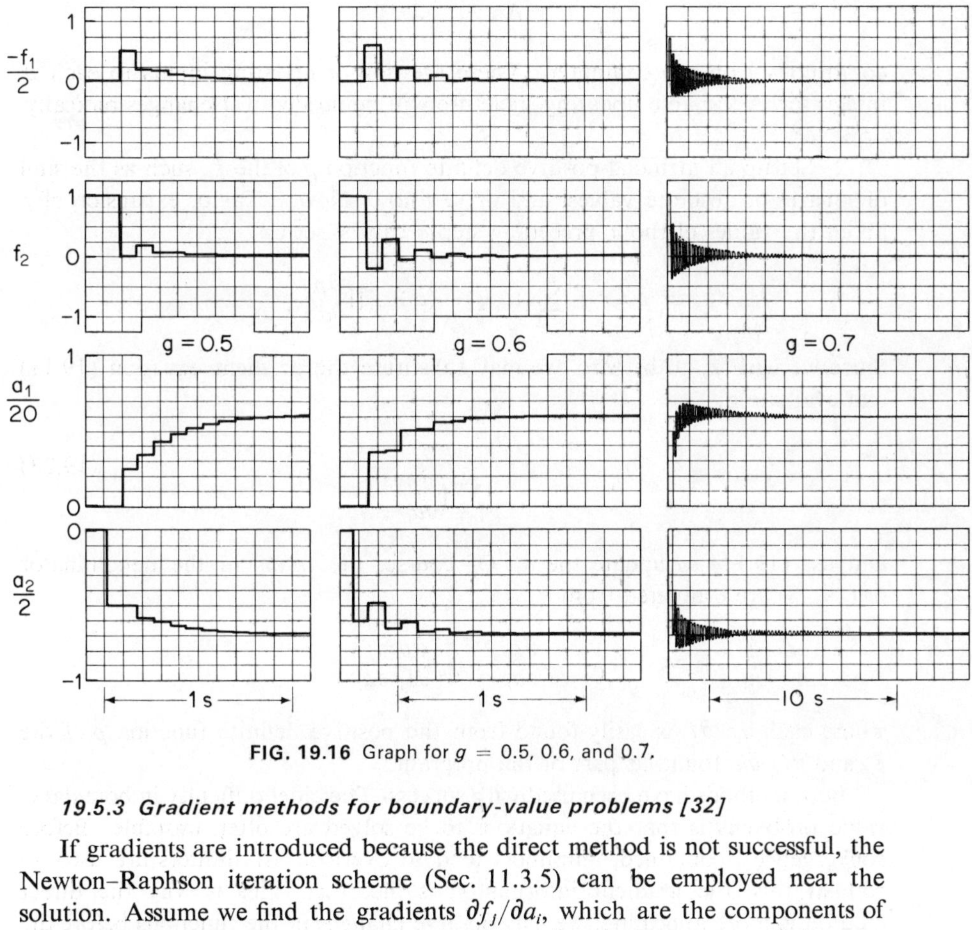

FIG. 19.16 Graph for $g = 0.5$, 0.6, and 0.7.

19.5.3 Gradient methods for boundary-value problems [32]

If gradients are introduced because the direct method is not successful, the Newton–Raphson iteration scheme (Sec. 11.3.5) can be employed near the solution. Assume we find the gradients $\partial f_j/\partial a_i$, which are the components of the Jacobian matrix $\mathbf{J}(\mathbf{f}, \mathbf{a})$, with $\mathbf{f}$ and $\mathbf{a}$ being the current vectors of elements f_j and a_i preceding step k. There are two simple ways to proceed:

1. Solve matrix equation

$$\mathbf{J}\,\Delta\mathbf{a} + \mathbf{f} = \mathbf{0}$$

and use

$$(a_i)_k = (a_i)_{k-1} + \Delta a_i$$

for the next guess. This procedure involves a matrix inversion that may be impractical on analog computers for large systems. Often, however, the elements of **J** are nearly constant near the solution. We may find $\mathbf{J}^{-1}$ off-line and use a modification of the direct method

$$\Delta\mathbf{a} = -\mathbf{J}^{-1}\mathbf{f}$$

continuously on the computer. There are now n updating inputs to each a_i integrator. A variable updating structure will be needed if **J** changes radically, however.

2. Set up an artificial positive definite function ρ of the f_j, such as the sum of squares or absolute values, and drive ρ to 0. Now, a Taylor expansion of ρ about the values of the a_i preceding step k can be set up:

$$\rho_k = \rho_{k-1} + \sum_{i=1}^{n} [(a_i)_k - (a_i)_{k-1}]\left(\frac{\partial \rho}{\partial a_i}\right)_{k-1} + \cdots$$

Since we wish ρ_k to be zero, we may substitute the gradient equation (19.11) and evaluate g:

$$g = \frac{\rho_{k-1}}{\sum_{i=1}^{n}\left(\frac{\partial \rho}{\partial a_i}\right)_{k-1}^{2}} \tag{19.23}$$

and use (19.11) to update the a_i. Of course, the terms in the denominator of (19.23) are evaluated from

$$\frac{\partial \rho}{\partial a_i} = \sum_{j=1}^{n} \frac{\partial \rho}{\partial f_j}\frac{\partial f_j}{\partial a_i}$$

where each $\partial\rho/\partial f_j$ is easily found from the positive definite function ρ of the f_j, and $\partial f_j/\partial a_i$ found as part of the program.

Both methods have been used with success. The chief difficulty in boundary-value problems is that the equations to be solved are often unstable. Before convergence is obtained, amplifiers tend to overload. If limiters are used to combat this, the gradient information is incorrect. This is why the direct method is more appealing; we may observe changes in the functions before the limit is reached to ascertain if convergence may eventually be reached.

19.6 Functional Optimization—Optimal Control Problems

When the parameters a_i of the system are controllable functions of time $a_i(t)$, other special techniques are used to formulate equations for solution. These techniques are embodied in the theories of the calculus of variations [23] and its extension for use in modern control applications, the maximum principle. Our intention here is to give a brief description of the maximum principle with a main emphasis on solving the resulting equations. Excellent references exist for an indepth discussion by Pontryagin, *et al.* (e. g., [24]), and for simple heuristic motivations [25] of the theory.

19.6.1 Problem statement

We consider minimizing the functional

$$\rho = \int_0^{t_f} f_0(x_1, x_2, \ldots, x_s, a_1, a_2, \ldots, a_n)\, dt \tag{19.24}$$

where the x_j vary according to the differential equations

$$\frac{dx_j}{dt} = f_j(x_1, x_2, \ldots, x_s, a_1, a_2, \ldots, a_n), \qquad x_j(0) = c_j \tag{19.25}$$

for $j = 1, 2, \ldots, s$, where the f_j are known and are continuously differentiable with respect to the x_j, and the a_i represent time-varying parameters (controls) which are bounded (usually to satisfy physical limitations such as a maximum thrust) and are piecewise continuous. Note that equations (19.25) and (19.24) are autonomous. If t appears explicitly, we can simply add the equation

$$\frac{dx_{s+1}}{dt} = 1, \qquad x_{s+1}(0) = 0$$

to make the system autonomous.

We must find the $a_i(t)$ to minimize ρ but usually under certain restrictions. We consider here only the following problem that occurs frequently in practice. Time t_f may or may not be given, and some (or all) of the final states $x_j(t_f)$ may or may not be given. Those that are given are individual values; i. e., we exclude boundary conditions other than the type

$$x_j(t_f) = b_j$$

where j takes on only those values between 1 and s inclusively to correspond with a given b_j. This is done to simplify the results of the maximum principle, and by no means includes all cases solvable by the theory. [For example, we may wish to minimize ρ while forcing the final state to reach any point on the circle $x_1^2(t_f) + x_2^2(t_f) = R^2$, but we do not consider such a case.] In addition, the $x_j(t)$ are unconstrained and are allowed to take on any real values.

Problems with none or all of the $x_j(t_f)$ given are called problems with free or fixed endpoints. If the a_i are found to satisfy the boundary conditions while minimizing ρ, they are called optimal controls, while the $x_j(t)$ are then called optimal trajectories. The special case of $f_0 = 1$ (with t_f not given) is called a time-optimal problem.

19.6.2 The maximum principle

To formulate the maximum principle for the problem stated, a new coordinate x_0 is defined by

$$\frac{dx_0}{dt} = f_0(x_1, x_2, \ldots, x_s, a_1, a_2, \ldots, a_n), \qquad x_0(0) = 0$$

so that x_0 corresponds to ρ at $t = t_f$. Now, define a new function, the Hamiltonian H, by introducing $s + 1$ *adjoint* functions p_j such that

$$H = \sum_{j=0}^{s} p_j f_j \tag{19.26}$$

where the p_j satisfy the equations

$$\frac{dp_j}{dt} = -\sum_{k=0}^{s} \frac{\partial f_k}{\partial x_j} p_k \tag{19.27}$$

The final system of $2s + 2$ equations can be conveniently written as

$$\frac{dx_j}{dt} = \frac{\partial H}{\partial p_j}, \qquad \frac{dp_j}{dt} = -\frac{\partial H}{\partial x_j}, \qquad j = 0, 1, \ldots, s \tag{19.28}$$

The maximum principle states essentially that if the $a_i(t)$ are to be optimal controls corresponding to optimal trajectories, it is necessary that there exist the continuous functions p_j defined by (19.27) and (19.26) such that

1. H is maximized at every instant of time t ($0 \leq t \leq t_f$) with respect to the control parameters.
2. p_0 is constant and ≤ 0.
3. H is constant and ≥ 0. If t_f is unspecified $H = 0$.
4. $p_j(t_f) = 0$ for all unspecified $x_j(t_f)$, for $j = 1, 2, \ldots, s$.
5. At least one $p_j(t) \neq 0$ for $0 \leq t \leq t_f$.

From the above, we see that the problem stated reduces to a two-point boundary-value problem. We must solve system (19.28) and satisfy all boundary conditions. If t_f is given, there are $s + 1$ initial conditions $p_j(0)$ to be found. Actually, one of them, which is nonzero, can be arbitrary (but of the right sign), since minimizing ρ is equivalent to minimizing $a\rho$, $a > 0$. If the necessary conditions are met for an optimal set of controls using a set $p_j(0)$, they will also be met using the initial values $ap_j(0)$. (The defining equations for dp_j/dt are linear.) Thus, this case becomes an s-parameter, two-point boundary-value problem.

If t_f is not given, it is an additional parameter. But then there is an additional condition, $H = 0$. This case is therefore an $(s + 1)$-parameter two-point boundary-value problem.

19.6.3 Mechanization of optimal control problems

After the Hamiltonian is formed, the scheme to generate the $a_i(t)$ (to maximize H) should be preselected before attempting to satisfy the boundary conditions. In this way, item 1 of the necessary conditions for optimality will always be satisfied throughout the search procedure. Either $a_i(t)$ will be at a boundary or it must satisfy

$$\frac{\partial H}{\partial a_i} = 0$$

Korn [26] and Steinmetz [27] suggest an implicit generation of each a_i at the output of a high-gain amplifier using

$$\dot{a}_i = g\frac{\partial H}{\partial a_i}, \qquad g \gg 0 \tag{19.29}$$

a scheme which should work most of the time. In using (19.29), the a_i amplifier should be limited to its known bounded values, if necessary.

With the above fundamentals, we illustrate a method of formulating and solving an optimal control problem on an iterative analog computer. The solution is based on the reordering method discussed in Sec. 19.5.

Example 19.6. Of the many examples of optimal control problems in the literature [27]-[31], we choose one, reported by Darcy and Hannen [28], to illustrate the procedures described above.

A maneuvering vehicle attempts rendezvous with a target. The target is in circular orbit with angular velocity ω about the origin of an inertial reference frame, and both vehicle and target are confined to the plane of orbit. The vehicle contains a thruster capable of producing accelerations up to $a_{\max}$ ft/s^2. It can be rotated to accelerate the vehicle in any direction in the plane. This system is describable [28] by the equations

$$\begin{aligned} \frac{d^2x}{dt^2} - 2\omega\frac{dy}{dt} &= a_1 \cos a_2 \\ \frac{d^2y}{dt^2} + 2\omega\frac{dx}{dt} - 3\omega^2 y &= a_1 \sin a_2 \end{aligned} \tag{19.30}$$

where a_1 is the thrust subject to $0 \leq a_1 \leq a_{\max}$, a_2 is the "steering" angle of the thruster, and x and y are rotating rectilinear coordinates in the plane, with origin at the target. Initial conditions of the coordinates and velocities are given. The problem is how to select $a_1(t)$ and $a_2(t)$ so that the integral

$$p = \int_0^{t_f} (1 + \lambda a_1)\, dt \tag{19.31}$$

is a minimum, while the vehicle is transported to the origin with zero velocity:

$$x(t_f) = \frac{dx(t_f)}{dt} = y(t_f) = \frac{dy(t_f)}{dt} = 0$$

This is a problem with fixed endpoints. If $\lambda = 0$ in (19.31), it becomes a time-optimal problem. If $\lambda > 0$, some restriction is placed on the amount of fuel expended.

A first-order set of equations for (19.30) is

$$\frac{dx_1}{dt} = x_2, \qquad \frac{dx_2}{dt} = 2\omega x_4 + a_1 \cos a_2$$

$$\frac{dx_3}{dt} = x_4, \qquad \frac{dx_4}{dt} = -2\omega x_2 + 3\omega^2 x_3 + a_1 \sin a_2$$

with $x_1 = x$, $x_2 = dx/dt$, $x_3 = y$, and $x_4 = dy/dt$. We add to the system

$$\frac{dx_0}{dt} = 1 + \lambda a_1$$

Now we form the Hamiltonian from (19.26)

$$H = p_0(1 + \lambda a_1) + p_1 x_2 + p_2(2\omega x_4 + a_1 \cos a_2) + p_3 x_4 + p_4(-2\omega x_2 + 3\omega^2 x_3 + a_1 \sin a_2) \tag{19.32}$$

From (19.27) we have

$$\frac{dp_0}{dt} = 0, \qquad \frac{dp_1}{dt} = 0, \qquad \frac{dp_2}{dt} = -p_1 + 2\omega p_4$$

$$\frac{dp_3}{dt} = -3\omega^2 p_4, \qquad \frac{dp_4}{dt} = -2\omega p_2 - p_3$$

Next, we preselect the scheme to generate a_1 and a_2. Since the only terms in H containing the parameters a_1 and a_2 are

$$a_1(\lambda p_0 + p_2 \cos a_2 + p_4 \sin a_2)$$

H is maximized with respect to a_1 if

$$a_1 = a_{\max}\, U(\lambda p_0 + p_2 \cos a_2 + p_4 \sin a_2) \tag{19.33}$$

The thruster must either be on full-blast or be off. Also

$$\frac{\partial H}{\partial a_2} = a_1(-p_2 \sin a_2 + p_4 \cos a_2) = 0 \tag{19.34}$$

or

$$\tan a_2 = \frac{p_4}{p_2}$$

which reduces to

$$\sin a_2 = \frac{p_4}{\sqrt{p_2^2 + p_4^2}}, \qquad \cos a_2 = \frac{p_2}{\sqrt{p_2^2 + p_4^2}} \tag{19.35}$$

Thus, we have also from (19.33)

$$a_1 = a_{\max}\, U(\sqrt{p_2^2 + p_4^2} + \lambda p_0) \tag{19.36}$$

These controls maximize H, since $a_1 > 0$ when the thruster is on. We mechanize (19.36) and (19.35) continuously to satisfy necessary condition 1 during the search for the $p_j(0)$. It was not necessary to use (19.29) in these cases; direct computations of the control schemes were easier.

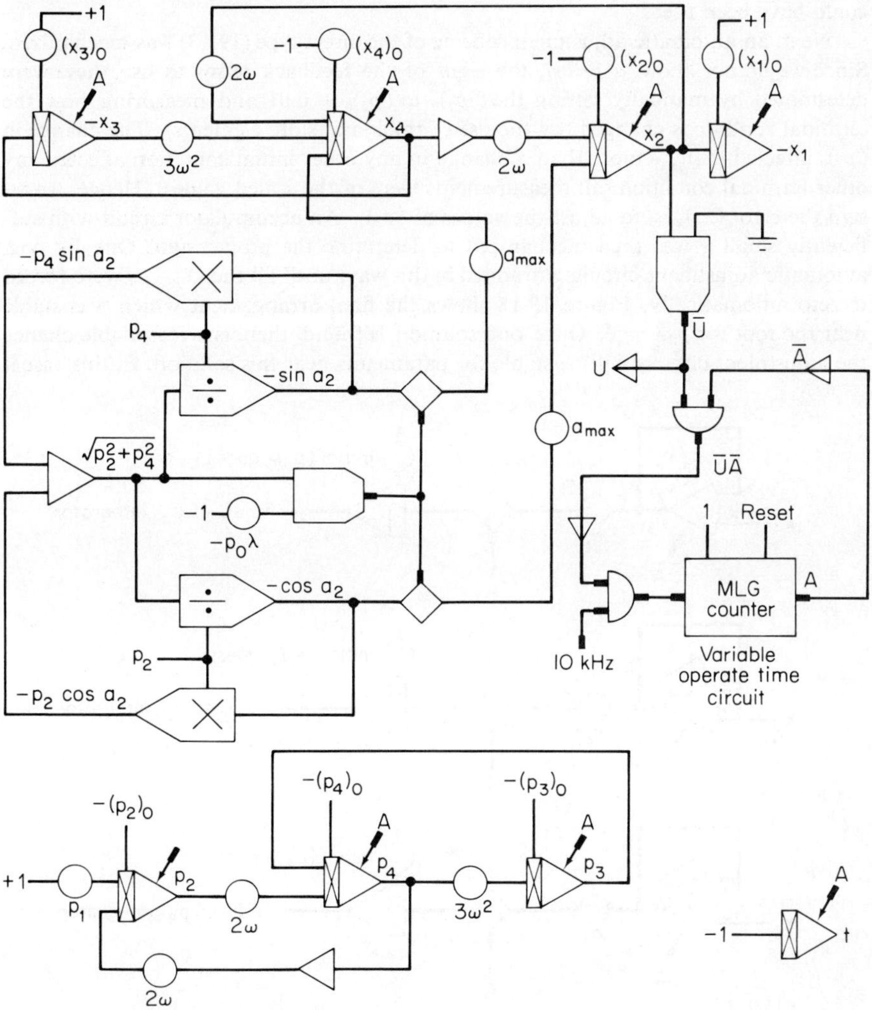

FIG. 19.17 Circuit for optimal control problem (Example 19.6).

Figure 19.17 shows the unscaled circuitry for the system without any automatic adjustments to satisfy boundary conditions and with an inverse resolver circuit used to generate sin a_2 and cos a_2. Since the main investigation centered on large initial negative values of x_1 and $x_2 > 0$ for the entire flight, a variable operate time circuit was used and the computation was ended when $x_2 = 0$. To start the search procedure, the $-p_0\lambda$ pot was set to 0, and the p_1 pot (p_1 is constant) was arbitrarily fixed. For any resemblance of a solution when $\lambda = 0$, we must have $p_1 > 0$, $(p_2)_0 > 0$, $(p_4)_0 < 0$ and $(p_3)_0 < 0$. The values of $(p_2)_0$, $(p_3)_0$, and $(p_4)_0$ were manually adjusted until the final states of x_1, x_3, and x_4 were reasonably close to the origin when $x_2 = 0$. Using a scope, this took about 20 min. with most of the time spent in getting a feel for how the $(p_j)_0$ affected the solution. A relaxation or random search procedure could have been used.

Next, an automatic adjustment scheme of the direct type (19.13) was mechanized. Since we do not know, *a priori*, the signs of the feedback terms to use, these were determined by manually setting the $(p_j)_0$ to $(p_j)_0 + 0.01$ and measuring how the terminal conditions changed (by the use of track-and-store elements). The change in $(p_3)_0$ affected $(x_4)_{x_2=0}$ more than a change in any other initial condition affected any other terminal condition (all measurements were of the scaled values). Hence, we assign the error $(x_4)_{x_2=0}$ to adjust the values of $(p_3)_0$. An accumulator circuit with sufficiently small g was then mechanized to determine the proper sign. One by one, automatic adjustment circuits are added in this way, until all the $(x_j)_{x_2=0}$ were forced to zero automatically. Figure 19.18 shows the final arrangement which was stable near the root for $p_0\lambda = 0$. Once one solution is found, there is a reasonable chance the adjustment circuits will be stable for parameters near this solution. In this case it

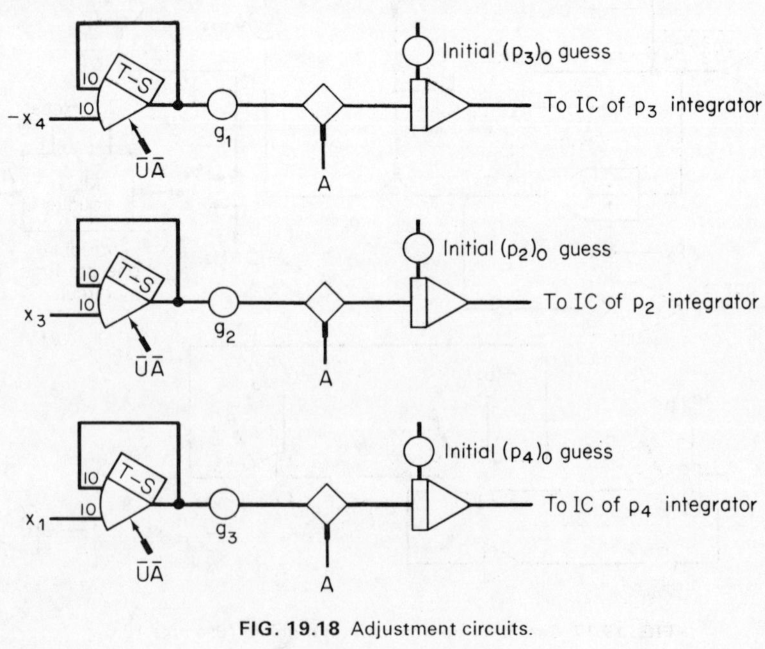

FIG. 19.18 Adjustment circuits.

was easy to establish that the system was stable for a wide range of λ as well as wide ranges in the $(x_j)_0$.

To find solutions for nonzero λ, the pot labeled $-p_0\lambda$ was increased (since $p_0 \leq 0$, $\lambda > 0$, $-p_0\lambda \geq 0$) to an arbitrary setting b:

$$b = -p_0\lambda \tag{19.37}$$

Necessary condition 3 requires $H = 0$ throughout the computation, and in particular at $t = t_f$. Since the thrust was on at this time and all $x_j = 0$, (19.32) reduces to

$$H_{t_f} = p_0(1 + \lambda a_{\max}) + a_{\max}\sqrt{p_2^2 + p_4^2} \tag{19.38}$$

For the given pot setting b, we store the value of $\sqrt{p_2^2 + p_4^2}$ at $t = t_f$. Say this value is

$$e = \sqrt{p_2^2 + p_4^2} \quad \text{at} \quad t = t_f$$

Now (19.38) reduces to

$$p_0 - ba_{\max} + ea_{\max} = 0$$

or

$$p_0 = ba_{\max} - ea_{\max}$$

which, from (19.37) results in

$$\lambda = \frac{b}{a_{\max}(e - b)} \tag{19.39}$$

For any given λ, the p_1 pot and/or the p_0 pot can be manually adjusted until (19.39) is satisfied, or the relationship can be automatically satisfied by using a fourth accumulator circuit that adjusts the value p_1. For this purpose, the error

$$\epsilon_0 = a_{\max}\lambda(e - b) - b \tag{19.40}$$

can be used. Although this is a five-parameter boundary-value problem, it is reduced to adjusting only 4 parameters (3 automatically and 1 manually). The parameter t_f is then read from an additional memory (not shown).

Figure 19.19 is a graph showing how the errors approach zero for a typical case where $\lambda = 0$. Before iteration, the errors were quite large, at least 0.3 for computer variables whose absolute value did not exceed 1. But convergence is quite rapid.

The general procedure is highly recommended, particularly if one solution (or near solution) is known, since variations of a constant parameter will most likely produce stable adjustments also. Previous methods can be used to get an initial approximate solution.

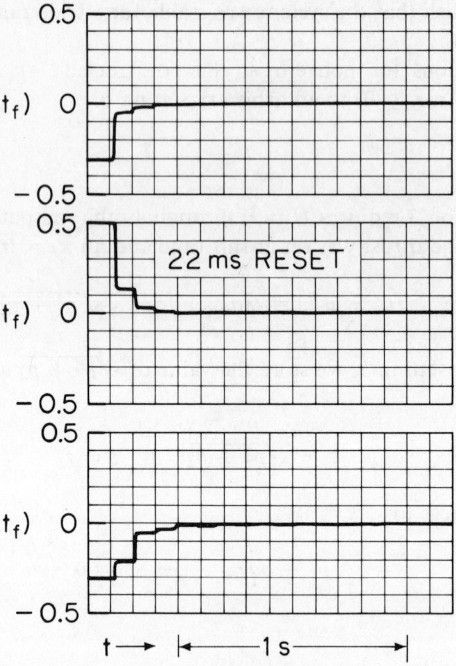

FIG. 19.19 Graphs of errors approaching zero.

It should be emphasized that the maximum principle provides a set of necessary conditions only. There may be more than one set of controls and trajectories satisfying these conditions. (Indeed, there may be none.) We must find them all and choose the one minimizing ρ.

PROBLEMS

19-1. Use the manual relaxation procedure to obtain a best quintic polynomial fit to the function sin (x^2), $0 \leq x \leq \pi/2$, with an absolute error norm.

19-2. Set up Example 19.1 using a least squares norm.

19-3. Set up Example 19.2 using a least squares norm.

19-4. Mechanize the best fifth-degree differential equation of the type (19.17) to produce a time delay, using an absolute error norm and the method in Sec. 19.4.2.

19-5. Set up an automatic iteration to solve

$$\ddot{x} + 0.2\dot{x} + bx = 0$$

for $0 \leq x \leq 10$, so that $x(0) = 50$ and $x(10) = 0$, for $0 \leq b \leq 0.01$, and graph the parameter $\dot{x}(0)$ as a function of b.

19-6. Find the solution of

$$\ddot{x} + x = 12.5 + 0.1 \int_0^\infty xe^{-0.2t}\, dt$$

given $x_0 = 0$, $\dot{x}_0 = b$, using the discrete reordering method, and graph the value of the integral as a function of b for $0 \leq b \leq 1$.

19-7. Find a such that the solution of

$$\ddot{x} + 2a\dot{x} + x = b$$

with $x_0 = 40$ and $\dot{x}_0 = 0$ minimizes the integral

$$\rho = \int_0^\infty x^2\, dt$$

using the discrete gradient method, and graph the function $a(b)$.

19-8. System (19.22) can be set up without a multiplier by using variable scaling. With y_i, $i = 0$–3, defined as $d^i y/dx^i$, the system can be written

$$\frac{dy_0}{dx} = y_1, \qquad (y_0)_0 = 0$$

$$\frac{dy_1}{dx} = y_2, \qquad (y_1)_0 = 0$$

$$\frac{dy_2}{dx} = y_3, \qquad (y_2)_0 = 1$$

$$\frac{dy_3}{dx} = a_1 y_0, \qquad (y_3)_0 = a_2$$

Now, let

$$Y_0 = a_1 y_0$$

$$Y_1 = y_1 (a_1^3)^{1/4}$$

$$Y_2 = y_2 (a_1^2)^{1/4}$$

$$Y_3 = y_3 (a_1)^{1/4}$$

so that the system becomes

$$\frac{dY_0}{dx} = Y_1 (a_1)^{1/4}, \qquad (Y_0)_0 = 0$$

$$\frac{dY_1}{dx} = Y_2 (a_1)^{1/4}, \qquad (Y_1)_0 = 0$$

$$\frac{dY_2}{dx} = Y_3 (a_1)^{1/4}, \qquad (Y_2)_0 = (a_1)^{1/2}$$

$$\frac{dY_3}{dx} = Y_0 (a_1)^{1/4}, \qquad (Y_3)_0 = a_2 (a_1)^{1/4}$$

Since the system is linear, $(Y_2)_0$ can be arbitrary and the problem reduces to finding $(Y_3)_0$. With $\dot{x} = 1/(a_1)^{1/4}$, no multipliers are needed. Mechanize this case.

19-9. If we fix t_f in Example 19.6, the error $x_2(t_f)$ can be used to adjust p_1. Since H is not necessarily 0, the four terminal conditions are then used to adjust the four unknown initial conditions. Describe the changes needed in Fig. 19.17 to mechanize this variation and indicate how to guarantee the stability of the adjustment.

19-10. With the variation mechanized in Problem 19-9, the time t_f must be varied to meet the condition $H = 0$. But this can also be done automatically using the ϵ_0 of (19.40). Describe the changes in Fig. 19.17 to fully mechanize this five-parameter system [42].

19-11. Set up a circuit to find the $a(t)$, $|a(t)| \leq 1$, so that the system

$$\frac{d^2x}{dt^2} = a(t)$$

is driven to the origin in minimum time. Use the discrete reordering method assuming the investigation is centered about varying initial conditions that nominally are $x_0 = (dx/dt)_0 = 0.5$.

19-12. Using the maximum principle, show that minimizing the integral

$$\rho = \int_0^b f(x, y, dy/dx)\, dx, \quad b \text{ constant}$$

with $y(0)$ and $y(b)$ given, reduces to finding the solution of the Euler differential equation

$$\frac{d}{dx}\left[\frac{\partial f}{\partial(dy/dx)}\right] - \frac{\partial f}{\partial y} = 0$$

which is a classical result of the calculus of variations.

19-13. A tubular reactor of specified length involves the reaction of two components x_1 and x_2 flowing at constant speed through the tube. The reaction equations are given by [27]

$$\frac{dx_1}{dt} = -k_1(T)x_1, \qquad \frac{dx_2}{dt} = k_1(T)x_1 - k_2(T)x_2$$

where the k_i are functions of the temperature T, which is controllable by heating and cooling units along the length of the tube. Thus, T can be made the controllable function of time. If the k_i are given by

$$k_1 = c_1e^{-b_1/T}, \qquad k_2 = c_2e^{-b_2/T}$$

set up the equations to determine how $T(t)$ should vary to *maximize* x_2. (*Hint*: Take

$$\rho = -\int_0^{t_f} \frac{dx_2}{dt}\, dt$$

where t_f is fixed. The adjoint variable p_0 can be combined with the adjoint variable p_2.)

19-14. Show that minimizing a function

$$\rho = \sum_{i=1}^{n} c_i x_i$$

can be set up without the p_0 adjoint variable provided the conditions

$$p_j(t_f) = -c_j, \qquad j = 1, 2, \ldots, s$$

are used. [All $x_j(t_f)$ are free endpoints.]

REFERENCES

[1] Bekey, G. A., "Optimization of Multiparameter Systems by Hybrid Computer Techniques, Part I," *Simulation*, vol. 2, Feb. 1964; Part II, vol. 2, March 1964.

[2] Harvey, R. A., *et al.*, "Analog Computer Methods for Parameter Optimization," *Simulation*, vol. 6, March 1966.

[3] Leitmann, G., ed., *Optimization Techniques With Applications to Aerospace Systems* (New York: Academic Press, Inc., 1962).

[4] Favreau, R. R., and R. Franks, "Random Optimization by Analog Techniques," *Proceedings of the 2nd Inter. Analogue Comp. Meetings* (Brussels: Presses Académiques Européennes, 1959).

[5] Munson, J. K., and A. I. Rubin, "Optimization by Random Search on the Analog Computer," *IRETEC*, vol. 8, June 1959.

[6] Mitchell, B. A., "A Hybrid/Analog-Digital One-Parameter Optimizer," *AICA*, vol. 6, Jan. 1964. See comment by W. Brunner in vol. 6, July 1964.

[7] Mitchell, B. A., Jr., "A Hybrid Analog-Digital Parameter Optimizer for Astrac II," *Proc. Spring Joint Comp. Conference*, 1964.

[8] Favreau, R. R., and W. Brunner, "Application of the Parameter Influence Coefficient Technique to Boundary Value Problems," *Princeton Computation Center Report No. 156*, Electronic Associates, Inc., Princeton, N. J., May 26, 1960.

[9] Bingulac, S., and P. Kokotovic, "Automatic Optimization of Linear Feedback Control Systems on an Analog Computer," *AICA*, vol. 7, Jan. 1965.

[10] Smith, D. A., and A. C. Soudack, "Analog Computer Gradient Techniques for Parameter Optimization," *AICA*, vol. 10, Oct. 1968.

[11] Howell, M., "Automatic Parameter Optimization as Applied to Transducer Design," *Proc. Spring Joint Comp. Conference*, 1963.

[12] Fel'dbaum, A. A., "Automatic Optimizer," *Automatika i Telemekhanika*, vol. 19, Aug. 1958.

[13] O'Grady, E. P., "Correlation Method for Computing Sensitivity Functions on a High-Speed Iterative Analog Computer," *IEEETEC*, vol. 16, April 1967.

[14] Gonzalez, R., and M. Muterspaugh, "Optimization by Parameter-Perturbation Correlation," *Simulation*, vol. 11, Aug. 1968.

[15] Janicke, M. T., and L. C. Just, "A Simulation of a Generalized Thermal Radiating Fin," *Simulation*, vol. 2, June 1964.

[16] Vichnevetsky, R. "One Dimension Iteration on the Analog Computer," *European Computation Center Report No. 43*, Electronic Associates, Inc., Brussels, Belgium, 1961.

[17] Fadden, E. J., "Determination of the Normal Modes of a Two-Point Boundary Value Problem," *AICA*, vol. 7, April 1965.

[18] Showalter, R. E., "The Solution of Boundary-Value Problems in Ordinary Differential Equations Employing Iterative Differential Analyzer Equipment," *Simulation*, vol. 5, Dec. 1965.

[19] Joss, E. J., "An Iterative Analogue Solution of a Beam Torsion Problem," *Proceedings of the 3rd Inter. Analogue Comp. Meetings* (Brussels: Presses Académiques Européennes, 1962).

[20] Cook, K. S., "An Iterative Analog Determination of Burning Time and Weight Distribution in a Two-Stage Rocket," *Instr. and Control Systems*, vol. 36, Dec. 1963.

[21] Lowe, W. W., "Fortune-Wheel Process for Generating Pseudo-Random Numbers," *Instr. and Control Systems*, vol. 35, Jan. 1962. (In the Analog Techniques section, T. C. Anderson, ed.)

[22] Tomović, R., *Sensitivity Analysis of Dynamic Systems*, transl. by D. Tornquist (New York: McGraw-Hill Book Co., 1963), pp. 57–78.

[23] Elsgolc, L. E., *Calculus of Variations* (Reading, Mass.: Addison-Wesley, 1962).

[24] Pontryagin, L. S., *et al.*, *The Mathematical Theory of Optimal Processes*, transl. by K. N. Trirogoff (New York: John Wiley & Sons, 1962).

[25] Witsenhausen, H. S., "A Heuristic Approach to the Maximum Principle," *Simulation*, vol. 2, June 1964.

[26] Korn, G. A., "Enforcing Pontryagin's Maximum Principle by Continuous Steepest Descent," *IEEETEC*, vol. 13, Aug. 1964.

[27] Steinmetz, H. L., "Using Pontryagin's Maximum Principle to Solve One-Dimensional Optimization Problems, With and Without Constraints, on an Iterative Analog Computer," *Simulation*, vol. 4, June 1965.

[28] Darcy, V. J., and R. A. Hannen, "An Application of an Analog Computer to Solve the Two-Point Boundary-Value Problem for a Fourth-Order Optimal Control System," *IEEEAC*, vol. 12, Feb. 1967.

[29] Lee, E. S., "Optimization by Pontryagin's Maximum Principle on the Analog Computer," *Proc. 4th Joint Automatic Control Conf.*, June 19–21, 1963.

[30] Anderson, M. D., and S. C. Gupta, "Analog Computer Solution of a Third-Order Pontryagin Optimum Control System," *Simulation*, vol. 5, Oct. 1965.

[31] Maybach, R. L., "Solution of Optimal Control Problems on a High-Speed Hybrid Computer," *Simulation*, vol. 7, Nov. 1966.

[32] Brunner, W., "An Iteration Procedure for Parametric Model Building and Boundary Value Problems," *Proc. Western Joint Comp. Conference*, 1961.

[33] Vichnevetsky, R., "Eigenvalues and Rate of Convergence in the Gradient Optimization Technique," *Princeton Computation Center Report No. 196*, Electronic Associates, Inc., Princeton, N. J., Dec. 16, 1964.

See also

[34] Greenspan, D., *et al.*, "On the Analog Solution of Two Point Boundary Value Problems," *Rev. Sci. Instr.*, vol. 28, Dec. 1957.

[35] Shubin, A. B., "10-Channel Automatic-Electronic Relay Optimizer," *Automatika i Telemekhanika*, vol. 21, May 1960.

[36] Hara, H. H., "Special Techniques for Two-Dimensional Air-to-Air Missile Simulation," *Simulation*, vol. 2, May 1964.

[37] Melsa, J. L., and M. J. Wozny, "Iterative-Differential-Analyzer Study of Prediction Networks," *Simulation*, vol. 3, Aug. 1964.

[38] Mathis, A., *et al.*, "Iterative Methods for the Analog Simulation of Physical Systems Containing Flowing Fluids," *AICA*, vol. 8, July 1966.

[39] Neustadt, L. W., "Synthesizing Time Optimal Control Systems," *J. Math. Analysis and Appl.*, vol. 1, Dec. 1960.

[40] Teraski, R. M., "Analog Computation of Green's Function for Integrating Two-Point Boundary Value Problems," *IRETEC*, vol. 11, Feb. 1962.

[41] Suryanarayanan, K. L., and A. C. Soudack, "Analog Computer Automatic Parameter Optimization of Nonlinear Control Systems with Specific Inputs," *IEEETC*, vol. 17, Aug. 1968.

[42] Hausner, A., "Simplified Analog/Hybrid Solutions to Algebraic Equations and Boundary-Value Problems," *Harry Diamond Labs. Report TR-1461*, Wash., D.C., Sept. 1969.

Chapter 20

ANALOG/HYBRID COMPUTERS III: CLOCKED LOGIC

As we have seen in the previous two chapters, point memory, mode control, and repetitive operation can handle a wide variety of iterative computations. But other auxiliary circuits are sometimes needed. This chapter describes equipment and uses of digital elements designed to make such auxiliary circuits easier to construct. The elements can be part of a modern analog/hybrid computer or a special unit slaved into the main analog computer console known as a digital expansion system (DES). Because digital states of these elements change only during master clock pulse signals, they are called *clocked* or *gated* elements.

With a digital expansion system, each run of a rep-op computation can be designed to serve a different purpose. This makes possible automatic sequential calculations hitherto confined to manual operator control. An exploratory run to approximate $\partial\rho/\partial a_1$, for example, can be made automatically by allowing a_1 to increase by a slight amount Δa_1 and storing the $\Delta\rho$ produced, without disturbing any of the system parameters, including a_1 Further, such a calculation automatically can be stepped to all a_i with an eventual storing of all components of the gradient vector of ρ. These can then be used in an ascent or descent process to increase or decrease ρ by changing the a_1. As we can readily appreciate, such a sequential set of calculations would be quite tedious to mechanize automatically with the components previously described, although it could be done. A digital expansion system or analog/hybrid computer offers a tremendous convenience for such mechanizations.

This discussion is based on the EAI digital expansion system, the DES-30 [1]. In the applications given, we assume the DES-30 is trunked to the MLG board of their 231R-V analog/hybrid computer. Many of the circuits shown have to be modified for other equipment, usually a modest exercise.

20.1 Clocked Digital Elements (Gated Logic) [1]

A clocked digital element is one that cannot change its digital state between periodic pulses emitted from a master clock. Changes may occur only during the pulse-time width (or very close to the trailing edge of the pulse), which is typically the order of nanoseconds. The clock period t_{CL} is much greater so that these are indeed pulses by comparison. Means to vary t_{CL} by preselecting one of a number of frequencies for the clock ($t_{CL} = 1$, 0.001, or 0.000001 s in the DES-30) are available within the unit, or an external oscillator can be used to provide any frequency. In addition, one clock pulse can be emitted by manual control to check sequences of digital operations. Elements are interconnected by patching a special board for the unit.

The operation of clocked elements differs slightly from that of AND gates or comparators that can change digital states almost instantly after commanded to do so. Outputs of all clocked elements are either 0 or 1 as usual, and inverted outputs are often automatically produced also.

A "blip" is a logic 1 for period t_{CL}.

A "bit" is a logic signal for period t_{CL}. (It is short for "binary digit.") A blip is therefore a "one bit."

A clocked element "sets" when its output becomes digital 1 and it "resets" when its output becomes digital 0. We omit the phrase "during or shortly after the next clock pulse" but this should be tacitly understood.

All clocked logic elements operate in parallel as do analog computer elements, but only at discrete times.

20.1.1 Flip-flop

The main element needed for counting and control is the flip-flop, a two-state clocked digital storage element that can change its state on command. It is an SRT (Set, Reset, Trigger) type with an enable and clear feature (Fig. 20.1). The state of the digital input variables R, S, E, and C at the beginning of the clock pulse determines the output state U at the end of the

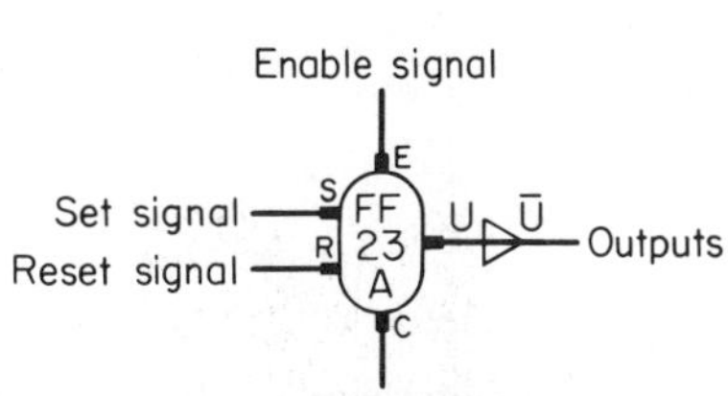

Flip-flop 23 A is used

(a) Symbol

With C=0 and E=1

S	R	U
1	0	1
0	1	0
0	0	Does not change
1	1	Changes state

If C=0 and E=0, U does not change state
If C=1, U=0

(b) Operation (state after next clock pulse)

FIG. 20.1 Flip-flop operation and symbolism.

clock pulse, as shown in the table. E is an enable signal, C a clear signal, S a set signal, and R a reset signal. If R and S are 1, the command is to "trigger" the flip-flop, i.e., change its state providing $E = 1$ and $C = 0$. All signals can come from any digital source whatever. The enable feature is equivalent to AND gating signals S and R before patching into the flip-flop; the clear signal likewise can be patched with AND gates. Its incorporation into flip-flop design is an added convenience which, at the same time, saves on the use of AND gates. Omission of a patching wire is equivalent to patching a 0 signal. This is different from AND gate operation where unused inputs are assumed at 1 levels.

Flip-flops by themselves are useful for storing digital signals for later use (Fig. 20.2).

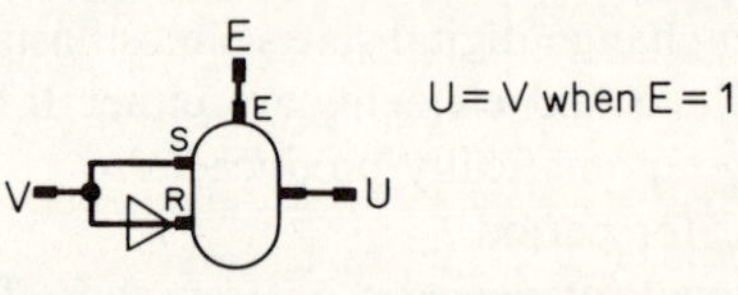

FIG. 20.2 Digital storage use of a flip-flop.

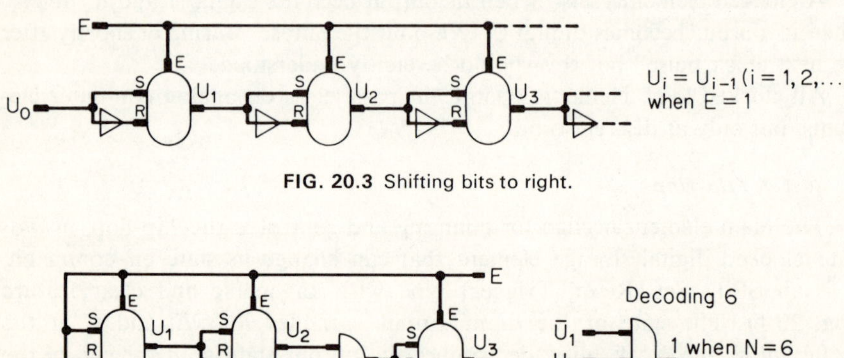

FIG. 20.3 Shifting bits to right.

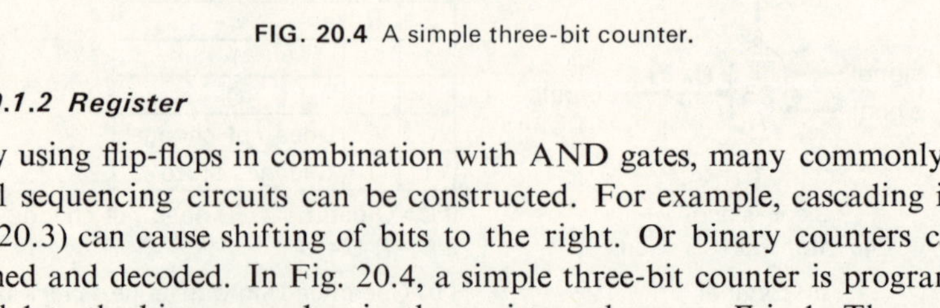

FIG. 20.4 A simple three-bit counter.

20.1.2 Register

By using flip-flops in combination with AND gates, many commonly used digital sequencing circuits can be constructed. For example, cascading inputs (Fig. 20.3) can cause shifting of bits to the right. Or binary counters can be designed and decoded. In Fig. 20.4, a simple three-bit counter is programmed to add 1 to the binary number in the register whenever E is 1. The number N may be represented algebraically by

$$N = U_1 + 2U_2 + 4U_3$$

and any number between 0 and 7 can thus be designated. Decoding requires inputting the proper signals into an AND gate ($N = 6$, for example, occurs only when $\bar{U}_1 U_2 U_3 = 1$). For n flip-flops in tandem there are, of course, 2^n possible states. Number states are in modulo arithmetic, mod 2^n.

Counting and shifting are used often enough that four-bit general-purpose registers are designed in lieu of four individual flip-flops, to facilitate programming. By simple patching, the register can be programmed for a counter mode, a shift register mode, or an individual flip-flop mode. Such a procedure also permits retaining the set-reset characteristics of an individual flip-flop in the register.

In the shift register mode, patching holes are provided for digital signals as follows:

S_i: The bit shifted into the first flip-flop (U_0 in Fig. 20.3)

SH: The shift signal. When $SH = 1$, all bits are shifted to the right. $U_1 = S_i$, and $U_{i+1} = U_i$. (The last one is destroyed.)

C: The clear signal. When $C = 1$, all flip-flops in the register are reset. (If $C = 1$, SH should be 0.)

E: An enabling signal for the first flip-flop. When $E = 1$, data could be entered into the first flip-flop through the normal S-R inputs. (If $E = 1$, SH should be 0.)

L: A ganged enabling signal for the last three flip-flops. When $L = 1$, data could be entered into these flip-flops through their normal S-R inputs. (If $L = 1$, SH should be 0.) This is the only enabling signal for these flip-flops. If used individually in the flip-flop mode, all must be enabled together. L stands for "LOAD." The process of overriding the shift signal to enter data into the register is called "loading the register." L and E are usually used together in this mode.

The signals E, L, R, and S for each flip-flop need not be patched; the register then behaves as a pure shift register. Figure 20.5 shows the symbol used for a four-bit shift register. The E, L, R_i, and S_i leads may be omitted if not used. Thus, in a pure shift mode only two inputs S_i and SH must be patched unless some clearing circuitry is needed.

In the counter mode, only two signals need be provided:

EC_i (enable carry in): When $EC_i = 1$, the counter will count only if
C_i (carry in): $C_i = 1$.

Thus, EC_i AND C_i must be high to count, a feature that eliminates an AND gate if counting depends on two signals. In addition, a carry-out signal C_o becomes high when all flip-flops in the register AND C_i are high. A clear provision, as well as parallel loading similar to the shift register mode, is also

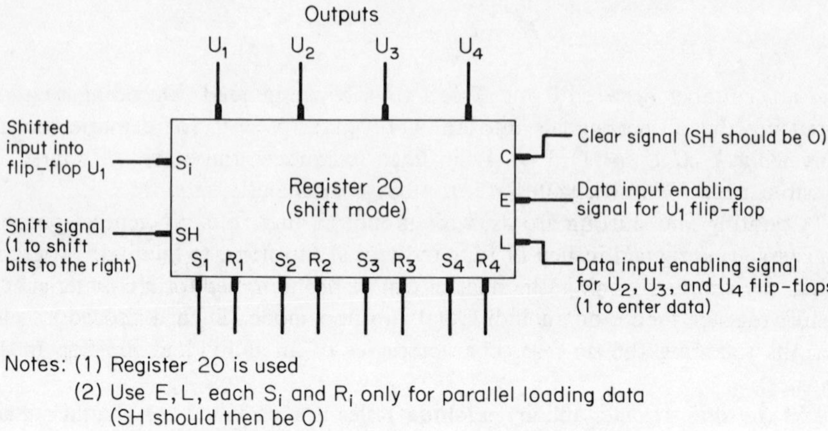

FIG. 20.5 Shift register symbol (four-bit).

optionally used. Figure 20.6 shows the symbol used for the register when in the counter mode.

Registers can be cascaded by properly patching the outputs of one unit into a succeeding one, thereby creating a register of any number of bits.

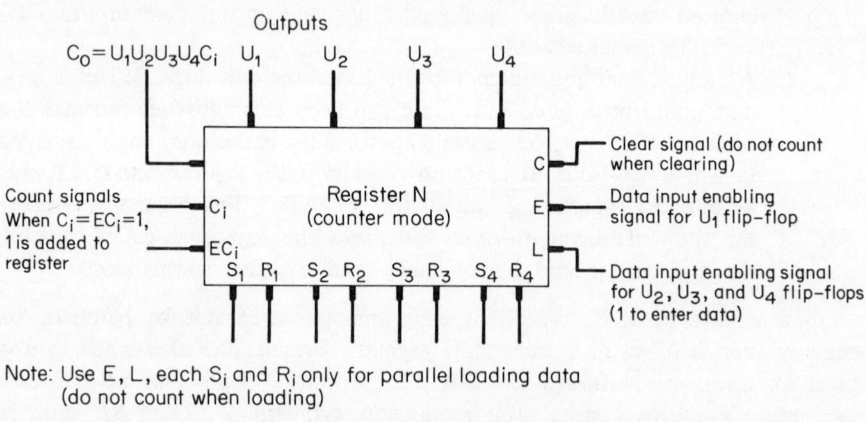

FIG. 20.6 Counter symbol (four-bit).

20.1.3 Simple register circuits

Ring shift register (*Fig. 20.7*). Useful for repeatedly circulating a one bit when controlling a sequence of n operations, one at a time. Initially, all four flip-flops are reset causing a one-bit to be loaded into the first flip-flop. Each time a shift signal is high, the one bit is shifted to its adjacent flip-flop to the right. After n shifts, U_1 is again high with all other flip-flops low. The shift

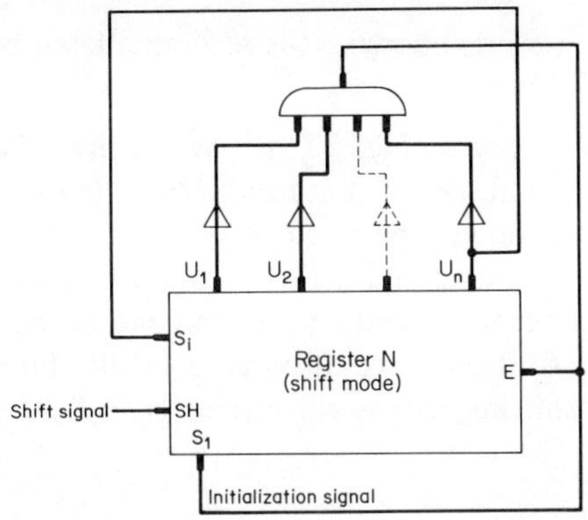

FIG. 20.7 Ring shift register programming (n-bit).

signal comes from a digital source which indicates that one operation is over and the next one should begin. The outputs U_i control switches, integrator modes, etc., allowing the operation to be performed.

Transposed ring shift (*Fig. 20.8*). Similar in purpose to a ring shift register. One four-bit register can produce eight possible states, each decoded with

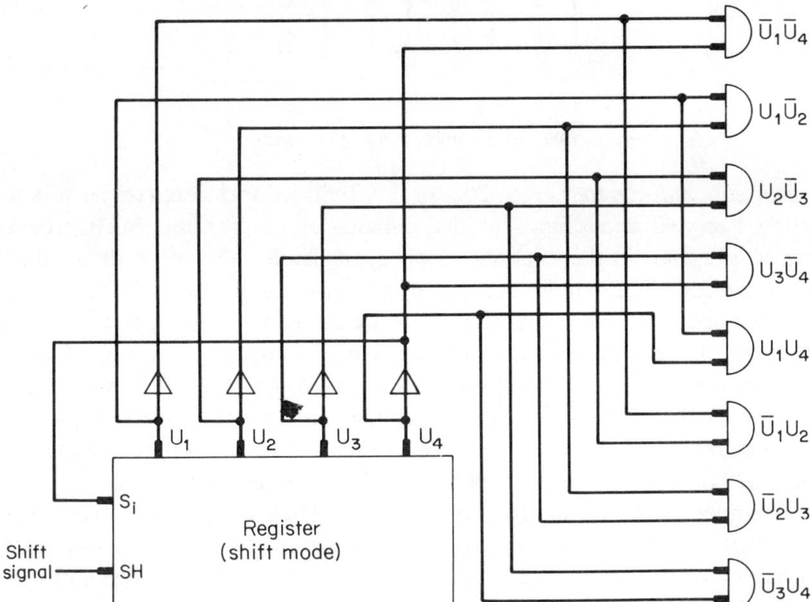

FIG. 20.8 Transposed ring shift.

a two-input AND gate. As shown, it is self-initializing because (0, 0, 0, 0) is one possible state.

Bidirectional shift register (*Fig. 20.9*). When the sequence of n operations is not fully unindirectional, i.e., if it is necessary to return to the previous state for any reason during the computation, then a bidirectional shift register can be programmed. Such a unit accepts commands to shift bits either right or left. The left shift is accomplished by loading the proper data on command through the enable and load inputs. If necessary, a flip-flop can be used to store U_4 for reentry when shifting left by the circuit of Fig. 20.2.

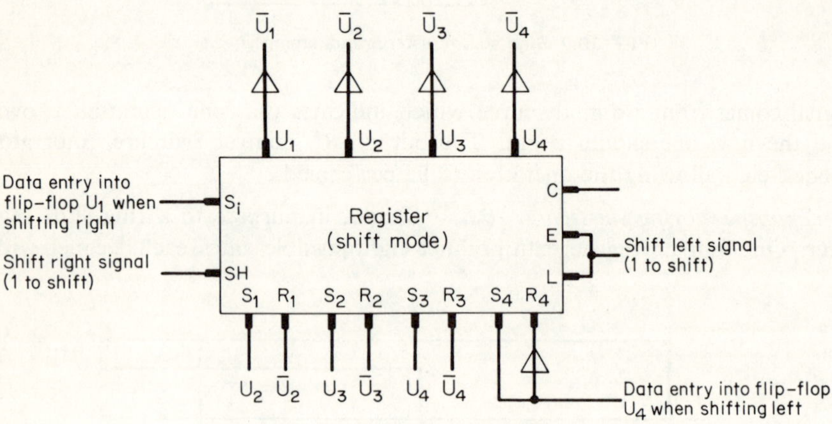

FIG. 20.9 Bidirectional shift register.

Bidirectional counter (*Fig. 20.10*). A bidirectional counter permits subtracting 1 as well as adding 1 to the contents of the register. Such programming of a register in the counter mode again loads the proper data into the register through the enable and load input commands instead of subtracting; the AND gates determine the number to be placed into the register. Upon the command to subtract 1 ($EB_i = 1$), a given flip-flop triggers only if all the preceding ones are reset. (The first one always triggers.) A "borrow-out" signal B_o [when (0, 0, 0, 0) is in the register] similar to carry-out signal C_o [when (1, 1, 1) is in the register], can be constructed with an AND gate if needed. As shown, a "borrow-in" signal B_i can be used if needed to force $B_o = \bar{U}_1\bar{U}_2\bar{U}_3\bar{U}_4B_i$, and is similar to the signal C_i when adding 1. Thus B_i AND EB_i must be high to subtract 1. To eliminate the borrow-out signal, AND gates 1 and 2 can be omitted with a high signal replacing B_i as a trigger of the first flip-flop.

The diagrams in Figs. 20.7 through 20.10 are working diagrams indicating all leads to be patched. For general schematics, simpler diagrams can be used.

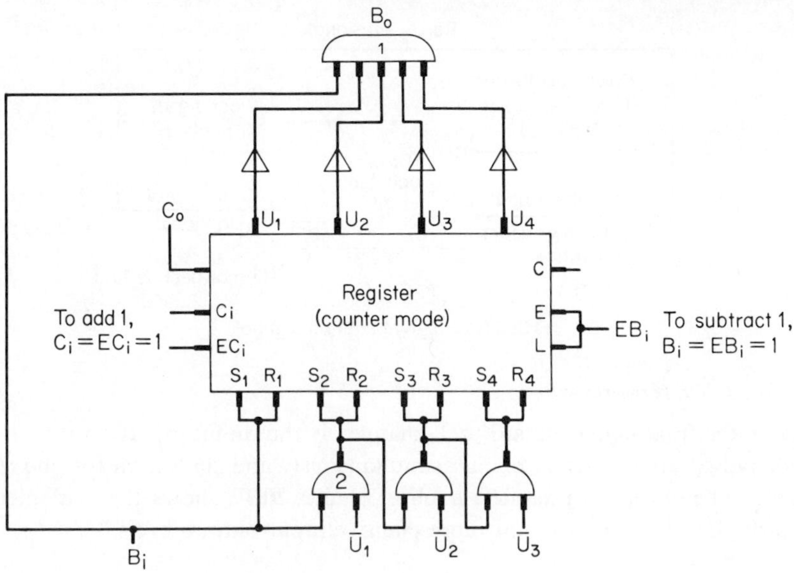

FIG. 20.10 Bidirectional counter.

20.1.4 Down counter

For obtaining large counts (0–99), specially designed down counters are available with the count presettable by thumbwheel switches. Two inputs C_i and EC_i are required for down counting and are similar in use to a register in the counter mode (except that EC_i is normally high if unused). In addition, clear and load inputs must also be patched for proper operation. The clear signal (1 to clear) must precede the load signal (1 to load) if loading is desired before the counter reaches 0. The number N selected on the thumbwheel switches is then loaded into the down counter register. EC_i must be 0 and L and C_i must be 1 to load properly. The register then down counts on command. When it reaches 0, a carry-out signal C_o is high provided C_i is high. This is the only output of the down counter; individual flip-flop outputs and trigger inputs are not available, saving much space on the patch panel. When 0 is in the register, it is necessary to reload it with N, otherwise a signal to down count places the number 99 into the register. Since the register contains 0 at the time of reloading, it is not necessary to clear it. The carry-out signal can be used to reload the register.

Figure 20.11 shows the working diagram of a down counter programmed to produce a carry-out blip after every 26 carry-in blips (one carry-in blip is needed to load the down counter). Down counters can be cascaded to produce very large counts.

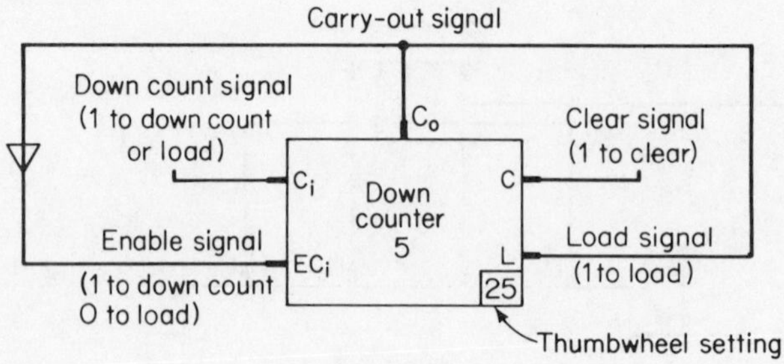

FIG. 20.11 Down counter symbol.

20.1.5 Differentiator

A differentiator detects a 0 to 1 change. If the input goes from 0 to 1 at a clock pulse, the output goes high simultaneously and stays high for one clock period. The output is therefore a blip. Figure 20.12 shows the symbol used for a digital differentiator and representative input-output levels.

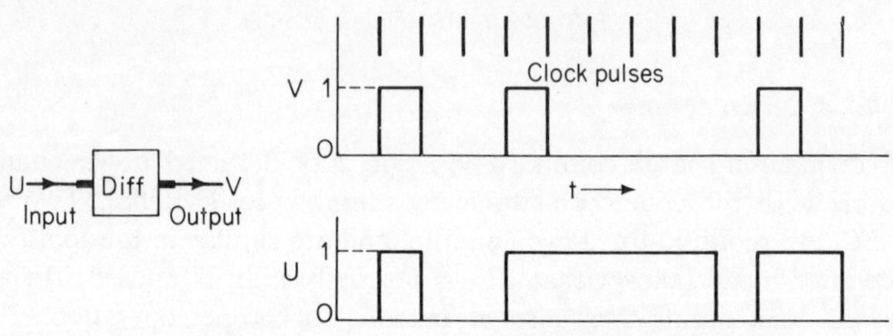

FIG. 20.12 Differentiator symbol.

Differentiators are useful for blip generation, which in turn can enable all elements through AND gates for one clock period. Suppose, for example, we wish to count the number of rep-op solutions produced by the variable drive voltage A, which goes from 0 to 1 at the start of the reset period. By feeding A into a differentiator, a blip is produced once during the reset time t_r $(A = 1)$, assuming many clock pulses occur during this time. The output signal of the differentiator can be used as an enable signal to the register in the counter mode as well as other clocked digital elements. Thus the entire digital expansion unit can be driven at the frequency of the computer solutions, even when variable. Such a circuit is shown in Fig. 20.13. Other enabling signals might include the OP signal of the MLG to avoid stepping the digital expansion unit while the computer is in the RESET or HOLD modes, or other clocked element outputs. During the reset period, other digital circuitry can also be programmed to do almost anything wished.

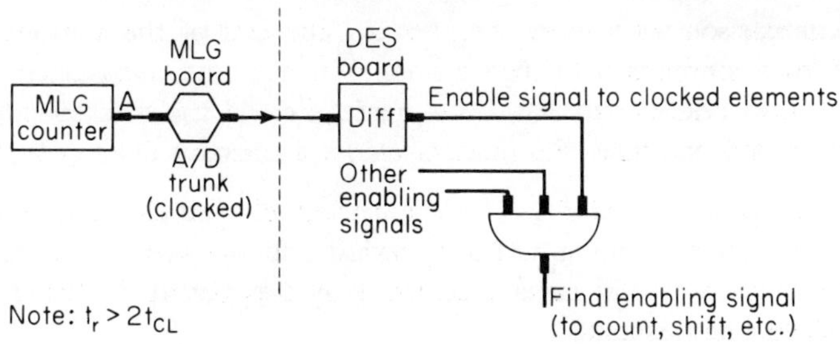

FIG. 20.13 Driving the digital expansion unit with the *A* drive voltage.

By constantly triggering a flip-flop, a square wave oscillator can be produced as an input to the master counter on the MLG unit producing *A*. Thus, the digital expansion system can also be used to drive the analog computer. There is little advantage to this, since the same circuitry for stepping clocked elements is required anyway. We shall always use the analog computer to drive the digital expansion unit as in Fig. 20.13. (A signal other than *A* can be used, provided it goes high once per solution cycle.)

A/D (Analog to DES-30) trunks, from the MLG board to the digital expansion board, are clocked, i.e., they contain flip-flops that change state only during clock pulses. Hexagon symbols are used for trunk lines as in Fig. 20.13. Care must be exercised in using unclocked signals for control when they are trunked from the analog portion to the digital expansion portion and back again. A comparator output, for example, may change state between clock pulses. If used through the digital expansion system to stop a track-and-store from tracking, a slight delay could be introduced. This could be made negligible by using a high (MHz) clock rate. But for smaller ones, do not use trunked signals to control tracking.

20.2 Programming with Parallel Logic Elements

Programming a digital expansion system is somewhat of art requiring some experience for proficiency. After deciding what is to be done, we must conceptually visualize how to do it in terms of opening and closing electronic switches (these include switches controlling track-and-store units and rep-op integrator modes). We must then program to get the necessary digital signals to operate the switches by interconnecting the proper digital sources so that they change state according to a predetermined time sequence. The timing sequence is crucial. Suppose we wish to make trial runs for some optimization scheme. In general, three periods are needed within a rep-op cycle:

1. Compute period t_c. This is the period (possibly variable from run to run) where we solve all differential equations necessary for determining a trial result for comparison with an existing result. The end of the compute period is signaled by a comparator changing state. At that time, another comparator indicates if a successful run has been made, i.e., if the trial somehow was "better" than previous runs. Parameters are held constant during this period.

2. *Digital updating period t_d.* In this period, clocked elements can change state to prepare for the analog updating period and the next compute period. Comparator outputs should remain constant in this period if they determine the state of clocked elements.

3. *Analog updating period t_a.* Here, parameters are updated when successful runs have been made, integrators are reset, and appropriate switches are opened or closed to prepare for the next run.

We use the timing sequence shown in Fig. 20.14. The normal operate period t_o is split into the compute period and the digital updating period by using the variable operate time circuitry of Fig. 18.9. In this way, the MLG

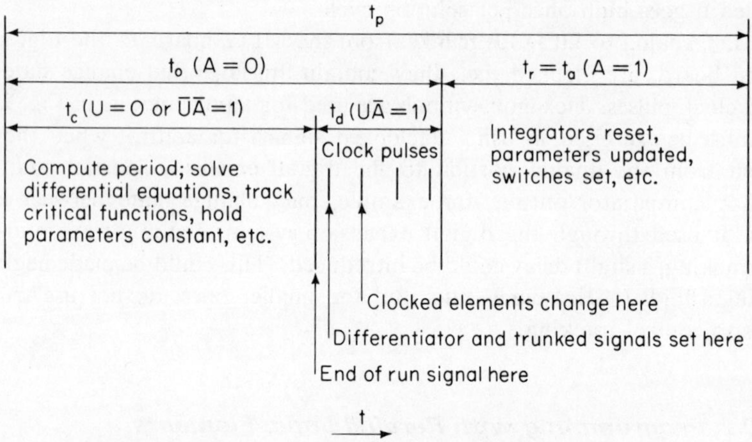

FIG. 20.14 Timing sequence used when computing with clocked digital elements.

counter operate setting denotes t_d and the MLG counter does not count during t_c. A differentiator then sets at the first clock pulse in the period t_d, which forces necessary changes in clocked element states at the next clock pulse (t_d should be at least two clock periods). The reset period t_r is used as the analog updating period t_a, changing parameters through accumulator circuits when necessary. Many variations to this sequence are possible; using one consistently merely aids the presentation of techniques.

"Initializing" the digital elements is just as important as initializing analog elements. The digital expansion system has pushbutton operate modes just as the main computer console:

CLEAR: All clocked elements are reset.
STOP : All clocked elements are inhibited from changing.
RUN : All clocked elements can change when enabled.

The digital expansion system can manually be placed in CLEAR, STOP, and then RUN. No elements will change because the differentiator remains reset. Provision is made to set any clocked element manually, if necessary, at any time. Thus, we adopt the following sequential steps as the initialization procedures.

(1) CLEAR the digital expansion unit
(2) Set any required flip-flop
(3) Place the analog computer in RESET and OPERATE
(4) Place the digital expansion unit in RUN

Step 2 can be omitted if all flip-flops initially are reset, or if the required flip-flops can be set automatically. We must also arrange the analog portion of the program so that no iteration takes place until step 4 is completed.

An end-of-problem signal can be generated, either by a counter, manual pushbuttons, or other digital logic. More often than not, the operator can determine when an iterative problem is over by monitoring certain analog voltages to determine if convergence has been attained. No end-of-problem signal is then needed.

The most difficult job in using gated logic is the design of the control circuitry. To make this easier, first draw a computer diagram showing all switches, track-and-store units, etc., that will be controlled by digital signals, and write out the conditions under which the signals should be set. Use the words AND, OR, or any Boolean function. Label the digital signals with names to help remember their function in a manner similar to labeling analog outputs. We always capitalize these names to remind us also that they are digital signals. Further, the names are chosen so that their functions correspond to their set state. This procedure is demonstrated in the examples of the next four sections.

20.3 Serial Solutions to Partial Differential Equations

Jury [2], [3] proposed using a continuous memory circuit for the serial solution of partial differential equations. As discussed in Sec. 14.6, an entire function u_i of one independent variable x produced at one value y_i of the other independent variable must be stored and used as an input to the equations when y is increased to the value $y_i + \Delta y = y_{i+1}$. Further complicating this process is the fact that a boundary-value problem must be solved to generate the new function $u_{i+1}(x)$. Since a simple two-point boundary-value problem is amenable to solution on an analog computer by iterative means, the entire solution of the partial differential equation can be done iteratively.

If the equations to be solved are unstable (and they usually are), the iterative boundary-value solution might be quite sensitive to parameter adjustments. In that case, noise and other errors might prohibit a satisfactory solution. Also, we must beware of using continuous memory circuits (Figs. 18.23–18.25) in situations where polynomial approximations are not justified. And we must not wait too long to update the memory; capacitor leakage causes the memorized function to deteriorate with time.

Example 20.1. We now solve the normalized heat conduction equation

$$\frac{\partial u}{\partial t} = \frac{\partial^2 u}{\partial x^2}$$

by the serial method as described in Chapter 14, by taking a backward finite difference approximation in time and integrating continuously in x. With $0 \leq x \leq 1$, we assume there is a heat sink maintaining $u(1, t) = 0$, while the temperature at the other end of the slab varies sinusoidally:

$$u(0, t) = \sin 2t$$

and the initial temperature profile $u_0 = u(x, 0) = 0$. The equation solved in rep-op is

$$\frac{d^2 u_i}{dx^2} = \frac{u_i - u_{i-1}}{\Delta t} \tag{20.1}$$

with the continuous memory used to store the $u_{i-1}(x)$ obtained with fixed t_{i-1}.

There are three distinct sections to this program.

(1) A boundary-value iteration as discussed in Sec. 19.5. Several cycles in rep-op are needed to insure convergence to the correct initial value of $(du_i/dx)_0$.

(2) A memory, to be updated only when u_i is found, and not after every cycle as in the case of integral equations (Sec. 18.5.2).

(3) A down counter to "step" t and update the memory after n iterations, where n may be determined by trial (we use $n = 10$). Such uniform stepping is very desirable to keep solutions uniform in t.

Figure 20.15 shows the circuitry to sample a slow time sin $2t$ oscillator, to solve the boundary-value problem in rep-op using an accumulator circuit for $\Delta t = 0.05$, to produce a timing sequence and an enabling signal E, and to count with a down counter that provides an UPDATE signal.

Each rep-op period of $t_p = 20$ ms is split into the subperiods of $t_c = 10$ ms, $t_d = 0.1$ ms, and $t_a = 9.9$ ms. In slow time, we let t continually increase using $\dot{t} = k$, but since k must be referenced to the Δt chosen, we have

$$k = \frac{\Delta t}{nt_p} = 0.2500$$

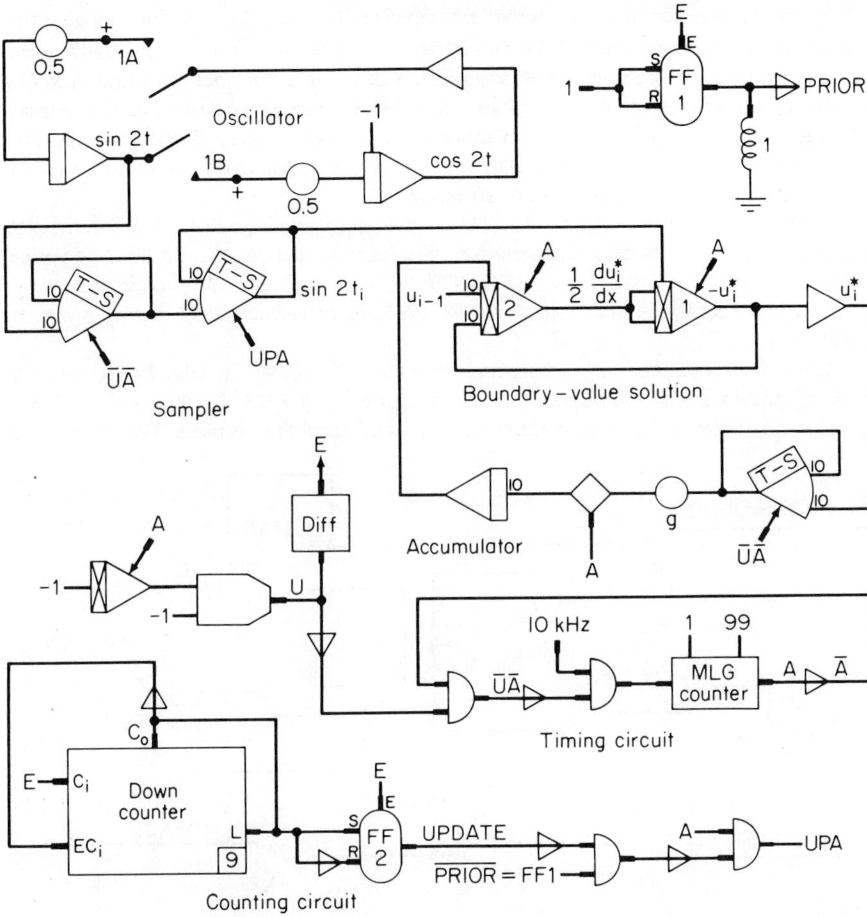

Notes: (1) $t_a = t_r = 9$ ms; $\Delta t = 0.05$
(2) Use 1 MHz clock
(3) Limit integrators 1 and 2 to ± 1
(4) DES-30 elements initially cleared
(5) UPA ≡ (UPDATE OR PRIOR) AND A
(6) EMC integrators use 0.01 μF capacitors.

FIG. 20.15 Control circuitry for Example 20.1.

The slow time oscillator pots are set to twice this value to produce sin $2t$, but the oscillator is inhibited by FF1 until the DES-30 is placed in RUN. The first track-store amplifier of the sampler tracks sin $2t$ during t_c (where $\bar{U}\bar{A} = 1$), and the second stores the value every tenth compute cycle, when the UPDATE signal goes high. Since we wish to force its value to zero PRIOR to placing the DES-30 in RUN, such a signal (which is just FF1) is also available. PRIOR is initially high but remains low after the very first enabling signal.

In the rep-op circuit, the symbol u_i^* replaces u_i since they are not equal until convergence has taken place. The circuit requires the function u_{i-1}, which will come from a continuous memory. The boundary value for u^* is met by automatically adjusting its initial derivative using an accumulator circuit as described previously. A value of 0.5 for g was found to produce convergence in about 5 cycles. Note that integrators 1 and 2 must be limited to prevent possible saturation due to the unstable solutions before convergent ones are obtained.

The timing circuit is standard. The end-of-compute signal U is differentiated digitally to provide the enabling signal E. This permits the down counter to step once during t_d, and the final signal, UPDATE AND A, is not only high during t_a once every 10 computer solutions, but also PRIOR to the iteration during t_a every cycle.

The continuous memory circuit that stores u_{i-1} is shown in Fig. 20.16, using a cubic approximation. Quantized values for x are 0, 0.3333, 0.6667, and 1. Since $u_{i-1}(0, t_{i-1}) = \sin 2t_{i-1}$, another track-store pair of amplifiers is used. For the middle

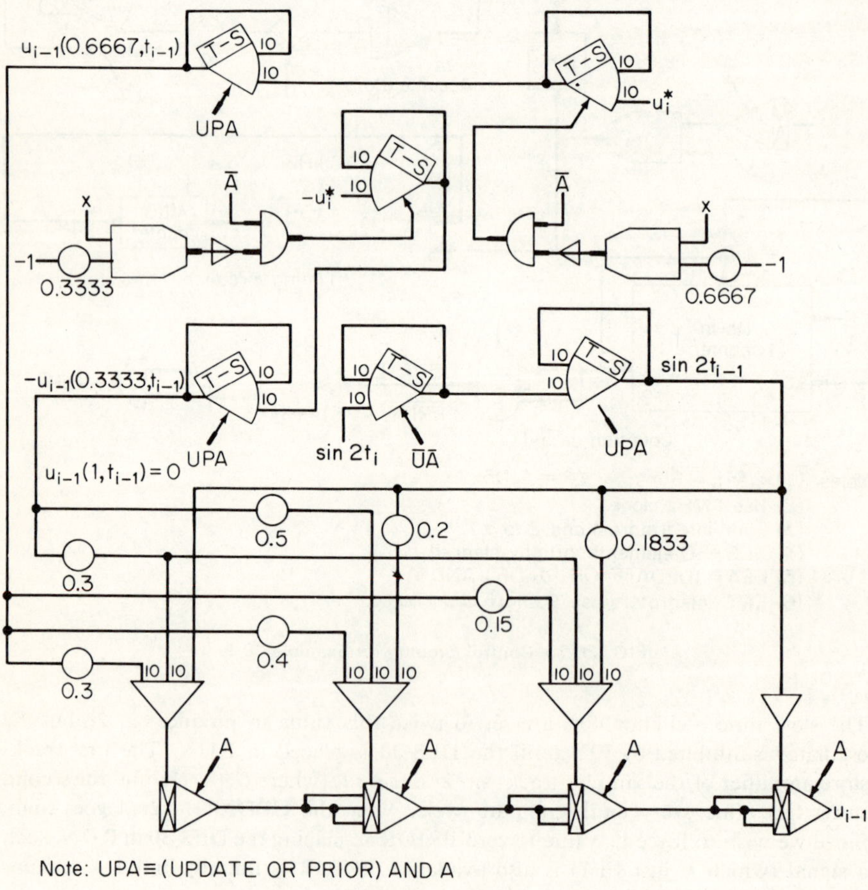

FIG. 20.16 Continuous memory circuit.

two points, tracking during t_c takes place every cycle until $x = 0.3333$ and 0.6667. The stored values, however, are updated during t_a after every 10 computer solutions, or after every solution PRIOR to placing the DES-30 into RUN. No inputs from $u_{i-1}(1, t_{i-1})$ are needed because this value is always 0.

When the computer is in OPERATE with the DES-30 inactive, $u_0 = 0$ is forced into the memory regardless of what value existed beforehand; the initial and final values of u_i^* are forced to be zero, so that the boundary-value solution will produce $u_i^* = 0$. Nothing happens the instant the DES-30 is placed into RUN because no enabling signal is obtained until the end of t_c. After the first enabling signal, the relays close enabling the oscillator to oscillate, and the down counter loads, setting both FF1 and FF2. After the second enabling signal, FF2 resets, forcing 0 to remain in the memory and in the sampled $\sin 2t_i$ and $\sin 2t_{i-1}$ amplifiers. After the tenth enabling signal, FF2 sets again, and the $\sin 2t_i$ amplifier stores the value of $\sin 0.1$, which then produces some iteration.

Output signals at any x and t can be used in another part of some larger problem. Figure 20.17 shows a typical set of outputs, $u_{i-1}(0, t)$ and $u_{i-1}(0.5, t)$. The former, of course, is just the sampled values of $\sin 2t$. Note the phase shift in the solutions, predictable by theory.

Throughout the solution, d^3u_i/dx^3 remained small, justifying a cubic polynomial approximation. But the error introduced by taking $\Delta t = 0.05$ needs analysis. A good method to ascertain these errors is to halve Δt and rerun to see if every other updat-

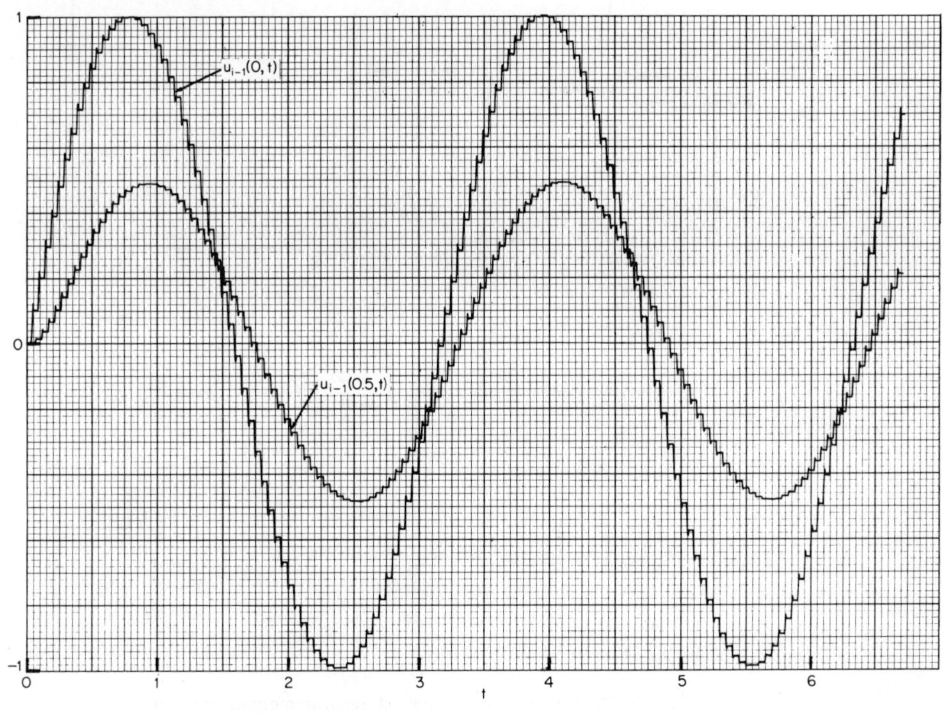

FIG. 20.17 Output $u_{i-1}(0, t)$ and $u_{i-1}(0.5, t)$.

ing resembles the previous solution. In this case, it does, assuring a reasonably good overall solution.

This is a rather simple example of a linear equation which demonstrates the techniques involved. The reader can appreciate the savings of equipment that can result when solving nonlinear equations. A finite difference scheme requires nonlinear equipment at all stations, but in this technique the equations need to be solved just once.

20.4 Time Sharing Integrators for Linear Systems

When gradient components for a general linear system of order s

$$\dot{\mathbf{y}} + \mathbf{A}(t)\mathbf{y} + \mathbf{g}(t) = \mathbf{0}, \qquad (\mathbf{y})_0 = \mathbf{y}_0 \tag{20.2}$$

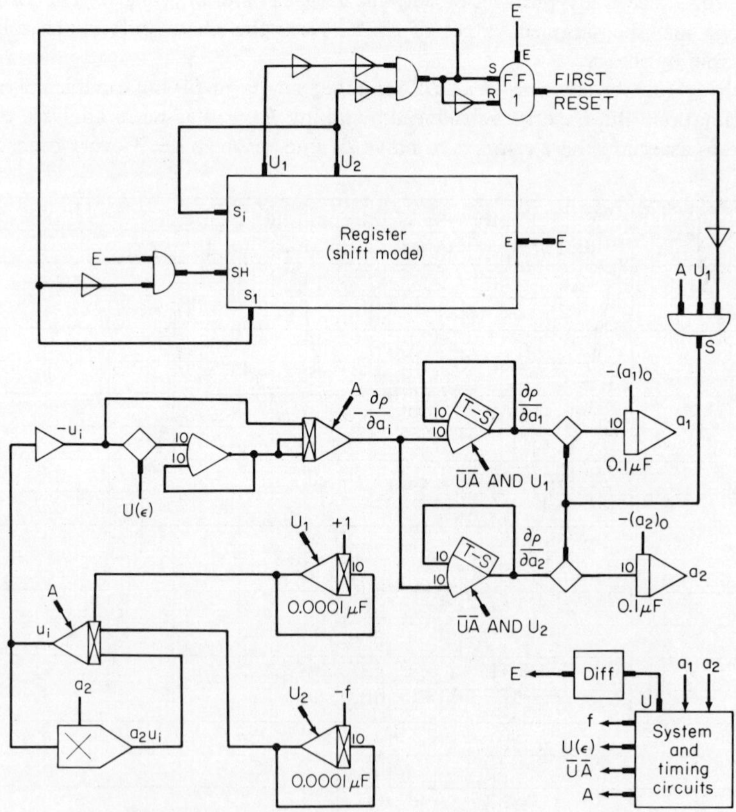

Note: All clocked elements initially cleared.

FIG. 20.18 Time sharing the partial derivative circuit for linear equations.

are needed for a discrete gradient optimization procedure, the method of Sec. 19.4.2 may require much more equipment than twice the components needed to mechanize (20.2). But the structures of the sensitivity equations and (20.2) are identical, the only difference being in the forcing function or the initial conditions. For, differentiating (20.2) with respect to a_i,

$$\frac{\partial \dot{\mathbf{y}}}{\partial a_i} + \mathbf{A}(t)\frac{\partial \mathbf{y}}{\partial a_i} + \frac{\partial \mathbf{A}}{\partial a_i}\mathbf{y} = \mathbf{0} \tag{20.3}$$

With $\mathbf{u} = \partial \mathbf{y}/\partial a_i$, (20.3) reduces to

$$\dot{\mathbf{u}} + \mathbf{A}(t)\mathbf{u} + \frac{\partial \mathbf{A}}{\partial a_i}\mathbf{y} = \mathbf{0}, \qquad (\mathbf{u})_0 = \frac{\partial \mathbf{y}_0}{\partial a_i} \tag{20.4}$$

In lieu of solving system (20.4) in parallel for each i needed, we may conveniently time share the $\mathbf{u}$ integrators, storing the partial derivatives as i is shifted every run, and updating the a_i every nth run. This saves much analog equipment for complicated systems, at the expense of simple logic and n times the convergence rate.

Example 20.2. Since (19.15) is linear with $s = 1$, we demonstrate this procedure by solving Example 19.3 with one set of circuitry to find $\partial p/\partial a_i$. Figure 19.9 contains the system and timing circuits. Figure 20.18 contains the new gradient circuit and the gated-element control.

Integrators used as switches provide the proper input or initial condition for the u_i integrator. An enabling signal E is obtained once again by differentiating U. The ring shift register cycles U_1 and U_2 high with the a_i-updating occurring every second reset period (not counting the FIRST RESET period).

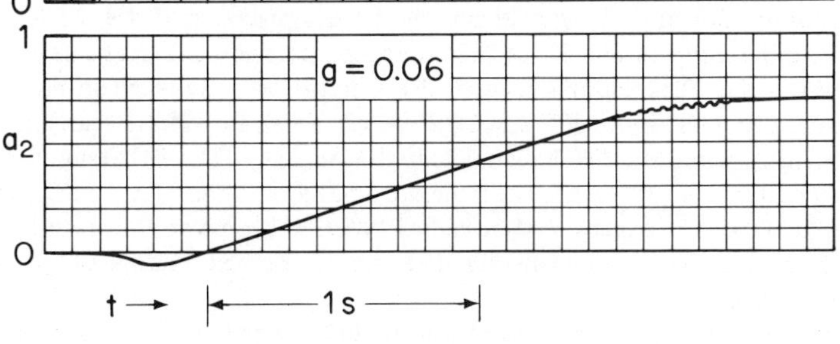

FIG. 20.19 Graph for $g = 0.06$.

With the computer in OPERATE and the DES-30 inactive, the a_i do not update because U_1 is cleared initially. After the first enabling signal, U_1 sets, but $\overline{\text{FIRST RESET}}$ becomes low, so that both updating switches still remain open in t_a. However, during the next compute period, $\partial p/\partial a_1$ is tracked and stored. After the second enabling signal, $U_2 = 1$, $U_1 = 0$, $\overline{\text{FIRST RESET}} = 1$. The switches remain open and f is switched into the u_i integrator, computing $\partial p/\partial a_2$. After the third enabling signal, and every other one thereafter, $U_1 = 1$, $U_2 = 0$, $\overline{\text{FIRST RESET}} = 1$, so that updating occurs when $A = 1$ (in reset).

As can be seen from the graph (Fig. 20.19), a convergence pattern identical with that of Fig. 19.10 is obtained with $g = 0.06$, taking exactly twice as long to converge.

20.5 Quasigradient Method by Relaxation

In this method, we do not actually find the gradients. A parameter is increased by Δa. If p decreases we are going in the right direction; if not, the initial increase was either too great or it is in the wrong direction. We assume the latter and decrease the parameter by Δa. Now if p decreases, we're heading correctly, otherwise the step size Δa is too large. Make Δa smaller (say $\Delta a/2$) and repeat. Any failure after two trials means the step size is too great, so Δa must be reduced. When Δa reaches a preset minimum (say 0.0001), switch to the next parameter. Note that this is a direct method requiring the generation of p.

To mechanize this scheme in a simple way we update the a_i and p after every trial, but note if a FAILURE or SUCCESS has occurred. We need

(1) A shift register to control the parameter to be investigated.
(2) A SUCCESS-FAILURE flip-flop.
(3) A Δa or $-\Delta a$ flip-flop, triggered when failure occurs.
(4) A two-bit counter to determine if TWO RUNS have been made after a shift to a new parameter.
(5) Means to cut Δa in half after TWO RUNS AND a FAILURE.

Example 20.3. Attacking the same problem as in Example 19.2, Fig. 20.20 is a circuit diagram showing all logic. When the DES is placed in RUN, FI is high. This controls the latching of the SUCCESS comparator, and enables the shift register to set U_1. After the FIrst enabling signal, FI remains low. Δa is cut in half by simple exponential decay during t_a if FAILURE AND TWO RUNS (at least) are high. It resets during t_a when the SHift signal is high ($\Delta a < \epsilon$). The SHift signal also shifts the register, clears the TWO RUNS indicator and latches the SUCCESS comparator. Finally, Δa TRGR is triggered after each FAILURE to reverse the switch states that update the a_i, and therefore reverse the direction of the next change of the parameter being worked on.

There are several shortcomings to this method. We allow a run after SH becomes high. Also, if the first run is a failure, a_i decreases by only Δa instead of $2\Delta a$. Both cases waste runs and possibly valuable computing time. We allow the a_i to change

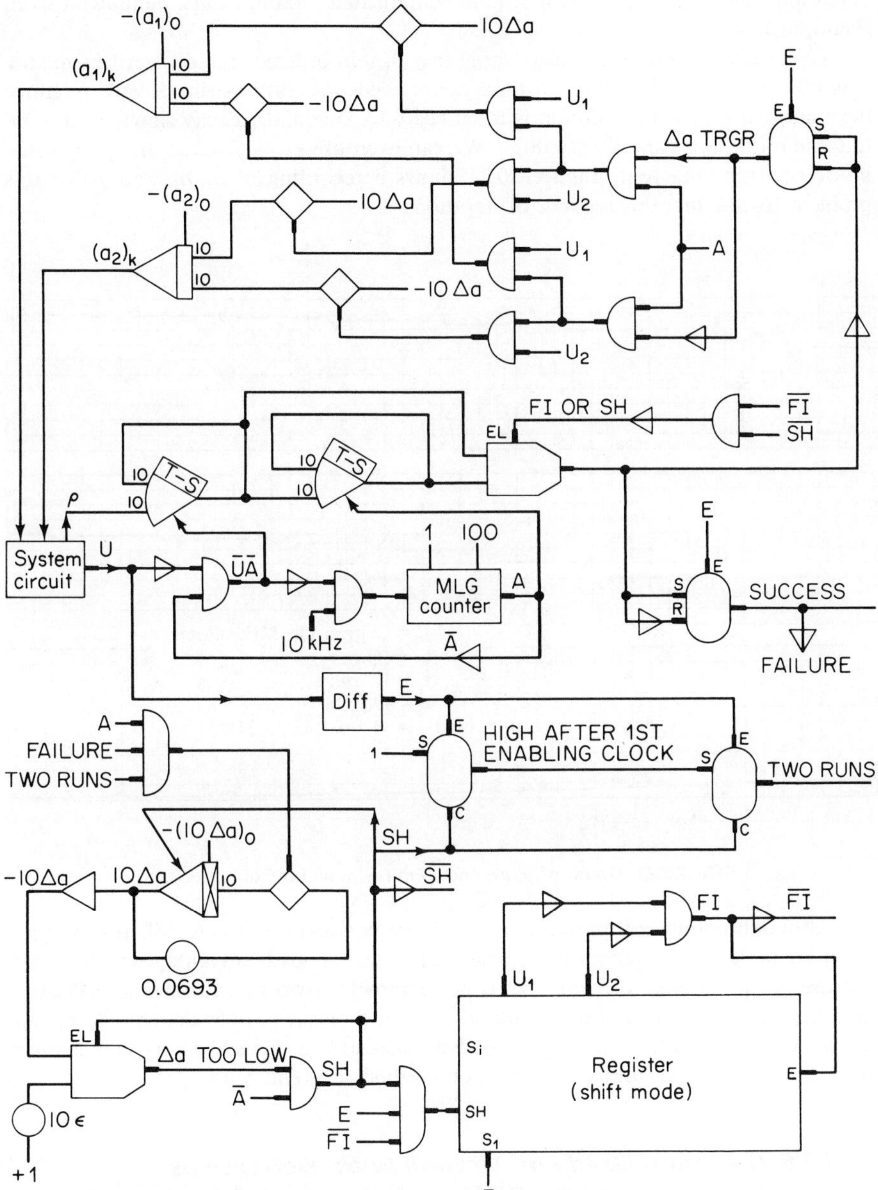

Note: All clocked elements initially cleared

FIG. 20.20 Circuit for the quasigradient relaxation method.

while Δa is changing exponentially. A failure after success reverses a_i by the amount $\Delta a/(2 \ln 2) \simeq 0.72\Delta a$ instead of $\Delta a/2$. (This is not particularly serious.) All of these difficulties can be corrected with a more complicated program to be demonstrated in Example 20.4.

To start, we should have $(\Delta a)_0$ large (say 0.1) in order to rapidly zero in on the a_i within 0.05. If $\epsilon = 0.0001$, 14 failures are needed to shift. A simple way to reduce the many runs near the solution is to let $(\Delta a)_0$ exponentially decay slowly with time to some minimum value (say 0.007). We can even use $\epsilon = 0.02(\Delta a)_0$ to speed shifting away from the root. Figure 20.21 shows a recording of a_1, a_2, and ρ for this problem for the first few seconds of iteration.

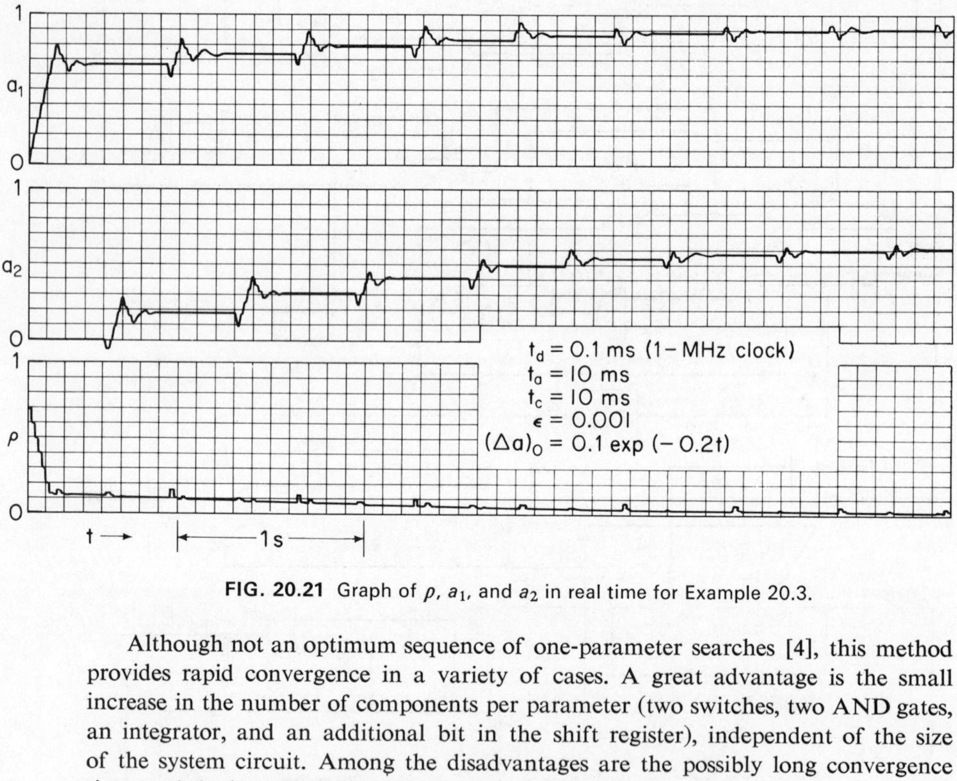

FIG. 20.21 Graph of ρ, a_1, and a_2 in real time for Example 20.3.

Although not an optimum sequence of one-parameter searches [4], this method provides rapid convergence in a variety of cases. A great advantage is the small increase in the number of components per parameter (two switches, two AND gates, an integrator, and an additional bit in the shift register), independent of the size of the system circuit. Among the disadvantages are the possibly long convergence time, and the inapplicability of relaxation methods in some cases.

20.6 Optimum Gradient Method with Derivatives Approximated by Differencing

To mechanize the optimum gradient method, we must find each $\partial\rho/\partial a_i$ and determine the g that minimizes ρ with this information. In lieu of this, we use several runs to zero in to minimum ρ in a way similar to Example 20.3, but attempt to correct the shortcomings of that method. Much of the ideas in the following example could be used to speed convergence of the quasigradient method in Example 20.3.

Example 20.4. (***Optimum Gradient Method***). We mechanize Example 20.3 in an elaborate way to demonstrate the versatility of logic circuitry and illustrate more sophisticated circuitry (at the expense of more equipment). The method consists in

(1) Determining the partial derivatives $\partial\rho/\partial a_i$ approximately with second-order error from (19.12).

(2) Mechanizing the discrete gradient method with arbitrary small g_0.

(3) Increasing g_k to $\simeq 1.24g_{k-1}$ as long as no failures occur, using the same derivative information.

(4) After the first failure, changing the sign of g_k and decreasing its magnitude to about one-fifth of its previous value.

(5) After a second failure, obtaining new derivative information and repeating the steps.

Only one failure is permitted because $\partial\rho/\partial a_i$ should be approaching 0. When near the optimum values of a_i, the first run always produces a failure with only small excursions. Because of noise, computer error, and approximation error, $\partial\rho/\partial a_i$ never is quite 0, but oscillations in the a_i become quite small.

To mechanize this scheme, digital and analog updating periods are further divided into two subperiods. In the digital updating period, the usual enabling signal E coming from the differentiator, which is set at the end of the compute period, is stored in a flip-flop. This signal E_1 is used as an enabling signal after the very next clock pulse, to set switches for updating g_k and the a_i in the next analog period. No run

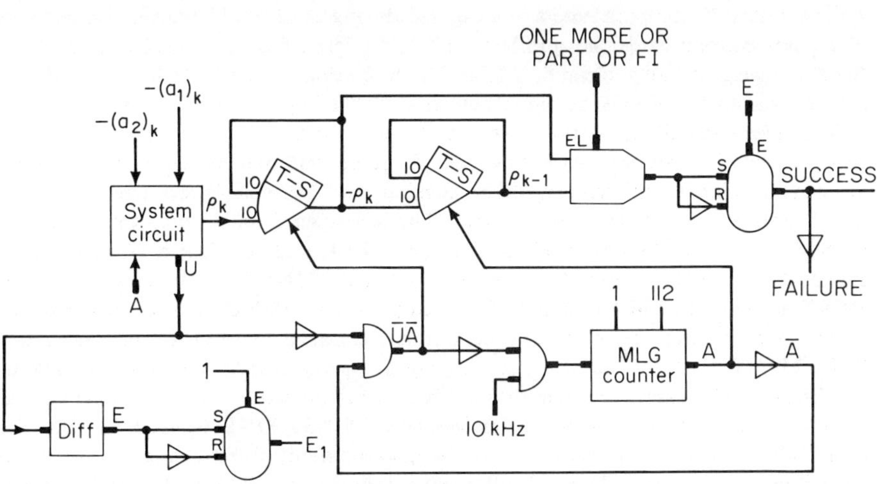

FIG. 20.22 Dual enabling signals for sequential operation in t_d.

is wasted if the SUCCESS flip-flop is enabled by E. Figure 20.22 shows this part of the circuitry. We update ρ every cycle so that a track-store pair of amplifiers can compute the SUCCESS signal. The success comparator is latched high under one of three circumstances: before the FIrst enabling signal sets the first flip-flop in the shift register, during the computation of the PARTial derivatives, or after the computation of the last partial derivative, when ONE MORE run is needed to store the current value of $\rho(a_1, a_2)$ into the ρ_{k-1} amplifier. This is necessary because we are time sharing these ρ amplifiers. During the partial derivative computation $\rho_{k-1} - \rho_k$ represents $\rho(a_i + \Delta a) - \rho(a_i - \Delta a)$, and during the periods where the a_i are updated (which we call the GRADient periods), these amplifiers produce the SUCCESS and FAILURE signals. An alternate procedure is to store the current best value of $\rho(a_1, a_2)$, thus not wasting a run after each partial derivative computation.

The MLG counter is set to produce an analog updating period t_a of 11.2 ms. We subdivide this period into two periods TA1 and TA2 as shown in Fig. 20.23. During TA1, g_k will be changed if necessary, and during TA2, the a_i will be updated. We have 1.2 and 10 ms to do these tasks.

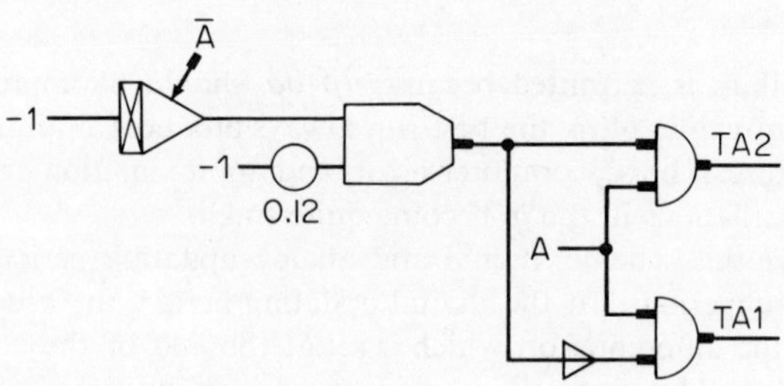

FIG. 20.23 Analog updating signals.

Figure 20.24 shows the logic used to count the failures during the GRADient periods and shift the parameters investigated during the PARTial derivative periods. The down counter loads the number 1 after the FIrst enabling pulse E_1. Since all flip-flops were initially cleared, FI was high, forcing U_1 and PART high. Since PART was low, MINUS remains low, and PLUS high. Thus Δa has a value of $+0.01$. Because GRAD is now low, the down counter is inhibited throughout this partial derivative routine. After the next E_1 pulse, the register still shows U_1 high since its shifting is inhibited by MINUS being low. MINUS is now triggered high, which produces -0.01 as a value of Δa. The following E_1 pulse shifts the register to U_2 high and MINUS low producing $\Delta a = +0.01$, and the fourth E_1 pulse again makes $\Delta a = -0.01$. After four cycles, U_2 AND MINUS are high, which resets PART to enter the GRAD phase of the computation. MINUS is also triggered low and U_1 high in preparation for the next time the partial derivatives are calculated. ONE MORE is set for use as a disabling signal for the extra run and Δa is made 0.

Figure 20.25 shows the remaining circuitry for updating the a_i and changing g. Two auxiliary control signals are needed. AT LEAST ONE SUCCESS and NO FAILURES in the GRADient part of the computation inhibit changes in g during TA1 if either are reset. These flip-flops are cleared when PART OR ONE MORE are high to insure a reset state after the first GRADient change. When both are high

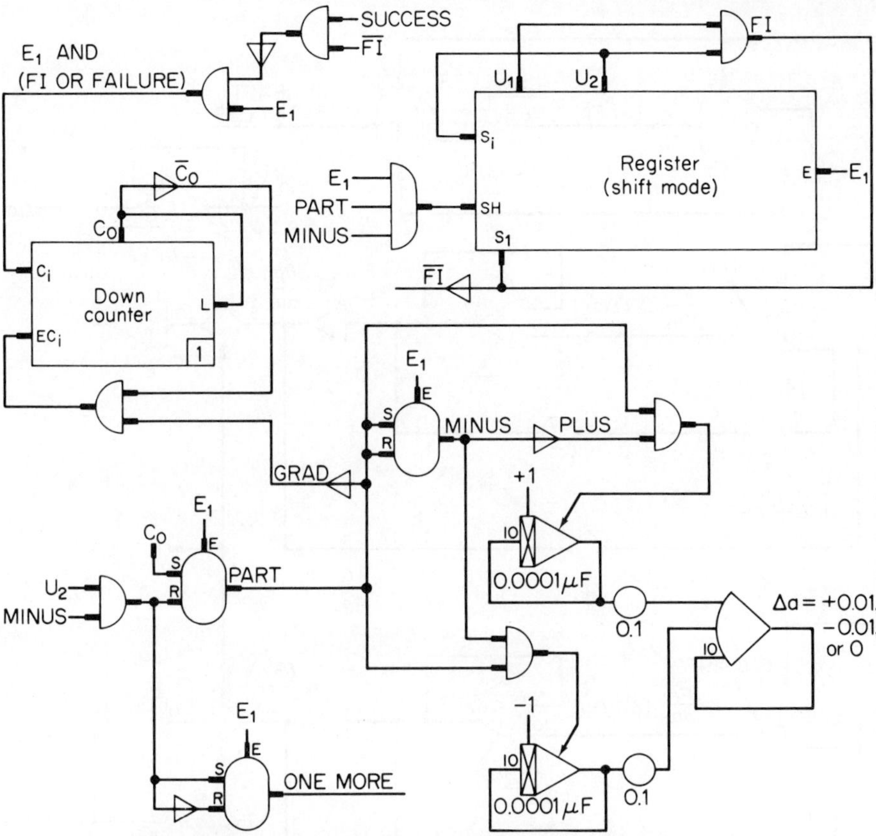

FIG. 20.24 Logic for PARTial derivative and GRADient routines and the generation of Δa.

(NO FAILURES after AT LEAST ONE SUCCESS), the value of each partial derivative integrator increases during TA1 by integrating the stored value (amplifiers 1 and 2) for 1.2 ms with positive feedback. Since the gain of the integrator is 200, its new value becomes 1.24 of its old. A FAILURE makes NO FAILURES low and inhibits this switch. Then, the switch with negative feedback is on. With a gain of 1000, the new value of the integrators becomes -0.2 of the old. Directions of a_i changes are reversed. The constants 1.24 and -0.2 were selected by trial to produce good results. Other constants between 1.00 and 1.5, and -0.1 and -0.5 all converged to the solution in different times. A second FAILURE sets PART and the entire process repeats.

In TA2, the a_i are always updated during the GRADient portion provided ONE MORE AND FI are low.

Figure 20.26 shows the results as obtained in real time for comparison with other methods. Convergence is more rapid than with the quasigradient method. The price paid is extensive logic and analog equipment.

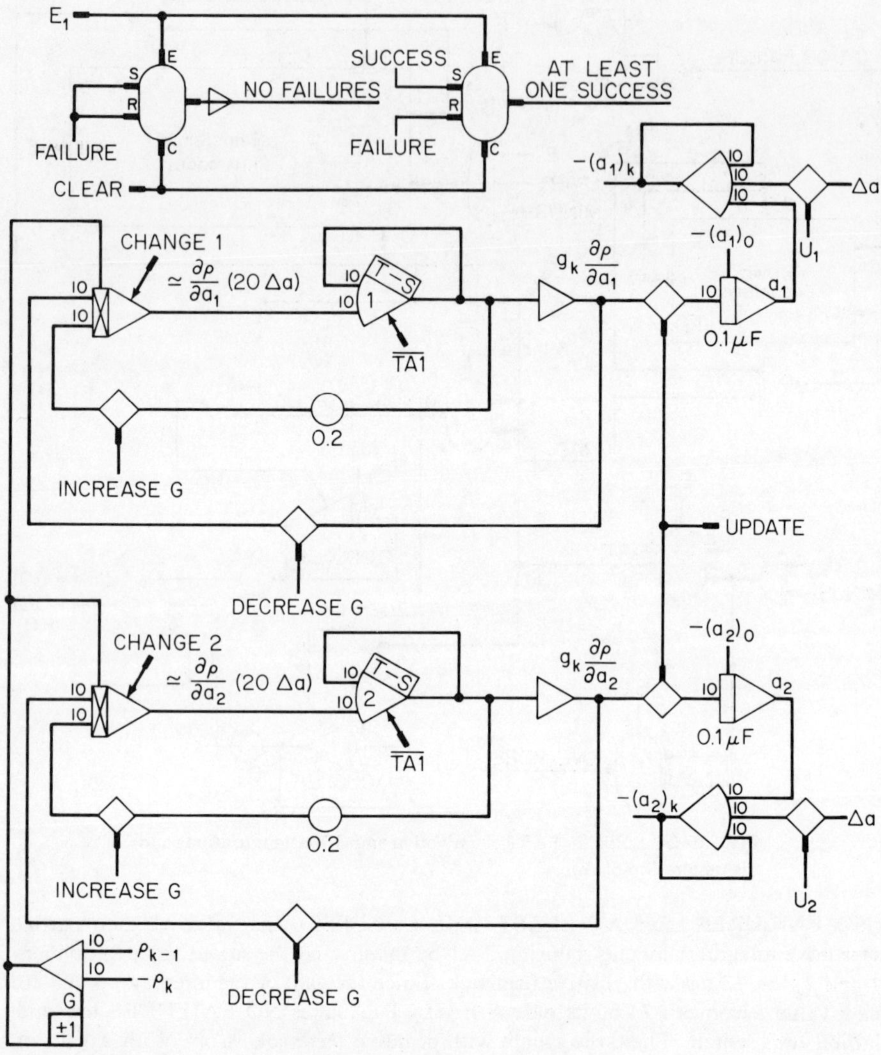

Notes: (1) CLEAR ≡ PART OR ONE MORE
(2) CHANGE 1 ≡ [(PART AND U_1) OR FI] AND $\overline{UA}$
(3) CHANGE 2 ≡ [(PART AND U_2) OR FI] AND $\overline{UA}$
(4) INCREASE G ≡ TA1 AND AT LEAST ONE SUCCESS AND NO FAILURES
(5) DECREASE G ≡ TA1 AND FAILURE
(6) UPDATE ≡ TA2 AND GRAD AND $\overline{\text{ONE MORE}}$ AND $\overline{\text{FI}}$

FIG. 20.25 Changing g and the a_i.

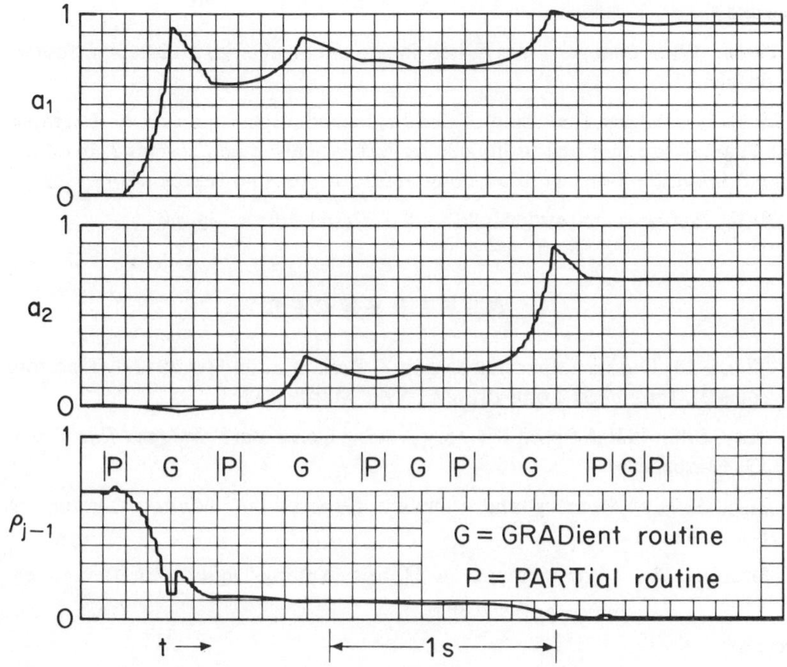

FIG. 20.26 Typical graph of the results of an optimum gradient method in real time.

PROBLEMS

20-1. Program a differentiator from a flip-flop and AND gate.

20-2. Cascade two four-bit registers into a single shift register of eight bits.

20-3. Initialize a ring shift register with a flip-flop instead of the AND gate as shown in Fig. 20.7.

20-4. Mechanize a four-bit bidirectional transposed ring shift register.

20-5. Cascade two down counters to produce a blip once every 3000 enabling cycles.

20-6. Program a general random search by relaxation for four parameters, assuming a general system circuit.

20-7. Show that $(\Delta a)_k = 0.72(\Delta a)_{k-1}$ in Fig. 20.20 when a FAILURE occurs after TWO RUNS.

20-8. Correct the difficulties listed for the scheme in Fig. 20.20 by using the ideas in Example 20.4.

20-9. Repeat Example 20.2 using a Chebyshev norm. (A maximum-value circuit is needed.)

20-10. What changes are required in the circuit of Fig. 20.24 for a four-parameter problem?

20-11. In the serial solution of the heat conduction equation of Example 20.1, what changes are required to include radiation heating and cooling? Insulating one end and fixing the other with a known temperature-time function?

20-12. Set up a first-order hold with a digital differentiator.

REFERENCES

[1] "DES-30, Digital Expansion System," Publ. No. 00 800 2042-1, Electronic Associates, Inc., West Long Branch, N.J., March 1966.

[2] Jury, S.H., "High Speed Memory Analog Computer," *Ind. and Eng. Chem.*, vol. 53, March 1961.

[3] Jury, S. H., "Memory and Function Generation," *Military Systems Design*, vol. 6, Jan.-Feb. 1962.

[4] Brunner, W., "Comment on 'A Hybrid/Analog-Digital One-Parameter Optimizer'," *AICA*, vol. 6, July 1964.

See also:

[5] Witsenhausen, H.S., "Hybrid Techniques Applied to Optimization Problems," *Proc. Spring Joint Comp. Conference*, 1962.

[6] Gilbert, E.G., "Hybrid Computer Solution of Time-Optimal Control Problems," *Proc. Spring Joint Comp. Conference*, 1963.

[7] Elgerd, O.I., "Hybrid Computation—A Means to Study Predictive Control System Behavior," *IEEE National Convention Record, Part 1*, 1964, pp. 30-36.

[8] Melanson, F., and P. Lanner, "Parameter Optimization by Hybrid Techniques," *Instr. and Control Systems*, vol. 37, Jan. 1964.

[9] Vichnevetsky, R., "Application of the Digital Expansion System (D. E. S.) to an Automatic-Iteration Problem: (Eigenvalue Search,)" *Princeton Computation Center Report No. 178*, Electronic Associates, Inc., Princeton, N.J., March 8, 1962.

[10] Brözel, R., "A Method of Analog Integration with Respect to a Dependent Variable," *Simulation*, vol. 8, Feb. 1967.

[11] Bullock, T.E., "An Improved Automated Nyquist Plotting Method Using Digital Logic," *Simulation*, vol. 11, Sept. 1968.

[12] Ung, M.T., and G. Klein, "The Use of Parallel Logic in Teaching Engineering and Science," *AICA*, vol. 10, Oct. 1968.

Appendix

GRAPHS OF STEP RESPONSE TO CHEBYSHEV AND PADÉ TIME-DELAY APPROXIMANTS

The graphs in Figs. A1–A10 were digitally computed and drawn to serve as accurate references when setting up Padé and Chebyshev approximants to a time delay. The functions plotted are the approximations to $U(t - T)$. These should be useful for determining whether a time delay circuit is operating properly. Isolate the time-delay circuit from the problem and let $f(t) = 1$ in Fig. 10.12. The output should agree with the graphs when set up properly, with peak values somewhat sensitive to the pot settings for larger n. Ordinates of the graphs have been shifted for clarity.

Note the poor response of the $C_{n,n}^{\phi,0.5}$ cases. Although oscillations in all of these cases eventually decay with the output approaching 1 as $t \to \infty$, they should probably never be used to delay discontinuous functions. Good results can be expected for bandwidth-limited signals.

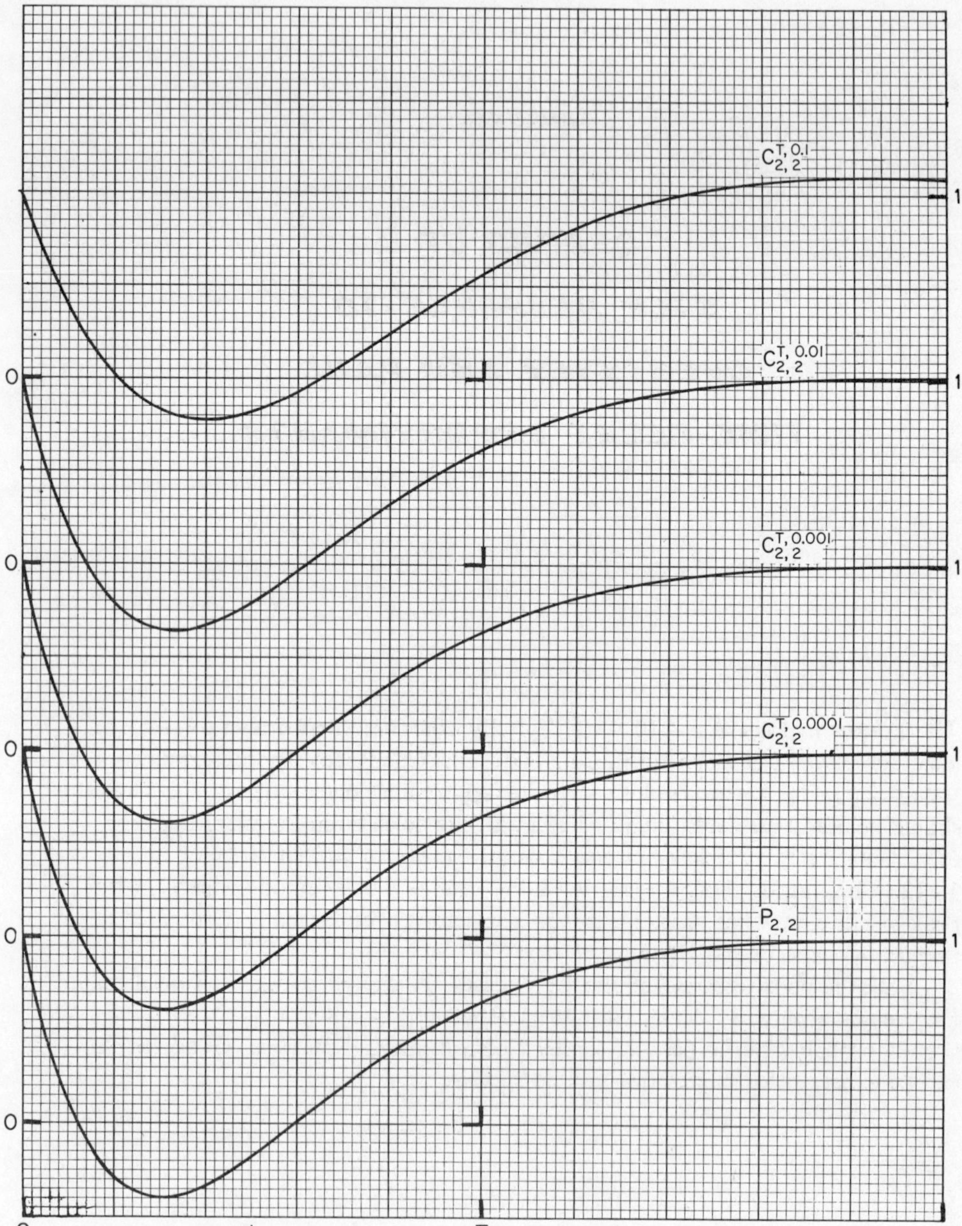

FIG. A1

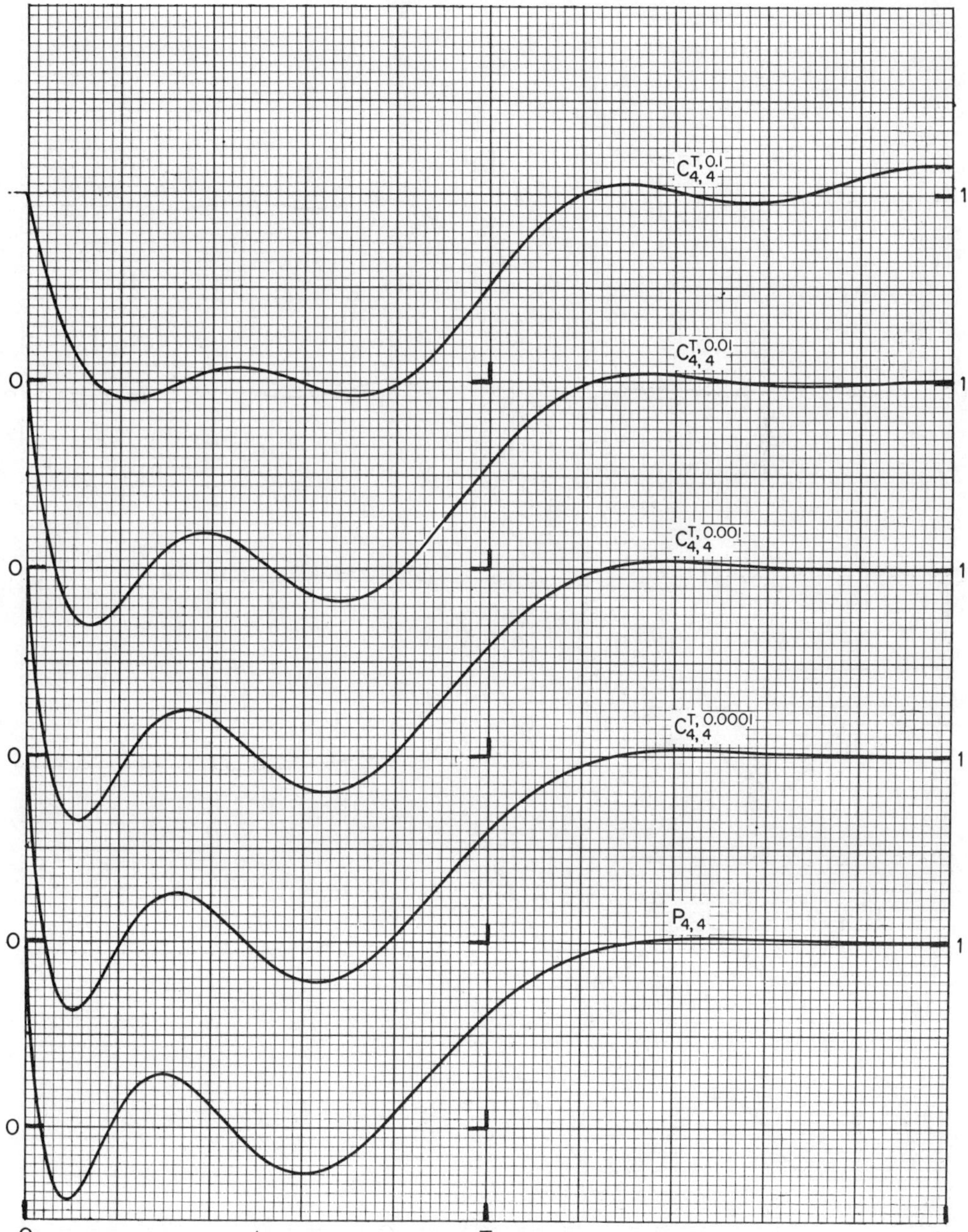

FIG. A2

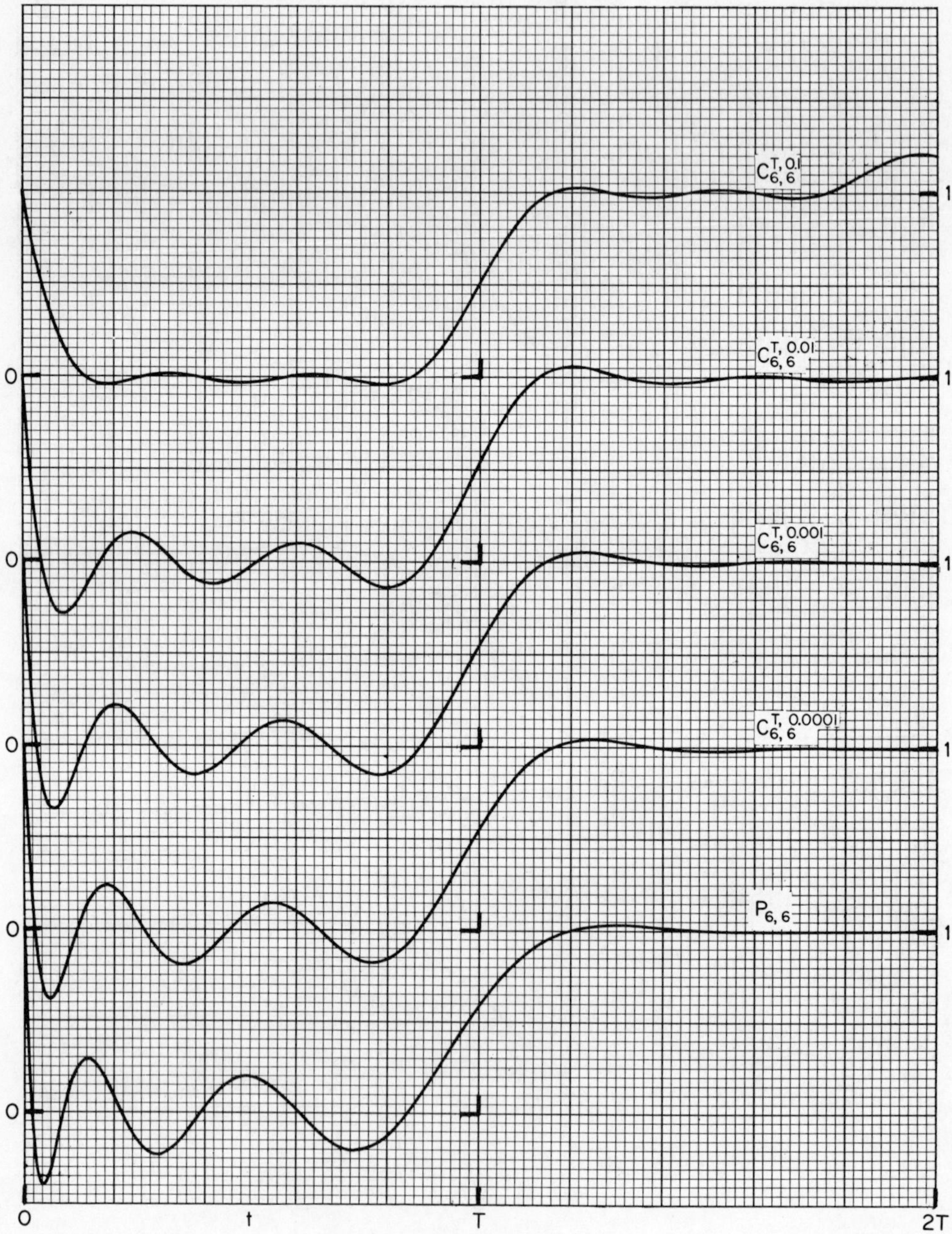

FIG. A3

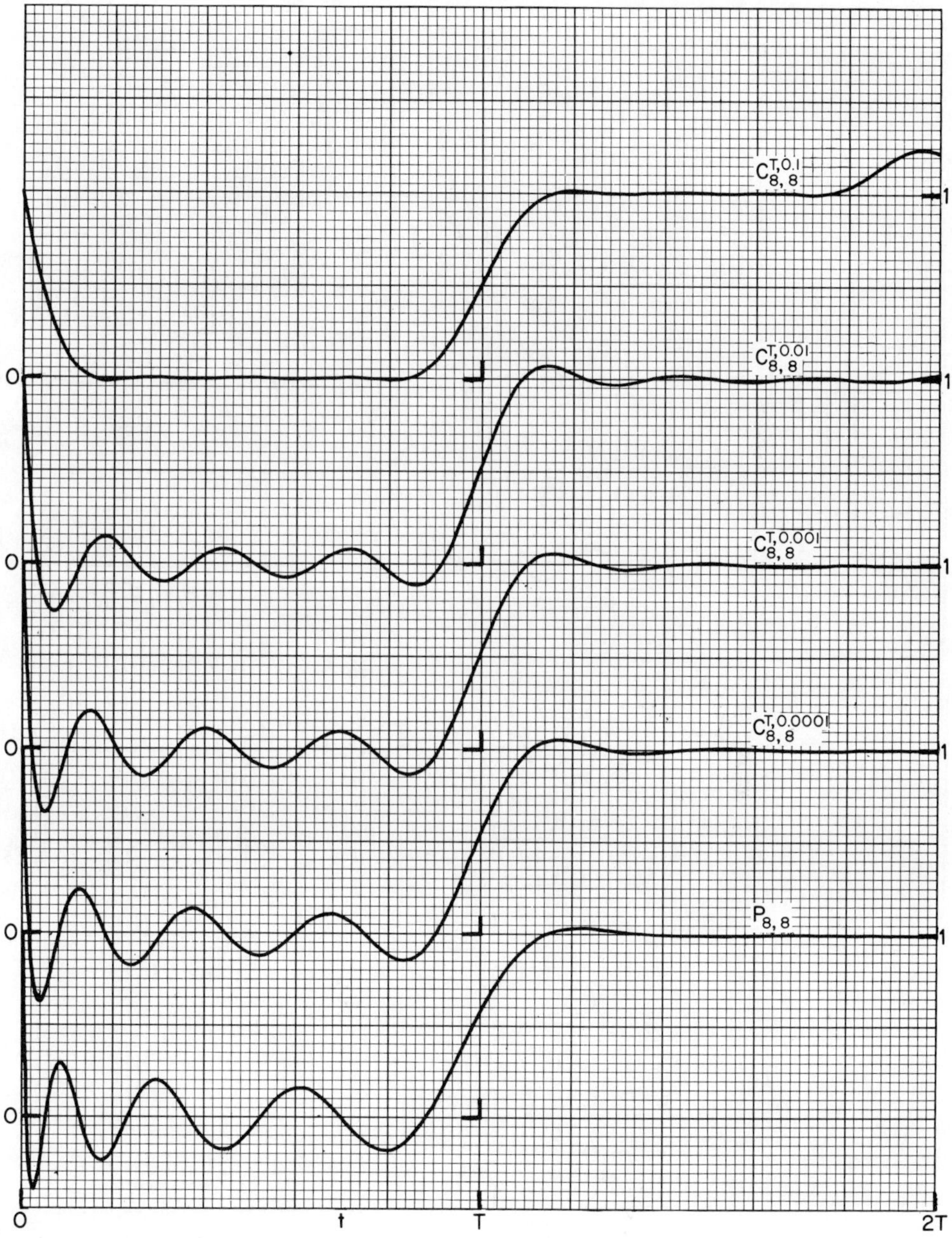

FIG. A4

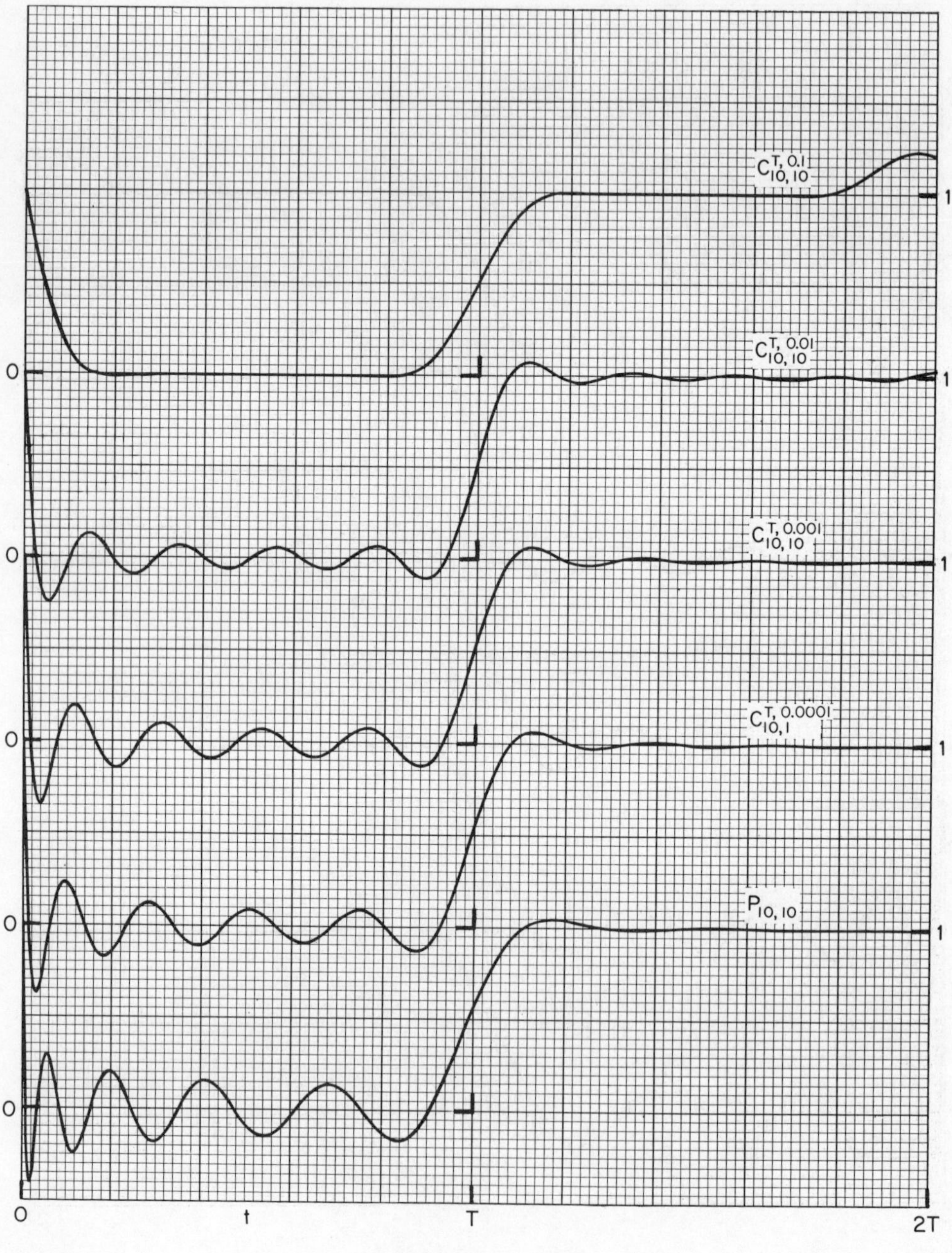

FIG. A5

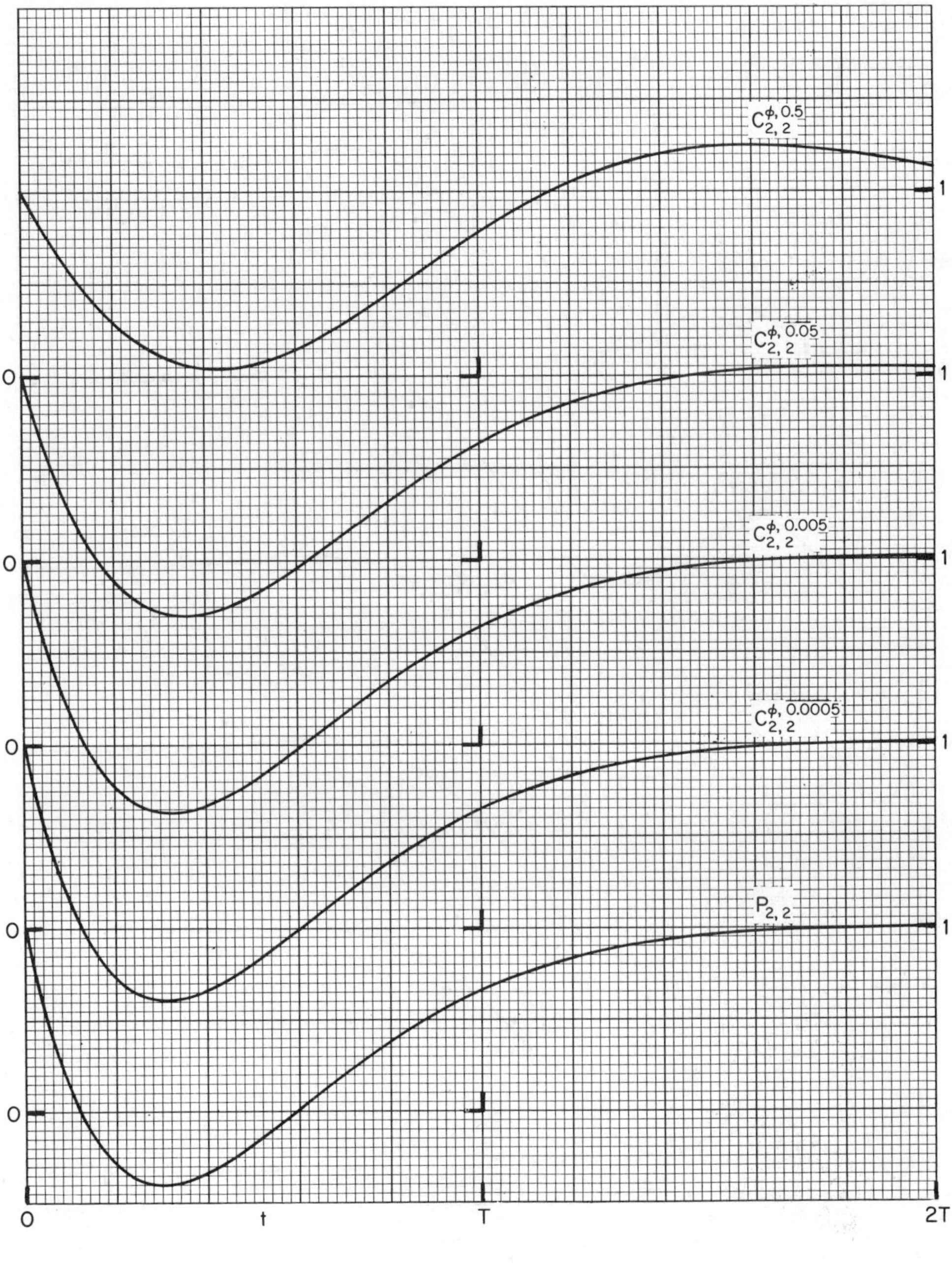

$C_{2,2}^{\phi,0.5}$
$C_{2,2}^{\phi,\overline{0.05}}$
$C_{2,2}^{\phi,0.005}$
$C_{2,2}^{\phi,\overline{0.0005}}$
$P_{2,2}$
0
1
0
t
T
2T

FIG. A6

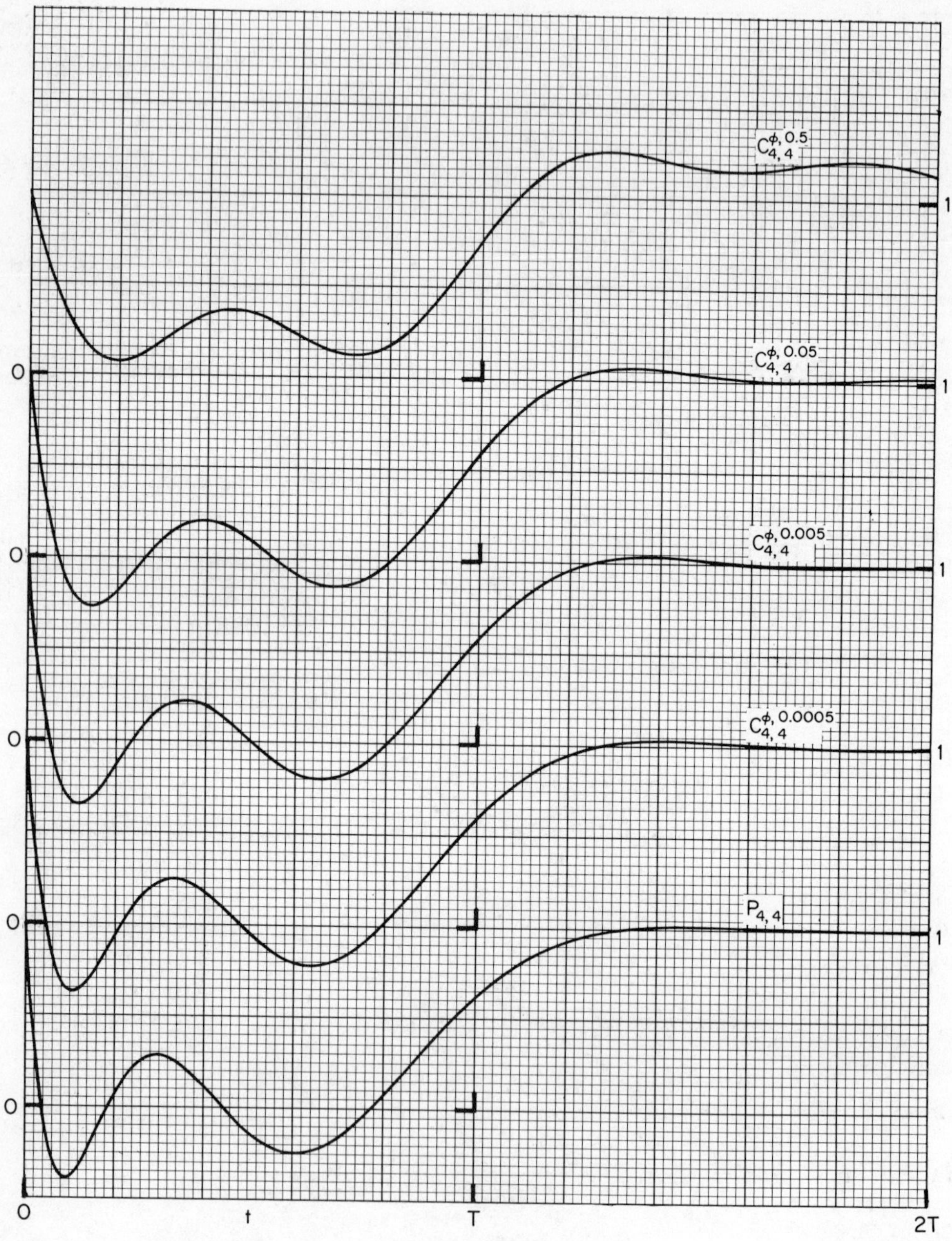

FIG. A7

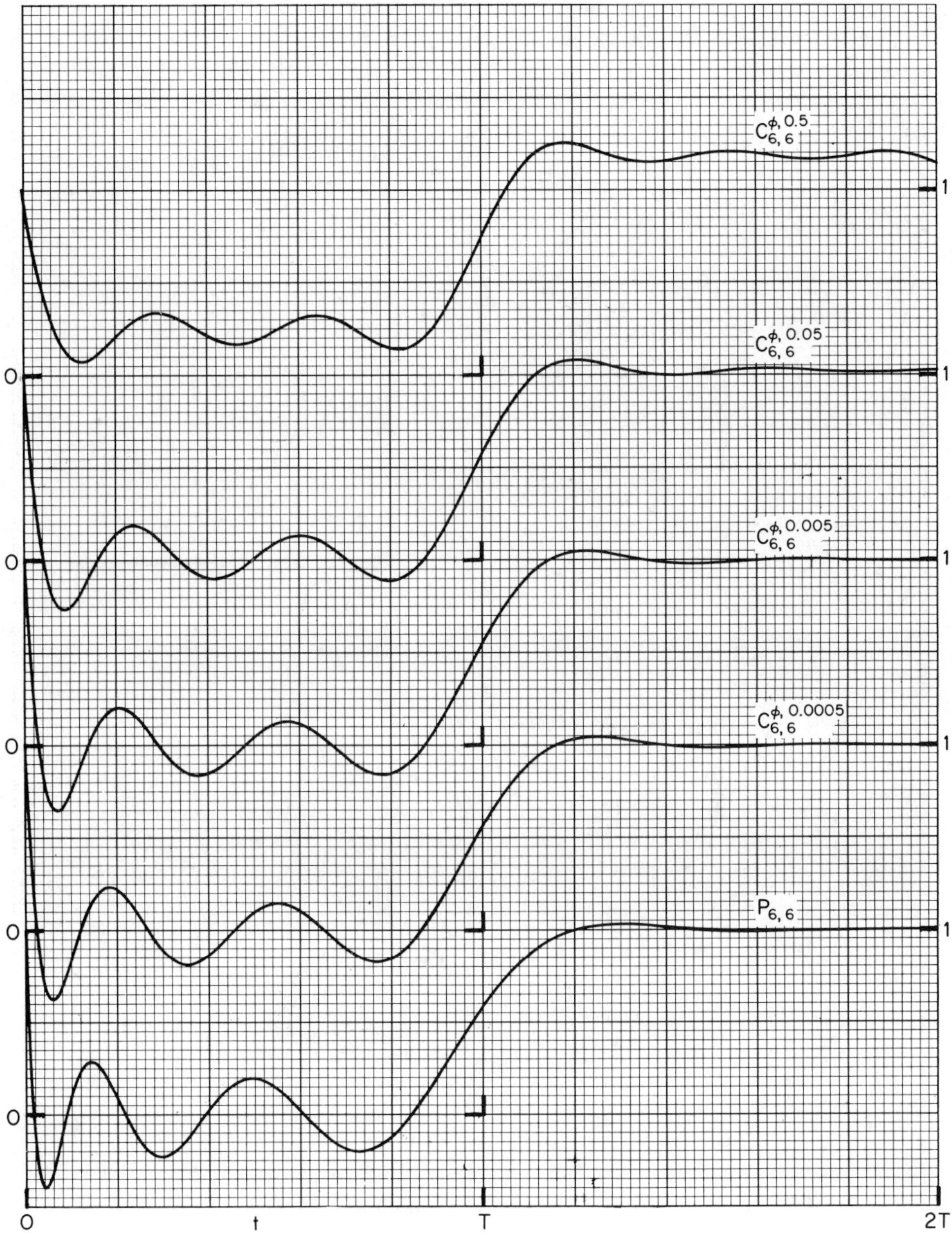

FIG. A8

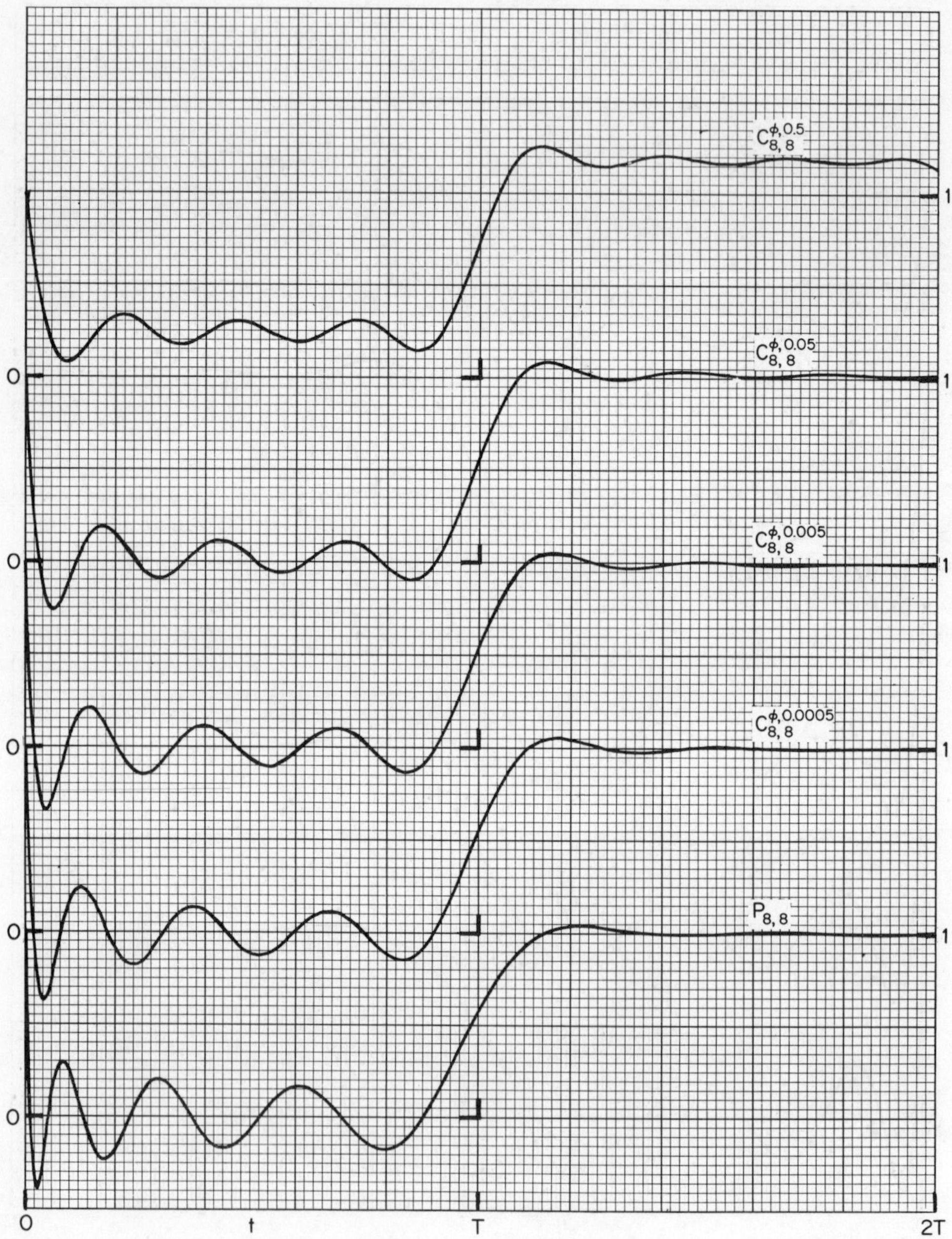

FIG. A9

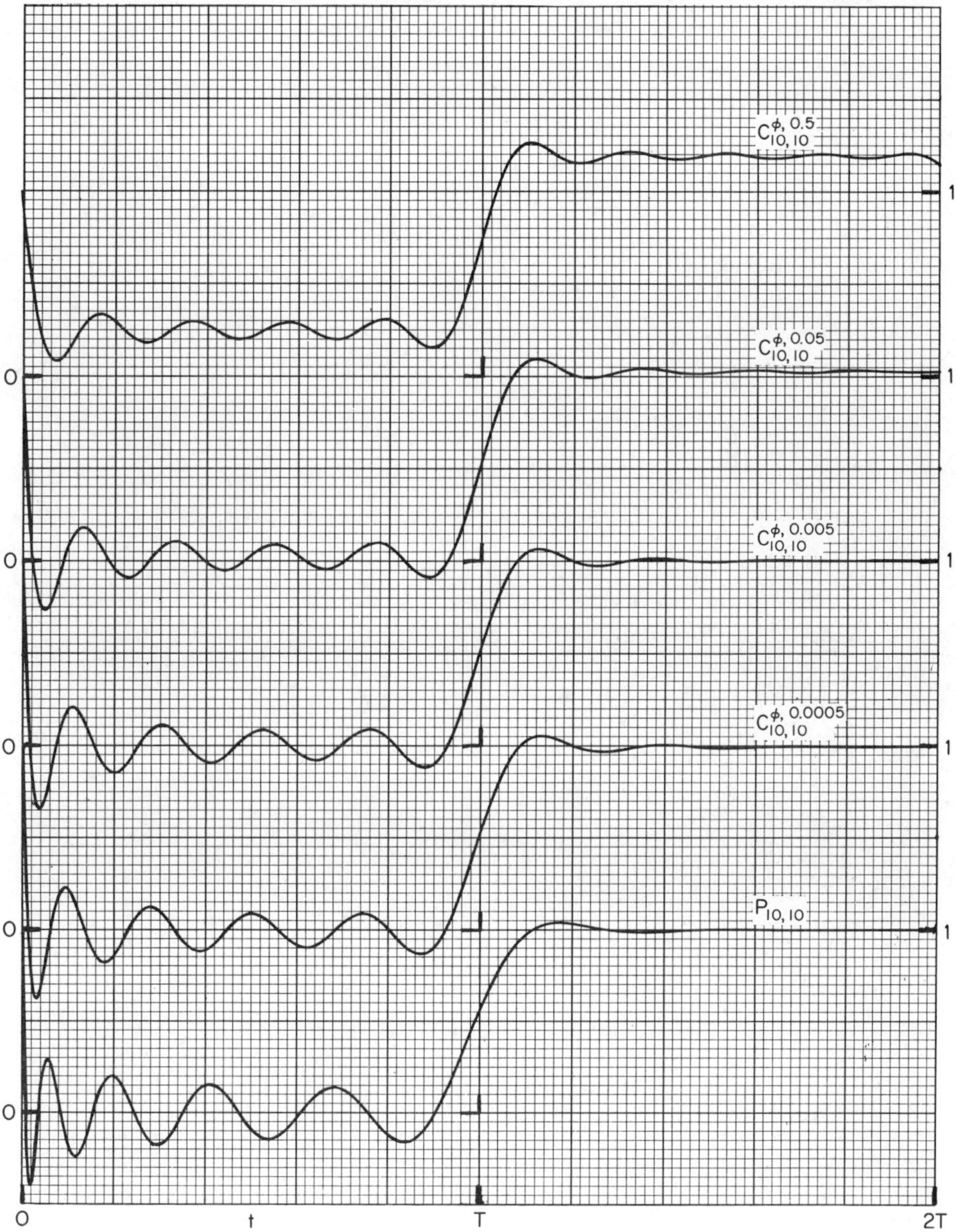

$C^{\phi,\,0.5}_{10,10}$
$C^{\phi,\,0.05}_{10,10}$
$C^{\phi,\,0.005}_{10,10}$
$C^{\phi,\,0.0005}_{10,10}$
$P_{10,10}$
0
1
t
T
2T

FIG. A10

INDEX